AF681196

NOUS·SOMMES·PRETS

SIMON FRASER UNIVERSITY
W.A.C. BENNETT LIBRARY

SB 608 F8 T75 2002

Tropical Fruit Pests and Pollinators

Biology, Economic Importance, Natural Enemies and Control

Dedication

To my nieces, Dalia Lucia and Carmen Yomaira, and nephews, Jario Alberto and Jorge Eduardo and to my adopted children Christina and Matthew. To Luz-Stella Cobo-Martinez for encouraging me to become an entomologist.

Acknowledgements

Extensive thanks to my co-editors, authors and co-authors for their willingness to participate in this endeavour. Dr Richard E. Litz for encouraging me to edit this book. Drs Richard Jones and William Brown for their support. Ms Elvia Esparza for the superb art work on the book cover. Ms Rita E. Duncan and Dr Diviina Amalin for their help. To the different reviewers of the book chapters.

J.E. Peña

Tropical Fruit Pests and Pollinators

Biology, Economic Importance, Natural Enemies and Control

Edited by

J.E. Peña

Tropical Research and Education Center
University of Florida, Florida
USA

J.L. Sharp

USDA-ARS, Miami,
Florida
USA

and

M. Wysoki

ARO, The Volcani Center
Bet Dagan
Israel

CABI *Publishing*

CABI *Publishing* is a division of CAB *International*

CABI Publishing
CAB International
Wallingford
Oxon OX10 8DE
UK

Tel: +44 (0)1491 832111
Fax: +44 (0)1491 833508
E-mail: cabi@cabi.org
Web site: www.cabi-publishing.org

CABI Publishing
10 E 40th Street
Suite 3203
New York, NY 10016
USA

Tel: +1 212 481 7018
Fax: +1 212 686 7993
E-mail: cabi-nao@cabi.org

©CAB *International* 2002. All rights reserved. No part of this publication may be reproduced in any form or by any means, electronically, mechanically, by photocopying, recording or otherwise, without the prior permission of the copyright owners.

A catalogue record for this book is available from the British Library, London, UK.

Library of Congress Cataloging-in-Publication Data
Tropical fruit pests and pollinators : biology, economic importance, natural enemies, and control / edited by J.E. Peña, J.L. Sharp, and M. Wysoki.
p. cm.
Includes bibliographical references (p.).
ISBN 0-85199-434-2 (alk. paper)
1. Tropical fruit--Diseases and pests. 2. Tropical fruit--Diseases and pests--Control. 3. Insect pollinators--Control. I. Peña, Jorge E., 1948- II. Sharp, Jennifer L. III. Wysoki, M.
SB608.F8 T75 2002
634'.6--dc21 2002001303

ISBN 0 85199 434 2

Typeset by AMA DataSet Ltd, UK
Printed and bound in the UK by Cromwell Press, Trowbridge

Contents

The colour plates can be found following p. 56

Contributors

E.L. Aguiar-Menezes, *Empresa Brasileira de Pesquisa Agropecuaria, Centro Nacional de Pesquisa de Agrobiologia, BR 465, Km 7, Caixa Postal 74505, Seropedica, RJ 23890-000 Brazil.*

M. Barbosa-Pereira, *Department of Entomology, ESALQ-USP, Piracicaba, São Paulo, Brazil.*

D.P. Bartholomew, *Department of Agronomy and Soil Science, University of Hawaii at Manoa, 1910 East-West Road, Honolulu, Hawaii 96822, USA.*

P.C.R. Cassino, *Universidade Federal Rural do Rio de Janeiro, Centro Integrado de Manejo de Pragas, 'Cincinnato Rory Goncalves', BR 465, Km 7, Seropedica, RJ 23890-000 Brazil.*

P.A. Follett, *USDA-ARS, Pacific Basin Agricultural Research Center, PO Box 4459, Hilo, Hawaii 96720, USA.*

S. Gazit, *Department of Horticulture, Faculty of Agricultural, Food and Environmental Quality Sciences, The Hebrew University of Jerusalem, PO Box 12, Rehovot 76100, Israel.*

C.S. Gold, *International Institute of Tropical Agriculture, ESARC, PO Box 7878, Kampala, Uganda.*

W.P. Gould, *USDA-ARS, Subtropical Horticulture Research Station, 13601 Old Cutler Road, Miami, FL 33158, USA.*

N.W. Heather, *School of Land and Food, The University of Queensland, Gatton College, Queensland 4345, Australia.*

J.S. Hwang, *Taiwan Agricultural Chemicals and Toxic Substances Research Institute, 189 Chung-Cheng Road, Wufeng, Taichung Hsien, Taiwan 413, Republic of China.*

G. Ish-Am, *Department of Horticulture, Faculty of Agricultural, Food and Environmental Quality Sciences, The Hebrew University of Jerusalem, PO Box 12, Rehovot 76100, Israel.*

E.B. Menezes, *Universidade Federal Rural do Rio de Janeiro, Centro Integrado de Manejo de Pragas, 'Cincinnato Rory Goncalves', BR 465, Km 7, Seropedica, RJ 23890-000 Brazil.*

H. Nadel, *International Centre for Insect Physiology and Ecology (ICIPE), PO Box 30772, Nairobi, Kenya.*

A. Pantoja, *Department of Crop Protection University of Puerto Rico, Mayaguez Campus, PO Box 9030 Mayaguez, Puerto Rico 00681-9030.*

J.E. Peña, *Tropical Research and Education Center, University of Florida, 18905 SW 280 Street, Homestead, FL 33031, USA.*

G.J. Petty, *Institute for Tropical and Subtropical Crops, Agricultural Research Council, Private Bag X1, Bathurst 6166, Republic of South Africa.*

B. Pinese, *Queensland Department of Primary Industries, PO Box 1054, Mareeba, Queensland 4880, Australia.*

P.A.C. Ooi, *FAO Regional Office for Asia and the Pacific, 39 Phra Atit Road, Bangkok 1022, Thailand.*

A. Raga, *Instituto Biologico, Caixa Postal 70, 13.001-970, Campinas, SP, Brazil.*

J.L. Sharp, *USDA-ARS, 13601 Old Cutler Road, Miami, FL 33158, USA.*

D. Smith, *Department of Primary Industries, Maroochy Horticultural Research Station, Nambour, Queensland 4560, Australia.*

M.A. Soares, *Universidade Federal Rural do Rio de Janeiro, Centro Integrado de Manejo de Pragas, 'Cincinnato Rory Goncalves', BR 465, Km 7, Seropedica, RJ 23890-000 Brazil.*

G.R. Stirling, *Biological Crop Protection (Pty) Ltd, 3601 Moggill Road, Moggil, Queensland 4070, Australia.*

M.A. van den Berg, *Institute for Tropical and Subtropical Crops, Private Bag X11208, Nelspruit 1200, South Africa.*

J.A. Villanueva-Jiménez, *Colegio de Postgraduados, Campus Veracruz, Apartado Postal 421, CP 91700 Veracruz, Veracruz, Mexico.*

G.K. Waite, *Queensland Horticulture Institute, Maroochy Research Station, PO Box 5083, SCMC, Nambour, Queensland 4560, Australia.*

A. Winotai, *Biological Control Group, Division of Entomology and Zoology, Department of Agriculture, Chatuchak, Bangkok 10900, Thailand.*

M. Wysoki, *Department of Entomology, ARO, Institute of Plant Protection, The Volcani Center, PO Box 6, Bet Dagan 50250, Israel.*

1 Introduction

Jorge E. Peña
Tropical Research and Education Center, University of Florida, 18905 SW 280 Street, Homestead, FL 33031, USA

Tropical Fruit

Tropical fruit form a large and diverse commodity group, that ranges from perennial to herbaceous species (Alexander and Possingham, 1984; Akora, 1998). The herbaceous group comprises important crop plants such as banana, pineapple, and papaya while the woody or perennial plants may include tree species, shrubs, and vines (Verheij and Coronel, 1992). In this book we have attempted to record and summarize what is known regarding pests and pollinators of tropical fruit from the major tropical fruit producing areas: South and South-East Asia, Australia, Africa, South and Central America and the Caribbean region. We have focused on tropical citrus, avocado, mango, pineapple, banana, passion fruit, litchi, guava, *Annona* spp., durian, mangosteen, acerola, and carambola. The reader will notice that the amount of information is extensive for citrus, avocado and pineapple, whereas relatively little has been compiled for durian, mangosteen, passion fruit, acerola, and carambola. This reflects the worldwide importance, consumer acceptance, and production margins of these fruit crops.

Tropical fruits are regularly grown in a variety of climates from latitude 23°27′ N to 23°27′ S of the equator, while some are grown up to approximately 37° N in Spain. Proximity to the sea, sea currents, altitude, direction of prevailing winds, rainfall, and humidity all have modifying effects on these crops. Tree phenology (flushing, quiescence, flowering, fruiting and leaf fall) is much more dramatic in the tropics than at high latitudes, where winter confines growth and development largely to one season each year (Verheij and Coronel, 1992). Day length in the tropics varies too little to noticeably affect the annual course of radiation or temperature. Thus, most climatic cues of plant development are not as firmly tied to calendar months as is the case at higher latitudes. The phenological cycle in the tropics may shift a few months from one year to the next. This complicates phenological studies, but should make it easier to detect which environmental factor (e.g. dry season) triggers a certain growth and development response (e.g. flowering). Because of their diversity, not all tropical fruit follow the same strategy for growth and development. Strategies are based on differences in tree form and branching habits. For instance, single-stemmed plants such as pineapple, papaya, and banana are contrasted with freely branching fruit trees (Verheij and Coronel, 1992). Most tropical fruit are considered as perennial plants which persist for several years without abrupt, major changes other than seasonal leaf formation, flowering, and fruit development (Bennett *et al.*, 1976). However, pineapple, papaya, and passion fruit are grown for shorter periods of time and arthropod management on these crops is influenced by their duration in the field (Ch. 6).

©CAB *International* 2002. *Tropical Fruit Pests and Pollinators* (eds J.E. Peña, J.L. Sharp and M. Wysoki)

Traditionally, tropical fruits were produced in the greatest quantity, and with the best quality, at or near the original centres of each species' distribution. As examples, the range and quality of mangoes is high in India, but other countries, such as Mexico, Guatemala and Israel, are seeking to optimize production. Litchis are traditional in southern China, but as a result of considerable production research, are now grown in the USA, Australia, and elsewhere. On the other hand, mangosteen, durian, and salak remain largely confined to their centres of origin, in the Malaya–Indonesia region. Pineapples, commonly those of the cayenne type, bananas, mainly of the dwarf Cavendish variety, and papayas of the 'Solo' type, are now almost ubiquitous in the tropics. A surprisingly wide range of tropical fruits including bananas, papayas, and pineapples are grown in high altitude areas (2000–2500 m) near the equator in countries such as Kenya, Colombia, Ecuador, Peru, Indonesia, and India. In nearly all cases, these fruits flourish in unusual environmental niches which are frost free and receive a high level of solar radiation.

According to FAO (2001), world production of tropical fruit (avocado, banana, citrus, mango, papaya, and pineapple) increased by 36–54% from 1975 to 2000 (Table 1.1), and production of minor crops (*Annona*, guava, passion fruit, etc.) increased by 24–29% during this period (Table 1.2). However, Verheij and Coronel (1992) caution that these figures could be misleading, because most fruit comes from trees scattered in home gardens, making it difficult to compile reliable statistical data. According to FAO (2001), India is the leading tropical fruit producing country, followed by Brazil, China, Mexico, and the Philippines. Major Asian producers are India, China, the Philippines, Indonesia, Thailand, and Pakistan. In the western hemisphere the leading producers of tropical fruit are Brazil, Mexico, Ecuador, Colombia, Costa Rica, and Cuba (FAO, 2001) (Table 1.3). Tropical fruit production in Africa is apparently concentrated in Nigeria, Burundi, South Africa, Congo, Cameroon, Egypt, and Angola.

Table 1.2. Trends in the global production of minor tropical fruit crops (*Annona*, guava, passion fruit, etc.) (source: FAO, 2001).

Years	Tropical fruit production (t)	Increase (%)
2000	15,331,309	24
1990	11,653,889	29
1980	8,374,428	29
1970	6,004,759	14
1961	5,165,192	

Table 1.3. Rank order of 15 countries with major tropical fruit production (source: FAO, 2001).

Country	Fruit production ('000 t)
1. India	35,142
2. Brazil	34,122
3. China	20,093
4. Mexico	9,800
5. Philippines	9,123
6. Indonesia	8,343
7. Thailand	7,941
8. Ecuador	5,000
9. Colombia	4,725
10. Pakistan	3,743
11. Costa Rica	2,101
12. Nigeria	1,629
13. Burundi	1,514
14. South Africa	1,433
15. Bangladesh	720

Table 1.1. Trends in global production of various tropical fruits (source: FAO, 2001).

Commodity	Production 1975 (t)	Production 2000 (t)	Increase (%)
Avocados	1,232,763	2,336,765	52
Bananas	31,688,547	58,687,214	54
Citrus	51,736,565	104,966,628	50
Mangoes	12,774,918	24,975,204	51
Papayas	1,944,494	5,363,167	36
Pineapples	7,219,688	13,455,362	53
Tropical fruit	6,894,951	15,331,309	45

Few tropical fruits are grown in Europe and only relatively small amounts in the USA, Australia, and Israel. In the USA, tropical fruits are grown in Florida and California, where the availability of suitable soils and occasional below-freezing temperature limit production. Hawaii grows a wide range of tropical fruits, but has an unusual maritime tropical climate that is modified by altitude. For this reason varieties of papaya and macadamia developed in Hawaii have complex environmental requirements, and are grown elsewhere with only limited success. Small quantities of tropical fruit for European markets are grown in Israel, generally with the support of irrigation in low-lying, saline desert areas with low relative humidity. Many equatorial tropical countries including those in Africa, the Americas, India, and South-East Asia are beginning to strengthen research programmes on propagation, genetic improvement, control of flowering, control of pathogens and arthropods, irrigation and nutrition, postharvest handling, and storage.

Integrated Pest Management (IPM) for Tropical Areas

Pest management has as its goal the control of agricultural pests to benefit society as a whole (Pimentel, 1981). Therefore, successful pest control strategies must take into account the complexities of society and the entire ecosystem (Pimentel, 1981). According to a definition proposed by FAO in 1972, IPM is

> a system which primarily takes into account the milieu and the population dynamics of the species under consideration and uses appropriate techniques and methods in a compatible manner in order to restrict the pest populations below the threshold of economic damage. Application of such a method mandates not only a thorough knowledge of the biology of the pest, i.e. all the biotic factors which can affect it and the effect of abiotic factors on the crop ecosystem, but also its market goals (export vs. internal consumption).

Fifty-one years have passed since the appearance of W. Ebeling's (1950) book, *Subtropical Entomology*, which continues to be cited as an essential reference for students and researchers of tropical and subtropical fruit entomology. However, tropical crop production has undergone major changes since the book was published. Several other texts dealing with pest management systems for tropical crops (Lamb, 1974; Hill, 1975; Swaine *et al.*, 1985; Braga *et al.*, 1998; Tandon, 1998; Mariu, 1999) have since been written. These publications are in response to substantial growth of tropical fruit production and increased concern for food safety. Many books dealing with integrated pest management have concentrated on field crop IPM with the occasional exception of temperate fruits, such as pear and apple (Boethel and Eikenbary, 1979), citrus (Anonymous, 1991) or nuts and almonds (Anonymous, 1985).

Because production of tropical fruit ranges from sophisticated plantation type crops to rudimentary backyard production, coupled with economic solvency and production purpose, internal vs. export production, it is difficult to generalize on a single method of crop protection. For instance, Verheij and Coronel (1992) maintain that tropical fruit crop protection or the use of commercial pesticides is limited to commercial cropping systems, i.e. orchards and corporate plantations. Entwistle (1972) supports the idea that the application of the concept of integrated control often marks a stage in the evolution of research where information accrued in fields of study originally treated largely as separate has expanded into a dynamic study of the interacting whole. This case may have happened for some crops such as pineapple, citrus, and banana but not for others. Again, the diversity of the tropical crops treated in this book makes it difficult to make a single assessment, as can be done for temperate fruit crops such as apple or pear. The diversity of tropical fruit farmers ranges from those with wide financial support to those with minimum resources. Small farmers in the tropics need an arsenal of various pest management tools (e.g. chemical, biological, cultural, etc.) in order to have a successful IPM programme (Goodell, 1984). Farmers in less developed countries lack resources and monetary support to build up an effective IPM farm-level implementation programme, while farmers in developed countries have well funded research and extension institutions (Goodell, 1984).

Integrated control of a pest or pest complex in a tropical crop requires a full investigation of relationships and interactions within the whole ecosystem (Entwistle, 1972). Because of the specifically complex society of plants and arthropods inhabiting the area where tropical fruits originate, tropical fruits may have a greater diversity of trophically associated arthropods than any temperate crop. For instance, Entwistle (1972) maintains that the 1400 arthropod species listed for cocoa can only be the skeleton crew of the total cocoa fauna. Thus, information on tropical fruit entomology is regionally uneven both in quantity and quality; this being the result of national policies or the absence of them, and the dominant nature of certain pest problems. Moreover, few authors have yet taken an integrated attitude to arthropods of tropical fruit crops.

Globalization of tropical fruit and rapid expansion of export markets has resulted in exotic plants grown outside of their centres of origin. For instance, citrus originated in Asia and is grown in the Neotropics while papaya and passion fruit originated in the Neotropics and now are grown in Asia, Australia, etc. Thus, upon arrival in their new area of domestication, exotic fruit plants are populated by an indigenous regional arthropod fauna, and by exotic arthropods that have also migrated from the crop's centre of origin. Thus, arthropod systems (pollinators, pests, and natural enemies) for introduced plant species will be somewhat different from those observed in the crop's area of origin. For instance, the European honeybee *Apis mellifera* did not co-evolve with avocado but it is used as a pollinator of avocado in Israel and Australia. Original pollinators from the area of origin of avocado in Mexico and Guatemala could be other more efficient hymenopteran species (see Ch. 8). In the Neotropics several insects have adapted to introduced plant species, such as citrus. For instance, the Neotropical, *Diaprepes abbreviatus* (Coleoptera: Curculionidae) has adopted citrus as one of its preferred host plants. *D. abbreviatus* is considered a key pest for citrus in Florida and the Caribbean region, while this pest is completely unknown in China and Asia (see Ch. 3).

Commercial production of tropical crops regularly starts on or near areas with little agricultural development. The constitution of arthropod fauna that exist in unexploited areas would be likely to invade new plants when any disturbance occurs, creating conditions for colonization of the new introduced plants. Such situations may occur when the fruit tree is planted on land prepared by clear felling or beneath the shade of secondary forest. In such circumstances, the arthropod fauna may still reflect that of the local primary and secondary plant communities of the area (Entwistle, 1972). Attempts should be made to bring together existing data into cohesive studies of insect communities in these important plant associations. Until these data are available, tropical fruit entomology will in many ways lack a fundamental perspective.

Tropical fruit crops provide a relatively long-term and stable environment, offering continuing habitats for both pests and natural enemies. This environment provides excellent opportunities for biological control and alternative pest management programmes (Bennett *et al.*, 1976). However, the extensive use of broad spectrum pesticides disrupts this stable environment, and leads to instability of arthropod densities (Hoyt and Burts, 1974). According to Hoyt and Burts (1974) any attempt to develop integrated control programmes in fruit crops must take into account the following: (i) knowledge of native or resident arthropod fauna; (ii) arthropod fauna affecting the tree crop in its area of origin or domestication; and (iii) the presence of natural enemies. The basis for integrated pest management includes the pest's biology and ecology, sampling and monitoring techniques, economic thresholds, and the application of management tactics, i.e. chemical, biological, autocidal, plant resistance, etc.

Biology and ecology

The botanical diversity of tropical fruit is enormous and their arthropod fauna is probably more complex than we suspect.

There is climatic and geographical variation within the range of tropical fruit cultivation which is known to affect the composition of insects on each tropical crop. For instance, *Bephratelloides* spp. (Hymenoptera: Eurytomidae) are the most common insects affecting *Annona* crops in the Neotropics. Within this group, *Bephratelloides cubensis*, a key pest species present in Florida, has a female-biased sex ratio, while a similar species, *B. pomorum*, dominant in *Annona*-growing regions of Brazil, has an almost 50 : 50 sex ratio. In the same way, the moth *Cerconota annonella* is found affecting *Annona* in the warmer regions of South America and the Caribbean but is absent from subtropical Florida. Insects can be classified as autochthonous species with a more local distribution, or as inter-regional species which may affect crops in different areas either by exceptional powers of dispersal coupled with the host and environmental tolerance, or as homophylous, i.e. tending to distribution by man. These classifications are not necessarily mutually exclusive and it may be difficult, as for example with such mealybugs as *Maconellicoccus hirsutus* and the aphid *Toxoptera aurantii*, to distinguish the main agents of dispersal (Entwistle, 1972).

Most arthropods affecting tropical fruit are polyphagous (mites, fruit flies, scales, whiteflies, and thrips) and studies on their biology have been published extensively. Thus readers requiring more detail should consult, among others, the books edited by White and Elson-Harris (1994) (fruit flies); Jeppson *et al.* (1975), Helle and Sabelis (1985a,b) (mites); Rosen (1990a,b) (armoured scales); and Parker *et al.* (1995) (thrips).

In order to apply IPM concepts to tropical fruit, we must better understand the biology and ecology of these arthropods particularly in their relationship with tropical fruit. This book reviews the biology of the most important pest species and pollinators affecting tropical fruit, but may not include some regionally important pests. Although baseline information is provided, this book demonstrates that there are several gaps of information pertaining to these arthropods.

Sampling and monitoring

Sampling populations to determine kinds and to estimate numbers of living species are the most fundamental research activities in ecology and the primary basis of integrated pest management (Pedigo, 1994). The basic aims of IPM are most effectively accomplished by employing preventive tactics such as biological control, plant resistance and cultural activities to maintain pest populations below the economic injury level. However, if these tactics fail, other curative methods, i.e. chemical or microbial control, are applied. The use of the latter methods calls for a practical sampling programme that will help to define pest density as well as the appropriate decision rules such as economic threshold and use of a curative method. Hare (1994) suggests that for pest management decisions it is necessary to rapidly and economically determine where the pest population is relative to its treatment threshold.

Sampling programmes for fruit trees are inherently more complex than for annual, herbaceous plants because of the wide variety of habitats that trees offer (Hare, 1994). For instance, in fruit trees substantial heterogeneity in pest populations can exist even within different parts of the leaf canopy, some pests are direct (fruit feeding) but can also be indirect (affecting shoot, wood, foliage, roots) (Hare, 1994). Beers *et al.* (1994) suggest that for fruit trees, sampling is better directed toward indirect pests because they have a greater probability of being present at non-damaging levels. Sampling and monitoring methods for tropical fruit are well established for some direct pests such as fruit flies, fruit borers (Ch. 4) as well for some indirect pests (i.e. banana weevil) (Ch. 2), leafminers (Ch. 3), mites (Keenan, 1997; see also Ch. 3) and defoliators (Ch. 8). However, the inability to correlate sampling data to infestation levels of the fruit at the time of harvest is still a problem (Chs 2 and 4). While attempts have been made to develop sampling techniques for several pests of citrus, mango, avocado, and banana (Smith *et al.*, 1997; see also Chs 2, 3 and 4), adequate sampling techniques are not available for some pests and crops (Chs 10 and 12).

Of course, the availability or allocation of resources for sampling, research and monitoring is a function of the place where the crop is grown, the purpose of fruit production (export vs. internal consumption), and the economic solvency of growers. Thus, grower organizations in Australia, the USA, South Africa, and Israel are able to fund research for tropical fruit, while tropical fruit growers in many developing countries cannot.

Economic thresholds

Pest management of fresh fruit focuses more upon preserving the appearance of the fruit rather than upon maximizing fruit production. While several additional factors may complicate determination of the actual economic status of pests, the level of damage tolerated by consumers and packers varies and complicates pest management decisions. Therefore, some confusion still exists regarding economic thresholds, economic injury levels, and economic damage with respect to temperate fruit (Hoyt and Burts, 1974), and this confusion is even greater for tropical fruit. For instance, there are insufficient data on the economic consequences of arthropods of tropical fruit and when available these data are often approximate. As export commodities, tropical fruit crops command consistently high prices, with highest prices for undamaged fruit of premium quality. Therefore the lack of knowledge of economic thresholds, and/or the non-existence of economic thresholds, prevent control programmes from focusing on the prevention of damage to tropical fruit by pests. This factor, together with the paucity of sampling techniques for tropical fruit pests, suggests that little effort has been focused on determining economic treatment levels (Chs 2, 3 and 6). Consequently, pesticides continue to be widely applied as prophylactic measures. Another source of insect-associated loss is pollination, which in several crops is thought to limit production. Objective studies are required to substantiate the value of pollinators and their effect on yield.

Chemical control

The choice of toxicants and availability of spraying equipment differs among tropical fruit producing countries. Verheij and Coronel (1992) maintain that in Asian countries, traditional tropical fruit crop protection measures such as chemical control are occasionally practised, being largely limited to small plants and nursery work, while in Australia, the USA, and Israel, there are few constraints regarding availability of equipment and pesticides to be used for tropical fruit. The use of pesticides in some areas is limited by tree size, since spraying of large trees is impossible with the equipment most small growers can afford. The scattered plantings in backyards, on field borders and along watercourses impede access of equipment. Moreover, the tree population may consist of many seedling trees, and factors such as biennial bearing would make it difficult to establish an annual spraying routine.

Because of a variety of issues, including toxic residues, etc., spraying should be based on sound investigation. *Ad hoc* spraying may well be uneconomic and counter-productive. For instance, in the case of defoliators that are usually only numerous in periods of maximal leaf production, application timing is critical for cost effective spraying. Secondly, spraying may disturb the balance between phytophagous insects and their natural enemies, leading to increasing numbers of secondary pests. Thirdly, the toxicants used may not always be those considered suitable to pollinating insects. Verheij and Coronel (1992) suggest that developed countries, which were the first to rely heavily on commercial biocides for fruit growing, are now adopting integrated crop protection systems that were implemented by small farmers in developing countries. Nevertheless, pesticides have been associated with significant successes in protecting tropical fruit from insect attack and increasing tropical fruit productivity. Thus, the control of fruit and foliar insect pests continues to depend on the use of chemical insecticides/acaricides (Chs 3 and 6).

Information regarding proper timing, spray volumes and knowledge of the pest complex differs among tropical fruit. While solid information is available for some crops like pineapple (Ch. 6), it is disregarded for others, e.g. papaya (Ch. 5). In Chapter 5 Pantoja *et al.* report that most papaya producers apply insecticides on a calendar basis. Such routine heavy use of sprays for controlling fruit flies and leafhoppers in papaya can cause heavy outbreaks of mites and other pests. Moreover, pesticidal procedures for controlling pests for some crops (e.g. citrus), when applied to other crops without reliable knowledge of the effects of these chemicals on the crop in question may exacerbate crop losses because of phytotoxicity, induced explosion of population densities of other pests, reduction of natural enemy populations, etc.

Widespread use of non-selective pesticides continues to be the rule (Chs 6, 10 and 12), but currently there is a trend towards evaluating a new generation of pesticides (Ch. 8), adoption of selective spraying (Ch. 7), proper timing of spray applications (Ch. 6), and determining the effect of pesticides on predators and parasitoids (Ch. 8).

In contrast to the extensive information on effects of pesticides on pests, there is very little (Ch. 12) information on their effects on pollinators (Chs 4, 7 and 8). For instance, Aguiar-Menezes *et al.* suggest in Chapter 12 that the deleterious effects on pollinators can be averted by proper timing of spray applications in passion fruit, according to the cultivar. Purple passion, whose flowers open during the morning hours, should be sprayed during late afternoon, while the yellow cultivar, whose flowers open in the afternoon, should be sprayed in the morning.

Attractants – pheromones

Non-selective traps such as black light and bait traps are commonly used in tropical fruit because they can provide useful information for monitoring purposes. Furthermore, development of selective traps (baited with attractants or pheromones) often to the pest species level, has been a tremendous boost to IPM of temperate fruit (Beers *et al.*, 1994) and it may provide the same benefit available for tropical fruit pests. Use of feeding and sexual attractants (secondary pheromones) in tropical fruit has been largely limited to fruit flies (Chs 5 and 9). Few species-specific attractants are available for other insect groups such as weevils (Curculionidae) (Gold *et al.*, 2002) and Lepidoptera (Ch. 8; see also Bailey *et al.*, 1988).

Biological control

Biological control has great potential as a major tactic for regulating pest populations in fruit orchards (Hoyt and Burts, 1974). The ability to apply biological control effectively has increased in recent decades because of greater knowledge of the arthropod fauna of some tropical fruits (citrus, avocado, pineapple, and mango) (Chs 3, 4, 8 and 9) but concentrated efforts to use biological control agents have rarely been made for most pests of various tropical fruit crops (Chs 2, 7, 10 and 12). For instance, efforts towards developing systems for biocontrol of the pink hibiscus mealybug, *Maconellicoccus hirsutus*, and the carambola fruit fly, *Bactrocera carambolae*, in the Caribbean were not initiated until after these pests had become major threats to valuable crops (Ch. 10). Generally, if a stenophagous insect (i.e. avocado weevils, papaya fruit fly, avocado thrips, and *Annona* fruit moths) is the major constraint to production of a single commodity, then efforts to develop biological control are pursued half-heartedly (Chs 5 and 8). The exceptions to this rule are the current strong efforts towards biocontrol of banana weevil (Ch. 2), and the search for natural enemies of avocado thrips (Ch. 8). Efforts also differ among countries. For instance, in Australia and Israel biological control is an important component of *Annona* pest management (Ch. 7), while other *Annona*-producing countries, i.e. Brazil, Ecuador, Venezuela, Colombia, and Mexico, rely on chemical pest control.

It is likely that with time the use of biological control agents for tropical fruit protection will increase extensively. The main constraints on this trend are the lack of commitment and money. Currently, very few people in tropical countries are researching these areas, and very few specialists receive adequate financial support. While major emphasis has been given to staple crops (i.e. rice and cassava), with the exception of banana, little or no emphasis has been given to fruit crops.

Scattered through this book are several references to microorganisms such as fungi, bacteria, and viruses pathogenic to insects. However, most pathogens of tropical fruit pests are virtually unknown. Apart from fungal and bacterial control of insects on citrus, avocado, and banana, there have been few studies into applied aspects of this very promising area. In general, it may be said that with the exception of citrus and avocado, the biological control of tropical fruit arthropods has been much neglected and that there is a great need for further work.

Host plant resistance

Host plant resistance offers considerable promise as a tactic in pest management. Even though it appears that most efforts are concentrated in developing genetic resistance to plant pathogens, use of this tactic merits attention for numerous crops against numerous arthropod pest species. Evaluation of tropical fruit germplasm collections for resistance to arthropod pests should become a high priority. Most efforts are directed at insect vectors (citrus, papaya, and pineapple), and pests of mango, avocado, and guava (Chs 4, 8 and 9). By contrast several valuable fruits, i.e. durian and acerola, continue to be largely unnaturalized and selections for improved traits have rarely been made.

Cultural practices

Cultural control concerns the employment of cultural management methods to minimize insect damage. The use of trap crops, host plant removal, removal of pests, and the reduction of pests' habitats are considered beneficial by some and controversial by others. For instance, Hansen and Amstrong (1990) reported that field sanitation did not reduce infestation by the mango nut weevil, *Sternochetus mangiferae* (Coleoptera: Curculionidae) in Hawaii, while in the Neotropics, removal of *Annona* spp. fruit infested with the seed borer *Bephratelloides cubensis* is considered to be a regular practice to avoid future infestations (Ch. 7). Untreated backyard trees and neglected plantings are considered to be major sources of pests such as fruit flies. Thus, in Australia, hygiene and attention to alternative host plants that can increase pest pressure on the *Annona* (custard apple) orchard, are important (Ch. 7). Mature fruit infested with yellow peach moth or with the Queensland fruit fly should be collected and destroyed. Preferred fruit fly hosts like guava and loquat should not be planted as trap crops in or near the *Annona* orchard.

According to Hoyt and Burts (1974), cultural practices generally do not offer a direct means for controlling pests but, when used properly, they can enhance natural enemy activity or retard pest population growth to a degree that is important in integrated control programmes. With polyphagous insects, the wild host plants may be so numerous that special eradication would be impracticable, yet removal of related wild host plants may be very beneficial where insect pests have restricted host ranges. Gold *et al.* in Chapter 2 describe crop sanitation as an important factor for maintaining low banana weevil densities in banana plantings. However, crop sanitation has not been firmly established for controlling important pests of avocados (i.e. weevils) (Ch. 8), *Annona* (annona seed borer) (Ch. 7), and mango (mango seed weevil) (Ch. 4).

Pollination

Interest in pollination is as ancient as civilized culture. Farmers have always shown a keen interest in the reproductive biology of plants and some mechanisms of fruit production (Real, 1983) and while most of the studies have focused on the temperate zone, there is an expanding emphasis on investigations of tropical forests and plant ecosystems. Several studies of tropical plant pollination have been conducted in diverse natural habitats (i.e. Beach, 1984; Valerio, 1984; Gottsberger, 1989) and this information has been applied to agricultural systems. Yet, it is necessary to remember that agroecosystems may lack some of the complex interactions observed in natural habitats. Young (1982) pointed out that for trees of tropical forests such as cacao, production of mature fruit is influenced by factors other than pollinator abundance. It is interesting to note that cacao originated as an understorey tree in Amazonian rainforests, where there was probably a well established equilibrium between pollinator abundance and abundance of flowers in the tree. Young (1982) concludes that successful pollination in a tropical crop depends in part upon the floristic complexity of the tree and the effects of surrounding habitats on the influx of insects influencing pollination within the agricultural habitat. The authors in this book provide different approaches to these issues. In Chapter 8, Wysoki *et al.* trace the different theories of pollination of avocado. While one school advocates the role of pollinators and the influence of native species of pollinators of avocado, another school asserts the theory that insect pollinators are not needed. A similar controversy exists with respect to mango (Ch. 4). Peña *et al.* in Chapter 7 review the flowering plant evolution of *Annona* spp., and discuss the methodologies for improving pollination mechanisms. In Chapter 12 Aguiar-Menezes *et al.* summarize studies on passion fruit pollination and discuss those insect species that could be effective pollinators and tactics that limit their effectiveness.

Conclusions

According to Aluja (1994) there are two major and opposing forces that drive the dynamics of pest management from a worldwide perspective: trends toward globalization of markets versus trends toward sustainability of agricultural practices and conservation of biodiversity. Aluja (1994) demonstrated that market forces compel several countries to comply with quality standards established in other countries. For example, Mexico and Brazil must respond to standards imposed by the USA or Japan that require fruit that is free of any cosmetic blemishes, and often this can be only be achieved by use of agrochemicals or quarantine treatments (Sharp and Heather, 2002).

Standard IPM methodologies, involving a number of control alternatives used alone or in combination (such as orchard sanitation, monitoring, early harvesting, parasitoid release and wild host removal) have achieved some success in controlling pests such as fruit flies for individual growers (Hendrichs, 1996). However, because these various tactics are usually implemented in a helter-skelter or uncoordinated manner, their effectiveness is compromised. Therefore, many commercial producers are forced to regularly apply insecticide-bait sprays to protect fruit in their orchards from flies dispersing from neighbouring orchards and other hosts.

Since very important polyphagous pests, i.e. fruit flies, weevils, lepidopterans, etc., affect similar tropical fruit species, the concept of area-wide pest control can be applied effectively in large geographical areas, where the same fruit crops are exploited (Klassen, 2000; Lindquist, 2000). When fruit growers pursue a concerted 'total pest population management strategy' over a substantial area, the number of individual pests produced is reduced progressively with time, and the number moving between neighbouring orchards is largely reduced. Under these area-wide conditions, which require an effective organization of growers, and in some instances increasing technical sophistication, IPM becomes much more effective. Malavasi *et al.* (1994) discuss this approach for the creation of pest-free areas to facilitate exporting of fruit. A pest-free area is one that lacks a quarantine-significant pest species, and is separated from infested areas by natural or artificial barriers. There are two types of pest-free areas: (i) pest-free zones are large geographic areas, such as the entire country of Chile, that are certified free of tropical fruit flies of economic importance; and (ii) pest-free production groves that require the demonstrated suppression of quarantine pests to non-detectable levels. Florida is able to export grapefruit to Japan by creating Caribbean fruit-fly-free grapefruit groves in about 22 counties. Requirements to establish pest-free fields of crop production include a sensitive detection programme, suppression of the quarantine-significant pest to non-detectable levels, strict control of the fields, and safeguards to prevent infestation during packing and transit to the port of export.

In conclusion, for better management of tropical fruit, pests' habits, importance of their damage and action levels should be known as the foundation for any programme. For the tropical crop in question, more emphasis should be devoted to improved knowledge of pollinators and ways to improve their effectiveness; these will be determined by research and by practical experience. After these steps are taken, development of better monitoring techniques for pests and natural enemies should follow as well as the evaluation of a feasible biological control (requirements and impact of biocontrol agents) programme. These are not merely research tools, but should also be a set of techniques and data with practical value for scouts and growers. Lastly, development of IPM packages for the whole pest complex should consider that all these arthropods inhabit the same universe. Some of them can be controlled with pesticides, others can be controlled by biological agents, cultural methods or by any means that are less harmful to the whole system.

Acknowledgements

I thank W. Klassen, C. Mannion and R. Giblin-Davis for comments to improve an earlier version of this manuscript.

References

Akora, R.K. (1998) Genetic resources of native tropical fruits in Asia: diversity, distribution and IPGRI's emphasis on their conservation and use. In: Akora, R.K. and Ramanatha Rao, V. (eds) *Tropical Fruits in Asia, Diversity, Maintenance, Conservation and Use.* Proceedings of the IPGRI-ICAR-UTANET Regional Training Courses, 18–31 May 1997, Bangalore, India, pp. 42–53.

Alexander, D.McE. and Possingham, J.V. (1984) Potential for fruit growing in tropical Australia. In: Page, P.E. (ed.) *Tropical Tree Fruits of Australia.* Queensland Department of Primary Industries, Brisbane, pp. 1–19.

Aluja, M. (1994) Bionomics and management of *Anastrepha. Annual Review of Entomology* 39, 155–178.

Anonymous (1985) *Integrated Pest Management for Almonds.* University of California, Division of Agriculture and Natural Resources, Oakland, California, Publication 3308, 156 pp.

Anonymous (1991) *Integrated Pest Management for Citrus*, 2nd edn. University of California, Division of Agriculture and Natural Resources, Oakland, California, Publication 3303, 144 pp.

Bailey, J., Hoffman, M. and Olsen, K. (1988) Black light monitoring of two avocado insect pests. *California Agriculture* 42, 26–27.

Beach, J.H. (1984) The reproductive biology of the peach or 'pejibaya' palm (*Bactris gasiapaes*) and a wild congener (*B. porschiana*) in the Atlantic lowlands of Costa Rica. *Principes* 28, 107–119.

Beers, E., Hull, L. and Jones, V. (1994) Sampling pest and beneficial arthropods of apple In: Pedigo, L. and Buntin, D. (eds) *Handbook of Sampling Methods for Arthropods in Agriculture.* CRC Press Boca Raton, Florida, pp. 383–415.

Bennett, F.D., Rosen, D., Cochereau, P. and Wood, B. (1976) Biological control of pests of tropical fruits and nuts. In: Huffaker, C.B. and Messenger, P.S. (eds) *Theory and Practice of Biological Control.* Academic Press, New York, pp. 359–387.

Boethel, D. and Eikenbary, R. (1979) *Pest Management Programs for Deciduous Tree Fruits and Nuts.* Plenum Press, New York, 256 pp.

Braga, R., Cardoso, J.E. and das Chagas, F. (1998) *Pragas de Fruteiras Tropicais de Importancia Agroindustrial.* Empresa Brasileira de Pesquisa Agropecuaria, Centro Nacional de Pesquisa Agrodindustria Tropical, Brasilia, 209 pp.

Ebeling, W. (1950) *Subtropical Entomology.* Lithotype Process Co, San Francisco, 747 pp.

Entwistle, P. (1972) *Pests of Cocoa.* Longmans, London, 779 pp.

FAO (Food and Agriculture Organization) (2001) FAO Statistical Data Base. http://appsl.fao.org/servlet/Xte.FAOStat Data Base.

Goodell, G. (1984) Challenges to international pest management research and extension in the third world: do we really want IPM to work? *American Entomologist* 30, 18–26.

Gottsberger, G. (1989) Beetle pollination and flowering rhythm of *Annona* spp. (Annonaceae) in

Brazil. *Plant Systematics and Evolution* 167, 165–187.

Hansen, J.D. and Armstrong, J.W. (1990) The failure of field sanitation to reduce infestation by the mango nut weevil, *Cryptorhynchus mangiferae* (F.) (Coleoptera: Curculionidae), in Hawaii. *Tropical Pest Management* 36, 359–361.

Hare, D.J. (1994) Sampling arthropod pests in citrus. In: Pedigo, L. and Buntin, D. (eds) *Handbook of Sampling Methods for Arthropods in Agriculture*. CRC Press, Boca Raton, Florida, pp. 417–432.

Helle, W. and Sabelis, M.W. (1985a) *Spidermites, Their Biology, Natural Enemies and Control*, Vol. 1A. Elsevier, Amsterdam, 405 pp.

Helle, W. and Sabelis, M.W. (1985b) *Spidermites, Their Biology, Natural Enemies and Control*, Vol. 1B. Elsevier, Amsterdam, 458 pp.

Hendrichs, J. (1996) Action programs against fruit flies of economic importance: session overview. In: McPheron, B. and Steck, G. (eds) *Fruit Fly Pests: a World Assessment of Their Biology and Management*. St Lucie Press, Delray Beach, Florida, pp. 513–519.

Hill, D. (1975) *Agriculture Pests of the Tropics and Their Control*. Cambridge University Press, New York, 516 pp.

Hoyt, S.C. and Burts, E.C. (1974) Integrated control of fruit pests. *Annual Review of Entomology* 19, 231–252.

Jeppson, L.R., Keifer, H. and Baker, E. (1975) *Mites Injurious to Economic Plants*. University of California Press, Berkeley, 614 pp.

Keenan, M. (1997) Foreword. In: Smith, D., Beattie, G. and Broadley, R. (eds) *Citrus Pests and Their Natural Enemies*. Department of Primary Industries, QI97030, Brisbane, Australia, 272 pp.

Klassen, W. (2000) Area-wide approaches to insect pest management: history and lessons. In: Tan, K.H. (ed.) *Proceedings of International Conference in Area-wide Control of Insect Pests*. 28 May–2 June 1998, Penang, Malaysia, Penerbit Universiti Sains Malaysia, pp. 21–38.

Lamb, K.P. (1974) *Economic Entomology in the Tropics*. Academic Press, New York, 1995 pp.

Lindquist, D.A. (2000) Pest management strategies: area-wide and conventional. In: Tan, K.H. (ed.) *Proceedings of International Conference in Area-wide Control of Insect Pests*. 28 May–2 June 1998, Penang, Malaysia, Penerbit Universiti Sains Malaysia, pp. 13–19.

Malavasi, A., Rohwer, G. and Campbell, D.S. (1994). Fruit fly free areas: strategies to develop them. In: Calkins, C.O., Klassen, W. and Liedo, P. (eds) *Fruit Flies and the Sterile Insect Technique*. CRC Press, Boca Raton, Florida, pp. 165–180.

Mariu, D. (1999) *Integrated Pest Management of Tropical Perennial Crops*. CIRAD, France, and Science Publishers, Enfield, New Hampshire, 167 pp.

Parker, B.L., Skinner, M. and Lewis, T. (1995) *Thrips Biology and Management*. Plenum Press, New York, 636 pp.

Pedigo, L. (1994) Introduction to sampling arthropod populations. In: Pedigo, L. and Buntin, D. (eds) *Handbook of Sampling Methods for Arthropods in Agriculture*. CRS Press, Boca Raton, Florida, pp. 1–14.

Pimentel, D. (ed.) (1981) *CRC Handbook of Pest Management in Agriculture*, Vol. I. CRC Press, Boca Raton, Florida, 597 pp.

Real, L. (1983) Introduction. In: Real, L. (ed.) *Pollination Biology*. Academic Press, New York, pp. 1–5.

Rosen, D. (1990a) *Armored Scale Insects: Their Biology, Natural Enemies and Control*, Vol. 1A. Elsevier, Amsterdam, 384 pp.

Rosen, D. (1990b) *Armored Scale Insects: Their Biology, Natural Enemies and Control*, Vol. 1B. Elsevier, Amsterdam, 687 pp.

Smith, D., Beattie, G. and Broadley, R. (1997) *Citrus Pests and Their Natural Enemies: Integrated Pest Management in Australia*. Queensland, Department of Primary Industries, Brisbane, Australia, 272 pp.

Swaine, G., Ironside, D.A. and Yarrow, W. H. (1985) *Insect Pests of Tropical Fruit and Vegetables*. Queensland Department of Primary Industries, Information Series QI83021, Brisbane, Australia, 350 pp.

Tandon, P.L. (1998) Management of insect pests in tropical fruit crops. In: Akora, R.K. and Ramanatha Rao, V. (eds) *Tropical Fruits in Asia, Diversity, Maintenance, Conservation and Use*. Proceedings of the IPGRI-ICAR-UTANET Regional Training Courses, 18–31 May 1997, Bangalore, India, pp. 235–245.

Valerio, C. (1984) Insect visitors to the inflorescence of the aroid *Diffenbachia oerstedii* (Araceae) in Costa Rica. *Brenesia* 22, 139–146.

Verheij, E.W. and Coronel, R.E. (1992) *Plant Resources of South-east Asia. Edible Fruits and Nuts*. Prosea Foundation, Bogor, Indonesia, 446 pp.

White, F. and Elson-Harris, M. (1994) *Fruit Flies of Economic Significance: Their Identification and Bionomics*. CAB International, Wallingford, UK, 601 pp.

Young, A. (1982) *Population Biology of Tropical Insects*. Plenum Press, New York, 511 pp.

2 Pests of Banana

C.S. Gold,[1] B. Pinese[2] and J.E. Peña[3]
[1]International Institute of Tropical Agriculture, ESARC, PO Box 7878, Kampala, Uganda; [2]Queensland Department of Primary Industries, PO Box 1054, Mareeba, Queensland 4880, Australia; [3]Tropical Research and Education Center, University of Florida, 18905 SW 280 Street, Homestead, Florida 33031, USA

Bananas and plantains (*Musa* spp.) are among the most important crops in tropical and subtropical climates. The genus *Musa* evolved in South-East Asia (Stover and Simmonds, 1987) where numerous undomesticated *Musa* species still grow as 'opportunistic weeds' (Price, 1995). Edible bananas (*Musa* spp., Eumusa series) originated within this region from two wild progenitors, *Musa acuminata* and *M. balbisiana*, producing a series of diploids, triploids and tetraploids through natural hybridization. Additionally, man has selected for parthenocarpy (development of fruit without pollination or seeds). Simmonds and Shepherd (1955) provided a key by which these naturally hybridized bananas may be divided into six genome groups (i.e. AA, AAA, AAB, AB, ABB, ABBB) based on the relative contributions of *M. acuminata* and *M. balbisiana*. Domesticated bananas include a wide range of dessert, cooking and brewing cultivars (Stover and Simmonds, 1987). The most extensively grown bananas are triploids.

Plantains (AAB) evolved in southern India with a secondary centre of *Musa* diversity in West Africa. East Africa has also evolved as a secondary centre of *Musa* diversity with numerous, locally evolved highland cooking and brewing clones (denoted AAA-EA). Cooking banana is the primary staple crop in the Great Lakes region of eastern Africa, while plantain is an important food in western Africa, Latin America and the Caribbean. Dessert bananas, especially Cavendish clones (AAA), are important export crops in Latin America and the Caribbean, Africa, Asia and the Pacific, and Australia.

In terms of gross value of production, bananas and plantains are the fourth most important global food crop. World production is calculated to be 87,934,558 t (INIBAP, 2001). Export bananas are the developing world's fourth most important commodity and most important fruit crop (INIBAP, 2001). Bananas grown for export are almost exclusively of one variety, 'Cavendish'; this cultivar accounts for little more than 10% of global production. Most banana production consists of a wide range of locally adapted clones that are consumed within the region (INIBAP, 2001).

Banana production systems range from low input kitchen garden and small-scale subsistence stands to large-scale, high input export banana plantations. Some commercial plantations of dessert bananas or plantains are replanted annually. By contrast, well-managed banana stands have produced stable yields for 30–100 years, even under low input conditions. At the same time, an extended harvest period ensures resource-poor farmers with food and income throughout the year. However, insect pests can cause reduced yields and shortened plantation life (Rukazambuga *et al.*, 1998; Gold *et al.*, 1999d).

©CAB *International* 2002. *Tropical Fruit Pests and Pollinators* (eds J.E. Peña, J.L. Sharp and M. Wysoki)

Phenology

Bananas are rhizomatous herbaceous plants ranging in height from 0.8 to 15 m (Turner, 1994). A *mat* (= banana stool) consists of an underground *corm* (rhizome) from which one or more plants (shoots) emerge. Adventitious roots spread extensively 4–5 m from the parent and down to 75 cm or more (Nakasone and Paull, 1998); however, most roots are near the soil surface. Plants represent a single shoot (pseudostem, stem, leaves, flower and bunch). Yield is normally expressed in kg/area/year reflecting both number and size of bunches harvested. The banana shoot consists of a *pseudostem* bearing the leaves and a true *stem* bearing the flower and fruit (bunch). The apparent stem or pseudostem is actually composed of leaf sheaths. New leaves emerge from the centre of the pseudostem. A single plant produces 25–50 leaves in its lifetime and normally supports 10–15 functional leaves at any one time (Nakasone and Paull, 1998). The true stem arises from the apical meristem after leaf production has terminated and grows through the centre of the pseudostem (Stover and Simmonds, 1987; Turner, 1994; Karamura and Karamura, 1995). One terminal inflorescence emerges from the true stem and bends downward after extrusion. After the fruit matures, the stem dies back to the corm. Farmers normally cut harvested plants between ground level and 1 m. Crop residues may be used for weevil traps (Gold, 1998).

New plants (*ratoons*) are produced by suckers emerging from lateral buds in the corm. These can be left *in situ* or serve as a source of planting material, in which case they are removed and planted elsewhere. Normally, a banana mat consists of three or more plant generations (= ratoons or crop cycles) at any one time. Plant density is controlled by the farmer through desuckering. As banana stands age, mats 'divide' and the relationship between plants (e.g. sharing of a common corm) becomes more tenuous; thus, in older stands, mat definition becomes unclear. Suckers used to establish new fields are called the 'mother plant or plant crop' (Stover and Simmonds, 1987; Turner, 1994).

Pests

Pest and disease pressures have increased considerably in recent years, and yield loss due to pests and/or disease attack is now considered the most significant factor affecting banana and plantain production worldwide (INIBAP, 2001). The status of specific pests and diseases reflects, in part, the banana clone and the management system. At a global level, diseases are considered the major threat, followed by nematodes and insects and mites. Nakasone and Paull (1998) suggest that the continuous nature of banana production makes pests such as nematodes more important than insects and mites; however, Pinese and Piper (1994) argue that management of insect pests is a key operation in banana production in Australia. Gold and Gemmill (1993) reported that pest problems in banana and plantain significantly reduce yields in all tropical regions, and particularly in Africa.

Detailed information on banana pests has been presented by authors in different parts of the world (Ostmark, 1974; Gold and Gemmill, 1993; Pinese and Piper, 1994; Fancelli and Martins, 1998). Integrated pest management tactics have been developed for several pests; in general, chemicals are widely used in some systems, although these are beyond the means of many growers. Environmental degradation and the development of insect resistance have highlighted the limitations of chemically based control strategies and the need for integrated pest management, including host plant resistance, cultural control and biological control.

In this chapter we will address management of the most common arthropod pests of banana in tropical areas.

Pests of Rhizome and Pseudostem

Banana weevil, *Cosmopolites sordidus* (Germar)

The banana weevil, *Cosmopolites sordidus* (Germar), is an important pest of banana, plantain and ensete (Plate 1). Damage to the banana plant is entirely the result of larval feeding. Weevil attack can prevent crop establishment, cause significant yield reductions in ratoon cycles and contribute to shortened plantation life. For example, the weevil has been implicated as a primary factor contributing to the decline and disappearance of East African highland cooking banana (*Musa* spp., genome group AAA-EA) from its traditional growing areas in central Uganda (Gold *et al.*, 1999b) and western Tanzania (Mbwana and Rukazambuga, 1999). Banana weevil pest status may reflect ecological conditions, banana types, cultivar selection, and management systems.

Taxonomy, morphology, distribution and host range

The banana weevil was first identified by Germar in 1824 from specimens collected in Java and given the name *Calandra sordida*. In 1885, Chevrolat changed this to its currently recognized name *Cosmopolites sordidus* (Germar). The genus *Cosmopolites* belongs to the subfamily Rhynchophorinae of the family Curculionidae (weevils and snout beetles). Taxonomic keys are presented by Zimmerman (1968a); adult morphology has been described by Moznette (1920), Beccari (1967), Zimmerman (1968b), Viswanath (1976) and Nahif *et al.* (1994); reproductive system morphology by Cuille (1950), Beccari (1967), Uzakah (1995) and Nahif (1998); and larval morphology by Moznette (1920) and Viswanath (1976). Males can be separated from females by curvature and punctuation of the rostrum and by curvature of the last abdominal sternite (Roth and Willis, 1965; Longoria, 1968). The sex ratio is 1 : 1.

The banana weevil originated within the Indo-Malayan region (Simmonds, 1966; Zimmerman, 1968b; Waterhouse, 1993), coincident with the area of origin of bananas (Stover and Simmonds, 1987). Spread of the weevil is presumed to have been through the movement of infested planting material. The weevil is currently found throughout Asia, Oceania, Australia, sub-Saharan Africa and the Americas (Cuille and Vilardebo, 1963). It is unclear how long the weevil has been present in most areas.

The banana weevil is an oligophagous pest with a narrow host range, attacking wild and cultivated clones in the genera *Musa* (banana, plantain, abaca) and *Ensete*. Reports of alternative hosts appear to be in error. Nevertheless, L. Traore (personal communication) was able to maintain larvae through several instars using a factitious host (processed *Xanthasoma saggittifolium*), while Pavis (1988) and Schmitt (1993) had modest success rearing larvae on artificial diets.

Banana weevil biology

ADULTS

Longevity, tropisms, and distribution. The banana weevil displays a classical 'K' selected life cycle (Pianka, 1970) with long life span and low fecundity. Adults may live up to 2 years (Froggatt, 1925; Treverrow *et al.*, 1992), although mean longevity under field conditions is not known. The adult is nocturnally active and characterized by negative phototropism, strong hygrotrophism, thigmotactism, gregariousness and death mimicry (Delattre, 1980; Ittyeipe, 1986; Tsai, 1986). The adults favour moist environments and are closely associated with banana mats (being primarily in the leaf sheaths, around the base of the mat or, occasionally, in larval galleries) or with detached residues (Gold *et al.*, 1999d) (Plate 2). The weevils feed on rotting banana tissue and are not pests (Budenberg *et al.*, 1993b). Under field conditions, the weevil can survive for 3 to 6 months once all banana material is removed (Froggatt, 1924; Peasley and Treverrow, 1986; Allen, 1989; Mestre and Rhino, 1997).

Daily and seasonal activity periods. Banana weevil adults are active between 1800 and 0600 h (Cuille, 1950; Uzakah, 1995) with greatest activity between 2100 and 0400 h (Uzakah, 1995). A substantial proportion of the population may be inactive for extended periods of time (S. Lux, personal communication). The weevils are sedentary during daylight hours.

Seasonal differences in trap captures of adult weevils have been reported by many authors. Trap captures may reflect activity patterns, but do not provide meaningful estimates of population density (Vilardebo, 1973) which require mark and recapture methods (Price, 1993; Gold and Bagabe, 1997). Higher trap captures have been found in dry seasons (Yaringano and van der Meer, 1975), rainy seasons (Cuille, 1950; C.S. Gold *et al.*, unpublished), or unrelated to climatic factors (Oliveira *et al.*, 1976; Delattre, 1980; Pavis, 1988). These conflicting results provide an unclear picture of when adults are most active and, hence, most vulnerable to control interventions.

Dispersal and movement. Dispersal of the weevil may be by walking, occasional flight and dissemination of infested planting material. The maximum observed distance moved by walking was 35 m in 3 days (Gold and Bagabe, 1997) and 60 m in 5 months (Delattre, 1980). Gold *et al.* (1999d) found only a small percentage of weevils moved > 25 m in 6 months. This suggests that dispersal by walking may contribute to invasion of neighbouring fields but not much beyond.

The nocturnal habit of the weevil has largely precluded direct observations on weevil flight under field conditions. Although the banana weevil has functional wings, most observers believe the weevil rarely flies (e.g. Gordon and Ordish, 1966; Pinese and Piper, 1994; Sponagel *et al.*, 1995). In contrast, Cuille and Vilardebo (1963) reported the weevil to be a good flyer and suggested that this may explain wider dissemination than a few kilometres.

The banana weevil's narrow host range and limited dispersal capability mitigate against immigration of adults into isolated or newly planted banana stands by walking or flying (Gold *et al.*, 1999d). It has been widely recognized that dispersal of banana weevil is primarily through infested planting material (e.g. Froggatt, 1925; Pinese and Piper, 1994). Banana suckers may contain adults in the leaf sheaths or immature stages within the rhizome. This suggests that the use of clean planting material is an important factor in establishing healthy banana stands.

Semiochemicals. Olfactory cues may be utilized by banana weevils in locating host plants, conspecifics and/or mates. Cuille (1950) was the first to propose and test 'chemotropisms' of banana weevil. He concluded that olfactory cues were most important in host location, while chemoreception played the predominant role in host acceptance. Cuille concluded that chemoreception is probably more important than olfaction for a sedentary insect like banana weevil. Further work by Budenberg and Ndiege (1993) and Budenberg *et al.* (1993b) found that males and females were both attracted to plant volatiles (i.e. kairomones); they suggested the weevils used these volatiles for attraction to food sources rather than oviposition sites. Budenberg *et al.* (1993a) also demonstrated the presence of an aggregation pheromone (sordidin) produced in the male hindgut and attractive to both sexes. It is likely that weevils are attracted to their host plants by plant volatiles and that the aggregation pheromone then serves as an arrestant from further movement.

The use of pheromones and attractive plant volatiles (i.e. kairomones) as a means of controlling banana weevil through mass trapping and/or in baits for delivery of *B. bassiana* was first proposed by Budenberg *et al.* (1993a) and Kaaya *et al.* (1993). Jayaraman *et al.* (1997) also suggested that semiochemical-enhanced mass trapping could overcome the low reproductive capacity of the insect and lead to successful control. Chemtica International SA in Costa Rica has begun commercial production of banana weevil pheromones (enhanced with kairomones) in a product called Cosmolure+ (C. Oehlschlager, personal communication).

Oviposition. Female weevils are sexually immature upon emergence with first oviposition occurring 4–6 weeks later (Uzakah, 1995). Mean oviposition rates of 1–11 eggs week^{-1} have been recorded in the laboratory (Cuille, 1950; Delattre, 1980; Koppenhofer, 1993a; Abera, 1997; Gold, Kagezi and Nemeye, unpublished) and 0.5–1.2 eggs week^{-1} in the field (Abera, 1997). Oviposition is believed to be greatly reduced in dry seasons (Cuille, 1950; Nonveiller, 1965).

Eggs (0.5 × 2 mm) are deposited singly in the base of the host plant (i.e. leaf sheaths, leaf scars, rhizome) in orifices (1–2 mm deep) excavated by the female weevil with her rostrum. Oviposition occurs on plants of any age and on crop residues (Abera *et al*., 1999). The weevils have been variously reported to prefer young plants (Cuille, 1950), preflowered plants (Treverrow and Maddox, 1993), flowered plants (Vilardebo, 1973; Mesquita and Caldas, 1986; Abera *et al*., 1999) and crop residues (Treverrow and Maddox, 1993; Gold and Bagabe, 1997; Abera *et al*., 1999).

IMMATURES Banana weevil developmental rates under ambient temperatures (reviewed by Schmitt, 1993; Traore *et al*., 1993) show wide variability in stage duration: 3–36 days for eggs, 12–165 days for larvae (Plate 3), 1–4 days for prepupae, 4–30 days for pupae and 24–220 days from egg to adult. The longest stage durations were found in Australia where seasons are pronounced and the range in weevil development times large; under these conditions, development rates were up to four times as long as recorded anywhere else. While temperature is certainly the most critical factor in determining developmental rates, relative humidity, cultivar, age of plant, food quality and population density may also be involved (Mesquita *et al*., 1984; Schmitt, 1993).

Traore *et al*. (1993) determined a developmental threshold for the egg stage of 12°C and a thermal requirement of 89 degree-days. Under tropical conditions, the egg stage is commonly 7–8 days. Field eclosion rates probably exceed 90% in the field. Koppenhofer and Seshu Reddy (1994) found lower hatchability for eggs in pseudostems, possibly due to higher water content or metabolites. Kiggundu (2000) suggested that viscosity and metabolites of plant sap in resistant clones may also reduce egg success.

Upon hatching, first instar larvae bore directly into the plant. The larvae feed throughout the rhizome and will also enter the true stem (i.e. after flowering). In severe attacks, the larvae may move from the mother plant into young suckers (Vilardebo, 1960; Champion, 1975). The number of instars has been variously reported to range from five to eight (reviewed by Gold *et al*., 1999c). Traore *et al*. (1996) found a developmental threshold for the larval stage of 8.8°C with a total thermal requirement of 538 degree-days. The non-feeding prepupal stage is 3–4 days (Mesquita *et al*., 1984; Gold *et al*., 1999c). Pupation is in a bare chamber excavated by the larvae near the rhizome surface of the host plant (Vilardebo, 1960). The developmental threshold for the pupal stage is 10.1°C with a thermal requirement of 121 degree-days (Traore *et al*., 1996). Under tropical conditions, the larval stage probably lasts 4–6 weeks, while the pupal stage is 1 week.

Sampling methods for estimating adult population density and larval damage

The most common sampling methods for banana weevil include trapping of adults and estimates of larval damage in recently harvested plants. Trapping as a means of assessing banana weevil population levels has been employed since 1912 (Knowles and Jepson, 1912) and continues to be widely used. A variety of traps (described by Castrillon, 1989) are made out of crop residues (i.e. recently harvested rhizomes and pseudostems). Those using rhizome material are generally more attractive to banana weevils than those made from pseudostems. Trap quality and climatic factors can influence trap catches, making interpretation of results difficult (Vilardebo, 1973; Bakyalire, 1992).

Weevil populations can be more accurately estimated using standard mark and recapture methods including trapping, marking, releasing and subsequent retrapping (Delattre, 1980; Price, 1993; Gold and Bagabe, 1997; Gold *et al*., 1997). In Cameroon, Delattre (1980) estimated weevil densities in two fields to be 2600 ha^{-1} and 15,600 ha^{-1}, respectively. In a single watershed in Ntungamo district, Uganda, Gold *et al*. (1997) found weevil densities to range from 1600 to 149,000 ha^{-1} with a median population of 9300 ha^{-1} (= 15 per mat). This within-site variability suggests that management plays an important role in regulating weevil populations.

Unfortunately, estimates of weevil populations are only poorly related to damage levels (Shillingford, 1988; Gowen, 1995). For example, Gold *et al*. (1997 and unpublished) found a weak relationship between estimates of weevil adult density and damage ($r = 0.11$). This suggests that, all other things being equal, control methods targeting adult weevils may be less effective than those against the immature stages, and may also require a considerable lag time before population reductions are translated into reduced damage and increased yields.

Vilardebo (1960) proposed that visual observations on larval damage are a better indicator of weevil pest status than trap counts and recommended counting the number of galleries exposed on the rhizome periphery. Vilardebo (1973) then proposed a coefficient of infestation (CI) in which the rhizome is pared below the collar to a depth of 1–2 cm and the proportion of rhizome surface (i.e. 0–100%) with weevil galleries is estimated. Although the CI has been widely used, the scoring system is not clearly defined and remains highly subjective. To standardize the scoring system, Mitchell (1978) developed a percentage coefficient of infestation (PCI) in which the upper rhizome is pared and presence/absence of peripheral weevil damage is recorded for each of ten sections, each covering 18° of the rhizome surface.

Internal damage within the rhizome probably has a greater impact on banana growth and bunch filling than damage to the rhizome periphery. Damage to the central cylinder is likely to affect nutrient transport and stem growth while damage to the cortex may adversely affect root development and lead to snapping and toppling (Taylor, 1991; Gold *et al.*, 1994b). Moreover, the ability of the weevil to penetrate the rhizome may be cultivar related; as such, the CI and PCI serve as poor estimates of internal rhizome damage (Ogenga-Latigo and Bakyalire, 1993; Gold *et al.*, 1994b). Therefore, Gold *et al.* (1994a,b) proposed a scoring method in which two cross-sections were made in the rhizome (at the collar and 10 cm below the collar). In each section they estimated the percentage of surface area taken up in galleries in the central cylinder and in the outer cortex. Rukazambuga (1996) reported a stronger relationship between damage to the central cylinder and yield, than with damage to the cortex and rhizome periphery.

Pest status

The pest status of banana weevil remains controversial, with yield loss estimates ranging from 0 to 100% (e.g. Sengooba, 1986; Sponagel *et al.*, 1995; Mestre and Rhino, 1997) (Table 2.1). It is often unclear how many yield loss estimates were derived and whether they purport to represent single fields or regional estimates. Moreover, some of these studies failed to partition damage effects of weevils and nematodes and yield loss estimates are, thus, confounded. Farmers are well aware that weevil damage is more important in older stands and on-station trials have shown increasing yield losses over time (Rukazambuga, 1996). Thus, single cycle yield loss trials may be misleading. In addition, the weevil's importance may be influenced by banana clone and management system. Weevil damage levels are likely to be very different on Cavendish bananas (AAA) grown in commercial plantations than on highland cooking bananas (AAA-EA) or plantains (AAB) grown in smallholdings. Pest status within genome groups is also in dispute. For example, recommended action thresholds on Cavendish banana vary from two weevils per trap in Brazil (Moreira, 1979) to 15–25 weevils per trap in Central America (Anonymous, 1989; Sponagel *et al.*, 1995).

Ecological factors and management practices may also influence yield loss. The weevil is absent (Lescot, 1988) or in low numbers (Castrillon, 2000; Gold and Okech, unpublished) above 1600 m above sea level. As a result, it is unimportant in much of the *Ensete* growing regions of Ethiopia (M. Bogale *et al.*, unpublished) and part of the highland banana growing region of eastern Africa. In Ghana, low levels of weevil damage in plantain may result from the short crop cycle (one or two ratoons) before replanting (Schill, 1996).

Yield losses to banana weevil have been associated with sucker mortality, plant loss, reduced bunch weights and shorter stand life. Newly planted stands in or near previously infested fields may suffer high levels of plant loss (Ambrose, 1984; Price, 1994) as a sucker can be killed by a single larva if it attacks the growing point (C. Gold, personal observation). Moreover, ovipositing weevils are

Table 2.1. Reported yield losses to banana weevil.

Continent/country	Yield loss	Clone	Methods	Reference
Latin America				
Brazil	30			Moreira, 1971
	20–50			Gallo *et al.*, 1978
	Abandoned		Observ.	Arleu and Neto, 1984
Colombia	up to 80[a]			Marcelino and Quintero, 1991
Cuba	22–34		Trials	Reinecke, 1976
	> 19		Trials	Calderon *et al.*, 1991
	20		Trials	Masso and Neyra, 1997
Ecuador	20–40			Champion, 1975
Honduras	8–26			Roberts, 1955
	0		Trials	Sponagel *et al.*, 1995
Peru	48[a]			Liceras *et al.*, 1973
Puerto Rica	30–70[a]			Ingles and Rodriguez, 1989
	90[b]		Trials	Roman *et al.*, 1983
Africa				
Cameroon	20–90			Lescot, 1988
Congo	up to 90			Ghesquiere, 1925
Ghana	25–90	AAB		Gorenz, 1963
Kenya	24–90	AAA-EA	Trials	ICIPE, 1991
	16–53%	AAA-EA	Trials	Ngode, 1998
	up to 100%	AAA-EA		Koppenhofer, 1993
	22–76%	AAA-EA	Trials	Musabyiamana, 1999
Tanzania				
Kagera	30	AAA-EA		Walker *et al.*, 1983
	30[ab]	AAA-EA		Sikora *et al.*, 1983
	15		Trials	Uronu, 1992
Uganda	5–44	AAA-EA	Trials	Rukazambuga *et al.*, 1998
	40–50[b]	AAA-EA	Trials	Gold *et al.*, 1998
Central	20–60	AAA-EA	Damage[c]	Gold *et al.*, 1999
Masaka	up to 100[a]	AAA-EA	Observ.	Sengooba, 1986
Rakai	up to 100[a]	AAA-EA	Observ.	Sengooba, 1986
	> 50%	AAA-EA	Observ.	Sebasigari and Stover, 1988
West Africa	35–40%	AAB		Sery, 1988
Asia/Pacific				
India	35[b]		Trials	Job *et al.*, 1986
Tonga	< 10		Damage[b]	Crooker, 1979
	30–60		Damage[b]	Englberger and Toupu, 1983
	up to 80[a]		Observ.	Pone, 1994

[a]Plant loss due to toppling and snapping attributable to weevils.
[b]Composite weevil and nematode.
[c]Estimated from Rukazambuga *et al.* (1998).

attracted to cut rhizomes, making newly detached suckers especially susceptible to attack. Toppling is often attributed to plant parasitic nematodes that attack the root system, thereby reducing anchorage (Gowen, 1995). However, it appears likely that weevil damage reduces root number and can also contribute to toppling. For example, in Ugandan field trials, Rukazambuga (1996) found extensive toppling in mats with high levels of weevil attack and low levels of nematode damage. Snapping (i.e. breaking of the rhizome) may also occur on plants with severe weevil damage (Gowen, 1995). In central Uganda, banana weevils have contributed to the decline in stand life from > 30 years to less than 4 years (Gold *et al.*, 1999b).

In a yield loss trial on highland banana in Uganda, Rukazambuga *et al.* (1998) related levels of weevil damage in the central cylinder (cf. Gold *et al.*, 1994b) to plant growth, maturation rates and yield over four crop cycles. (Nematode damage was low (root necrosis index < 3%) suggesting that effects were attributable only to weevils.) Damage rose from 4% in the plant crop to 17% in the third ratoon (N.D.T.M. Rukazambuga, unpublished), while yield loss to weevils increased from 5% in the plant crop to 44% in the third ratoon. In the third ratoon, plant loss and reduced bunch size were equally responsible for lower yields.

Weevil pressure has been widely believed to be associated with 'poor' (i.e. low levels of) management, stressed plants, bad drainage, acid or low fertility soils, weedy fields, inadequate sanitation, extended droughts and nematode infestations (e.g. Froggatt, 1925; Ostmark, 1974; Speijer *et al.*, 1993; Gowen, 1995; Sponagel *et al.*, 1995). In Uganda, higher levels of weevils have been attributed to reduced crop sanitation and other management practices (Gold *et al.*, 1999b).

In contrast, Rukazambuga (1996) found similar percentages of yield loss (27%) in stressed (i.e. intercropped with finger millet) and vigorous (i.e. mulched monoculture) banana. This translated into a 2.5 t ha^{-1} loss in the intercrop and a loss of 6.3 t ha^{-1} in the mulch. These data suggest that banana weevil can be an important constraint in well-managed banana stands.

Integrated pest management of banana weevil

Current research results suggest that no single control strategy will be likely to provide complete control for banana weevil. Therefore, a broad integrated pest management (IPM) approach might provide the best chance for success in controlling this pest. The components of such a programme include habitat management (cultural control), biological control, host plant resistance and (in some cases) chemical control.

Habitat management (cultural control)

Habitat management can reduce herbivore levels by creating an environment that reduces pest movement, promotes host plant tolerance of pest attack and/or is unfavourable to pest build-up. For banana weevil, habitat management includes the use of clean planting material, management of crop residues (i.e. sanitation) and trapping.

CLEAN PLANTING MATERIAL IN CLEAN FIELDS The use of clean planting material can reduce initial banana weevil infestations and retard pest build-up for several crop cycles. At the same time, it can protect new banana stands against nematodes and some diseases. Suckers used as planting propagules often contain weevil eggs, larvae and, occasionally, adults. This provides the principal entry point of banana weevils into newly planted fields. Removing these weevils from planting material eliminates the most important source of infestation in new plantations. The insect's low fecundity and slow population growth further suggest that a reduction in initial infestation level will impede population build-up and provide extended protection to newly planted fields. As a result, the use of clean planting has been widely recognized and promoted.

A number of methods have been proposed for freeing planting materials from weevils. These include tissue culture, paring and hot-water treatment. With all methods, reinfestation remains a critical concern. Froggatt (1925) advocated the use of clean planting material and recommended against planting near infested fields or in previously infested sites. Previously infested fields can be rid of weevils by crop rotation or fallowing.

The use of tissue culture plantlets for banana weevil control has been recommended by Peasley and Treverrow (1986). Unlike other methods, tissue culture plants are likely to be 100% free of banana weevils and nematodes at the time of planting. However, tissue culture plantlets are not universally available or affordable (Seshu Reddy *et al.*, 1998).

Paring, or removal of the outer surface of the rhizome, has also been widely recommended (e.g. Froggatt, 1925; Sein, 1934; Seshu Reddy *et al.*, 1998). Paring can expose weevil galleries and allow the farmer to reject heavily damaged suckers. Removal of all of the leaf sheaths and paring of the entire rhizome will eliminate most weevil eggs and first instar larvae. Many later instar larvae are likely to be deeper within the rhizome and not removed by paring.

Hot water treatment to kill weevil eggs and larvae was first recommended in the 1920s (Ghesquiere, 1924) and continues to be promoted (Seshu Reddy *et al.*, 1993, 1998). Ordinarily, the rhizomes are pared and then completely submerged in hot water. Sein (1934) reported that placing suckers in boiling water for 1 min killed all weevil eggs and surface larvae. The use of some hot water treatment regimes (e.g. 52°C for 27 min or 54°C for 20 min) is also a highly effective control against banana nematodes (Seshu Reddy *et al.*, 1993, Speijer *et al.*, 1995). These temperatures have been suggested for concurrent management of weevils and nematodes (Seshu Reddy *et al.*, 1993, 1998). However, Gold *et al.* (1998a) found less than 33% mortality of banana weevil larvae for similar temperature regimes. Larval survival was greater in the central cylinder than in the cortex. Arroyave (1985) also reported that hot water baths are not effective at killing larvae deep within the rhizome.

Gold *et al.* (1998b) found lower weevil numbers for 11–27 months in plots planted from (i) pared or (ii) pared and hot water treated rhizomes than in plots planted with untreated suckers (controls). Weevil damage levels in controls were 70–200% higher than in plots grown from treated planting material for the plant crop. However, all treatments displayed similar levels of weevil damage in the first ratoon. Hot water treatment had little advantage over paring for controlling weevil but afforded excellent nematode control for the duration of the trial.

However, paring to remove weevil eggs and expose larval damage has not been widely adopted by farmers in East Africa. Many farmers believe that suckers will not perform well following removal of most or all of the root system. In Tanzania, for example, Taylor (1991) reported that farmers viewed the recommendation of rhizome paring with 'extreme disbelief'. Implementation of hot water baths for the control of banana weevils and nematodes requires investment in a hot water tank and a heating source (e.g. electricity, gas burner, wood). As a result, adoption by resource-poor farmers may be limited (c.f. Ssennyonga *et al.*, 1999).

CROPPING SYSTEMS AND CROP MANAGEMENT

The use of multiple cropping systems for the control of banana weevil may be limited. Mixed cropping systems often result in lower insect pressure by reducing immigration rates, interfering with host plant location and increasing emigration rates (Altieri and Letourneau, 1982; Risch *et al.*, 1983). However, banana weevils are sedentary insects that live in perennial systems in the presence of an abundant supply of hosts. Kehe (1985) found much lower incidence of weevil attack (CI = 6%) in plantains mixed with older coffee stands (i.e. > 5 years) than in those mixed with younger coffee plants (CI = 91%), with cacao (CI = 88%) or with annual crops (CI = 79%). He postulated that caffeine produced by older coffee plants served as an insecticide or feeding inhibitor. By contrast, Uronu (1992) tested a series of intercrops and failed to find a viable crop mixture that would both reduce weevil numbers and produce satisfactory banana yields. Gold and McIntyre (unpublished) found no effect of green manures with reported insecticidal properties (i.e. *Canavalia*, *Mucuna*, *Tephrosia*) either on weevil adult numbers or on damage.

Weeding, removal of trash from the base of the mat, deleafing, and desuckering have all been reported as means of eliminating shelters for weevils or making the environment at the base of the mat less favourable (Wallace, 1938; Seshu Reddy *et al.*, 1998). However, few data are available to demonstrate possible effects of these practices on weevil levels. Recent work has demonstrated that grass mulches may increase weevil damage by creating a more favourable environment (i.e. cool, moist conditions) for adult weevils (Price, 1994; Rukazambuga, 1996; Braimah, 1997). In Tanzania and Uganda, some farmers mulch away from the base of the mat as a means of reducing weevil infestations (Varela, 1993; Ssennyonga *et al.*, 1999).

Deep planting and earthing up have been recommended to render the rhizome inaccessible to ovipositing females and prevent high mat. Seshu Reddy *et al.* (1993) planted cooking bananas at depths of 15, 30, 45, and 60 cm in drums and reported that shallow planted suckers were more prone to attack, although some weevils were able to find the deepest planted suckers. However, Abera (1997) showed that weevils freely oviposit in leaf sheaths, while M. Masanza (unpublished) found greater levels of subterranean oviposition during dry seasons.

CROP SANITATION Following harvest, crop residues may serve as shelters for adults (Gold *et al.*, 1999d) and oviposition sites for females. For example, Gold *et al.* (1999d) found > 35% of adult weevils to be associated with crop residues. Banana weevils readily oviposit on residues for extended periods after harvest (Abera, 1997; M. Masanza, unpublished). For some clones, attack on residues may be more extensive than that against growing plants (e.g. Gros Michel in Ecuador; Cavendish in Australia and Latin America, Kisubi in Uganda (Vilardebo, 1960; Treverrow and Bedding, 1993; Gold and Bagabe, 1997; H.E. Ostmark, personal communication).

Crop sanitation has been widely recommended to eliminate weevil refuges and breeding sites (Ghesquiere, 1924; Hargreaves, 1940; Waterhouse and Norris, 1987; Smith, 1995). Methods include cutting residues at or below the soil surface and chopping or splitting old rhizomes and pseudostems. However, the value of sanitation as a means of weevil control has been disputed. For example, Peasley and Treverrow (1986) and Treverrow *et al.* (1992) suggest that crop hygiene (i.e. sanitation) is the long-term key to weevil control and that without it all other control measures are pointless. Nanne and Klink (1975) report that sanitation can drastically reduce weevil populations. Farmers in central Uganda felt that abandonment of sanitation practices was an important factor in increasing weevil problems on their farms (Gold *et al.*, 1999b). In contrast, Jones (1968) felt that sanitation required too much labour, while Gold (1998) suggested that it is possible that crop residues might act as 'trap crops' drawing gravid female weevils away from growing plants. Much of this debate is speculative, based on perceptions of weevil pest status, intuitive beliefs on population dynamics and on-farm observations. Unfortunately, there have been virtually no data from controlled studies on the role of crop sanitation in weevil population dynamics and related damage.

TRAPPING ADULT WEEVILS The use of trapping adults to control banana weevils has been controversial. Knowles and Jepson (1912) noted the attraction of banana weevils to crop residues (i.e. harvested rhizomes and pseudostems) and proposed trapping adults with these materials. Since then, trapping has been widely recommended as a banana weevil control by many workers (e.g. Sein, 1934; Hargreaves, 1940; Ndege *et al.*, 1995; Seshu Reddy *et al.*, 1995). Enhanced trapping with the addition of chemicals, biopesticides and/or semiochemicals has also been proposed (Veitch, 1929; Yaringano and van der Meer, 1975; Schmitt *et al.*, 1992; Budenberg *et al.*, 1993a; Kaaya *et al.*, 1993). For example, Jayaraman *et al.* (1997) and Alpizar *et al.* (1999) suggested that mass trapping with semiochemicals could overcome the weevil's low fecundity and slow population build-up, and lead to successful control. Similarly, Braimah (1997) suggested that the use of pseudostem traps, enhanced by semiochemicals and combined with other compatible control methods (e.g. entomopathogens), holds the key to banana weevil control. In contrast, Mestre (1997) concluded that the weevil is a poor candidate for mass trapping with semiochemicals because it is soil dwelling, sedentary and rarely flies.

The effect of trapping on weevil populations will, in part, reflect the intensity of trapping (i.e. frequency and density) and the types of materials used. Inclusion of rhizome material increases a trap's attractivity to weevils. Thus disc-on-stump traps tend to have higher weevil catches than pseudostem traps (see Castrillon, 1989, for description). However, pseudostem trapping is most often recommended for systematic trapping studies because a single harvested plant can only support one disc-on-stump trap (fixed in space), while the same plant can provide material for many pseudostem traps (placed where the farmer deems most useful. In addition, it is likely that trapping in established fields will result in a gradual decline in weevil numbers with a lag time required before effects are manifested in reduced damage.

Weevil reductions due to trapping have been reported by Masanza (1995), Ndege *et al.* (1995), Seshu Reddy *et al.* (1995), Vilardebo (1950), Yaringano and van der Meer (1975), Arleu and Neto (1984), Arleu *et al.* (1984), Koppenhofer *et al.* (1994), and Ngode (1998). For example, Yaringamo and van der Meer (1975) reported a 50% population reduction from 4 months of rhizome trapping, but the means by which this reduction was determined is not clear. Seshu Reddy *et al.* (1995) also found a 50% reduction in weevils captured following systematic trapping. Koppenhofer *et al.* (1994) implemented three trapping studies and found reductions of 33–67% over periods of time ranging from 7 weeks to 1 year.

In each of these studies, comparisons were made with initial populations and, thus, the trials lacked proper controls, making the results inconclusive. For example, reported weevil reductions in the Seshu Reddy *et al.* (1995) study and two Koppenhofer *et al.* (1994) trials were interpreted from trap capture rates, which may have reflected weather conditions and trap efficiency (Vilardebo, 1973). In a third trial, Koppenhofer *et al.* (1994) released a known number of weevils and then estimated populations using mark and recapture methods. However, weevil population declines of the same magnitude as that reported in Koppenhofer *et al.*'s (1994) third trial have been found for field populations of marked and released weevils in trials where trapping was not conducted (Rukazambuga, 1996; Gold and Night, unpublished).

Controlled studies to determine the efficacy of pseudostem trapping in reducing weevil populations were conducted under farmer conditions in Ntungamo district, Uganda. Twenty-seven farms were then stratified on the basis of weevil population density and divided among three treatments: (i) researcher-managed trapping (one trap/mat/month); (ii) farmer-managed trapping (trap intensity at discretion of farmer); and (iii) controls (no trapping). After one year, weevil populations had declined by 61% in researcher-managed fields, by 43% where farmers managed trapping and by 23% in controls (Gold *et al.*, 2002). Effects were highly variable among farms within treatments and there were no significant treatment effects. Moreover, there was only a weak relationship between the number of weevils removed and the change in population density. The results suggest that intensive trapping can, but does not always, reduce weevil numbers.

The use of enhanced trapping with semiochemicals could result in higher rates of weevil removal at lower trap densities and with reduced labour. Chemtica International SA, in Costa Rica, tested lures with male aggregation pheromones and found that a formulation, Cosmolure+ (containing a mixture of the four sordidin isomers plus plant volatiles) was most attractive to both male and female banana weevils (C. Oehlschlager, personal communication). Using interference studies by collecting weevils from pheromone-baited pitfall traps placed at different distances, Oehlschlager (personal communication) determined that the optimum spacing of traps was 20 m. Current recommendations are to place 4 traps ha^{-1}. These traps are replaced monthly and systematically moved through the field.

Alpizar *et al.* (1999) reported that pitfall traps with Cosmolure+ collected 12 times as many weevils as unbaited sandwich traps. Through interference studies, they estimated the effective radius of trap attractivity at 2.5–7.5 m. They then tested Cosmolure+ in three plantain fields and in one Grand Enano (AAA) stand using Chemtica recommendations on trap placement. In the plantain systems, weevil capture rates in treated plots remained at initial levels for 9 months and steadily decreased thereafter, while trap captures remained steady in control plots. Over 18 months, damage levels, measured by Vilardebo's (1973) coefficient of infestation, decreased from 15% to 12% in the treated plots, while increasing from 15% to 34% in controls. This resulted in a 25% yield gain for the first 18 months. Similar results were found in the Grand Enano field, where treated plots had less damage and a 32% yield advantage.

Biological control with arthropods

Biological control efforts against banana weevil have included the use of exotic natural enemies (classical biological control), endemic natural enemies, secondary host associations and microbial control (e.g. entomopathogens, endophytes, entomophagous nematodes). Microbial control agents may require repeated applications as biopesticides, although they lack the toxic side effects of chemical insecticides. As such, they may entail periodic application costs on the part of the farmer.

CLASSICAL BIOLOGICAL CONTROL Classical biological control of banana weevil may be possible. Introduced pests, unimportant in their native habitats, often reach damaging levels when released from the control of coevolved natural enemies. The banana weevil appears to fit this pattern. Therefore, exploration for banana weevil natural enemies in Asia followed by selection, quarantine and release of suitable species could establish an herbivore equilibrium below economic thresholds.

The first searches for natural enemies in Asia were undertaken by Muir in 1908 (Froggatt, 1925), Jepson (1914) and Froggatt (1928). They identified a histerid *Plaesius javanus* Erichson, the staphylinids *Belonuchius ferrugatus* Erichson and *Leptochirus unicolor* Lepeletier, a cucujid *Canthartus* sp. and a leptid (rhagionid) fly *Chrysophila ferruginosa* (Wied) as being predacious on the banana weevil and banana stem weevil (*Odoiporus longicollis* Oliv.). Later searches revealed the presence of other histerids (e.g. *Hololepta* spp.) and a hydrophilid (*Dactylosternum hydrophiloides* MacLeay). All of these are opportunistic predators that feed on a range of prey. The most important appeared to be *P. javanus* whose larvae and adults both attack banana weevil immatures. This predator is most commonly found in deteriorating banana residues and rarely enters weevil galleries in living plants.

Between 1913 and 1959, 45 attempts were made to introduce eight natural enemies from Asia to other banana growing regions in the world. These introductions tended to be done with small predator consignments. In most cases, the natural enemies either failed to establish following introduction or were ineffective (Waterhouse and Norris, 1987). Only in Fiji and Jamaica has there been any suggestion of even partial control (Waterhouse and Norris, 1987). Although these results are discouraging, Neuenschwander (1988) recommended additional searches that might focus on parasitoids (especially of the relatively vulnerable egg stage). Such natural enemies tend to more host specific and more effective biological control agents than opportunistic predators such as *P. javanus*.

An intensive search for natural enemies of banana weevil carried out in 2001 in Sumatra, Indonesia failed to reveal the presence of either egg or larval parasitoids (A. Aberu, personal communication).

ENDEMIC NATURAL ENEMIES Endemic natural enemies reported in Latin America, Africa and Asia (reviewed by Beccari, 1967; Schmitt, 1993; Koppenhofer, 1993b,c; Seshu Reddy *et al.*, 1998) include nabids, cydnids, capsids, reduviids, mirids, thrips, rhagionids, sarcophagids, histerids, carabids, hydrophilids, staphylinids, dermaptera, curculionids, scarabaeids, tenebrionids, and formicids. Very little information is available on the efficacy of these natural enemies and most appear to be of limited importance.

Koppenhofer *et al.* (1992) listed 12 predators of banana weevil in western Kenya. These included three staphylinids, three histerids, one hydrophilid, one carabid, one tenebrionid, two labiids, and one carcinophorid earwig. In laboratory studies, these predators variously searched rhizomes of living plants and pseudostem and rhizome residues. Eleven predators attacked the banana weevil egg stage, ten attacked the first two larval instars, nine attacked the third and fourth instar, while four attacked later stages. Using high predator densities under experimental conditions, two of these predators reduced weevils by up to 50% in suckers, 39% in stumps and 40–90% in residue pseudostems, *T. interocularis* reduced weevil densities in spent pseudostems by 42%, while the other predators were unimportant (Koppenhofer and Schmutterer, 1993; Koppenhofer, 1995). However, the number of natural enemies used in these experiments far exceeded field densities, suggesting that the impact of these predators in banana stands is likely to be limited (Koppenhofer and Schmutterer, 1993).

Hargreaves (1940) was the first to suggest that ants might have potential as biological control agents of banana weevil in Africa, although no studies were ever conducted. During the 1970s, Cuban researchers began a biological control programme using the myrmicine ant *Pheidole megacephala* (Fabricius) against sweet potato weevils (Perfecto, 1994). Roche (1975) suggested that *Tetramorium guineense* (Mayr) might be capable of suppressing banana weevil populations. Both *P. megacephala* and *T. guineense* will enter crop residues and remove eggs and larvae (S. Rodriguez, personal communication). *T. guineense* was also observed to nest in leaves and crop residues.

Roche and Abreu (1982, 1983) began the propagation and dissemination of *T. guineense* colonies. Colonies with up to 22 queens and 62,000 workers and immatures were collected and liberated in new fields (Perfecto and Castineiras, 1998). Ant establishment was followed by the rapid appearance of new colonies. At the onset of one trial, weevil trap catches were greater than ten per mat. Liberation of ants on 8% and 50% of the mats provided total field coverage in 6 and 2 months, respectively (Roche and Abreu, 1983). Eighteen months later, weevil trap catches were 56–65% lower. Based on these results, Roche and Abreu (1983) recommended releasing ants on 25–30% of the area for 'complete control' in 3–4 months.

The potential of myrmicine ants to control banana weevil has also been demonstrated by Castineiras and Ponce (1991). They released 9 and 15 *P. megacephala* colonies ha^{-1} into plantain plots (separated by 200 m alleys) 6 months after planting. During the first crop cycle, weevil trap captures and damage indices were similar in plots where ants had been released, in carbofuran-treated plots and controls. In the second cycle, ants reduced weevil trap captures by 54–69% and damage by 64–66%, with a corresponding yield increase of 15–22%. The level of control provided by ants was similar to that of the pesticide.

Myrmicine ants in the genera *Tetramorium* and *Pheidole* are widespread (e.g. Walker and Dietz, 1979; Varela, 1993). Their control potential in other banana systems is unknown. One concern, however, is that these ants may also protect homopteran pests of other crops (A.M. Varela, personal communication).

SECONDARY HOST ASSOCIATION Neuenschwander (1988) suggested that natural enemies of closely related hosts offer the promise for efficient secondary associations with banana weevil. Traore (1995) investigated the possible use of the mymarid egg parasitoid *Anaphes victus* Huber against banana weevil in Benin. *A. victus* is an important parasitoid of weevil eggs in the Americas. This parasitoid was selected for study because it searches near the soil level, is habitat- rather than species-specific and because it effectively suppresses populations of carrot weevil (*Listronotus oregonensis* (LeConte)) (Boivin, 1993).

In the laboratory, *A. victus* readily accepted banana weevil eggs, but parasitoid emergence was negligible (0–2%) (Traore, 1995). In contrast, *A. victus* immatures successfully emerged from water hyacinth weevil, *Neochetina eichorniae* Warner, demonstrating that the parasitoid could successfully complete its development within a new host. Traore (1995) attributed failure to emerge from banana weevil eggs to its larger size and the fact that the larvae of *A. victus* failed to consume all of the banana weevil egg contents, with decomposition of uneaten material contributing to pupal failure.

Microbial control

Microbial agents tested against the banana weevil include entomopathogenic fungi (e.g. *Beauveria bassiana* and *Metarhizium anisopliae*) (Plate 4), entomopathogenic nematodes (e.g. *Steinernema* spp. and *Heterorhabditis* spp.) (Plate 5) and endophytes (e.g. non-pathogenic *Fusarium* spp.). Entomopathogenic fungi and nematodes are mostly used to kill adult weevils, while endophytes target the immature stages. Although a number of strains have shown promise in the laboratory and in preliminary field studies, efficient and economically viable delivery systems still need to be developed. Only in a few sites have entomopathogens been reported to establish following applications in banana fields. Without adequate establishment, entomopathogens will require repeated applications as a biopesticide. This will entail continued production, distribution, and storage costs that will be passed on to the farmer.

ENTOMOPATHOGENIC FUNGI Entomopathogenic fungi have been tested against banana weevil since the 1970s (Ayala and Monzon, 1977; Delattre and Jean-Bart, 1978). Numerous strains have been screened against adult banana weevils in Africa and the Americas, employing diverse formulations, spore concentrations and application methods (reviewed in part by Nankinga, 1999; Nankinga *et al.*, 1999). These strains, isolated from a wide range of insects, as well as from *Galleria* baits, were predominantly *B. bassiana* and, secondarily, *M. anisopliae*. Additional research has been conducted on mass production on a range of substrates (e.g. maize, rice), spore viability and storage, formulations (powders, water solutions, mineral oils), doses, application methods and mortality rates after varying time intervals. In spite of the variability in methods, laboratory studies conducted in many different banana-growing regions have consistently demonstrated high levels of weevil mortality (often exceeding 95%) to a large number of strains.

For example, Brenes and Carballo (1994) screened 24 isolates of *B. bassiana* (from Hemiptera, Lepidoptera, ants and other weevils) by shaking the insects in conidial powder. The six most promising isolates were selected for further testing. Mortality of weevils dipped in water suspensions containing 1×10^9 spores ml^{-1} ranged from 73 to 100% with a LT_{50} of 7–10 days. Using a range of spore concentrations, an LT_{90} of 2.67×10^9 spores ml^{-1} was calculated for the most promising isolate. Carballo and de Lopez (1994) then found 31–63% adult mortality when *B. bassiana* conidial powder or spores on rice substrate were applied to pseudostem traps.

Contreras (1996) screened five strains of *B. bassiana* in the laboratory in oil-based formulations and found 65–95% weevil mortality in 15 days, with an LT_{50} of 2.5–8 days. Carballo (1998) tested water-based and oil-based formulations of *B. bassiana*. Formulations with > 20% oil caused high levels of mortality in the weevil (independent of fungal effects), while weevil mortality was negligible in solutions with 10% oil. Using a 15% oil solution, he found mortality to range from 10% at 1×10^7 spores ml^{-1} to 97% at 5×10^8 ml^{-1}.

Nankinga (1994) allowed weevils to walk on PDA cultures and found five isolates of *B. bassiana* produced > 96% mortality after 21 days. Topical applications of the same isolates in water suspensions produced 73–100% mortality. Nankinga (1994) also found mortality rates to be directly related to spore dose for three strains of *B. bassiana*. Higher doses killed almost all weevils, while females were more susceptible than males to lower doses of the pathogen. Topical applications by dispersion or immersion caused much higher rates of mortality than spraying pathogen solutions on to soil or pseudostem traps.

Although it is difficult to compare the results of different studies because of the wide range of methods used, it is clear that the most effective strains were capable of causing high mortality in the laboratory at lower spore concentrations and in shorter periods of time. In general, promising isolates of *B. bassiana* were more effective than those of *M. anisopliae* (Delattre and Jean-Bart, 1978; Batista Filho *et al.*, 1987; Mesquita, 1988; Busoli *et al.*, 1989; Kaaya *et al.*, 1993; Nankinga, 1994). Mineral oils used in formulations often affected mortality of weevils independent of entomopathogens, but tended to be more expensive than other substrates.

Under field conditions, naturally occurring infection rates (determined for trapped weevils) tended to be between 0 and 6% (Mesquita *et al.*, 1981; de Souza *et al.*, 1981; Van den Enden and Garcia, 1984; Gomes, 1985; Peña *et al.*, 1993; C.M. Nankinga, unpublished; C.S. Gold, personal observation). This suggests that entomopathogens may have to be periodically applied as biopesticides. However, few studies have addressed delivery systems and efficacy under field conditions.

Delattre and Jean-Bart (1978) were unsuccessful in reducing weevils by spraying *B. bassiana* spores on to the base of mats, while Mesquita (1988) obtained 5% infection after immersing pseudostem traps in *B. bassiana* spore solutions. Following applications of *B. bassiana* spores to pseudostem traps in rice paste and oil-based formulations, Batista

Filho *et al.* (1991, 1995, 1996) found 61% and 20% fewer weevils, respectively, in treated traps. Contreras (1996) suggested applications might be better placed on disc-on-stump traps that, in his studies, captured 3–4 times as many weevils as pseudostem traps. He found higher infection rates in oil formulations than in rice-based substrates of *B. bassiana* immediately after application, but the reverse was true 8 days after application. Nankinga (1999) found 50–60% infection rates for weevils collected in disc-on-stump and pseudostem traps treated with *B. bassiana* spores in maize culture, 55–61% for traps treated with spores in oil suspension, 23–44% for traps treated with spores in water suspension and 0% in oil and water controls. Pathogenicity decreased in treated soils after 2 weeks, although 15% of weevils collected from treated soils showed signs of infection 5 months after treatment. Nankinga (1999) also monitored weevil trap catches for 8 months in a trial with two *B. bassiana* applications. Mean weevil counts were lowest in plots treated with maize formulation (40) followed by plots receiving soil formulation (54), oil formulation (68) and controls (81). The incidence of field mortality of weevils observed in traps was low, with a maximum of 5% and often under 1%. Maize-based formulations also tended to reduce weevil damage levels in the central cylinder and cortex. These reductions in weevil damage suggest that infected weevils may not have been attracted to pseudostem traps and that actual mortality was higher.

Disc-on-stump and pseudostem traps may aggregate weevils at delivery sites for entomopathogens (Kaaya *et al.*, 1993; Contreras, 1996; Nankinga, 1999). Budenberg *et al.* (1993a) further suggested that semiochemicals might increase weevil attractivity of entomopathogen-baited traps. Currently, pheromones are exploited in pitfall traps. This would require a modification of the current pitfall trap design for use of pheromone such that the weevils become infected rather than drown. Such a method would only be advantageous over standard pitfall trapping if infected adults were able to transmit the pathogen to other weevils. Preliminary laboratory experiments suggest that such transmission does occur (Nankinga, 1994; Schoeman and Schoeman, 1999), but it is unclear how important this is under field conditions, where weevil density is relatively low.

ENDOPHYTES A wide variety of endophytic fungi have been isolated from nearly all examined plants (ranging from grasses to trees) and plant tissues (Carroll, 1991). Many of these have developed mutualistic relationships with plants and some act as antagonists to pests and diseases. Endophytes can enhance resistance to specialist herbivores that have evolved mechanisms to circumvent the plant's normal defences (Carroll, 1991; Breen, 1994).

The use of endophytes in banana to give extended protection to tissue culture planting material is currently under study. Griesbach (2000) obtained 200 isolates of endophytic fungi from 64 recently harvested highland bananas (AAA-EA) and Pisang awak (ABB). Spore suspensions of 12 isolates (eight *Fusarium* spp., three *Acremonium* spp., one *Geotrichium* sp.) caused 80–100% mortality in weevil eggs, while 74 additional isolates caused 60–79% mortality. Screening of the 12 most promising isolates against banana weevil larvae gave 0–48% mortality, with the best two strains being *F.* cf *concentricum* (48%) and *F. oxysporum* (32%). Griesbach (2000) was also able to successfully inoculate tissue culture plants with endophytes. Preliminary pot trials on the effects of inoculated endophytes on weevil damage in tissue culture plants produced some positive but largely inconsistent results. Current research efforts concern pathogenicity testing of candidate isolates, distribution and persistence within the host plant and efficacy in reducing weevils in potted plants.

ENTOMOPATHOGENIC NEMATODES The use of entomopathogenic nematodes (EPNs) for insect control and their potential use against banana weevil has been reviewed by Treverrow *et al.* (1991), Parnitzki (1992) and Schmitt (1993). The most commonly used species are within the genera *Steinernema* and *Heterorhabditis*. These have received wide attention as biological control agents because of their wide host range and ability to kill the host rapidly with no adverse effects on the environment (Schmitt, 1993). Five species of *Xenorhabdus* bacteria are mutualistically associated with EPNs. Infective juvenile nematodes enter through the host's natural orifices (*Steinernema*) or interskeletal membrane (*Heterorhabditis*) (Treverrow *et al.*, 1991). After entering the host, the nematodes penetrate mechanically into the haemocoel and release *Xenorhabdus* which causes septicaemia and insect death within 1–2 days (Schmitt, 1993).

EPNs may be more effective against banana weevil larvae than against weevil adults (Figueroa, 1990; Peña and Duncan, 1991). However, the cryptic habitat of weevil larvae within living plants makes delivery difficult (Treverrow and Bedding, 1993; Treverrow, 1994). In addition, it is difficult to know which plants are infected with larvae. Therefore, applications cannot be restricted to those plants with high weevil numbers.

Treverrow *et al.* (1992) and Treverrow and Bedding (1993) developed a delivery system for EPNs, capitalizing on the weevil's attraction to cut rhizomes and damaged plants. Two conical shaped cuts were made in residual rhizomes. These cuts attracted adult weevils and provided thigmotactic stimuli that encouraged them to remain at the infection sites. The holes also buffered the delivery site against temperature extremes and provided excellent conditions (high humidity, moderate temperatures, protection against ultraviolet light) for nematode persistence. The nematodes were released at a density of 250,000 per hole in a formulation including a polyacrylic gel (to reduce water build-up and incidence of nematode drowning) with an adjuvant of 1% paraffin oil (to encourage the weevils to raise their elytra, exposing the first spiracle for nematode entry). The nematodes persisted for up to 50 days and attacked both adults and larvae (Treverrow *et al.*, 1991; Treverrow, 1994). At moderate weevil infestation levels, nematode baits performed as well as or better than insecticides (Treverrow, 1993; Treverrow and Bedding, 1993), but were not as effective as pesticides in heavily infested fields (Treverrow, 1994). However, controls based on EPNs were not economically competitive with pesticides (Treverrow, 1993, 1994).

Host plant resistance

The literature on susceptibility of *Musa* clones to banana weevil attack is largely fragmentary, with highly variable and often contradictory findings (Pavis and Lemaire, 1997; Kiggundu *et al.*, 1999). Most often, reported results reflect comparisons among a small number of clones used in field trials. Fogain and Price (1994), Ortiz *et al.* (1995), Anitha *et al.* (1996) and Kiggundu (2000) conducted screening trials to identify existing clones displaying resistance to banana weevil.

The variability in findings may reflect, in part, differences in sampling methods in assessing weevil damage. For example, in a screening trial, Kiggundu (2000) found that Nsowe (AAA-EA) scored highest among highland banana clones in damage to the rhizome surface but among the lowest in internal rhizome damage. Pavis and Lemaire (1997) and Mestre (1997) noted the need for standard screening methods and reference cultivars. Kiggundu (2000) recommended the use of total cross-section damage (cf. Gold *et al.*, 1994a) as this measure has a high level of heritability and is well correlated with other indices of weevil damage. In contrast, Rukazambuga *et al.* (1998) suggested the use of damage to the central cylinder, as this damage appeared to have the greatest impact on plant growth and yield.

RESISTANCE ACROSS GENOME GROUPS Edible bananas (*Musa* spp., Eumusa series) originated within South-East Asia from two wild diploid progenitors, *Musa acuminata* and *M. balbisiana*, producing a series of parthenocarpic diploids and triploids through natural hybridization. Both *M. acuminata* and *M. balbisiana* escaped attack in a screening trial in Cameroon (Fogain and Price, 1994).

Plantains (AAB) appear to be more susceptible to banana weevil attack than other clonal groups (Ghesquiere, 1925; Fogain and Price, 1994; Gold *et al.*, 1994a; Price, 1994; Kiggundu, 2000). Highland cooking bananas (AAA-EA) also appear susceptible to banana weevil (Sikora *et al.*, 1989; Gold *et al.*, 1994a; Kiggundu, 2000). Reports on susceptibility of AAA dessert bananas (e.g. Gros Michel, Cavendish, Williams, Valery) have ranged from resistant to susceptible (e.g. Mesquita *et al.*, 1984; Fogain and Price, 1994; Gold *et al.*, 1994a; Stanton, 1994; Sponagel *et al.*, 1995; Kiggundu, 2000). Sponagel *et al.* (1995) suggest that weevils favour crop residues of AAA dessert bananas over developing plants and are thus unimportant. AB and ABB bananas are often considered among the most resistant *Musa* clones to banana weevil (Mesquita *et al.*, 1984; Seshu Reddy and Lubega, 1993; Gold *et al.*, 1994a; Ortiz *et al.*, 1995; Abera *et al.*, 1999; Kiggundu, 2000). In germplasm collections in Cameroon and Nigeria, ensete appeared to be highly susceptible to banana weevil (Pavis and Lemaire, 1997; C.S. Gold, personal observation). However, in Ethiopia, where the crop is most widely grown, ensete largely escapes attack because most production is above the weevil's upper elevational threshold (M. Bogale *et al.*, unpublished).

CLONAL RESISTANCE In screening trials in Cameroon (Fogain and Price, 1994) and Nigeria (Ortiz *et al.*, 1995), all plantain clones appeared susceptible to banana weevil. In contrast, Chavarria-Carvajal (1998) evaluated eight plantain clones and found the common dwarf plantain and a Lacknau clone to have less than 20% of the damage occurring in Sin Florescencia and Rhino Horn plantains. Irizarry *et al.* (1988) and Fogain and Price (1994) also found Lacknau clones less susceptible than other plantains.

As part of a larger screening trial, Kiggundu (2000) evaluated 26 highland bananas for susceptibility to banana weevil. Cluster analysis suggested that 7 clones were highly susceptible to weevil attack (mean damage 9%), 13 clones were intermediate in susceptibility (6%) and 6 clones were resistant (4%). Among the more resistant cooking clones were Mbwazirume and Nakyetengu, which are both widely grown. One brewing clone was considered susceptible, three were intermediate in susceptibility and two appeared relatively resistant.

MECHANISMS CONFERRING RESISTANCE Successful attack of bananas by banana weevils involves host plant location, host plant acceptance (oviposition) and host plant suitability (larval survival, developmental rate and fitness). Host plant resistance may be attributed to antixenosis (non-preference), antibiosis and/or host plant tolerance (Painter, 1951). For banana weevil, available data suggest that antibiosis is the most important factor conferring host plant resistance, while antixenosis is of little importance. Little has been reported on host plant tolerance to banana weevil damage, as such work would require yield loss studies over several crop cycles.

Antixenosis. Antixenosis suggests that resistant clones avoid pest attack by reducing rates of host plant location (i.e. attraction) and/or host plant acceptance; the combined effects of these two processes would be reduced oviposition. Pavis and Lemaire (1997) suggested that antixenotic factors might also deter adult feeding.

The data on the relative attraction of banana weevils to susceptible and resistant clones is equivocal. Budenberg *et al.* (1993b) and Pavis and Minost (1993) reported that females were equally attracted to cut rhizomes and volatiles from resistant and susceptible cultivars. In contrast, Minost (1992) found that Burmanica (AA) was most attractive to adult weevils, followed by Pisang awak (ABB), Borneo (AA) and French Clair (AAB), while Petit Naine (AAA) and Rose (AA) were much less attractive. However, clonal attraction was not related to subsequent weevil damage. Similarly, Musabyimana (1995), Abera (1997) and Kiggundu (2000) found some differences among trap captures at the base of different clones, but these differences were not related to damage at harvest. These studies suggest that resistance mechanisms must be related to oviposition and/or larval development rather than to host plant attraction.

Little work has been done on host plant acceptance. Abera (1997) found field oviposition on Kayinju (ABB) to be similar to that on highland banana clones, even though the latter displayed much higher levels of weevil damage. Kiggundu (2000) looked at oviposition on resistant and susceptible clones in both choice and no-choice experiments. There was very little mean separation and the lower levels of oviposition occurred on clones

(Atwalira, Nakyetengu, Muvubu) that were not considered resistant.

The banana weevil is a relatively sedentary insect living in perennial systems with an abundance of host plants. Kiggundu (2000) argued that it is unlikely that banana weevils might walk far looking for a suitable host. Most likely, tenure time at the base of any given mat and oviposition may be more related to environmental factors such as soil moisture. Available data indicate that the weevil will freely oviposit on both susceptible and resistant clones, suggesting that antibiosis plays the most important role in host plant resistance (Abera, 1997; Kiggundu, 2000).

Antibiosis. Antibiotic factors are those which negatively influence larval performance (i.e. poorer survivorship, slower development rates, reduced fitness). These factors may range from physical (e.g. sticky sap and latex, rhizome hardness) to nutritional quality and deficiencies to toxic secondary plant substances.

In a mixed cultivar trial, Abera (1997) found weevil damage to the interior of the rhizome to be 5–25 times higher in five highland banana clones than in Kayinju. Banana weevil attraction and oviposition on Pisang awak was similar to that on the highland bananas, while larval survivorship was estimated as 10–23 times higher in highland bananas than in Kayinju. From these data, Abera (1997) concluded that antibiosis explained why Kayinju was resistant to banana weevil. Mesquita and Alves (1983), Mesquita *et al.* (1984) and Mesquita and Caldas (1986) found that banana weevil immatures developed faster and had fewer ecdyses on some clones than on others. Lemaire (1996) reported slower larval development and higher larval mortality on the resistant clone Yangambi-Km5. Kiggundu (2000) found that two resistant clones, FHIA-03 and Kayinju, increased larval developmental time. Larval mortality ranged from 5 to 100%, with highest levels occurring in resistant clones such as Kayinju (100%) and Kabula (AAA-EA) (90%). Larvae reared on Mbwazirume (AAA-EA), FHIA-03, Ndiizi (AB), and Yangambi-Km5 (AAA) also had high mortality rates. Rhizome extracts from Kayinju applied to susceptible rhizome material inhibited larval feeding, while extracts from susceptible clones did not.

Pavis and Minost (1993) found a negative correlation ($r = -0.47$) between rhizome hardness and infestation rate and hypothesized mechanical resistance to oviposition or larval development. However, Ortiz *et al.* (1995) found no relationship between rhizome hardness and weevil damage scores in segregating progenies, suggesting that other resistance mechanisms may be more important.

Tolerance. Tolerance suggests that the host plant can sustain high levels of insect damage without yield reduction. Cuille and Vilardebo (1963) argued that Gros Michel (AAA) was resistant because the large size of the rhizome conferred tolerance to weevil attack. Kiggundu (2000) also found that rhizome size can reduce the proportion of damaged tissue. Pavis (1993) suggested that the vigorous growth of Pisang awak allowed it to tolerate moderate levels of attack. However, no studies have compared damage thresholds and related yield losses for different *Musa* clones.

BREEDING FOR RESISTANCE To date, there have been no attempts to breed bananas or plantains for resistance to banana weevil. Breeding for resistance depends upon sound knowledge of resistance mechanisms, resistance markers and the genetics of resistance (Kiggundu *et al.*, 1999). As a foundation, it is important to determine whether there are useful sources of resistance within the available germplasm. Kiggundu (2000) observed that the wild diploid Calcutta-4 and the clones Yangambi-Km5 and FHIA-03 showed high levels of resistance and might be exploited in breeding programmes. Lemaire (1996) and Mestre and Rhino (1997) also found Yangambi-Km5 to be highly resistant to banana weevil. Calcutta-4 has already been successfully used in conventional breeding programmes in Nigeria and Uganda, while the male/female fertility of Yangambi-Km5 and FHIA-03 still needs to be determined (Kiggundu, 2000).

Pesticides

Chemical pesticides for control of banana weevil may be applied to protect planting material (through dipping of suckers or applications in planting holes), periodically applied at the base of the mat after crop establishment, and/or applied to pseudostem traps to increase trap catches. Since the first recommendation in 1907 for the use of chemicals (i.e. Bordeau mixture) to control banana weevil (Gravier, 1907), there have been numerous studies on the relative efficacy of different insecticides under different formulations and application rates, persistence and the appearance of insecticide resistance in banana weevils. Chemicals remain an important part of banana weevil control although costs often make them prohibitive for subsistence farmers.

In 1951, the use of chemicals gained further importance with the advent of synthetic insecticides that largely replaced labour-demanding cultural controls such as trapping or sanitation. As with many other pests, the introduction of chemicals in the 1950s for the control of banana weevil was greeted with optimism. Braithwaite (1958) suggested that eradication of the banana weevil might be achieved with aldrin and dieldrin.

A wide range of chemicals, encompassing all major classes of insecticides, have been tested and recommended as effective for the control of banana weevil (reviewed, in part, by Sponagel *et al.*, 1995 and Seshu Reddy *et al.*, 1998). Many once recommended chemicals have been banned or otherwise fallen out of favour because of their high levels of mammalian toxicity, environmental concerns and/or the development of resistance. Nevertheless, insecticides can often achieve high levels of control in short periods of time.

Insecticide resistance in banana weevil has been documented in Australia, Latin America and Africa (reviewed by Gold *et al.*, 1999a) for a range of chemicals including cyclodienes (aldrin, BHC, heptachlor, dieldrin), organophosphates (chlorpyrifos, ethoprophos, pirimiphos-ethyl and prothiophos), and carbamates (carbofuran). Cross-resistance has also been demonstrated (Edge, 1974; Collins *et al.*, 1991). In Uganda and Tanzania, outbreaks of banana weevil in the mid-1980s were attributed to pest resurgence following development of resistance to dieldrin (Sengooba, 1986; Sebasigari and Stover, 1988; Gold *et al.*, 1999a) leading to loss of confidence in chemical control by some farmers.

Botanicals

In Kenya, Musabyimana (1999) conducted a detailed study on the effects of neem (*Azadirachta indica*) seed derivatives on banana weevil adult activity, success of immatures and resulting damage. This research, including both laboratory and field trials, employed different formulations of neem seed powder (NSP), neem kernel powder (NKP), neem cake (NC) and neem oil (NO). The azadirachtin content was determined as 4000 p.p.m. for NSP, 5500 p.p.m. for NKP, 5800 p.p.m. for NC and 850 p.p.m. for NO. Musabyimana's (1999) results suggest that neem derivatives can reduce weevil damage by interfering with each stage of attack: (i) fewer adults will locate or remain at the host plant; (ii) females locating the host plant will have reduced oviposition; (iii) eclosion rates will be lower; (iv) an antifeedant effect will delay and reduce larval feeding; and (v) larval fitness (developmental rates, size, survivorship) will be reduced.

Direct application of NC and NSP to the soil was much more cost effective than applications of aqueous solutions. Overall, NSP appeared to be the preferred derivative as it was easier to produce and had better effects. From these results, Musabyimana (1999) recommended application rates of 60–100 kg ha^{-1} once every 4 months.

In laboratory studies in Cameroon, Messiaen (2000) had results consistent with those of Musabyimana (1999): neem had a repellent effect on adults and reduced oviposition levels and eclosion rates. However, in field studies, Messiaen *et al.* (2000) found limited advantage in weevil control from applications of neem dips (i.e. aqueous solution of concentrated NSP) and no benefits from granular applications of NSP (30–100 g per plant). In two trials, neem treatments did not influence weevil adult populations, although damage was reduced in one experiment. However, neem dips reduced sucker mortality by 73–85% and total plant mortality by 50%.

From his results in Kenya, Musabyimana (1999) concluded that NSP and NC soil applications are effective enough to do away with paring and hot water treatment of suckers to be used as planting material. His data suggest that extended protection under field conditions is possible. However, the largely negative results obtained by Messaien *et al.* (2000) in Cameroon were inconsistent with those of Musabyimana (1999) and show that it would be useful to conduct further studies at additional sites. Also, the availability of neem products, their economic viability and their acceptance by farmers need to be determined.

Banana pseudostem borer, *Odoipurus longicollis* (Olivier)

The banana pseudostem borer or pseudostem weevil is considered a minor to important pest of banana and Manila hemp in parts of India, Nepal, Burma, Sri Lanka, Thailand, Indonesia, China and elsewhere in Asia (Froggatt, 1928; Pinto, 1928; Kung, 1955; Dutt and Maiti, 1972; Waterhouse, 1993). The weevil bears a superficial resemblance to the banana weevil, *Cosmopolites sordidus* (Germar), but is slightly larger and its elytra do not completely cover the abdomen (Pinto, 1928; Dutt and Maiti, 1972). The morphology of the adult and the immature stages have been described by Dutt and Maiti (1972). Sexing of adults is based on punctuation of the rostrum. Males also tend to be smaller than females. Observations on sex ratio range from 1 : 1.45 (male : female) (Kung, 1955) to 1 : 1.17 (Dutt and Maiti, 1972).

Life history, habits, pest status

The life cycle of the banana pseudostem borer has been described by Pinto (1928), Kung (1955) and Dutt and Maiti (1972). The adult weevils feed on living and decomposing banana leaf tissues, but eat little and are not considered pests. Damage is done by the larval stage. The larvae attack the pseudostem and stem of banana plants, although they will occasionally feed within the rhizome. This contrasts with the banana weevil, which attacks the rhizome and will rarely enter the pseudostem. As such, pseudostem borer damage may be clearly visible, while banana weevil damage can only be observed by dissection of the rhizome. The pseudostem borer will attack both living plants and harvested stumps (Pinto, 1928). In some sites, the weevil may show a preference for crop residues (B. Pinese, personal observation).

The colour of the adult weevil varies with age from reddish brown to black (Dutt and Maiti, 1972). The weevil is characterized by long life span, negative phototropism, thigmotropism, gregariousness, hydrotropism, and death mimicry (Kung, 1955). Most adults live 6–10 months, although some can survive for more than 2 years (Pinto, 1928). In contrast to the banana weevil, the adult readily flies, although it has been described as a 'poor' flyer (Dutt and Maiti, 1972). In spite of being negatively phototropic, diurnal flight may occur (C.S. Gold, personal observation).

Oviposition is in the leaf sheaths of living plants or residues (Pinto, 1928). The weevils are especially attracted to and readily oviposit in cuts in banana material (Kung, 1955). Kung (1955) suggested that ovipositing females prefer stressed plants, while Pinto (1928) observed greater oviposition in residues than in living plants. Dutt and Maiti (1972) found greatest oviposition in pseudostems with a girth of 25–50 cm, with little oviposition in plants < 25 cm or > 75 cm in girth. The eggs are placed singly in chambers made with the female's rostrum. Pinto (1928) found oviposition rates of 1–6 eggs day^{-1} and observed two weevils to produce 103 and 185 eggs, respectively, in 4.5 months. The egg stage has been reported as 3–4 days (Pinto, 1928) and 5–12 days (Kung, 1955).

The larvae pass through four (Kung, 1955) or five (Dutt and Maiti, 1972) instars. The first instar remains in the same leaf sheath where the egg was placed. Subsequent instars bore into the inner leaf sheaths or pseudostem (Dutt and Maiti, 1972). The mature larvae may be twice the size of those of banana weevil (Pinto, 1928). The larval stage has been reported as 14 days (Pinto, 1928), 3–6 weeks

(Kung, 1955) and 26–68 days (Dutt and Maiti, 1972).

The prepupa forms a pupal chamber within the host plant. Pupation is within a cocoon formed out of plant fibres. The prepupal stage lasts 3–9 days, while the pupal period has been reported as 3–20 days (Pinto, 1928; Kung, 1955; Dutt and Maiti, 1972). Following emergence, the adults may pass extended periods and even mate within the host plant. Normally, there is a preoviposition period of 1 month (Pinto, 1928; Dutt and Maiti, 1972).

Tunnelling by the weevil larvae can lead to the rotting of pseudostem tissues and breakage in the wind (Dutt and Maiti, 1972). Although the weevil has been described as an important pest, there has been no quantification of plant loss and/or yield reductions. Further work in this area seems a prerequisite for the development of any integrated pest management programme.

Control measures

Little information is available on the control of banana pseudostem weevil. Pinto (1928) recommended selecting new and clean sites (Pinto, 1928), the use of clean planting material, crop sanitation (e.g. burying infested residues), rogueing of infested plants, crop rotation (to rid fields of weevils), and trapping by placement of residue slices on the ground. Zhou and Wu (1986) and Mathew *et al.* (1997) suggested removing dead leaves to eliminate hiding places for the adults. Whereas the weevil flies and its dispersal capacity has not been determined, it is unclear how effective the use of these cultural methods might be. Kung (1955) collected more than 2 million weevils in traps over a 4-month period, but did not provide information on the effect of this trapping programme on weevil populations.

Chemical control and host plant resistance have also been suggested. Mathew *et al.* (1996, 1997) found swabbing of the pseudostem and borer holes to be effective in reducing weevil infestations. In India, Ishaque (1978) found one variety to be completely free of the weevil, while two others appeared resistant; Charles *et al.* (1996) screened banana germplasm and found six varieties to be severely infested, three heavily infested, four moderately infested and 27 lightly infested.

West Indian sugarcane borer, (Olivier) *Metamasius hemipterus sericeus*

The West Indian sugarcane borer, *Metamasius hemipterus sericeus* (Olivier), also known as 'the rotten stalk borer of sugarcane' in Puerto Rico and 'picudo rayado' in South America, is an important pest of palms, sugarcane, pineapple, and bananas in the West Indies, Mexico, Central South America and Florida, USA (Vaurie, 1966; Woodruff and Baranowski, 1985; Giblin-Davis *et al.*, 1994).

Life history and habits

The adult stage of *M. hemipterus sericeus* is free-living within banana pseudostems, palm fronds and sugarcane sheaths. This weevil is distinguished from other species by the colour (red to yellow and black) and size (9–14 mm), but can be confused with *Metamasius callizona*, which infests bromeliads. Sometimes the colour pattern of *M. hemipterus sericeus* is variable, causing some confusion. The colours of the elytra are one-half to one-third red and the remainder mostly black; the pronotum and venter are black to red and black. Adult females of *M. h. sericeus* lay eggs in cracks or damaged areas of the host plant or in petioles or crown shafts of certain species of healthy palms. The adult can live up to 60 days. The female can deposit approximately 500 eggs (Castrillon and Herrera, 1980). The egg is 1.5 mm in length, oval, and hatches between 3 and 7 days after oviposition.

The yellowish larvae are typical legless weevil grubs and similar in most aspects to other members of the Rhynchophorinae. The average size of the larvae is 1.3 to 1.9 cm. The larvae bore and feed in the host tissue, causing extensive physical damage which can lead to the death of the host (Giblin-Davis *et al.*, 1994). The larval stage lasts between 50 and 60 days. Then a fibrous pupal case is constructed (similar to that of the giant palm weevil,

Rhynchophorus palmarum (L.)). The pupal stage is spent in this cocoon for about 10–20 days.

Damage

Generally, the West Indian cane weevil is considered a secondary pest of sugarcane (Simmonds, 1966; Sosa *et al.*, 1997) and it is attracted to damaged or rotting plant material. In sugarcane, Sosa *et al.* (1997) consider that the weevil is attracted to cane damaged by either mechanical cultivation, harvesting equipment, rats, borers, disease or natural growth cracks. Raigosa (1974) considers that *Metamasius* females prefer to deposit eggs on canes that have been damaged by *Diatraea*. In Colombia this type of damage is known as the complex *Diatraea-Metamasius*. *Metamasius* has also been observed infesting canes used as seedpieces (Raigosa, 1974). In Florida, Sosa *et al.* (1997) observed infestation levels ranging from 8% to 32% of the stalks of CP 85–1382. Sosa *et al.* (1997) found little or no infestation in other cultivars growing next to CP 85–1382. Larvae of *M. hemipterus sericeus* can seriously affect ornamental palms, *Phoenix canariensis, Ptychosperma macarthurii, Ravenia rivularis, Roystonia regia*, destroy banana and plantain, *Musa* spp., and interspecific hybrids of *Saccharum* (Vaurie, 1966; Giblin-Davis *et al.*, 1994; Peña *et al.*, 1995). Larval tunnelling in palms starts in the petioles, wounds in petioles, crown, stem and then extends into healthy leaf or stem tissue. Affected palms are often characterized by the production of an amber-coloured and gummy exudate in the stem, crown shaft or petioles, and galleries in the leaves, petioles, and stems (Giblin-Davis *et al.*, 1994).

In banana the larvae feed in the upper part of the pseudostem. The first symptoms are yellowing of lower leaves and consequent rotting of the pseudostem.

Trapping

Raigosa (1974) captured an average of 30 weevils per trap using pieces of bamboo filled with pieces of fermenting cane. Peña *et al.* (1995) captured a maximum of ten weevils per trap per week using banana pseudostem disc traps. Giblin-Davis *et al.* (1994) tested different trap designs and the response of adults to semiochemicals and concluded that weevil counts increased with combinations of ethyl acetate, sugarcane and/or the aggregation pheromone metalure, compared with the use of any compound alone (ethyl acetate, sugarcane or metalure).

Population assessment of *M. h. sericeus* is problematic. Trapping of adults is often used to monitor weevil numbers. Viladerbo (1973) and later Castrillon and Herrera (1980) suggested the use of a 'sandwich trap' using banana pseudostem as an attractant. Using the same trapping method, Peña *et al.* (1995) determined that population build-up of the weevil increased during spring, summer and early autumn in Florida, but warned that the number of weevils collected at these traps was consistently low. Giblin-Davis *et al.* (1994) have determined major aggregation pheromone compounds that can be used for trapping and for monitoring weevil populations in the field.

Sampling

Sosa *et al.* (1997) surveyed sugarcane fields in Florida, by two or three people walking the field looking for infested stalks for 30 min. If damaged cane was found, the field was considered infested. Later Sosa *et al.* (1997) sampled by walking along the aisle for ~80 m, moving inside the field and marking off the first 3 m of row, alternating between 2 centre rows and by striping leaves of the bottom half of stalks for external symptoms of infestation such as lodging or cracked rinds with frass extruding. Then they determined the number of larvae, pupae or adults per sample. Sosa *et al.* (1997), determined a mean of 2.4 weevils per stalk, suggesting that this pest should be monitored closely.

Seasonality

Densities of *M. h. sericeus* were low in Florida when a banana disc trap was used to assess population density. Population build-up was observed from April through to June and between November and December.

Chemical control

Chemically based pest controls, currently recommended by some control programmes, represent a short-term and questionable strategy for resource farmers in Florida and the Caribbean region with accompanying health and environmental concerns for the entire area. Giblin-Davis *et al.* (1996) demonstrated that adults of *M. h. sericeus* were killed by labelled rates of acephate, carbofuran, chlorpyrifos, cyfluthrin, disulfoton, imidachloprid, isofenphos, lindane and vydate. Raigosa (1974) and Rossignoli (1972) examined the use of sugarcane poisoned traps and determined that this method was adequate for control of *M. h. sericeus* while Nogueira (1976) and Sarah (1990) demonstrated that chemical control is not always possible against *M. h. sericeus*.

Natural enemies

ENTOMOPATHOGENS The use of entomopathogens provides a promising but expensive means of control of *M. h. sericeus*. Entomogenous fungi, *Beauveria bassiana* (Balsamo) Vuillemin and *Metarhizium anisopliae* (Metchnikoff) Sorokin, have gained considerable attention as potential control for weevils (Mesquita *et al.*, 1981; Peña *et al.*, 1995; Giblin-Davis *et al.*, 1996). For example, a study undertaken by Peña *et al.* (1995) demonstrated that naturally occurring *B. bassiana* was an important mortality factor to adults of *M. h. sericeus* in Florida. *B. bassiana* infection increased up to 70% between March and April 1991, when more than ten weevils were captured per trap (Peña *et al.*, 1995). However, more information was needed on the effect of this fungus before pest management decisions could be made. Giblin-Davis *et al.* (1996) demonstrated that the nematode *Steinernema carpocapsae* was efficacious against larvae but not against adults of *M. h. sericeus* and concluded that because of the high potential for high weevil production in Florida and the cyptic habitat of the boring stages of this weevil, chemical insecticides and entomopathogenic nematodes will need to be applied frequently and over a long period of time for effective management.

PARASITOIDS Surveys for biological control agents (predators, parasites) of *M. h. sericeus* have been unsuccessful (J.E. Peña, unpublished). Thus, very little is known about effective biocontrol agents of *M. h. sericeus* in the Americas and the Caribbean. Siqueira *et al.* (1996) identified predators of *Metamasius* at the family level and stated that they were more abundant in Brazil than parasitoids. The predacious families included Labiduridae, Histeridae, Staphylinidae, Carabidae, Cicindelidae, Formicidae, Reduviidae, and Tachinidae. Search of host-specific parasitoids in one of the areas of origin has not provided any positive results (J.E. Peña, unpublished). However, Cave and Alvarez del Hierro (1997) found that the tachinid *Admontia* spp., which was observed in Honduras parasitizing *Metamasius quadrilineatus*, could be tested as a possible parasitoid of other species of *Metamasius*.

Lixophaga sphenophori (Villeneuve), a successful parasitoid of a related sugarcane weevil species (*Rhabdoscelus obscurus* (Boisduval)), is currently considered as a candidate for biological control of *M. h. sericeus*. *L. sphenophori* was collected in New Guinea parasitizing the sugarcane weevil *Rhabdoscelus obscurus* and was successfully introduced into Hawaii for control of the same weevil species in sugarcane in 1910 (Waggy and Beardsley, 1972). The biology and behaviour of *L. sphenophori* is similar to another successfully introduced tachinid, *Lixophaga diatreae*, which is a parasite of lepidopterous borers in sugarcane. *L. diatreae* is larviparous and the complete life stage lasts 20.5 to 32 days (Scaramuza, 1930). The mode of entry by the parasitoid into the weevil larva and the 'symptoms' of parasitized larvae have been described by Olson (1970a,b). Topham and Beardsley (1973) have studied the behaviour of *L. sphenophori* flies in the field. These authors found a high correlation between presence of flies and number of nectar source plants. Leeper (1972) reported that *Euphorbia hirta*, *Ricinus communis*, *Crotalaria incana* and *C. mucromata* were the most important sources for adult fly feeding. The first two plant species are widely found in Florida and could provide the nutritional requirements of the fly upon its introduction. Ota and Mitchell (1971) reported that application of several insecticides in sugarcane did not adversely affect parasitism of *L. sphenophori* on *Rhabdoscelus* larvae.

Other pests of roots and rhizomes

Roots and rhizomes are also attacked by scarab beetle larvae, mealybugs, *Chavesia* sp., and cydnid bugs of the genus *Scaptcoris* whose exudations inhibit growth of soil fungi (Ostmark, 1974).

Pests of Flowers and Fruits

Thrips: *Chaetanaphothrips orchidii* (Moulton), *C. signipennis* (Bagnall), *Caliothrips bicinctus* Bagnall, *Frankliniella parvula* Hood, *Heliothrips haemorrhoidalis* (Bouché), *Hercinothrips bicinctus* (Bagnall), *Thrips hawaiiensis* (Morgan), *Tryphactothrips* sp.

Thrips cause superficial skin blemishes on immature and developing banana fruit. Damage is primarily cosmetic although severe attacks may result in splitting of the peel with subsequent development of secondary roots. Usually only fruit grown commercially requires treatment with prophylactic pesticides. In Australia, chemical treatments are routinely applied to prevent the rusty brown discoloration caused by the pantropical banana rust thrips *Chaetanaphothrips signipennis*. The closely related *C. orchidii* (Moulton) causes similar damage in Central and South America. The banana flower thrips *Thrips hawaiiensis* (Morgan), a widespread and polyphagous flower feeder in Oriental and Pacific regions, damages fruit at flowering as it oviposits and feeds on fruit during and immediately after emergence of the inflorescence. The slightly raised silvery grey lesions caused by this thrips are locally referred to as 'corky scab'.

Banana thrips *Hercinothrips bicinctus* (Bagnall), *Caliothrips bicinctus, Chaetanaphothrips orchidii, C. signipennis, Thrips hawaiiensis* and *Tryphactothrips lineatus* (Ostmark, 1974; Reis and Souza, 1986; Pinese and Piper, 1994) (Plates 6 and 7) are the most important peel-blemishing insects, producing a range of damage symptoms on immature fruit. Most of these species are found in the inflorescences or between fruits (Gallo *et al.*, 1988). The skin of severely infested fruit may crack, allowing secondary invasion of pathogens. The banana rust thrips *Chaetanaphothrips signipennis*, apparently native to north Queensland, Australia, was originally described from Sri Lanka. It has also been recorded in Fiji, Panama, Trinidad, Brazil, Honduras, Costa Rica, and Florida. In Brazil, injury by *Tryphactothrips lineatus* is regularly observed on 30-day-old fruits, or on fruits > 32 mm in diameter (Martinez and Palazzo, 1971).

In Mexico, *Frankliniella parvula* prefers to oviposit in the epidermis of young banana fruits and less frequently in the flower parts. In Yemen, *Scirtothrips aurantii* and *Thrips pusillus* cause fruit spotting on bananas. Small circular spots first appear on the surface of the fruit, gradually enlarge, blacken and develop into oily, water-soaked lesions (Childers and Achor, 1995).

In Australia, *T. hawaiiensis* causes a superficial skin injury locally referred to as 'corky scab'. Adults are attracted to the emerging inflorescence. Female oviposition and subsequent nymphal and adult feeding cause damage on the developing fruit while the bunch is wrapped closely in the bracts. Oviposition punctures result in localized raised 'pimples' which disappear as the fruit develops, while the superficial grazing by the thrips develops into the slightly raised silvery grey areas of 'corky scab'. This damage is more prevalent during dry periods and is more commonly associated with fruit fingers on the lower bunch hands, the rachis and attacked cushion (Pinese and Piper, 1994).

Typical life history

Eggs are inserted into the plant tissues including fruit, pseudostem and leaf petioles, depending on species. Surfaces that are in close contact are preferred for oviposition and development. The eggs hatch in 1 to 2 weeks. Nymphs are clear to straw-coloured and, like the adults, shun sunlight, quickly dispersing when disturbed from their cryptic hiding places between adjacent fruit or from under leaf bracts on the pseudostem. Pupation takes place on the plant or in the soil near the base of the plants, depending on species. Banana rust thrips may spend part of their life cycle in the soil, while banana flower thrips stages all occur on the host plant (Pinese and Piper, 1994).

For example, in Australia, the entire life cycle of *T. hawaiiensis* is spent on the fruit or other parts of the plant. During summer months, the period from egg to adult for this species is 3 weeks.

Monitoring

Monitoring methods for rust and flower thrips and their damage have been developed and are used in Australia (Pinese and Piper, 1994).

Management

Cultural methods of control, such as clean cultivation and removal of trash, promote the exposure of pupae to desiccation, but do not provide effective control (Simmonds, 1966). In Brazil, Gallo *et al.* (1988) recommended the use of chemical control as soon as the flowers are formed, elimination of flowers after the fruits are formed, removal of alternative host plants, and covering banana bunches with a bag impregnated with insecticides. Martinez and Palazzo (1971) recommend harvesting fruits with 34 mm diameter. According to Simmonds (1966) no useful natural enemies of the banana thrips are known. Pinese (2001) also reported a lack of beneficials for *C. signipennis* in north Queensland, Australia, although Pinese and Piper (1994) found that a number of generalist predatory bugs, coccinellids and chrysopids feed on flower thrips and can reduce their numbers.

Chemical control methods consist of enclosing the bunch inside an insecticide-treated bag. This practice, once widespread, is still recommended in South America. A single pesticide injection into the emerging inflorescence, a treatment specifically aimed at the banana scab moth, is also efficacious against *T. hawaiiensis* and helps protect from early *C. signipennis* infestation. Attaching a piece of chlorpyrifos-impregnated ribbon to the upper bunch stalk also provides extended protection against *C. signipennis*. Untreated polythene bunch covers significantly reduced damage compared to uncovered fruit while bunch covers impregnated with 1% chlorpyrifos provided almost total protection (Pinese, 1987).

Banana fruit scarring beetles, *Colaspis hypochlora* Lefevre

Colaspis hypochlora has been reported from Mexico, Central America, Colombia and British Guyana (Simmonds, 1982). It appeared first as a banana pest in Colombia in 1922; the outbreak reached a peak 3 years later and declined in subsequent years. Eggs are most commonly placed in cavities gnawed by the female beetle in the sheath or root of the banana plant; however, they may also be laid in the soil to a depth of up to 1 cm. Eggs may be placed singly or in groups of up to 45. The larvae remain in the soil, feeding on the roots of grasses, especially those of *Paspalum conjugatum*. The larvae are sensitive to soil moisture and will move downwards in dry soils. If the soil remains dry for long periods, the larvae may die. Pupation occurs in the soil, the depth depending on the moisture present.

On emergence, the adults leave the soil feed on the leaves of various weeds as well as on the young leaves and fruits of banana. The adult beetle is nocturnal in habit. Following emergence, the beetles live for 9 to 12 days. In Colombia, breeding occurs only during the wet season, which allows production of four broods (April, June, August and October/ November).

The adult beetles feed on the unfurled leaves of the banana, on the skin of the fruits, marking the fruit in such a way as to render it unsaleable. Most of the scarring occurs on the lower proximal surfaces of the fingers, reflecting the fact that the beetle chooses the most sheltered spots for feeding. The scars are mostly oval in shape, and their damage can be confused with that caused by the fruit-scarring bee *Melipona* (*Trigona*) *amalthea* (Olivier). Scarring is worst on banana fruits growing in close proximity to drainage canals, or grassy-weedy patches in the fields and or near grassy roadsides, thus reflecting the larval ecology of the insect.

Control

Cutting grasses selectively around infested areas has been recommended for reducing *Colaspis* infestations. Natural enemies of the beetle in Colombia do little to reduce the population, though predation by frogs, lizards and spiders has been observed (Simmonds, 1966). Chemical control is not recommended.

Irapua bee, *Trigona spinipes* Fabricius

The 'Irapua' bee *Trigona spinipes* Fabricius visits the inflorescences of banana and sometimes causes damage to young fruits. Injury appears as dark, very well defined spots on the fruit and, in cases of heavy bee infestation, the damage appears along the angles of the fruit (Silva and Fancelli, 1998). Silva and Fancelli (1998) recommend as control measures to harvest fruits early or to protect fruits with polythene bags impregnated with insecticides.

Banana moths, *Opogona glycyphaga* Meyrick, *Opogona sacchari* Bojer, *Tirathaba rufivena* (Walker)

The banana moth *Opogona sacchari* Bojer (= *O. subcervinella* (Walker); *Tinea subcervinella* (Walker), *Gelichia sanctae-helenae* Mellis) has been recognized as a pest of bananas and row crops in both tropical and temperate environments. *O. sacchari* has been reported from Mauritius, Canary Islands, Madagascar, Italy, Belgium, The Netherlands, Great Britain, Brazil, Peru, Barbados, and the United States. *O. sacchari* is highly polyphagous and attacks at least 42 plant species, among them *Musa paradisiaca* and *Musa sapientum*. The sugarcane bud moth, *O. glycyphaga* is a significant banana fruit pest in Australia (Pinese and Piper, 1994) and *Tirathaba rufivena* occassionally inflicts damage to young fruit adjacent to rainforest.

O. sacchari belongs to a group of species that are detritus feeders and rarely feed on living plant tissue. The eggs are minute and laid singly on the banana fruit (Pinese and Piper, 1994). The egg stage is 10 days at 18°C (Martin, 1983). The larva has seven instars with total stage duration of 50–90 days, while the pupal stage is 21 days (Martin, 1983; Davis and Peña, 1990). The lowest larval temperature threshold is –5°C. The pupa is found under a tough silken cocoon and adults rest with their wings folded on banana leaves. The life cycles of *O. glycyphaga* and *T. rufivena* have not been studied.

Damage

O. sacchari prefers to oviposit on fresh inflorescences as opposed to dry flowers (Moreira, 1979a,b). The larvae may feed on bunches, pseudostems and sometimes even attacks the rhizome (Martin, 1983). The most serious damage occurs to the banana inflorescence. The larvae burrow into the substratum and seldom feed on exposed material. Their presence is usually indicated by the accumulation of frass and other debris entangled in larval silk over the surface of the injury. In Brazil, larval galleries cause rotting of fruits (Moreira, 1979a,b). In Florida, there are ten generations per year with adult populations highest between the spring and early autumn months. Adult activity is observed between 01:00 h and 04:00 h (12L : 12D) before the end of scotophase. Martin (1983) recomended the use of black light traps for collection of adults.

O. glycyphaga causes superficial scarring predominantly near the flower ends or at contact points between the lower flowers and the fruits. Thus damage is mostly limited to areas where larvae can feed under the protection of spent floral parts. Presence of copious black frass pellets attached to webbing is indicative of attack by this pest.

T. rufivena larvae feed deep into the pulp, in contrast to the superficial feeding by *O. glycyphaga*. This moth is most damaging in bunches near rainforests, suggesting that rainforest plants are their preferred food and bananas are an incidental host.

Control

The best control method for banana fruit moths is prevention. This can be done by cleaning debris from banana plantations (Martin, 1983). In Brazil, Cintra (1975) recommends manual removal of affected inflorescences. Chemical control has also been recommended (Moreira, 1979b) and the larvae of *O. sacchari* are known to be susceptible to carbamates and organophosphates. However, insecticide efficacy in controlling fruit moths drops rapidly after 8 days (Peña *et al.*, 1990a,b).

The entomopathogenic nematodes *Steinernema carpocapsae* and *Heterorhabditis heliothidis* provided 16–23% larval mortality during the first 2 weeks following application (Peña *et al.*, 1990a,b). In Australia, treatment of bunches with entomopathogenic nematodes for of control banana scab moth and rust thrips provided effective control of both *O. glycyphaga* and *T. rufivena*.

Banana scab moth, *Nacoleia octasema* (Meyrick)

The banana scab moth, *Nacoleia* (*Notarcha, Lamprosema*) *octasema* is known from Indonesia, New Guinea, the Solomon Islands, New Caledonia, Fiji, Tonga, Samoa, and Queensland (Simmonds, 1982) (Plate 8). In Queensland its distribution is confined to coastal areas north of Townsville. Sibling species or biotypes may be present as host preference varies markedly in different locations between what appear to be morphologically similar insects. For example, bananas are not attacked on mainland New Guinea or parts of Indonesia or Malaya where the moth feeds on *Pandanus* and *Nipa*. This may indicate the presence of non-banana-feeding races being present in these areas (Paine, 1964). While a valid explanation for this anomaly has not been put forward, taxonomic clarification to determine if distinct races are present in geographically separate areas is required before meaningful attempts at the introduction of biological controls can progress.

Damage

Larval feeding on young banana fruits, primarily on the outer curve, causes superficial scarring. As the fruit develops, lesions turn into black scabs from which the insect's common name is derived. Severe feeding can extend to cover the majority of the finger, preventing normal fruit development and causing fruit distortion. The scab moth is most active during wet, humid and hot conditions when, if left untreated, total bunch damage is possible. Banana is the preferred commercial host but feeding has been observed on *Heliconia*, *Nipa* and *Pandanus* and these hosts can provide suitable breeding sites during periods when bananas are not available.

Biology

The adult female lays dorsoventrally flattened scale-like eggs singly or, more commonly, in small clusters of up to 20 overlapping eggs. Eggs are placed on the smooth bracts of the inflorescence, the bases of the adjacent leaves and, uncommonly, the upper pseudostem. Females live for 4–5 days and lay 80–120 eggs (Simmonds, 1966). Egg laying is confined to the period immediately prior to and during emergence of the inflorescence, a period spanning about 7 days. After emergence, the young larvae migrate under the upper bunch bracts to feed on the immature fruit of the basal (upper) hands. As the bracts and underlying fruit hands lift, the larvae migrate progressively to the nearest lower unlifted hand to feed in the protected area between the fruit and the bunch stalk until they reach the distal (bottom) hands. Although not naturally gregarious, limitations on suitable feeding sites can lead to crowding (Paine, 1964). By the time the last hand lifts, the larvae have completed five instars and are usually fully grown.

The larvae grow and feed more voraciously as they move down the bunch. Thus, damage tends to increase in severity from the top to the bottom hands. During hot summer conditions, when fruit and bunch development is rapid, the larvae complete their development under the bracts of the male bud among the upper male flowers or migrate back up the bunch, feeding between fruit fingers. Pupation occurs primarily within a frail silken cocoon covered by frass on the bunch or among the trash near the base of the plant, or within dry leaf petioles on the lower pseudostem. The entire life cycle from egg hatch to egg laying is completed in aproximately 28 days during summer. Cooler and drier winter periods are not suitable for scab moth, when development periods are greatly increased. In Australia, its restricted climatic range has prevented its spread into the drier and cooler environments outside of the wet coastal tropical regions of north Queensland.

All adult behaviour is crepuscular with adult emergence and mating occurring from 30 to 60 min after sunset (B. Pinese, personal observation). Adults are very cryptic by day and are seldom observed and, if disturbed, fly short distances before seeking shelter.

Control

Chemical treatments remain the most important control tactic for commercial production. As the entire life cycle is completed on the plant, cultural control is not practical against this pest (Simmonds, 1982; Waterhouse and Norris, 1987) and biological control has very limited potential (Paine, 1964). Franzmann (1979) recorded only nine larval parasites and no egg parasites from extensive field collection of eggs and larvae. These included two species of tachinid fly parasite, *Bacromyiella ficta* and *Argyrophylax proclinata* Crosskey, and one elasmid wasp, *Elasmus* sp. Although Waterhouse and Norris (1987) list 12 species which were introduced into Fiji, Java, and Western Samoa from 1929 to 1964, only one, *Chelonus* sp., established in Fiji where it exerts variable control. Poor efficiency has also been noted from naturally occurring parasites. In north Queensland, high populations of the ants *Tetramorium bicarinatum* (Nylander) and *Pheidole megacephala* appeared to significantly reduce the incidence of damage and *T. bicarinatum* was observed attacking larvae. However, the generally poor level of biological control of *N. octasema* in bananas led Paine (1964) to the pessimistic conclusion that limitations of food and climate are more important than parasites in the regulation of this pest.

In commercial plantations, pesticide applications to the 'throat' of plants which are approaching inflorescence emergence by aerial or ground-based cover sprays or by dusting individual plants have been replaced by a single targeted injection into the inflorescence. Diluted organophosphates or carbamates (20–40 ml) are injected into the upper third of the inflorescence to prevent damage to fruit (Pinese and Piper, 1994). This treatment has to be applied to the vertical inflorescence, requiring at least weekly selection and treatment of inflorescences.

Fruit flies: Banana fruit fly, *Bactrocera musae* (Tryon); Queensland fruit fly, *Bactrocera tryoni* (Froggatt)

Both the banana fruit fly, *Bactrocera musae* (Tryon), and the Queensland fruit fly, *Bactrocera tryoni* (Froggatt), are pests of banana grown in north Queensland (Australia) the Torres Strait, Papua New Guinea and nearby islands. *B. musae* is primarily a pest of cultivated and wild banana and will sting and oviposit in very immature fruit, but its eggs may fail to develop unless oviposition occurs in fruit approaching maturity (climacteric). *B. tryoni* has an extensive host range but will only sting ripe or ripening banana fruit. Commercial bananas harvested at the mature 'hard green' stage (preclimacteric) are not considered hosts of either species for quarantine purposes.

Eggs are laid in the pulp just below the skin. As fruit ripens, they hatch and the maggots feed on the soft flesh and, when fully developed, fall to the ground to pupate. The complete life cycle takes about 2.5 weeks during the hot summer months.

Control

Control relies on good plantation management to prevent 'mixed ripe' bunches which can attract and act as breeding sites for fruit fly. Harvesting at the 'hard green' stage, prior to fruit becoming susceptible, is the main method of ensuring damage-free fruit. Insecticide treatments are not necessary.

Pests of Foliage

Banana skipper, *Erionota thrax* (Linnaeus)

The banana skipper, *Erionota thrax* (Linnaeus), is a minor to severe pest of bananas and *Musa textilis* in South-East Asia (Kalshoven, 1981) and, since 1986, Papua New Guinea (Sands *et al.*, 1988). Damage is also recorded from bamboo, coconut and a range of palms including oil and nipa (Waterhouse and Norris, 1989) although it has been suggested that other species may be responsible for the records on palms and bamboo (Sands *et al.*, 1988; Waterhouse and Norris, 1989). Banana clumps in isolated villages in Java had a very patchy damage distribution, ranging from severe defoliation to nil on clumps growing within close proximity (B. Pinese, personal observation). Heavy rainfall and strong winds are unsuitable for banana skipper. Entry of water into the leaf rolls drowns the larva (particularly the first instar) and wind-torn leaf laminae are unsuitable for the production of leaf roll shelters. For these reasons, outbreaks in Malaysia and Indonesia are more common after a drought and in wind-protected areas (Kalshoven, 1981).

Direct fruit production losses would only be significant following heavy defoliation, since banana plants can withstand at least 20% leaf lamina loss before production is affected (Ostmark, 1974). None the less, bananas in South-East Asia are grown for aesthetic value (Hoffmann, 1935) and for culinary purposes where even minor infestations would be detrimental.

The adult butterfly lays bright yellow eggs singly or in groups of up to 25. These are laid at dusk or at night, preferentially on the lower leaf lamina midway between the midrib and the outer edge. Eggs turn bright red and the pale green larvae hatch after about 5 to 8 days. The larvae move to the outer leaf lamina where they commence feeding and then produce loose rolls by cutting the leaf and rolling the lamina towards the midrib. Larvae feed and grow within the rolls, commencing a new roll once the midrib is reached. The second and subsequent three instars are covered in a white waxy powder that provides protection from drowning during high rainfall. The larval stage lasts between 20 and 30 days depending on temperature. Pupation occurs within the leaf roll and lasts from 8 to 12 days. Adults emerge in the afternoon and are most active in the evening and early morning when they actively fly around banana plants to mate and oviposit.

Control

The banana skipper is adequately controlled by a range of beneficial insects such that other control measures are seldom required. If unusually heavy outbreaks occur, the collection and destruction of leaf rolls is helpful. In Indonesia, egg parasitoids, including *Ooencyrtus erionotae*, *Agiommatus* sp. and *Anastatus* sp., can parasitize 50–70% of the eggs (Kalshoven, 1981). Young larvae are attacked by *Apanteles erionotae* while older third instar larvae are preferred by *Scenocharops* sp. (Waterhouse and Norris, 1989). The pupal parasitoids *Brachymeria* sp., *Xanthopimpla* sp. and *Pediobius* sp. also contribute to biological suppression of *E. thrax*.

Bagworm, *Oiketicus kirbyi* Guilding

The bagworm *Oiketicus kirbyi* Guilding is a polyphagous insect (Costa Lima, 1945; Martorell, 1945; Ebeling, 1959) that became a leaf-feeding pest of bananas in Costa Rica in 1958 (Stephens, 1962) and later caused some damage in plantations of banana in Colombia (Garcia, 1987). The maggot-like female deposits eggs in the posterior portion of the pupal case that remains as a secured 'bag'. Eggs hatch in 27 to 32 days and the larval stage may range from 207 to 382 days (Stephens, 1962). The female pupal period lasts 10 to 33 days while the male pupal stage lasts 11 to 39 days. The larvae-female adult can live 14 days during which time she may deposit up to 6700 eggs. The males live only 3 to 5 days.

Control

Two dipterous parasitoids, *Sarcophaga* (*Sarcodexia*) *lambens* Wiedemann and *Achaotoneura* sp., attack medium and large bagworms. A braconid, *Iphiaulax* sp. parasitizes all larval sizes, whereas *Psychidosmicra* sp. (Hymenoptera: Chalcidae) attacks only small larvae. Other hymenopterous parasites found by Stephens (1962) are *Casinaria* sp., *Phobetes* sp., and *Carinodes* sp. Predacious ants frequently attack larvae of all sizes. A microsporidiam, *Nosema* sp., and a fungus, *Beauveria bassiana* (Balsamo) Vuillmen, infect all larval stages (Stephens, 1962) in Costa Rica. In Colombia, small larvae are attacked by *Psychidosmicra* sp., *Spilochalcis* sp., *Iphiaulax* sp., and *Brachymeria* sp. (Garcia, 1987).

Cultural control is recommended by removing all larvae from infested plantations. Microbial control, using *Bacillus thuringiensis*, is recommended in Colombia as an effective control (Garcia, 1987).

Caterpillars, *Caligo* spp., *Opsiphanes* spp., *Antichloris* spp.

Several caterpillars sporadically, cause damage to the foliage of bananas. Often their presence is the result of disruptions to the ecosystem that surround banana plantations (Mesquita and Alves, 1984). For example, *Caligo* spp., *Opsiphanes* spp., and *Antichloris* spp. can become pests of banana when their natural enemies are reduced. At such times, these have been reported as key pests of banana in Brazil (Tourner and Viladerbo, 1966; Tourner *et al.*, 1966).

Control

In Brazil larvae of *Caligo* spp. are parasitized by *Hemimasipoda* sp. (Diptera: Tachinidae), *Spilochalcis* spp. (Hymenoptera: Chalcidae); *Opsiphanes* spp. is parasitized by *Apanteles* spp. (Hymenoptera: Braconidae), *Horismenus* spp. (Hymenoptera: Eulophidae), *Spilochalcis* spp. (Hymenoptera: Chalcidae) and by *Xanthozoma melanopyga* (Diptera: Tachinidae); *Anthocloris* spp. is parasitized by *Telenomus* spp. (Hymenoptera: Scelionidae), *Calocarcelia* sp. (Diptera: Tachinidae), *Meteorus* spp. (Hymenoptera: Braconidae) (Fancelli and Martins, 1998).

Sucking insects – Banana aphid, *Pentalonia nigronervosa* Coquerel

The banana aphid, *Pentalonia nigronervosa* Coquerel, is the sole vector of banana bunchy top disease, one of the serious viral diseases of *Musaceae* (Magnaye, 1959). Although the disease was observed as early as 1915 and became serious in 1923 (Ocfemia, 1930), this aphid gained importance only when the first proof of its ability to transmit bunchy top virus from diseased to healthy abaca was established (Ocfemia, 1930). Other researchers claim that only this species and the subspecies, *O. nigronervosa* f. *caladii*, are capable of transmitting the disease (Espino and Ocfemia, 1948). The banana aphid is a widely distributed tropical and subtropical species. Distribution records include the Philippines, Ellice Island, Malaya, Fiji, Samoa, Papua, Sabah, India, Ceylon, Mauritius, Africa (including Malawi and Rwanda), Egypt, Sierra Leone, Zanzibar, Australia, Central America, and Bermuda. In some cases (e.g. Uganda) the aphid is present, but the bunchy top virus is not.

Both alate and apterous forms of *Pentalonia nigronervosa* coexist in dense colonies on banana plants and breed continuously throughout the year (Varma and Capoor, 1958). *P. nigronervosa* populations increase at moderate humidity and temperature and decrease during drought and heavy rainfall (Kolkaila and Soliman, 1954; Varma and Capoor, 1958; Kung, 1963; Menon and Christudas, 1967; Gavarra, 1968). In Egypt, the greatest aphid activity occurs from December to February, with the population being at its lowest from March to May (Kolkaila and Soliman, 1954). In the warmer regions the aphid is most abundant during winter from May to July (Goddard, 1929; Ayyar, 1954; Varma and Capoor, 1958; Kung, 1963; Menon and Christudas, 1967). Besides the *Musaceae*, this species can be found feeding on species in the families

Araceae, Zingiberaceae, Cannaceae; however, their main hosts are abaca and banana.

Biology

The banana aphid reproduces parthenogenetically throughout the year and no males have been observed. It is also entirely viviparous as eggs are unknown. From 21 to 26 overlapping generations have been recorded throughout the year with no hibernation (Kung, 1963). The average fecundity of one apterous female is 31.5 (Johnson, 1963; Kung, 1963; Magnaye, 1959). Threshold development in degree-days ranged from 287 to 330°d at 22–31°C and 69–72% relative humidity (Lomerio and Calilung, 1993). There are four to five nymphal instars and the total developmental period ranges from 8 to 21 days with an average of 14.5 days (Kung, 1963).

The preferred sites for *P. nigronervosa* are within the whorl of the growing shoot and between the leaves on young suckers and growing plants and on immature fruit. During rainy months, the aphids are found around the base of the pseudostems at soil level or several centimetres below (Calilung, 1978). Colonies sometimes infest flowers and fruits (Kolkaila and Soliman, 1954). The aphids colonize the roots and corms of abaca but are rarely found on roots of banana (Magee, 1927; Ocfemia and Garcia, 1947; Wardlaw, 1972). In Australia, heavy bunch infestations, which result in sooty mould development on fruit, are infrequent due to effective biological control from a complex of beneficial insects. Outbreaks are associated with cooler weather in autumn and spring and disruption of natural control from broad specrum pesticides.

Dispersal can be by flight (Magee, 1927; Goddard, 1929) or assisted by the ants *Ragiolepsis longipes* Gord. and *Dolichoderus bituberculatus* Mayr.

Host plant resistance

Facundo and Sumalde (1998) determined that the abaca cultivars Itoalus × Magsarapong No. 7 and Pacol × CES III-2 were resistant to *P. nigronervosa*, compared with the development observed in the resistant cultivar Tinauagan Pula (TP). They consider that resistance of the hybrids to the aphid vector may partly explain their resistance to banana bunchy top disease.

Mites

Mites are considered minor pests of banana, but can erupt into epidemics, causing damage to leaves (Ostmark, 1974) and less often to fruit. Mite outbreaks are usually due to insecticide-induced disruption and some mite epidemics have been triggered by road dust inhibiting the action of predators (Ostmark, 1974).

Most of the mites are oligophagous species with restricted distribution. For instance, the banana rust mite *Phyllocoptruta musae* Keifer is reported to cause spotting on bananas in Queensland (Jeppson *et al.*, 1975), but it is not reported in other parts of the world. Mites of the genus *Tetranychus* are most commonly reported as pests of banana. The banana spider mite (also known as the strawberry spider mite) (Plate 9) *T. lambi* Pritchard and Baker and the twospotted spider mite *T. urticae* Koch are common pests of a broad range of crops and are widely distributed (Pinese and Piper, 1994).

Damage

Mite damage in banana is confined to the underside of the leaves; however, in severe outbreaks the mites can move to bunches and damage the fruit. Leaf damage appears at first as isolated bronzed rusty patches, which later coalesce along the leaf veins as the infestation increases. Fruit damage by *T. lambi* is found mainly at the cushion end of the fingers. Feeding in this area causes a red to purple-black discoloration of the fruit surface that may later dry out and crack (Pinese and Piper, 1994). *T. urticae* damages the tips of fingers mostly on the top hands, giving the affected fruit a dull silvery grey appearance. The main impact of mite damage is on yield, where loss of phytosynthetic capacity results in slower plant growth and fruit filling. The impact is most noticeable if plants are stressed by other factors such as drought or nematode and banana weevil borer damage to roots and corms. In experimental trials, where mite pressure and plant response were assessed, an arbitrary 'medium' leaf damage, where bronzing damage was readily visible on leaves, caused a 2 week extension to the fruit filling period from bunch emergence to harvest (B. Pinese, unpublished).

Biology

The life cycles and appearance of *T. lambi* and *T. urticae* are similar. The main distinguishing feature between them is the relative lack of fine webbing in infestations of *T. lambi*, whereas high populations of *T. urticae* are always associated with webbing (Pinese and Piper, 1994).

Monitoring and control

Pinese and Piper (1994) recommended fortnightly monitoring during periods of hot dry weather when mite development and activity is highest. Five plants are selected and mite damage to leaves is assessed using three categories: (i) low or few scattered discrete mite colonies; (ii) medium or mite colonies scattered but numerous and coalescing between the interveinal areas; and (iii) high, or mite colonies coalescing, with bronzing damage over most of the leaves.

Miticide treatments are required only when mean damage to leaves reaches the 'medium' level. Other considerations relate to the presence or absence of mite predators, particularly the mite-eating ladybird beetle, *Stethorus fenestralis* (Coleoptera: Coccinellidae), plant vigour and anticipated weather conditions. To achieve maximum benefit from miticide treatments, high volume cover sprays using a minimum of 500 l ha^{-1} must be applied when leaves are turgid to ensure good spray cover under the leaves where the mites are present. A second application, 14 days later, is usually required to control mite carryover.

Acknowledgement

This is Florida Agricultural Experiment Station's Journal Series No. R-08408.

References

Abera, A.M.K. (1997) Oviposition preferences and timing of attachment by the banana weevil (*Cosmopolites sordidus* Germar) in East African highland banana (*Musa* spp.). MSc thesis, Makerere University, Kampala, Uganda.

Abera, A.M.K., Gold, C.S. and Kyamanywa, S. (1999) Timing and distribution of attack by the banana weevil (Coleoptera: Curculionidae) in East African highland banana (*Musa* spp.). *Florida Entomologist* 82, 631–641.

Allen, R.N. (1989) *Control of Major Pests and Diseases of Bananas*. Department of Agriculture New South Wales, 10 pp.

Alpizar, D., Fallas, M., Oehlschlager, A.C., Gonzalez, L. and Jayaraman, S. (1999) Pheromone-based mass trapping of the banana weevil, *Cosmopolites sordidus* (Germar) and the West Indian sugarcane weevil *Metamasius hemipterus* L. (Coleoptera: Curculionidae) in plantain and banana. *Memorias XIII Reunion ACORBAT*, 23–27 November 1998. Guayaquil, pp. 515–538.

Altieri, M.A. and Letourneau, D.K. (1982) Vegetation management and biological control in agroecosystems. *Crop Protection* 1, 405–430.

Ambrose, E. (1984) Research and development in banana crop protection (excluding Sigatoka) in the English speaking Caribbean. *Fruits* 39, 234–247.

Anitha, N., Rajamony, L. and Radhakrishnan, T.C. (1996) Reaction of banana clones against major biotic stresses. *Planter* 72, 315–321.

Anonymous (1989) *Guia Educativa No. 1: Trampeo Para el Picudonegro en Platano*. Fundacion Hondureña Investigaciones Agricolas, La Lima, 14 pp.

Arleu, R.J. and Neto, S.S. (1984) Broca da bananeira *Cosmopolites sordidus* (Germ., 1824) (Coleoptera: Curculionidae). *Turrialba* 34, 359–367.

Arroyave, F.P. (1985) Control del picudo negro *Cosmopolites sordidus* Germar en semilla vegatativa de platano (*Musa* AAB Simmonds). Ingeniero Agronomo thesis, Universidad de Caldas, Manizales, Colombia.

Ayala, J.L. and Monzon, S. (1977) Ensayo sobre diferentes dosis de *Beauveria bassiana* para el control del picudo negro del platano (*Cosmopolites sordidus*) (Germ.). *Centro Agricultura, Revista Ciencias de Facultad de Ciencias Agricolas* 4, 19–24.

Ayyar, G.R. (1954) Unrecorded host plant of *Pentalonia nigronervosa* Coq. *Current Science* 23, 408–409.

Bakyalire, R. (1992) A study of the life cycle and behaviour of the banana weevil *Cosmopolites sordidus* Germar, in Uganda. MSc thesis, Makerere University, Uganda, 118 pp.

Batista Filho, A., Paiva Castro, L.M., Myazaki, I., Bastos Cruz, B.P. and Oliveira, D.A. (1987) Controle biolgico do 'moleque' da bananeira *Cosmopolites sordidus* (Germ., 1824) pelo uso de fungo entomogenos, no laboratorio. *Biologico, Sao Paulo* 53, 1–6.

Batista Filho, A., Sata, M.E., Leite, L.G., Raga, A. and Prada, W.A. (1991) Utilizacao de *Beauveria bassiana* (Bals.) Vuill., no controle do moleque da bananeira *Cosmopolites sordidus* (Germ., 1824) (Coleoptera: Curculionidae). *Revista Brasileira Fruticultura, Cruz das Almas* 13, 35–40.

Batista Filho, A., Leite, L.G., Raga, A., Sato, M.E. and Oliveira, J.A. (1995) Utilizacao de *Beauveria bassiana* (Bals.) Vuill. no manejo de *Cosmopolites sordidus* Germ., 1824, em Miracatu, Sao Paulo. *Biologico* 57, 17–19.

Batista Filho, A., Leite, L.G., Alves, E.B. and Aguiar, J.C. (1996) Controle de *Cosmopolites sordidus* (Coleoptera: Curculionidae) por fipronile e seu efeito sobre *Beauveria bassiana*. *Arquivos do Instituto Biologico* 63, 47–51.

Beccari, F. (1967) Contributo alla conoseenza del *Cosmopolites sordidus* Ger. (Coleoptera: Curculionidae), Parte I-II. *Rivista di Agricultura Subtropicale e Tropicale* 61, 51–93; 131–150.

Boivin, G. (1993) Les parasitoides des oeufs de Curculionidae. In: Gold, C.S. and Gemmill, B. (eds) *Biological and Integrated Control of Highland Banana and Plantain Pests and Diseases. Proceedings of a Research Coordination Meeting*. International Institute of Tropical Agriculture, Cotonou, Benin, pp. 97–106.

Braimah, H. (1997) Laboratory studies on the host plant searching behaviour and chemical ecology of the banana weevil, *Cosmopolites sordidus* (Germ. 1824), (Coleoptera: Curculionidae). PhD thesis, University of Reading, Reading, UK.

Braithwaite, B.M. (1958) Ground spray treatments for control of the banana beetle borer (*Cosmopolites sordidus*) (Germ.). *Journal of the Australian Institute of Agricultural Science* 24, 27–34.

Breen, J.P. (1994) *Acremonium* endophyte interactions with plant resistance to insects. *Annual Review of Entomology* 39, 401–423.

Brenes, S. and Carballo, M. (1994) Evaluacion de *Beauveria bassiana* (Bals.) para el control biologico del picudo del platano *Cosmopolites sordidus* Germar. *Manejo Integrado de Plagas* 31, 17–21.

Budenberg, W.J. and Ndiege, I.O. (1993) Volatile semiochemicals of the banana weevil. In: Gold, C.S. and Gemmill, B. (eds) *Biological and Integrated Control of Highland Banana and Plantain Pests and Diseases. Proceedings of a Research Coordination Meeting*. International Institute of Tropical Agriculture. Cotonou, Benin, pp. 75–86.

Budenberg, W.J., Ndiege, I.O. and Karago, F.W. (1993a) Evidence for volatile male-produced pheromone in banana weevil *Cosmopolites sordidus*. *Journal of Chemical Ecology* 19, 1905–1915.

Budenberg, W.J., Ndiege, I.O., Karago, F.W. and Hansson, B.S. (1993b) Behavioral and electrophysiological responses on the banana weevil *Cosmopolites sordidus* to host plant volatiles. *Journal of Chemical Ecology* 19, 267–277.

Busoli, A.C., Fernandes, O.A. and Tayra, O. (1989) Controle da broca da bananeira *Cosmopolites sordidus* Germar 1824 (Coleoptera: Curculionidae) atraves dos fungos entomopatogenicos *Beauveria bassiana* (Bals) Vuill. e *Metarhizium anisoplae* (Metschn.) Sorok. (Hyphomycetes). *Annais Sociedade Entomologica Brasileira* 18, 33–41.

Calderon, A., Castineiras, A. and Lopez, M. (1991) Efecto de los biocidas y fertilizantes empleados en el cultivo del platano en Cuba sobre los hongos entomopatogenos. *Proteccion de Plantas* 1, 21–31.

Calilung, V.J. (1978) Aphid vectors of plant viruses in the Philippines. *Philippine Entomology* 4, 219–232.

Carballo, M. (1998) Mortalidad de *Cosmopolites sordidus* condiferentes formulaciones de *Beauveria bassiana*. *Manejo Integrado de Plagas* 48, 45–48.

Carballo, M. and de Lopez, M.A. (1994) Evaluacion de *Beauveria bassiana* (Bals.) para el control biologico del *Cosmopolites sordidus* Germar y *Metamasius hemipterus* en condiciones de campo. *Manejo Integrado de Plagas* 31, 22–24.

Carroll, G.C. (1991) Fungal associates of woody plants as insect antagonists in leaves and stems. In: Barbosa, P., Krischik, V.A. and Jones, C.G. (eds) *Microbial Mediation of Plant–Herbivore Interactions*. John Wiley & Sons, New York, pp. 253–271.

Castineiras, A. and Ponce, E. (1991) Efectividad de la utilizacion de *Pheidole megacephala* (Hymenoptera: Formicidae) en la lucha biologica contra *Cosmopolites sordidus* (Coleoptera: Curculionidae). *Proteccion de Plantas* 1, 15–21.

Castrillon, C. (1989) *Plagas del cultivo del platano. Curso de Actualizacion sobre Problemas Sanitarios en Platano.* Instituto Colombiano Agropecuario, La Dorada, Colombia, 54 pp.

Castrillon, C. (2000) *Distribucion de las Especies de Picudo del Platano. Evaluacion de sus Entomopatogenos Nativos en el Departamento de Risaralda.* Corporacion Instituto Colombiano Agropecuario, Manizales, Colombia, 72 pp.

Castrillon, C. and Herrera, J.G. (1986) *Los Picudos Negro y Rayado del Platano y Banano.* ICA-Informa Separata, 4 pp.

Cave, R. and Alvarez del Hierro, D. (1997). *Admontia* sp: a hope for biological control of *Metamasius callizona* in Florida. *Annual Meeting Florida Entomological Society,* 4–7 August, Daytona Beach, Florida, p. 7.

Champion, J. (1975) Productions bananieres et recherché scientifique. *Fruits* 30, 11–17.

Charles, J.S.K., Thomas, M.J., Menon, R., Premalatha, T. and Pillai, S.J. (1996) Field susceptibility of banana to pseudostem borer *Odoiporus longicollis* Oliver. In: Nayar, N.K. and George, T.E. (eds) *Technological Advancement in Banana/Plantain Production and Processing.* Kerala Agricultural University, Mannuthy, India, p. 32.

Chavarria-Carvajal, J.A. (1998) Response of eight plantain clones to nematodes and the corm-weevil (*Cosmopolites sordidus* Germar) in Puerto Rico. *Memorias XIII Reunion ACORBAT,* 23–27 November 1998. Guayaquil, pp. 539–546.

Childers, C. and Achor, D. (1995) Thrips feeding and oviposition injuries to economic plants, subsequent damage and host responses to infestation. In: Parker, B.L., Skinner, M. and Lewis, T. (eds) *Thrips Biology and Management.* Plenum Press, New York, pp. 31–51.

Cintra, A.F. (1975) *Opogona* sp. nova praga da bananicultura em Sao Paulo. *O Biologico Sao Paulo* 41, 223–231.

Collins, P.J., Treverrow, N.L. and Lambkin, T.M. (1991) Organophosphorus insecticide resistance and its management in the banana weevil borer, *Cosmopolites sordidus* (Germar) (Coleoptera: Curculionidae), in Australia. *Crop Protection* 10, 215–221.

Contreras, T. (1996) Evaluacion de trampas de pseudotallo y formulaciones de *Beauveria bassiana* (Bals) en el combate del picudo negro del platano *Cosmopolites sordidus* (Germar) en Costa Rica. Unpublished thesis, Mag. Sci. CATIE, 68 pp.

Costa Lima, A.M. (1945) Psychidae (Boisduval, 1892). *Insectos do Brazil* 5, 186–194.

Cuille, J. (1950) Recherches sur le charancon du bananier. Institut de Fruits et Agrumes Coloniaux. *Serie Technique No. 4,* Paris, 225 pp.

Cuille, J. and Vilardebo, A. (1963). Les calandrini nuisibles au bananier. In: Balachowsky, A.S. (ed.) *Entomologie appliquee a l'agriculture.* Masson et Cie, Paris, pp. 1099–1114.

Davis, D. and Peña, J.E. (1990) Biology and morphology of the banana moth, *Opogona sacchari* (Bojer) and its introduction into Florida (Lepidoptera: Tineidae). *Proceedings of the Washington Entomological Society* 92, 593–618.

Delattre, P. (1980) Recherche d'une methode d'estimation des populations du charancon du bananier, *Cosmopolites sordidus* Germar (Col.: Curculionidae). *Acta Oecologica: Oecologia Applicata* 1, 83–92.

Delattre, P. and Jean-Bart, A. (1978). Activités des champignons entomopathogenes (Fungi imperfecti) sur les adultes de *Cosmopolites sordidus* Germ. (Coleoptera: Curculionidae). *Turrialba* 28, 287–293.

de Souza, A.F., Warumby, J.F., de Moura, R.J.M., de Almeida, J.L. and Dantas, A.P. (1981) Dinamica populacional de *Cosmopolites sordidus* (Germar, 1824) e *Metamasius hemipterus*, e ocorrencia de epizootias por *Beauveria bassiana* em plantios de bananeira 'Prata' situados em topografia de varzea e de serra, no estado de Pernambuco. *IPA Divulga* 3, 252–268.

Dutt, N. and Maiti, B.B. (1972) Bionomics of the banana pseudostem weevil, *Odoiporus longicollis* Oliv. (Coleoptera:Curculionidae). *Indian Journal of Entomology* 34, 20–30.

Ebeling, W. (1959) *Subtropical Fruit Pests.* University of California, Berkeley, 436 pp.

Edge, V.E. (1974) Cyclodiene-BHC resistance in *Cosmopolites sordidus* (Germ.) (Coleoptera: Curculionidae) in New South Wales, Australia. *Bulletin of Entomological Research* 64, 1–7.

Espino, R.C. and Ocfemia, G.O. (1948) An additional insect vector of bunchy top of abaca and Manila hemp plant. *Philippine Agriculture* 31, 231–232.

Facundo, H.T. and Sumalde, A. (1998) Feeding preference of *Pentalonia nigronervosa* Coq. (Hemiptera: Aphididae) for and its comparative development on bunchy top susceptible abaca variety and resistant hybrids. *Philippine Entomology* 12, 169–177.

Fancelli, M., and Martins, A.L. (1998) Pragas da bananeira. In: Braga, R., Cardoso, J.E. and Chagas, F. (eds) *Pragas de fruteiras tropicais de importancia agroindustrial.* Empresa Brasileria de Pesquisa Agropecuaria, pp. 41–51.

Figueroa, W. (1990) Biocontrol of the banana root borer weevil, *Cosmopolites sordidus* (Germar) with steinernematid nematodes. *Journal of Agriculture of the University of Puerto Rico* 74, 15–19.

Fogain, R. and Price, N.S. (1994) Varietal screening of some *Musa* cultivars for susceptibility to the banana weevil, *Cosmopolites sordidus* (Coleoptera: Curculionidae). *Fruits* 49, 247–251.

Franzmann, B.A. (1979) Parasitism of banana scab moth *Nacoleia octasema* (Meyrick) (Lepidoptera: Pyralidae) in north Queensland. *Journal of the Australian Entomological Society* 18, 135–136.

Froggatt, J.L. (1924) Banana weevil borer (*Cosmopolites sordidus* Chev.). *Queensland Agricultural Journal* 21, 369–378.

Froggatt, J.L. (1925) The banana weevil borer (*Cosmopolites sordidus*). *Queensland Journal of Agriculture* 24, 558–593.

Froggatt, J.L. (1928) The banana weevil borer in Java, with notes on other crop pests. *Queensland Journal of Agriculture* 6, 530–541.

Gallo, D. (1978) *Manual de Entomologia Agricola.* Agronomia Ceres, Sao Paulo, 531 pp.

Gallo, D., Nakano, N., Silvera-Neto, S., Carvalho, R.P., de Batista, G.C., Berti Filho, E., Parra, J.R.P., Zucchi, R.A., Alves, S.B. and Vendramin, J.D. (1988) Pragas das fruitiferas-Bananeira. *Manual de Entomologia Agricola.* Sao Paulo, Agronomica Ceres, pp. 428–434.

Garcia, F. (1987) Aspectos biologicos y manejo del gusano canasta *Oiketicus kirbyi. Instituto Colombiano Agropecuario, Boletin Tecnico* 149, 23 pp.

Ghesquiere, M.J. (1924) La maladie des bananiers dans le Bas-Congo. *Bulletin Agriculture Congo Belge, Bruxelles* 15, 171–175.

Ghesquiere, M.J. (1925) La maladie du bananier au Congo Belge. *Bulletin Agriculture Congo Belge, Bruxelles* 3–4, 556–560.

Giblin-Davis, R.M., Peña, J. E. and Duncan, R.E. (1994) Lethal pitfall trap for evaluation of semiochemical-mediated attraction of *Metamasius hemipterus sericeus* (Coleoptera: Curculionidae). *Florida Entomologist* 77, 247–255.

Giblin-Davis, R., Peña, J. and Duncan, R. (1996) Evaluation of an entomopatholgenic nematode and chemical insecticides for control of *Metamasius hemipterus sericeus* (Coleoptera: Curculionidae). *Journal of Entomological Science* 31, 240–251.

Goddard, E.J. (1929) Bunchy-top of banana. *Journal for Australian Science and Industry* 11, 21–27

Gold, C.S. (1998) Banana weevil: ecology, pest status and prospects for integrated control with emphasis on East Africa. In: Saini, R.K. (ed.) *Proceedings of the Third International Conference on Tropical Entomology,* 30 Oct.–4 Nov. 1994, Nairobi. ICIPE Science Press, Nairobi, pp. 47–71.

Gold, C.S. and Bagabe, M.I. (1997) Banana weevil, *Cosmopolites sordidus* Germar (Coleoptera: Curculionidae), infestation of cooking and beer bananas in adjacent stands in Uganda. *African Entomology* 5, 103–108.

Gold, C.S. and Gemmill, B. (1993) *Biological and Integrated Control of Highland Banana and Plantain Pests and Diseases.* Proceedings of a Research Coordination Meeting, Cotonou, Benin.

Gold, C.S., Speijer, P.R., Karamura, E.B. and Rukazambuga, N.D. (1994a) Assessment of banana weevils in East African highland banana systems and strategies for control. In: Valmayor, R.V., Davide, R.G., Stanton, J.M., Treverrow, N.L. and Roa, V.N. (eds) *Proceedings of Banana Nematode/Borer Weevil Conference,* Kuala Lumpur, 18–22 April 1994, pp. 170–190.

Gold, C.S., Speijer, P.R., Karamura, E.B., Tushemereirwe, W.K. and Kashaija, I.N. (1994b) Survey methodologies for pest and disease assessment in Uganda. *African Crop Science Journal* 2, 309–321.

Gold, C.S., Okech, S.H. and Ssendege, R. (1997) Banana weevil population densities and related damage in Ntungamo and Mbarara districts, Uganda. In: Adipala, E., Tenywa, J.S. and Ogenga-Latigo, M.W. (eds) *African Crop Science Conference Proceedings,* Pretoria 13–17 January 1997. Makerere University, Kampala, Uganda, pp. 1207–1219.

Gold, C.S., Night, G., Abera, A. and Speijer, P.R. (1998a) Hot-water treatment for control of banana weevil, *Cosmopolites sordidus* Germar (Coleoptera: Curculionidae) in Uganda. *African Entomology* 6, 215–221.

Gold, C.S., Night, G., Speijer, P.R., Abera, A.M.K. and Rukazambuga, N.D.T.M. (1998b) Infestation levels of banana weevil, *Cosmopolites sordidus* Germar, in banana plants established from treated propagules in Uganda. *African Entomology* 6, 253–263.

Gold, C.S., Bagabe, M.I. and Ssendege, R. (1999a) Banana weevil, *Cosmopolites sordidus* (Germar): (Coleoptera: Curculionidae) tests for suspected resistance to carbofuran and dieldrin in Masaka District, Uganda. *African Entomology* 7, 189–196.

Gold, C.S., Karamura, E.B., Kiggundu, A., Bagamba, F. and Abera, A.M.K. (1999b) Geographic shifts in highland cooking banana (*Musa* spp., group AAA-EA) production in Uganda. *International Journal of Sustainable Development and World Ecology* 6, 45–59.

Gold, C.S., Nemeye, P. and Coe, R. (1999c) Recognition and duration of larval instars of banana weevil, *Cosmopolites sordidus* Germar, in Uganda. *African Entomology* 7, 49–62.

Gold, C.S., Rukazambuga, N.D.T.R., Karamura, E.B., Nemeye, P. and Night, G. (1999d)

Recent advances in banana weevil biology, population dynamics and pest status with emphasis on East Africa. In: Frison, E., Gold, C.S., Karamura, E.B. and Sikora, R.A. (eds) *Mobilizing IPM for Sustainable Banana Production in Africa. Proceedings of a Workshop on Banana IPM Held in Nelspruit, South Africa,* 23–28 November 1998. INIBAP, Montpellier, pp. 35–50.

Gold, C.S., Okech, S.H. and Nokoe, S. (2002) Evaluation of pseudostem trapping as a control measure against banana weevil, *Cosmopolites sordidus* (Coleoptera: Curaulionidae). *Bulletin of Entomological Research* 92, 35–44.

Gomes, C. (1985) Estudo do comportamento da broca da bananeira *Cosmopolites sordidus* (German, 1824) (Coleoptera: Curculionidae), visando seu controle. Unpublished doctorate thesis, University of Sao Paulo, 82 pp.

Gordon, J. and Ordish, G. (1966). Insect pests of banana. *Bananas* PANS *Manual No. 1*. Ministry of Overseas Development, London, pp. 33–35.

Gorenz, A.M. (1963) Preparation of disease free planting materials of banana and plantain. *Ghana Farmer* 7, 15–18.

Gowen, S.R. (1995) Pests. In: Gowen, S. (ed.) *Bananas and Plantains*. Chapman & Hall, London, pp. 382–402.

Gravier, C. (1907) Sur un coleoptore (*Sphenoporous striatus* Fahr) qui attaque les bananiers a Sau Thome (Golfe de Guinee). *Bulletin Museum Natural d'Histoire, Paris* 13, 30–32 (cited in Viswanath, 1976).

Griesbach, M. (2000) Occurrence of mutualistic endophytes in bananas (*Musa* spp.) and their potential as biocontrol agents of the banana weevil *Cosmopolites sordidus* (Germar) (Coleoptera: Curculionidae) in Uganda. PhD dissertation, University of Bonn, Bonn, 131 pp.

Hargreaves, H. (1940) Insect pests of bananas. In: Tothill, J.D. (ed.) *Agriculture in Uganda*. Oxford University Press, Oxford, pp. 121–124.

Hoffmann, W.E (1935) Observations on an hesperid leafroller and a lace bug, two pests of banana in Kwangtung. *Lingnan Science Journal* 14, 634–649.

ICIPE (1991) *Annual Report: 1990*. ICIPE, Nairobi, Kenya.

INIBAP (2001) http://www.inibap.org/presentation/inibap_eng.htm

Irizzary, H., Rivera, E., Rodriguez, J., Beauchamp de Caloni, I. and Oramas, D. (1988) The Lacknau plantain: a high yielding cultivar with field resistance to the corm weevil, *Cosmopolites sordidus* (Germar). *Journal of the Agricultural University of Puerto Rico* 72, 353–363.

Ishaque, N.M.M. (1978) A note on the incidence of *Odoiporus longicollis* (Oliv.) in Assama. *Pesticides* 12, 22–24.

Ittyeipe, K. (1986) Studies on the host preference of banana weevil borer *Cosmopolites sordidus* Germ. (Curculionidae: Coleoptera). *Fruits* 41, 375–379.

Jayaraman, S., Ndiege, I.O., Oehlschlager, A.C., Gonzalez, L.M., Alpizar, D., Falles, M., Budenberg, W.J. and Ahuya, P. (1997) Synthesis, analysis, and field activity of sordidin, a male-produced aggregation pheromone of the banana weevil, *Cosmopolites sordidus*. *Journal of Chemical Ecology* 23, 1145–1161.

Jepson, F.P. (1914) A mission to Java in quest of natural enemies for a coleopterous pest of bananas (*Cosmopolites sordidus*, Chevr.). *Department of Agriculture of Fiji, Bulletin No. 7*, 23 pp.

Jeppson, L.R., Keifer, H. and Baker, E. (1975) *Mites Injurious to Economic Plants*. University of California Press, Berkeley, USA, 614 pp.

Johnson, J. (1963) Biology of the bunchy top aphid *Pentalonia nigronervosa* Coq. *Agricultural Research in Kerala* 2, 45–51.

Jones, D.E. (1968) Attraction of banana rhizome volatiles to the banana root borer, *Cosmopolites sordidus* Germar. United Fruit Company, La Lima, Honduras, 7 pp.

Kaaya, G.P., Seshu Reddy, K.V., Kokwaro, E.D. and Munyinyi, D.M. (1993) Pathogenicity of *Beauveria bassiana*, *Metarhizium anisoplae* and *Serratia marcescens* to the banana weevil *Cosmopolites sordidus*. *Biocontrol Science and Technology* 3, 177–187.

Kalshoven, L.G.E. (1981) *Pests of Crops in Indonesia*. P.T. Ichtiar Baru-Nun Hoeve, Jakarta, 701 pp.

Karamura, E. B. and Karamura, D.A. (1995) Banana morphology – part II: The aerial shoot. In: Gowen, S. (ed.) *Bananas and Plantains*. Chapman & Hall, London, pp. 190–205.

Kehe, M. (1985) Les principaux insectes depredateurs du plantain en Cote D'Ivoire: importance des infestations et incidence agroeconomique. In: *Proceedings of the Third Meeting: International Association for Research on Plantain and Bananas*, 27–31 May 1985, Abidjan, pp. 94–101.

Kiggundu, A. (2000) Host plant reactions and resistance mechanisms to banana weevil, *Cosmopolites sordidus* (Germar) in Ugandan *Musa* germplasm. MSc thesis, Orange Free State University, 98 pp.

Kiggundu, A., Vuylsteke, D. and Gold, C.S. (1999) Recent advances in host plant resistance to banana weevil, *Cosmopolites sordidus* Germar. In: Frison, E., Gold, C.S., Karamura, E.B. and Sikora, R.A. (eds) *Mobilizing IPM for Sustainable*

Banana Production in Africa. Proceedings of a Workshop on Banana IPM Held in Nelspruit, South Africa, 23–28 November 1998. International Network for the Improvement of Banana and Plantain, Montpellier, pp. 87–96.

Knowles, L.H. and Jepson, F.P. (1912) Department of Agriculture. *Fiji Bulletin*. 17 pp. (cited by Cuille, 1950).

Kolkaila, A.M. and Soliman, B. (1954). A study of the banana aphid *Pentalonia nigronervosa* Coq. (Hemiptera: Homoptera: Aphididae) *Bulletin de la Societé Fouad 1er d'Entomologie* 38, 231–250.

Koppenhofer, A.M. (1993a) Observations on egg-laying behaviour of the banana weevil, *Cosmopolites sordidus* (Germar). *Entomologia Experimentalis et Applicata* 68, 187–192.

Koppenhofer, A.M. (1993b) Egg predators of the banana weevil, *Cosmopolites sordidus* (Germar) (Col.: Curculionidae) in western Kenya. *Journal of Applied Entomology* 116, 352–357.

Koppenhofer, A.M. (1993c) Search and evaluation of natural enemies of the banana weevil. In: Gold, C.S. and Gemmill, B. (eds) *Biological and Integrated Control of Highland Banana and Plantain Pests and Diseases. Proceedings of a Research Coordination Meeting*. International Institute of Tropical Agriculture, Cotonou, Benin, pp. 87–96.

Koppenhofer, A.M. (1995) Bionomics of *Euborellia annulipes* in western Kenya (Dermaptera: Carcinophoridae). *Entomologia Generalis* 20, 81–86.

Koppenhofer, A.M. and Schmutterer, H. (1993) *Dactylosternum abdominale* (F.) (Coleoptera: Hydrophilidae): a predator of the banana weevil. *Biocontrol Science and Technology* 3, 141–147.

Koppenhofer, A.M. and Seshu Reddy, K.V. (1994) A comparison of rearing methods for the banana weevil, *Cosmopolites sordidus* (Germar) (Coleoptera: Curculionidae) on its natural host. *Insect Science and its Application* 15, 191–195.

Koppenhofer, A.M., Seshu Reddy, K.V., Madel, G. and Lubega, M.C. (1992) Predators of the banana weevil, *Cosmopolites sordidus* (Germar) (Col.: Curculionidae) in western Kenya. *Journal of Applied Entomology* 114, 530–533.

Koppenhofer, A.M., Seshu Reddy, K.V. and Sikora, R.A. (1994) Reduction of banana weevil populations with pseudostem traps. *International Journal of Pest Management* 4, 300–304.

Kung, K.S. (1955) The banana stem-borer weevil (*Odoiporus longicollis* Oliv.) in Taiwan. *Journal of Agriculture and Forestry (Taiwan)* 4, 83–111 (Chinese: English summary).

Kung, K.S. (1963) Ecological studies of the banana aphid, *Pentalonia nigronervosa* Coq. *Plant Protection Bulletin* 5, 224–246.

Leeper, J.R. (1972). Adult feeding behavior of *Lixophaga sphenophori*, a tachinid parasitoid of the New Guinea sugarcane weevil. *Proceedings of the Hawaiian Entomological Society* 21, 403–412.

Lemaire, L. (1996) Les relations semiochimiques chez le charancon *Cosmopolites sordidus* (Germar) (Coleoptera: Curculionidae) et la resitance de sa plante-hote, le bananier. PhD thesis, University of Montpellier, Montpellier, France.

Lescot, T. (1988) Influence de l'altitude sur les populations du charancon des bananiers (*Cosmopolites sordidus* Germar). *Fruits* 43, 433–437.

Liceras, L., Urrelo, G. and Beltran, F. (1973) Ensayo para el control del gorgojo negro del platano, *Cosmopolites sordidus* Germar (Coleoptera: Curculionidae), al momento de la siembra. *Revista Peruana de Entomologia* 16, 50–54.

Lomerio, E.O. and Calilung, V.J. (1993) Comparative development of *Pentalonia nigronervosa* on five host plants. *Philippine Agriculture* 9, 101–151.

Longoria, A. (1968) Diferencias sexuales en la morfologia externa de *Cosmopolites sordidus* Germar (Coleoptera: Curculionidae). *Ciencias Biologicas, La Habana* 1, 1–11.

Magee, C.J. (1927) Investigations on the bunchy-top disease of bananas. *Bulletin of the Council of Science Industry Research, Australia* 30, 64 pp.

Magnaye, L.V. (1959) Studies on the identity and relationship of the abaca and banana bunchy-top virus in the Philippines. MS thesis. University of the Philippines, College of Agriculture, College Laguna, Philippines.

Marcelino, L. and Quintero, J.A. (1991) Influencia del picudo negro (*Cosmopolites sordidus*) y de la precipitacion en los platanares de cuatro localidades de Baru, Chiriqui. *Revista Ciencia Agropecuaria* 7, 49–57.

Martin, R. (1983) Morfologia y ecobiologia de *Opogona* (*Hieroxestis*) *subcervinella* (Wilk.) (Lepidoptera: Heterocera), plaga de los platanares canarios. PhD thesis. Universidad de la Laguna, La Laguna, Spain, 353 pp.

Martinez, J.A. and Palazzo, D.A. (1971) Ferrugem da banana ocasionada per tripes. In: *Congreso Brasileiro de Fruiticultura* 1, 1971. Campinas, SP, Anais Campinas: Sociedade Brasileira Frutais, pp. 201–206.

Martorell, L.F. (1945) A survey of the forest insects of Puerto Rico. *Journal of Agriculture, University of Puerto Rico* 29, 544–545.

Masanza, M. (1995) Integrating pseudostem trapping, chemical and biological control for the management of the banana weevil (*Cosmopolites sordidus* Germar). MSc thesis. Makerere University, Kampala, Uganda, 93 pp.

Masso, E. and Neyra, M. (1997) Danos y perdidas causadas por *Cosmopolites sordidus* en el cultivo del platano. *Agrotecnia de Cuba* 27, 86–88.

Mathew, M.P., Thomas, J., Pauol, T.S., Sivaraman, S., Nair, S.R. and Ibraham, K.K. (1996) Chemical control of pseudostem borer of banana *Odoiporus longicollis* Oliver. In: Nayar, N.K. and George, T.E. (eds) *Technological Advancement in Banana/Plantain Production and Processing.* Kerala Agricultural University, Kerala, India.

Mathew, M.P., Nair, S.R. and Sivarman, S. (1997) Management of pseudostem borer *Odoiporus longicollis. Indian Journal of Entomology* 59, 269–273.

Mbwana, A.S.S. and Rukazambuga, N.D.T.M. (1999) Banana IPM in Tanzania. In: Frison, E., Gold, C.S., Karamura, E.B. and Sikora, R.A. (eds) *Mobilizing IPM for Sustainable Banana Production in Africa. Proceedings of a Workshop on Banana IPM held in Nelspruit, South Africa,* 23–28 November 1998. INIBAP, Montpellier, pp. 237–245.

Menon, M.R. and Christudas, S.D. (1967) Studies on the population of the aphid, *Pentalonia nigronervosa* Coq., on banana plants in Kerala. *Agricultural Research Journal Kerala* 5, 84–85.

Mesquita, A.L.M. (1988) Controle biologico das brocas da bananeira *Cosmopolites sordidus* (Germar, 1824) e *Metamasius hemipterus* (Linne, 1764) com fungos entomogenos. *Reunion de la Associacion para cooperaction en investigatciones de banano en America Tropical (ACORBAT),* 8 Santa Marta, Colombia 1987, pp. 311–324.

Mesquita, A.L.M. and Alves, E.J. (1983) Aspectos da biologia da broca-do-rizoma em diferentes cultivares de bananeira (*Cosmopolites sordidus, Musa acuminata*). *Pesquisa Agropecuaria Brasileira* 18, 1289–1292.

Mesquita, A.L. and Alves, E.J. (1984) Lepidopteros desfolhadores de banana e seus inimigos naturais Cruz das Almas, *Empresa Brasileira Pesquisa Agropecuaria, Comunicado Technico, 3.*

Mesquita, A.L.M. and Caldas, R.C. (1986) Efeito da idade e da cultivar de bananeira sobre a biologia e preferencia do *Cosmopolites sordidus* (Germar, 1824) (Coleoptera: Curculionidae). *Fruits* 41, 245–249.

Mesquita, A.L.M., Lucchini, F., Alves, E.J. and Caldas, R.C. (1981) Influencia dos fatores ambientais no grau de parasitismo de *Beauveria bassiana* sobre *Cosmopolites sordidus* e *Metamasius hemipterus,* em cultivo da bananeira. *Pesquisa Andamento* No. 14, 4 pp.

Mesquita, A.L.M., Alves, E.J. and Caldas, R.C. (1984) Resistance of banana cultivars to *Cosmopolites sordidus* (Germar 1824). *Fruits* 39, 254–257.

Messiaen, S. (2000) Neem (*Azadirachta indica*), wood ashes, coffee husk and hot pepper (*Capsicum* spp.) for controlling the banana weevil (*Cosmopolites sordidus*): investigations into their effect and mode of action. Unpublished manuscript. Centre de Recherches Regionales sur Bananiers et Plantains. Njombe, Cameroon, 13 pp.

Messiaen, S., Fogain, R., Ysenbrandt, H. and Sama Lang, P. (2000) *In situ* efficacy of neem (*Azadirachta indica* A. Juss) for controlling banana weevil *Cosmopolites sordidus* Germar (Coleoptera: Curculionidae). Unpublished manuscript. Centre de Recherches Regionales sur Bananiers et Plantains. Njombe, Cameroon, 13 pp.

Mestre, J. (1997) Les recherches recentes sur le charancon des bananiers, *Cosmopolites sordidus* (Germar, 1824) (Coleoptera: Curculionidae). *Fruits* 52, 67–82.

Mestre, J. and Rhino, B. (1997) Les etudes sur le charancon des bananiers, *Cosmopolites sordidus* (Germar, 1924): Bilan sommaire – Neufchateau 1995–1997. *CIRAD-FHLOR Document JM-97–04.* Neufchateau, 20 pp.

Minost, C. (1992) Etude de la communication semiochemique chez le charancon du bananier, *Cosmopolites sordidus* Germar (1824) (Coleoptera: Curculionidae). DAA thesis, Institut National Agronomique, Paris.

Mitchell, G. (1978) The estimation of banana borer population and resistance levels. *Technical Bulletin* 2, Windward Island Banana Growers Association, St Lucia, 34 pp.

Moreira, R.S. (1971) A broca das banaeiras. *Correio Agricola (Sao Paulo)* 1, 10–12.

Moreira, R.S. (1979a) Bananis livres de broca produzem o dobro. *Correio Agricola* 2, 202–206.

Moreira, R.S. (1979b) Pragas. In: Moreira, R.S. (ed.) *Banana: teoria e practica de cultivo.* Campinas, Fundacao Gargill, 335 pp.

Moznette, G.F. (1920) Banana root-borer. *Journal of Agricultural Research* 19, 39–46.

Musabyimana, T. (1995) Studies on the banana weevil (*Cosmopolites sordidus*) and nematode complex in western Kenya: March–October 1995. *Final Report, International Center of Insect Physiology and Ecology,* 23 pp.

Musabyimana, T. (1999) Neem seed for the management of the banana weevil, *Cosmopolites sordidus* Germar (Coleoptera: Curculionidae) and banana parasitic nematode complex. PhD thesis, Kenyatta University, Nairobi.

Nahif, A.A. (1998) Morphology, histology and histochemistry of the reproductive system of *Cosmopolites sordidus* (Coleoptera: Curculionidae). Part 2. The female internal genitalia.

Unpublished manuscript, University of Bonn, 13 pp.

Nahif, A.A., Koppenhofer, A. and Madel, G. (1994) Morphologie, biologie und bedeuntung von *Cosmopolites sordidus*, Germar, 1824 (Coleoptera: Curculionidae. *Zeitschirt fur Angewandt Zoologie* 4, 435–447.

Nakasone, H.Y. and Paull, R.E. (1998) *Tropical Fruits*. CAB International, Wallingford, UK, 445 pp.

Nankinga, C.M. (1994) Potential of indigenous fungal pathogens for the biological control of the banana weevil, *Cosmopolites sordidus* (Germar), in Uganda. MSc thesis. Makerere University, Kampala, Uganda.

Nankinga, C.M. (1999) Characterization of entomopathogenic fungi and evaluation of delivery systems for the biological control of the banana weevil, *Cosmopolites sordidus*. PhD dissertation. University of Reading, UK.

Nankinga, C.M., Moore, D., Bridge, P. and Gowen, S. (1999) Recent advances in microbial control of banana weevil. In: Frison, E., Gold, C.S., Karamura, E.B., and Sikora, R.A. (eds) *Mobilizing IPM for Sustainable Banana Production in Africa. Proceedings of a Workshop on Banana IPM Held in Nelspruit, South Africa*, 23–28 November 1998. INIBAP, Montpellier, pp. 73–85.

Nanne, H.W. and Klink, J.W. (1975) Reducing banana root weevil adults from an established banana plantation. *Turrialba* 25, 177–179.

Ndege, L.J., Ndile, M.R. and Bosch, C.H. (1995) Farmers' assessment of a weevil trapping trial in Ntoija village, Bukoba district. Lake Zone Farming Systems project, Agricultural Research Institute, Maruku, Tanzania. *Progress Report No. 9*, 10 pp.

Neuenschwander, P. (1988) Prospects and proposals for biological control of *Cosmopolites sordidus* (Germar) (Coleoptera: Curculionidae) in Africa. In: *Nematodes and the Borer Weevil in Bananas.. Proceedings of a Workshop Held in Bujumbura, Burundi*, 7–11 December 1987. INIBAP, pp. 54–57.

Ngode, L. (1998) Management of banana weevil *Cosmopolites sordidus* Germar (Coleoptera: Curculionidae) for improved banana yield in western Kenya. PhD thesis, Kenyatta University, Kenya.

Nogueira, S.B. (1976) Efeito de alguns inseticidas, alcois, e esteres aplicados em iscas contra as brocas da bananeira, *Cosmopolites sordidus, Metamasius ensirostris* e *Metamasius unaequalis* (Coleoptera: Curculionidae). *Congreso Brasilerio Entomologia* 3, 90–91.

Nonveiller, G. (1965) Comment proteger les bananiers contre les attaques du charancon. *Cameroun Agriculture Pastoral Forestier* 87, 32–43.

Ocfemia, G.O. (1930) Bunchy top of abaca or Manila hemp. I. A study of the cause of the disease and its method of transmission. *Annual Journal of Entomology* 17, 1–18.

Ocfemia, G.O. and Garcia, F.P. (1947) Further studies on the transmission of bunchy top and mosaic of abaca. Separation of two diseases and mechanics of inoculation by *Pentalonia nigronervosa* Coq. *Philippine Agriculture* 31, 87–97.

Ogenga-Latigo, M.W. and Bakyalire, R. (1993) Use of pseudostem traps and coefficient of infestation (PCI) for assessing banana infestation and damage by *Cosmopolites sordidus* Germar. *African Crop Science Journal* 1, 39–48.

Oliveira, A.M. de, Sudo, S., Barcellos, D.F., Mendes, S.G., Maiolino, W. and Meneguelli, N. do A. (1976) Fluctuacao da populacao de *Cosmopolites sordidus* e *Metamasius* spp. em bananais de Agra dos Reis, Estado do Rio de Janeiro. *Pesquisa Agropecuaria Brasileira, Serie Agronomica* 11, 37–41.

Olson, F.J. (1970a) Mode of entry by the parasitoid maggot of the cane weevil tachinid, *Lixophaga sphenophori* into the New Guinea sugarcane weevil larvae, *Rhabdoscelus obscurus*. *Proceedings of the Hawaiian Entomological Society* 21, 109–112.

Olson, F.J. (1970b) Pathologies of the New Guinea sugarcane weevil grub, *Rhabdoscelus obscurus* caused by the parasitoid cane weevil tachinid. *Hawaiian Entomological Society* 20, 579–581.

Ortiz, R., Vuylsteke, D., Dumpe, B. and Ferris, R.S.B. (1995) Banana weevil resistance and corm hardness in *Musa* germplasm. *Euphytica* 86, 95–102.

Ostmark, H.E. (1974) Economic insect pests of bananas. *Annual Review of Entomology* 19, 161–176.

Ota, A. and Mitchell, W. (1971) An insecticide test to control the sugarcane beetle borer (Coleoptera: Curculionidae). *Hawaiian Planters Record* 58, 147–158.

Paine, R.W. (1964) The banana scab moth, *Nacoleia octasema* (Meyrick): its distribution, ecology and control. *Technical Papua South Pacific Communication*, No 145, 70 pp.

Painter, R.H. (1951) *Insect Resistance in Crop Plants*. Macmillan, New York, 512 pp.

Parnitzki, P. (1992) Biologische Bekampfund des Russelkafers *Cosmopolites sordidus* (Germar) mit entomopathogenen Nematoden der Gattungen *Heterorhabditis* und *Steinernema* sowie Untersuchungen zur Biologie des Schadlings. PhD dissertation. University of Bonn, Bonn.

Sikora, R.A., Bafokuzara, N.D., Mbwana, A.S.S., Oloo, G.W., Uronu, B. and Seshu Reddy, K.V. (1989) Interrelationship between banana weevil, root lesion nematode and agronomic practices, and their importance for banana decline in the United Republic of Tanzania. *Food and Agriculture Organization Plant Protection Bulletin* 37, 151–157.

Silva, C.G. and Fancelli, M. (1998) Banana insect pests. In: Galan S.V. (ed.) *Proceedings of an International Symposium of Banana in the Subtropics*. Tenerife, Spain, pp. 385–393.

Simmonds, N.W. (1966) *Bananas*. Longman, London, UK, 512 pp.

Simmonds, N.W. (1982) *Bananas*, 2nd edn. Longman, Harlow, UK.

Simmonds, N.W. and Shepherd, K. (1955) The taxonomy and origins of the cultivated bananas. *Linnean Society of Botany* 55, 302–312.

Simmonds, N.W. and Simmonds, F.J. (1953) Experiments on the banana borer, *Cosmopolites sordidus* in Trinidad, BWI. *Tropical Agriculture* 30, 216–223.

Siqueira, H.A., Barreto, R., Cavalcante, T. and Picanco, M. (1996) Conrole biologico de *Cosmopolites sordidus* e *Metamasius* sp. (Coleoptera: Curculionidae) em bananeoira por predadores e parasitoides. *V Siconbiol, Faz de Iguacu, Brasil*, 9–14 June, p. 161.

Smith, D. (1995) Banana weevil borer control in south-eastern Queensland. *Australian Journal of Experimental Agriculture* 35, 1165–1172.

Sosa, O., Shine, J. and Tai, P. (1997) West Indian cane weevil (Coleoptera: Curculionidae): a new pest of sugarcane in Florida. *Journal of Economic Entomology* 90, 634–638.

Speijer, P.R., Budenberg, B. and Sikora, R.A. (1993) Relationships between nematodes, weevils, banana and plantain cultivars and damage. *Annals of Applied Biology* 123, 517–525.

Speijer, P.R., Gold, C.S., Kajumba, C. and Karamura, E.B. (1995) Nematode infestation of 'clean' banana planting material in farmers fields in Uganda. *Nematologica* 41, 344.

Sponagel, K.W., Diaz, F.J. and Cribas, A. (1995) *El Picudo Negro del Platano, Cosmopolites sordidus* Germar. Fundacion Hondureña Investigacion Agricola. La Lima, Honduras, 35 pp.

Ssennyonga, J.W., Bagamba, F., Gold, C.S., Tushemereirwe, W. K., Ssendege, R. and Katungi, E. (1999) *Understanding Current Banana Production with Special Reference to Integrated Pest Management in Southwestern Uganda*. International Center of Insect Physiology and Ecology, Nairobi, Kenya, 47 pp.

Stanton, J.M. (1994) Status of nematode and weevil borer problems in Australia. In: Valmayor, R.V., Davide, R.G., Stanton, J.M., Treverrow, N.L. and Roa, V.N. (eds) *Proceedings of Banana Nematode/Borer Weevil Conference*, Kuala Lumpur, 18–22 April 1994. INIBAP, Los Baños, pp. 48–56.

Stephens, C.S. (1962) *Oiketicus kirbyi* (Lepidoptera: Psychidae) a pest of bananas in Costa Rica. *Journal of Economic Entomology* 55, 381–386.

Stover, R.H. and Simmonds, N.W. (1987) *Bananas*, 3rd edn. John Wiley & Sons, New York, 469 pp.

Taylor, B. (1991) Research field work on upland bananas, *Musa* spp., principally *acuminata* triploid AAA types in the Kagera region of Tanzania. With observations on growth and causes of decline in crop yield. *Rivista di Agricultura Subtropicale e Tropicale* 85, 349–392.

Topham, M. and Beardsley, J. (1973) Influence of nectar source plants on the New Guinea sugarcane weevil parasite, *Lixophaga sphenophori* (Villeneuve). *Proceedings of the Hawaiian Entomological Society* 22, 145–154.

Tourner, J.C. and Viladerbo, A. (1966) Lepidopteres defoliateur du bananaier en Equateur: morphologie et biologie. II *Opsiphanes tamarindi*, var. *corrosus* Stitchel. *Fruits* 21, 159–166.

Tourner, J.C., Viladerbo, A., and Stomayor, B. (1966) Lepidopteres defoliateurs du bananier en Equateus: morphologie et biologie. I *Caligo eurilochus* Stich. (Brassolidae). *Fruits* 21, 55–56.

Traore, L. (1995) Facteurs biologiques de mortalite de Curculionidae en mileux tempere et tropical. PhD thesis, McGill University, Montreal, Canada.

Traore, L., Gold, C.S., Pilon, J.G. and Boivin, G. (1993) Effects of temperature on embryonic development of banana weevil, *Cosmopolites sordidus* Germar. *African Crop Science Journal* 1, 111–116.

Traore, L., Gold, C.S., Boivin, G. and Pilon, J.G. (1996) Developpement postembryonnaire du charancon du bananier. *Fruits* 51, 105–113.

Treverrow, N. (1993) An integrated management program for banana weevil borer. *Final Report. HRDC. Project Number: Fr/0012/RO*. Wollongbar Agricultural. Institute, Wollongbar NSW, Australia, 40 pp.

Treverrow, N. (1994) Control of the banana weevil borer, *Cosmopolites sordidus* (Germar) with entomopathogenic nematodes. In: Valmayor, R.V., Davide, R.G., Stanton, J.M., Treverrow, N.L. and Roa, V.N. (eds) *Proceedings of Banana Nematode/Borer Weevil Conference*, Kuala Lumpur, 18–22 April 1994. International Network for Improvement of Banana and Plantain, Los Baños, pp. 124–138.

Treverrow, N. and Bedding, R.A. (1993) Development of a system for the control of the banana weevil borer, *Cosmopolites sordidus*, with entomopathogenic nematodes. In: Bedding, R., Akhurst, R. and Kaya, H. (eds) *Nematodes and the Biological Control of Insect Pests*. CSIRO, Melbourne, pp. 41–47.

Treverrow, N. and Maddox, C. (1993) The distribution of *Cosmopolites sordidus* (Germar) (Coleoptera: Curculionidae) between various types of banana plant material in relation to crop hygiene. *General and Applied Entomology* 23, 15–20.

Treverrow, N., Bedding, R., Dettmann, E.B. and Maddox, C. (1991) Evaluation of entomopathogenic nematodes for the control of *Cosmopolites sordidus* Germar (Coleoptera: Curculionidae), a pest of bananas in Australia. *Annals of Applied Biology* 119, 139–145.

Treverrow, N., Peasley, D. and Ireland, G. (1992) *Banana Weevil Borer: A Pest Management Handbook for Banana Growers*. Banana Industry Committee. NSW Agriculture, 28 pp.

Tsai, Y.P. (1986) Major insect pests of banana. In: *Plant Protection in the Republic of China (1954–1984)*, 6 pp.

Turner, D.W. (1994) Bananas and plantains. In: Schaffer, B. and Andersen, P.C. (eds) *Handbook of Environmental Physiology of Fruit Crops, Vol. III, Subtropical and Tropical Crops*. CRC Press, Boca Raton, Florida, pp. 37–61.

Uronu, B.E.M.A. (1992) The effect of plant resistance and cultural practices on the population densities of banana weevil *Cosmopolites sordidus* (Germar) and on banana yield. PhD thesis, Kenyatta University, Kenya.

Uzakah, R.P. (1995) The reproductive biology, behaviour and pheromones of the banana weevil, *Cosmopolites sordidus* Germar (Coleoptera: Curculionidae). PhD thesis, University of Ibadan, Nigeria.

Van den Enden, H. and Garcia, E.A. (1984) Reconocimiento del control natural del picudo negro del platano (*Cosmopolites sordidus*) Germar en la zona del gran Caldas. *Universidad Nacional Facultad de Agronomia, Manizales, Colombia*, 114 pp.

Varela, A.M. (1993) Report on a trip to Bukoba and Muleba districts (Kagera region) to assess the possibility for biological control of banana weevil with ants. Unpublished manuscript. Agriculture Research Institute, Maruku, Tanzania, 15 pp.

Varma, P.M. and Capoor, S.P. (1958) Mosaic disease of cardamon and its transmission by the banana aphid. *Pentalonia nigronervosa* Coq. *Indian Journal of Agricultural Science* 28, 97–107.

Vaurie, P. (1966) A revision of the Neotropical genus *Metamasius* (Coleoptera: Curculionidae, Rhynchophorinae). Species groups I and II. *Bulletin American Museum Natural History* 131, 213–337.

Veitch, R. (1929) The banana weevil borer. In: Veitch, R. and Simmonds, J.H. (eds) *Pests and Diseases of Queensland Fruits and Vegetables*. Octaro, Brisbane, pp. 255–263.

Vilardebo, A. (1950) Conditions d'un bon rendement du peigeage de *Cosmopolites sordidus*. *Fruits* 5, 399–404.

Vilardebo, A. (1960) *Los insectos y nematodos de las bananeras del Ecuador*. Instituto Franco-Euaotiano de Investigaciones Agromicas, Paris, 78 pp.

Vilardebo, A. (1973) Le coefficient d'infestation, critere d'evaluation du degre d'attaques des bananeraies par *Cosmopolites sordidus* Germ. le charancon noir du bananier. *Fruits* 28, 417–431.

Viswanath, B.N. (1976) Studies on the biology, varietal response and control of banana rhizome weevil, *Cosmopolites sordidus* (Germar) (Coleoptera: Curculionidae). PhD thesis, Bangalore, India.

Waggy, S. and Beardsley, J. (1972) Biological studies on two sibling species of Lixophaga (Diptera: Tachinidae) parasites of the New Guinea sugarcane weevil, *Rabdoscellus obsucrus* (Boisduval). *Proceedings of the Hawaiian Entomological Society* 21, 485–494.

Walker, A.K. and Dietz, L.L. (1979) A review of entomophagous insects in the Cook Islands. *New Zealand Entomology* 7, 70–82.

Wallace, C.R. (1938) Measurement of beetle borer migration in banana plantations. *Journal of the Australian Institute of Agricultural Sciences* 4, 215–219.

Wardlaw, C.W. (1972) *Diseases of the Banana and of the Manila Hemp Plant*. Macmillan, London, 615 pp.

Waterhouse, D.F. (1993) *The Major Arthropod Pests and Weeds of Agriculture in Southeast Asia*. Australian Centre for International Agricultural Research, Canberra, 141 pp.

Waterhouse, D.F. and Norris, K.R. (1989) *Cosmopolites sordidus* (Germar). In: Waterhouse, D.F. and Norris, K.R. (eds) *Biological Control: Pacific Prospects*. Inkata Press, Melbourne, Australia, pp. 152–158.

Woodruff, R.E. and Baranowski, R.M. (1985) *Metamasius hemipterus* (Linnaeus) recently established in Florida (Coleoptera: Curculionidae). *Florida Department of Agriculture and Consumer Services Division of Plant Industry, Entomology* Circular No. 272, 4 pp.

Yaringano, C. and Van der Meer, F. (1975) Control del gorgojo del platano, *Cosmopolites sordidus* Germar, mediante trampas diversas y pesticidas granulados. *Revista Peruana de Entomologia* 18, 112–116.

Zhou, S.F. and Wu, X.Z. (1986) Monitoring and control of banana borer *Odoiporus longicollis* (Olivier). *Acta Phytophylactica Sinica* (China) 13, 195–199.

Zimmerman, E.C. (1968a) The *Cosmopolites* banana weevils (Coleoptera: Curculionidae; Rhynchophorinae). *Pacific Insects* 10, 295–299.

Zimmerman, E.C. (1968b) Rhynchophorinae of southeastern Polynesia. *Pacific Insects* 10, 47–77.

1.

2.

4.

3.

5.

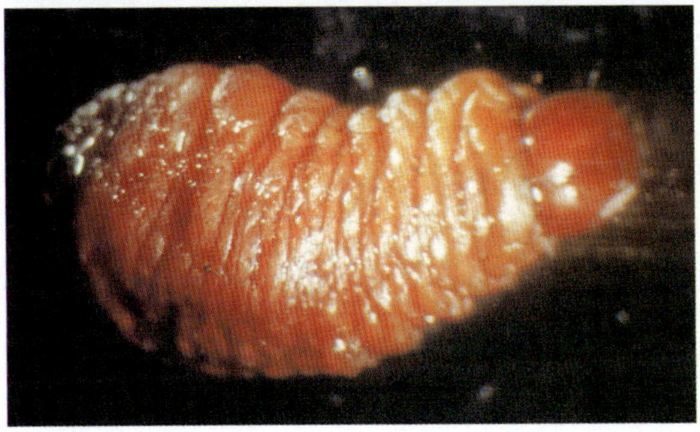

6.

Banana

Plate 1. Damage by banana weevil to banana corm (J. Peña).

Plate 2. *Cosmopolites sordidus*, adult (R. Duncan).

Plate 3. Banana weevil, larva (R. Duncan).

Plate 4. Banana weevil adult infected with *Beauveria bassiana* (R. Duncan).

Plate 5. Banana weevil larva infected by nematodes (R. Duncan).

Plate 6. Severe damage to banana by banana rust thrips (B. Pinese).

7.

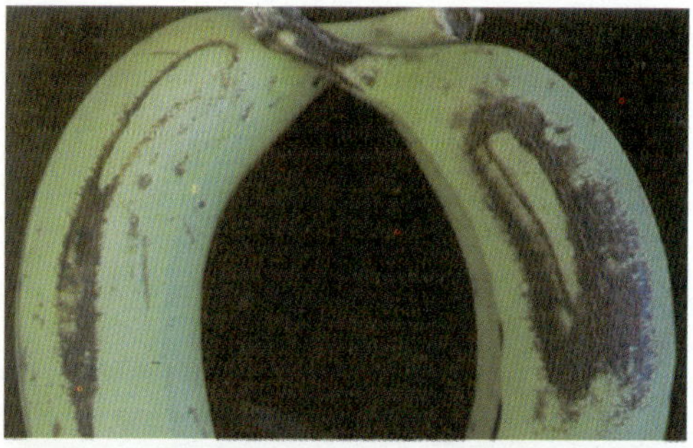

8.

9.

10.

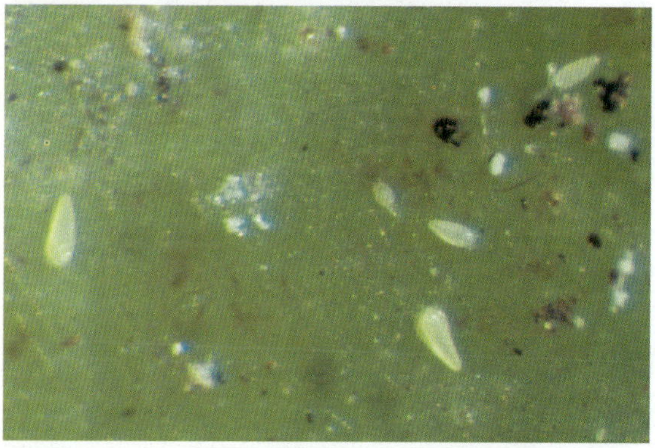

12.

11.

Banana

Plate 7. Typical damage to mature banana by rust thrips (B. Pinese).
Plate 8. Banana scab moth larva feeding on immature fruit (B. Pinese).
Plate 9. Damage to banana by banana spider mite (B. Pinese).

Citrus

Plate 10. Citrus rust mite, *Phyllocoptruta oleivora* (J. Peña).
Plate 11. Broad mite, *Polyphagotarsonemus latus* (DPI-Queensland).
Plate 12. Oriental spider mites (DPI-Queensland).

13.

14.

15.

16.

18.

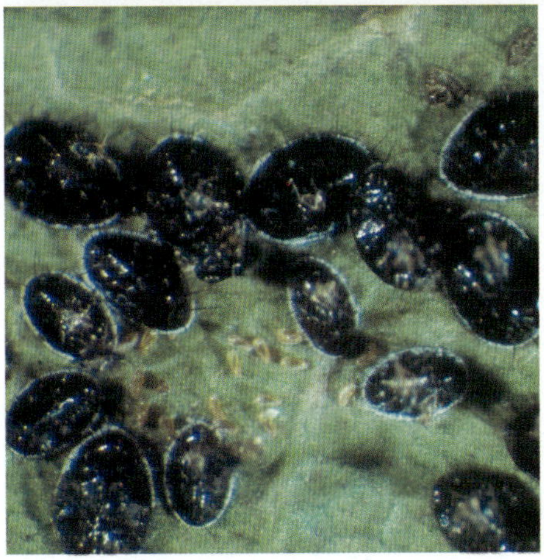

17.

Citrus

Plate 13. Citrus leaves damaged by oriental spider mite (DPI-Queensland).
Plate 14. Red scale *Aonidiella aurantii* on fruit (DPI-Queensland).
Plate 15. Green coffee scale along midrib (DPI-Queensland).
Plate 16. Longtailed mealybug (DPI-Queensland).
Plate 17. *Anagyrus fusciventris*, a wasp parasite of longtailed mealybug (DPI-Queensland).
Plate 18. Citrus blackfly *Aleurocanthus woglumi* (R. Duncan).

19.

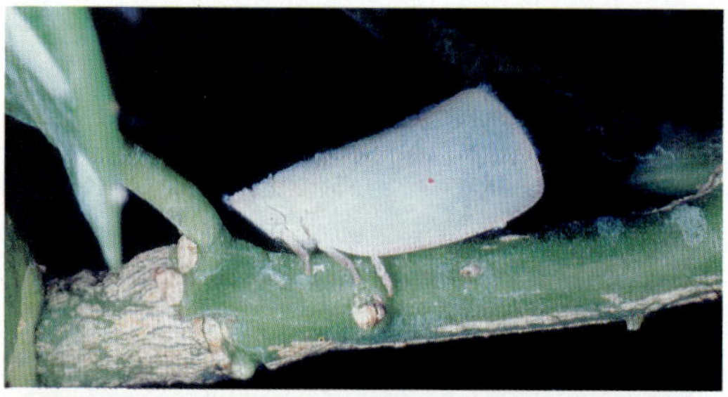

20.

21.

22.

23.

24.

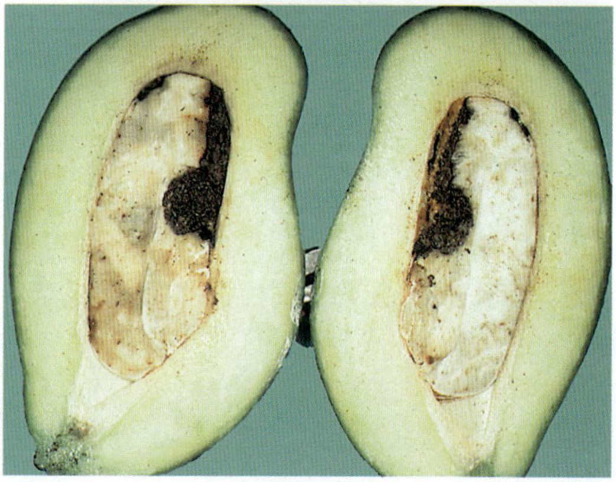

Citrus
Plate 19. Citrus planthopper (DPI-Queensland).
Plate 20. Citrus leafminer, adult (R. Duncan).
Plate 21. Citrus leaves damaged by citrus leafminer (R. Duncan).
Plate 22. Fruit piercing moth (DPI-Queensland).

Mango
Plate 23. Queensland fruit fly (G. Waite).
Plate 24. Damage by mango seed weevil (G. Waite).

25.

26.

27.

28.

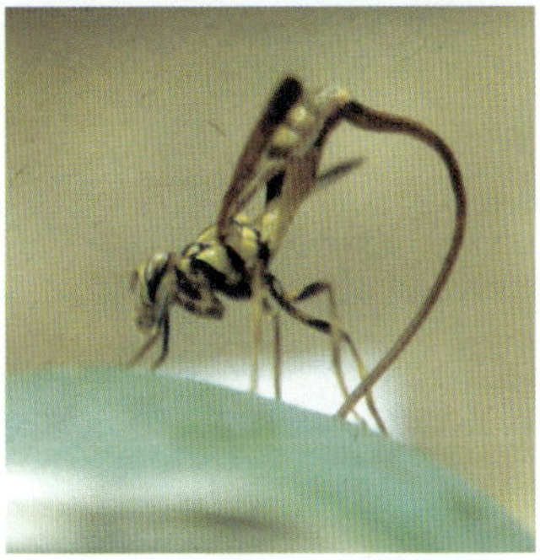

29.

30.

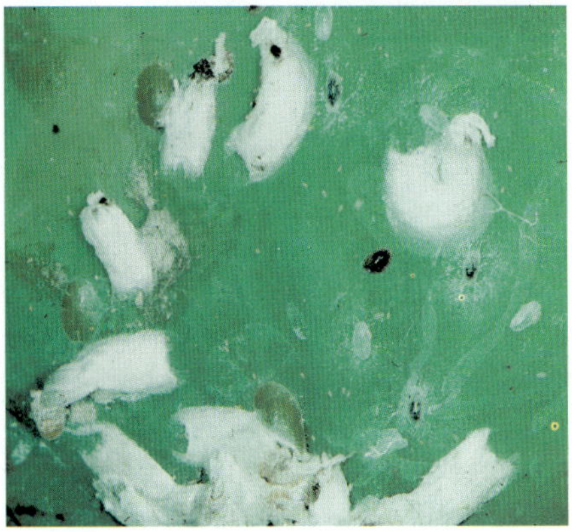

Mango

Plate 25. Red-banded thrips (G. Waite).
Plate 26. Mango scale (G. Waite).
Plate 27. Pink wax scale, *Ceroplastes rubens* (G. Waite).

Papaya

Plate 28. Papaya fruit fly, *Toxotrypana curvicauda*, female (R. Swanson).
Plate 29. Damage to green fruit by papaya fruit fly (J. Peña).
Plate 30. Papaya scale, *Philephedra tuberculosa* (M. Shepard).

31.

32.

33.

36.

35.

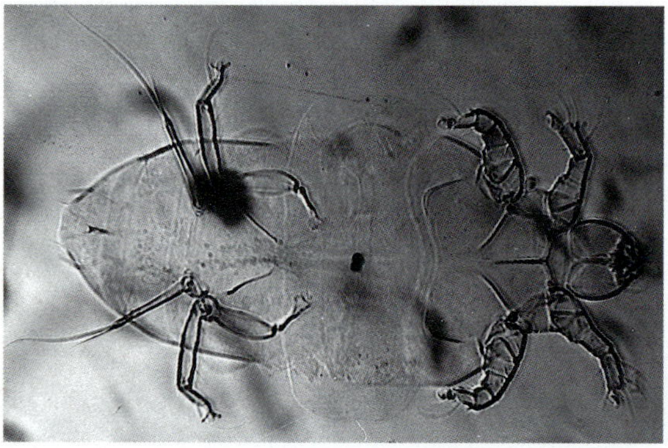

34.

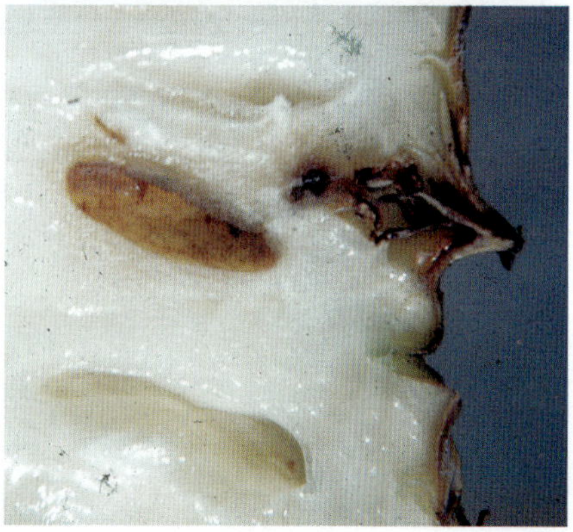

Papaya

Plate 31. Damage to papaya leaves by two-spotted mite (J. Peña).
Plate 32. Papaya leafhopper (R. Duncan).

Pineapple

Plate 33. Damage to pineapple roots by *Meloidogyne* (R. Bradley).
Plate 34. Damage to pineapple by leathery pocket mite (G. Petty).
Plate 35. Leathery pocket mite (G. Petty).
Plate 36. Thrips affecting pineapple (G. Petty).

37.

38.

41.

39.

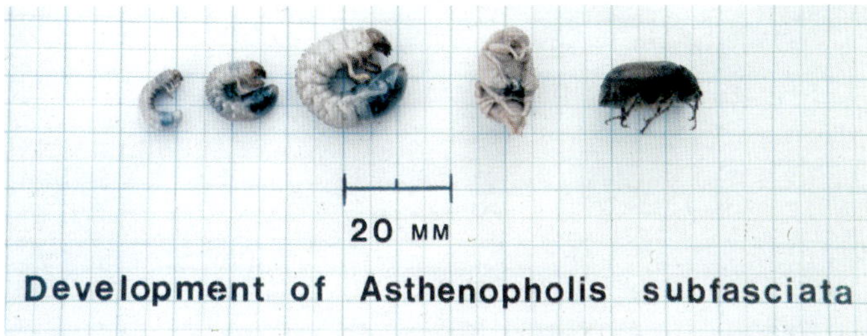

40.

42.

Pineapple
Plate 37. Thrips damage to pineapple (G. Petty).
Plate 38. Mealybug wilt to pineapple (G. Petty).
Plate 39. *Asthenopholis subfasciata*, stages of white grub damaging pineapple (G. Petty).
Plate 40. Symptoms of white grub injury to pineapple.

***Annona* sp.**
Plate 41. Atemoya flower (H. Nadel).
Plate 42. *Carpophilus freemani*, pollinating nitidulid (H. Nadel).

43.

46.

45.

44.

47.

48.

***Annona* sp.**
Plate 43. *Annona montana* flower (J. Peña).
Plate 44. *Annona muricata* pollinating scarabs (J. Peña).
Plate 45. *Annona* seed borer *Bephratelloides cubensis* (H. Nadel).
Plate 46. *Annona* seed borer damage to atemoya (J. Peña).
Plate 47. *Cerconota anonella*, adult (F. Garcia).
Plate 48. Damage to fruit by *Cerconota anonella* (J. Peña).

49.

50.

51.

52.

53.

54.

***Annona* sp.**
Plate 49. *Cratosomus bombinus,* adult (Monica Barbosa-Pereira).
Plate 50. Damage to branches by *Heilipus velamen* (Monica Barbosa-Pereira).

Avocado
Plate 51. *Protopulvinaria pyriformis* (M. Wysoki).
Plate 52. Avocado lacebug, *Pseudacysta perseae* (J. Peña).
Plate 53. *Retithrips syriacus* (M. Wysoki).
Plate 54. *Heliothrips haemorrhoidalis* (M. Wysoki).

55.

56.

57.

58.

59.

60.

Avocado

Plate 55. *Heliothrips haemorrhoidalis* damage (M. Wysoki).

Plate 56. *Conotrachelus perseae* (N. Bautista).

Plate 57. *Heilipus laurii* (N. Bautista).

Plate 58. 'Hass' male flower at first dehiscence (stage D_2). The stigma is dry, the nectaries secreting nectar (E. Lahav).

Plate 59. 'Reed' female flower (stage B_2). Field picture (A. Shuv).

Plate 60. Honeybee collecting nectar from 'Ettinger' male flower (stage D_2). Pollen accumulates on the ventral thorax. Field picture x 3.4 (A. Shuv).

61.

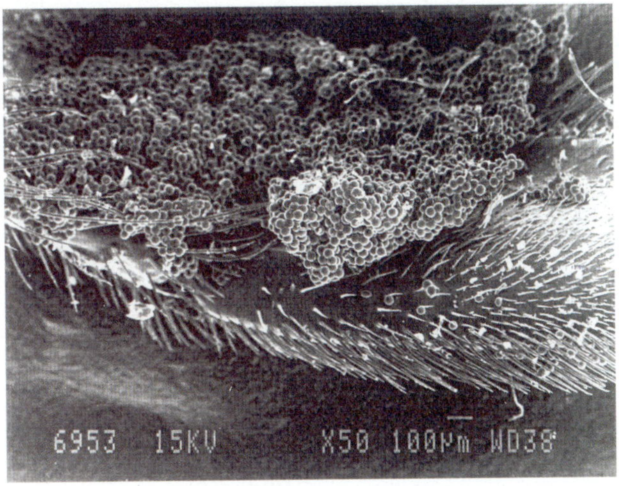

62.

64.

63.

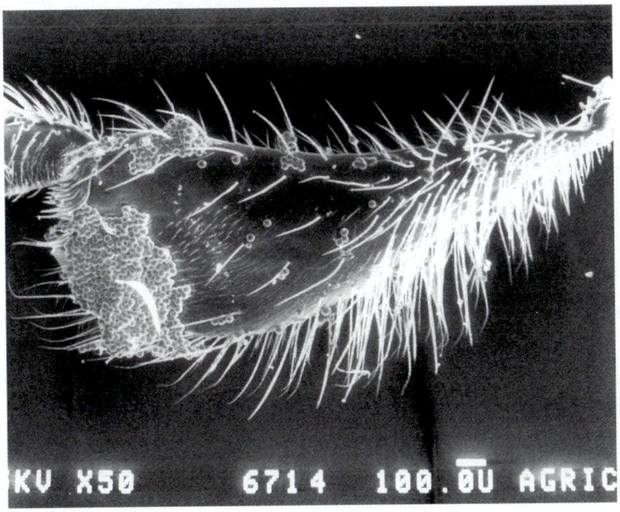

65.

Avocado

Plate 61. 'Ettinger' pollen grains organized into a honeybee's pollen load. The pollen is not well packed and the pollen load tends to disintegrate. SEM x 50, the bar represents 100 µm (G. Ish-Am).

Plate 62. The stingless bee *Geotrigona acapulconis* collecting nectar from 'Hass' female flower stage (stage B_2), and holding the stigma by a leg. Field picture x 6.2 (G. Ish-Am).

Plate 63. Avocado pollen packed to form pollen load on a *Geotrigona acapulconis* pollen basket. SEM x 50, the bar represents 100 µm (G. Ish-Am).

Plate 64. The Mexican honey wasp *Brachygastra mellifica* collecting nectar from a male flower before dehiscence (stage D_1). The anthers touch the wasp's lateral thorax, ventral abdomen and legs. Field picture x 6.0 (G. Ish-Am).

Plate 65. The fly *Lucilla sericata* (Calliphoridae) collecting nectar from 'Reed' female flower (Stage B_2). Field picture x 7.0 (G. Ish-Am).

66.

67.

69.

71.

68.

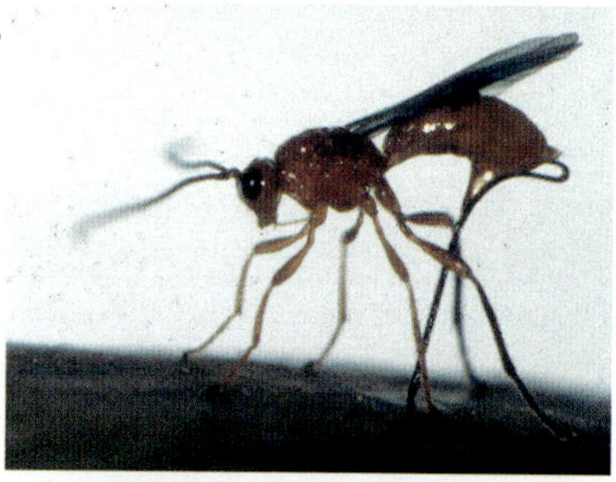

70.

Guava

Plate 66. *Anastrepha suspensa*, adult (R. Duncan).
Plate 67. McPhail trap (W. Gould).
Plate 68. *Diachasmimorpha longicaudata* (H. Glenn).
Plate 69. Guava fruit protected against Caribbean fruit fly (R. Duncan).

Minor fruits

Plate 70. *Hypomeces squamosus* (A. Winotai).
Plate 71. *Conogethes punctiferalis* (A. Winotai).

72.

73.

74.

75.

76.

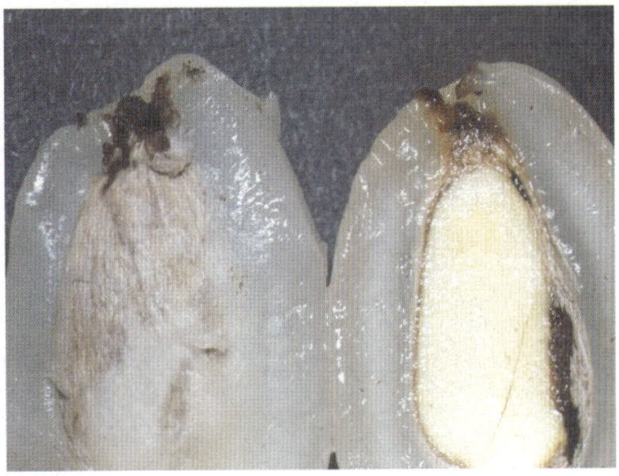

Minor fruits

Plate 72. *Mudaria luteileprosa* (A. Winotai).
Plate 73. Durian mealybug infestation (A. Winotai).
Plate 74. *Phyllocnistis* sp. mining rambutan leaf (A. Winotai).
Plate 75. *Conopomorpha* spp. (A. Winotai).
Plate 76. *Conopomorpha* damage (A. Winotai).

77.

78.

80.

79.

81.

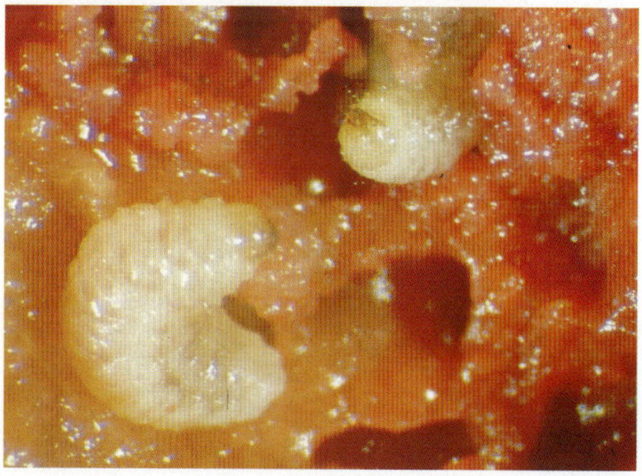

Minor fruits
Plate 77. *Hyperaeschrella* sp. (A. Winotai).
Plate 78. *Bactrocera* sp. (A. Winotai).
Plate 79. Aphids infesting carambola fruitlets (J. Peña).
Plate 80. Acerola weevil (A. Hunsberger).
Plate 81. Acerola weevil larva (A. Hunsberger).

82.

83.

84.

85.

86.

87.

Litchi, longan

Plate 82. Damage to litchi by nut borer (G. Waite).
Plate 83. *Cryptophlebia* adult (G. Waite).
Plate 84. Green shield scale (G. Waite).
Plate 85. Fruitspotting bug causing fruit fall (G. Waite).
Plate 86. Erinose mite (G. Waite).
Plate 87. Damage by erinose mite to flowers (G. Waite).

88.

89.

90.

91.

92.

93.

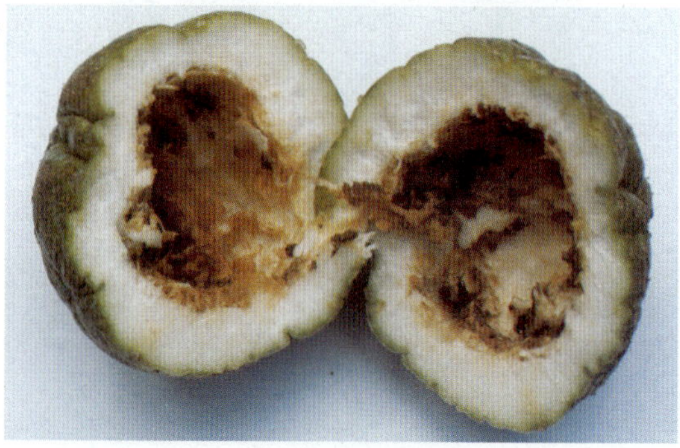

Passion fruit

Plate 88. Pollination of *Passiflora edulis* yellow passion fruit by the carpenter bee, *Xyolocopa* spp. (E. Aguiar-Menezes).

Plate 89. Adult of *Agraulis vanillae vanillae* (Lepidoptera: Nymphalidae) (E. Aguiar-Menezes).

Plate 90. *A. vanillae* caterpillar feeding on leaves of yellow passion fruit (E. Aguiar-Menezes).

Plate 91. Coreidae. Left: adult of *Veneza zonatus*; right: adult of *Leptoglossus gonagra* (E. Aguiar-Menezes).

Plate 92. *Anastrepha* spp., a pest of passion fruit (E. Aguiar-Menezes).

Plate 93. Immature yellow passion fruit damaged by fruit fly larvae (E. Aguiar Menezes).

3 Tropical Citrus Pests

D. Smith[1] and J.E. Peña[2]

[1]Department of Primary Industries, Maroochy Horticultural Research Station, Nambour, Queensland 4560, Australia; [2]Tropical Research and Education Center, University of Florida, 18905 SW 280 Street, Homestead, FL 33031, USA

Introduction

Citrus is host to a large number of pests worldwide. Ebeling (1959) lists about 875 insects and mites, albeit less than 10% are of major importance. Talhouk (1975) lists 144 well known species. Warmer temperatures and higher humidities usually result in there being up to three times as many pests in the tropics in comparison with the higher latitude and Mediterranean climatic areas. Southern China, for example, has over 120 species of some significance (Anonymous, 1975) as has India (Ghosh, 1990), while Japan has about 40 (Anonymous, 1981). Warmer areas of South Africa have about 100 species (Bedford *et al.*, 1998) while Morocco has about 30 (Papacek, 1997). Northern and north eastern Australia also have about 100 species. Dry southern inland regions of Australia have about 30 (Smith *et al.*, 1997). Hemipterous pests (scales, mealybugs, aphids, whiteflies, leafhoppers) constitute 30–60% of the pests, followed by lepidoptera (fruit boring and piercing moths, leafrollers, and leafminers), mites, beetles, bugs, flies, and thrips. Citrus pests are classified according to the severity and frequency of attack and whether they attack fruit or other parts of the tree (Nasca *et al.*, 1981). Some pests have a greater potential for control by natural enemies. Talhouk (1975) classes the pests as major, occasional or of little importance and a similar system (major, occasionally important or minor) is used by Smith *et al.* (1997). Every citrus region has an even more select group of key pests – three or four major pests dominating the whole pest monitoring programme. Some insects are key pests because they are vectors of serious diseases like greening or tristeza. Major pests are those that occur most frequently (usually each season) and can seriously affect fruit yield or quality or tree health. They can have limited potential for biological control. A grading of occasionally important means the pest occurs more sporadically, or causes less damage, while a pest of minor importance has the potential to cause problems but is usually present in small numbers. Secondary or induced pests are normally kept at a minor level by natural enemies but flare up when there is too much disruption from broad spectrum pesticides.

Differences between tropical and higher latitude citrus growing areas and practices

A major difference between tropical citrus and citrus grown in high latitudes is the effect of climate on tree phenology. Citrus growth in semitropical (latitude 26° N) to intertropical (23° N) areas is mostly governed by the amount of precipitation between winter and summer and the alternance between precipitation between the wet and the dry season

©CAB *International* 2002. *Tropical Fruit Pests and Pollinators* (eds J.E. Peña, J.L. Sharp and M. Wysoki)

rather than by changes in temperature, such as those observed in the Mediterranean and California, USA (Cassin, 1984). Tropical and intertropical citrus growing areas show major differences among themselves in citrus varieties (Murata, 1997). In South-East Asia sweet orange, grapefruit and lemon are of little importance, while tangerines play a relatively large role in the region (Verheij and Coronel, 1992). In other tropical areas, there are more oranges (many produced for juice) and pomelos and limes produced for both processing and fresh fruit (Ghosh, 1990). Tropical citrus production has grown, from 71 million t in 1980 to approximately 78 million t year^{-1} (FAO, 2000). In the tropics, flowering and fruiting is less seasonally defined than in subtropical and temperate areas. In the latter, citrus species (other than lemons) usually flower once a year except when stressed (e.g. by lack of water). Shoot growth of most varieties in cooler areas occurs in distinct flushes with 50% of more in the spring flush. Autumn flushes produce vigorous shoots with large, wide leaves. In summer these shoots do not carry flowers except on lemons, or when it results in an out-of-season crop that is often of no economic value. Management practices such as irrigation and fertilizer application can be used to minimize summer/autumn vegetative growth. In the tropics, flushing is less defined and is almost continual on some varieties. In some tropical areas, trees are even encouraged to have several flowerings per season and the constant presence of young fruit can make pest control much more difficult. Citrus farming practices also differ between the tropics and higher latitudes. Often there are more small orchards of 1–10 ha managed by a single family. Subsequently there is less mechanization and spraying equipment is less efficient. Financial risk can pressure smaller growers into using chemical rather the biological options for pest control. Propagation techniques for nursery stock tend to be less stringent with regard to rootstock and budwood selection and nursery production. Pest monitoring is practised on an increasing number of larger orchards but less so in the many small orchards. While larger orchards need daily care to prevent losses, small farms cannot afford this type of management (Verheij and Coronel, 1992). Postharvest treatments, waxing, grading and packing are less developed and more fruit is sold on domestic markets. Disease control has an important role in all citrus areas but even more so in the tropics and particularly the Asian tropics where Asian greening is transmitted by the citrus psylla *Diaphorina citri* Kuwayama. For instance, because of the marcotting system of propagation, many trees are already infested with greening even before planting out. Integrated pest management (IPM) systems in the tropics have to contend with control of important vector pests such as citrus psylla.

Key Pests

Mites

At least a dozen species of mites attack tropical citrus (Jeppson *et al.*, 1975; Cermeli, 1983; Atachi, 1991; Quiros-Gonzalez, 1996) (Table 3.1). The species are classified within the Eriophyidae, Tarsonemidae, Tetranychidae and Tenuipalpidae.

Eriophyidae

DESCRIPTION, BIOLOGY, DAMAGE The citrus rust mite (CRM), *Phyllocoptruta oleivora* (Ashmead), is the most important arthropod citrus pest worldwide (Browning *et al.*, 1995). Knapp *et al.* (1996a) report that this mite is a serious pest of citrus in most humid regions of the world. Other significant eriophyids on citrus are: citrus bud mite, *Eriophyes sheldoni* (Ewing) (worldwide); citrus grey mite, *Calacarus citrifolii* Keifer (Transvaal, South Africa); brown citrus rust mite, *Tegolophus australis* Keifer (Queensland, Australia); and the pink citrus rust mite, *Aculus pelekassi* Keifer (Japan).

According to Knapp (1983), the adult citrus rust mite is a yellow or light brown mite and has an elongated, wedge-shaped body about three times as long as wide (Plate 10). The mite has two pairs of short anterior legs and a pair of lobes on the posterior end which assist in moving and clinging to plant

Table 3.1. Important insect and mite pests of citrus in subtropical and tropical areas of the world.

Subtropical/tropical region	Key pests	Major or occasionally important pests
Southern Africa – Natal, Transvaal, Swaziland, Zimbabwe, Mozambique Important diseases – greening (South African), alternaria brown spot, blackspot, phytophthora, tristeza, leprosis	• Red scale *Aonidiella aurantii* • Citrus thrips *Scirtothrips aurantii* • Citrus psylla *Trioza erythreae* • Mediterranean fruit fly *Ceratitis capitatia* • Natal fruit fly *Ceratitis rosa* • False codling moth *Cryptophlebia leucotreta*	• Bud mite *Eriophyes sheldoni* • Citrus rust mite *Phyllocoptruta oleivora* • Citrus red mite *Panonychus citri* • Oriental spider mite *Eutetranychus orientalis* • Broad mite *Polyphagotarsonemus latus* • Citrus mealybug *Planococcus citri* • Spherical mealybug *Nipaeoccus viridis* • Florida red scale *Chrysomphalus aonidium* • Cottony cushion scale *Icerya purchasi* • Spiny blackfly *Aleurocanthus spiniferus* • Black citrus aphid *Toxoptera citricidus* • Citrus leafhopper *Penthimeola bella*
Australia – Queensland and Northern Territory Important diseases – blackspot, melanose, alternaria brown spot, lemon scab, phytophthora, tristeza	• Red scale • Queensland fruit fly *Bactrocera tryoni* • Citrus rust mite • Brown citrus rust mite *Tegolophus australis* • Oriental spider mite	• Broad mite • Bud mite • Citrus snow scale *Unaspis citri* • Long soft scale *Coccus longulus* • Green coffee scale *Coccus viridis* • Pink wax scale *Ceroplastes* rubens • Florida wax scale *Ceroplastes floridensis* • Purple scale *Lepidosaphes becki* • Citrus mealybug • Spherical mealybug • Planthoppers *Siphanta* spp., *Colgaroides acuminata* • Citrus leafhopper *Empoasca smithi* • Spined citrus bug *Biprorulus bibax* • Citrus leafminer *Phyllocnistis citrella* • Corn earworm *Helicoverpa armigera* • Fruit piercing moths *Othreis* spp. • Citrus gall wasp *Bruchophagus fellis* • Scirtothrips *Scirtothrips albomaculatus* • Citrus rust thrips *Chaetanaphothrips signipennis* • *Iridomyrmex* spp.

continued

Table 3.1. *Continued.*

Subtropical/tropical region	Key pests	Major or occasionally important pests
		• *Achaea* spp. • Fruit borer *Citripestis sagittiferella* • Oriental fruit fly • Papaya fruit fly *Bactrocera philippinensis* • Leaf eating weevil *Hypomyces squamosus* • Citrus bark borer *Agrilus occipitalis* • Citrus stink bug *Rhynchocoris humeralis*
North Africa – Mediterranean Important diseases – phytophthora, septaria spot, mal secco, xyloporosis	• Mediterranean fruit fly • Red scale • Citrus red mite • Citrus leafminer • Citrus rust mite	• Bud mite • Broad mite • Two-spotted mite • Dictyospermum scale *Chrysomphalus dictyospermi* • Yanone scale • Oleander scale *Aspidiotus neri* • Purple scale • Chaff scale • Black scale • Florida wax scale • Soft brown scale • Cottony cushion scale • Citrus mealybug • Citrus mealybug *Pseudococcus cryptus* • Fig wax scale *Ceroplastes rusci* • Citrus whitefly • Citrus woolly whitefly • Brown citrus aphid • Black citrus aphid • Spiraea aphid • Bud moth *Prays citri* • *Cryptoblabes gnidiella*

surfaces. Egg deposition begins within a day or two of the female reaching maturity and continues throughout her life, about 20 days. The eggs hatch in about 3 days during summer months and immature mites undergo two moults, with each nymphal stage lasting from 1 to 3 days during the summer (McCoy *et al.*, 1988).

The citrus rust mite is found on leaves, fruits and young branches. Affected leaves exhibit shallow yellow punctures, giving them a pale appearance similar to that produced by *Eutetranychus banksi* McGregor (Ochoa *et al.*, 1994). Various types of damage can be seen in fruits according to whether they were attacked early or late (McCoy and Albrigo, 1975). Attacks of *P. oleivora* on *Citrus sinensis* result in blackening or darkening of the fruit. When the attack is early the fruit is also smooth and opaque, whereas in late attack the blackening is shiny. The surface of attacked fruit in *C. limon*, *C. limettoides* and *C. aurantifolia* takes on a smooth yellow brown to white appearance which is opaque in early attacks and shiny in late attacks. Occasionally, *C. limon* shows fine cracking. Effects of citrus mite on fruit drop and tree growth have been reported by Allen (1979). *P. oleivora* population densities can fluctuate depending on weather conditions or phenological tree conditions, but some studies in the Neotropics indicate that temperature around 24.5°C and 30–90% RH favour its development.

MONITORING AND ECONOMIC THRESHOLDS Monitoring methods for CRM have been developed by different authors (Yothers and Miller, 1934; Allen and Stamper, 1979; Hall *et al.*, 1991, 1994; Peña and Baranowski, 1992; Smith *et al.*, 1997). Action levels depend on fruit size, percentage of fruit infested and predatory mite activity (Smith *et al.*, 1997). Nascimento *et al.* (1982) suggest that *P. oleivora* population levels can be estimated by collection of five leaves per plant. They can be assessed by counting the total number of mites per leaf, counting the number of mites on 2.5 cm^2 or using a brushing machine. The efficacy of these three methods is similar (Oliveira *et al.*, 1982). In Florida, CRM monitoring should be initiated as soon as populations are detected and continue every 2–3 weeks throughout the fruit season. Fruit should be sampled at random representing the four quadrants of the tree (Bullock *et al.*, 1999). In Brazil, Gravena and Trevizoli (1984) found that a density of more than 70 mites per cm^2 was the threshold injury level for the citrus rust mite. An action level of 5–10% of young and old fruit infested with CRM could be considered the action level in Australia (Smith *et al.*, 1997). Knapp *et al.* (1996a) reported that the amount of citrus rust mite blemish acceptable for fresh fruits varies in Florida, generally it must not exceed 5% of the surface area. However, citrus grown for processing can withstand up to 75% surface area damage before reduced fruit size and juice content and increased fruit drop become evident. In Florida, USA, six CRM cm^{-2} is considered the threshold where pesticide intervention would be required and ten CRM cm^{-2} would be the action threshold where treatment would be required as soon as possible (Bullock *et al.*, 1999).

BIOLOGICAL CONTROL According to Browning *et al.* (1995) citrus rust mite is attacked by a complex of natural enemies, but generally under Florida, USA conditions, beneficial organisms are unable to consistently maintain rust mites below damaging levels. In Mexico, an unidentified cecidomyiid appears to be the major biological control agent (Ruiz, 1997) of *P. oleivora* and densities of the phytoseiid, *Euseius mesembrinus* (Dean) appear also to be correlated with increased densities of *P. oleivora* and *E. banksi* (Ruiz-Cancino *et al.*, 1996). In Benin, the exotic phytoseiids (*Amblyseius aerealis* (Muma), *Euseius concordis* (Chant), *Galendromus annectens* (De Leon)) introduced against the cassava mite, *Mononychelus tanajoa* (Bondar), are thought to have potential against *P. oleivora* and *E. sheldoni* (Ewing) (Atachi, 1991). In Australia, Smith *et al.* (1997) and Smith and Papacek (1991) report as the most important predators of CRM the phytoseiids *Euseius victoriensis* (Womersley), *Euseius elinae* Schicha, *Amblyseius herbicolus* (Chant) and *A. lentiginosus* Denmark and Schicha as well as the pathogen, *Hirsutella* sp. Samson *et al.* (1980) described the taxonomy of *Hirsutella thompsonii* Fisher in Florida. Applications of

H. thompsonii at 10^{10} conidia per plant are recommended in Cuba, when more than 25% of the fruits are infested (Otero *et al.*, 1994).

Tarsonemidae

The broad mite, *Polyphagotarsonemus latus* (Banks) is an important pest of tropical citrus (Jeppson *et al.*, 1975; Gerson, 1992) (Plate 11). The broad mite has a longevity of 13.4 ± 1.0 days for females and 12.0 ± 2.4 days for males. The intrinsic rate of increase (r_m) is 0.359, with a mean generation time of 10.34 and net reproductive rate (*Ro*) 41.0 (Jones and Brown, 1983; Vieira and Chiavegato, 1999). The potential for damage caused by *P. latus* has become evident during the last two decades (Gerson, 1992). Attacks tend to be concentrated on the young leaves, and sometimes cause damage to specific plant parts, e.g. stems, flowers, fruitlets or tips of shoots (Nucifora, 1961; Costilla, 1980). Hot, humid weather during exposure to broad mite feeding seems to intensify the symptoms of damage (Brown and Jones, 1983). The principal symptoms of attack consist of deformation of the leaves and suberization of the floral buds, growing tips and fruit. Apparently, *P. latus* mouthparts are unsuitable for effective penetration of renitent tissues (Jeppson *et al.*, 1975). Broad mite damage can be similar in appearance to that caused by citrus rust mite. The broad mite damage on citrus appears as a thin, silver-grey skin that can be easily scratched off (Smith *et al.*, 1997). Its feeding appendages are suitable for penetrating succulent tissues, but are quite incapable of penetrating thick-walled, lignified, and often varnished tissues, such as are found in mature stems and leaves. Toxins injected during feeding, presumably of salivary origin, cause alteration of normal tissue ontogeny in the host plant (Aubert *et al.*, 1980). On limes, larger numbers of *P. latus* are found on shaded parts of the fruit compared to the stylar, peduncle and sunlit regions. This pattern may be due to the propensity of the mites to avoid sunlight, or to avoid parts of the fruit with low relative humidity (Peña and Baranowski, 1992).

As a consequence of damage to plant tissue and disturbance of plant physiological processes, changes in growth intensity, flowering and yield may be observed. The most common change is a retardation of the growth of the organs of damaged plants. *P. latus* reduced the total leaf area and leaf water content of damaged lime and sour orange plants in Florida (Peña and Bullock, 1994).

INJURY LEVELS TO FRUIT Under greenhouse conditions, limes began to show damage 4–6 days after infestation and severe damage to the fruit epidermis appears 12 days after infestation (Peña, 1990) or when lime fruits are two-thirds mature (Smith *et al.*, 1997).

ECONOMIC INJURY LEVELS Despite the economic importance of *P. latus*, very few authors have determined the relationship between *P. latus* density and injury to citrus. Estimates of economic injury levels can be obtained using equations that describe the relationship between lime fruit surface damaged and broad mite days, and between percentage of fruits damaged per tree and broad mite days. Smith *et al.* (1997) recommend an action level of 5% of the fruit infested, particularly in coastal areas of Australia, or where predators are absent. Peña (1990) uses the term broad mite days and reported that in Florida the economic injury level per lime tree will fluctuate between 42 and 45 broad mite days for the spring and summer harvest, respectively.

CHEMICAL CONTROL The economic injury level, as mentioned earlier, is extremely low. In particular, the rapid injury caused to fruits, leaves and flowers necessitates a treatment at an early stage in population development to prevent excessive injury. Several acaricides, including cyclocompounds (e.g. propargite), diphenyl compounds (e.g. dicofol), organic phosphates (e.g. carbophenothion) and natural avermectins have been found to be effective against broad mites (Bullock, 1978; Schoonhoven *et al.*, 1978; Peña, 1988).

BIOLOGICAL CONTROL

Predators. Few studies have been conducted to investigate the suitability of broad mites as targets for biological control. The potential of phytoseiid mites as predators has been reported for different areas and crops (Moutia, 1958; Badii and McMurtry, 1984; McMurtry *et al.*, 1984; Hariyappa and Kulkarni, 1989). In Florida, six species of predacious mites (four Phytoseiidae, one Bdellidae and one Ascidae) were observed on lime fruits (Peña *et al.*, 1989a; Peña, 1992a,b). *Typhlodromalus peregrinus* (Muma) accounted for 72.4% of the predacious mites and outnumbered *Typhlodromis dentilis* (DeLeon), *Amblyseius aerelis* (Muma), *Galendromus helveolus* (Chant), *Bdella distincta* Baker and Balock and *Asca muma* Hurlbutt. In Australia, *Euseius victoriensis* is considered effective in subcoastal areas as well as the ladybird *Scymnus* sp. The effectiveness of mite predators for controlling broad mite populations was demonstrated by Peña *et al.* (1989a). In an exclusion experiment, population densities of *P. latus* increased in plots treated with pyrethroids immediately after the first insecticide application. During the dry season, the percentage of damaged fruits per tree was 3.2 and 3.65 times higher in predator-free plots than in plots with predators. During the humid season, however, there were no significant differences between the percentage of fruits injured per tree in the predator-free plots and plots with predators. Several factors could be responsible for this. Since most of these mites are facultative predators, the presence of other preferred prey species or food substrates might influence the predator mite response. Also, since the broad mite has a short generation time (Jones and Brown, 1983), and fruit injury is observed in 4 to 6 days (Peña, 1990), the ratio of predator to broad mite populations may need to be higher than that observed.

Pathogens. Pathogens have potential as control agents of phytophagous mites, or contribute to the natural regulation of mite populations. Fungi infecting Tetranychidae and Eriophyidae have been documented by different researchers, but field applications of fungi against mites have been carried out by few investigators, which indicates that the major constraints are the germination of spores and penetration of the fungus into the mite, which are very poor at humidities below 100%. Since development of *P. latus* is positively related to relative humidities between 75% and 90%, and development of entomopathogenic fungi require between 90 and 100% RH, fungi might offer another possible way to reduce broad mite populations. Peña *et al.* (1996a) compared the toxicity of *Beauveria bassiana*, *Paecilomyces fumosoroseus* and *Hirsutella thompsonii* and confirmed that all isolates tested were able to infect *P. latus* under laboratory conditions. Under greenhouse conditions, despite significant mortality of *P. latus*, a fungal epizootic was variable among performed tests. Failures of this nature are not uncommon when attempts are made to use fungi as mycoacaricides. In this instance, the failure is most probably related to fluctuation of relative humidity and temperature differences between environments where fungi are tested. The relative humidities were very high (approaching 100%) in the Petri dish bioassay, but variable (50–90%) in the whole-plant experiments.

Tenuipalpids

Brevipalpus spp. feed on the exposed surface of citrus fruit, but are also found on the undersides of leaves and green twigs (Smith *et al.*, 1997). In *Citrus aurantifolia*, the fruit shows off-white, irregular cracking, covering up to 80% of the epidermis. In Honduras, this mite is associated with the citrus scab pathogen *Elsinoe fawcettii* in the same host (Ochoa *et al.*, 1994). Ochoa *et al.* (1994) found attacks by *Brevipalpus* in conjunction with *P. oleivora* and the fungus *Sphaceloma fawcettii*.

Brevipalpus phoenicis (Geijskes) is considered a key pest of citrus in Brazil (Gravena *et al.*, 1995), particularly for the transmission of virus-like leprosis to fruits and twigs. Gravena *et al.* (1995) considered that adult mites are the best transmitters of the virus, while young forms are less efficient. Action thresholds for tenuipalpids are higher than for other mites. For instance, Smith *et al.* (1997) recommended an action threshold of 20% fruit infested.

BIOLOGICAL CONTROL The predator *Euseius citrifolius* is considered effective against *B.*

phoenicis in Brazil (Gravena *et al.*, 1995) and *E. victoriensis* is important in Australia. *Amblyseius citri* Vande-Merme and Ryke is a predator in South Africa (Bedford *et al.*, 1998).

Other species are citrus flat mite *Brevipalpus californicus* (Banks), ornamental flat mite *Brevipalpus obovatus* Donnuadic and *Brevipalpus lewisi* (McGregor) (Bedford *et al.*, 1998).

Tetranychids

Citrus red mite *Panonychus citri* (McGregor) is a cosmopolitan species. Injury is characterized by light coloured (etched) areas called stippling, which give a greyish or silvery appearance to the leaves or fruit (Knapp *et al.*, 1996a). *P. citri* densities on limes grown in southern Florida increase during spring and early autumn (Peña and Baranowski, 1992). In Florida, the combined influence of undetermined numbers of citrus red mites and weather may result in heavy leaf drop, twig dieback and fruit drop (Knapp *et al.*, 1996a). However, the citrus red mite is not considered a problem during extremely hot, dry weather in Australia (Smith *et al.*, 1997). *P. citri* occurs throughout eastern and South-East Asia (Beattie, 1997; Beattie and Watson, 1997). In China, damage levels to the leaf of 20–30% resulted in yield losses. In the Guangzhou area, the economic injury level is reached in spring when 21 mites are recorded per tender leaf (Tan *et al.*, 1989). The predators, *Amblyseius newsami* Evans and *A. nicholsi* Evans are considered to respond to population build-up of *P. citri* in China. Phytoseiids are vital for biocontrol of *P. citri* wherever it occurs.

Other significant tetranychids – Texas citrus mite, *Eutetranychus banksi* (McGregor), African citrus mite, *Eutetranychus africanus* (Tucker), the oriental spider mite, *Eutetranychus orientalis* Klein) (Plates 12 and 13) – are important pests of citrus in the Americas, Africa and the Near East, Asia, Australia and South Africa, respectively (Charanasri *et al.*, 1988). The two-spotted mite *Tetramychus urticae* Koch occurs on citrus in some areas, e.g. Australia (Smith *et al.*, 1997). In the Philippines, *Eutetranychus cendani* Rimando appears to be the most common tetranychid species (Cendaña *et al.*, 1984).

E. africanus produces fine stippling on the leaves, causing them to drop prematurely without turning brown. Heavy rain is a limiting factor in the distribution of the species. Injury by the Texas citrus mite is similar to that produced by *P. citri* (Jeppson *et al.*, 1975). Eggs of this species are found throughout the year in Texas and low relative humidity and temperatures above 27°C are favourable for the development of the species (Jeppson *et al.*, 1975). The oriental red mite feeds on the upper leaf surfaces, producing grey spots, and later leaves have a chlorotic appearance. Infested leaves weaken and finally drop; twigs die, which results in bare trees. Injury is more severe in the autumn, especially under low soil water and humidity conditions. The longevity of adults is about 12 days in summer, 14–18 days in the spring and autumn, and up to 21 days during the winter (Jeppson *et al.*, 1975). The time of year populations peak and most injury occurs is largely determined by prevailing temperatures and humidity. This species seems to prefer sour lemon stock to sweet lemon, mandarin and orange. As with *P. citri*, phytoseiids are vital in controlling many tetranychid species. In Australia the dominant phytoseiid is *E. victoriensis* (Smith *et al.*, 1997). In South-East Asia species such as *Amblyseius largoensis* (Muma) and *Amblyseius longispinus* (Evans), *Amblyseius deleoni* (Muma and Denmark), *Amblyseius cinctus* Corpuz and Rimando and *Phytoseius hongkongensis* Swirski and Shecter, are important.

Scales

Scales are the most abundant and variable group of citrus pests (Bennett and Alam, 1985; Quezada, 1989; Rose, 1990). They have a tremendous host range and can settle on citrus twigs, branches, trunk and fruits. The families, Coccidae (soft scales) and Diaspididae (armoured scales) are the most important. Armoured scales tend to be more abundant on trees growing under adverse conditions, or in situations unfavourable to their natural enemies (see Tables 3.2 and 3.3).

Table 3.2. Common wasp parasitoids of soft scales on citrus (Prinsloo, 1984; Malipatil *et al.*, 2000).

Parasitoids	Main hosts on citrus	Origin of parasitoid
Aphelinidae		
Coccophagus ceroplastae	*Ceroplastes rubens*, *Coccus viridis*, *Coccus elongatus* and *Pulvinaria polygonata*	Japan Taiwan
C. lycimnia	*Coccus hesperidum* and *Coccus pseudomagnoliarum*	Unknown
C. semicircularis	*C. hesperidum* and *C. pseudomagnoliarum*	(?) Europe
C. pulvinariae	*C. hesperidum*	South Africa
C. cowperi	*C. hesperidum*	Uganda
C. catherinae	*Ceroplastes brevicauda*	South Africa
Euryischomyia flavithorax	*C. hesperidum* and *C. viridis*	Unknown
Encyrtidae		
Anicetus beneficus	*C. rubens*	Japan
A. communis	*Ceroplastes destructor*	South and East Africa
Anicetus (= *Paraceraptrocerus*) *nyasicus*	*C. destructor*	South Africa
Diversinervus elegans	*C. hesperidum*, *C. destructor*, *Ceroplastes floridensis*, *Saissetia nigra* and *Saissetia oleae*	Ethiopia
D. stramineus	*C. viridis*	Kenya
Encyrtus aurantii (= *lecaniorum*)	*C. hesperidum*	(?) Europe
E. infelix	*S. coffeae*	(?) Ethiopia
Metaphycus lounsburyi (= *bartletti*)	*S. oleae*	South Africa
M. helvolus	*C. hesperidum*, *S. oleae* and *C. pseudomagnoliarum*	South Africa
Metaphycus sp. n. A (*Guerrieri & Noyes*) (*fomerly lounsburyi*)	*S. oleae*	South Africa
M. varius	*C. rubens*	(?) Australia
M. luteolus	*C. hesperidum*	California
M. stanleyi	*C. hesperidum*	South Africa
M. galbus	*C. hesperidum*	South Africa
Microterys nietneri	*C. hesperidum* *Ceroplastes floridensis*	South Africa Unknown
Eulophidae		
Aprostocetus ceroplastae	*C. destructor* *C. floridensis* *C. sinensis*	South Africa
Pteromalidae		
Scutellista caerulea	*S. oleae* *S. coffeae* *S. nigra* *C. rubens* *C. floridensis* *C. destructor* *C. sinensis*	(?) South/East Africa
Moranila california	*C. floridensis* *C. rubens*	(?) Australia
Cryptochetidae		
Cryptochetum iceryae	*Icerya purchasi*	Australia

Table 3.3. Common wasp parasitoids of armoured scales on citrus (Rosen and Debach, 1979; Prinsloo, 1984; Rosen, 1994; Noyes, 1998; Malipatil *et al.*, 2000).

Parasitoids	Main hosts on citrus	Origin of parasitoid
Aphelinidae		
Aphytis chrysomphali	*Aonidiella aurantii* and *Aonidiella citrina*	Mediterranean
A. columbi	*Chrysomphalus aonidum* and *Lepidosaphes beckii*	Australia
A. holoxanthus	*C. aonidum*	China
A. lepidosaphes	*Lepidosaphes beckii*	(?) China
A. lingnanensis	*A. aurantii, L. beckii* *L. gloverii* and *Unaspis citri*	China/eastern Asia
A. africanas	*A. aurantii*	South Africa
A. yanonensis	*Unaspis yanonensis*	China
Aphytis roseni	*Selenaspidus articulatus*	Kenya
A. hispanicus	*Parlatoria pergandii*	? Mediterranean
A. proclia	*Chrysomphalus dictyospermi*	? Mediterranean
Encarsia citrina	*A. aurantii* *A. citrina* *A. orientialis* *C. aonidum* *U. citri* and *L. beckii* *Chrysomphalus dictyospermi*	China/California
E. perniciosi	*A. aurantii* *U. citri*	(?) China
Pteroptrix smithi	*C. aonidium*	China
Encyrtidae		
Comperiella bifasciata	A. aurantii *A. citrina*	China China
C. lemniscata	*A. orientalis*	China/South-East Asia
Habrolepis rouxi	*A. aurantii*	South Africa

Armoured scales

The most important armoured scale species attacking citrus have been listed by Ebeling (1959). They are red scale *Aonidiella aurantii* (Maskell) (Plate 14), purple or mussel scale *Lepidosaphes beckii* (Newman) and citrus snow scale *Unapis citri* (Comstock), also Florida red scale *Chrysomphalus aonidium* (L.), dictyospernum scale *Chrysomphalus dictyospermi* (Morgan), Glovers scale *Lepidosaphes gloverii* (Packard), yanone scale *Unaspis yanonensis* Kuwana, West Indian red scale *Selenaspidus articulatus* (Morgan), black parlatoria *Parlatoria zizyphi* (Lucas), and chaff scale *Parlatoria pergandii* (Comstock). Similarity of armoured species caused some confusion and misdirection at times in biological control (Rose, 1990). The species of armoured scales in the genera *Aonidiella, Chrysomphalus, Lepidosaphes, Parlatoria, Pinnaspis* and *Unaspis* are believed to be native to the same Asian regions as citrus (Rose, 1990). Perhaps this is substantiated by the fact that many effective parasitoids used against these scales are from those areas (DeBach, 1976).

Red scale, Aonidiella aurantii *(Maskell)*

Red scale is a key pest in many subtropical and temperate parts of the world, e.g. South Africa, North Africa, Australia, California, Texas, Mexico and parts of South America (Bennett *et al.*, 1976; Anonymous, 1984; Gill, 1988; Smith *et al.*, 1997; Bedford *et al.*, 1998). In the Asian tropics and subtropics (where the scale originated) it is only occasionally important and is usually controlled by parasitoids – *Aphytis* spp., *Comperiella bifasciata* Howard and *Encarsia* spp. The main *Aphytis* species in red scale are *Aphytis melinus* DeBach and *A. lingnanensis* Compere and also *A. africanus* Quednow in South Africa (Luck *et al.*, 1982; Dahms and Smith,

1994; Smith *et al.*, 1997). The key to obtaining control in these other regions has been the introduction of these parasitoids. *Chilocorus negritus* has been a valuable predator in South Africa (Bedford *et al.*, 1998).

LIFE CYCLE The adult female gives birth to 100–150 mobile young, called crawlers, at a rate of two or three a day over a 6–8 week period. The crawlers emerge from under their mother's scale cover, and search for a suitable feeding site on leaves, shoots or fruit. Crawlers wandering on the tree canopy can be blown by the wind into neighbouring trees or orchards. Once a crawler settles, it inserts its mouthparts into the plant and starts feeding. It secretes a white waxy covering, and at this stage is called a 'whitecap'. After a period of feeding and growth, the insect moults. The cast skin is attached to the scale cover, giving the cover its typical red colour. The development stage and sex of red scale can be determined by the shape and size of the scale cover. After the second stage, scales can be identified as male or female. The scale cover of males is elongated, while the scale cover of females is circular.

The male develops through a prepupal and pupal stage under a scale cover, before emerging as a delicate, winged insect. It is attracted to the female by a pheromone, and dies after mating. In first-stage and second-stage females, and in third-stage (i.e. adult) unmated females, the scale cover is not attached to the body of the scale. When first-stage and second-stage females are moulting, and when third-stage females have the scale cover it is attached to the body of the scale.

In the tropics there are six or seven generations per year in comparison with two to four generations in higher latitudes.

Florida red scale, Chrysomphalum aonidum *(L.)*

Florida red scale is also normally a minor pest in eastern Asia (the centre of its origin) being controlled by *Aphytis holoxanthus* (DeBach) and also *Pteroptrix smithi* Compere. Transported around the world it became a major pest in many areas, e.g. Israel, Australia, Florida, Mexico, until one or both of these parasitoids were introduced (Smith, 1978; Rosen and DeBach, 1979; Rosen, 1994; Browning *et al.*, 1995).

Purple scale, Lepidosaphes beckii *(Newman)*

Purple scale was the most abundant and injurious insect in Florida citrus prior to 1960 (Knapp *et al.*, 1996a). It is troublesome where it has spread throughout the world, e.g. South Africa, Australia, California, Texas, Central and South America (Nasca *et al.*, 1981), but is a relatively minor problem in eastern Asia, its probable centre of origin. *Aphytis lepidosaphes* Compere is the dominant parasitoid and it appears also to have spread with its host worldwide (DeBach, 1971). It prefers a dense canopy, with infestations being heavier at the centre of the tree and on the north and northeast quadrants.

Citrus snow scale, Unaspis citri *(Comstock)*

Citrus snow scale is the most economically important pest in Florida since 1960 (Knapp *et al.*, 1996b), but is considered of low importance in South America (Mosquera, 1979). It had a major pest ranking in Queensland, Australia (Smith *et al.*, 1997). Citrus snow scale attacks the trunk and large limbs of the tree, leaves and fruits are infested only after the main branches. Natural enemies including *Telsimia* sp., *Chilocorus cacti* L. (Coccinellidae), *Encarsia* spp., *Aphytis lignanensis* Comp. (Hong Kong strain) are mentioned by several authors (Brooks, 1964; Quezada, 1989; Browning *et al.*, 1995). Coronado and Ruiz (1996) list *Encarsia* spp. as the most important parasitoid of *U. citri* in Mexico. In southern China, *Encarsia inquerenda* (Silvestri) is an important parasitoid and *Aphytis debachi* De Bach also occurs. Good biocontrol of the scale has been achieved in Queensland with the predator *Chilocorus circumdatus* Gylenhall (of Chinese origin) (Smith *et al.*, 1995).

Yanone scale, Unaspis yanonensis *Kuwana*

Yanone scale replaces *U. citri* as an occasionally important pest in China, Taiwan and

Japan but is normally controlled by *Aphytis yanonensis* (Rosen and DeBach, 1979).

Glovers scale, Lepidosaphes gloveri *(Packard)*

Glovers scale is reported as a serious pest of limes in Cuba, particularly in areas with an average temperature of 25–29°C and relative humidity fluctuating between 70 and 75%. In Mexico, the introduction of *A. lepidosaphes* Compere in the 1950s reduced the levels of infestation of this species (Ruiz, 1997).

Rufous scale is a serious pest of citrus in Peru and occurs in the West Indies (Guerrero, 1969; Beingolea, 1994). The parasitoid *Aphytis roseni* DeBach and Gordy was successfully introduced from Kenya in 1971.

Soft scales

The main soft scale pests occur in the following genera: (i) *Coccus* – soft brown scale *Coccus hesperidum* L, green coffee scale *Coccus viridis* (Green) (Plate 15), citricola scales, *Coccus pseudomagnoliarum* (Kuwana), long soft scale *Coccus longulus* (Douglas); (ii) *Saissetia* – black scale *Saissetia oleae* (Olivier) and hemispherical scale *Saissetia coffeae* (Walker); (iii) *Ceroplastes* – pink or red wax scale *Ceroplastes rubens* Maskell, white wax scale *Ceroplastes destructor* Newstead, Florida wax scale *Ceroplastes floridensis* Comstock, hard wax scale *Ceroplastes sinensis* Del Guercio, fig wax scale *Ceroplastes rusci* L., citrus wax scale *Ceroplastes brevicacuda* Hall, barnacle scale *Ceroplastes cirripediformis* Comstock; (iv) *Pulvinaria* – soft green scale *Pulvinaria aethiopica* (De Lotto), cottony citrus scale *Pulvinaria polygonata* Cockerell, green shield scale *Pulvinaria psidii* Maskell and orange pulvinaria *Pulvinaria aurantii* Cockerell; (v) *Protopulvinaria* – pyriform scale *Protopulvinaria pyriformis* Cockerell; and (vi) *Icerya* – cottony cushion scale *Icerya purchasi* Maskell (Ben-Dov, 1993; Browning *et al.*, 1995; Smith *et al.*, 1997).

Soft brown scale, Coccus hesperidum *L.*

Soft brown scale typifies a species that has become cosmopolitan in its distribution.

LIFE CYCLE Male soft brown scales are rare. Females can reproduce without mating, and give birth to a total of about 200 live young (crawlers). The life cycle takes approximately 2 months in summer. Development within generations is not synchronized, i.e. various stages are present at any given time, and the generations overlap.

SEASONAL HISTORY In eastern Australia, there are four or five generations per year but three or four generations per year in southern areas (Smith *et al.*, 1997).

HABITS Adult soft brown scales infest leaves and twigs, and occasionally green fruit. The scales produce large amounts of honeydew, resulting in the growth of sooty mould. The honeydew also attracts ants, which can interfere with biological control. Newly hatched crawlers move on to younger growth and the stalks of fruit. Young scales move around until they are half-grown, and then migrate towards leaves and small green twigs where they settle and reproduce.

The main natural enemies in Australia are *Metaphycus* spp., *Coccophagus* spp., *Microterys flavus* and *Diversinervus elegans*. Predators include ladybirds like *Rhyzobius* spp. and lacewings (Smith *et al.*, 1997).

Details on life cycle, characteristics, distribution, economic importance and damage caused by these scales are reported in Browning *et al.* (1995), Hammon and Williams (1984) and Smith *et al.* (1997). Soft scales are soft-bodied and have no separate protective covering, but the body is usually protected with a thick waxy or mealy secretion. Soft scales produce large amounts of sugary honeydew. The main economic damage caused by soft scales is from downgrading of fruit quality due to sooty mould fungus growing on the honeydew (Smith *et al.*, 1997). Soft scales usually infest fruits, leaves and young twigs. Heavily infested plants can lose vigour and become unproductive as foliage becomes covered in sooty mould. The honeydew attracts ants which defend the scales against natural enemies causing disruption and aggravating the scale problem.

Natural enemies (particularly encyrtid parasitoids) are extremely important in biological control of soft scales. The main encyrtid genera involved are *Metaphycus*, *Anicetus*, *Microterys* and *Diversinervus*. The pteromalids *Scutellista caerulea* and *Moranila* spp., the eulophid *Tetrastichus ceroplastae* and the aphelinid *Coccophagus* spp. are also important (Smith *et al.*, 1997; Malipatil *et al.*, 2000). Coccinellids and chrysopids also play an important role, the most notable case being the ladybird *Rodolia cardinalis* Mulsant for cottony cushion scale (Bennett *et al.*, 1976).

There are one or two orthezid citrus pests including *Orthezia praelonga* Douglas that are found on the leaves, branches, flowers and trunk occurring in Central and South America including Colombia (Garcia Roa, 1995). The most effective natural control agents are the predators *Hyperaspis* sp. (Coccinellidae), *Ambracius dufouri*, *Proba vittiscutis* (Miridae) and *Chrysopa* sp. Use of oils and soap helps to reduce damaging populations.

Mealybugs

The main mealybug pests of citrus include citrus mealybug *Planococcus citri* (Risso), citrophilous mealybug *Pseudococcus calceolarie* (Maskell), longtailed mealybug *Pseudococcus longispinus* (Targioni and Tozzetti) (Plate 16), *Pseudococcus cryptus* Hempel, *Pseudococcus citriculus* Green, spherical mealybug *Nipaecoccus viridis* (Newstead), tuber mealybug *Pseudococcus viburni* (*affinis*) (Maskell), comstock mealybug *Pseudococcus comstocki* Kuwana and oleander mealybug *Paracoccus burnerae* (Brain) (Knapp *et al.*, 1996; Smith *et al.*, 1997; Bedford *et al.*, 1998). The life cycle of mealybugs is similar to that of soft scales, feeding on plant sap, and excreting honeydew.

The citrus mealybug is often encountered in sheltered locations on the citrus tree, such as cracks, folds in leaves, and in fruit clusters or under fruit buttons (Browning *et al.*, 1995). Citrus mealybug feeding on fruit surfaces may result in discoloration and severe infestations may cause fruit drop.

In Florida, increases of *P. citri* are observed during summer and autumn, often corresponding to periods when pesticides are frequently applied to lime orchards (Peña and Baranowski, 1992). In Australia, the spring generation of crawlers of *Pseudococcus calceolariae* migrate to young fruit in late November and early December (Smith *et al.*, 1997). *P. citri* occurs at low levels due to the combined actions of the ladybeetle predator, *Cryptolaemus montrouzieri* (Mulsant), several parasitic wasps and a fungal disease (Browning *et al.*, 1995). Besides the mealybug ladybird, the encyrtid *Leptomastix dactylopii* Howard and the lacewings, *Oligochrysa lutea* (Walker), *Micromus* sp., and *Mallada signata* are considered important as natural enemies of the citrus mealybug in Australia (Waterhouse, 1988; Smith *et al.*, 1997). The encyrtid wasps *Tetracnemoidea brevicornis* (Girault) and *Anagyrus fusciventris* (Girault) are major parasites of *P. calceolariae* in southeastern Australia (Table 3.4), while the aphelinid wasp *Coccophagus gurneyi* Compere is more important on the east coast of Australia (Smith *et al.*, 1997). Members of the encyrtid genus *Anagyrus* (Plate 17) are important parasitoids of mealybugs, e.g. *Anagyrus pseudococci* (Girault) in *P. citri* and *Anagyrus agraensis* Saraswat in *N. viridis*. *Pseudaphycus flavidulus* (Brethes) and *Leptomastix epona* Noyse parasitize *P. viburni* (Ripa and Rodriguez, 1999). *Coccidoxenoides peregrinus* (Timberlake) attacks first instar *P. citri*.

Citrus whiteflies and blackflies

About 30 species of citrus whiteflies and blackflies have been reported throughout the world, out of which a number of species like citrus spiny whitefly *Aleurocanthus spiniferus* Quantaince, citrus blackfly *Aleurocanthus woglumi* Ashby (Plate 18), cloudy winged whitefly *Dialeurodes citrifolii* (Morgan), citrus whitefly *Dialeurodes citri* (Ashmead), *Dialeurodes elongata* Dozier and citrus woolly whitefly *Aleurothrixus floccosus* (Maskell) are considered important in several tropical citrus areas (Ghosh, 1990). The Australian citrus whitefly *Orchamoplatus citri* (Takahashi) is a minor pest in Australia. Details on the life history and damage of *Dialeurodes citri* (Ashmead), *D. citrifolii* and *Paraleyrodes* spp.

Table 3.4. Common wasp parasitoids of mealybugs on citrus (Prinsloo, 1984; Noyes, 1998; Malipatil *et al.*, 2000).

Parasitoids	Main hosts on citrus	Origin of parasitoid
Aphelinidae		
Coccophagus gurneyi	*Pseudococcus calceolariae*	Australia
Encyrtidae		
Anagyrus agraensis	*Nipaecoccus viridis*	(?) North Africa
A. fusciventris	*Pseudococcus longispinus*	Australia
	P. calceolariae	
A. pseudococci	*P. citri*	Mediterranean
Coccidoxenoides peregrinus	*P. citri*	China
Leptomastidea abnormis	*P. citri*	Mediterranean
Leptomastix dactylopii	*P. citri*	Brazil (or Africa?)
Tetracnemoidea brevicornis	*P. calceolariae*	Australia
	P. longispinus	
T. peregrina	*P. longispinus*	(?) South America
	P. calceolariae	
T. sydneyensis	*P. longispinus*	Australia
	P. calceolariae	
Pseudaphycus flavidulus	*Pseudococcus viburni (affinis)*	Chile
P. maculipennis	*P. viburni*	Europe
Leptomastic epona	*P. viburni*	Europe
Pteromalidae		
Ophelosia spp.	*P. calceolariae*	Unknown
Platygasteridae		
Allotropa sp.	*P. calceolariae*	China

are given by Browning *et al.* (1995). Most adults of citrus whiteflies and blackflies congregate on young leaves, where they lay eggs on the underside of the leaves; nymphs are regularly found on older leaves on the middle and bottom tree canopy. The life history, seasonal abundance and host range of *A. floccosus* has been determined by Salinas *et al.* (1996) in the Philippines. This species is known as the woolly whitefly, a name derived from the wool-like filaments that develop in the third nymphal instar of the insect (Watson, 1915). The nymphs of this species are smaller than *Aleurodicus* sp. More *D. citrifolli* (Morgan) were observed in Florida on new flushes than on the old ones. Common density peaks are observed from early autumn through early winter (Peña and Baranowski, 1992).

DAMAGE Whiteflies inflict their damage by sucking sap, excreting honeydew, which favours the growth of sooty moulds that interfere with photosynthesis, and producing a dense mat of woolly material that persists for months on the leaves, which, in turn, serves as an excellent refuge for scales (Watson, 1915). *A. floccosus* is also considered to be a potential vector of plant viruses and virus-like organisms that cause plant diseases and associated disorders (Bird and Maramorosch, 1978).

BIOLOGICAL CONTROL The main parasites of whitefly occur in the Aphelinidae – *Encarsia* spp. and *Eretmocerus* spp. also *Cales* spp. and *Amitus* spp. (Platygasteridae) (Clausen, 1978) (Table 3.5). Ladybirds, lacewings and fungi (*Aschersonia*) are also important. *Encarsia opulenta* (Silvestri) is important on citrus blackfly, *E. lahorensis* (Howard) on citrus whitefly, *E. variegata* on nesting whitefly (Browning *et al.*, 1995). The introduction of *Amitus hesperidum* Silvistri and *E. opulenta* (Silvestri) to Florida is probably responsible for the reduction of *A. woglumi* (Ashmead) densities (Browning *et al.*, 1995).

A. spiniferus is a minor pest of citrus in Taiwan, but it can also cause the reduction of yield when population is high. There are six generations of the spiny blackfly in a year.

Table 3.5. Common wasp parasitoids of whiteflies and blackflies on citrus (Clausen, 1978; Browning *et al.*, 1995).

Parasitoids	Main hosts on citrus	Origin of parasitoid
Aphelinidae		
Eretmocerus series	*Aleurocanthus woglumi* and *Aleurocanthus spiniferus*	Malaysia
E. portoricensis	*Aleurothrixus floccosus*	
E. haldemani	*A. floccosus*	
Cales noaki	*A. floccosus*, *Paraleyrodes* sp. and *Orchamoplatus citri*	
Encarsia smithi	*A. woglumi*	Japan
	A. spiniferus	
E. clypealis	*A. woglumi* and *A. spiniferus*	India
E. opulenta	*A. woglumi* and *A. spiniferus*	South Asia
E. merceti	*A. woglumi*	
Encarsia spp.	*Orchamoplatus citri* and *Paraleyrodes* sp.	Australia
E. lahoriensis	*Dialeurodes citri*	Pakistan
Platygasteridae		
Amitus hesperidum	*A. spiniferus* and *A. woglumi*	India, Pakistan
A. spiniferus	*A. floccosus*	

Three natural enemies of this blackfly are a predacious drosophilid, *Acletoxenus* sp., and two parasitic wasps, *Encarsia smithi* (Silv.) and *A. hesperidum*; however, they were not considered very effective. Several entomopathogenic fungi infecting this blackfly, e.g. *Aschersonia* spp. and *Aegerita webbneri* Fawc., were found in the wet season. The number of eggs laid per female averaged 19 (Chien and Chiu, 1986).

Citrus psyllids

The psyllid vectors of Asian greening disease *Diaphorina citri* Kuway and African greening *Trioza eritreae* (Del Guercio) are considered a most important threat to citrus in different tropical areas of the world (Aubert *et al.*, 1980; Nariani, 1981; Van den Berg *et al.*, 1991). The citrus psyllid *D. citri* is widely distributed and important in Asia (Atwal *et al.*, 1970; Broadbent *et al.*, 1980; Aubert and Quilici, 1983; Beingolea, 1988; Tandom, 1997) while *T. eritreae* is indigenous to Africa, where it is found in many countries as well as on the Islands of St Helena, Mauritius, Madagascar and Réunion (Mamet, 1955; Breniére and Dubois, 1965; CIE, 1967; Moran and Brown, 1973; Etienne, 1978). *Diaphorina citri* are brown to light brown (black in *Trioza erytreae*) (Mead, 1977) actively flying insects and measure 3–4 mm in length. While at rest, they raise their body upward. The nymphs are orange-yellow in colour, flattened and circular in shape. The eggs are anchored by means of a short stalk embedded in the plant tissues. Females can lay more than 800 eggs. Nymphs pass through five instars. Total life cycle requires from 15 to 47 days, depending upon the season. Adults may live for several months. For more information on life history see Husain and Nath (1927), Atwal *et al.* (1970), Catling (1970), Capoor *et al.* (1974).

The eggs of *T. eritreae* are elongate pear shape, and about 0.3 mm long; yellow when first laid, turning brownish. They are usually laid on the edges, or main veins, of very young leaves, being anchored to the leaf blade by a short appendage. Hatching takes about 5–6 days. There are five nymphal instars, and the whole nymphal period occupies 2–3 weeks. Females may live for a month and lay about 600 eggs (Hill, 1975). The historical review of infestations of *T. erytreae* is provided by Van Den Berg and Fletcher (1988).

DAMAGE The damage is caused by the nymphs and adults sucking sap from buds and leaves (Tandom, 1997) and transmitting the organism that causes greening disease (Mead, 1977). Greening was first considered a viral infection, but later the causal agent was identified as a phloem-restricted intracellular bacterium (Evers and Grisoni, 1991). The two strains of greening are associated with each vector, the African and the Asian, and are not adapted to the same range of temperature (Bove *et al.*, 1974). The mechanical damage by these psyllids leaves the leaves conspicuously pitted, the pits opening to the lower leaf surface. In severe attacks the leaf blades are cupped or otherwise distorted and yellow in colour, especially when young (Hill, 1975). There is no systematic data available on extent of damage; however, *D. citri* has been reported causing loss of US$1.04 million in India (Tandom, 1997). Trees severely affected by the South African or heat-sensitive race of the greening disease are badly stunted and produce crops of predominantly greened, worthless fruit. These fail to ripen and cannot be used for processing as they impart an objectionable flavour which may taint large volumes of juice (Van Den Berg and Fletcher, 1988). The importance of greening in South Africa was emphasized when approximately 100,000 sweet orange trees were rendered commercially unprofitable by the disease. The South African greening disease was also partly responsible for the collapse of citrus production in Réunion and Mauritius (Van Den Berg and Fletcher, 1988).

MONITORING *T. erytreae* was found to be highly attracted to yellow surfaces, particularly fluorescent yellow-green (about 530 nm) in citrus orchards in South Africa (Samways, 1987). Water pan traps painted with yellow are also used successfully to trap other psyllids (Omole, 1980). Yellow cards were useful to detect immigrant infestations of *T. erytreae* from adjacent vegetation into citrus orchards (Van den Berg *et al.*, 1991).

BIOLOGICAL CONTROL Several species of predators have been reported feeding on eggs and nymphs of citrus psylla. The spider, *Marpissa tigrina* Tikader is considered one of the most important predators of this pest in India (Sadana, 1991). Besides these, four species of parasitoids have been reported from India (Tandom, 1991a) and up to 15 from the Asian Pacific region (Etienne, 1978; Waterhouse, 1998; Tang and Aubert, 1990) (Table 3.6).

In Réunion, the impact of *T. erytreae* and *D. citri* was reduced through releases of the parasites *Tetrastychus dryi* Waterston, *Tetrastychus radiatus* Waterston, and the encyrtid *Diaphorencyrctus aligarhensis* (Shafee, Alam and Agarwal) (Aubert and Quilici, 1983).

PLANT RESISTANCE In India, of 20 species and cultivars tested for their reaction to greening, sweet lime was resistant alone or as rootstock for mosambi orange. Italian, Eureka and Lisbon lemon were tolerant and showed mild symptoms (Nariani, 1981).

IPM STRATEGY In India, at the initiation of new flush, sprays of monocrotophos (0.025%) or dimethoate (0.03%) are delivered at a 10–12 day interval (Bindra *et al.*, 1974; Nariani, 1981; Tandom, 1991b). For effective management, pruning infested shoots and their destruction, maintaining orchard sanitation, and application of 1% pongamia oil and 4% neem oil at 21 and 7 day intervals is recommended (Tandom, 1997). Careful integrated biological and chemical suppression of the psylla vector is considered the most practical entomological method for containing greening disease on mature trees (Samways, 1987).

Nariani (1981) suggests an integrated control schedule incorporating use of disease-free or heat-treated budwood, injection with tetracycline antibiotics, insecticide sprays and use of tolerant rootstocks to reduce disease losses. On the other hand, Aubert *et al.* (1980) suggested the use of disease-free nursery stock, introduction of hymenopterous parasitoids and use of antibiotics as the best action against psyllids and the greening disease.

Aphids

Worldwide, 16 species of aphids are reported to feed regularly on citrus (Halbert and Brown, 1996). Aphids infest citrus blossom and young growth. Of these species, four are found consistently in Florida groves, including cowpea aphid *Aphis craccivora* Koch, cotton or melon aphid *Aphis gossypii* Glover, spiraea aphid *Aphis spiraecola* Patch, black citrus aphid *Toxoptera aurantii* (Boyer de Fonscolombe) and more recently brown citrus aphid *Toxoptera citricida* (Kirkaldy).

The brown citrus aphid (BrCA), is the most important vector of citrus tristeza virus (CTV) because of its high vector efficiency, especially for severe strains (Meneghini, 1946; Costa and Grant, 1951; Yokomi *et al.*, 1994). An extensive bibliographical review of BrCA is provided by Michaud (1998). It is currently distributed in South-East Asia, Africa south of the Sahara, Australia, New Zealand, the Pacific Islands, South America, the Caribbean Islands, Central America and, most recently, Florida (Carver, 1978; Yokomi *et al.*, 1994; Halbert and Brown, 1996). One of the most devastating citrus crop losses ever reported

Table 3.6. Common wasp parasitoids of psyllids, aphids, planthoppers and leafhoppers on citrus (Clausen, 1978; Prinsloo, 1984; Tang and Aubert, 1990; Smith *et al.*, 1997; Waterhouse, 1998).

Parasitoids	Main hosts on citrus	Origin of parasitoid
Aphelinidae		
Aphelinus gossypii	*Toxoptera citricidus*, *T. aurantii*	Australia, Hawaii
	Aphis spireacola	
	Aphis gossypii	
	A. spireacola	
	A. gossypii	
Centrodora scolypopae	*Scolypopa australis*	Australia
C. penthimiae	*Penthimiola bella*	South Africa
Braconidae		
Lysiphlebus testaceipes	*A. spireacola*	California
	T. aurantii	
	A. gossypii and *T. citricidus*	
L. japonica	*T. citricidus*	Japan
L. fabarus	*T. citricidus*, *T. aurantii* and *A. gossypii*	Turkey
		India
Aphidus colemani	*T. citricidus*, *T. aurantii* and *A. gossypii*	
Lepolexis scutellaris	*A. gossypii* and *A spireacola*	India
Binodoxiys spp.	*A gossypii*	India
Aphidius matricariae	*T. aurantii*	
Platygasteridae		
Aphanomerus spp.	*Colgar peracutum*, *Colgaroides acuminata* and *Siphanta* spp.	Australia
Eulophidae		
Tamarixia radiata	*Diaphorina citri*	India
T. dryi	*Trioza erytreae*	South Africa
Encyrtidae		
Achalcerinys spp.	*C. peracutum*	Australia
Psyllaephagus pulvinatus	*T. erytreae*	South Africa
Ooencyrtus spp.	*C. peractum*	Australia
	C. acuminata	
	Siphanta spp.	
Diaphorencyrtus aligarhensis	*D. citri*	Southeastern Asia
Mymaridae		
Stethynium nr. empoascae	*E. Smithi*	Australia
Anagrus baeri	*Empoasca smithi*	Australia
Chaetomymar lepidius	*P. bella*	South Africa
C. gracile	*P. bella*	South Africa
Elasmidae		
Elasmus zehntneri	*P. citrella*	Philippines, Java
Braconidae		
Microbracon phyllocnistoides	*P. citrella*	Indonesia
Bracon sp.	*P. citrella*	Philippines

followed the introduction of BrCA into Brazil, Argentina and Venezuela (Carver, 1978; Roca-Peña *et al.*, 1995). BrCA is considered larger than the other species of aphids occurring in citrus (Halbert and Brown, 1996) and its identification is difficult because most adult wingless forms of aphids are shiny black, and nymphs are dark reddish brown. A field key to adult wingless forms of BrCA is provided by Halbert and Brown (1996). The life cycle of BrCA is anholocyclic, meaning that there is no sexual cycle in the autumn, and thus no males, no oviparae and no eggs (Komazaki, 1987); nymphs mature in 6–8 days at temperatures of 20°C or higher (Komazaki, 1987) and the total life cycle can take as little as 1 week (Smith *et al.*, 1997).

Citrus is propagated vegetatively, which greatly increases the possibility for spreading the phloem-limited virus because CTV is graft transmissible. Thus, the first step in any management programme should be to ensure that budwood and nursery stock are free of disease (Halbert and Brown, 1996). Another aspect of cultural control is inoculum suppression, or destruction of abandoned or volunteer crop plants that become reservoirs of pests and disease (Bishop *et al.*, 1992; Plumb and Johnstone 1995).

NATURAL ENEMIES Aphid parasitoids include *Aphidius* spp. and *Aphelinus* spp. A wide range of predators occur, e.g. in Australia (Smith *et al.*, 1997). *Chilomenes lunata* (F) is important in South Africa (Bedford *et al.*, 1998). A range of predators attack citrus aphids. These include ladybirds, e.g. the transverse ladybird (*Coccinella transversalis*), the common spotted ladybird (*Harmonia conformis*), *Harmonia testudinaria*, the variable ladybird (*Coelophora inaequalisi*), and the yellow-shouldered ladybird (*Scymnodes lividigaster*); syrphid flies, e.g. the common hoverfly (*Simosyrphus grandicornis*); and lacewing larvae.

Planthoppers and leafhoppers

These include flatids and jassids (Plate 19). The flatids or planthoppers, e.g. *Siphanta* spp., citrus planthopper *Colgar peracutum* (Walker) and mango planthopper *Colgaroides acuminata* (Walker) produce copious honeydew with resulting sooty mould (Smith *et al.*, 1997). The jassids or leafhoppers, particularly *Empoasca* spp. and also (in South Africa) citrus leafhopper *Penthimiola bella* (Stal.), suck maturing fruit causing multiple oleocellosis-like spots (Smith *et al.*, 1997; Anonymous, 1984; Bedford *et al.*, 1998).

Mymarid wasp egg parasitoids are important natural enemies of leafhoppers and the aphelinid *Centrodora penthimiae* Annecke parasitizes *P. bella*. The encyrtids *Alchalcerinys* sp., *Ooencyrtus* spp. and *Aphanomerus* spp. are important egg parasitoids of planthoppers and dryinid wasps and strepsiptera parasitize the nymphs and adults (Smith *et al.*, 1997).

Lepidoptera pests

The main groups of lepidopterous pests are: fruit borers (mainly tortricids but also pyralids and blastobasid larvae); leafminers notably citrus leafminer *Phyllocnistis citrella* Stainton; leafrollers (mainly tortricids); fruit piercing or sucking moths (mainly *Othreis* spp. and *Oraesia* spp.); citrus butterflies (*Papilio* spp.); and flower or bud moths (*Prays* spp.). The false codling moth *Cryptophlebia leucotreta* Megre is important in South Africa (Bedford *et al.*, 1998). The carob moth *Ectomyclois ceratoniae* (Zeller) is a minor pest in the Mediterranean and in Africa. American bollworm or corn ear worm *Helicoverpa armigera* (Hubner) is a sporadic pest, e.g. in South Africa and eastern Australia and South-East Asia. Citrus flower moth *Prays citri* (Milliere) is distributed worldwide. *Adoxophyes* spp. are occasional pests in South-East Asia and Australia. Table 3.7 lists the important parasitoids of lepidopteran citrus pests.

Citrus leafminer

The citrus leafminer *Phyllocnistis citrella* was first noted in South-East Asia in 1856 and slowly dispersed to Japan (1927), Korea, Philippines (1915), Australia (1918), East Africa (1967) and West Africa (1970). Since 1993, this insect has infested most other citrus

Table 3.7. Common wasp parasitoids of other lepidoptera on citrus (Smith *et al.*, 1997).

Parasitoids	Main hosts on citrus	Origin of parasitoid
Encyrtidae		
Ooencyrtus spp.	*Papilio* spp. and *Othreis* spp.	Worldwide
Braconidae		
Apanteles spp.	*Cryptophlebia leucotreta, Tortrix* spp., *Papilio* spp.	Worldwide
Dolichogenidea arisanus	*Epiphyas postvittana*	Worldwide
Microplitis spp.	*Helicoverpa armigera* and *H. punctigera*	Worldwide
Trichogrammatidae		
Trichogramma spp., e.g. *pretiosum*	*H. armigera, Tortrix* spp., *C. leucotreta* and *E. postvittana*	Worldwide
Chalcididae		
Brachymeria spp.	*Tortrix* spp.	Worldwide
Pteromalidae		
Pteromalus puparium	*Papilio* spp.	Worldwide
Eulephidae		
Euplectrus kurandaensis	*Tiracola plagiata*	Worldwide
Scelionidae		
Telenomus spp.	*H. armigera* and *Othreis* spp.	Australia
Ichneumonidae		
Phytodietus sp.	*Isotenes miserana*	Australia
Xanthopimpla sp.	*E. postvittana*	Australia

producing areas of the world. The first infestation of the Americas occurred in Florida, USA, followed by Mexico and the Caribbean region, the Mediterranean region, Central America, the Near East countries and lately most countries of South America. It is a common pest on citrus in South-East and eastern Asia and in some areas is heavily sprayed. It tends to aggravate injury from citrus canker.

BIOLOGY The biology of the leafminer is typical of other species of leafmining moths (Plate 20). Small eggs are laid on young leaves, and hatching larvae produce serpentine mines beneath the leaf epidermis, where they feed upon the liquid contents of leaf cells. Life history of the leafminer has been studied by Badawy (1969), Batra *et al.* (1988) and Garrido (1995). Total generation time can fluctuate between 13 and 52 days depending on temperature.

Species, and hybrids of the genus *Citrus* and other species of the family *Rutaceae* appear to be the primary host plant group of *P. citrella*. Published accounts (Sandhu and Batra, 1978a,b) indicate that some cultivars are more susceptible than others. The most resistant species was mandarin and the most susceptible was lime.

RELATIONSHIP BETWEEN *P. CITRELLA* INJURY AND PLANT PHENOLOGY *P. citrella* injury to the plant causes reduction of the leaf surface area responsible for capturing energy for tree growth. In heavy populations, leafminer larvae prevent newly emerging leaves from expanding fully, causing leaves to remain curled and twisted (Plate 21). Very few studies have addressed the relationship between *P. citrella* levels, damage and, ultimately, its effect on yield reduction (Binglin and Mingdu, 1996; Hunsberger *et al.*, 1996). Most important, measurement of leafminer damage remains an undefined term. Studies conducted by Schaffer *et al.* (1996) in lime determined that visual damage estimates in percentage of area damaged (underside and upperside of the leaves) tended to overestimate leaf damage by about 30%. Number of larvae per leaf and the number of days mining were correlated with leaf damage.

Hunsberger *et al.* (1996) found that lime yield was significantly reduced (37.7%) by CLM damage levels in south Florida. The same authors established relationships between fertilized trees and stressed fertilized trees and concluded that there was a positive correlation ($r^2 = 0.57$) between leaf abscission and leaf damage.

Binglin and Mingdu (1996) reported that no influence on growth, development of flushes or yield would occur with under 20% damaged area in tender leaves. These authors report that in southern China the functional relationship between the number of larva (x) and the percentage of damaged area (y) in a tender leaf is: $y = -1.005 + 23.246x$ ($R = 0.98$), giving an economic threshold of 0.74 larva/tender sweet orange leaf for the autumn generation. Peña *et al.* (2000) reported economic injury levels in Florida to fluctuate between 16 and 23% and between 18 and 85% leaf area damaged for 15-year-old and 5-year-old Tahiti lime trees, respectively.

SEASONALITY Field observations from China, Australia and some areas of Florida indicate that in subtropical regions spring flushes are the least damaged of seasonal growth periods, while summer and autumn flushes suffer the most serious infestations (Knapp *et al.*, 1995). Different seasonality is observed for subtropical areas (Peña *et al.*, 1996b), tropical areas, the Mediterranean, North Africa and the Near East. For instance, Smith and Beattie (1996) demonstrated that infestation of the new growth begins in early December in Queensland and 4–6 weeks later in southern Australian states. Up to 100% of young leaves in a flush can be attacked from January to April when infestation levels drop rapidly with increasing parasitism, reduced new growth and cooler temperatures. Peak numbers of adults were caught throughout the year with peak numbers in spring, summer and autumn in south Florida (Peña, 1998). The total number of moths trapped was 2.5 times higher in 1994 than in 1995. Relationships were established between adult catch and egg deposition during the year in south Florida. Results also suggested that numbers of moths trapped are not influenced by the amount of rainfall but might be influenced by temperature (Peña, 1998). Huang and Huang (1989) and Peña and Schaffer (1996, 1997) have made some studies of egg and larva distribution patterns and sampling methodology.

The sex attractant (Z,Z)-7,11-hexadecadienyl acetate and (Z,E)-7,11-hexadecadien-1-ol discovered originally from *Pectinophora gossypiella* (Bierl *et al.*, 1974) has been used to trap citrus leafminer. This commercially available lure (Ando *et al.*, 1985) failed to indicate male activity in Florida, Spain, China and Korea (Du *et al.*, 1989; Tongyuan *et al.*, 1989; Caleca and Lo Verde, 1996; Malausa *et al.*, 1996; Ortu, 1996; Jacas and Peña, 2002). Male catches in pheromone traps may help to predict leafminer density in orchards before reaching outbreak densities.

CHEMICAL CONTROL Many growers have relied on chemical control of *P. citrella* but effective chemical control of CLM is difficult to achieve and total reliance on chemical control has been demonstrated to be ineffective (Knapp *et al.* 1995, 1996a). Application of chemicals should take into consideration flushing pattern, actual population present, tree age (mature trees vs. young trees), production system (orchard vs. nursery), application method (foliar vs. soil application), and presence of natural enemy complex.

Some selective materials such as avermectin are useful to check infestations and petroleum oil sprays (at 0.25–0.5%) applied three or four times per flush are also effective (Rae *et al.*, 1996; Liu *et al.*, 1999, 2001).

BIOLOGICAL CONTROL The most important aspect of citrus leafminer management is biological control. While in many cases the diversity of natural enemies of the leafminer (hymenopterous parasitoids, predacious arachnids, ants, lacewings) accounts for significant reduction of the leafminer population (Amalin *et al.*, 1996; Binglin and Mingdu, 1996; Browning, 1996; Browning *et al.*, 1996; Cano *et al.*, 1996; Castaño *et al.*, 1996; Cave, 1996; Hoy and Nguyen, 1997; Morakote and Nanta, 1996; Peña *et al.*, 1996b; Smith and Neale, 1996; Waterhouse, 1998), in other cases, their presence and activity are low.

In the area of origin of the citrus leafminer, the parasitoid complexes are varied (LaSalle and Schauff, 1996) (Table 3.8) and their value as mortality factors differs among geographical areas. For example, in China,

Table 3.8. Common wasp parasitoids of citrus leafminer on citrus (LaSalle and Schauff, 1996; Hoy and Nguyen, 1997).

Parasitoids	Origin of parasitoid
Encyrtidae	
Ageniaspis citricola	Thailand, Philippines Indonesia
Asecodes delucchi	Thailand
Eulophidae	
Cirrospilus quadristriatus	China
Citrostichus phyllocnistoides	China, Thailand
Semielacher petiolatus	Australia
Zaommomentedon brevipetiolatus	Thailand, Japan
Closterocerus trifasciatus	Thailand
Stenomesius japonicus	Japan, China
Sympiesis striatipes	Thailand, Japan
Kratosyma citri	Thailand, PNG
Elachertus sp.	China
Chrysocharis pentheus	Japan, Taiwan
Quadrastichus sp.	Thailand
Pnigalio flavipes	Florida
Diglyphus begini	Florida

Binglin and Mingdu (1996) demonstrated that larvae and pupae are attacked by four or five species of parasitoids in Guangdong Province and at least seven species against larvae and two species against pupae in Fujian Province. These authors considered that the dominant species *Tetrastychus phyllocnistoides* Narayanan and *Cirrospilus quadristriatus* (Subba Rao and Ramamani) in Guangdong province, and *Elachertus* sp. in Fujian Province played an important role in controlling the population development of the leafminer. In Thailand, 13 species of hymenopterous parasitoids were found. Parasitism during 1991 through 1993 fluctuated between 36.73% and 72.05%. The average percentage parasitism by *Ageniaspis citricola* Logvinovskaya, *Quadristriatus* sp., *Cirrospilus ingenuus* Gohan, *Teleopterus* sp., *Eurytoma* sp., *C. itrifasciatus*, *Citrostichus phyllocnistoides* (Narayanan), *Sympiesis striatipes* Ashmead and undetermined species was 11.39, 8.77, 4.84, 2.25, 0.23, 0.19, 0.18, 0.13, and 29.32%, respectively.

In other Asiatic areas, e.g. Japan, where *P. citrella* was introduced 60 years ago, the dominant parasitoids, *Sympiesis striatipes*, *Chrysocharis pentheus* (Walker), *C. phyllocnistoides* and *Zaommomentedon brevipetiolatus* Kamijo produced high levels of parasitism. However, Ujiye (1996) considered that these native parasitoids cannot give an adequate biological control of the leafminer.

In the western hemisphere, the current biological control situation has been studied by different researchers (Hoy and Nguyen, 1994; Browning *et al.*, 1996; Cano *et al.*, 1996; Castaño *et al.*, 1996; Cave, 1996; French and Legaspi, 1996; Peña *et al.*, 1996b; Perales *et al.*, 1996). The seasonal occurrence of parasitoids of *P. citrella* was studied from 1993 through 1995 in Florida (Peña *et al.*, 1996b). The parasitoids collected were identified as *Pnigalio minio*, *Cirrospilus* sp., *Closterocerus* sp., *Sympiesis* sp., *Horismenus* sp., *Zagrammosoma multilineatum* (Ashmead), *Oncophanes* sp., and *Elasmus tischeriae* Howard. Most of these parasitoids are eulophids. Levels of parasitism in Florida indicated that, through parasitoid conservation, biological control of *P. citrella* by its native parasitoids could play an important role in leafminer management.

The importance of more detailed studies to determine the role of the indigenous parasitoids has been emphasized by Browning (1996). Perales *et al.* (1996) reported that *Cirrospilus* sp. preferred second and third instar larvae, while *Horismenus* sp. was observed in prepupa or pupa and *Elasmus* sp., was observed in pupa. Duncan and Peña (2000) reported fecundity, and host stage preferences of *Pnigalio minio*. Much research is still needed to determine the effect of the different chemicals on the beneficial entomofauna of Florida. However, preliminary data (Peña *et al.*, 1999) show that Agrimek might be the least harmful pesticide to the native parasitoid.

CLASSICAL BIOLOGICAL CONTROL Introduction of exotic species (e.g. *A. citricola*) from the area of origin has proven to be successful for Australia and Florida (Hoy and Nguyen, 1994; Smith and Beattie, 1996). There is a clear need for other parasitoid species adapted to different climatic areas. In Queensland *A. citricola* can kill 80–90% of its host during January–April, resulting in much less attack on the late summer–autumn flushes. Long cool winters and extreme dry heat (up to

45°C) in summer have a detrimental effect on parasitism levels. In southern Australian states, *Semielacher petiolatus* Girault can reach parasitism levels of 20–60% but not until March–April. High rates of parasitism in March–April appear to have an important influence on the rate of reinfestation in the following season, possibly by reducing the size of the overwintering leafminer populations (Smith and Beattie, 1996). A classical biological control project was initiated in Florida in February 1994 with the introduction of the parasitoid *A. citricola* from Australia (Hoy *et al.*, 1996). Rearing methods were developed (Smith and Hoy, 1995) and releases began in May 1994. *A. citricola* established, dispersed and overwintered in Florida during 1994/95. Maximum dispersal of *A. citricola* was documented 25 km from an initial release site (Pomerinke and Stansly, 1996). In some sites, parasitism of pupae was found to be as high as 99% and parasitism rates of 60–80% were common (Hoy *et al.*, 1995). However, the success of *A. citricola* in the humid Florida climate (Hoy *et al.*, 1996) has not been repeated in the dry areas of Texas (V. French, personal communication).

Parasitoid rearing has been improved and modified by different researchers (Argov and Rossler, 1996; Hoy *et al.*, 1996; Johnson *et al.*, 1996; Peña and Duncan, 1996; Smith and Neale, 1996). Because there is no artificial diet available for citrus leafminer, rearing of parasitoids requires an understanding of the biology of the leafminer and the parasitoid as well as the phenology and flushing patterns of the host plant (Smith and Hoy, 1995).

PREDATORS Binglin and Mingdu (1996) reported that the predators *Chrysopa boninensis*, *C. sinica* and the predacious bug *Orius insidiosus* and some ants and spiders are reported to prey on larvae of the leafminer. Amalin *et al.* (1996) reported that arachnids, *Chiracanthium inclusum*, *Clubiona* sp., *Trachelas volutus*, *Hivana velox*, *Hentzia palmarum*, were considered the most dominant predators feeding on larvae of citrus leafminer in south Florida, followed by lacewings, coccinellids and ants. In Florida, Amalin *et al.* (2002) reported that a higher percentage mortality of *P. citrella* was obtained by predation (Amalin *et al.*, 2001) by the indigenous natural enemies than by the introduced parasitoid, *A. citrella*.

Fruit piercing moths

In India, fruit sucking or piercing moths belong to the genera *Othreis*, *Achaea*, *Calpe*, *Anua*, *Ophideres*. In Brazil, *Gymnadrosoma aurantium* Costa Lima (Lepidoptera: Grapholitidae) (Cermeli, 1983) causes damage by sucking juice of citrus. Quezada (1989) reports that *Othreis scabellum* (Guenee) and *O. serpentifera* Walker (Noctuidae) are a serious problem for citrus in Honduras. Ripe fruits of sweet orange, mandarin orange and sweet limes are commonly attacked. The insects pierce the fruit rind and expose it to secondary infections (Plate 22). The fruit usually drops within a few days and thus becomes unmarketable. It is claimed that through elimination of alternative host plants from the vicinity of citrus orchards the insect infestation could be reduced (Ghosh, 1990). *O. fullonia* (Clerck), *O. materna* (L.) and *Eudocima salaminia* (Cramer) are significant pests in northeastern Australia.

Coleopteran pests

Beetle pests of citrus include the branch and trunk borers of the family Cerambycidae (e.g. *Anoplophora chinensis* Forster in China), leaf, fruit and root eating weevils of the family Curculionidae, e.g. Fuller's rose weevil *Asynonychus cervinus* Crotch and leaf eating chrysomelids (e.g. monolepta beetle *Monolepta australis* Jacoby, in Australia).

Root weevils

Several species of weevils (Curculionidae) affect citrus in Florida and the Caribbean (Schroeder and Beavers, 1977). These are the West Indian sugarcane rootstalk borer weevil, *Diaprepes abbreviatus* (L.); Fuller's rose beetle, *Asynonychus cervinus* (Boheman), the citrus root weevil, *Pachneus litus* (Germar); the northern citrus root weevil *Pachneus opalus* (Olivier); and the little leaf notcher *Artipus floridanus* Horn (Otero *et al.*, 1995; Knapp *et al.* 1996b). Other species of weevils considered important citrus pests in Central and South America are *Exopthalmus* spp. and *Compsus* spp. (Coto *et al.*, 1995). *Hypomeces squamasus* Fabricus occurs in South-East Asia and citrus snout weevil *Sciobius marshalls* Schoeman in South Africa. All weevil species are similar in biology and the damage they do to citrus trees (Knapp *et al.*, 1996b).

The adults feed on leaves, often leaving a characteristic pattern of notches around the edges. Except for Fuller's rose weevil, which deposits its eggs beneath the calyx of the fruit (Knapp *et al.*, 1999), female root weevils lay their eggs in clusters on leaves and, by secreting a sticky substance, cement leaves together around egg clusters. A single female may lay as many as 5000 eggs during her life of 3 to 4 months. In 7 or 8 days the eggs hatch; larvae leave the cluster and fall to the soil. After they enter the soil, they may stay there for 8–12 months, feeding on plant roots. They pupate in the soil, and then adults emerge, come up out of the soil and start the cycle all over again eating leaves and laying eggs. The costs of fighting an infestation of the insect in a citrus grove are considerable. Currently, it involves an autumn application of nematodes for larval control, use of spring and autumn sprays to kill adults (McCoy and Simpson, 1994) and use of fungicides in the spring, summer and autumn to control *Phytophthora*. Lesions caused by the larvae increase susceptibility to root rot. Adding up these costs could reach \$912 per hectare. Factoring a theoretical loss of 10% of the citrus trees and a 25% decrease in harvests, the estimated loss reaches \$3140 ha^{-1} year^{-1}.

Whitwell (1990) reports that *Diaprepes famelicus* causes losses of \$30,000 ha^{-1} in citrus nurseries in the Caribbean. This author reported that the use of soil-applied insecticides did not reduce root damage to citrus nursery beds in the Caribbean island of Dominica, probably because larvae were too deep in the soil. Whitwell (1990) considers that pesticides may have exacerbated the problem by reducing natural enemies.

The current and past research on *Diaprepes abbreviatus* (Coleoptera: Curculionidae) was summarized by Hall (1995) and analysed by McCoy and Simpson (1994). Its biology has been revised by Fennah (1942), Woodruff (1968) and McCoy and Simpson (1994). Current research for *Diaprepes* has focused on trapping, kairomones and pheromones, use of mortality factors, e.g. entomopathogenic nematodes and fungi (Beavers and Schroeder, 1980; Schroeder 1990; Adair 1994; McCoy and Simpson, 1994; J.E. Peña *et al.*, unpublished), determination of proteins to reduce larval feeding, plant resistance (Shapiro *et al.*, 1996) and use of audiographic methods to determine larvae in the soil (Shapiro and Gottwald, 1995). For a review of the biology of *Diaprepes abbreviatus* see Schroeder and Beavers (1977). Distribution and sampling have been investigated by Beavers and Selhime (1975), while Adair *et al.* (1998) determined *D. abbreviatus* oviposition preferences. A. Hunsberger *et al.* (unpublished) observed that *D. abbreviatus* prefers to deposit eggs on sorghum rather than other host plants.

INJURY CAUSED BY LARVAE The larvae girdle the taproot, destroying the tree's ability to take up nutrients and causing it to die. This type of injury provides an avenue for root rot infections. One larva can kill a young tree, and several larvae can cause serious decline to established trees.

BIOLOGICAL CONTROL A new species of entomogenous nematode, *Steinernema riobravis*, offers improved control (Schroeder, 1990). Bullock (1995) found that *S. riobravis* offers 89–90% control of *Diaprepes*. The beneficial nematodes enter weevil larvae and release a plug of bacteria which multiply in the larva's body cavity and digest its insides, creating a material which the nematodes feed upon. This process kills the larvae within 24–48 h, and the nematodes continue to grow and mate. The female nematode releases infective juveniles which then go out and find other host larvae (Anonymous, 1995). In central Florida citrus areas, nematodes are applied August–October and March and May (Rhodes, 1990).

EGG PARASITOIDS The introduction into continental USA of some species of parasitic wasps, e.g. *Tetrastichus* (*Quadrastichus*) *haitiensis* Gahan (Armstrong, 1987) collected from

Caribbean countries with a heavy *Diaprepes* population has not been considered effective (Anonymous, 1996; Beavers *et al.*, 1980). However, preliminary results by R. Franqui (personal communication) report higher parasitism rates of *Quadrastichus haitiensis* on eggs of *D. abbreviatus*. For several years, Etienne *et al.* (1990) reported that the egg parasite *Ceratogramma etiennei* (Hymenoptera: Trichogrammatidae) appeared to be specific to eggs of the genus *Diaprepes* and recommended its introduction into other Caribbean islands and into the USA. *C. etiennei* has been successfully reared in the laboratory and its biology has been determined by Etienne *et al.* (1990). The same authors observed a parasitism level of 30% of the egg population in the island of Guadeloupe. Moreover, the population of *Diaprepes* sp. in Guadeloupe appears to be lower after parasitoid augmentation, compared to the pest situation in other Caribbean islands such as Dominican Republic and Puerto Rico (J. Etienne, personal communication). *C. etiennei* is now cultured at TREC-Homestead (Peña *et al.*, 1998).

PREDATORS Eggs and larvae of *Diaprepes abbreviatus* have been reported as prey of nine predator species in Florida (Whitcomb *et al.*, 1982; Tryon, 1986). Predation by ants is reduced during the afternoon, but increases during the evening. Richman *et al.* (1983) concluded that predation by ants was less important in neonate larvae than for eggs and late instar larvae. Later Jaffe *et al.* (1990) demonstrated that first instar larvae appear to produce repellents that reduce predation. Tryon (1986) reported that earwigs are effective night predators of first instar larvae in Florida.

Dipteran pests

The main dipterous pests are fruit flies and blossom midges. Fruit flies can be major pests, capable of causing near total fruit loss and can be of major concern from a quarantine standpoint in exported fruit (Vijaygasegaran, 1993).

Fruit flies affecting citrus belong to the genera *Anastrepha, Bactrocera, Ceratitis, Dirioxa, Monacrostichus* and *Rhagoletis*. The biology of the most important fruit flies has been summarized by Christensen and Foote (1960), their ecology has been summarized by Bateman (1972) and their management has been assessed by McPheron and Steck (1996).

Bactrocera *spp.*

Bactrocera spp. are pests of major importance in the eastern hemisphere. The *Bactrocera* spp. reported on citrus by White and Elson-Harris (1992) include: *B. aquilonis* (May), *B. caryeae* (Kapoor), *B. caudata* (Fabricius), *B. cilifer* (Hendel), *B. correcta* (Bezzi), *B. cucurbitae* (Coquillett), *B. curvipennis* (Froggatt), *B. diversa* (Coquillett), *B. dorsalis* (Hendel), *B. facialis* (Coquillett), *B. frauenfeldi* (Schiner), *B. jarvisi* (Tryon), *B. halfordiae* (Tryon), *B. kriki* (Froggatt), *B. latifrons* (Hendel), *B. malgassa* Munro, *B. melanota* (Coquillett), *B. melas* (Perkins & May), *B. minax* (Enderlein), *B. neohumeralis* (Hardy), *B. passiflorae* (Froggatt), *B. psidii* (Froggatt), *B. quinaria* (Bezzi), *B. tau* (Walker), *B. tsuneonis* (Miyake), *B. trilineola* Drew, *B. trivialis* (Drew), *B. tryoni* (Froggatt), *B. zonata* (Saunders), *B. umbrosa* (Fabricius), and *B. xanthodes* (Broun). Among these, *B. minax* is considered one of the most destructive pests of citrus in China (Huasong *et al.*, 1998), being particularly damaging to *Citrus sinensis* and *C. aurantium*. *B. tryoni* is the major fruit fly pest of citrus in eastern Australia. *B. papayae* and *B. dorsalis* are important throughout South-East Asia and the western Pacific – Philippines, Indonesia, Pakistan (Allwood and Drew, 1996).

LIFE CYCLE AND DAMAGE *Bactrocera* spp. female fruit flies insert their eggs beneath the skin of citrus, especially in ripening fruit. White banana-shaped eggs are deposited in clusters of about a dozen eggs. A total of 200–400 eggs are laid by *Anastrepha fraterculus*, 1200–1500 eggs by *B. dorsalis* and 50 eggs by *B. minax*. The larvae tunnel into the fruit, contaminating the fruit with frass and providing entry for fungi and bacteria. Fully grown larvae measuring *ca.* 7 mm drop to the ground and enter the soil where they pupate. The egg period lasts from 2 to 20 days. There are three larval instars. Larval and pupal periods are each 2–4 weeks. Pupal stage of *B. minax* in China is 180 days (Huasong *et al.*, 1998). Studies conducted by Singh (1991) with *B. dorsalis* in India indicated that pupal period was longest (18 days) at 15°C and shortest (6 days) at 35°C. After emergence, the females require a protein source for egg maturation. Warm, humid weather is considered to be favourable for *Bactrocera* fruit flies and pest populations build up as citrus ripening occurs. *Bactrocera* populations decrease during dry periods.

MONITORING Fruit fly activity has been monitored in Australia using Dakpot® fruit fly traps hung beneath the tree canopy. Methyl eugenol is considered the most powerful male lure for oriental fruit flies. Methyl eugenol was used successfully for control and eradication of *B. dorsalis* in Oahu (Steiner and Lee 1955), Rota Island (Steiner *et al.*, 1965), Okinawa, Kume, Miyako and Uaekama Islands (Iwahashi, 1984) and has also been used for monitoring *B. umbrosus* in the Philippines (Umeya and Hirao, 1975). Fruit flies of both species were controlled by mass trapping of males with methyl eugenol and infestations were brought to sub-economic levels in Pakistan (Mohyuddin and Mahmood, 1993).

Protein or yeast baits containing a toxicant such as malathion are commonly used to control *B. tryoni* in Australia (Smith *et al.*, 1997). The bait mixture is applied as a course spray on the tree skirt in quantities of 15 l ha^{-1} at 7–14 day intervals while fruit are susceptible and flies active. The baits are very effective and relatively non-disruptive to natural enemies.

BIOLOGICAL CONTROL A good number of braconid parasitoids are recorded from *Bactrocera* spp. (Waterhouse, 1993); however, they give a low level of control. The parasitoids of *B. zonata* that have been found in Pakistan include *Opinus longicaudatus* (Ashm.), *Dirhinus giffardii* Silv., and *Bracon* sp. The parasitoids, *O. longicaudatus*, *D. giffardii* and *Spalangia grotiusi* Girault are commonly reported from *B. dorsalis* (Syed *et al.*, 1970); however, their incidence is extremely low. *Opius* spp. introduced from Malaysia into Hawaii became established against *B. dorsalis* (Clancy *et al.*, 1952); however, fruit flies directly damage produce that is to be marketed. As a result, a small fruit fly population can cause economic damage, reducing the success of classical biological control programmes. *Biosteres arisanus* (Sonan) has recently been considered to have some promise for control of *B. dorsalis* in Hawaii (Harris *et al.*, 1998).

Ceratitis *spp.*

OCCURRENCE AND DAMAGE Seven species of *Ceratitis*, *C. capitata* (Wiedemann), *C. catoirii* Guerin-Méneville, *C. cosyra* (Walker) *C. discussa* (Munro), *C. malgassa* Munro, *C. quinaria* (Bezzi) and *C. rosa* Karsch attack citrus species (White and Elson-Harris, 1992). The Mediterranean fruit fly *C. capitata* is the most important and a common polyphagous pest in citrus growing areas of Hawaii, western Australia, Mexico, Réunion and South America (Etienne, 1966; Morin 1967). *Ceratitis cosyra* occurs in Africa whereas *C. catoirii* Guer occurs in Réunion (Etienne, 1968). The female can oviposit all over the fruit, with no preference for any part. Later, when fruit becomes suitable for maggot development, the oviposition sites become light in colour and the tissue softens. The fully grown maggots leave the fruit and pupate in the soil. The developmental period is approximately 3–4 weeks and 8–10 generations $year^{-1}$ can occur depending on temperature and other factors intrinsic to the fly population (Hill, 1975).

BIOLOGICAL CONTROL Several parasitoids, e.g. *Opius fullawayi* (Silvestri), *O. humilis* Silvestri, *O. incisi* Silvestri, *O. kraussi* Fullaway, *O. tryoni* Cameron, *O. bellus* Gahan, *Biosteres longicaudatus* Ashmead, *B. tryoni* (Couron) and *B. oophilus* (Fullaway) have been reported parasitizing *C. capitata* (Beardsley, 1961; Wharton and Marsh, 1978). Bess *et al.* (1961) reported that the most important parasitoids collected from *C. capitata* in Hawaii were *O. vandenboschi*, *O. oophilus* and *B. longicaudatus*. The fungus *Beauveria tenella* is reported from pupae of *B. manix* in China (Huasong *et al.*, 1998). Again control is usually inadequate.

Anastrepha *spp.*

LIFE CYCLE AND OCCURRENCE *Anastrepha* spp. are endemic to the western hemisphere and their range extends from the southern United States to northern Argentina and includes the Caribbean islands (Aluja, 1994). Six *Anastrepha* species (*A. fraterculus* (Wiedemann), *A. ludens* (Loew), *A. ocresia* (Walker), *A. serpentina* (Wiedemann), *A. suspensa* (Loew), and *A. striata* (Schiner) are associated with citrus (White and Elson-Harris, 1992). Mexican fruit fly, *Anastrepha ludens*, is reported to be the most common fruit fly pest attacking citrus in the Americas. Abundance of *Anastrepha* populations has been positively correlated with temperature and negatively correlated with relative humidity. However, a study by Aluja *et al.* (1989) demonstrated the lack of a clear relationship between rainfall and *Anastrepha* fly captures in mango orchards in Mexico. The same authors caution that differences in population fluctuations can occur within the same orchard. Most of what is known today is based on basic biology studies carried out between 1900 and 1944 (Aluja, 1994). The basic life cycle is very similar among all *Anastrepha* spp. for which the biology is known. Caribbean fruit fly eggs are laid in small groups just below the rind of citrus fruit (Browning *et al.*, 1995). Egg incubation of Mexican fruit fly (*A. ludens*), requires 3.8 days; larval development 14.2 days and pupal development 14.2 days at 27 ± 2°C (Leyva *et al.*, 1991). In the majority of *Anastrepha* species, the females deposit their eggs, e.g. *c.* 15–19 eggs per *A. ludens* female, in either the epicarp or mesocarp of ripening fruit. Depending on the species, eggs are laid either singly or in clusters. Larvae pass through three instars before emerging from the fruit and burrowing into the ground to pupate.

Damage to grapefruit in Honduras can reach 87%, and each fruit could yield 5.7 to 11 larvae (Sponagel *et al.*, 1996).

TRAPPING McPhail traps has been a standard procedure for controlling *Anastrepha* spp. for 35 years even though such methods have been ineffective. The most widely used traps during the last 35 years to monitor and in some cases control *Anastrepha* populations are glass and plastic versions of the McPhail trap, which is baited with a mixture of protein (occasionally hydrolysed cottonseed together with borax, molasses or fermented juices) and water (Balock and Lopez 1969). The McPhail trap, however, has several drawbacks. It is expensive, breaks easily and is cumbersome to service. Colours such as yellow and orange, reflecting maximally within a narrow spectral region, i.e. 500–590 nm, have proven to be effective for capturing several *Anastrepha* species, e.g. *A. fraterculus*, *A. ludens* and *A. suspensa*, when used in spherical, rectangular or cylindrical traps. In commercial grapefruit orchards in Honduras, monitoring is practised in combination with chemical and cultural control practices (Sponagel *et al.*, 1996).

Use of broad-spectrum pesticides with hydrolysate protein baits has been the rule for control of *Anastrepha* species (Hentze *et al.*, 1993). Tests using cyromazine, a highly specific growth regulator against Diptera, against various *Anastrepha* species resulted in reduction of infestation of fruits (Diaz *et al.*, 1996).

BIOLOGICAL CONTROL Both classical biological control and repeated augmentative releases of mass-reared parasitoids have been used to suppress *Anastrepha* populations. Parasitoid species such as *Diachasmimorpha longicaudata*, *Doryctobracon crawfordi*, *Ganapis pelleranoi*, *Biosteres vandenboschi* and *Aceratoneuromya indica* have been imported and released in the USA, Mexico, Costa Rica, Brazil and Peru for the control of *A. suspensa*, *A. ludens* and *A. fraterculus*.

CULTURAL CONTROL Orchard sanitation, i.e. collection and destruction of all unwanted fruit on the trees and on the ground, contributes significantly towards reducing damaging fruit fly populations (Vijaysegaran, 1993). However, orchard sanitation is difficult to implement and enforce (Vijaysegaran, 1993). Early harvesting is a technique that is successfully practised on other tropical fruits. Uncontrolled breeding of fruit flies in poorly managed or abandoned orchards should be avoided.

PHYSICAL CONTROL Wrapping or bagging individual fruits is a method not recommended for citrus (Vijaysegaran, 1993). Moreover the shortage of labour appears to be the major constraint to the widespread use of bagging for production of selected fruits. At maturity, oranges and grapefruits are susceptible to *A. suspensa*, whereas immature fruit are not susceptible to attack (Browning *et al.*, 1995). The plant growth regulator gibberellic acid (GA_3) reduces attack of *A. suspensa* on grapefruits (Greany *et al.*, 1987). The success of this strategy has prompted the application of GA_3 into a broadly based pest-free zone in Florida.

Citrus midges

The citrus midge, *Prodiplosis longifila* Gagne, a neotropical species, is present throughout most of the lime production areas of Florida and has gained pest status since 1984 (Peña and Mead, 1988; Peña *et al.*, 1989b). Damage by *P. longifila* normally affects the flower parts, including the ovary, stamens and petals. After larval feeding, fungal infection by *Colletotrichum gloeosporioides* Penz, *Cladosporium herbarum* LK ex Fr. var. *citricola* and *Penicillum* sp., follows, causing the death of flowers. *P. longifila* has been reported feeding on wild cotton, lucerne, beans, tomatoes and potatoes.

Eggs of this species are deposited individually or in groups of 12–59 on stamens or on the ovary. Lime flowers are susceptible to egg laying from the time the flower is approximately 4 mm in diameter until the flower is near petal break. The eggs hatch in *c.* 1–2 days depending on the prevailing temperature conditions. Upon hatching, the larvae find their way to the stamens and to the lower portion of the ovary on which they feed. Larvae mature in about 10 days and drop from the flower and burrow into the soil to pupate (Peña *et al.*, 1989b). If undisturbed by insecticide applications, the parasitoid *Synopeas* spp. can maintain the population density at non-economic levels (Peña *et al.*, 1990). The citrus blossom midge *Cecidomyia* sp. is a minor pest in eastern Australia (Smith *et al.*, 1997).

Thrips

There are about six important thrips pests of citrus worldwide. *Scirtothrips* is the most damaging genus. Citrus thrips *Scirtothrips citri* Moulton is a key pest of areas like California, *Scirtothrips aurantii* Favre is a key pest in South Africa and chilli thrips *Scirtothrips dorsalis* is a key pest of South-East Asia. Other significant species are citrus rust (or orchid) thrips *Chaetanaphothrips orchidii* Moulton and greenhouse thrips *Heliothrips haemorrhoidalis* (Bouchel) (worldwide) *Taenothrips* sp. (South Africa) *Scirtothrips albomaculatus* Bianchi (eastern Australia) and *Megaleurothrips kellyanus* (Bagnall) (Australia).

Scirtothrips damage which occurs mostly at and shortly after fruit set, is characterized by grey scar tissue around the stem and/or between touching fruit. One of the major causes of fruit blemish and downgrading throughout the world is wind rub (40%) and it is easy to confuse the two (Bedford, 1943). Bedford *et al.* (1998) describe the life and seasonal history of *S. aurantii* in South Africa.

LIFE HISTORY The biology and habits of *S. aurantii* in Zimbabwe were described in detail by Hall (1930). Bedford (1943) described the different stages at Rustenburg, South Africa as follows:

- *Egg*. The egg is bean-shaped when first laid and swells with the development of the embryo. It is translucent with a smooth thin shell. The eggs are laid separately within the soft tissues of small green fruit and the tender leaves and shoots of new growth, and are invisible to the naked eye. The egg measures 0.22 mm by 0.11 mm. Most eggs are found just beneath the upper surface of the leaves.
- *First instar larva*. When newly hatched the larva is colourless and translucent but soon becomes translucent white. As it matures and before moulting the larva becomes yellowish. Wing pads are not present. The mean larval length is 0.32 mm.
- *Second instar larva*. The body colour of the second instar larva progresses from

white when newly moulted to a deep orange-yellow. The mean length before pupation is 0.75 mm.

At maturity the larvae drop to the ground after sunset and pupate beneath the citrus trees. In this way lethal soil temperatures are avoided. Most larvae drop close to the tree trunk. Sometimes a larva will pupate under the calyx of a citrus fruit but this appears to occur only very rarely. Like the first instar larva, wing pads are not present.

- *Prepupa*. The pale yellow prepupa possesses short wing pads and can walk about if disturbed. It is a non-feeding stage. The eyes are reddish while the antennae and legs are translucent. Mean length is 0.57 mm. Moulting to the pupa takes place within the soil or leaf litter.
- *Pupa*. In this stage the translucent wing pads are longer and extend back to approximately segment VIII of the abdomen. The antennae are folded back over the head. The colour is similar to that of the prepupa. The overall body length is 0.56 mm. Mortality during pupation is estimated at 56.3% (Gilbert, 1992).
- *Adult female*. The female is pale, orange-yellow, and like all thrips the wings are folded longitudinally over the body when at rest.
- *Adult male*. The male is smaller than the female at 0.6–0.7 mm in length and the abdominal markings are sometimes inconspicuous.

LIFE CYCLE The egg stage takes from 6 to 24 days, depending on the season. The combined duration of the two larval stages is an average of 7.6 days in midsummer (December to February) and 13.5 days in August. Hall (1930) gives the larval stage as 5 days in Zimbabwe.

The duration of the prepupal stage from October to December is 1 day and that of the pupal stage 3 days, making a total of 4 days. The total duration of the life cycle is 18.4 days during summer and 44 days during winter. Hibernation does not take place in any stage. The mean number of eggs per female per day is 1.2 during the summer and 0.43 during July. Hall (1930) estimates the pre-ovipositional period at about 2.5 days.

SEASONAL HISTORY The number of generations per year is estimated to be 9.4 at Rustenburg (Bedford, 1943). *Scirtothrips aurantii* numbers are generally highest during the 6 months from September to February. Diapause does not occur, presumably due to the mild climatic conditions which prevail in the citrus-producing areas of southern Africa. Larvae and adults may therefore be found at any time of the year if suitable feeding areas are available.

The adults lay eggs on the first growth flush of the new season. First generation larvae develop to maturity on this flush and then pupate in the soil beneath the tree. The subsequent emergence of adults tends to coincide with the hardening-off of this flush and the end of blossoming. The adults then begin to infest the small fruitlets as they begin to swell and become dark green, especially when they grow up against the protecting sepals.

NATURAL ENEMIES The two most common predators of the citrus thrips at Rustenburg (Bedford, 1943) are *Haplothrips bedfordi* Jacot-Guillarmod, which is only common from January to April, and an anthocorid bug, *Orius* (= *Thriphleps*) *thripoborus* (Hesse), which is only common on citrus trees from April to July. These and other predators had no effect in controlling thrips populations which cause severe damage (up to 75% culls or more) in orchards left unsprayed for several years.

Grout and Richards (1992) and Grout and Stephen (1993), investigated the value of indigenous phytoseiid mites as predators of *S. aurantii*. Two important species are *Euseius addoensis* (Van der Merwe and Ryke) and *E. citri* (Van der Merwe and Ryke). The former is especially important as an effective predator in the Eastern Cape where it is very common. Table 3.9 lists the important parasitoids of citrus thrips.

True bugs

There are at least six significant bugs on citrus: coreid bugs like the leaf-footed bug *Leptoglossus phyllopus* (L.), shield bugs like green vegetable bug *Nezara viridula* (L.),

Table 3.9. Common wasp parasitoids of bugs and thrips on citrus (Smith *et al.*, 1997).

Parasitoids	Main hosts on citrus	Origin of parasitoid
Aphelinidae		
Centrodora darwini	*Biprorulus bibax*	Australia
Scelionidae		
Trissolcus basalis	*Nezara viridula*	North Africa, USA
T. oenone	*B. bibax*	Australia
T. ogyges	*B. bibax*	Australia
T. mitsukurii	*N. viridula*	Japan
T. nakagawai	*N. viridula*	Japan
Gryon spp.	*Amblypelta nitida* and *A. lutescens*	Australia
	Anoplocnemis curvipes	South Africa
Pteromalidae		
Acroclisoides tectacorisi	*B. bibax*	Australia
Eupelmidae		
Anastatus biproruli	*B. bibax*	Australia
Anastatus spp.	*Musgraveia sulciventris*	Australia
Thripobius semiluteus	*Heliothrips haemorrhoidalis*	Australia
Goethana shakespeari	*H. haemorrhoidalis*	Australia
Trichogrammatidae		
Megaphragma mymaripenne	*H. haemorrhoidalis*	Australia

antestia bug *Antestiopsis* spp. and spined citrus bug *Biprorulus bibax* Breddin (Browning *et al.*, 1995; Smith *et al.*, 1997; Bedford *et al.*, 1998). The fruit sucking bug *Rhynchocoris humeralis* Thunber occurs commonly throughout South-East Asia.

Egg parasitoids

The scelionid genus *Trissolcus* and the eupelmid genus, *Anastatus* are important (Table 3.9). Assassin bugs like *Pristhesancus plagipennis* are significant predators.

Other pests

Cicadas

A few cicadas are recorded as minor pests, e.g. bladder cicadas *Cytosoma schmeltzi* Distant in Australia and *Cryptotympana atrata* Fabricius in China.

Orthopterans

Orthopteran pests include katydids, crickets and grasshoppers – all minor pests (Browning *et al.*, 1995; Smith *et al.*, 1997; Bedford *et al.*, 1998).

Termites

The giant northern termite *Mastotermes darwinensis* Froggatt is a formidable pest of citrus in the northern tropics of Australia (Smith *et al.*, 1999) attacking even living tissue. Sub-Saharan Africa has several genera attacking citrus but mostly involving dead branches (Bedford *et al.*, 1998). Subterranean termite *Reticulitermes flavipes* (Koler) occurs in Florida (Browning *et al.*, 1995).

Ants and wasps

Paper wasps *Polistes* spp. occasionally pose a problem to pickers and pruners. The citrus gall wasp *Bruchophagus fellis* (Girault) of eastern Australia is an unusual pest, attacking the young twigs and causing severe galling. It is heavily parasitized by *Megastigmus* spp. The main threat from Hymenoptera comes from ants and their attraction to honeydew producing pests – soft scales, mealybugs, planthoppers, whiteflies and aphids. Serious pests include Argentine ant (Ripa and Rodriguez, 1999), *Linepithema humilis* (Mayer), and other *Iridomyrmex* spp., leafcutting ants *Atta* spp. and fire ants *Solenopsis* spp.

Other pests

These include spiders, snails, slugs and nematodes. Spiders are more a problem for pruners and pickers. Several snail species are important, e.g. the common brown garden snail *Helix aspersa* (Miller) (Smith *et al.*, 1997), the dune snail *Theba pisana* (Muller) (South Africa) and the slugs *Urocyclis* spp. (Bedford *et al.*, 1998) and *Deroceras* spp. (Chili) (Ripa and Rodriguez, 1999).

The most important and widespread nematode which parasitizes citrus is the citrus nematode (*Tylenchulus semipenetrans*). Others which sometimes cause problems are the root lesion nematode (*Pratylenchus* spp.) and the stubby root nematode (*Paratrichodorus* sp.). Several other types of plant parasitic nematodes have been found in association with citrus trees, but they have not been found to be economic pests. These include *Paratylenchulus* sp., *Xiphinema* sp. and *Hemicriconemoides* sp. (Smith *et al.*, 1997).

Pest Management Practice

Because of the greater number of pest and disease problems in the tropics and the greater number of small farms with less emphasis on monitoring, pesticide use can be heavy and up to 24 or more sprays are applied per season. Application, particularly for sedentary scales and mealybugs and also mites, can be inefficient with poor coverage. Pesticides tend to be broad spectrum and very disruptive to natural enemies. Citrus is a long-term crop offering continuity of overwintering sites and habitats for both pests and natural enemies, and some degree of stability. However, extensive use of disruptive pesticides negates these factors. Hand-operated spraying machinery common on small farms also gives little protection to the user.

Integrated pest management relies heavily on the biological control of pests by their natural enemies (parasitoids, predators or pathogens) and integrates into the system any useful cultural techniques (to reduce pests and their damage or improve the environment for natural enemies) and selective use of pesticides. IPM needs as its basis a sound knowledge of the pests and their natural enemies, action levels and management strategies. Systematic monitoring of both the pests and the levels of their natural enemies is vital (Smith and Papacek, 1993; Papacek, 1997).

The record of IPM implementation worldwide is currently more impressive on paper than in practice. However, fruit tree pest control systems based purely on pesticides are faced with major problems of resistance development, development of secondary pests, rising costs and threats to environmental and human health. Properly managed IPM systems in citrus can cost a fraction of full chemical programmes and give better quality and production of fruit (Hardman *et al.*, 1993).

In developing a successful IPM programme for citrus there are a number of important components:

1. *Sound knowledge of the pests* – their habits, the importance of the damage caused and action levels. An action level is the point at which action should be taken to avoid economic loss. It is determined by research (looking at pest densities, the impact of natural enemies, damage levels, economics of production, costs of spraying and expected market returns) and by practical experience. Often the best means of determining an action level is to make a 'good estimate' based on experience and then to test the estimate in commercial practice. It is very important to determine the real economic impact of a pest. Citrus leafminer, for example, is extensively sprayed in the tropics and subtropics but there is increasing evidence of a low economic impact except on young trees (Schaffer *et al.*, 1996). There is a tendency for growers to apply repetitive sprays (up to two dozen a year) in South-East Asia because the damage to the leaves is readily seen. Unless workable action levels for pests in a complex are proposed and adhered to, 'management' will soon degenerate into 'calendar spraying'.

2. *Sound knowledge of biological control.* The natural enemies of pests, with some measure of their significance, must be known. Natural enemies may include native or introduced species with both often having a vital role. Where exotic pest species have been introduced, researchers should if necessary have access to 'classical biological control' allowing the introduction of natural enemies from the pest's centre of origin to redress the imbalance. Excessive bureaucratic impediments to the introduction of natural enemies is counter-productive to good biological control.

Mass rearing and augmentative or inundative release of natural enemies is a proven useful technique in many citrus areas. It is of most use where a natural enemy is checked by adverse conditions e.g. cold winters, very hot dry periods in summer, discontinuity between host generations or disruption from pesticide application. Mass rearing and augmentation can be costly, however, and should not be used unless clearly necessary and effective. *Aphytis* spp. for red scale, are some of the most commonly used mass reared beneficials in citrus (Papacek and Smith, 1985).

Attention also must be paid to cultural practices that can be contrary to biological control, e.g. excessive dust from roads, bare soil inter-rows. Phytoseiid mite populations in Queensland orchards are increased sixfold by allowing some inter-row growth and flowering of the grass *Chloris gayana kunth* (Smith and Papacek, 1991).

3. *Simple and effective monitoring procedures* – both for pests and their natural enemies. Monitoring is a vital component of IPM. It is best done by a trained scout who is dedicated to the task. Orchardists can perform it themselves but are often too busy and too involved with the crop to make impartial decisions on pest management. The large number of small orchardists in the tropics, however, practically demands that individual growers learn to regularly count their pests and natural enemies. This means that IPM and monitoring must be extended and demonstrated to such growers and backed up on a long-term basis with help on quality control and research needs.

Monitoring is not a research tool but must be quick and practical for the scout or grower. An average block of trees (1 ha) should be assessed in no more than 30 min and sampling repeated at regular intervals (7 to 28 days) throughout the season. The data collected need to be accurate and repeatable. A random number of trees e.g. ten per hectare, are assessed per block and fruit, leaves and/or twigs assessed (Smith *et al.*, 1997). It is important to record the data, initially on a small orchard sample card and then, if possible, on a computer spreadsheet. The data then recorded allow a decision to be made on any actions to take.

4. *Development of complete IPM packages for a whole pest complex*. Some key pests may have poor biological control and must be treated with pesticides. Biocontrol studies have been made on *Scirtothrips* spp. by Grout and Richards, 1992 and Grout and Stephen, 1993. Serious transmitters of disease are also a major problem in the tropics, the most notable being *D. citri* (transmitter of Asian greening throughout South-East Asia). Control of such a pest can dominate the whole programme. Biocontrol studies have intensified with *D. citri* in recent years (Mercado *et al.*, 1991; Osman and Quilici, 1991; Tang and Huang, 1991; Tang and Wu, 1991) and the use of selective chemicals particularly petroleum spray oils is also an option (Rae *et al.*, 1997). Other pests are very important because of quarantine status, e.g. fruit flies. Fortunately, good selective baiting techniques exist for fruit flies and also postharvest treatments. It is important for researchers to develop IPM packages that are complete and to demonstrate them in display blocks or in key orchards within a region. In spite of the seriousness of some key pests and the variety of pest species occuring in the tropics, successful IPM has been implemented in many areas worldwide.

5. *Careful use of pesticides*. Selective pesticides can be a powerful tool in IPM programmes. The main problems in pesticide use are ignorance of the effect of the pesticides on natural enemies, inefficient spraying machinery and practices, rapid development of resistance and off target spray drift, which can disrupt natural enemies.

There is considerable information on the toxicity of pesticides to natural enemies (Broadley and Thomas, 1995; Hattingh and Tate, 1995; Smith *et al.*, 1997). Synthetic pyrethroids such as deltamethrin are persistent and very disruptive when used on citrus, and most organophosphates and carbamate insecticides are also disruptive. The last decade has seen the development of a range of new pesticide groups with some characteristics that showed promise for selectivity (Sparkes, 1998). These include systemic neo-nicotinoids, insect growth regulators, phenylpyrazols, avermectins and spinosyns. There are still unfortunately some problems with these materials, with the systemics causing mite outbreaks (G.A.C. Beattie, personal communication, 1999) and some of the IGRs showing persistant toxicity to coccinellids (Hattingh and Tate, 1995; Smith *et al.*, 1999).

Inefficient spraying is a chronic problem in citrus worldwide. Scale, mealybug and most mite infestations require spraying to point of runoff with a high percentage of coverage to gain control. Carmen (1975, 1977) details operations of the oscillating boom, the preferred sprayer for large citrus trees. Air blast sprayers with and without towers are probably the most used sprayers in higher latitude areas. Rotary atomizers and high velocity sprayers with air shear nozzles have also been tried (Smith *et al.*, 1997). Air blast sprayers decrease rapidly in efficiency when tree height increases above 4 m and when spray volumes drop below 5000 l ha^{-1}. Mature trees 5 m high normally require about 10,000 l ha^{-1} for thorough wetting for sedentary pests. The top centre of the tree is the most difficult to reach with spray. Air blasters with towers can cover trees this high but lower spray volumes will reduce efficiency. In the many small orchards in the tropics, spraying is done more by hand-held wands driven by small pumps on a spray cart, or where there are canals, in a boat in the canal. There tends to be considerable exposure of operators to spray drift. Resistance development to pesticides is a continual problem, notably with pests like red scale, citrus rust mite, tetranychid mites, psyllids and citrus thrips (Bedford *et al.*, 1998). Spray drift can easily occur between adjacent blocks of trees or between neighbouring orchards or (in some circumstances) between orchards or farms several kilometres apart. Levels of drift may be extremely low but if the pesticide is persistent and highly toxic to key parasitoids, e.g. *Aphytis* spp., serious disruptions will occur. Careful use of pesticides is not only important on individual farms but throughout continuous farming areas.

References

Adair, R.C. (1994) A four-year field trial of entomopathogenic nematodes for control of *Diaprepes abbreviatus* in a flatwoods citrus grove. *Proceedings of the Florida State Horticultural Society* 107, 63–68.

Adair, R.C., Nigg, H., Simpson, S. and Lefebre, L. (1998) Ovipositional preferences of *Diaprepes abbreviatus* (Coleoptera: Curculionidae). *Florida Entomologist* 81, 225–234.

Allen, J.C. (1979) The effect of citrus rust mite damage on citrus tree growth. *Journal of Economic Entomology* 72, 327–330.

Allen, J.C. and Stamper, J.H. (1979) Frequency distribution of citrus rust mite damage on citrus fruit. *Journal of Economic Entomology* 72, 746–750.

Allwood, A.J. and Drew, R.A.I. (1996) Management of fruit flies in the Pacific. A regional symposium, Nadi, Fiji 28–31 October 1996. *ACIAR Proceedings No. 76*, 267 pp.

Aluja, M. (1994) Bionomics and management of *Anastrepha*. *Annual Review of Entomology* 39, 155–178.

Aluja, M., Cabrera, M., Guillen, J., Celedonio, H. and Azora, I. (1989) Behavior of *Anastrepha ludens* and *A. serpentina* (Diptera: Tephritidae) on wild mango tree (*Mangifera indica*) harbouring three McPhail traps. *Insect Science and its Application* 10, 309–318.

Amalin, D.M., Peña, J.E. and McSorley, R. (1996) Abundance of spiders in lime groves and their potential role in suppressing the citrus leafminer population. In: Hoy, M. (ed.) *Proceedings of International Conference, Managing Citrus Leafminer*. University of Florida, p. 72.

Amalin, D.M., Reiskind, J., Peña, J.E. and McSorley, R. (2001) Predatory behavior of three species of sac spiders attacking citrus leafminer. *Journal of Arachnology* 29, 72–81.

Amalin, D.M., Peña, J.E., Duncan, R.E., Browning, H.W. and McSorley, R. (2002) Natural mortality factors acting on citrus leafminer, *Phyllocnistis citrella*, in lime orchards in South Florida *BioControl* 47, 327–347.

Ando, T., Taguchi, K.Y., Uchiyama, M., Ujiye, T. and Kuroko, H. (1985) (7Z-11Z)-7,11-hexadecadienal: sex attractant of the citrus leafminer moth, *Phyllocnistis citrella* Stainton (Lepidoptera: Phyllocnistidae). *Agricultural Biological Chemistry, Tokyo* 49, 3633–3653.

Anonymous (1975) *Illustration of Citrus Pests and Diseases*. Chinese Academy of Agricultural Sciences, Sichuan People's Press, 185 pp.

Anonymous (1981) *Major Citrus Insect Pests in Japan*. Japan Plant Protection Association, Tokyo, 16 pp.

Anonymous (1984) *IPM for Citrus*. University of California Publication, 3303.

Anonymous (1995) Nematodes play increased role in controlling weevils. *Florida Grower and Rancher* (September) 88, 6–7.

Anonymous (1996) *Detections of Diaprepes abbreviatus in Florida*. Florida Department of Agriculture and Consumer Services Division Plant Industry, Bureau of Eradication and Control, 2 pp.

Argov, Y. and Rossler, Y. (1996) Introduction, release and recovery of several exotic natural enemies for biological control of the citrus leafminer, *Phyllocnistis citrella* in Israel. *Phytoparasiticae* 24, 33–38.

Armstrong, A. (1987) Parasitism of *Tetrastichus haitiensis* Gahan on egg masses of *Diaprepes abbreviatus* in Puerto Rico. *Journal of the Agricultural University Puerto Rico.* 71, 407–409.

Atachi, P. (1991) Récents dégats d'acariens des Citrus au Benin. *Fruits* 46, 577–579.

Atwal, A.S., Chaudhary, J.P. and Ramzan, M. (1970) Studies on the development and field population of citrus psylla, *Diaphorina citri* Kuwayama (Psyllidae: Homoptera). *Journal of Research of Punjab Agricultural University* 7, 333–338.

Aubert, B. and Quilici, S. (1983) Nouvel équilibre biologique observé á la Réunion sur les populations de psyllidés aprés l'introduction et l'etablissement d'hymenoptéres chalcidiens. *Fruits* 38, 771–779.

Aubert, B., Bové, J.M. and Etienne, J. (1980) La lutte contre la maladie du greening des agrumnes á l'ile de la Réunion. Resultats and perspectives. *Fruits* 35, 605–624.

Badawy, A. (1969) The morphology and biology of *Phyllocnistis citrella* Staint., a citrus leafminer in the Sudan. *Bulletin Société Entomologique Egypte* 51, 95–103.

Badii, M.H. and McMurtry, J.A. (1984) Feeding behavior of some phytoseiid predators on the broad mite, *Polyphagotarsonemus latus* (Acari: Phytoseiidae, Tarsonemidae). *Entomophaga* 29, 49–53.

Balock, J. and Lopez, D.F. (1969) Trapping for control of the Mexican fruit fly in mango and citrus groves. *Journal of Economic Entomology* 62, 53–56.

Bateman, M.A. (1972) The ecology of fruit flies. *Annual Review of Entomology* 17, 493–518.

Batra, R.C., Sandhu, G.S., Sharma, S.C. and Singh, R. (1988). Biology of the citrus leafminer on some rootstocks and its relationship with abiotic factors. *Punjab Horticultural Journal* 28, 30–35.

Beardsley, J.W. (1961) A new review of Hawaiian Braconidae (Hymenoptera). *Proceedings of the Hawaiian Entomological Society* 27, 333–366.

Beattie, G.A.C. (1997) Cheju Island, South Korea 13–18 Oct. 1997. Overseas trip report.

Beattie, G.A.C. and Watson, D.M. (1997) Report on overseas travel to China and Southeast Asia 28 March to 24 April 1997, ACIAR China and Southeast Asia trip report.

Beavers, J.B. and Schroeder, W. (1980) Sugarcane rootstalk borer weevil, *Diaprepes abbreviatus*: insect growth regulators. *Insecticide and Acaricide Tests* 5, 234.

Beavers, J.B. and Selhime, A.G. (1975) Futher attempts to establish the weevil egg parasite, *Tetrastichus haitiensis* in Florida. *Florida Entomologist* 58, 29–31.

Beavers, J.B., Lovestrand, S.A. and Selhime, A.G. (1980) Establishment of the exotic parasite *Tetrastichus haitiensis* (Hym: Eulophidae) and recovery of a new *Trichogramma* (Hym: Trichogrammatidae) from root weevil egg masses in Florida. *Entomophaga* 25, 91–94.

Bedford, E.C.G. (1943) The biology and economic importance of the South African citrus thrips, *Scirtothrips aurantii* Faure. *Publications of the University of Pretoria, South Africa, Natural Science* 7, 1–68.

Bedford, E.C.G., Van den Berg, M.A. and De Villiers, E.A. (1998). *Citrus Pests in the Republic of South Africa*. Institute for Tropical and Subtropical Crops and Outspan, 288 pp.

Beingolea, G. (1988) *Integrated Pest Control (IPC) in Citrus*. (Indonesia) Research and Development for Citrus Rehabilitation, FAO, 46 pp.

Beingolea, O.G. (1994) Role of *Aphytis roseni* DeBach and Gordh in the biological control of the rufous scale, *Selanaspidis articulatus* (Morgan). In: Rosen, D. (ed.) *Advances in the Study of Aphytis (Hymenoptera: Aphelinidae)*. Intercept, Andover, pp. 105–118.

Ben-Dov, Y. (1993) *A Systematic Catalogue of the Soft Scale Insects of the World*. Flora and Fauna Handbook, No. 9. Sandhill Crane Press, 536 pp.

Bennett, F.D. and Alam, M.M. (1985) An annotated check list of the insects and allied terrestrial arthropods of Barbados. *Caribbean Agricultural Research and Development Institute*, 81 pp.

Bennett, F.D., Rosen, D., Cocheaureau, P. and Wood, B.J. (1976) Biological control of pests of tropical fruits and cuts. In: Huffaker, C.B. and Messenger, P.S. (eds) *Theory and Practice of*

Biological Control. Academic Press, New York, pp. 359–395.

Bess, H. A., van den Bosch, R. and Haramoto, F. H. (1961) Fruit fly parasites and their activity in Hawaii. *Proceedings of the Hawaiian Entomological Society* 27, 367–378.

Bierl, B.A., Beroza, M., Staten, R.T., Sonnet, P.E. and Adler, V. (1974) The pink bollworm sex attractant. *Journal of Economic Entomology* 67, 211–216.

Bindra, O.S., Sohi, B.S. and Batra, R.C. (1974) Note on the comparative efficacy of some contact and systemic insecticides for the control of citrus psylla in Punjab. *Indian Journal of Agricultural Sciences* 43, 1087–1088.

Binglin, T. and Mingdu, H. (1996) Managing the citrus leafminer in China. In: Hoy, M. (ed.) *Proceedings International Conference Managing Citrus Leafminer*. University of Florida, pp. 49–52.

Bird, J. and Maramorosch, K. (1978) Virus and virus diseases associated with whiteflies. *Virus Research* 22, 55–111.

Bishop, G.W., Kleinschmidt, G.D., Knutson, K.W., Mosley, A.R., Thornton, R.E. and Voss, R.E. (1992) *Integrated Pest Management for Potatoes in the Western United States*. University of California, Division of Agriculture and Natural Resources. Oakland, CA. Publication 3316, 146 pp.

Bove, J.M., Calavan, E.C., Capoor, S.P., Cortez, R.E. and Schwarz, R. (1974) Influence of temperature on symptoms of California stubborn, South African greening, Indian Citrus decline, and Philippines leaf mottling diseases. *Procedings 6th Conference, International Organization of Citrus Virologists*, pp. 12–15.

Breniére, J. and Dubois, J. (1965) Catalogue des insectes nuisibles aux cultures malgaches. *Review of Applied Entomology* A53, 640.

Broadbent, P., Beattie, G.A.C., Van Velsen, R.J. and Freeman, B. (1980) Citrus cultivation in China. *Agricultural Gazette of NSW* 91 (141), 32–35.

Broadley, R. and Thomas, M. (eds) (1995) *The Good Bug Book: Beneficial Insects and Mites Commercially Available in Australia for Biological Pest Control*. Australian Biological Control Inc., Queensland Department of Primary Industries and Rural Industries Research and Development Corporation, 53 pp.

Brooks, R.F. (1964) Control of citrus snow scale, *Unaspis citri* (Comst.) Florida. *Proceedings Florida State Horticultural Society* 77, 66–70.

Brown, R. and Jones, V.P. (1983) The broad mite on lemons in southern California. *California Agriculture* 37, 21–22.

Browning, H.W. (1996) Parasite communities associated with citrus leafminer, *Phyllocnistis citrella*. In: Hoy, M. (ed.) *Proceedings of International Conference on Managing Citrus Leafminer*. University of Florida, p. 74.

Browning, H., McGovern, R., Jackson, L., Calvert, D. and Wardoski, W. (1995) *Florida Citrus Diagnostic Guide*. Florida Science Source, 244 pp.

Browning, H., Peña, J.E. and Stansly, P. (1996) Evaluating impact of indigenous parasitoids on populations of citrus leafminer. In: Hoy, M. (ed.) *Proceedings of International Conference on Managing Citrus Leaf Miner*. University of Florida, pp. 14–15.

Bullock, R.C. (1978) Greenhouse evaluation of foliar miticides for control of broad mite on citrus. *Florida Entomologist* 61, 62.

Bullock, R.C. (1995) Suppression of *Pachnaeus litus* and *Diaprepes abbreviatus* (Coleoptera: Curculionidae) adult emergence with *Steinernenma carpocapsae* (Rhabditidae: Steinernematidae) soil drenches in field evaluations. *Proceedings of the Florida State Horticultural Society* 107, 90–92.

Bullock, R.C., Childers, C.C., Hall, D.G., Knapp, J., McCoy, C., Rogers, J. and Stansly, P.A. (1999) *Florida Citrus Pest Management Guide: Citrus Rust Mites. Fact Sheet ENY-603, 1999*. Florida Cooperative Extension Service, Institute of Food and Agricultural Sciences, University of Florida, 5 pp.

Caleca, V. and Lo Verde, G. (1996) Prove sull' attrattivit, di z7,z11-esadecadeinale su maschi di *Phyllocnistis citrella* Stainton (Lep: Gracillariidae) in Sicilia: dati preliminari. *Phytophaga (Italy)* 6, 183–191.

Cano, E., de la Llana, A., Hernandez, J., Ruiz, F., Peña, J.E. and Evans, G. (1996) Dynamics and biological control of the citrus leafminer in Nicaragua. In: Hoy, M. (ed.) *Proceedings of International Conference on Managing Citrus Leafminer*. University of Florida, p. 76.

Capoor, S.P., Rao, D.G. and Viswanath, S.M. (1974) Greening disease of citrus in the Deccan trap country and its relationship with the vector, *Diaphorina citri* Kuwayama In: Weathers, L.G. and Cohen, M. (eds) *Proceedings of the 6th Conference of the International Organization of Citrus Virologists*. University of California, Division of Agricultural Sciences, pp. 43–49.

Carmen, G.E. (1975) Spraying procedures for pest control in citrus. In: *Citrus. Ciba Geigy Agrochemicals Technical Monograph* no. 4, pp. 28–34.

Carmen, G.E. (1977) Evaluation of citrus sprayer units with air towers. *Citrograph* 62, 134–139.

Carver, M. (1978) The black citrus aphids, *Toxoptera citricidus* (Kilkardy) and *T. aurantii* (Boyer de Fonscolombe) (Homoptera: Aphididae).

Journal of the Australian Entomological Society 17, 263–270.

Cassin, P.J. (1984) Comportment des variétés d'agrumes dans les differentes regions de production. *Fruits* 39, 263–276.

Castaño, O., Garcia, F., Trochez, A., Rojas, L., Peña, J.E., and Evans, G. (1996) Biological control of the citrus leafminer, *Phyllocnistis citrella* in Colombia. In: Hoy, M. (ed.) *Proceedings of International Conference on Managing Citrus Leafminer*. University of Florida, p. 76.

Catling, H.D. (1970) Distribution of the psyllid vectors of citrus greening disease, with notes on the biology and bionomics of *Diaphorina citri*. *FAO Plant Protection Bulletin* 18, 8–15.

Cave, R. (1996) Biological control of the citrus leafminer in Honduras. In: Hoy, M. (ed.) *Proceedings of International Conference on Managing Citrus Leafminer*. University of Florida, p. 78.

Cendaña, S.M., Gabriel, B.P. and Magallona, E.D. (1984) *Insect Pests of Fruit Plants in the Philippines. Citrus*. Department of Entomology, University of the Philippines at los Baños, pp. 27–37.

Cermeli, M. (1983) Plagas y su control. *Citricas, Serie Petroleo y Agricultura, Fusagri, Venezuela* 1, 71–109.

Charanasri, V., Saringkaphaibul, C., Kongchuensin, M., Kulpiyawat, T. and Wongsiri, N. (1988) Taxonomic study on mites as pests of tangerine, *Citrus reticulata* Blanco, in Thailand. In: *Entomology and Zoology Division. Annual Report*. Department of Agriculture, Bangkok (in Thai), 133–177 pp.

Chien, C. and Chiu, S. (1986) Biology of the citrus spiny blackfly, *Aleurocanthus spiniferus*. *Journal of Agricultural Research China* 35, 222–229.

Christensen, L.D. and Foote, R.H. (1960) Biology of fruit flies. *Annual Review of Entomology* 5, 171–192.

CIE (1967) *Distribution Maps of Pests, Series A (Agricultural). Map no. 234*. Commonwealth Institute of Entomology, London.

Clancy, D., Marucci, P. and Dresner, E. (1952) Importation of natural enemies to control oriental fruit fly in Hawaii. *Journal of Economic Entomology* 45, 85–90.

Clausen, C.P. (ed.) (1978) *Introduced Parasites and Predators of Arthropod Pests and Weeds: a World Review*. CSDA Agriculture Handbook No. 480, 545 pp.

Coronado, J.M. and Ruiz, E. (1996) Escama nevada de los citricos *Unaspis citri* (Comstock) (Homoptera: Diaspididae). *Folleto Entomologico, UAM Agronomia y Ciencias, UAT, Mexico*, 2, 2 pp.

Costa, A.S. and Grant, T.J. (1951). Studies on transmission of the tristeza virus by the vector, *Aphis citricidus*. *Phytopathology* 41, 105–113.

Costilla, M.A. (1980) Bioecological aspects of the white citrus mite. *Revista Industrial, Agricola, Tucuman* 57, 15–21.

Coto, D., Saunders, J.L., Vargas, C. and King, A.C. (1995) *Plagas invertebradas de cultivos tropicales con emfasis en America Central*. Centro Agronomico Tropical de Investigacion y Ensenanza, Manual Tecnico 12, Costa Rica, 66 pp.

Dahms, E.C. and Smith, D. (1994) The *Aphytis* fauna of Australia. In: Rosen, D. (ed.) *Advances in the Study of* Aphytis *(Hymenoptera: Aphelinidae)*. Intercept, Andover, pp. 245–255.

DeBach, P. (1969) Biological control of diaspine scale insects on citrus in California. *Proceedings of an International Citrus Symposium, Riverside, California* 2, 801–815.

DeBach, P. (1971) Fortuitous biological control from ecesis of natural enemies. In: *Entomological Essays to Commemorate the Retirement of Profession*, pp 293–307.

DeBach, P.H. (1976) The *Aphytis* story: an illustration of principals and practice in biological control. *25th Annual Faculty Research Lecture*. University of California, Riverside.

Diaz, F., Toledo, J., Enkerlin, W. and Hernandez, J. (1996) Cyromazine: effects on three species of *Anastrepha*, In: McPheron, B.A. and Steck, G.J. (eds) *Fruit Fly Pests: a World Assessment of Their Biology and Management*. St Lucie Press, Delray Beach, Florida, USA, pp. 333–337.

Du, T.Y., Xiong, J.J. and Wang, Z. (1989) (7Z, 11Z)-7, 11-hexadecadienal: sex attractant of *Phyllocnistis citrella* Liu and Zeng. *Kungchong Zhishi* 26, 147–149.

Duncan, R.E. and Peña, J.E. (2000) Fecundity, host stage preferences and the effects of temperature on *Pnigalio minio* (Hymenoptera: Eulophidae), a parasitoid of *Phyllocnistis citrella* (Lepidoptera: Gracillariidae) *Proceedings of the Florida State Horticultural Society* 113, 20–24.

Ebeling, W. (1959) *Subtropical Fruit Pests*. University of California, Division of Agricultural Sciences, 436 pp.

Etienne, J. (1966) *Lutte Contre les Mouches des Fruits*. Institute Recherches Agronomiques Tropicales Réunion, pp. 112–113.

Etienne, J. (1968) *Lutte Contre les Mouches des Fruits*. Rapporte de Institute Recherches Agronomiques Tropicales Réunion, pp. 117–124.

Etienne, J. (1978) Controle biologique á la Réunion de *Trioza erytreae* (Hompt.: Psyllidae) au moyen de *Tetrastichus dryri* (Hymenoptera: Eulophidae) *Fruits* 33, 877–882.

Etienne, J., Mauleon, H. and Pintureau, B. (1990) Biologie et dynamique de *Ceratogramma etiennei* (Hymenoptera: Trichogrammatidae) parasite de *Diaprepes abbreviatus* (Coleoptera: Curculionidae) en Guadeloupe. In recontres Caraibes en lutte biologique, INRA, Paris, 1991. *Les colloques, no. 58*, pp. 458–468.

Evers, G. and Grisoni, M. (1991) Present situation of the citrus greening disease in Tanzania and proposal for control strategies. *Fruits* 46, 171–177.

FAO (2000) http://appsl.fao.org/serlet/xte.FAO statdatabase

Fennah, R.G. (1942) *The Citrus Pests Investigation in the Windward and Leeward Islands British West Indies 1937–1942.* Agricultural Advisory Department, Imperial College of Tropical Agriculture, Trinidad, BWI, 66 pp.

Flint, M.L. (1991) *Integrated Pest Management for Citrus,* 2nd edn. University of California, Statewide Integrated Pest Management Project, Division of Agricultural and Natural Resources, Publication 3303, 144 pp.

French, J.V. and Legaspi, J. (1996) Citrus leafminer in Texas: population dynamics, damage and control. In: Hoy, M. (ed.) *Proceedings of International Conference on Managing Citrus Leafminer.* University of Florida, p. 80.

Garcia Roa, F. (1995) Manejo de *Orthezia praelonga,* plaga de citricos. Instituto Colombiano Agropecuario, Programa de choque tecnologico, Palmira 1995, 9 pp.

Garrido, A. (1995) El minador de los citricos (*Phyllocnistis citrella* Stainton): morfologia, biologia, comportamiento, daños, interaccion con factores foraneos. *Phytoma* 72, 84–92.

Gerson, U. (1992) Biology and control of the broad mite, *Polyphagotarsonemus latus* (Banks) (Acari: Tarsonemidae). *Experimental and Applied Acarology* 13, 163–178.

Ghosh, S.P. (1990) Citrus. In: Bose, T.K. and Mitra, S.K. (eds) *Fruits: Tropical and Subtropical.* Naya Prokash, Calcutta, India, pp. 105–110.

Gilbert, M.J. (1992) The ecology of the South African citrus thrips *Scirtothrips aurantii* Faure, and its economic implications. PhD thesis, Rhodes University, Grahamstown, South Africa, 193 pp.

Gill, R.J. (1988) *The Scale Insects of California.* Parts 1 and 2. Californian Department of Agriculture.

Gravena, S. and Trevizoli, D. (1984) Injury levels for the citrus rust mite *Phyllocoptruta oleivora* (Ashm.). *Proceedings of the International Society Citriculture* 1, 457–460.

Gravena, S., Yamamoto, P. and Paiva, P.E. (1995) Integrated pest management of citrus insect/mite pests (Brazil): present status. In: Gravena, S. and Donadio, L.C. (eds) *Citrus Integrated Management of Insect and Mite Pests.* 3rd International Citrus seminar, 1994, São Paulo, Brazil, pp. 197–212.

Greany, P.D., McDonald, R.E., Shaw, P., Schroeder, W., Howard, D., Hatton, T., Davis, P. and Rasmussen, G. (1987) Use of gibberellic acid to reduce grapefruit susceptibility to attack by the Caribbean fruit fly *Anastrepha suspensa* (Diptera: Tephritidae). *Tropical Science* 27, 261–270.

Grout, T.G. and Richards, G.I. (1992) *Euseius addoensis addoensis,* an effective predator of citrus thrips, *Scirtothrips aurantii,* in the eastern Cape Province of South Africa. *Experimental and Applied Acarology* 15, 1–13.

Grout, T.G. and Stephen, P.R. (1993) Predation of alate citrus thrips, *Scirtothrips aurantii* Faure (Thysanoptera: Thripidae), by *Euseius citri* (Van der Merwe and Ryke) (Acari: Phytoseiidae) at low temperatures. *African Entomology* 1, 264–265.

Guerrero, O.B. (1969) Biological control of citrus pests in Peru. In: Chapman, H. (ed.) *Proceedings of the First International Citrus Symposium,* Vol. 2, University of California, Riverside, pp. 781–789.

Halbert, S.E. and Brown, L.G. (1996) *Toxoptera citricida* (Kirkaldy), brown citrus aphid identification, biology and management strategies. Florida Department Agriculture and Consumer Services, Division of Plant Industry, *Entomology Circular* 374, 4 pp.

Hall, D.G. (1995) A revision to the bibliography of the sugarcane rootstalk borer weevil, *Diaprepes abbreviatus* (Coleoptera: Curculionidae) *Florida Entomologist* 78, 364–377.

Hall, D.G., Childers, C.C. and Eger, J.E. (1991) Estimating citrus rust mite (Acari: Eriophyidae) levels on fruit and individual citrus trees. *Environmental Entomology* 20, 382–390.

Hall, D.G., Childers, C.C. and Eger, J.E. (1994) Spatial dispersion of sampling of citrus rust mites (Acari: Eriophyidae) on fruit in Hamlin and Valencia orange groves in Florida. *Journal of Economic Entomology* 87, 687–698.

Hall, W.J. (1930) The South African citrus thrips in Southern Rhodesia. *Mazoe Citrus Experimental Station Publication* 1, pp. 1–55.

Hammon, A.B. and Williams, M.L. (1984) The soft scale insects of Florida (Homoptera: Coccoidea: Coccidae). In: *Arthropods of Florida and Neighboring Land Areas.* Florida Department of Agriculture and Consumer Services 11, 194 pp.

Hardman, J.R., Papacek, D. and Smith, D. (1993) The use of integrated pest management in citrus orchards in Queensland: an economic perspective. In *Aggregated Papers of the 37th Annual Conference of the Australian Agricultural Research and Economics Society*. University of Sydney, 9–11 February 1993.

Hariyappa, A.S. and Kulkarni, K. (1989) Interaction between the predatory mite, *Amblyseius ovalis* (Evans) and chili mite, *Polyphagotarsonemus latus* (Banks). *Journal of Biological Control* 3, 31–32.

Harris, E.J., Bautista, R. and Spencer, J.S. (1998) Utilization of the egg-larval parasitoid *Biosteres arisanus* for augmentative biological control of tephritid fruit flies. *Proceedings Fifth International Symposium on Fruit Flies of Economic Importance*, University Sans Malaysia, 1–5 June, Penang, Malaysia, p. 41.

Hattingh, V. and Tate, B. (1995) Effects of field weathered residues of insect growth regulators on some Coccinellidae (Coleoptera) of economic importance as biocontrol agents. *Bulletin of Entomological Research* 85, 489–493.

Hentze, F., Mata, R. and Urbina, R. (1993) A Central American program for fruit fly control In: Aluja, M. and Liedo, P. (eds) *Fruit Flies: Biology and Management*. Springer-Verlag, New York.

Hill, D. (1975) *Agricultural Insect Pests of the Tropics and Their Control*. Cambridge University Press, New York, 516 pp.

Hoy, M. and Nguyen, R. (1994) Classical biological control of the citrus leafminer in Florida. *Citrus Industry* 75, 22–25.

Hoy, M. and Nguyen, R. (1997) Classical biological control of the citrus leafminer – *Phyllocnistis citrella*. Stainton. *Tropical Lepidoptera* 8 (Suppl. 1), 19 pp.

Hoy, M.A., Nguyen, R., Hall, D., Bullock, R., Pomerinke, M., Peña, J., Browning, H. and Stansly, P. (1995) Establishment of citrus leafminer parasitoid, *Ageniaspis citricola* in Florida. *Citrus Industry* 76, 12–17.

Hoy, M., Johnson, S. and Nguyen, R. (1996) Classical biological control of the citrus leafminer: in theory and practice. In: Hoy, M. (ed.) *Proceedings of International Conference on Managing Citrus Leafminer*. University of Florida, pp. 16–20.

Huang, M. and Huang, M. (1989) Study on the spatial pattern of larvae of *Citrus* leafminer. In *Studies on the Integrated Pest Management of Citrus Insect Pests*. Academic Book and Periodical Press, Guanzhou, pp. 90–95.

Huasong, W., Jianguo, H., Daguang, L., Wen, K. and Yuanying, L. (1998) Area wide control of chinese citrus fly, *Bactrocera* (*Tetradacus*) *minax* and studies on mating characteristics. *Proceedings of the Fifth International Symposium on Fruit Flies of Economic Importance*. University Sans Malaysia, 1–5 June, Penang, Malaysia, p. 41.

Hunsberger, A., Peña, J.E. and Schaffer, B. (1996) Relationships of citrus leafminer density to citrus damage and yield. In: Hoy, M. (ed.) *Proceedings of International Conference on Managing Citrus Leafminer*, University of Florida, p. 86.

Husain, M.A. and Nath, D. (1927) The citrus psylla (*Diaphorina citri*, Kuw.) (Psyllidae: Homoptera). *Memoirs of the Department of Agriculture of India* 10, 1–27.

Iwahashi, O. (1984) The control project of the oriental fruit fly in Okinawa. *Chinese Journal of Entomology* 4, 107–120 (in Chinese).

Jacas, J. and Peña, J. (2002) Calling behavior of two different field populations of *Phyllocnistis citrella* Stainton: effect of age and photoperiod. *Florida Entomologist* (in press).

Jaffe, K., Mauleon, H. and Kermarrec, A. (1990) Quantitative evaluation of ants as biological control agents with special reference to predators on *Diaprepes* spp. (Coleoptera: Curculionidae) on citrus groves in Martinique and Guadeloupe. In: *Recontres Caraibes en lutte biologique, INRA, Paris, 1991. Les colloques, no. 58*, pp. 405–416.

Jeppson, L.R., Keifer, H. and Baker, E.W. (1975) *Mites Injurious to Economic Plants*. University of California Press, Berkeley, 614 pp.

Johnson, S., Vaughn, A. and Bourgeois, W.J. (1996) Rearing and release methods for *Ageniaspis citricola* for a classical biocontrol program of the citrus leafminer in Lousiana. In: Hoy, M. (ed.) *Managing the Citrus Leafminer. Proceedings International Conference, Orlando, Florida*. University of Florida, Gainesville, p. 63.

Jones, V.P. and Brown, R.D. (1983) Reproductive responses of broad mite, *Polyphagotarsonemus latus* (Acari: Tarsonemidae) to constant temperature-humidity regimes. *Annals of the Entomological Society America* 76, 466–469.

Knapp, J.L. (1983) *Florida Citrus Integrated Pest and Crop Management Handbook*. Florida Cooperative Extension Service, Institute of Food and Agricultural Services, University of Florida, Gainesville, SP-14, 81pp.

Knapp, J., Albrigo, L.G., Browning, H.W., Bullock, R.C., Heppner, J.B., Hall, D.G., Hoy, M., Nguyen, R., Peña, J.E. and Stansly, P. (1995) Citrus leafminer, *Phyllocnistis citrella* Stainton: current status in Florida – 1995. In: Knapp, J.L. (ed.) *Citrus Leafminer Workshop*. 8–9 February 1995, Orlando, Florida. Institute of Food and

Agricultural Science, University of Florida, Gainesville, 35 pp.

Knapp, J.L., Peña, J.E., Stansly, P.A., Bullock, R.C. and Shapiro, J. (1996a) Chemical control of the citrus leafminer: what are the options? In: Hoy, M. (ed.) *Proceedings of the International Conference on Managing the Citrus Leafminer, Orlando, Florida, 23–25 April*. University of Florida, Gainsville, Florida, pp. 21–24.

Knapp, J., Noling, J.W., Timmer, L.W. and Tucker, D.P. (1996b) Florida citrus IPM. In: Rosen, D., Bennett, F. and Capinera, J. (eds) *Integrated Management: a Florida Perspective*. Intercept, Andover, UK, pp. 317–348.

Knapp, J., Bergh, J., Oswalt, T., Salyani, M., Parsons, L., Tucker, D., Wheaton, T., Bullock, R., Childers, C., Hall, D., McCoy, C., Rogers, J., Stansly, P., Browning, H., Graham, J., Timmer, L., Peña, J., Futch, S., Duncan, L., Noling, J., Roberts, P., McMillan, R., Sonoda, R., Brown, G., Davies, F., Ismail, M. and Stover, E. (1999) *Florida Citrus Pest Management Guide*. University of Florida, Gainesville, Institute of Food and Agricultural Sciences, Florida Cooperative Extension Service. SP_43.

Komazaki, S. (1987) Growth and reproduction in the first two and summer generations of two citrus aphids, *Aphis citricola* van der Goot and *Toxoptera citricidus* (Kilkardy) (Homoptera: Aphididae), under different thermal conditions. *Applied Entomology and Zoology* 23, 220–227.

LaSalle, J. and Schauff, M.E. (1996) The genera of chalcid parasitoids (Hymenoptera: Chalcidoidea) of citrus leafminer, *Phyllocnistis citrella* Stainton (Lepidoptera: Gracillaridae): aworkshop presentation. In: Hoy, M. (ed.) *Proceedings of International Conference on Managing the Citrus Leafminer*. 23–25 April 1996, Orlando, Florida, University of Florida, Gainesville, 119 pp.

Leyva, J.L., Browning, H. and Gilstrap, F. (1991) Development of *Anastrepha ludens* (Diptera: Tephritidae) in several host fruit. *Environmental Entomology* 20, 1160–1165.

Liu, Z.M., Beattie, G.A.C., Jiang, L. and Watson, D.M. (1999) Volumes of petroleum spray oil required for control of *Phyllocnistis citrella* Stainton (Lepidoptera: Gracillariidae) in mature citrus orchards. *Australian Journal of Entomology*, 38(2), 141–144.

Liu, Z.M., Beattie, G.A.C., Hodgkinson, M., Rose, H.A. and Jiang, L. (2001) Influence of petroleum-derived spray oil aromaticity, equivalent *n*-paraffin carbon number and emulsifier concentration on oviposition by citrus leafminer, *Phyllocnistis citrella* Stainton (Lepidoptera: Gracillariidae). *Australian Journal of Entomology* 40, 193–197.

Luck, R.F., Podoler, H. and Kifir, R. (1982) Host selection and egg allocation behaviour by *Aphytis melinus* and *A. lingnanensis*: comparison of two facultatively gregarious parasitoids. *Journal of Ecological Entomology* 7, 397–408.

Malausa, J.C., Quilici, S. and Brun, P. (1996) Status and first studies on the citrus leafminer in France: Azur Coast, Corsica and Reunion Island, In: Hoy, M. (ed.) *Proceedings of International Meeting: Managing the Citrus Leafminer*. 22–25 April 1996, Orlando, Florida. University of Florida, Gainesville, Florida, p. 88.

Malipatil, M.B., Dunn, K.L. and Smith, D. (2000) *An Illustrated Guide to the Parasitic Wasps Associated with Citrus Scale Insects and Mealybugs in Australia*. Department of Natural Resources and Environment publications, Agriculture Victoria, Horticultural Research Development Corporation.

Mamet, J.R. (1955) A revised food-plant catalogue of the insects of Mauritius. *Bulletin Department of Agriculture Mauritius 90 [Rev. appl. Ent.* (1956) A44,252*]*

McCoy, C. and Albrigo, G. (1975) Feeding injury to the orange caused by the citrus rust mite, *Phyllocoptruta oleivora* (Prostigmata: Eriophyodea). *Annals of the Entomological Society of America* 68, 289–292.

McCoy, C. and Simpson, S.E. (1994) Past and current IPM strategies to combat the spread of *Diaprepes abbreviatus* in Florida citrus. *Proceedings of the Caribbean Food Crops Society, St Thomas* 30, 247–255.

McCoy, C., Albrigo, G. and Allen, J.C. (1988) The biology of citrus mites and its effects on fruit quality. *Citrus Industry* (September), 44–49.

McMurtry, J.A., Badii, M.H. and Johnson, H.G. (1984) The broad mite, *Polyphagotarsonemus latus*, as a potential prey for phytoseiid mites in California. *Entomophaga* 29, 83–86.

McPheron, B.A. and Steck, G. (1996) *Fruit Fly Pests: A World Assessment of Their Biology and Management*. St Lucie Press, Delray Beach, 587 pp.

Mead, F. (1977) The asiatic citrus psyllid, *Diaphorina citri* Kuwayama (Homoptera: Psyllidae). Florida Department of Agriculture and Consumer Services, *Entomology of Circular* 80, 4 pp.

Meneghini, M. (1946) Sobre a natureza e transmissibilidade do doenca 'tristeza' dos citrus. *Biologico (Sao Paulo)* 12, 285–287.

Mercado, B.G., Pableo, F., Gavarra, M.R. and Gonzales, C.I. (1991) Population studies and biological control of *Diaphorina citri* Kuwayama, the insect vector of citrus greening disease in the Philippines. In: Chung, Ke and Osman, S.B. (eds) *Rehabilitation of Citrus Industry in the Asia Pacific Region. Proceedings of the 6th International Asia Pacific Workshop on Integrated Citrus Health Management*. 24–30 June 1991, Kuala Lumpur, Malaysia.

Michaud, J.P. (1998) A review of literature on *Toxoptera citricida* (Kirkaldy) (Homoptera: Aphididae). *Florida Entomologist* 81, 37–61.

Mohyuddin, A.I. and Mahmood, R. (1993) Integrated control of mango pests in Pakistan. *Acta Horticulturae* 341, 467–483.

Morakote, R. and Nanta, P. (1996) Natural enemies of citrus leafminer, *Phyllocnistis citrella* Stainton in Thailand. In: Hoy, M. (ed.) *Proceedings of International Conference on Managing Citrus Leafminer*. University of Florida, p. 90.

Moran, V.C. and Brown, R.P. (1973) The antennae, host plant chemoreception and probing activity of the citrus psylla. *Journal of the Entomological Society South Africa* 36, 191–202.

Morin, C. (1967) *Cultivo de frutales tropicales*. Librerias ABC, Lima, Peru, pp. 225–229.

Mosquera, F. (1979) Algunas consideraciones sobre la importancia economica del piojo blanco de los citricos, *Unaspis citri*. *Revista Colombiana de Entomologia* 5, 53–56.

Moutia, L.A. (1958) Contribution to the study of some phytophagous Acarina and their predators in Mauritius. *Bulletin of Entomological Research* 49, 59–75.

Murata, T. (1997) Citrus. In: Mitra, S.K. (ed.) *Postharvest Physiology and Storage of Tropical and Subtropical Fruits*. CAB International, Wallingford, UK, pp. 21–46.

Nariani, T.K. (1981) Integrated approach to control citrus greening disease in India. *Proceedings of the International Society of Citriculture* 1, 471–472.

Nasca, A.J., Teras, A.C., Fernandez, R.V. and Pasqualini, A.J. (1981) *Animals Perjudicials and Beneficos a los Citruicos en el Noroeste Argentino*. Centro de Investigaciones Sobre Regulaciones de Poblaciones de Organismos no ciuos Argentina.

Nascimento, A.S., Caldas, R.C. and Da Silva, L.M.S. (1982) Amostrage, do acaro ferrugem *Phyllocoptruta oleivora* (Ashmead, 1879) (Acari: Eriophyidae). *Anais Sociedade Entomologica Brasileira* 11, 157–162.

Noyes, J.S. (1998) *Catalogue of Chalcidoidea of the World*. ETI/The Natural History Museum (CD ROM).

Nucifora, A. (1961) Observazioni sulla reproduzioni di *Hemitarsonemus latus* (Banks) (Acarina: Tarsonemidae). *Academia Nazionale Italiana Entomologica* 10, 142–153.

Ochoa, R., Aguilar, H. and Vargas, C. (1994) *Phytophagous Mites of Central America: an Illustrated Guide*. Centro Agricultura Tropical Internacional y Enseñanza, Turrialba, Costa Rica, Serie Tecnica. Manual tecnico no. 6, 234 pp.

Oliveira de, C.A.L., Mauro, A.O. and Kronka, S.N. (1982) Comparacao de metodos para estimativas da populacao do acaro *Phyllocoptruta oleivora* (Ashmead, 1879) na cultura do citrus. *Anais da Sociedade Entomologica Brasileira* 11, 101–104.

Omole, M. (1980) Attraction to colour traps and height of flight in *Trioza urticae* L. (Hem.: Psyllidae). *Entomologist's Monthly Magazine* 116, 173–175.

Ortu, S. (1996) Field observations on sexual attraction of *Phyllocnistis citrella* Stainton (7Z, 11Z)-7,11-hexadecadienal. *Proceedings XX International Congress of Entomology*. 25–31 August, Florence, Italy, p. 496.

Osman, S. and Quilici, S. (1991) Studies of citrus greening vector, *Diaphorina citri* Kuwayama, natural enemies and alternate hosts in Malaysia. In: Chung, Ke and Osman, S.B. (eds) *Rehabilitation of Citrus Industry in the Asia Pacific Region. Proceedings of the 6th International Asia Pacific Workshop on Integrated Citrus Health Management*. 24–30 June 1991, Kuala Lumpur, Malaysia.

Otero, O., Montes, M., Arteaga, E., Rodriguez, N., Gonzalez, C., Cabrera, R., Broche, R., Castellanos, A. and Fernandez, O. (1994) Manual de orientaciones para el manejo fitosanitario de las principales plagas y enfermedades de los citricos. *Departamento de Proteccion de Plants, Instituto de Investigaciones de Citricos*, 21 pp.

Papacek, D. (1997) *IPM for Citrus*. Report on short term consultancy to Morocco 20 May–24 June 1997. IPM P/L Mundubbera.

Papacek, D.F. and Smith, D. (1985) *Aphytis lingnanensis*. In: Singh, P. and Moore, R.F. (eds) *Handbook of Insect Rearing*, Vol. 1. Elsevier, Amsterdam.

Peña, J.E. (1990) Relationships of broad mite (Acari: Tarsonemidae) density to lime damage. *Journal of Economic Entomology* 83, 2008–2015.

Peña, J.E. (1992a) Predator–prey interactions between *Typhlodromalus peregrinus* and *Polyphagotarsonemus latus*: effects of alternative prey and other food resources. *Florida Entomologist* 75, 241–427.

Peña, J.E. (1992b) Use of phytoseiid mites in the biological control of broad mites *Polyphagotarsonemus latus*. *75th Annual Meeting, Florida Entomological Society*, 10–13 August, p. 17 (Abstract).

Peña, J.E. (1998) Population dynamics of citrus leafminer (Lepidoptera: Gracillariidae) as measured by interception traps and egg and larva sampling in lime. *Journal of Entomological Science* 33, 90–96.

Peña, J.E. and Baranowski, R.M. (1992) Spatial dispersion and sampling of lime pests in Florida. *Proceedings of the Florida State Horticultural Society* 105, 289–294.

Peña, J.E. and Bullock, R.C. (1994) Effects of broad mite, *Polyphagotarsonemus latus* (Acari: Tarsonemidae) feeding on vegetative plant growth. *Florida Entomologist* 77.

Peña, J.E. and Duncan, R. (1996) Rearing methods for citrus leafminer and *Pnigalio minio* using Tahiti limes In: Hoy, M. (ed.) *Managing the Citrus Leafminer. Proceedings of an International Conference, Orlando, Florida.* University of Florida, Gainesville, p. 65.

Peña, J.E. and Mead, F. (1988) Citrus gall midge, *Prodiplosis longifila* (Diptera: Cecidomyiidae). Florida Department of Agriculture and Consumer Services. Division of Plant Industry. *Entomology Circular* 312, 2 pp.

Peña, J.E. and Schaffer, B. (1996) Sampling and monitoring for the citrus leafminer in Florida, USA. In: Hoy, M. (ed.) *Proceedings of International Conference on Managing Citrus Leafminer.* University of Florida, p. 91.

Peña, J.E. and Schaffer, B. (1997) Intraplant distribution and sampling of the citrus leafminer (Lepidoptera: Gracillariidae) on lime. *Journal of Economic Entomology* 90, 458–464.

Peña, J.E., Baranowski, R. and Denmark, H. (1989a) Developments toward biological control of the broad mite on limes in southern Florida. *Florida Entomologist* 72, 373–377.

Peña, J.E., Cagné, R. and Duncan, R. (1989b) Biology and characterization of *Prodiplosis longifila* (Diptera: Cecidomyiidae) on lime in Florida. *Florida Entomologist* 72, 444–450.

Peña, J.E., Duncan, R. and Torres, V. (1990) Control of the citrus midge *Prodiplosis longifila* Cagné in southern Florida limes. *Proceedings of the Interamerican Society for Tropical Horticulture* 34, 159–161.

Peña, J.E., Osborne, S. and Duncan, R.E. (1996a) Potential of fungi as biocontrol agents of *Polyphagotarsonemus latus* (Acari: Tarsonemidae) *Entomophaga* 41, 27–36.

Peña, J.E., Duncan, R. and Browning, H. (1996b) Seasonal abundance of *Phyllocnistis citrella* (Lepidoptera: Gracillariidae) and its parasitoids in south Florida citrus. *Environmental Entomology* 25, 698–702.

Peña, J.E., Etienne, J., Duncan, R. and Pinto, J. (1998) Introduction of *Ceratogramma etiennei* (Hymenoptera: Trichogrammatidae) for biological control of *Diaprepes abbreviatus* in Florida, USA. In: Hassan, S. (ed.) *Egg Parasitoids. 5th International Symposium, IIOBC, Cali, Colombia*, March 1998, pp. 145–148.

Peña, J.E., Duncan, R., Klema, E. and Hunsberger, A. (1999) Evaluation of direct and indirect action of insecticides and acaricides for control of lime and avocado pests. *Proceedings of the Florida State Horticultural Society* 112, 213–217.

Peña, J.E., Hunsberger, A. and Schaffer, B. (2000) Citrus leafminer (Lepidoptera: Gracillariidae) density: effect on yield of 'Tahiti' lime. *Journal of Economic Entomology* 93, 374–379.

Perales, M., Arredondo, H. and Garza, E. (1996) Parasitoids of the citrus leafminer in Colima, Mexico. In: Hoy, M. (ed.) *Proceedings of International Conference on Managing Citrus Leafminer.* University of Florida, p. 93.

Plumb, R.T. and Johnstone, G.R. (1995) Cultural, chemical, and biological methods for the control of barley yellow dwarf. In: d'Arcy, C.J. and Barnett, P.A. (eds) *Barley Yellow Dwarf – 40 Years of Progress.* APS Press, St Paul, Minnesota, pp. 307–319.

Pomerinke, M.A. and Stansly, P.A. (1996) Establishment and dispersal of *Ageniaspis citricola* in southwest Florida, In: Hoy, M. (ed.) *Managing the Citrus Leafminer. Proceedings of International Conference, Orlando, Florida.* University of Florida, Gainesville, p. 94.

Prinsloo, G.L. (1984) An illustrated guide to the parasitic wasp associated with citrus pests in the Republic of South Africa. *Science Bulletin* no. 402. Department of Agriculture.

Quezada, J.R. (1989) Citricos. In: Andrews, K.L. and Quezada, J.R. (eds) *Manejo Integrado de Plagas Insectiles en la Agricultura.* Escuela Agricola Panamericana, El Zamorano, Honduras, pp. 471–487.

Quiros-Gonzalez, M. (1996) Plant feeding mites on orchards under irrigation of Tahiti lime, *Citrus latifolia* Tanaka, in Zulia state, Venezuela. In: Mitchell, R., Horn, D., Needham, G. and Welbourn, C. (eds) *Acarology IX, Volume 1, Proceedings Ohio Biological Survey.* Columbus, Ohio, pp. 203–205.

Rae, D.J., Beattie, G.A.C., Watson, D.M., Liu, Z.M. and Jiang, L. (1996) Effect of petroleum spray oils without and with copper fungicides on the control of citrus leafminer, *Phyllocnistis citrella* Stainton (Lepidoptera: Gracillaridae). *Australian Journal of Entomology* 35, 247–251.

Rae, D.J., Liang, W.G., Watson, D.M., Beattie, G.A.C. and Huang, M.D. (1997) Evaluation of petroleum spray oils for control of the Asian citrus psylla, *Diaphorina citri* (Kuwayama)

(Hemiptera: Psyllidae), in China. *International Journal of Pest Management* 43, 71–75.

Rae, D.J., Watson, D.M., Huang, M.D., Cen, Y.J., Wang, B.Z., Beattie, G.A.C., Liang, W.G., Tan, B.L. and Liu, D.G. (2000) Efficacy and phytotoxicity of multiple oil sprays on sweet orange (*Citrus sinensis* (L.)) and pummelo (*Citrus grandis* (L.)) in Southern China. *International Journal of Pest Management* 46, 125–140.

Rhodes, L.F. (1990) Investigations into the potential of entomogenous nematodes for the control of the citrus root weevil, *Diaprepes famelicus* Oliver. In: Pavis, C. and Kermarrec, A. (eds) *Recontres Caraibbes en Lutte Biologique, INRA, Paris. Les colloques no. 58*, 161–165.

Richman, D.B., Buren, W.F. and Whitcomb, W.H. (1983) Predatory arthropods attacking the eggs of *Diaprepes abbreviatus* (L.) (Coleoptera: Curculionidae) in Puerto Rico and Florida. *Journal of Georgia Entomological Society* 18, 335–342.

Ripa, S. and Rodriguez, A. (eds) (1999) *Plagas de citricos, sus enemigos naturales y manejo*. Instituto de Investigaciones Agropecuarias, Ministerio de Agricultura Coleccion Libros (NIA No 3). Santiago, Chile.

Roca-Peña, M.A., Lee, R.F., Lastra, R., Niblett, C.L., Ochoa-Corona, F.M., Garnsey, S.M. and Yokomi, R.K. (1995). Citrus tristeza and its aphid vector, *Toxoptera citricida*. Threats to citrus production in the Caribbean and North America. *Plant Disease* 79, 437–445.

Rose, M. (1990) Citrus: diaspidid pest problems and control in crops. In: *Armored Scale Insects: Their Biology, Natural Enemies and Control, Vol. B.* Elsevier, New York, pp. 535–540.

Rosen, D. (ed.) (1994) *Advances in the Study of* Aphytis *(Hymenoptera: Aphelinidae)*. Intercept, Andover, 362 pp.

Rosen, D. and DeBach, P. (1979) *Species of* Aphytis *of the World (Hymenoptera: Aphelinidae)*. Israel Universities Press, Jerusalem, and W. Junk, The Hague.

Ruiz, E. (1997) Control natural y biologico de plagas citricolas en Tamaulipas, Mexico. II Curso Internacional de Citricultura, Manejo Integral Fitosanitario, Cd. Victoria, Tamaulipas, Mexico, pp. 131–135.

Ruiz-Cancino, E., Escamilla, O., Briones, F. and Van Driesche, R. (1996) *Euseius mesembrinus* (Dean) (Phytoseiidae) in Tamaulipas, Mexico. In: Mitchell, R., Horn, D., Needham, G. and Welbourn, C. (eds) *Acarology IX, Vol. 1, Proceedings Ohio Biological Survey*. Columbus, Ohio, pp. 221–222.

Sadana, G.L. (1991) Mode of hunting and functional response of the spider *Marpissa tigrina* Tikader (Salticidae: Arachnidae) to the density of its prey, *Diaphorina citri*. *Entomon* 16, 279–282.

Salinas, M.D., Sumalde, A.C., Calilung, V.J. and Bajet, N.B. (1996) Life history, seasonal abundance, host range and geographical distribution of the wooly whitefly, *Aleurothrixus floccosus* (Maskell) (Homoptera: Aleyrodidae). *Philippine Entomology* 10, 67–89.

Samson, R.A., McCoy, C. and O'Donnell, K. (1980) Taxonomy of the acarine parasite. *Hirsutella thompsonii*. *Mycologia* 72, 359–377.

Samways, M.J. (1987) Phototactic response of *Trioza erytreae* (Del Guercio) (Hemiptera: Triozidae) to yellow-coloured surfaces, and an attempt at commercial suppression using yellow barriers and trap trees. *Bulletin of Entomological Research* 77, 91–98.

Sandhu, G.S. and Batra, R.C. (1978a) Incidence of citrus leafminer *Phyllocnistis citrella* injury on different cultivars and rootstocks of citrus in the Punjab. *Punjab Horticultural Journal* 18, 92–96.

Sandhu, G.S. and Batra, R.C. (1978b) Sources of resistance as observed under field infestation against citrus leafminer. *Science Culture* 44, 122–124.

Schaffer, B., Peña, J., Colls, C.A. and Hunsberger, A. (1996) Accuracy of visual estimates of leaf damage from citrus leafminer and the relationship between larval density and mining period on leaf damage and photosynthesis of 'Tahiti' lime leaves. In: Hoy, M. (ed.) *Proceedings of International Conference on Managing Citrus Leafminer*. University of Florida, p. 97.

Schoonhoven, A.V., Piedrahita, J., Valderrama, R. and Galvez, G. (1978) Biologia, dano y control del acaro tropical *Polyphagotarsonemus latus* (Banks) (Acarina: Tarsonemidae) en frijol. *Turrialba* 28, 77–80.

Schroeder, W. (1990) Suppresion of *Diaprepes abbreviatus* (Coleoptera: Curculionidae) adult emergence with soil application of entomopathogenic nematodes (Nematoda: Rhabditida). *Florida Entomologist* 73, 680–683.

Schroeder, W.J. and Beavers, J.B. (1977) Citrus root weevils in Florida: identification, biology and control. *Proceedings of the International Society of Citriculture* 2, 498–500.

Shapiro, J. and Gottwald, T. (1995) Resistance of eight cultivars of citrus rootstock to larval root weevil, *Diaprepes abbreviatus* L. (Coleoptera: Curculionidae). *Journal of Economic Entomology* 88, 148–154.

Shapiro, J.P., Schroeder, W. and Stansly, W. (1996) Bioassay and field tests of *Bacillus thuringiensis* and an organosilicone surfactant against the citrus leafminer, *Phyllocnistis citrella* Stainton

(Lepidoptera: Gracillariidae). *Florida Entomologist* 81(2), 201–210.

Singh, G. (1991) Loss assessment, ecology and management of mango fruit fly, *Dacus* sp. *Acta Horticulturae* 291, 425–436.

Smith, D. (1978) Biological control of scale insects in south-eastern Queensland. 2. Control of circular black scale, *Chrysomphalus ficus* Ashmead, by the introduced parasite, *Aphytis holoxanthus* DeBah. *Journal of the Australian Entomological Society* 17, 373–377.

Smith, D. and Beattie, G.A.C. (1996) Australian citrus IPM and citrus leafminer. In: Hoy, M. (ed.) *Proceedings of International Conference on Managing the Citrus Leafminer*. University of Florida, pp. 34–46.

Smith, D. and Neale, C. (1996) Rearing techniques for citrus leafminer parasitoids in Queensland. In: Hoy, M. (ed.) *Proceedings of International Conference on Managing Citrus Leafminer*. University of Florida, pp. 66–68.

Smith, D. and Papacek, D.F. (1991) Studies of the predatory mite *Amblyseius victoriensis* (Womersley) (Acarina: Phytoseiidae) in citrus orchards in south-east Queensland: control of *Tegolophus australis* Kiefer and *Phyllocoptruta oleivora* (Ashmead) (Acarina: Eriophyidae), effect of pesticides, alternate host plants and augmentative release. *Experimental and Applied Acarology* 12, 195–217.

Smith, D. and Papacek, D.F. (1993) *Report on Short Term Consultancy Mission to Thailand: IPM in Citrus*. 22 August–12 September, QDPI, Thai-German. 78 pp.

Smith, D., Papacek, D.F. and Smith, N.J. (1995) Biological control of citrus snow scale in South East Queensland. *Israel Journal of Entomology* 29, 253–260.

Smith, D., Beattie, G.A.C. and Broadley, R. (1997) *Citrus Pests and their Enemies: Integrated Pest Management in Australia. Queensland Department of Primary Industries Series QI97030*, 272 pp.

Smith, J. and Hoy, M. (1995) Rearing methods for *Ageniaspis citricola* (Hymenoptera: Encyrtidae) and *Cirrospilus quadristriatus* (Hymenoptera: Eulophidae) released in a classical biological control program for the citrus leafminer *Phyllocnistis citrella* (Lepidoptera: Gracillariidae). *Florida Entomologist* 72, 600–608.

Smith, K.M., Smith, D. and Lisle, A.T. (1999) Effect of field-weathered residues of pyriproxyfen on the predatory coccinellids *Chilocous circumdatus* Gyllenhal and *Cryptolaemus montrouzieri* Mulsant. *Australian Journal of Experimental Agriculture* 39, 995–1000.

Sparkes, T.C. (1998) New insecticides and acariades: mode of action and potential role in IPM. In: *Integrated Pest Management – Future Challenges. 6th Australasian Applied Entomological Research Conference*. University of Queensland, Brisbane Australia, 21 September–20 October. P998, pp. 124–130.

Sponagel, K., Diaz, F. and Cribas, A. (1996) *Las moscas de la fruta (Diptera: Tephritidae) y su importancia en la agricultura de Honduras.* Fundacion Hondureña de Investigacion Agricola, La Lima, Honduras, 75 pp.

Steiner, L.F. and Lee, R.K. (1955) Large area tests of male annihilation method against oriental fruit fly control. *Journal of Economic Entomology* 48, 311–317.

Steiner, L.F., Mitchell, W.C., Harris, E., Kozuma, T. and Fujimoto, M. (1965) Oriental fruit fly eradication by male annihilation. *Journal of Economic Entomology* 58, 961–964.

Syed, R.A., Ghani, M.A. and Murtaza, M. (1970) Studies on trypetids and their natural enemies in West Pakistan III. *Dacus (Strumeta) zonatus* (Saunders). *Technical Bulletin 13*, Commonwealth Institute of Biological Control, pp. 1–16.

Talhouk, A.S. (1975) Citrus pests throughout the world, distribution and importence. In: *Citrus*. Ciba-Geigy Agrochemicals Technical Monograph 4, 21–28.

Tan, B., Huang, M. and Jialum, L. (1989) Studies on the damage and economic threshold of citrus red mite, *Panonychus citri* (McGregor) to citrus. *Studies on the Integrated Management of Citrus Insect Pests*. Academic Book and Periodical Press. Guangzhou, China, pp. 15–26.

Tandom, P.L. (1991a) Major achievements in pest management in tropical fruit crops, deficiencies in the present programme and new thrust for future. In: *A Decade of Research on Pests of Horticultural Crops (1980–1990)*. Group discussion of entomologists working in the coordinated projects of horticultural crops, 28–29 January 1991, pp. 36–37.

Tandom, P.L. (1991b) Problems and prospects of insect pest management in fruit trees. In: Dhaliwal, G.S. and Arora, R. (eds) *Trends in Agricultural Pest Management*. Commonweath Publishers, New Delhi.

Tandom, P.L. (1997) Management of insect pests in tropical fruit crops. In: Arora, R.K. and Ramanatha Rao, V. (eds) *Tropical Fruits in Asia: Diversity, Maintenance, Conservation and Use.* Proceedings of the IPGRI-ICAR-UTFANET Regional Course on the Conservation and Use of Germplasm of Tropical Fruits in Asia, 18–31 May, Bangalore, India, pp. 235–245.

Tang, Y.Q. and Aubert, B. (1990) An illustrated guide to the identification of parasitic wasps associated with *Diaphorina citri* Kuwayama in the Asian Pacific region. In: *Proceedings of the 4th International Asian Pacific Conference on Citrus Rehabilitation*. 4–10 February 1970, Chiang Mai, Thailand, pp. 228–245.

Tang, Y.Q. and Huang, Z.P. (1991) Studies on the biology of two primary parasites of *Diaphorina citri* Kuwayama (Homoptera: Psyllidae). In: Chung, Ke and Osman, S.B. (eds) *Rehabilitation of Citrus Industry in the Asia Pacific Region. Proceedings of the 6th International Asia Pacific Workshop on Integrated Citrus Health Management*. Kuala Lumpur, Malaysia, 24–30 June 1991.

Tang, Y.Q. and Wu, M.X. (1991) Interspecific host discrimination between two primary parasites of Asian citrus psyllid *Diaphorina citri* Kuwayama. In: Chung, Ke and Osman, S.B. (eds) *Rehabilitation of Citrus Industry in the Asia Pacific Region. Proceedings of the 6 International Asia Pacific Workshop on Integrated Citrus Health Management*. Kuala Lumpur, Malaysia, 24–30 June 1991.

Tongyuan, D., Jinjun, X., Zeshua, W. and Fanlei, K. (1989) 7Z, Z11-hexadecadienal:sex attractant of *Phyllocnistis wampella* Liu & Zeng. *Kunchong Zhishi* 3, 147–149.

Tryon, H. (1986) The striped earwing and ant predators of sugarcane rootstock borer in Florida citrus. *Florida Entomologist* 69, 336–343.

Ujiye, T. (1996) Parasitoid fauna of the citrus leafminer in Japan, Taiwan and Thailand. In: Hoy, M. (ed.) *Proceedings of International Conference on Managing Citrus Leafminer*. University of Florida, p. 100.

Umeya, K. and Hirao, J. (1975) Attraction of the jackfruitfly, *Dacus umbrosus* F. (Diptera: Tephritidae) and lacewing, *Chrysopa* sp. (Neuroptera: Chrysopidae) to lure traps baited with methyl eugenol and cue-lure in the Philippines. *Applied Entomology and Zoology* 10, 60–62.

Van den Berg, M.A. and Fletcher, C.D. (1988) A bibliography of the citrus psylla, *Trioza erytreae* (Del Guercio) (Hemiptera: Triozidae), up to 1987. *Phytoparasitica* 16, 47–61.

Van den Berg, M.A., Deacon, V.E. and Steenekamp, P.J. (1991) Dispersal within and between citrus orchards and native hosts, and nymphal mortality of citrus psylla. *Trioza erytreae* (Hemiptera: Triozidae). *Agriculture, Ecosystems and Environment* 35, 297–309.

Verheij, E.W. and Coronel, R.E. (1992) *Plant Resources of South East Asia: Edible Fruits and Nuts*. Prosea Foundation, Bogor, Indonesia, 446 pp.

Vieira, M.R. and Chiavegato, L.G. (1999) Biology of *Polyphagotarsonemus latus* (Banks) (Acari: Tarsonemidae) on lemon (*Citrus limon* Burm). *Annais Sociedade Entomologia Brasileira* 28, 27–33.

Vijaysegaran, S. (1993) Control of fruit flies in the tropical regions of Asia. In: Aluja, M. and Liedo, P. (eds) *Fruit Flies: Biology and Management*. Springer-Verlag, New York.

Waterhouse, D.F. (1993) *Biological Control: Pacific Prospects*. Supplement 2. Australian Centre for International Agriculture Research, Canberra, 138 pp.

Waterhouse, D.F. (1998) *Biological Control of Insect Pests: Southeast Asian Prospects*. Australian Centre for International Agriculture Research, Canberra, 118 pp.

Watson, J.R. (1915) The woolly whitefly. *Florida Agricultural Experimental Station Bulletin* 126, 79–101.

Wharton, R.A. and Marsh, P.M. (1978) New world Opinae (Hymenoptera: Braconidae) parasitic on Tephritidae (Diptera). *Journal of the Washington Academy of Sciences* 68, 147–165.

Whitcomb, W H., Gowan, T. and Buren, W. (1982) Predators of *Diaprepes abbreviatus* larvae sugarcane rootstalk borer weevil in Florida. *Florida Entomologist* 65, 150–158.

White, I.M. and Elson-Harris, M. (1992) *Fruit Flies of Economic Significance*. CAB International, Wallingford, UK.

Whitwell, A.C. (1990) *Diaprepes* spp. problems in Dominica and some possible solutions. In: Pavis, C. and Kermarrec, A. (eds) *Recontres Caraibbes en Lutte Biologique. INRA, Paris. Les colloques no. 58*. pp. 529–541.

Woodruff, R.E. (1968) The present status of a West Indian weevil (*Diaprepes abbreviatus*) in Florida (Coleoptera: Curculionidae). Florida Department of Agriculture and Consumer Services, *Plant Industry Entomology* 77, Gainesville, Florida.

Yokomi, R.K., Lastra, R., Stoetzel, M.B., Damsteegt, V.D., Lee, R.F., Garnsey, S.M., Gotwald, T.R., Rocha-Peña, M.A. and Niblett, C.L. (1994) Establishment of the brown citrus aphid (Homoptera: Aphididae) in Central America and the Caribbean basin and transmission of citrus tristeza virus. *Journal of Economic Entomology* 87, 1078–1085.

Yothers, W.W. and Miller, R. (1934) Methods for determining rust mite abundance. *Florida State Horticultural Society* 47, 53–55.

4 Pests and Pollinators of Mango

G.K. Waite
Queensland Horticulture Institute, Maroochy Research Station, PO Box 5083, SCMC, Nambour, Queensland 4560, Australia

The mango was supposed to have originated in the Indo-Burma region (Verheij, 1991), but Mukherjee (1997) concluded that it most likely originated in South-East Asia, particularly in the Malay Archipelago. However, it was probably first grown as a food crop in India, where its cultivation is thought to have commenced at least 4000 years ago. Small fruit with thin flesh can still be found in parts of India and these were probably the ancestors of today's selections, which offer the greater size and flavour found in scores of cultivars grown worldwide. The arrival of the Portuguese in India in the late 15th century apparently benefited the Indian mango industry, as the knowledge concerning vegetative propagation was put to good use to plant large orchards of superior selections. The global spread of mango cultivation followed as European explorers transported the fruit to Africa, North and South America and the Caribbean. The 19th and 20th centuries have seen its range expand even further. Mangoes are now grown as an important crop all over the world, in both tropical and subtropical areas.

Importance of the Crop

The mango is a significant cultural and religious symbol in India and is much revered there (Mukherjee, 1997). It is the most important fruit of Asia and ranks fifth worldwide in production after bananas, citrus, grapes and apples. Mangoes form a regular part of the diet of people in areas where the fruit is easily grown; it may be eaten ripe or green. In addition, a proportion of the fruit is juiced or processed into preserves, chutney, frozen purée and dried mango. The seeds yield a starch and the skins are a source of anacardic acid. World production of mangoes in 2000 was estimated to be about 24.5 million t (FAOSTAT Database, 2001).

In its spread around the world, the mango has been accompanied by some of its original insect pests. The seed weevils have an easy and free ride in fruit which escapes quarantine restrictions. Other pests have followed mangoes to their new homes, but such occurrences may not always be linked to mangoes in the first instance, e.g. the Mediterranean fruit fly, *Ceratitus capitata* (Wiedemann), and some scale insects. Because the pests concerned are polyphagous, sooner or later, if mango is one of their hosts, they will find it in the new land. Often, in the absence of the natural enemies they left behind, these pests assume greater importance in their new range than at home.

Every production area has its unique complex of pests, many individuals of which may be shared with other areas. Generally, there are a few key pests against which specific management techniques must be applied. Often, because there are no effective

©CAB *International* 2002. *Tropical Fruit Pests and Pollinators*
(eds J.E. Peña, J.L. Sharp and M. Wysoki)

natural enemies present, or the pests breed on alternative hosts and are very mobile and fly into orchards to damage the crop, chemical controls must be applied. There are usually several minor pests which only occasionally cause a problem under normal circumstances, but which may become serious pests for various reasons. The indiscriminate use of pesticides that disrupt natural enemies; the introduction of new cultivars that are more susceptible to particular pests; and changing international export protocols, which may impose a nil tolerance on certain pests thus creating an artificial pest status, may elevate previously minor pests to the top rank.

Pests of Flowers

Mango flowers are attacked by a variety of insects such as midges, leafhoppers, caterpillars and thrips, as well as mites. Excessive damage at this critical time has the potential to limit production even before the fruit is set, though it should always be remembered that most tree crops, including mangoes, produce an excess of flowers which usually set more fruit than the tree can mature. Consequently, some allowance for pest damage can be made without detriment to the final crop.

Mango hoppers (Homoptera: Cicadellidae)

DISTRIBUTION AND BIOLOGY Numerous species of leafhoppers are reported infesting mango trees throughout Asia, but the most important are *Idioscopus clypealis* Lethierry, *Idioscopus niveosparsus* Lethierry, *Idioscopus magpurensis* Pruthi and *Amritodius atkinsoni* Lethierry (Soomroo *et al.*, 1987; Veeresh, 1988). Often all species occur together as in Bihar and south India, or with one or more being the major pests in particular areas e.g. *A. atkinsoni* and *I. clypealis* in the Punjab, *I. niveosparsus* in Gujarat, Maharashtra and Karnataka and *A. atkinsoni* in north India (Veeresh, 1988). In the Philippines the mango blossom hopper, *I. clypealis,* was found to constitute about 98% of the hopper population present on the flowers, with *I. niveosparsus* and occasionally *Bakera* (*Typhlocyba*) *nigrobilineata* Melichar accounting for the balance (Alam, 1994). The leafhopper causing most problems in Guangxi Province, China, is *Idioscopus incertus* Baker.

In the Philippines, *I. clypealis* females lay 100–200 eggs on flower buds and panicles, and in the midrib of young leaves (Anonymous, 1994). Eggs take 4–7 days to hatch and the nymphal period, occupied by only four instars, takes about 8–10 days. Adults live for up to 315 days and reproduce only during the mango flowering period (Alam, 1994). Mango hoppers in India, including *I. clypealis,* are reported to pass through five nymphal instars, occupying from 5 to 20 days, and the adults survive from one flowering to the next (Sohi and Sohi, 1990). Populations are at a minimum in December–January and reach a peak in March–April, when the mangoes flower (Tandon *et al.*, 1983). Peña *et al.* (1998) report that the species complete varying numbers of generations per year, depending on the location e.g. 4–5 generations of *A. atkinsoni* in the Punjab of Pakistan and 1–6 generations throughout India, while *I. clypealis* has 1–4 generations in the Philippines and 5–6 in India. Mohyuddin and Mahmood (1993) found that *I. niveosparsus* and *A. atkinsoni* in Pakistan moved up and down the tree at different times of the year and were concentrated in the lower canopy during May.

In 1998, *I. niveosparsus* was recorded for the first time on mangoes in Queensland, Australia. The infestation was confined to an area around a remote sea port through which the pest apparently gained entry, and has not yet spread to commercial production areas (B. Pinese, Mareeba, 1998, personal communication).

DAMAGE Adult and nymphal leafhoppers damage the flowers by piercing the tissues with their proboscis and sucking on the plant sap. The concentrated feeding of hundreds, and sometimes thousands, of individuals per panicle, results in the flowers withering and dropping, and no fruit is set. In addition to the direct effect of their feeding, the hoppers produce honeydew on which sooty mould thrives. This interferes with photosynthesis and pollination, if the flowers have not already been destroyed by the hoppers' feeding (Anonymous, 1994; Peña and Mohyuddin, 1997). Differential cultivar susceptibility has been reported in the Philippines, with *I. clypealis* causing the most severe damage to cv. 'Carabao', while *B. nigrolineata* prefers the cv. 'Kathmitha' (Alam, 1994).

BIOLOGICAL CONTROL Little information is available from the Philippines with respect to natural control of mango hoppers. However, Alam (1994) reported that the entomophagous fungi *Metarhizium* sp. and *Beauveria* sp. caused 100% mortality in caged populations of *I. clypealis*. *B. bassiana* Balsamo (Vuilllemin) and *Verticillium lecanii* (Zimmerman) Viegas are known to infect *I. clypealis* in India (Srivastava and Tandon, 1986). In Pakistan, Mohyuddin and Mahmood (1993) recorded the parasitoids *Gonatocentrus* sp., *Mirufens* sp. nr. *mangiferae* Viggiani and Hayat, *Centrodora* sp. nr. *scolypopae* Valentine, *Aprostocetus* sp. and *Quadrastichus* sp. from hopper eggs, and the ectoparasitoid, *Epipyrops fuliginosa* Tames, from adults. In India, Fasih and Srivastava (1990) noted the egg parasitoids *Aprostocetus* sp., *Gonatocerus* sp. and *Polynama* sp. In addition, five species of predators, *Chrysopa lacciperda* (Kimmins), *Mallarda boninensis* (Okomote), *Bochartia* sp., a mantid and a lygaeid were found to prey on the nymphs. *Aprostocetus* sp., *Platygaster* sp., *Synopeas* sp. and *Zatropis* sp. were recorded as egg parasitoids in Dominica (Whitwell, 1993). *E. fuliginosa* parasitizes *Idioscopus* spp. in Thailand.

MONITORING AND CONTROL Because mango hoppers attack the flowers and are capable of completely preventing fruit set, there are few options available for spray decisions based on monitoring. Most recommendations involve the application of effective insecticides before hopper populations become too intense, and as the trees commence to flower (Khanzada and Naqvi, 1985). Generally, the flowers are the target for the sprays applied, but Mohyuddin and Mahmood (1993) recommended that one spray be applied in May, and that it need only be applied to the lower 5 m of the trunk and canopy where the hoppers rested. A sequential sampling plan for monitoring hopper populations was suggested by Verghese and Rao (1985) while others have set specific infestation levels at which control should be initiated (Serrano and Palo, 1933; Shah *et al.*, 1984).

Many insecticides have been used against mango hoppers including a variety of botanical concoctions, chlorinated hydrocarbons, organophosphates and synthetic pyrethroids (Dakshinamurthy, 1984; Datar, 1985; Sohi and Sohi, 1990). Most of these have been applied as cover sprays on a schedule basis (Cendana *et al.*, 1983). In recent years the synthetic pyrethroids have given excellent control, as has imidacloprid, a systemic compound belonging to the nitroguanadine group of chemicals (Verghese, 1998). The technique of trunk injection with systemic organophosphates such as monocrotophos, dimethoate and vamidothion provided good control of hoppers for up to 7 weeks in India (Thontadarya *et al.*, 1978; Shah *et al.*, 1983), and imidacloprid used in the same way in the Philippines has shown promising results (G.K. Waite, 1998, unpublished).

Several non-chemical procedures are recommended to control mango hoppers. Planting resistant varieties is an obvious tactic and Nachiappan and Baskaran (1984) noted that in India, the cultivars 'Chinnarasam', 'Bangalora', 'Khader' and 'Beneshan' were resistant to *A. atkinsoni*, while Singh (1997a) listed a further ten Indian cultivars that are resistant. Smoking orchards to deter hoppers was suggested by Otanes and Toquero (1927), and Serrano and Palo (1933) recommended that light traps set up with kerosene lanterns placed in the middle of a basin containing soapy water could be used to reduce hopper populations just before flowering. Hoppers are noted for their habit of seeking secluded and sheltered areas to rest. A common recommendation is to prune large and dense trees to open them up to sunlight and so discourage the pests from remaining in the trees (Bondad, 1989; Anonymous, 1994). For cultivars that respond to artificial flower induction, peak hopper infestations can be avoided, for instance on cv. 'Carabao' in the Philippines, by inducing the trees to flower early in October–December (Anonymous, 1994).

The literature on mango hoppers is extensive, especially from India and Pakistan, and the reader is referred to Veeresh (1988) and Sohi and Sohi (1990), for a listing of references concerning the pests in those countries.

Midges (Diptera: Cecidomyiidae)

DISTRIBUTION AND BIOLOGY The mango blister midge or mango gall midge, *Erosomyia mangiferae* Felt, infests mangoes in the West Indies. According to Barnes (1948) it was first recorded in St Vincent and later in Trinidad and St Lucia. Female midges lay their eggs in the developing flower and leaf buds. When the eggs hatch after 2–3 days, the larvae cause the tissue in which they are feeding to swell and form galls. Up to 70% of small fruit may also be affected (Whitwell, 1993). After 10–14 days the larvae leave the galls and pupate in the soil. Adults emerge about 7 days later (Callan, 1941; Barnes, 1948). The abundance of the pest has been noted to be favoured by dry seasons that are unusually wet (Murray, 1991).

Five cecidomyiid species including *Erosomyia indica* Grover and Prasad, are reported to attack mango flowers and to cause severe damage in India (Kulkarni, 1955; Prasad, 1966, 1972; Butani, 1979), while *Dasyneura mangiferae* (Felt) does the same in Hawaii (USDA, 1981).

BIOLOGICAL CONTROL *Platygaster* sp., *Systasis dasyneurae* and *Eupelmus* sp. parasitize *Dasyneura* sp. in India, while *Tetrastichus* sp., *Aprostocetus* sp. and *Mirufens longifunculata* Viggiani and Hayat attack *Erosomyia indica* (Grover, 1986; Fasih and Srivastava, 1990). *Pirene* sp. has been recorded as an ectoparasitoid of *Procystiphora mangiferae* Felt.

Lepidoptera

DISTRIBUTION AND BIOLOGY Numerous species of Lepidoptera have been found to infest mango flower panicles in all production areas, but in few cases have the species involved been identified. In Florida, *Pococera atramentalis* Lederer (Pyralidae) and *Platynota rostrana* (Walker) (Tortricidae) are the two most damaging species of a complex that also includes *Pleuroprucha insulsaria* (Guneeé) (Geometridae), *Tallula* spp. (Pyralidae) and *Racheospilla gerularia* (Hubner) (Peña and Mohuddin, 1997). In the Philippines, the tip borer, *Chlumetia transversa* Walker (Noctuidae), is second only to mango hoppers as a pest of flowers (Bondad, 1989). Whitwell (1993) mentions that 14 lepidopterous species infest mango flowers in Dominica, with the most common being the geometrids, *Eupithecia* sp., *Chloropteryx glauciptera* Hampson and *Oxydia vesulia* (Cramer). Caterpillars belonging to the families Geometridae, Lymantriidae, Noctuidae, Pyralidae and Tortricidae infest mango flowers in Australia (Cunningham, 1989). The larvae of all of these species feed on the florets and the flower stems. Many of them construct shelters by webbing the destroyed flower parts together (Cunningham, 1989; Peña, 1993). *Penicillaria jocosatrix* Gueneé (Noctuidae) is a major pest in Guam, where it can consume all of the flowers on a panicle (Schreiner, 1987). The female moth deposits eggs on or near new leaves and inflorescences and frequently on spider webs. The larvae develop more quickly on these younger plant parts than on older leaves (Nafus *et al.*, 1991).

DAMAGE Feeding by the larvae on the various flower parts destroys the potential for maximum fruit set, and some of the species involved also damage young fruit (Peña, 1993). In many cases infestation levels and consequent damage are insignificant compared to the total number of flowers produced, and the capacity of the tree to set and carry a full crop to maturity (Cunningham, 1989).

BIOLOGICAL CONTROL In general, because of the lack of knowledge concerning the individual species attacking mango flowers, little information is available regarding their natural control. Peña (1993) noted that *Macrocentrus delicatus* Cresson parasitizes *P. atramentalis* in Florida. The hymenopterous parasitoids, *Aleiodes* sp. nr. *circumscriptus* (Nees) (Braconidae) and *Euplectrus* sp. nr. *parvulus* Ferriere (Eulophidae), along with the tachinid *Blepharella lateralis* Macquart, were introduced into Guam from India in 1986–1987, for the control of *P. jocosatrix*. Although *Aleiodes* sp. failed to establish, the other parasitoids successfully reduced *P. jocosatrix* populations by about 75%, and fruit production increased substantially (Nafus, 1991).

MONITORING AND CONTROL Cunningham (1989) recommended weekly monitoring of flower panicles for the presence of damaging caterpillars in Australia. Webbed shelters should be pulled apart in order to record the presence of larvae, with endosulfan being applied when caterpillar infestation is significant. No actual infestation levels were stated for the initiation of such action. Peña (1993) made similar recommendations for mangoes in Florida, so that potential problems can be detected early and chemical controls applied if they are necessary. On Guam, Schreiner (1987) found that *Bacillus thuringiensis* Berliner (Bt) could be used to control *P. jocosatrix*, but only if populations were carefully monitored, or regular sprays were applied to cover the overlapping periods of larval emergence.

Thrips (Thysanoptera: Thripidae)

DISTRIBUTION AND BIOLOGY *Frankliniella bispinosa* (Morgan) and *Frankliniella kelliae* (Sakimura) frequently infest mango blossoms in Florida, where they feed on the nectaries and anthers, presumably affecting fruit set (Peña, 1993). Sakimura (1981) described *F. kelliae* as a species distinct from *F. difficilis* Hood, and noted that it had been collected from a range of host plants from Florida and the Caribbean area. The western flower thrips, *Frankliniella occidentalis* (Pergande), damages flowers and fruit in Israel (Wysoki *et al.*, 1993). This species made a dramatic appearance in countries all over the world during the 1990s, and given the Israeli experience it would not be surprising to find that it has become a pest of mango flowers in at least some of these countries. Van Lenteren *et al.* (1995) report that the life cycle of *F. occidentalis* takes from 14.8 to 16.6 days at 25°C. The thrips *Frankliniella cubensis* attacks mango flowers in Costa Rica during the dry season (Jirón, 1993).

Scirtothrips dorsalis Hood, chilli thrips, is regarded as being an important pest of mango flowers in Thailand. The thrips attack the flowers, fruit, young shoots and new flush. Populations reach a peak during hot dry weather. In very dry springs in Australia the plague thrips, *Thrips imaginis* Bagnall, may heavily infest mango flowers, but they apparently cause little if any damage (G.K. Waite, 1996, unpublished). Tandon and Verghese (1987) recorded *Thrips palmi* Karny infesting mango flowers in India, leaving scab-like marks where they had fed, and retarding growth.

DAMAGE Thrips feed on all parts of the mango flower panicle, including the petals, anthers and pollen. They may destroy the florets so that no fruit sets, and also reduce the vigour of the panicle so that growth is retarded.

BIOLOGICAL CONTROL Loomans and van Lenteren (1995) reviewed the biological control agents of thrips and considered that the predators *Orius* sp., *Anystis agilis* Banks and *Hypoaspis aculifer* (Canestrini) showed good potential in the biocontrol of *F. occidentalis*. *Ceranisus menes* Walker attacks that species in Israel (Rubin and Kuslitzky, 1992).

MONITORING AND CONTROL Thrips populations can be monitored by regular inspection of flower panicles and control decisions made, based on predetermined economic injury levels if they have been set (Peña, 1993). Monitoring for the presence of the chilli thrips in Thailand is carried out by shaking shoots, flush leaves or inflorescences over a container, and counting the number of insects dislodged. Sampling units are selected from all four sides of the tree. When five thrips per sample are found, a spray of carbaryl is recommended. A predatory mite, *Amblyseius* sp., and a *Stethorus* sp. have been found in association with the thrips and are thought to prey on them (S. Krairiksh, Thailand, 1999, personal communication).

Verghese *et al.* (1988) studied the spacial distribution of the pests on the tree, and considered that a sample of 55 panicles from the lower canopy provides a good estimate of the thrips population present on the whole tree. However, they did not recommend at what infestation level controls should be applied. Peña (1993) quoted a formula for calculating the number of thrips in a panicle, based on the number caught on sticky traps. In the absence of reliable estimates of the true population size, the trap method has not been shown to give an unbiased estimate of numbers.

Pests of Fruit

Mango fruit are a very valuable commodity in most areas of production, whether they are consumed locally, sent to central markets in large cities, or exported. There are numerous alternative uses for inferior fruit such as juice, purée and dried mango, but the price paid for good quality fruit without imperfections of any kind, especially insect-induced, usually provides the farmer with a better return. For this reason, the fruit needs to be protected from a range of insects that may cause physical damage or loss, or merely affect its outward appearance.

Fruit flies (Diptera; Tephritidae)

Flies of the family Tephritidae, commonly known as fruit flies, are pests of mangoes in most parts of the world (Hill, 1975; Veeresh, 1988; Bondad, 1989; Cunningham, 1989; Aluja, 1994). The female flies oviposit in maturing fruit and the larvae burrow through the flesh, feeding on it and initiating rots. Fruit may fall prematurely or, if eggs are laid just before the fruit is harvested, spoilage in the package will occur as the fruit ripens and the eggs hatch. Ideally, adult flies should be prevented from laying in the fruit. The biology and behaviour of fruit flies has been studied in detail by Christenson and Foote (1960), Bateman (1972), and Aluja *et al.* (1997).

DISTRIBUTION AND ECOLOGY *Anastrepha* spp. range from the southern USA through Central America and the Caribbean, to Argentina in South America (Aluja, 1994). White and Elson-Harris (1992) reported that eight species of *Anastrepha* have been found in association with mangoes. *Anastrepha obliqua* (Macquart) is a pest of mango and other soft fruits in most Caribbean countries except Grenada, St Vincent and the Grenadines. The Mexican fruit fly *Anastrepha ludens* (Loew) is a serious pest in Belize, *Anastrepha serpentina* (Wiedemann) and *Anastrepha striata* Schiner occur in Guyana and Trinidad and Tobago, while *A. suspensa* (Loew) and *Anastrepha ocresia* Walker infest fruit in Jamaica (Rhodes, 1992). Segarra (1988) determined that only *A. obliqua* attacked mangoes in Puerto Rico, despite a long-held belief that the only other fruit fly species present in the country, *A. suspensa*, was also a problem. *A. ludens* is the major pest of mangoes grown at high altitudes in Mexico, while *A. obliqua* dominates at lower altitudes (Aluja *et al.*, 1996). *A. obliqua* has been shown to be the major fruit fly pest in Costa Rica and Guatemala (Jirón, 1996) and is responsible for up to 94% of damage attributed to fruit flies (Jirón and Hedström, 1988; Jirón *et al.*, 1988). In Peru, the important species attacking mango are *Anastrepha distincta* Greene, *Anastrepha fraterculus* Wiedemann, *A. serpentina*, *A. striata* and *Anastrepha chiclayae* Greene (Morin, 1967; Tijero, 1992).

Bactrocera tryoni (Froggatt), *Bactrocera neohumeralis* (Hardy), *Bactrocera jarvisi* (Tryon), *Bactrocera zonata* (Saunders), *Bactrocera frauenfeldi* Schiner and *Bactrocera dorsalis* (Hendel) are all reported to attack mango (Umeya and Hirao, 1975; Cunningham, 1989). The name oriental fruit fly, *B. dorsalis*, had for many years unwittingly been given to a complex of flies attacking a variety of fruits throughout South-East Asia. Drew and Hancock (1994) separated the complex into 52 separate species. Many of these were described from specimens caught in pheromone traps and have no host records associated with them, but *B. dorsalis*, *Bactrocera carambolae* Drew and Hancock, *Bactrocera occipitalis* (Bezzi), *Bactrocera papayae* Drew and Hancock and *Bactrocera philippinensis* Drew and Hancock, have all been reared from mango fruit (Plate 23).

The Mediterranean fruit fly is the most widespread of a number of *Ceratitis* spp. reported to attack mangoes throughout the world. *C. capitata* occurs in Israel, Spain, Mexico, Réunion, Mexico, South America and South Africa (Morin, 1967; Galan-Sauco, 1990; Vayssieres, 1997; De Villiers and Steyn, 1998). *Ceratitis cosyra* (Walker), the marula fruit fly, and *Ceratitis rosa* Karsch, the Natal fruit fly, also attack mangoes in South Africa (De Villiers and Steyn, 1998) while *C. rosa* also occurs in Réunion (Vayssieres, 1997), along with *Ceratitis catoirii* Guer (Etienne, 1968). *C. cosyra* is by far the most important fruit fly pest of mangoes in South Africa, accounting for about 99% of individuals emerging from infested fruit in a survey conducted by Labuschagne *et al.* (1996a).

The life cycle of all *Anastrepha* spp. studied is basically the same. Eggs are laid singly or in groups, mostly into the pulp of the developing fruit. The larvae develop through three instars, after which they leave the fruit to pupate in the soil. The rate of development is influenced by temperature and the food medium. Adult flies spend much time exploring the surface of the leaves of plants, even those which are not reproductive hosts. In doing this they appear to obtain food in the form of yeasts and bacteria (Christenson and Foote, 1960). Leyva *et al.* (1991) found that the eggs of *A. ludens* took 4.6 days to hatch in mango, but only 3.8 days in grapefruit. For the larval stage, this was reversed, with development taking 27 days in citrus, but only 23 days in mango.

Bactrocera spp. have a similar biology and behave in a similar way to *Anastrepha* spp. Warm, humid weather favours fruit fly activity and in Queensland, Australia, populations increase rapidly as the summer rains increase and the mango maturation season progresses (Cunningham, 1989).

DAMAGE Female fruit flies often probe immature fruit with their ovipositor, causing blind stings in which no eggs are laid. These stings may lead to later infection by disease organisms, but more often the sap which exudes from the puncture is of prime concern, since it burns the skin of the fruit (Wysoki *et al.*, 1993). Eggs laid closer to maturity hatch, and the larvae proceed to burrow through the flesh, making it inedible, and eventually causing the fruit to rot and fall.

BIOLOGICAL CONTROL Numerous parasitoids are known to attack fruit flies in all parts of the world, and while they undoubtedly have some effect on fly populations, they are unable to reduce numbers to levels that prevent economic damage. Syed *et al.* (1970) indicated that *Biosteres longicaudatus* (Ashmead), *Dhirinus giffardii* Silvestri and *Spalangia grotiusi* Girault parasitize *B. dorsalis*, while *B. longicaudatus*, *D. giffardii* and *Bracon* sp. attack *B. zonata* in Pakistan.

Opius and *Biosteres* spp., notably *Opius humilis* Silvestri, *Opius fullawayi* (Silvestri), *Opius kraussi* Fullaway, *Opius incisi* Silvestri, *Opius tryoni* (Cameron), *Opius bellus* Gahan, *Biosteres vandenboschi* Fullaway, *Biosteres oophilus* (Fullaway), *B. longicaudatus* and *Biosteres tryoni* (Couron), are recorded as parasitoids of *C. capitata* (Beardsley, 1961; Bess *et al.*, 1961; Wharton and Marsh, 1978).

Diachasmimorpha longicaudata, *Ganaspis pelleranoi*, *B. vandenboschi*, *Doryctobracon crawfordi* and *Aceratoneuromya indica* have been released for the biocontol of *Anastrepha ludens*, *A. suspensa* and *A. fraterculus* in Mexico, Brazil, Costa Rica, Peru and USA (Peña and Mohyuddin, 1997).

MONITORING AND CONTROL Fruit fly numbers are easily monitored with pheromone traps, although trap catches have not been related to the risk to the crop. In Australia, cuelure is used in Dakpot™ traps which are monitored weekly to detect sustained increases in fly numbers, rather than threshold catches (Cunningham, 1989). Sprays of dimethoate or fenthion applied every 2 weeks, or bait sprays (protein autolysate plus chlorpyrifos) applied at least weekly, are recommended. These should commence 6 weeks before the anticipated harvest date, or as trapping indicates.

In South Africa, the use of traps for monitoring fly numbers is recommended, with bait sprays being applied accordingly. Protein hydrolysate combined with the insecticides mercaptothion or trichlorfon is used as the bait, and the mixture is applied to small areas of the canopy in volumes of 250–1000 ml per tree. The elimination of potential breeding hosts of the flies such as wild tobacco, *Solanum mauritianum* Scop., wild guavas and bramble berries, as well as orchard sanitation, are also measures that minimize the fruit fly risk (De Villiers and Steyn, 1998).

McPhail traps are used to monitor *Anastrepha* spp. numbers in Peru, and control measures are applied when individual trap catches average two flies per week (Herrera and Viñas, 1977). Bait sprays are applied preferably from the ground, so that they can be restricted to small areas of individual trees, thus minimizing their impact on natural enemies. If applied from the air, the risk of upsetting natural controls is increased because of the reduced control over where the chemical lands, and other pest problems may flare (Soto-Manatiu *et al.*, 1987). Aluja *et al.* (1996) found that more flies were caught in McPhail traps baited with hydrolysed protein, placed around the edges of orchards than within orchards, and large trap catches were not correlated with high infestation levels in the crop.

In other countries, methyl eugenol has been used to control fruit flies and even eradicate them. *B. dorsalis* was reported to have been eradicated from Oahu (Steiner and Lee, 1955), Rota Island (Steiner *et al.*, 1965) and Okinawa, Kume, Miyako and Uaekama Islands (Iwahashi, 1984). Mass trapping of male flies with methyl eugenol reduced infestations to subeconomic levels in Pakistan (Mohyuddin and Mahmood, 1993). Balock and Lopez (1969) report the trapping of *A. ludens* in McPhail traps using a pelletized mixture of cottonseed hydrolysate and borax dissolved in water as the bait. Infestations in mangoes were reduced by 98% when one to five traps per tree were deployed, depending on tree size.

Jirón (1996) detailed a management strategy for fruit flies infesting mangoes in Central America. This included cultural techniques such as attempting to induce mangoes to flower and fruit during the dry season when fruit fly numbers were minimal, planting of cultivars with synchronous flowering, 'live fence' management (avoiding known fruit fly hosts in the species composition of bordering vegetation), parasitoid introductions, monitoring fly numbers through trapping, and appropriate insecticide applications.

The cv. 'Carabao' in the Philippines can be induced to flower by the application of potassium nitrate sprays. This ensures that a uniform flowering will occur in individual orchards, but not all farmers in an area may decide to induce their trees at the same time. Nevertheless, there is the possibility of reducing exposure to fruit fly and mango hopper infestation through the manipulation of flowering to produce the major crop during the dry season. In addition, a common practice in the Philippines is to enclose the developing fruit in newspaper envelopes, which are applied by hand when the fruit is about 80 mm long. The envelopes are secured with a skewer of bamboo. Reasonable protection from fruit flies is provided, but as the fruit expands the end of the envelope often opens to allow entry of the flies (G.K. Waite, 1998, unpublished). While Manoto *et al.* (1984) considered this approach to be the most effective way of preventing fruit fly damage, Bondad (1985) showed that even with the most durable paper, only 46% of the fruit was protected. Trapping, fruit wrapping and the application of bait sprays are all recommended procedures in Thailand (S. Krairiksh, Thailand, 1999, personal communication).

In the search for fruit fly controls that might be less damaging to the orchard ecosystem, Robacker *et al.* (1996) examined the effect of 55 isolates of *Bacillus thuringiensis* (Berliner) on adults and larvae of *A. ludens*. While flies which fed on agar pellets carrying the bacterium died, aqueous formulations, which are necessary for application to trees in the field, did not kill flies.

In areas that produce a variety of fruit crops with differing maturity times, the fruit fly problem in the late maturing crops will generally be worse than for the early maturing fruit. In Israel, mango orchards adjacent to citrus groves are often infested by flies which have bred in the citrus fruits (Wysoki *et al.*, 1993). The application of pesticides to mangoes to control fruit flies is effective in Pakistan, but it has sometimes led to the development of damaging scale infestations due to the disruption of natural enemies (Mohyuddin and Mahmood, 1993).

Cultivars which are resistant to fruit fly attack have been identified in some areas. Carvalho *et al.* (1996) found that cv. 'Espada' was unaffected by *A. obliqua*, but cv. 'Carlota' was extremely susceptible. Not only did the flies not attack the fruits extensively, but adult flies which did develop from larvae feeding on the resistant cultivar lived for a shorter time.

Seed weevils (Coleoptera: Curculionidae)

The mango seed weevil, *Sternochetus mangiferae* (F.) (Coleoptera: Curculionidae), is widely distributed in Africa, Asia, Australia, the Pacific Islands (Cunningham, 1989) and the Caribbean (McComie and De Chi, 1994). *Sternochetus gravis* (F.) and *Sternochetus frigidus* (F.) occur in India (Dey and Pande, 1987) and Bangladesh (Alam, 1972). *S. frigidus* was first detected in the Philippines in 1987 (Basio *et al.*, 1994) and has caused extensive damage to mangoes on the island of Palawan to which it is presently restricted (Anonymous, 1994). These weevils are important pests of mangoes partly because they infest the seeds and may be present in otherwise sound fruit, which makes them a quarantine concern (Yee, 1958; Dey and Pande, 1987; Cunningham, 1989) (Plate 24). They may also be a nuisance to nursery staff wishing to propagate trees from infested seeds. In India, the seeds are used as a source of starch and for animal and poultry food, oils, seed meal and organic manure, so that weevil infestation also affects production of these products (Bagle and Prasad, 1984). Follett and Gabbard (2000) found that mango seeds can withstand substantial damage by seed weevils and still germinate successfully, especially the polyembryonic cultivars. Germination of damaged seeds in the single-seeded cultivar 'Haden' still exceeded 70%, suggesting that concerns with respect to a deleterious impact on nursery industries are exaggerated. Follett and Gabbard (2000) also conclude that mango seed weevils do not seriously affect mango yields or marketability. Nevertheless, it is still regarded as a significant quarantine threat and unsuccessful attempts have been made to kill it using heat, cold and fumigation treatments (Balock and Kozuma, 1964). Irradiation may be a useful alternative disinfestation treatment (Heather and Corcoran, 1992).

Females of *S. mangiferae* generally lay eggs in small green fruit, and the larvae tunnel to the seed, where they feed and develop. Because the fruit are small and green when the egg hatches, the tunnel made by the small larva disappears as the fruit expands and ripens. Most infested mangoes are consumed in ignorance of the presence of the insect in the seed. Shukla and Tandon (1985) found that in India during March–April, eggs took 6–7 days to hatch and the larva passed through five instars and a pupal stage over a period of 43.7 days. Female weevils lived for about 302 days and males for 267 days. Only one generation occurred per year. Adult weevils emerged in late May–early June and remained in a facultative diapause until the following season. In contrast to *S. mangiferae*, the larvae of *S. gravis* and *S. frigidus* live and feed in the pulp as the fruit grows, and the former species has been known to infest up to 100% of fruit in some orchards (Dey and Pande, 1987). Females of *S. frigidus* lay eggs on mature green fruit and cover them with a brown exudate. They then chew a notch in the fruit, causing the sap to flow and cover the eggs with an opaque white film. A female lays about 15 eggs day^{-1} and perhaps 175–300 eggs in a lifetime. The period from egg to adult occupies 40–50 days (Anonymous, 1994).

No natural enemies of *S. mangiferae* were recorded in India, and all commercial cultivars were susceptible to infestation (Bagle and Prasad, 1984; Shukla and Tandon, 1985). Field sanitation has been recommended as a control measure (Van Dine, 1906), but Hansen and Armstrong (1990) found that it failed to reduce the incidence of infestation because the weevils are apparently able to fly further than had been thought. The best insecticidal treatment was deltamethrin applied twice in 3 weeks, commencing in early March. In South Africa, the organophosphate fenthion, the pyrethroids deltamethrin, fenvalerate and esfenvalerate, along with the insect growth regulator triflumuron, are registered for weevil control. Recent trials have found that endosulfan also gives good control (Joubert, 1997). Dispersal of the weevils may be by flying to trees or crawling up trunks from overwintering sites, but humans play a major role in the movement of the pest in infested fruit over longer distances (Hansen, 1993). Newly matured weevils may remain in the seed for long periods before they emerge, and because of this, movement into new areas is facilitated in exported fruit. Quarantine procedures, which prohibit the movement of mango plants and fruit from the island of Palawan to other parts of the Philippines, have been enforced to protect the valuable industry in the rest of the country from *S. frigidus*, so that exports to foreign countries can be maintained (Anonymous, 1994).

Mango seed borer (Lepidoptera: Pyralidae)

DISTRIBUTION AND BIOLOGY *Deanolis sublimbalis* Snellen, the mango seed borer, which was commonly referred to as *Noorda albizonalis* Hampson in much of the literature (Waterhouse, 1998), is an important pest of mangoes in the Philippines (Anonymous, 1994), Vietnam (Nguyen *et al.*, 1998a; Van Mele *et al.*, 2001), China (Li *et al.*, 1997), Thailand, Indonesia and Papua New Guinea (Cunningham, 1984). The oval white eggs are laid in groups at the fruit apex and take 3–4 days to hatch. The larvae develop through five instars in 14–20 days and they pupate in cocoons in the soil. The period from egg to adult takes from 28 to 40 days. The insect apparently prefers mango, but other species of *Mangifera* have been recorded as hosts (Waterhouse, 1998).

DAMAGE The distinctive red-banded larvae feed on and bore through the mango pulp to the seed, which is consumed. Up to 11 larvae have been recorded in a single fruit, but usually there is only one. Infested fruit split and rot, and fall to the ground (Anonymous, 1994). In Guimaras, Philippines, Golez (1991) recorded 12.5% fruit infestation and in serious outbreak years, 40–50% yield reductions are possible. Waterhouse (1998) considered that, since *D. sublimbalis* is capable of causing such levels of damage, it may be a more important pest of mangoes than has generally been realized. Among several suggestions for this status are that it may have been overlooked as a pest or that it has only recently spread to new areas and so has only just become evident as a pest there. Van Mele *et al.* (2001) suggested that damage caused by *D. sublimbalis* in the Mekong Delta has been wrongly attributed to fruit flies.

BIOLOGICAL CONTROL In the Guimaras Islands of the Philippines, the vespid wasp, *Rychium attrisimum*, preys on the larvae as they leave the fruit to pupate. Larvae are used to stock the wasps' nests as food for their young. The egg parasitoids *Trichogramma chilonis* Ishii and *Trichgramma chilotreae* have been recorded attacking the pest in Luzon (Golez, 1991).

MONITORING AND CONTROL Mango fruit become susceptible to the seed borer at about 60 days post-induction, and insecticide applications should commence then. Further sprays at 75, 90 and 105 days post-induction are required to fully protect the fruit. The most effective chemicals were found to be deltamethrin and cyfluthrin (Golez, 1991). Other recommendations are to remove infested fruit from the tree before the larvae can leave them to attack neighbouring fruit, to wrap the fruit in protective bags at 55–65 days after induction, and to destroy fallen fruit (Anonymous, 1994).

Fruitspotting bugs (Hemiptera: Coreidae)

The yellowish green coreid bugs, *Amblypelta lutescens lutescens* (Distant) and *Amblypelta nitida* Stål occur along the coast of Queensland, and attack most of the tropical and subtropical fruit crops grown there (Waite and Huwer, 1998). They prefer to feed on young, green fruit, but *A. l. lutescens* also damages the terminals of a number of hosts. In tropical north Queensland, *A. l. lutescens* is the dominant species and feeds on the young fruit, causing black lesions to develop and the fruit to fall. It also feeds on the terminals and leaf petioles, causing wilting and dieback (Cunningham, 1989). In the subtropical south, both species attack mango, but *A. nitida* confines its attention to the fruit, while

A. l. lutescens attacks fruit as well as terminal growth (G.K. Waite, 1995, unpublished). The bugs breed in natural rainforest areas, and fly into the orchards to feed on the fruit and terminals. Female bugs lay individual, opalescent green eggs under leaves. There are five nymphal instars and a generation takes about 40 days.

The main predators of fruitspotting bugs are spiders, particularly members of the family Thomisidae. Several species of egg parasitoids have been recorded. In north Queensland, *Ooencyrtus* sp. (Encyrtidae), *Anastatus* sp. (Eupelmidae) and *Gryon* sp. (Scelionidae) parasitized 37.5–91.6% of the eggs of *A. l. lutescens* (Fay and Huwer, 1994). In south Queensland, *Anastatus* sp. and *Gryon meridianum* (Dodd) have been found to parasitize eggs of *A. nitida* and *A. l. lutescens* to a similar degree (Waite and Petzl, 1994).

Because fruitspotting bugs continuously migrate into orchards, more than one insecticide spray may be required to protect the young fruit. However, the fruit are safe from attack once they have grown to a length of about 50 mm, and two or three sprays of endosulfan at intervals of 2 weeks are generally sufficient to protect them from the bugs.

The coconut bug, *Pseudotheraptus wayi* Brown, was first recorded on mangoes in South Africa in 1977, and now also attacks guavas, pecans, macadamias, avocados and loquats. It causes damage similar to that of *Amblypelta* spp. (De Villiers, 1990).

Helopeltis sp. (Miridae) are minor pests of mango fruit in the Philippines and in northern Australia, where they feed on the fruit and cause superficial corky blemishes that detract from the fruit's appearance. Cashew and cacao are alternative hosts. Insecticides can be used to control them but in the Philippines bagging, which is carried out for protection from fruit fly, is also effective (Anonymous, 1994).

Tip wilters (Hemiptera: Coreidae)

The tip wilter, *Anoplocnemis curvipes* (Fabricius), can be a serious pest of young mango trees in South Africa. The bugs feed on the new flush, leaf veins and flower stalks, causing the tissue to die where they have fed. They breed on many alternative hosts such as weeds, vegetables, ornamentals, granadillas and citrus, as well as mangoes. The eggs are laid in rows on the leaves, and adults and nymphs numbering ten or more may kill all of the new growth on a small mango tree, leading to retardation of growth. As no chemicals are registered for its control, the insects must be collected by hand and destroyed (De Villiers, 1998).

Thrips (Thysanoptera: Thripidae)

The South African citrus thrips, *Scirtothrips aurantii* (Sign.), has become an increasingly damaging pest in recent years. Nymphs and adults feed on small green fruit and cause a blemish on the skin. The damage is only cosmetic, but it is sufficient to downgrade fruit and make it unsuitable for export. The lesions vary from silvering to skin cracking, and if high populations are present when the fruit is very small, the whole surface may be scarred. The thrips can be monitored with yellow sticky traps. Methamidophos is registered for control of the pest and is used as an undiluted trunk treatment, applied with a paint brush (Grové, 1998a).

Fruit piercing moths (Lepidoptera: Noctuidae)

The noctuid moths *Eudocima* (*Othreis*) *fullonia* (Clerck), *Eudocima materna* (Linnaeus) and *Eudocima salaminia* (Cramer) are common pests of ripening fruit in Queensland. Litchis, carambolas and citrus are particularly susceptible (Fay and Halfpapp, 1999), but mangoes may also be attacked. The moths possess a barbed proboscis with which they bore into the fruit and suck the juice. No satisfactory control method has been found apart from totally netting orchards to exclude them.

Pests of Leaves and Buds

Gall midges (Diptera: Cecidomyiidae)

Gall midges infesting the leaves of mango have been recorded from the Caribbean, Brazil, India, South Africa (*Procontarinia matteiana* Kieffer and Cecconi), China (*Erosomyia* spp.) and Guam (*Procontarinia schreineri* Harris). Larvae of the latter species form blister galls on the young leaves. When the larvae mature and leave the galls, the tissue is invaded by the anthracnose fungus, *Colletotrichum gloeosporioides* Penz. The galls dry up and fall out, leaving a typical 'shot-hole' effect that was originally thought to be caused solely by the fungus. Similar symptoms have been noted in relation to infestation by this species on mangoes in Saipan (Marianas Islands), Belau and Yap (Caroline Islands), and in the Philippines (Harris and Schreiner, 1992). Heaviest infestations of the midge were recorded during the wet season, possibly because of the increased survival of the immature stages encouraged by the high humidity.

Amradiplosis echinogalliperada Mabi and *P. matteiana* have been reported from Uttar Pradesh (Mani, 1943) and the Punjab (Rao, 1956). *P. matteiana* is considered to be indigenous to India where Jhala *et al.* (1987) studied the susceptibility to it of 17 mango cultivars in south Gujarat. The cultivars 'Alphonso' and 'Kesar' were most heavily infested, while 'Deshi Malgoba' and 'Benisan' were least affected. Phosphamidon and monocrotophos were found to give the best chemical control of the pest (Jhala *et al.*, 1990). In Africa, the cultivar 'Heidi' is very susceptible to *P. matteiana* while 'Sensation' is resistant (Githure *et al.*, 1997). The eulophid, *Chrysonotomyia pulcherrima*, parasitizes the larvae heavily, and De Villiers *et al.* (1987) considered that the midge was never a serious pest. However, Grové (1998b) noted that in some areas, parasitism was less effective and the midge can cause severe damage. Apart from India and South Africa, *P. matteiana* now occurs in Mauritius, Kenya, Réunion, Oman and Malaysia, having spread on imported mango plants (Githure *et al.*, 1997).

Thrips (Thysanoptera: Thripidae)

The red-banded thrips, *Selenothrips rubrocinctus* (Giard), has been recorded as a pest of mango leaves from the Caribbean, Hawaii, Australia, South Africa, Brazil and Florida (Plate 25). The thrips prefer to feed on the leaf, especially adjacent to the midrib, where they cause a silvering that develops into necrosis, leading to eventual leaf drop. When infestations are severe, fruit may also be attacked and the silvering turns yellow to brown and is speckled with dark, dried faeces (Cunningham, 1989). Eggs are inserted into the leaf tissue and covered with a drop of fluid that dries to form a black, disc-like cover. The adult thrips are almost black, with a red band on the first abdominal segment. The nymphs are pale orange with the first two abdominal segments and the anal segment coloured bright red (Moznette, 1922). Yee (1958) recommended malathion as a control when necessary in Hawaii, while in Australia, Cunningham (1989) recommended endosulfan. Bartlett (1938) stated that the hymenopterous parasitoid, *Goethana parvipennis* Gahan, attacks the thrips in Puerto Rico.

The Mediterranean mango thrips, *Scirtothrips mangiferae* Priesner, is recognized as a major pest of mangoes in Israel. The thrips feed on the young shoots and leaves, causing the shoots to become stunted and the leaves to curl and drop (Wysoki *et al.*, 1993). Low overwintering populations form the basis for an increase in spring numbers that peak in summer. The pest's abundance can be monitored using yellow sticky traps. Good control has been achieved with the application of fluvalinate or acephate when the economic threshold of ten thrips per shoot had been reached (Ganz *et al.*, 1990).

Bud mites (Acari: Eriophyidae)

The mango bud mite, *Eriophyes mangiferae* (Sayed) is reported to attack the buds of terminals. The normal symptom is a proliferation of shoots on the terminal, giving rise to a witches' broom effect. However, if the fungus *Fusarium* sp. is present the tree develops floral and foliar galls because the fungus causes necrosis of the hypertrophia that would normally result in witches' brooming (Ochoa *et al.*, 1994).

In Florida, *E. mangiferae* has been found in association with malformed mango flowers (Peña, 1993) and there has been conjecture that it may be implicated in the vectoring of diseases that could be the real cause of the malformation. There has been difficulty proving this hypothesis, and some consider that tissue damage caused by the mite allows infection by *Fusarium* sp. (Denmark, 1983). The life history of the mite has been described by Abou-Awad (1981). The life cycle is completed in 15 days at 25–27°C, and high summer temperatures appear to have an adverse effect on the mite. The mango bud mite occurs in Australia but has never been a problem except occasionally in backyard trees (Cunningham, 1989). Orozco Santos and Núñez (1988) reported that *E. mangiferae* had been found in Mexico, and that it was associated with a disorder termed 'achaparramiento', the witches' broom effect described above. The mite could be controlled with sulphur or dimethoate. Doreste (1981) reports that in Venezuela *E. mangiferae* is increasing in distribution and intensity.

E. mangiferae is considered to be a major pest in the Punjab, Delhi and Uttar Pradesh in India, where it has threatened the very existence of the mango industry (Singh and Mukherjee, 1989). Agrawal and Singh (1988) reported that it becomes more abundant in April and gradually reaches a peak in June.

Control of the mango bud mite has been achieved in Egypt with four sprays of dichlorvos (Osman, 1979). As with many eriophyid pests, e.g. litchi erinose mite, infestations should be anticipated through monitoring and a knowledge of the phenology and flushing behaviour of the trees, so that sprays can be applied to apparently uninfested buds (Rai *et al.*, 1966). The phytoseiid, *Amblyseius swirskii* Athias-Henriot, has been found in association with *E. mangiferae* (Abou-Awad, 1981).

Spider mites (Acari: Tetranychidae)

A number of mites belonging to the genus *Oligonychus* feed on the upper surface of mango leaves. The mango mite, *Oligonychus mangiferae* Rahman and Sapra, is a common pest in India, Egypt, Mauritius, Peru, Israel and some parts of Asia. Its biology has been studied by Rai *et al.* (1988). The tea red spider mite, *Oligonychus coffeae* (Nietner), is an important pest of avocados and a sporadic pest of mangoes in Australia, where high populations may cause leaf bronzing if excessive pesticide use eliminates its main predator, the coccinellid, *Stethorus* sp. (Cunningham, 1989). A similar situation occurs in Central America and the USA where the avocado brown mite, *Oligonychus punicae* (Hirst), and the avocado red mite, *Oligonychus yothersi* McGregor, cause leaf bronzing and, in severe cases, necrosis and leaf fall (Andrews and Poe, 1980). It is considered there that the problem is exacerbated by sprays directed at other pests. *O. punicae* also attacks mango in Puerto Rico (Comroy, 1958).

Other mites (Acari: Tarsonemidae)

Broad mite, *Polyphagotarsonemus latus* (Banks), may occasionally infest trees in the field but is more often associated with nursery trees (Wolfenbarger, 1956). Symptoms of broad mite attack include stunting and crinkling of new leaves and rolling of leaf margins. The mites can be controlled with applications of dicofol or sulphur (Yee, 1987). (See also Chapter 3.)

Mango scale (Homoptera: Diaspididae)

Scale insects, especially diaspidid scales, are among the most important pests of mangoes in some parts of the world. In Australia and especially in South Africa, the mango scale, *Aulacaspis tubercularis* (Newstead), is regarded as a key pest mostly because when it infests the fruit, even though the scale can be brushed off, it may leave blemishes on the fruit (Plate 26). This results in their rejection for the lucrative export market (Labuschagne *et al.*, 1995; Joubert, 1997). As well as infesting fruit, *A. tubercularis* infests mango leaves, causing discoloured and necrotic areas which, in severe cases, may result in leaf fall. In Queensland, the diaspidid *Phenacaspis dilitata* (Green) produces similar symptoms (Cunningham, 1989). The biology of *A. tubercularis* has been described by Van Halteren (1970). In north Queensland, breeding occurs throughout the year. Females lay about 50 eggs under the protective scale covering. On hatching, female crawlers become sedentary and secrete a circular scale cover, while the male crawlers congregate, and each secretes a fine white filament that curls over the body. Older instars secrete a white rectangular cover that has two distinct grooves. The adult male is winged and is capable of flight. Labuschagne *et al.* (1995) studied the population dynamics of the scale in South Africa and found that it was more abundant on the southern shaded sector of the mango tree where temperatures are more moderate.

Sprays of methidathion, chlorpyrifos and Lo-Vis™ oil have been recommended for the control of *A. tubercularis* in Queensland. The insect growth regulator, buprofezin, has given better control in field trials (De Faveri and Brown, 1995). In South Africa, Labuschagne *et al.* (1995) recorded the parasitoid, *Encarsia citrina* (Crawford) and the predators *Rhizobius lophanthae* (Blaisdell), *Chilocorus nigritus* (Fabricius) and *Aleurodothrips fasciapennis* (Franklin) attacking the scale. An aphelinid parasitoid, *Aphytis* sp., and the predatory nitidulid beetle, *Cybocephalus binotatus* Grouvelle, were introduced into South Africa from Thailand in 1995. They have been released in the field and appear to have established (Labuschagne *et al.*, 1996b; Daneel and Dreyer, 1997).

Soft scales (Homoptera: Coccidae)

Swirski *et al.* (1997) noted that 63 species of soft scales have been recorded infesting mango, but of these only about six are considered to be economically important. Heavy infestations of *Ceroplastes pseudoceriferus* Green in Bangladesh cause wilting of leaves, malformation of flowers and general malaise of the tree, so that flowering is reduced (Ali, 1978). It is also a serious pest in Taiwan where it undergoes three generations a year (Wen and Lee, 1986). A number of parasitoids control the pest in India (Sankaran, 1955), Bangladesh (Ali, 1978) and Taiwan (Wen and Lee, 1986).

Milviscutulus mangiferae (Green), mango shield scale, is reported to be a pest of the crop in a number of countries. Many detrimental effects are attributed to its feeding, including leaf fall, failure of buds to open, reduced tree vigour, fruit drop and even death of trees. Outbreaks of the pest in Israel have been attributed to interference with biological control agents by exhaust fumes emitted by vehicles using nearby highways (Gerson, 1975; Swirski *et al.*, 1997). Many natural enemies have been reported to attack *M. mangiferae*. In Israel, *Microterys flavus* (Howard) and *Coccophagus eritreaensis* Compere are the most common parasitoids (Avidov and Zaitzov, 1960) while in South Africa, *Coccophagus pulvinariae* Compere, *Tetrastichus* sp. and *Marietta javensis* (Howard) have been recorded (Kamburov, 1987). Oil and malathion have been commonly recommended as chemical control measures (Ebeling, 1959; Avidov and Zaitzov, 1960).

Pulvinaria (*Cloropulvinaria*) *psidii* (Maskell) is rated as a serious pest in India (Gopalakrishnan and Narayanan, 1989), but it is apparently only a minor pest in Pakistan where it is kept under control by natural enemies (Mahmood and Mohyuddin, 1986). If these are disrupted by chemical sprays, then problems may be encountered. *C. psidii* can be a serious pest of litchis in Australia (Waite and Elder, 1996) but does not affect mangoes there.

Pink wax scale, *Ceroplastes rubens* Maskell, is a common pest of mangoes in Australia, where severe infestations may develop as a result of natural enemy disruption through spraying (Plate 27). Heavy films of sooty mould cover the leaves and may also contaminate the fruit. The introduced parasitoid, *Anicetus beneficus* Ishii, can provide good control under unsprayed conditions, but corrective applications of oil may be required in spring or late summer to coincide with the presence of young scales (Cunningham, 1989). This pest may also become a problem under similar circumstances in Florida, USA (Peña, 1993).

Mealybugs (Homoptera: Pseudococcidae)

DISTRIBUTION AND BIOLOGY *Rastrococcus invadens* Williams is a polyphagous mealybug which, in addition to mango, attacks a range of hosts including citrus, breadfruit, banana, frangipani and *Ficus* spp. (Agounké *et al.*, 1988). The mealybug has been noted as a serious pest of mango in Ghana since 1982, and it has since spread to other West African countries (Willink and Moore, 1988; Ivbijaro *et al.*, 1992). It has been the subject of intensive study with respect to its biology and biological control. The adult females are pale green-yellow and covered with white wax, which extends in long filaments at the anterior and posterior ends. The first instars are yellow, and they prefer to settle along the midrib of the leaf. Development takes 28–30 days, with females living for 80–90 days and producing 160–190 young. Males are winged and they outnumber females by about six to one (Willink and Moore, 1988).

Rastrococcus spinosus (Robinson) was reported to be a pest in southern Pakistan (Mahmood *et al.*, 1980) and in the Philippines (Morrill and Otanes, 1947), but more recent literature from the Philippines (Bondad, 1989; Anonymous, 1994) does not mention it as an economic pest.

DAMAGE The insect may infest leaves, flowers and fruits from which it sucks the sap, and when it is present in large numbers, heavy deposits of honeydew encourage the growth of a thick film of sooty mould on the foliage, which interferes with photosynthesis and reduces fruit yield. In Ghana, losses of more than 80% have been reported and farmers, desperate to rid themselves of the pest, have resorted to cutting mango trees down (Willink and Moore, 1988). Apart from the loss of production of an important fruit, the mealybug has affected the amenity aspect of community mango trees by spoiling their appearance and changing social attitudes towards them (Agounké *et al.*, 1988).

BIOLOGICAL CONTROL In Benin, the coccinellid *Chilocorus nigritis* (F.) was the most common predator of *R. invadens*. Several other predators were found but there was no native parasitoid which attacked the newcomer. Narasimham and Chacko (1988) noted that *R. invadens* was scarce in India. They determined that *Anagyrus* sp., later described as *Anagyrus mangicola* Noyes (Noyes, 1990), and an encyrtid, *Gyranusoidea tebygi* Noyes (Noyes, 1988), were worthy of further investigation for the purpose of biological control of *R. invadens* in West Africa. Detailed investigations, including host specificity studies, were carried out at the International Institute for Biological Control (IIBC) in the UK. These showed that *G. tebygi* was unlikely to compete with *Epidinocarsis lopezi* (De Santis), a parasitoid being used for the biocontrol of *Phenacoccus manihoti* Matile-Ferrero on cassava. It was introduced into Togo in 1987, and subsequently into Benin, Ghana, Nigeria, Gabon and Zaire, where it established easily and effected good control (Neuenschwander, 1989). The dilemma of whether to introduce *A. mangicola* into this successful biological control situation was solved when attempts to establish that parasitoid in the field failed (Moore and Cross, 1993).

MONITORING AND CONTROL The spatial distribution of the mealybug on mango trees was studied by Boavida *et al.* (1992) who developed binomial sampling plans for estimating the pest's population and the effect of natural enemies. However, they considered the plans were suitable only for estimating medium to high populations of mealybugs, and that for low populations, selective sampling of infested units would assist in determining what was happening with respect to natural enemy populations. The only practical solution to the mealybug problem seems to rest with biological control, which is making good progress. The application of chemical controls including neem extract and pirimiphos-ethyl has been to no avail, since the pest rapidly reinfests trees (Agounké *et al.*, 1988) and, of

them to see at first hand how ecosystems work, and what relationships exist between pests, natural enemies and the crop. In addition, they are able to inform the researchers in the early stages of development of particular management approaches, what is and what is not acceptable from the practical farming point of view. The 'farmer first' approach has been found to provide a good basis not only for conducting research, but also for having the findings of that research adopted (Chambers *et al.*, 1989). A whole crop approach to technology transfer in mangoes that includes information on pests has been adopted in Queensland. Crop-specific information kits that are part of the Agrilink Series address damage symptoms, the likely cause of the symptoms and the most appropriate control strategies (Meurant *et al.*, 1999).

An examination of the mango industries throughout the world suggests that in most there are key pests for which the only practical control is to apply insecticides. When this is the case, monitoring for the presence of pests assumes high importance. Strategies that rely on calendar sprays can be replaced with managed spraying programmes if monitoring and threshold levels are well developed. While there are very few effective pesticides available that are target-specific or generally 'soft' on beneficial species, it may be possible, through the adjustment of dosage rates and timing, to ameliorate the side effects of necessary sprays.

References

Abou-Awad, B. (1981) Ecological and biological studies on the mango bud mite, *Eriophyes mangiferae* (Sayed), with description of the immature stages (Eriphyoidea: Eriophyidae). *Acaralogia* 22, 145–150.

Agounké, D., Agricola, U. and Bokonon-Ganta, H.A. (1988) *Rastrococcus invadens* Williams (Hemiptera: Pseudococcidae), a serious exotic pest of fruit trees and other plants in West Africa. *Bulletin of Entomological Research* 78, 695–702.

Agrawal, L. and Singh, J. (1988) Present status of *Eriophyes mangiferae* in eastern Uttar Pradesh. In: ChannaBasavanna, G.P. and Virakta-math, C.A. (eds) *Progress in Acarology*, Vol. 2. Oxford-IBH, New Delhi, pp. 109–114.

Alam, M.Z. (1972) *Some Insect and Mite Pests of Fruit Trees in East Pakistan and Their Control* (revised edition). Bangladesh Agricultural Research Institute, Dacca, 120 pp.

Alam, S.N. (1994) Population dynamics, varietal reactions and microbial control of different species of mango leafhoppers. MSc thesis, University of the Philippines, Los Baños.

Ali, M. (1978) A report on the wax scales, *Ceroplastes ceriferus* Green and *Chloropulvinaria polygonata* (Ckll.) (Homoptera: Coccidae) on mango and their natural enemies. *Bangladesh Journal of Zoology* 6, 69–70.

Aluja, M. (1994) Bionomics and management of *Anastrepha*. *Annual Review of Entomology* 39, 155–178.

Aluja, M., Celedonio-Hurtado, H., Liedo, P., Cabrera, M., Castillo, F., Guillén, J. and Rios, E. (1996) Seasonal population fluctuations and ecological implications for management of *Anastrepha* fruit flies (Diptera: Tephritidae) in commercial mango orchards in southern Mexico. *Journal of Economic Entomology* 89, 667.

Aluja, M., Jiménez, A., Piñero, J., Camino, M., Aldana, L., Valdés, M.E., Castrejón, V., Jácome, I., Dávila, A.B. and Figueroa, R. (1997) Daily activity patterns and within-field distribution of papaya fruit flies (Diptera: Tephritidae) in Morelos and Veracruz, Mexico. *Journal of Economic Entomology* 90, 505–520.

Anantanarayanan, K.P. and Subramanian, T.R. (1955) A note on the unusual occurrence of the mango leaf-weevil (*Rhynchaenus mangiferae*, Mshll.) at Coimbatore. *Madras Agricultural Journal* 42, 373–376.

Anderson, D.L., Sedgeley, M., Short, J.R.T. and Allwood, A.J. (1982) Insect pollination in mango in northern Australia. *Australian Journal of Agricultural Research* 33, 541–548.

Andrews, K. and Poe, S. (1980) Spider mites of El Salvador, Central America (Acari: Tetranychidae). *Florida Entomologist* 63, 502–505.

Angeles, N., Dedordy, N., Paredes, P. and Requena, J. (1971) Citrus blackfly in Venezuela. *Agronomia Tropical* 21, 71–75.

Anonymous (1994) Major insect pests and diseases of mango and their control. In: *The Mango Technical Committee. The Philippines Recommends for Mango*. Los Baños, Laguna: PCARRD and PARRFI, 124 pp.

Avidov, Z. and Zaitzov, A. (1960) On the biology of the mango shield scale, *Coccus mangiferae* (Green), in Israel. *Ktvim* 10, 125–137.

Bagle, B.G. and Prasad, V.G. (1984) Varietal incidence and control of stone weevil *Sternochetus mangiferae* Fabricius. *Indian Journal of Entomology* 46, 389–392.

Balock, J.W. and Kozuma, T.T. (1964) Notes of the biology and economic importance of the mango weevil, *Sternochetus mangiferae* (Fabricius), in Hawaii. *Proceedings of the Hawaii Entomological Society* 3, 353–364.

Balock, J.W. and Lopez, D. (1969) Trapping for control of the Mexican fruit fly in mango and citrus groves. *Journal of Economic Entomology* 62, 53–56.

Barnes, H.F. (1948) *Gall Midges of Economic Importance*, Vol. 3, *Gall Midges of Fruit*. Crosby Lockwood and Sons, London, 184 pp.

Bartlett, K. (1938) The introduction and colonization in Puerto Rico of *Dasycapus parvipennis* Gahan, a parasite of thrips. *Agricultural Notes, Puerto Rico Experiment Station*. United States Department of Agriculture No. 87.

Basio, R.G., Johnson, P.J., Pua, D.R., Bergonia, H.T., Diloy, C.C. and Villegas, E.L. (1994) Mango pulp weevil (*Sternochetus frigidus* (Fabr.)) (Curculionidae: Coleoptera) found in Palawan. *Philippine Entomologist* 9, 350–351.

Bateman, M.A. (1972) The ecology of fruit flies. *Annual Review of Entomology* 17, 493–518.

Beardsley, J.W. (1961) A new review of the Hawaiian Braconidae (Hymenoptera). *Proceedings of the Hawaiian Entomological Society* 27, 333–366.

Bess, H.A., van den Bosch, R. and Haramoto, F.H. (1961) Fruit fly parasites and their activities in Hawaii. *Proceedings of the Hawaiian Entomological Society* 27, 367–378.

Bhatia, R., Gupta, D., Chandel, J.S. and Sharma, N.K. (1995) Relative abundance of insect visitors on flowers of major subtropical fruits in Himachal Pradesh and their effect on fruit set. *Indian Journal of Agricultural Sciences* 65, 907–912.

Boavida, C., Neuenschwander, P. and Schulthess, F. (1992) Spatial distribution of *Rastrococcus invadens* Williams (Hom., Pseudococcidae) in mango trees. *Journal of Applied Entomology* 114, 381–391.

Bondad, N.D. (1985) Mango bagging. *Animal Husbandry and Agricultural Journal* (June), 11–14.

Bondad, N.D. (1989) *The Mango (especially as observed in the Philippines)*. Rex Printing Company, Quezon City, Philippines, 402 pp.

Butani, D.K. (1979) *Insects and Fruits*. Periodical Expert Book Agency, Delhi.

Callan, E. McC. (1941) The gall midges (Diptera: Cecidomyidae) of economic importance in the West Indies. *Tropical Agriculture, Trinidad* 17, 117–127.

Carvalho, R. da S., Nascimento, A., Morgante, J. and Fonseca, N. (1996) Susceptibility of different mango varieties to the attack of the fruit fly, *Anastrepha obliqua*. In: McPheron, B. and Steck, G. (eds) *Fruit Fly Pests: a World Assessment of their Biology and Management*. St Luce Press, Delay Beach, pp. 325–331.

Cendana, S.M., Gabriel, B.P. and Magallona, E.D. (1983) Mango. In: *Insect Pests of Fruit Plants in the Philippines*. Department of Entomology, College of Agriculture, University of the Philippines, Los Baños, pp. 53–60.

Chambers, R., Pacey, A. and Thrupp, L.A. (eds) (1989) *Farmer First – Farmer Innovation and Agricultural Research*. Intermediate Technology Publications, London, 219 pp.

Chahal, B.S. and Singh, D. (1977) Bionomics and control of mango shoot borer, *Chlumetia transversa* Walk. (Noctuidae: Lepidoptera). *Indian Journal of Horticulture* 34, 188–193.

Christenson, L.D. and Foote, R.H. (1960) Biology of fruit flies. *Annual Review of Entomology* 5, 171–192.

Comroy, H.L. (1958) A preliminary survey of the plant mites of Puerto Rico. *Journal of the Agricultural University of Puerto Rico* 42, 39–144.

Cunningham, I. (1989) Pests. In: Bagshaw, J. (ed.) *Mango Pests and Disorders*. Department of Primary Industries Information Series, Q189007, pp. 10–21.

Dag, A. and Gazit, S. (1996) Mango pollinators in Israel. *Alon HaNotea* 50, 458–465.

Dakshinamurthy, A. (1984) The mango hopper and its control in *Tripura*. *Pesticides* 18, 13–14.

Daneel, M.S. and Dreyer, S. (1997) Further studies on the establishment of *Aphytis* sp. and *Cybosephalus binotatus* in mango orchards for the control of mango scale. *South African Mango Growers' Association Yearbook* 17, 144–146.

Datar, V.V. (1985) Synthetic pyrethroids in the control of mango hoppers (*Amritodius atkinsoni* Leth.) 1982. *Pesticides* 19, 13–14.

Davenport, T.L. and Núñez-Elisea, R. (1997) Reproductive physiology. In: Litz, R.E. (ed.) *The Mango – Botany, Production and Uses*. CAB International, Wallingford, UK, pp. 69–146.

De Faveri, S.G. and Brown, J.D. (1995) Effect of buprofezin, fluvalinate, Lo-Vis oil and imidacloprid against the mango scale *Aulacaspis tubercularis* (Newstead). In: Holmes, R. (ed.) *Mango 2000. Proceedings of Marketing and Production Workshop*. Townsville, Queensland, Australia, pp. 153–159.

Denmark, H.A. (1983) *Eriophyes mangiferae* (Sayed), a pest of mango (Acarina: Eriophyidae). Florida Department of Agriculture and

Conservation Services, Division of Plant Industry, *Entomology Circular* 254, 2 pp.

De Villiers, E.A. (1990) Coconut bugs on avocados. *Farming in South Africa*, Avocados H.5. 2 pp.

De Villiers, E.A. (1998) Pests – tip wilter. In: De Villiers, E.A. (ed.) *The Cultivation of Mangoes*. Institute for Tropical and Subtropical Crops, Nelspruit, South Africa, pp. 164–165.

De Villiers, E.A. and Steyn, W.P. (1998) Pests – fruit flies. In: De Villiers, E.A. (ed.) *The Cultivation of Mangoes*. Institute for Tropical and Subtropical Crops, Nelspruit, South Africa, pp. 135–141.

De Villiers, E.A., Fourie, P.F. and Petty, G.T. (1987) Miscellaneous pests. In: Myburgh, A.C. (ed.) *Crop Pests in Southern Africa. Bulletin 411, Citrus and Other Subtropicals*, p. 109.

Dey, K. and Pande, Y.D. (1987) Evaluation of certain non-insecticidal methods of reducing infestation of the mango nut weevil, *Sternochetus gravis* (F.) in India. *Tropical Pest Management* 33, 27–28.

Doreste, E. (1981) Informacion sobre el eriofido del mango, *Eriophyes mangiferae* (Sayed), en Venezuela. *Revista Facultad de Agronomia* 23, 91–100.

Drew, R.A.I. and Hancock, D.L. (1994) The *Bactrocera dorsalis* complex of fruit flies (Diptera: Tephritidae: Dacinae) in Asia. *Bulletin of Entomological Research Supplement Series*, No. 2, 68 pp.

Eardley, C.D. and Mansell, M.W. (1994) Report on the natural occurrence of insect pollinators in a mango orchard. *Yearbook of the South African Mango Growers' Association* 14, 65–66.

Ebeling, W. (1959) *Subtropical Fruit Pests*. University of California, Division of Agricultural Sciences, Los Angeles, 436 pp.

Etienne, J. (1968) Lutte contre les mouches des fruits. *Rapporte de Institut de Récherches Agronomiques Tropicales, Réunion*, pp. 117–124.

FAOSTAT Database (2001) Food and Agriculture Organization of the United Nations, Rome. http://apps.fao.org/lim500/nphwrap.pl?Production.Crops.Primary&Domain=SUA

Fasih, M. and Srivastava, R.P. (1990) Parasites and predators of insect pests of mango. *International Pest Control* 32, 39–41.

Fay, H.A.C. and Halfpapp, H.K. (1999) Activity of fruit-piercing moths, *Eudocima* spp. (Lepidoptera: Noctuidae), in north Queensland crops: some effects of ripe fruit type, locality and season. *Australian Journal of Entomology* 38, 16–22.

Fay, H.A.C. and Huwer, R.K. (1994) Egg parasitoids collected from *Amblypelta lutescens lutescens* (Distant) (Hemiptera: Coreidae) in north Queensland. *Journal of the Australian Entomological Society* 32, 365–367.

Follett, P.A. and Gabbard, Z. (2000) Effect of mango weevil (Coleoptera: Curculionidae) damage on mango seed viability in Hawaii. *Journal of Economic Entomology* 93, 1237–1240.

Free, J.B. and Williams, I.H. (1976) Insect pollination of *Anacardium occidentalis* L., *Mangifera indica* L., *Blighia sapida* Koenig and *Persea americana* Mill. *Tropical Agriculture* 53, 125–139.

Galán-Saúco, V. (1990) *Los Frutales Tropicales en los Subtropicos*. Ediciones MundiPrensa, Madrid, pp. 83–84.

Galán-Saúco, V., Fernández Galván, D., Hernández Conde, J.C. and Morales Navarro, A. (1997) Preliminary studies on fruit-set of mango cultivar Tommy Atkins under greenhouse cultivation in the Canary Islands. *Acta Horticulturae* 455, 530–537.

Ganz, S., Hamieri, Y. and Nakach, Y. (1990) Control trials on the mango thrips. *Annual Report of the Eden Experiment Station*, Israel.

Gerson, U. (1975) A soft scale as an urban pest. *Israel Journal of Entomology* 10, 25–28.

Githure, C.W., Schoeman, A.S. and McGeoch, M.A. (1997) Differential susceptibility of mango cultivars in South Africa to galling by the mango gall fly, *Protocontarinia matteiana* Keiffer and Cecconi (Diptera: Cecidomyiidae). *South African Mango Growers' Association Yearbook* 17, 138–143.

Golez, H.G. (1991) Bionomics and control of the mango seed borer, *Noorda albizonalis* Hampson (Pyralidae: Lepidoptera). *Acta Horticulturae* 291, 418–424.

Gopalakrishnan, C. and Narayanan, K. (1989) Occurrence of *Fusarium oxysporum* Schlecht and its pathogenicity on guava scale *Chloropulvinaria psidii* Maskell (Hemiptera: Coccidae). *Current Science* 58, 92–93.

Grové, T. (1998a) Pests – thrips. In: De Villiers, E.A (ed.) *The Cultivation of Mangoes*. Institute for Tropical and Subtropical Crops, Nelspruit, South Africa, pp. 167–170.

Grové, T. (1998b) Mango gall fly. In: De Villiers, E.A. (ed.) *The Cultivation of Mangoes*. Institute for Tropical and Subtropical Crops, Nelspruit, South Africa, p. 171.

Grover, P. (1986) Population fluctuation of *Erosomyia indica* and *Dasyneura amaramanjarae* and co-related extent of damage. *Cecidologia Internationale* 7, 1–28.

Hansen, J.D. (1993) Dynamics and control of the mango seed weevil. *Acta Horticulturae* 341, 415–420.

Hansen, J.D. and Armstrong, J.W. (1990) The failure of field sanitation to reduce infestation by the mango weevil, *Cryptorhynchus mangiferae* (F.)

(Coleoptera: Curculionidae). *Tropical Pest Management* 36, 359–361.

Harris, K.M. and Schreiner, I.H. (1992) A new species of the gall midge (Diptera: Cecidomyiidae) attacking mango foliage in Guam, with observations on its pest status and biology. *Bulletin of Entomological Research* 82, 41–48.

Heather, N.W. and Corcoran, R.J. (1992) Effects of ionizing energy on fruit flies and seed weevil in Australian mangoes. In: *Panel Proceedings of the Final Research Coordination Meeting on 'Use of Irradiation as a Quarantine Treatment of Food and Agricultural Commodities'*. Kuala Lumpur, Malaysia, August 1990, IAEA, Vienna.

Herrera, A.J. and Viñas, L.E. (1977) Fruit flies (Diptera: Tephritidae) on mangoes in Chulucanas, Piura. *Revista Peruana de Entomologia* 20, 107–114.

Hill, D. (1975) *Agricultural Insect Pests of the Tropics and their Control*. Cambridge University Press, New York.

Ivbijaro, M.F., Udensis, N., Ukwela, U.M. and Anno-Nyako, F.V. (1992) Geographical distribution and host range in Nigeria of the mango mealybug, *Rastrococcus invadens* Williams, a serious exotic pest of horticulture and other crops. *Insect Science Applications* 13, 411–416.

Iwahashi, O. (1984) The control project of the oriental fruit fly in Okinawa. *Chinese Journal of Entomology* 4, 107–120 (in Chinese).

Jhala, R.C., Patel, Z.P. and Shah, A.H. (1987) Studies on the relative occurrence of leaf-gall midge (*Procontarinia matteiana* Keiffer and Cecconi) on different varieties of mango in south Gujarat. *Tropical Pest Management* 33, 277–279.

Jhala, R.C., Patel, K.G., Patel, C.B. and Shah, A.H. (1990) Field efficacy of different insecticides for the control of mango leaf-gall midge *Protocontarinia matteiana* Keiffer and Cecconi. *International Pest Control* 32, 40–41.

Jirón, L.F. (1993) Plagas del follaje y de paniculas asociadas con el mango en Costa Rica. *Memorias I Seminario Nacional del Cultivo del Mango* Punta Arenas, Costa Rica, pp. 22–26.

Jirón, L.F. (1996) Management guidelines for *Anastrepha obliqua* associated with mango in Central America. *Fruits* 52, 25–30.

Jirón, L.F. and Hedström, I. (1985) Pollination ecology of mango (*Mangifera indica* L.) (Anacardiaceae) in the neotropic region. *Turrialba* 35, 269–277.

Jirón, L.F. and Hedström, I. (1988) Occurrence of fruit flies of the genera *Anastrepha* and *Ceratitus* (Diptera: Tephritidae) and their host plant availability in Costa Rica. *Florida Entomologist* 71, 62–73.

Jirón, L.F. and Lobo, J.A. (1995) Las abejas del genero *Trigona* y su asociacion con cultivo del mango (*Mangifera indica* L.). *Boltec* 28, 66–68.

Jirón, L.F., Soto-Manitiu, J. and Norrbom, A.L. (1988) A preliminary list of the fruit flies of the genus *Anastrepha* (Diptera, Tephritidae) in Costa Rica. *Florida Entomologist* 71, 130–137.

Joubert, P. (1997) Pest management – mango. In: *Biennial Report*, April 1995–March 1997. Institute for Tropical and Subtropical Crops, Nelspruit, South Africa, p. 25.

Kamburov, S. (1987) The mango shield scale *Protopulvinaria mangiferae* (Green) (Homoptera: Coccidae), a new pest on mango (*Mangifera indica*). *Citrus and Subtropical Fruit Journal* 635, 10–11.

Khanzada, A.G. and Naqvi, K.M. (1985) The optimum time for the control of mango hopper *Idioscopus* spp. (Hemiptera: Cicadellidae). *Proclamation of the Entomological Society of Karachi* 14/15, 149–156.

Kulkarni, H.L. (1955) Incidence of mango flower galls in Bombay and Karnataka. *Journal of the Bombay Natural History Society* 53, 147–148.

Labuschagne, T.I., van Hamburg, H. and Froneman, I.J. (1995) Population dynamics of the mango scale, *Aulacaspis tubercularis* (Newstead) (Coccoidea: Diaspididae), in South Africa. *Israel Journal of Entomology* 29, 207–217.

Labuschagne, T.I., Brink, T., Steyn, W.P. and de Beer, M.S. (1996a) Fruit flies attacking mangoes – their importance and post harvest control. *South African Mango Growers' Association Yearbook* 16, 17–19.

Labuschagne, T.I., Daneel, M.S. and De Beer, M. (1996b) Establishment of *Aphytis* sp. (Hymanoptera: Aphelinidae) and *Cybocephalus binotatus* Grouvelle (Coleoptera: Nitidulidae) in mango orchards in South Africa for control of the mango scale, *Aulacaspis tubercularis* Newstead (Homoptera: Diaspididae). *South African Mango Growers' Association Yearbook* 16, 20–22.

Lakra, R.K., Kharub, E. and Singh, Z. (1980) Pest management systems for the mango mealybug *Drosicha mangifera* Green – a polyphagous pest of fruit trees in Haryana. *Indian Journal of Entomology* 42, 153–165.

Levya, J.L., Browning, H.W. and Gilstrap, F.E. (1991) Development of *Anastrepha ludens* (Diptera: Tephritidae) in several host fruit. *Environmental Entomology* 20, 1160–1165.

Li, L.Y., Wang, R. and Waterhouse, D.F. (1997) The distribution and importance of major insect pests and weeds of agriculture and plantation forests in southern China. *Chinese Academy of Agricultural Sciences and ACIAR*, 185 pp.

Litsinger, J.A. (1993) A farming systems approach to insect pest management for upland and lowland rice farmers in tropical Asia. In: Altieri, M.A. (ed.) *Crop Protection Strategies for Subsistence Farmers*. Intermediate Technology Publications, London, pp. 45–101.

Loomans, A.J.M. and van Lenteren, J.C. (1995) Biological control of thrips pests: a review on thrips parasitoids. Wageningen University Papers 95.1, 92–95.

Mahmood, R. and Mohyuddin, A.I. (1986) Integrated control of mango pests. Commonwealth Institute of Biological Control, Pakistan Agricultural Research Council, Islamabad, 11 pp.

Mahmood, R., Mohyuddin, I. and Kazimi, S. (1980) *Rastrococcus spinosus* (Robinson) (Homoptera: Pesudococcidae) and its natural enemies in Pakistan. *Proceedings of the 1st Pakistan Congress of Zoology*. Tandojam, pp. 291–294.

Mani, M.S. (1943) Studies on Indian Honididae (Cecidomyidae: Diptera). *Indian Journal of Entomology* 5, 151–164.

Manoto, E.C., Vijayasegaran, S. and Shamsudin, M.O. (1984) Postharvest entomology of mango. In: Mendoza, D.B. Jr and Wills, R.B.H. (eds) *Mango: Fruit Development, Postharvest Physiology and Marketing in ASEAN*. ASEAN Food Handling Bureau, Kuala Lumpur, Malaysia, pp. 62–68.

McComie, L. and De Chi, W. (1994) Review of the mango seed weevil, *Sternochetus mangiferae* (F.), and its distribution and population density in Trinidad and Tobago. *Tropical Fruits Newsletter, IICA* 10, 12.

Meurant, N., Holmes, R., MacLeod, N., Fullelove, G., Bally, I. and Kernot, I. (1999) *Mango Information Kit*. Agrilink Series, Queensland Department of Primary Industries, Brisbane, 211 pp.

Mohyuddin, A.I. (1981) A review of biological control in Pakistan. *Proceedings of the 2nd Pakistan Congress of Zoology*. Tandojam, Pakistan, pp. 31–79.

Mohyuddin, A.I. and Mahmood, R. (1993) Integrated control of mango pests in Pakistan. *Acta Horticulturae* 341, 467–483.

Moore, D. and Cross, A.E. (1993) Biological control of the fruit tree mealybug, *Rastrococcus invadens* Williams; single or multiple introduction? *Acta Horticulturae* 341, 433–440.

Morin, C. (1967) Plagas. In: *Cultivo de Frutales Tropicales*. Librerias A.B.C., S.A., Lima, Peru, pp. 225–229.

Morrill, A.W. and Otanes, F.Q. (1947) DDT emulsion to control mealybugs and scales. *Journal of Economic Entomology* 40, 599–600.

Moznette, G. (1922) Insects injurious to the mango in Florida and how to combat them. *Farmers Bulletin* 1257.

Mukherjee, S.K. (1997) Introduction: botany and importance. In: Litz, R.E. (ed.) *The Mango: Botany, Production and Uses*. CAB International, Wallingford, UK, pp. 1–19.

Murray, R.C. (1991) Arthropod pests of mango in the Caribbean Islands. *Review Paper, Crop Protection*. University of the West Indies, Kingston.

Nachiappan, R.M. and Baskaran, P. (1984) Quantum of feeding and survival of mango leafhopper adults on inflorescence of certain varieties of mango. *Indian Journal of Agricultural Science* 54, 312–314.

Nafus, D. (1991) Biological control of *Penicillaria jocosatrix* (Lepidoptera: Noctuidae) on mango in Guam with notes on the biology of its parasitoids. *Environmental Entomology* 20, 1725–1731.

Nafus, D., Schreiner, I. and Dumaliang, N. (1991) Survival and development of the mango shoot caterpillar (Lepidoptera: Noctuidae) in relation to leaf age, host and distribution on the host in tropical Anacardiaceae. *Environmental Entomology* 20, 1619–1626.

Narasimham, A.U. and Chacko, M.J. (1988) *Rastrococcus* spp. (Hemiptera: Pseudococcidae) and their natural enemies in India as potential biocontrol agents for *R. invadens* Williams. *Bulletin of Entomological Research* 78, 703–708.

Neuenschwander, P. (1989) Biocontrol of mango mealybug. *IITA Research Briefs* 9, 5–6.

Nguyen, T.T.C., Nguyen, V.H. and Le van, D. (1998a) Insect pests on citrus, durian, longan and mango in selected provinces of the Mekong Delta of Vietnam. In: Van Mele, P. and Nguyen, V.H. (eds) *Proceedings of the First Symposium on Fruit Production in the Mekong Delta Focussing on Integrated Pest Management*. Cantho, Vietnam, pp. 71–76 (in Vietnamese).

Nguyen, T.N., Nguyen, V.H. and Tran, T.H. (1998b) Study on the symptoms of 'branch drying and tree dying of mango at Can Tho and Tien Giang Province. In: Van Mele, P. and Nguyen, V.H. (eds) *Proceedings of the First Symposium on Fruit Production in the Mekong Delta Focussing on Integrated Pest Management*. Cantho, Vietnam pp. 77–85 (in Vietnamese).

Noyes, J.S. (1988) *Gyranusoidea tebygi* sp. n. (Hymenoptera: Encyrtidae), a parasitoid of *Rastrococcus* (Hemiptera: Pseudococcidae) on mango in India. *Bulletin of Entomological Research* 78, 313–316.

Noyes, J.S. (1990) A new species of *Anagyrus* (Hymenoptera: Encyrtidae) from India attacking *Rastrococcus invadens* (Homoptera:

Pseudococcidae), a pest of mango and citrus in West Africa. *Bulletin of Entomological Research* 80, 203–207.

Ochoa, R., Aguilar, H. and Vargas, C. (1994) Phytophagous mites of Central America: an illustrated guide. *Centro Agronomico de Investigacio y Enseñanza, Technical Series* 6. Turrialba, Costa Rica.

Orozco Santos, M. and Nuñez, E.R. (1988) 'Achaparramiento', un desorden del mango en el estado de Colima. *Revista Mexicana de Fitopatologia* 6, 131–135.

Osman, A.A. (1979) Notes on the control of the bud mite *Aceria mangiferae* (Sayed) in Egypt (Acarina: Eriophyidae). *Bulletin of the Entomological Society of Egypt, Economic Series 9*, pp. 119–126.

Otanes, F.Q. and Toquero, A.G. (1927) Notes on the mango twig borer (*Euclea capito* Pasc.). *Philippine Agricultural Review* 20, 249–250.

Peña, J.E. (1993) Pests of mango in Florida. *Acta Horticulturae* 341, 395–406.

Peña, J.E. and Mohyuddin, A.I. (1997) Insect pests. In: Litz, R.E. (ed.) *The Mango: Botany, Production and Uses*. CAB International, Wallingford, UK, pp. 327–362.

Peña, J.E., Mohyuddin, A.I. and Wysoki, M. (1998) A review of the pest management situation in mango agroecosystems. *Phytoparasitica* 26, 129–148.

Popenoe, W. (1917) *The Pollination of the Mango.* US Department of Agriculture Bulletin, No. 542.

Prasad, S.N. (1966) Studies on gall midges infesting mango inflorescence. *Cecidologia Indica* 1, 1–19.

Prasad, S.N. (1972) On the control of mango midge pests. *Cecidologia Indica* 7, 51–63.

Prasad, V. and Singh, R.K. (1976) Prevalence and control of mango mealy bug *Drosicha stebbingii* (Green) in Bihar. *Indian Journal of Entomology* 38, 214–224.

Rai, B., Verma, S. and Kumar, K. (1966) Evaluation of pesticides for control of mango bud mite, *Aceria mangiferae* Sayed (Acarina: Eriophyidae). *Indian Journal of Entomology* 28, 176–180.

Rai, B., Shah, A. and Patel, R. (1988) Biology of *Oligonychus mangiferus* (Tetranychidae: Acarina), a pest of mango in Gujarat. *Gujarat Agricultural University Research Journal* 14, 5–10.

Rao, S.N. (1956) Notes and biology of gall midges (Homididae: Diptera) from India. *Indian Journal of Entomology* 18, 79–81.

Rhodes, L.F. (1992) Major arthropod pests of mango in the Caribbean (Part II). *IICA Tropical Fruits Newsletter* 4, 3–4.

Robacker, D.C., Martinez, A.J., Garcia, J.A., Diaz, M. and Romero, C. (1996) Toxicity of *Bacillus thuringiensis* to Mexican fruit fly (Diptera: Tephritidae). *Journal of Economic Entomology* 89, 104–109.

Rubin, A. and Kuslitzky, W. (1992) First record of *Ceranisus menes* (Hymenoptera: Eulophidae) in Israel. *Phytoparasitica* 20, 123–124.

Sakimura, K. (1981) A review of *Frankliniella bruneri* Watson and description of *F. kelliae*, n. sp. (Thysanoptera: Thripidae). *Florida Entomologist* 64, 483–491.

Sankaran, T. (1955) The natural enemies of *Ceroplastes pseudoceriferus* Green (Hemiptera: Coccidae). *Journal of Scientific Research of the Banaras Hindu University* 5, 100–119.

Schreiner, I. (1987) Mango shoot caterpillar control. *Insecticide and Acaricide Tests* 12, 94.

Segarra, A.E. (1988) Identity and economic importance of fruit flies attacking mango in Puerto Rico: a two-year survey. *The Journal of Agriculture of the University of Puerto Rico* 72, 325.

Serrano, F.B. and Palo, M.A. (1933) Blossom blight of mangoes in the Philippines. *Philippine Journal of Science* 1, 211–277.

Shah, A.H., Jhala, R.C., Patel, G.M. and Patel, S.H. (1983) Evaluation of effective dose of monocrotophos for the control of mango hopper *Amritodius atkinsoni* Lethierry (Cicadellidae: Homoptera) by injection method and its comparison with foliar application. *Gujarat Agricultural University Research Journal* 9, 14–18.

Shah, A.H., Jhala, R.C., Patel, G.M. and Patel, C.B. (1984) Studies on the need based chemical control of the mangohopper, *Amritodius atkinsoni* (Leth.). *International Pest Control* 26, 15–17.

Shahjahan, M. and Ahmad, M. (1978) Occurrence of the mango shoot borer *Chlumetia transversa* (Lepi.-Noct.) in the Agricultural University area, Mymemsing, Bangladesh. *Bangladesh Journal of Zoology* 6, 193–206.

Sharma, P.L. and Tandon, P.L. (1972) Incidence and control of fig borer, *Batocera rufomaculata* Degeer (Damiidae: Coleoptera) in Himachal Pradesh. *Himachal Horticultural Journal* 12, 29–30.

Sharma, S., Abbas, S.R., Shukia, R.P. and Sharma, S. (1998) An easy and quick method of breeding flies for pollination of mango blossoms. *Insect Environment* 4, 76–77.

Shukla, R.P. and Tandon, P.L. (1985) Bio-ecology and management of the mango weevil, *Sternochetus mangiferae* (Fabricius) (Coleoptera: Curculionidae). *International Journal of Tropical Agriculture* 3, 293–303.

Singh, G. (1997a) Resistance against hoppers in mango. *Phytoparasitica* 25, 1.

Singh, G. (1997b) Pollination, pollinators and fruit setting in mango. *Acta Horticulturae* 455, 116–123.

Singh, J. and Mukherjee, I.N. (1989) Pest status of phytophagous mites in some northern states of India. In: *Proceedings of The First Asia-Pacific Conference of Entomology*. Chiang Mai, Thailand, pp. 192–203.

Singh, K.P. (1999) Evaluation of the pollination efficiency of different insects on mango (*Mangifera indica*). *Indian Forester* 125, 333–335.

Sohi, A.S. and Sohi, A.S. (Snr.) (1990) Mango leafhoppers (Homoptera: Cicadellidae) – a review. *Journal of Insect Science* 3, 1–12.

Soomroo, A., Khuhro, R. and Ansari, H. (1987) Status of insects associated with mango blossom. *Proceedings V Pakistan Congress of Zoology*. Karachi, Pakistan, pp. 123–125.

Soto-Manatiu, J., Jirón, L.F. and Hernandez, R. (1987) Chemical control and ecological observations of fruit flies of the genus *Anastrepha* Schiner (Diptera: Tephritidae) on mango. *Turrialba* 37, 145–251.

Srivastava, R.P. (1981) Comparative efficacy of various insecticidal dusts against mango mealybug eggs. *Indian Journal of Entomology* 43, 225–229.

Srivastava, R.P. and Tandon, P.L. (1986) Natural occurrence of two entomophagous fungi pathogenic to mango hopper, *Idioscopus clypealis* Leth. *Indian Journal of Plant Pathology* 4, 121–123.

Steiner, L.F. and Lee, R.K. (1955) Large area tests of male annihilation method for oriental fruitfly control. *Journal of Economic Entomology* 48, 311–317.

Steiner, L.F., Mitchell, W.C., Harris, E., Kozuma, T. and Fujimoto, M. (1965) Oriental fruit fly eradication by male annihilation. *Journal of Economic Entomology* 58, 961–964.

Swirski, E., Ben-Dov, Y. and Wysoki, M. (1997) Mango. In: Ben-Dov, Y. and Hodgson, C.J. (eds) *Soft Scale Insects – Their Biology, Natural Enemies and Control (7B)*, pp. 241–254.

Syed, R.A., Ghani, M.A. and Murtaza, M. (1970) Studies on the trypetids and their natural enemies in west Pakistan. IV. Further observations on *Dacus* (*Strumeta*) *dorsalis* Hendel. *Technical Bulletin of the Commonwealth Institute of Biological Control* 13, 17–30.

Tandon, P.L. and Verghese, A. (1987) New insect pests of certain fruit crops. *Indian Journal of Horticulture* 44, 120–121.

Tandon, P.L., Krishnaiah, K. and Prasad, V.G. (1975) Chemical control of mango shoot borer *Chlumetia transversa* Walker (Noctuidae: Lepidoptera). *Indian Journal of Horticulture* 32, 154–155.

Tandon, P.L., Lal, B. and Rao, G.S.P. (1983) Prediction of mangohopper *Idioscopus clypealis* (Leth.) population in relation to physical environmental factors. *Entomon* 8, 257–261.

Thontadarya, T.S., Viswanath, B.N. and Hiremath, S.C. (1978) Preliminary studies on the control of mango hoppers by stem injection with systemic insecticides. *Current Science* 47, 379.

Tigvattnannont, S. (1988) Biological and autoecological studies of the mango leaf-cutting weevil, *Deporaus marginatus* Pascoe (Coleoptera: Attelabidae). *Khon Kaen Agricultural Journal* 16, 51–62.

Tijero, R.F. (1992) El cultivo del mango en el Perú. *Ediciiones Fundeagro* 3, 53–54.

Umeya, K. and Hirao, J. (1975) Attraction of the jackfruitfly, *Dacus umbrosus* F. (Diptera: Tephritidae) and lace wing, *Chrysopa* sp. (Neuroptera: Chrysopidae) to lure traps baited with methyl eugenol and cue-lure in the Philippines. *Applied Entomology and Zoology* 10, 60–62.

United States Department of Agriculture (USDA) (1981) *Plant Pest News* 1.

Van Dine, D.L. (1906) The mango weevil, *Cryptorhynchus mangiferae* (Fabr.). *Hawaii Agricultural Experiment Station Bulletin* No. 17, 1–11.

Van Halteren, P. (1970) Notes on the biology of the scale insect *Aulacaspis tubercularis* Newst. (Diaspididae, Hemiptera) on mango. *Ghana Journal of Agricultural Science* 3, 83–85.

Van Lenteren, J.C., Loomans, A.J., Tommasini, M.G., Maini, S. and Riudavets, J. (1995) *Biological Control of Thrips Pests*. Wageningen Agricultural University Press, Wageningen, 201 pp.

Van Mele, P., Cuc, N.T.T. and Van Huis, A. (2001) Farmers' knowledge, perceptions and practices in mango pest management in the Mekong Delta, Vietnam. *International Journal of Pest Management* 47, 7–16.

Van Whervin, L.W. (1968) Citrus weevils of Jamaica and some of their parasites. *Technical Bulletin No. 1*, Citrus Research Unit, UWI, St Augustine.

Vayssieres, J.F. (1997) Enquête sur les arthropodes ravageurs et auxiliaires du litchi à la Réunion. *Réport au Département des Productions Fruitières et Horticoles, CIRAD-FLHOR, Ile de la Rèunion*, 40 pp.

Veeresh, G.K. (1988) Pest problems in mango – world situation. *Acta Horticulturae* 123, 551–565.

Verghese, A. (1998) Effect of imidacloprid on mango hoppers, *Idioscopus* spp. (Homoptera: Cicadellidae). *Pest Management in Horticultural Ecosystems* 4, 70–74.

Verghese, A. and Rao, G.S.P. (1985) Sequential sampling plan for leafhoppers, *Idioscopus clypealis* Lethierry. *Entomon* 10, 285–290.

Verghese, A., Tandon, P.L. and Rao, G. (1988) Ecological studies relevant to the management of *Thrips palmi* Karny on mango in India. *Tropical Pest Management* 34, 55–58.

Verheij, E.W.M. (1991) *Mangifera indica* L. In: Verheij, E.W.M. and Coronel, R.E. (eds) *Plant Resources of South-East Asia, No.2, Edible Fruits and Nuts*. Pudoc, Waginingen, pp. 211–216.

Waite, G.K. and Elder, R.J. (1996) Lychee/longan insect pests and their control. In: *Proceedings of the Fourth National Lychee Seminar Including Longans*. Yeppoon, Queensland, pp. 102–109.

Waite, G.K. and Huwer, R.K. (1998) Host plants and their role in the ecology of the fruitspotting bugs *Amblypelta nitida* Stål and *Amblypelta lutescens lutescens* (Distant) (Hemiptera: Coreidae). *Australian Journal of Entomology* 37, 340–349.

Waite, G.K. and Petzl, N. (1994) A search for egg parasites of *Amblypelta* spp. (Hemiptera: Coreidae) in south-east Queensland. In: Menzel, C. and Crosswell, A. (eds) *Maroochy Research Station Report No. 7*, pp. 13–14.

Waterhouse, D.F. (1998) Biological control of insect pests: Southeast Asian prospects. *ACIAR Monograph* No. 51, 548 pp.

Wen, H.C. and Lee, H.S. (1986) Seasonal abundance of the ceriferous wax scale (*Ceroplastes pseudoceriferus*) in southern Taiwan and its control. *Journal of Agricultural Research of China* 35, 216–221.

Wester, P.J. (1920) The Mango. *Philippine Bureau of Agriculture Bulletin* 18, 3–70.

Wharton, R.A. and Marsh, P.M. (1978) New world Opiinae (Hymenoptera: Braconidae) parasitic on Tephritidae (Diptera). *Journal of the Washington Academy of Sciences* 68, 147–165.

White, I.M. and Elson-Harris, M. (1992) *Fruit Flies of Economic Significance: Their Identification and Bionomics*. CAB International, Wallingford, UK 601 pp.

Whitwell, A.C. (1993) The pest/predator/parasitoid complex on mango inflorescences in Dominica. *Acta Horticulturae* 341, 421–432.

Willink, E. and Moore, D. (1988) Aspects of the biology of *Rastrococcus invadens* Williams (Hemiptera: Pseudococcidae), a pest of fruit crops in West Africa and one of its primary parasitoids, *Gyranusoidea tebygi* Noyes (Hymenoptera: Encyrtidae). *Bulletin of Entomological Research* 78, 709–715.

Wolfenbarger, D.O. (1956) Controlling mango insect pests. *Florida Agricultural Extension Service Circular* 147, 1–14.

Wongsiri, S. and Chen, P.P. (1995) Effects of agricultural development on honey bees in Thailand. *Bee World* 76, 3–5.

Woodruff, R.E. (1985) Citrus weevils in Florida and the West Indies: preliminary report on systematics, biology and distribution (Coleoptera: Curculionidae). *Florida Entomologist* 68, 370–379.

Wysoki, M., Ben-Dov, Y., Swirski, E. and Izhar, Y. (1993) The arthropod pests of mango in Israel. *Acta Horticulturae* 341, 452–466.

Yee, W. (1987) The mango in Hawaii. *University of Hawaii Agricultural Extension Service Circular* 388, 19–22.

5 Pests of Papaya

Alberto Pantoja,[1] Peter A. Follett[2] and Juan A. Villanueva-Jiménez[3]
[1]*Department of Crop Protection, University of Puerto Rico, Mayaguez Campus, PO Box 9030 Mayaguez, Puerto Rico, 00681-9030;* [2]*USDA-ARS, Pacific Basin Agricultural Research Center, PO Box 4459, Hilo, Hawaii 96720, USA;* [3]*Colegio de Postgraduados, Campus Veracruz, Apartado Postal 421, CP 91700 Veracruz, Veracruz, Mexico*

Papaya *Carica papaya* L. is a major tropical fruit cultivated in frost-free areas. Papaya has culinary, medical and industrial uses but is mainly cultivated for its edible fruit. As with most tropical fruits grown in diverse geographical regions, papaya is affected by several arthropods. Fruit quality for the fresh market is important. Therefore, arthropods that blemish the fruit skin, enter the fruit, or feed on the pulp or seeds can cause high losses and are considered economically important.

There are 134 species of arthropods that affect papaya (Table 5.1). Most of the species belong to the Hexapoda, while 12 belong to the Acarina. Twenty-six species are fruit flies in the family Tephritidae. Eighty-seven species can potentially attack or damage the fruit, but are mainly associated with the foliage or the trunk. One species is a seed borer. Five species affect the flowers, and three species are root feeders. At least 12 species are known vectors of important papaya diseases.

In different papaya growing areas, fruit flies, leafhoppers, mites, and scale insects are considered key pests requiring frequent pesticide applications. Fruit flies are the most important papaya pests either due to their direct effect on the fruit or for quarantine related issues. Mites are usually secondary pests causing economic damage especially after human intervention to control other pests, mainly fruit flies and leafhoppers. Aphids and leafhoppers are important pests due to their vector capacity, but rarely cause direct damage to the trees.

All species reported from papaya are presented in Table 5.1. Species that rarely attack papaya, or species with unknown pest status or economic importance are only mentioned briefly or are tabulated. Information on geographical distribution of the arthropod and the type of damage is provided.

Origin and Distribution of the Crop

Papaya is cultivated in the tropical and neotropical regions of the world between 32° North and South. Its origin in Central America and current distribution of the crop have been documented by Storey (1984), Campbell (1984), and Morton (1987). Although papaya was probably cultivated by early civilizations, no botanical records are available prior to the arrival of Columbus in America (Morton, 1987). The wide and rapid dissemination of the crop from Central America to the Caribbean region, Asia, Africa, and the Pacific in the 16th century is associated with its propagation by seed, aggressive growth, long-term

©CAB *International* 2002. *Tropical Fruit Pests and Pollinators* (eds J.E. Peña, J.L. Sharp and M. Wysoki)

Table 5.1. Arthropods affecting papaya.

ORDER/Family	Geographical distribution[a]	References	Plant part[b]
HEMIPTERA			
Coreidae			
Amblypelta cocophaga China	PI	FAO, 2000	FR
Amblypelta costalis szentivanyi Van Duzee	PI	FAO, 2000	FR
Amblypelta gallegonis Lever	PI	FAO, 2000	FR
Amblypelta lutescens papuensis (Distant)	PI	FAO, 2000	FR
Amblypelta theobromae Brown	PI	FAO, 2000	FR
Brachylybas variegatus Le Guillou	PI	FAO, 2000	FR
Miridae			
Fulvius angustatus Usinger	PI	FAO, 2000	FR
Pentatomidae			
Nezara viridula (L.)	PI, WI	Martorell, 1976	FR
Tingidae			
Corythucha gossypii (F.)	WI	Martorell, 1976	FL
HOMOPTERA			
Coccidae			
Coccus discrepans (Green)	AS, ME	Ben-Dov *et al.*, 2000	FO, FR, TR
Coccus hesperidum hesperidum L.	AF, AS, AU, CA, EU, ME, NA, PI., SA, WI	FAO, 2000	FO, FR, TR
Coccus longulus (Douglas)	AF, AS, AU, CA, EU, ME, NA, PI, SA, WI	FAO, 2000	FO, FR, TR
Drepanococcus chiton (Green)	AS, PI	FAO, 2000	FO, FR, TR
Eucalymnatus tessellatus (Signoret)	AF, AS, AU, EU, NA, PI, SA, WI	FAO, 2000	FO, FR, TR
Milviscutulus mangiferae (Green)	AF, AS, CA, EU, NA, PI SA, WI	Ben-Dov, 1993	FO, FR, TR
Parasaissetia nigra (Nietner)	AF, AS, AU, CA, EU, ME, NA, PI, SA, WI	FAO, 2000	FO, FR, TR
Philephedra tuberculosa Nakahara & Gill	CA,.NA, SA	Nakahara and Gill, 1985	FO, FR, TR
Protopulvinaria pyriformis (Cockerell)	AF, AS, CA, EU, NA, PI, SA, WI	Ben-Dov, 1993	FO, FR, TR
Saissetia oleae oleae (Olivier)	AF, AS, AU, CA, EU, ME, NA, PI, SA, WI	Ben-Dov *et al.*, 2000	FO, FR, TR
Saissetia coffeae (Walker)	AF, AS, AU, CA, EU, ME, NA, PI, SA, WI	FAO, 2000	FO, FR, TR
Conchaspididae			
Conchaspis angraeci Cockerell	AF, AS, AU, CA, EU, NA, PI, SA, WI	FAO, 2000	FO, FR, TR

Pseudococcidae			
Dysmicoccus nesophilus Williams & Watson	PI	FAO, 2000	FO, FR, TR
Ferrisia virgata (Cockerell)	AF, AS, AU	FAO, 2000	FO, FR, TR
Nipaecoccus viridis (Newstead)	AF, AS, AU	Ben-Dov *et al.*, 2000	FO, FR, TR
Paracoccus marginatus Williams & Granara de Willink	CA, SA, WI	Williams and Granara, 1992	FO, FR, TR
Planococcus citri (Risso)	PI	FAO, 2000	FO, FR, TR
Pseudococcus jackbeardsleyi Gimpel & Miller	AS, CA, NA, PI, SA, WI	Ben-Dov *et al.*, 2000	FO, FR, TR
Pseudococcus longispinus (Targioni Tozzetti)	AF, AS, AU	Ben-Dov *et al.*, 2000	FO, FR, TR
Pseudococcus viburni (Signoret)	AF, AS, AU	Ben-Dov *et al.*, 2000	FO, FR, TR
Diaspididae			
Aonidiella aurantii (Maskell)	PI	FAO, 2000	FO, FR, TR
Aonidiella comperei Mckenzie	PI	FAO, 2000	FO, FR, TR
Aonidiella inornata Mckenzie	PI	FAO, 2000	FO, FR, TR
Aonidiella orientalis (Newstead)	AF, ME, PI	FAO, 2000	FO, FR, TR
Aspidiotus destructor Signoret	AS, PI	FAO, 2000	FO, FR, TR
Aspidiotus excisus Green	PI	FAO, 2000	FO, FR, TR
Aspidiotus macfarlanei Williams & Watson	PI	FAO, 2000	FO, FR, TR
Chrysomphalus dictyospermi (Morgan)	AF, AS, AU, CA, EU, ME, NA, PI, SA, WI	FAO, 2000	FO, FR, TR
Howardia biclavis (Comstock)	PI	FAO, 2000	FO, FR, TR
Morganella longispina (Morgan)	PI	FAO, 2000	FO, FR, TR
Pseudaonidia trilobitiformis (Green)	PI	FAO, 2000	FO, FR, TR
Pseudaulacaspis cockerelli (Cooley)	PI	FAO, 2000	FO, FR, TR
Pseudaulacaspis pentagona (Targioni-Tozzetti)	WI	Medina and Franqui, 1999a,b	FO, FR, TR
Pseudoparlatoria ostreata Cockerell	AF, AS, AU, CA, EU, ME, NA, PI, SA, WI	Medina and Franqui, 1999a,b	FO, FR, TR
Asterolecaniidae			
Asterolecanium pustulans (Cockerell)	AS, PI	FAO, 2000	FO, FR, TR
Margarodidae			
Icerya aegyptiaca (Douglas)	PI	FAO, 2000	FO, FR, TR
Icerya purchasi Maskell	AF, AS, AU, CA, EU, ME, NA, PI, SA, WI	FAO, 2000	FO, FR, TR
Icerya seychellarum (Westwood)	AF, AS, PI	FAO, 2000	FO, FR, TR
Steatococcus samaraius Morrison	PI	FAO, 2000	FO, FR, TR

continued

Table 5.1. *Continued.*

ORDER/Family	Geographical distribution[a]	References	Plant part[b]
Cicadellidae			
Empoasca papayae Oman	SA, WI	Medina and Franqui, 1999a,b	FO, VE
Empoasca canavalia Long	WI	Medina and Franqui, 1999a,b	FO
Empoasca dilataria DeLong & Davidson	WI	Medina and Franqui, 1999a,b	FO
Empoasca fabalis Harris	WI	Medina and Franqui, 1999a,b	FO, VE
Emposaca insularis Oman	WI	Medina and Franqui, 1999a,b	FO, VE
Empoasca stevensi Young	WI	Medina and Franqui, 1999a,b	FO, VE
Empoasca solana DeLong	PI, WI	Medina and Franqui, 1999a,b	FO
Poeciloscarta laticeps Metcalf & Bruner	WI	Medina and Franqui, 1999a,b	FO
Sanctanus fasciatus (Osborn)	WI	Medina and Franqui, 1999a,b	FO
Cixiidae			
Oliarus complectus Ball	WI	Medina and Franqui, 1999a,b	FO
Derbidae			
Omolicna puertana Caldwell	WI	Medina and Franqui, 1999a,b	FO
Aleyrodidae			
Trialeurodes variabilis (Quaintance)	WI	Medina and Franqui, 1999a,b	FO
Aleurocanthus woglumi Ashby	AF, AS, AU, CA, EU, ME, NA, P., SA, WI	FAO, 2000	FO
Aleurodicus destructor (Mackie)	PI	FAO, 2000	FO
Aleurodicus dispersus Russell	CA, EU, NA, PI, SA, WI	FAO, 2000	FO
Tetraleurodes acaciae (Quaintance)	CA, NA, PI., SA, WI	FAO, 2000	FO
Aphididae			
Aphis coreopsidis (Thomas)	WI	Medina and Franqui, 1999a,b	FO
Aphis craccivora Koch	PI	Higa and Namba, 1971	FO, VE
Aphis gossypii Glover	PI, WI	Medina and Franqui, 1999a; Abreu, 1994	FO, VE
Aphis middletonii (Thomas)	PI	Higa and Namba, 1971	FO
Aphis nerii Boyer de Fonscolombe	WI	Medina and Franqui, 1999a; Abreu, 1994	FO, VE
Aphis spiraecola Patch	WI	Medina and Franqui, 1999a; Abreu, 1994	FO, VE
Macrosiphum euphorbiae (Thomas)	PI	Higa and Namba, 1971	FO, VE
Myzus persicae (Sulzer)	PI, WI	Medina and Franqui, 1999a; Abreu, 1994	FO, VE
Rhopalosiphum maidis (Fitch)	PI	Higa and Namba, 1971	FO, VE
Toxoptera aurantii (Boyer de Fonscolombe)	WI	Medina and Franqui, 1999a; Abreu, 1994	FO

THYSANOPTERA			
Thripidae			
Selenothrips rubrocinctus (Giard)	PI	FAO, 2000	FO, FL, FR
Thrips tabaci Lindeman	PI	Yee *et al.*, 1970	FO, FL, FR
Frankliniella occidentalis (Pergande)	PI	Sakimura, 1972	FO, FL, FR, VE
Frankliniella fusca (Hinds)	PI	Sakimura, 1972	FO, FL, FR, VE
COLEOPTERA			
Curculionidae			
Acicnemis crassiusculus Fairmaîre	PI	FAO, 2000	FO
Diaprepes abbreviatus (L.)	NA, WI	Abreu, 1994; Medina *et al.*, 1999	FO, RO
Rhabdoscelus obscurus (Boisduval)	PI	FAO, 2000	FO
Metamasius hemipterus (Linnaues)	CA, NA, SA, WI	FAO, 2000	RO
Anthribidae			
Araecerus vieillardi Montr.	PI	FAO, 2000	TR
Nitidulidae			
Carpophilus maculatus Murray	PI	FAO, 2000	FR
Scarabaeidae			
Protaetia orientalis (Gory & Percheron)	PI	FAO, 2000	RO
DIPTERA			
Tephritidae			
Anastrepha ludens (Loew)	CA, NA	White and Elson-Harris, 1992	FR
Anastrepha suspensa (Loew)	NA, WI	White and Elson-Harris, 1992	FR
Bactrocera bryoniae (Tyron)	PI	FAO, 2000	FR
Bactrocera cucumis (French)	AU	White and Elson-Harris, 1992	FR
Bactrocera cucurbitae (Coquillett)	AF, AS, ME, PI	White and Elson-Harris, 1992	FR
Bactrocera dorsalis (Hendel)	AS, PI	White and Elson-Harris, 1992	FR
Bactrocera diversa (Coquillett)	AS,	White and Elson-Harris, 1992	FR
Bactrocera facialis (Coquillett)	PI	FAO, 2000	FR
Bactrocera frauenfeldi (Schiner)	AU, PI	White and Elson-Harris, 1992	FR
Bactrocera jarvisi (Tryon)	AU	White and Elson-Harris, 1992	FR
Bactrocera kirki (Froggatt)	PI	FAO, 2000	FR
Bactrocera melanota (Coquillett)	PI	FAO, 2000	FR
Bactrocera musae (Tryon)	PI	White and Elson-Harris, 1992	FR
Bactrocera neohumeralis (Hardy)	PI	FAO, 2000	FR
Bactrocera passiflorae (Froggatt)	PI	White and Elson-Harris, 1992	FR

continued

Table 5.1. *Continued.*

ORDER/Family	Geographical distribution[a]	References	Plant part[b]
Bactrocera trilineola Drew	PI	White and Elson-Harris, 1992	FR
Bactrocera tryoni (Froggatt)	AU, PI	White and Elson-Harris, 1992	FR
Bactrocera xanthodes (Broun)	PI	White and Elson-Harris, 1992	FR
Bactrocera zonata (Saunders)	AS, AU, PI	White and Elson-Harris, 1992	FR
Ceratitis capitata (Wiedemann)	AF, AS, AU, CA, EU, ME, NA, PI, SA, WI	White and Elson-Harris, 1992	FR
Ceratitis catoirii Guérin-Méneville	AF	White and Elson-Harris, 1992	FR
Ceratitis rosa Karsch	AF, AS	White and Elson-Harris, 1992	FR
Dacus bivittatus (Bigot)	AF	White and Elson-Harris, 1992	?
Euphranta lemniscata (Enderlein)	AS, PI	White and Elson-Harris, 1992	FR
Myoleja nigroscutellata (Hering)	AS, ME	White and Elson-Harris, 1992	NR
Toxotrypana curvicauda Gerstaecker	CA, NA, SA, WI	Weems, 1969	FR, SE
LEPIDOPTERA			
Noctuidae			
Epitomiptera orneodalis (Guenee)	WI	Medina and Franqui, 1999a,b	FO
Agrotis ipsilon Hufnagel	PI, WI	Medina and Franqui, 1999a,b	FO
Eudocima fullonia (Clerck)	PI	FAO, 2000	FO, FR
Tiracola plagiata (Walker)	PI	FAO, 2000	FO
Sphingidae			
Erynnis alope (Drury)	SA, WI	Clavijo and Chacín, 1992	FO
Erynnis ello (L.)	CA, NA, SA, WI	Medina and Franqui, 1999a,b	FO
Erynnis lassauxi merianae Grote	WI	Medina and Franqui, 1999a,b	FO
Phycitidae			
Davara carica Dyar	WI	Medina and Franqui, 1999a,b	FO, FR, TR
Tortricidae			
Amorbia emigratella Busck	CA, NA	FAO, 2000	FO
Adoxophyes fasciculana Walker	PI	FAO, 2000	FO, FR
Decadarchis minuscula Wals	PI	FAO, 2000	FO

ACARI			
Eryophidae			
Calacarus citrifolli Keifer	WI	Abreu, 1994	FO
Calacarus brionese Keifer	WI	Medina and Franqui, 1999a; Abreu, 1994	FO, FR
Tarsonemidae			
Polyphagotarsonemus latus (Banks)	AU, AS, AF, NA, SA, PI, WI	Abreu, 1994	FO, FR
Tydeidae			
Tydeus spp	WI	Medina and Franqui, 1999a; Abreu, 1994	FO, FR
Tetranychidae			
Eutetranychus banski (McGregor)	PI	Yee *et al.*, 1970	FO, FR
Tetranychus cinnabarinus (Boisduval)	AF, ME, PI, WI	FAO, 2000	FO, FR, VE
Tetranychus urticae Koch	SA, WI	BIOSIS, 2000	FO, FR
Tretanychus tumidus (Banks)	WI	Medina and Franqui, 1999a; Abreu, 1994	FO, FR
Tetranychus truncatus Ehara	PI	FAO, 2000	FO, FR
Tenuipalpidae			
Brevipalpus phoenicis (Geijskes)	SA, PI, WI	BIOSIS, 2000	FO, FR
Tuckerellidae			
Tuckerella ornata (Tucker)	PI, WI	Yee *et al.*, 1970	FO
Tuckerella pavoniformis Ewing	CA, SA, PI, WI	Yee *et al.*, 1970; Abreu, 1994	FO

[a]Geographical distribution
AF = Africa, AS = Asia, AU = Australia, CA = Central America, EU = Europe, ME = Middle East, NA = North America, PI = Pacific Islands, SA = South America, WI = West Indies & Caribbean.
[b]Part affected
FR = fruits, FO = foliage, FL = flowers, RO = roots, SE = seed, TR = trunk, VE = vector, NR = not recorded, ? = doubtful host.

seed viability, high economic returns, and its adaptation to wide and diverse soil conditions and climates (Harkness, 1967; Seelig, 1970; Campbell, 1984).

This herbaceous plant is also known as papaw, paw paw, kapaya, kepaya, lapaya, tapaya, papayao, papaya, papaia, papita, lechosa, fruta bomba, mamon, mamona, mamao, and tree melon, and belongs to the dicotyledonous family *Caricaceae*. This small botanical family, indigenous to tropical and subtropical America, is represented by 31 species. *Carica papaya* is the most economically important and widely cultivated species, while *Carica pubescens* (A.D.C.) Solms-Laumb, known as chamburro and babaco (*C. heilbornii* Heilborn) are available in Latin American markets, but are of little economic importance. Even though *C. pubescens* and *C. heilbornii* have been studied for commercial production and as a potential source of resistance genes to papaya viruses, their exploitation requires further investigation and development (Campbell, 1984, 1996; Duke, 1985).

Uses of Papaya

Papaya is mainly cultivated for its edible fruit, but medical and industrial uses have been documented (Seelig, 1970; Morton, 1977, 1987; Duke, 1983, 1984; Yadava *et al.*, 1990). The intended use is important for pest control determinations, as fruit quality is a key factor for fresh market consumption. Ripe fruits are mostly eaten fresh, but green fruits can be cooked as a vegetable or candied by cooking on sugar. The leaves are used as greens in tropical America, the flowers are eaten in Java, the bark is used for making ropes in Africa, and the tree is used as an ornamental plant (Duke, 1983). The foliage has also been tested as green foliage to feed small mammals (Aduku *et al.*, 1989).

The latex of ripe fruits contains a proteolytic enzyme, papain, used in industry and for medical treatments. The industrial and medical uses of papain have been reviewed by Duke (1984, 1985, 1990), Morton (1977, 1987), Seelig (1970), Poulter and Cagygill (1985), Nuñez (1982), and Yadava *et al.* (1990). Papain is used to tenderize meat, clarify beer, treat digestive disorders, degum natural silk, extract fish oil; it is also used in shampoos and face-lifting preparations, in leather and rayon industries, in the manufacture of rubber and chewing gum, in photography, etc. The value of papaya as a medicinal plant is well known (Quisumbing, 1951; Chopra, 1958; Nuñez, 1982). The United States Food and Drug Administration (USDA-FDA) approved the use of chymopapain for treatment of lumbar hernia in humans (Duke, 1983, 1984). Seed extracts have bactericidal potential (Emeruwa, 1982). Sharma and Ogbeide (1982) suggested the use of papaya as a renewable energy source for the production of alcohol fuels.

Production

According to FAO (2000), during 1998 more than 5.2 million t of papaya were produced in 47 countries. Brazil is the largest papaya producer with 32% of the total production (1,700,000 t). Nigeria (751,000 t), Mexico (498,000 t), Indonesia (489,948 t), and India (450,000 t) are among the top papaya producers. Brazil is the third country in area harvested (35,000 ha), but has the highest yields (485,714 kg ha^{-1}). During 1998, Nigeria harvested the largest area (90,000 ha) followed by India (40,000 ha) (FAO, 2000). Most of the papain and papaya for industrial use come from Africa.

In the USA, papaya is cultivated on all the major islands of Hawaii and in Florida. The principal area of commercial production is Kapoho in the Puna district of the island of Hawaii (Yee *et al.*, 1970). Since the mid-1930s, the cultivar 'Solo' has been grown in Hawaii. The papaya ringspot virus-resistant variety 'Rainbow' is now replacing 'Solo' as the predominant variety in Hawaii and Oahu. Hawaii currently grows 1500 ha of papayas with a farm gate value of approximately US$28 million in 1999 and production is increasing.

Depending on the market and use, papayas are harvested green, ripe or unripe. In Hawaii papaya is harvested when the skin is 80% green. In Puerto Rico papaya to be candied in sugar is harvested green, while papaya for the local market is harvested when the skin is 20–25% yellow. Fruit quality for the market is very important. Fruits must be handled with extreme care to avoid scratching and bruising.

Arthropods Associated with Papaya

As with most tropical fruits grown in varied geographical regions, papaya is affected by several arthropods that can be considered key or secondary pests. No comprehensive worldwide list of papaya pests is available, but arthropods from specific regions have been reviewed by a few authors. Morton (1987) reported 11 key arthropod pests affecting papaya in a wide geographical area. FAO-GPPIS (2000) recorded 159 pests for papaya (71 species in the Arthropoda), but the list does not include many insects commonly found on papaya in the Americas and the Caribbean region. Martorell and Adsuar (1952) reviewed the insects associated with papaya in the Antilles and Florida, while Wolcott (1933, 1948), Martorell (1976), Medina *et al.* (1999), Medina and Franqui (1999a,b), and Abreu (1994) listed 32 species of arthropods associated with papaya in Puerto Rico. In Hawaii there are 26 species of insects and mites that attack papaya (Yee *et al.*, 1970; Anonymous, 2000; Follett, 2000). In Australia fruit flies are a predominant concern to papaya producers (Monzu *et al.*, 2000).

Table 5.1 provides a list of arthropods associated with papaya worldwide. The Hexapoda represent 91% of the total number of species, and those remaining are in the Acarina. Tephritid flies (Diptera) are the only direct fruit and seed feeders with 26 species (19%). Eighty-seven species of arthropods can attack or damage the fruit under heavy infestations. The majority of the species (93) attack the foliage or the trunk (42), five species affect the flowers, and three are root feeders. At least 12 species are vectors of important papaya diseases and one mite species is vector of a fungal pathogen. Homoptera is the largest group with 65 species (49% of the total) in 11 families.

Pests Attacking the Fruit

Fruit flies

Fruit flies (Diptera: Tephritidae) are the only group of insects that actually penetrate the pulp or seeds. In spite of their worldwide distribution, their economic status remains unsolved. Twenty-six species from seven genera, *Anastrepha* (two species), *Bactrocera* (17 species), *Ceratitis* (three species), *Dacus*, *Euphranta*, *Myoleja*, and *Toxotrypana* (one species each) attack papaya fruits (Martorell, 1976; Morton, 1987; White and Elson-Harris, 1992; Abreu, 1994; Medina and Franqui, 1999a,b; FAO-GPPIS, 2000; Monzu *et al.*, 2000).

White and Elson-Harris (1992), McPheron and Steck (1996), and Thompson (1998) provided information on the geographical distribution, biology, natural enemies, and synonyms of fruit flies. Most of the *Dacus* spp. affecting papaya are now placed in the genus *Bactrocera*. Only known from Africa, *D. bivittatus* (Bigot) is the only species in this genus reported from papaya, but is considered a doubtful host (Wilson and Elson-Harris, 1992; Thompson, 1998).

Tephritid fruit flies have been serious pests in Hawaii since the first species was found around 1895 (Harris, 1989). They are widespread, occurring from sea level to 2100 m elevation, and attack hundreds of plants, including many crop species. Three species feed on papaya. The oriental fruit fly, *Bactrocera dorsalis* (Hendel), is the most common species and is found at all elevations. The Mediterranean fruit fly, *Ceratitis capitata* (Wiedemann), was once common at lower elevations where papaya is primarily grown but was displaced when the oriental fruit fly appeared in 1945, and is now more frequent at higher elevations (Bess, 1953; Vargas *et al.*, 1995). Melon fly, *Bactrocera curcurbitae* (Coquillett), also feeds on papaya at low elevations.

BIOLOGY Eggs of fruit flies are regularly laid below the skin of the ripening papaya fruit and typically hatch in 1–4 days. Medflies and melon flies lay 10–15 eggs day^{-1}, singly or in clusters, whereas oriental fruit flies lay 130 eggs per day usually in groups of ten, but sometimes as many as 100 or more (Messing, 1999). Larvae will feed for 1–4 weeks, depending on temperature, and drop from the fruit to pupariate in the soil under the papaya plant. Adults emerge in 1–2 weeks. Damage to papaya fruits is caused primarily by larval feeding.

Bactrocera *spp.*

This genus, native to Asia, Australia, and the South Pacific, can be found in Africa and temperate Europe (White and Elson-Harris, 1992). Guiana, French Guiana, and Surinam are the only papaya producing areas of the neotropics where *Bactrocera* spp. are reported (Harris, 1989; White and Elson-Harris, 1992). Seventeen species of *Bactrocera* are known to affect papaya. Monzu *et al.* (2000) and Huxham (2000) list *Bactrocera papayae* (Drew and Hancock) as one of the most threatening pests to papaya in Australia. The Asian papaya fruit fly *B. papayae*, a major polyphagous species recorded from 193 species (Allwood *et al.*, 1999), is native to South-East Asia, invaded Queensland in 1995 (Hancock *et al.*, 2000) and was subsequently eradicated. Females can lay eggs in green papayas and citrus and young bananas. Female *B. papayae* has a long ovipositor, allowing it to penetrate past the sap layer of green fruits.

Anastrepha *spp.*

This genus is native to the neotropics and has not established itself outside the Americas (White and Elson-Harris, 1992). Two species from this genus attack papaya (Swanson and Baranowski, 1972; Nguyen *et al.*, 1993): the Caribbean fruit fly, *Anastrepha suspensa* (Loew) (Weems, 1966; Swanson and Baranowski, 1972; Nguyen *et al.*, 1993) can develop on fully ripe papayas (Lara *et al.*, 1989), but green papayas are resistant to attack. The reduced preference for green papaya fruits is associated with the presence of benzyl isothiocyanate in the latex of green fruits (Seo *et al.*, 1982a,b, 1983; Liquido *et al.*, 1989; Liquido, 1991c; Nguyen *et al.*, 1993). Although *A. suspensa* is a pest of several crops in Puerto Rico (Martorell, 1976) it has not been reported as a pest of papaya on the island (Martorell, 1976; Abreu, 1994; Medina and Franqui, 1999a,b; Medina *et al.*, 1999).

Toxotrypana curvicauda *Gerstaecker*

The papaya fruit fly, *T. curvicauda*, is considered to be the most damaging insect pest of papaya (Abreu, 1994; Heath *et al.*, 1996; Medina *et al.*, 1999) (Plate 28). This species is present in Central America, the Caribbean, and in tropical and subtropical areas of North and South America, including Florida, USA (Wolcott, 1933; Peña, 1986; Peña *et al.*, 1986; Abreu, 1994; Landolt, 1994a,b; Medina *et al.*, 1999). Although it was originally reported as an exclusive host of *C. papaya* (Knab and Yothers, 1914; Wolfenbarger and Walker, 1974; Castrejón and Camino, 1991), recent studies have demonstrated a broader host range, including other species of *Caricaceae* (*Jacavita mexicana* (A.D.C.) = *Pileus mexicanus* Johnston), four species of the family *Asclepiadaceae*, and occasionally mango, *Mangifera indica* L. (Weems, 1969; Castrejón, 1987; Castrejón and Camino, 1991; Peña, 1993; Landolt, 1994a).

The adult *T. curvicauda* fly resembles a common *Polistes* wasp in behaviour, size, form and general coloration. A long, slender, curved ovipositor, exceeding the length of its body, is a distinctive characteristic of this fly (Knab and Yothers, 1914; Medina *et al.*, 1999). The long ovipositor allows the female to lay the eggs (about ten per fruit) inside young fruits. The preferred fruit size for oviposition is between 5 and 12 cm in diameter (Aluja *et al.*, 1997b). Upon emergence, the larva feeds on the seed mass and lining of the seed cavity, damaging the fruit (Plate 29). In Puerto Rico, 84% fruit loss has been observed in semi-commercial plots (Abreu, 1994). The grown larva exits the fruit to pupate in the soil. Depending on the temperature and soil humidity, larval and pupal development requires 15–17 days and 2–6 weeks respectively (Weems, 1969; Landolt *et al.*, 1985; Peña *et al.*, 1986; Aluja *et al.*, 1997b).

Ceratitis *spp.*

Three species, *C. capitata* (Wiedemann), *C. catoirii* Guerin-Meneville, and *C. rosa* Karsch, are known from papaya. The Mediterranean fruit fly, *C. capitata*, is a polyphagous pest in almost all papaya producing areas of the world. The distribution of *C. catoirii*, and *C. rosa* is restricted to Asia and Africa (White and Elson-Harris, 1992; Thompson, 1998).

FRUIT FLY SAMPLING AND MONITORING Most reports on the papaya fruit fly *T. curvicauda* are from the USA and Mexico. Knab and Yothers (1914) reviewed records and distribution of the pest. Detailed descriptions of eggs, larvae, pupae, and adults were provided in this early work. Further studies have concentrated on adult behaviour and the male sex pheromone (Landolt and Hendrichs 1983; Landolt, 1984a; Landolt *et al.*, 1985, 1988, 1991; Landolt and Heath, 1988, 1990; Landolt and Reed, 1990; Aluja *et al.*, 1997a,b). Peña (1986) and Peña *et al.* (1986) studied oviposition and feeding behaviour on papaya seeds. Aluja *et al.* (1997a) quantified daily activity patterns and within-field distribution of the papaya fruit fly, and advised that if papaya plantations are mixed (papaya, mango, avocado, soursop), both within-orchard distribution and daily movement patterns differ when compared with those observed in papaya planted as a monocrop. Sampling methods and the use of pheromone traps for *T. curvicauda* have been studied by Chuman *et al.* (1987), Landolt and Heath (1988, 1990), Landolt and Reed (1990), Landolt *et al.* (1985, 1991), and Heath *et al.* (1996).

In Hawaii, fruit flies are monitored using traps baited with male lures. Methyl eugenol is used to attract oriental fruit flies, cuelure is used for melon fly, and trimedlure (most widely used) or ceralure for medfly (CABI and EPPO, 1997). Fay *et al.* (1997) reported that methyl eugenol trapping and regular host fruit surveys are recommended in Australia to monitor *B. papayae*.

CONTROL Several methods have been reported for papaya fruit fly control. Cultural control methods were reported by Aluja *et al.* (1997a,b) and Landolt (1984b), while chemical control measures were provided by Mason (1922), Wolfenbarger (1962), Conover and Waddill (1981), Peña and Nagel (1988), Aluja (1993, 1994), and Abreu (1994). Traditional measures involve insecticides and toxic baits. Unlike other tephritids attracted to protein baits, *T. curvicauda* does not require protein for ovarian maturation, therefore baits based on brown sugar and insecticides are recommended for control (Landolt, 1984b; Sharp and Landolt, 1984). Other control measures include destruction of infested fruits and removal of wild hosts.

Population suppression in papaya fields can be achieved by several methods. Sanitation is a first step; fruits should be removed as they ripen, and all fallen or infested fruit should be destroyed (Liquido, 1991b, 1993). In Hawaii, sanitation is usually insufficient by itself because fruit flies are abundant on alternative host plants and can fly in from outside areas. Insecticide protection is possible using cover or bait sprays (Messing, 1999). Malathion is the most commonly used insecticide but the microbe-derived toxin spinosad may soon become a widely accepted alternative to malathion (Peck and McQuate, 2000). Malathion can be combined with protein hydrolysate to form a bait spray (Roessler, 1989). With bait sprays, male and female fruit flies are attracted to a protein source from which ammonia emanates. Bait sprays are preferred to cover sprays because they are applied as spot treatments and the impact on natural enemies is minimal.

Biological control has been tried on fruit flies with little success, but the potential of inundative parasitoid releases (Bautista *et al.*, 1998), alone or with bait sprays, is being studied (Wood and Hardin, 2000). The parasitoid, *Doryctobracon toxotrypanae* (Marsh) from Costa Rica has been reported from *T. curvicauda*. The most effective parasitoid enemy of medfly and oriental fruit fly in Hawaii is *Fopius arisanus* (Vargas *et al.*, 1993). Petcharat (1997) reports that *Diachasmimorpha longicaudata* Ashmead (Hymenoptera: Braconidae) is responsible for 42% reduction of *B. papayae* densities in Thailand. Male annihilation, using attraction of males to insecticide-laced lures (Vargas *et al.*, 2000), and sterile insect techniques, using releases of large numbers of sterile flies to disrupt reproduction, have been used elsewhere to eradicate fruit flies but these tactics are not presently considered feasible in Hawaii.

Harvesting early is an effective means to avoid fruit fly damage. Papayas are usually fruit-fly-free when picked less than one-quarter ripe (Seo *et al.*, 1982a,b; Liquido *et al.*, 1989; Liquido 1990, 1991a,c). Although fruit continues to ripen after harvest, harvesting too early to avoid fruit fly infestation can result in diminished fruit flavour, as fruit will not ripen fully.

In Mexico, *T. curvicauda* incidence is affected by altitude and precipitation. In the State of Veracruz, a low incidence of papaya fruit flies is reported on 'Cera' type and the 'Maradol' variety (Machain, 1983). However, at higher altitudes in Morelos State (1400–1800 m), *T. curvicauda* is a serious problem. Differences in varietal susceptibility have been documented for the 'Hawaiian' and 'Cera' varieties (Aluja *et al.*, 1994). It is not clear whether the effects of aggressive predators, unfavourable meteorological conditions, or specific *T. curvicauda* strains are responsible for such differences in papaya susceptibility in Mexico. The combination of orchard design, trap crops, and border trappings have been proposed as a means to reduce *T. curvicauda* damage in commercial orchards in Mexico (Aluja *et al.*, 1997a,b). A trap crop of 10 m surrounding the main block of papaya trees can reduce the incidence of *T. curvicauda* (Aluja *et al.*, 1997a,b).

ERADICATION A continuous effort to eradicate and prevent *C. capitata* from invading new areas of southern Mexico, northern Central America, and the USA has been successful based on the strategic geographic location, and a permanent cooperative effort among phytosanitary authorities in Mexico, Guatemala, and the USA. Most agricultural regions of Mexico have been free of the medfly for decades. However, with a highly diversified agroecological system and heterogeneous social and cultural population structure, the confronting political, military, migratory and civil disputes make possible recurrent medfly invasions from southern regions into Mexico and the USA.

QUARANTINE Hawaii serves as a reservoir for the introduction of tephritid fruit flies and other regulatory pests into the USA mainland. Quarantine regulations require papaya to be treated before export from Hawaii to the USA mainland and Japan. Currently, papayas for export from Hawaii receive a single-temperature vapour heat treatment developed to disinfest fruit of fruit flies. This treatment requires fruits to be heated to a fruit centre temperature of 47.2°C during a treatment duration of not less than 4 h (Armstrong *et al.*, 1995). A recent study showed that all stages of white peach scale on the surface of papaya fruit are also killed by this vapour heat treatment (Follett and Gabbard, 1999).

Arthropods Affecting the Foliage and the Trunk

Most of the arthropods associated with papaya affect the foliage, the trunk or both (Table 5.1). Under heavy infestations some species of scale insects, thrips and mites can affect flowers and fruits, but are usually associated with the trunk or the foliage. The orders Hemiptera, Homoptera, Thysanoptera, Coleoptera, Lepidoptera, and Acarina are associated with the foliage or the trunk and represent 7, 49, 3, 5, 9, and 9% of the total number of species reported for the crop, respectively (Table 5.1). A few species of Hemiptera, Thysanoptera and Acari can also attack the flowers and fruits.

Scale insects

Thirty-eight species (28%) from 24 genera and six families of scale insects affect papaya. Two families, Diaspididae and Coccidae, represent 66% of the scale insects reported and 19% of the total number of arthropods related to papaya.

White peach scale, Pseudaulacaspis pentagona *(Targioni-Tozetti)*

Pseudaulacaspis pentagona (Targioni-Tozetti) (Homoptera: Diaspididae) has a cosmopolitan distribution and is one of the most economically important scale insects in southeastern USA where it is a serious pest of peaches and other fruit and ornamental crops (Nakahara, 1982; Gill, 1997). The white peach scale was collected for the first time in Hawaii in September 1997 on papaya. In Hawaii, its distribution is presently limited to the east side of the island of Hawaii, but is expected to expand rapidly.

Females begin laying eggs about 2 weeks after mating and lay their full complement of eggs within 8–9 days. Eggs hatch in 3–4 days after oviposition (Bennett and Brown, 1958). Crawlers settle and begin feeding within 2 days after hatching and complete development in about 1 week. Two subsequent moults requiring about 3 weeks produce adult females. Second instar males construct an oblong armour and after three moults become winged adults. Progeny is produced only through mating. The white peach scale initially attacks the trunks of papaya plants near the base. Overcrowding causes spread of the scale up the trunk, and in heavily infested trees scales move up on to fruit, preferring to settle in the calyx and peduncle regions.

White peach scale is a potential threat to the Hawaiian papaya industry as a source of tree stress and fruit downgrading, and as a quarantine pest on fruit for export. California, an important destination for Hawaiian papayas, has given *P. pentagona* a 'Q' rating, meaning that the pest is not found in the state, yet has economic pest status where it is known to occur. If live white peach scales are detected during inspection in California, plant quarantine officials could take action, and this scenario could also occur at other overseas destinations for Hawaiian papayas.

CONTROL Elsewhere, *P. pentagona* is attacked by parasites and predators (Bennett, 1956; Collins and Whitcomb, 1975), but chemical control is often required to prevent severe crop injury. Control in the field in Hawaii has been attempted using Sunoil sprays to the trunk of the papaya tree with limited success.

Philephedra tuberculosa *Nakahara and Gill*

This scale insect is a pest of papaya, sugar apple, *Annona squamosa* L., soursop, *Annona muricata* L. and several species of ornamentals (Nakahara and Gill, 1985; Peña *et al.*, 1987; Abreu, 1994) (Plate 30). Three types of damage by this scale insect have been documented: distortion of the apical point during the seedling stage; flower and leaf drop occuring after heavy infestations (Peña *et al.*, 1987); and cosmetic damage occuring if the females attach to the fruits (Peña *et al.*, 1986; Abreu, 1994; Medina and Franqui, 1999a). Damage to fruits is of considerable importance, as affected fruits are unmarketable; however, the pest status of this insect remains unsolved.

The life history, natural enemies and behaviour of *P. tuberculosa* have been studied by Peña and McMillan (1986) and Peña *et al.* (1987). Life span is 24 and 59 days for males and females, respectively. Females produce 87 crawlers per day in a 7-day oviposition period. Important natural enemies include ten arthropods and *Verticillium lecanii* (Zimmenn.).

Mealybugs

Papaya mealybug, Paracoccus marginatus *Williams and Granara de Willink*

Paracoccus marginatus Williams and Granara de Willink is a pest of papaya, cassava (*Manihot esculenta* Crantz), *Hibiscus* spp., aubergine (*Solanum melongena* L.), avocado (*Persea americana* Mill.), annona (*Annona* spp.), and sweet potato (*Ipomoea batatas* (L.) Lam.). The insect has been reported from papaya in Baja, California and from cassava in the central valleys of Mexico. The papaya mealybug occurs in tropical and subtropical climates, principally in the coastal states of Mexico (Williams and Granara, 1992). The papaya mealybug has been reported since 1994 in the Caribbean islands of Antigua, Belize, British Virgin Islands, the Dominican Republic, Guatemala, Haiti, Nevis, St Kitts, Puerto Rico, the US Virgin Islands and Costa Rica, and from continental USA (Florida) since 1998 (Miller *et al.*, 1999).

The insects feed on leaves, stems, fruits and even on seedlings. Mealybugs cause deformation, wrinkling and rolling of the leaf edges and early leaf drop (Miller *et al.*, 1999). Attack to unripe fruits causes sap running and blemishes, a source of fruit downgrading. Papaya fruits can be heavily infested with mealybugs, becoming white and essentially inedible. Under heavy infestations, the abaxial side of the lower leaves can be covered with insects that congregate near the main vein. In Mexico, heavy infestations occur rarely in commercial orchards, probably due to the presence of natural enemies. However, major infestations in Veracruz, Mexico, occur in June prior to the summer rainy season (Valencia, 1975).

Diagnostic characters are a yellowish body colour, with a series of short waxy filaments around the margin, of less than one-quarter the length of the body. The ovisac is produced ventrally only, but can be two or more times the length of the body of the adult female. When specimens are placed in alcohol, they become blue-black, which is characteristic of the species in this genus (Williams, 1986). Miller *et al.* (1999) reviewed the morphological characters of the adult female.

CONTROL Biological control appears to be the main factor keeping the species under control in Mexico, where the most important natural enemies are *Anagyrus* sp., *Acerophagus* sp., near *texanus* Howard, and *Apoanagyrus* sp. (González *et al.*, 1999). Common predators are *Chrysopa* sp. and *Chilocorus cacti* L. which are usually found in low densities. Due to its potential pest status in the Caribbean region, a classical biological control programme against *P. marginatus* was initiated, involving introduction of parasites from Mexico into the Bahamas and Florida, USA (González *et al.*, 1999).

Mites

Twelve species of mites in seven genera affect papaya. Mites are probably the most persistent arthropod pests of papaya (Plate 31). The lack of basic information on mite biology and ecology on papaya has prevented the development of effective management practices. Naturally occurring predators can suppress mite populations after pesticides are removed from the system. However, most papaya producers apply insecticides on a calendar basis, disrupting the natural pest balance. In Hawaii, during early spring, when natural enemies are low and plants are susceptible, mite populations can reach densities that trigger the use of disruptive acaricides (i.e. beneficial predatory mites and pest mites are killed), and a pesticide treadmill begins for the rest of the season.

Three species of pest mites feed on papaya in Hawaii: the carmine mite, *Tetranychus cinnabarinus* (a key pest); the red and black flat mite, *Brevipalpus phoenicis* (an occasional pest), and a newly invading species, the papaya leaf edge roller mite, *Calacarus brionese* (an eriophyid mite introduced into Hawaii in 1990 and currently spreading between islands). Most recently the papaya leaf edge roller mite has become a primary concern.

Otherwise considered secondary pests, present mainly during the dry season, mites are currently considered the most persistent papaya pest in Mexico. As a result of a production expansion during the 1990s, Mexican growers relied on chemical control to produce high quality fruits and triggered an outbreak of mites.

Carmine spider mite, T. cinnabarinus *(Boisduval)*

T. cinnabarinus is a cosmopolitan species considered a papaya pest in several countries (Hernández *et al.*, 1995). Females have an oval shape with a red or green coloration and a body from 0.4 to 0.5 mm in length. In a 3-week life span the female oviposits 200 eggs on the abaxial side of the leaf. Eggs are about 0.1 mm long with an incubation period of 4–7 days. Males are smaller than females. As with most mites, larvae have six legs, and their bodies are light pink. Nymphal development requires 3–5 days. The protonymphal and deutonymphal stages, characterized by a red to green coloration and the presence of four pairs of legs, require 6–10 days to complete development.

Carmine mites feed on epidermal cells, causing leaf curling, chlorosis, and plant stunting. Mites can also cause premature yellowing and leaf drop as well as direct damage on small fruits. Mites feeding on developing fruits can cause fruit downgrading. In Puerto Rico up to 79% of commercial papaya trees can be affected by *T. cinnabarinus* (Abreu, 1994). In addition to direct feeding damage, the presence of *T. cinnabarinus* on Puerto Rican papayas is associated with infection by *Oidium* spp. (Abreu, 1994). The combined effect of mites and fungal damage causes severe chlorosis.

Broad mite, Polyphagotarsonemus latus *(Banks) (Tarsonemidae)*

Polyphagotarsonemus latus (Banks) is a cosmopolitan polyphagous species, with a wide host range. Affected leaves curl up and break easily. Under severe infestations the foliar area is reduced and leaves are severely deformed and chlorotic. Abundant webbing on the foliage is characteristic of this species. The damage caused by this mite, sometimes confused with symptoms of papaya ringspot virus, can be distinguished by the absence of aqueous spots and ring forms on stems and fruits, which are characteristic of the virus infection.

In Puerto Rico and Brazil, *P. latus* is an important pest, attacking papaya seedlings in greenhouses, while in Chiapas, Mexico, damage by *P. latus* is severe under field conditions (M. de Coss, personal communication). Severe stunting to seedlings was documented by Abreu (1994) in Isabela, Puerto Rico. The damage caused by this mite in Brazil has been confused with papaya bunchy top (Fletchman, 1983).

CONTROL Sulphur dusting and mineral oils are commonly used for mite control in Mexico; however, pesticide resistance is suspected in the dry Pacific regions, where *T. cinnabarinus* is the most important pest of papaya. Hawaii growers typically apply sulphur or a synthetic acaricide on a calendar basis to suppress mites.

The most effective predator in Hawaii and Puerto Rico appears to be the predatory mite, *Phytoseiulus macropilis* (Banks) (Prasad, 1966; Abreu, 1994). Two other predatory mites also commonly occur in Hawaii, *Euseius* sp. and *Phytoseiulus hawaiiensis*, but their dynamics are poorly understood at present (Prasad, 1966). According to Abreu (1994) the presence of *P. macropilis* is common in papaya, but densities are too low to effectively control mites. In Mexico a coccinellid, *Stethorus* sp., feeds on spider mites, but their impact on controlling natural populations has not been established.

Leafhoppers (Homoptera: Cicadellidae)

Nine cicadellid species from three genera (*Empoasca*, *Poeciloscarta* and *Sanctanus*) can affect papaya (Table 5.1). Leafhoppers cause two types of damage: direct feeding and secondary damage as vectors (Plate 32). Symptoms of leafhopper feeding include tip burn, wrinkling and cupping of the leaves, burning of leaf margins in large trees, and stunting of smaller plants (Ebesu, 1985; Medina *et al.*, 1999). Leafhoppers are more important for their vectoring ability than for the mechanical damage.

In the Caribbean region, papaya production is severely limited by papaya bunchy top disease, transmitted by *Empoasca papayae* Oman, *E. stevensi* Young, and *E. insularis* Oman (Baker, 1936; Adsuar, 1946; Sein and Adsuar, 1947; Martorell and Adsuar, 1952; Nielson, 1968; Haque and Parasram, 1973; Fletchman, 1983; Web and Davis, 1987; Davis, 1994; Brunner *et al.*, 1996; Davis *et al.*, 1996, 1997). There has been controversy on the causal agent of papaya bunchy top (Bird and Adsuar, 1952; Pontis, 1953; Eugene *et al.*, 1968; Nielson, 1968; Story and Halliwell, 1969). Recently Davis *et al.* (1996, 1997) reported that a rickettsia and not a phytoplasm is associated with papaya bunchy top disease. Although attempts to culture the rickettsia in axenic culture have been unsuccessful, if this rickettsia causes the disease it will be the first report of a leafhopper-transmitted, lactifer-inhabiting, plant pathogenic rickettsia (Davis *et al.*, 1996, 1997).

Empoasca *spp.*

In Puerto Rico, where the papaya bunchy top was first reported (Cook, 1931), seven leafhopper species, *Empoasca canavalia* DeLong, *E. dilitaria* Delong and Davidson, *E. papayae* Oman, *E. insularis* Oman, *E. fabalis* DeLong, *Poeciloscarta laticeps* Metcalf and

Bruner, and *Sanctanus fasciatus* (Osborn) are associated with papaya (Martorell, 1976; Ramirez, 1997); however, only two species, *E. papayae* (the predominant species) and *E. fabalis* (secondary species), are commonly collected from papaya and are associated with papaya bunchy top (Ramirez, 1997).

The principal vector of papaya bunchy top, *E. papayae*, is not known to occur in Florida or Hawaii where another species, *E. stevensi* Young, is present. However, *E. stevensi* is reported to be a papaya bunchy top vector in Trinidad (Young, 1953; Haque and Parasram, 1973). This species was identified for the first time in Hawaii in 1980 as the cause of phytotoxaemia to papaya (Ebesu, 1985).

Little is known about the population dynamics and natural enemies of *E. fabalis* and *E. papayae* (Martorell, 1976; Ramirez, 1997). Furthermore the biology and development of both species are unknown on papaya plants. In Puerto Rico, *E. papayae* is the principal vector of papaya bunchy top (Brunner *et al.*, 1996) representing 90% of the *Empoasca* spp. associated with papaya (Ramirez, 1997). Heavy *E. papayae* infestations on papaya are related to the presence of weeds in the field. The critical period of susceptibility to papaya bunchy top includes the first 90 days after transplanting (Brunner *et al.*, 1996; Ramirez, 1997). This suggests that management practices for *E. papayae* should focus on the early period of plant establishment to reduce the contact between the insect vector and the crop. The natural parasitoid fauna of *E. papayae* have not been established.

Symptoms of *E. stevensi* feeding are similar to the familiar 'hopperburn' associated with feeding by other *Empoasca* species (Ebesu, 1985). In Hawaii, egg incubation requires 7–14 days and the nymphal period is approximately 12 days. Adult feeding results in more severe injury than nymphal feeding (Ebesu, 1985).

Aphids (Homoptera: Aphididae)

Aphids do not colonize papaya plants and are considered minor pests, but several species, *Aphis coreopsidis* (Thomas), *Aphis nerii* Boyer de Fonscolombe, *Aphis gossypii* Glover, *Aphis spiraecola* Patch, *Myzus persicae* (Sulzer) and *Toxoptera aurantii* (Boyer de Fonscolombe) can be found on papaya plants (Namba and Higa, 1981; Abreu, 1994; Medina *et al.*, 1999) or collected on water pan traps in papaya fields (Villanueva-Jiménez and Peña, 1991; Rabara *et al.*, 1996). Aphids are considered a serious threat to papaya production due to their ability to transmit diseases, in particular papaya ringspot virus (PRSV) and the papaya mosaic virus (Adsuar, 1947a,b; Pontis, 1953; Nariani, 1956; Ishii and Holtzmann, 1963; Khurana and Bhargava, 1971; Becerra, 1987). Papaya production is limited by PRSV in many areas of the world including Brazil, Hawaii, India, Philippines, Puerto Rico, Mexico, and Malaysia. In a 7-year study, Abreu (1994) found no aphids colonizing papaya plants in Puerto Rico.

Several aphid species (*M. persicae, M. euphorbiae, A. spiraecola* (= *A. citricola*), *A. gossypii, A. craccivora, A. nerii, Rhopalosiphum maidis,* and *T. auranti* (Boyer de Fosc) are capable of transmitting PRSV to papayas in Hawaii (Ishii and Holtzmann, 1963; Namba and Kawanishi, 1966; Higa and Namba, 1971; Namba and Higa, 1981; Labonne *et al.*, 1992), Mexico (García, 1987; Villanueva and Villanueva-Jiménez, 1994), and Puerto Rico (Adsuar, 1946; Martorell and Adsuar, 1952; Schaefers, 1969).

PRSV produces distinct ringspots on fruits, stunting of plants, and leads to a reduction in fruit production. Control is difficult because the virus is transmitted by aphids in a non-persistent manner (Namba and Higa, 1981). This means that the virus can be acquired during brief probes, retained for several hours, and transmitted to a new host within minutes. Thus, PRSV can be transmitted by non-colonizing species, and insecticides may not be effective at preventing transmission (Namba and Higa, 1981).

In the early 1960s on the island of Oahu and in the late 1980s on the island of Hawaii, the chief papaya production areas at the time were abandoned due to infestation by a highly virulent strain of PRSV (Ishii and Holtzmann, 1963). Recently, through genetic transformation with PRSV, a PRSV-resistant variety of papaya, 'Rainbow', has been developed (Fitch *et al.*, 1992). Areas previously abandoned on both islands are now in cultivation with the transgenic variety.

CONTROL Refined oil sprays have been suggested as physical barriers for viral transmission. However, inconsistent results and high control costs prevent wide adoption of this technology (Bhargava and Khurana, 1969; Hernández *et al.*, 2000a). Organophosphate and carbamate insecticides are not recommended, as sublethal doses can accelerate viral transmission by exciting probing activity in aphids. Applications of neem (*Azadirachta indica* A. Juss.) extracts affect aphid probing, but do not prevent PRSV transmission (Hernández *et al.*, 2000b; Pérez *et al.*, 2000).

Integrated crop management strategies for papaya aphids have been developed in Veracruz, Mexico (Flores *et al.*, 1995), and in the Philippines (Opina and Tomines, 1996) to manage the viral disease and the vectors. Barrier crops have been proposed as a way to interfere with aphid landing and searching behaviour. The use of companion crops such as sorrel (*Hibiscus sabdariffa* L.) planted around the fields 2 months before transplanting papaya may reduce the virus incidence (Becerra, 1989). The deep red to purple *H. sabdariffa* coloration is believed to repel aphids from landing on papaya fields. Intercropping barriers of maize or sorghum are used as intermediate landing crops in the Philippines (Andrade *et al.*, 1994; Opina and Tomines, 1996). Aphids, carrying non-persistent virus, clean their stylets after feeding on the companion crops, reducing the viral infection (Andrade *et al.*, 1994). Destruction of plants with viral symptoms reduces the source of primarily inoculum and delays development of the infection in the field (Becerra, 1987; Arenas *et al.*, 1992). Protecting the seedlings under polypropylene or anti-aphid covers is recommended to reduce rapid field infestations (Andrade *et al.*, 1994).

Little is known about natural enemies of aphids on papaya. A coccinellid *Cycloneda sanguinea* L. found on papaya plants feeds on aphids (Machain, 1983), but the effect of this predator on natural aphid populations still has to be determined.

Defoliators (Lepidoptera, Coleoptera)

Papaya has few defoliators. Grasshoppers, hornworms and a weevil are reported as defoliators. Little information is available on these defoliators.

Hornworms

Hornworms (Lepidoptera: Sphingidae), are papaya and cassava (*Manihot esculenta*) pests (Clavijo and Chacín, 1992; Abreu, 1994; Medina and Franqui, 1999b). In Veracruz, Mexico, *Erinnys ello* (L.) is considered the most important papaya defoliator. In North, Central, and South America the larvae can be found year round, but it is more abundant during the summer months. Three species, *E. alope* (Drury), *E. ello*, and *E. lassuxi merianae* Grote are reported from papaya in Puerto Rico (Martorell, 1976; Medina and Franqui 1999b), but damage is sporadic (Abreu, 1994).

Last instar *E. ello* larvae are 90–120 mm in length. Body colour varies from green, red, grey, yellow, to dark green or black. The larva feeds on leaves, destroying seedlings and small trees and defoliating larger trees in a short time, then drops to the ground for pupation. The larva has a dorsal horn at the end of the abdomen. The pupae have a reddish brown to black coloration. Moths are nocturnal and present reddish brown to grey forewings. Males have longitudinal dark spots on the forewings. The hindwings are reddish brown with a dark brownish red marginal band with two dorsal black dotted lines on the abdomen. Wing span is 80–90 mm.

CONTROL Weed control practices and soil preparation can reduce adult and pupal populations, respectively. In Mexico, manual removal of last instar larvae is a common practice in smallholdings; however, carbaryl and malathion have been used on small larvae in large plantings (Machain, 1983). The gregarious and polyembryonic braconid wasp, *Apanteles* sp., parasitizes hornworn larvae. After completing its development, the parasitic larvae emerge to pupate by covering hornworms with white cocoons. These cocoons are very often hyperparasitized as well. According to Abreu (1994) natural control keeps hornworm under control in Puerto Rican papaya orchards.

Minor Pests of Papaya

Yee *et al.* (1970) listed all the insect pests known to occur on papaya in Hawaii at the time and labelled them as major, minor, or occasional pests. Minor pests on the list included the cotton aphid, *Aphis gossypii*; potato aphid, *Macrosiphum euphorbiae*; southern green stink bug, *Nezara viridula*; southern garden leafhopper, *Empoasca solana*; black cutworm, *Agrotis ipsilon*; coconut scale, *Aspidiotus destructor*; mining scale, *Howardia biclavis*; obscure mealybug, *Pseudococcus obscurus*; onion thrips, *Thrips tabaci*; the Texas citrus mite, *Eutranychus banksi*; and two tukerellid mites, *Tukerella ornata*, and *T. pavoniformis*. The spiralling whitefly, *Aleurodicus dispersus* Russell, was discovered in Hawaii in 1978 and reportedly became a serious pest soon after its introduction. Two thrips, *Frankliniella occidentalis* and *F. fusca*, are present in Hawaii and are known to be vectors of tomato spotted wilt virus on papaya (Sakimura, 1972). The impact these minor pests have on papaya production in Hawaii is unknown.

Diaprepes abbreviatus (L.) (Coleoptera: Curculionidae)

Puerto Rico and Florida are the only papaya producing areas reporting *D. abbreviatus* defoliating papaya. The diaprepes weevil is a known sugarcane and citrus pest (Wolcott, 1948; Martorell, 1996) that has been documented as causing defoliation to papaya (Abreu, 1994). Although the pest status or economic importance of *D. abbreviatus* on papaya has not been defined, up to 99 *D. abbreviatus* adults per papaya tree have been observed in northern Puerto Rico in orchards where 51% of the trees were defoliated by the weevil (Abreu, 1994). The diaprepes weevil is present in Florida, USA, on citrus and ornamentals (Schroeder and Beavers, 1977) and lately it has been observed on papaya, feeding on leaves, fruit and roots (J.E. Peña, personal observation).

Fruitspotting bugs

Six species of coreids from the genera *Amblypelta* and *Brachylybas* affect papaya in Australia, New Guinea, and the South Pacific Islands (Brown, 1958a,b; Waite, 1990). The genus *Amblypelta* includes 15 species that attack several fruits, nuts, and papaya. The most damaging species are *A. cocophaga* China, *A. theobromae* Brown and *A. lutescens papuensis* Brown (Brown, 1958b; Waite, 1990; Waite and Huwer, 2000). Waite and Huwer (2000) reviewed the host records and the pest status of *Amblypelta* in Australia.

A few species of the orders Coleoptera and Lepidoptera are reported from papaya in the Pacific islands (Table 5.1), but little is known about their importance and damage to papaya production. Furthermore, little information is available on papaya pests from the Pacific islands, except for Hawai and Australia.

Discussion

Papaya is cultivated in semi-permanent stands in frost-free areas in the tropics and subtropics. A large number of arthropods are associated with papaya. Fruits for fresh consumption and quarantine issues are important marketing factors. Therefore, farmers implement some type of pest management tactic to harvest high quality, blemish-free fruits. Traditionally, most papaya producers apply insecticides on a calendar basis. Public pressure regarding pesticide regulation, the disruption of the natural pest balance, contamination, resurgence of secondary pests, and the need for the least possible disruption of the environment encouraged the search for natural pest control mechanisms.

Human intervention to control fruit flies and leafhoppers on papaya has triggered outbreaks of mites and scales that can become persistent pests of papaya. Naturally occurring predators can suppress secondary pest populations after pesticides are removed from the system, but the lack of basic information on secondary pest biology and ecology on papaya has prevented the development of effective management practices, thereby encouraging producers to use pesticides.

A clear understanding of the pest biology, behaviour, population dynamics and pest status is the foundation for the development of integrated pest management (IPM) strategies. Unfortunately, in spite of the economic importance and wide geographical distribution of the crop, papaya pest control, with the exception of fruit flies, has been poorly studied. Currently, no IPM programme is available, even for an insect complex such as fruit flies, where abundant information on behavioural responses to pheromone and host finding, trapping systems, habitat manipulation, orchard design, and sanitation practices is available. Current and recent developments in the integration of sampling, and the use of food attractants and insecticides has allowed a reduction in the use of broad spectrum pesticides; however, farmers rely heavily on insecticide use and postharvest treatments to manage fruit flies. Research is needed on biologically and culturally based practices to manage indirect pests and to integrate all available tactics for insects damaging the fruit.

Aphids and leafhoppers are important pests of papaya in the Americas and the Caribbean mainly due to their vectoring capacity. Factors affecting host finding and colonization by aphids and leafhoppers need to be studied and integrated to existing cultural practices for other pests, mainly fruit flies. Virus-resistant papaya varieties are available, but further work is needed on aphid sampling, host finding, colonization and insect–pathogen relationships. Papaya bunchy top is still a limiting factor for papaya production in the Caribbean, but little work has been conducted on the vector biology, sampling, natural enemies and the pathogen–insect–plant relationship. Only poor to modest relationships have been shown between aphids and leafhopper vectors and the number of affected plants in a field. It is therefore unclear how chemical control of adults will reduce damage.

Biological and cultural control tactics on papaya-based systems need further attention. Culturally based practices can provide a first line of defence against secondary pests and such practices are available for other crops in countries producing papaya. Research and extension protocols should emphasize integrating cultural and biologically based practices in order to develop IPM and integrated crop management programmes.

References

Abreu, E. (1994) Insectos de la papaya y prácticas de control. In: *Memorias Foro: Cultivo, Producción y Manejo*. Estación Experimental Agrícola de Puerto Rico, Recinto de Mayagüez, Univ. Puerto Rico, Aguadilla, Puerto Rico, 20 May 1994, pp. 58–62.

Adsuar, J. (1946) Transmission of papaya bunchy top by a leafhopper of the genus *Empoasca*. *Science* 103, 316.

Adsuar, J. (1947a) Studies on virus diseases of papaya (*Carica papaya*) in Puerto Rico. I. Transmission of papaya mosaic. *Journal of the Agriculture University of Puerto Rico* 31, 248–256.

Adsuar, J. (1947b) Studies on virus diseases of papaya (*Carica papaya*) in Puerto Rico. II. Transmission of papaya mosaic by green citrus aphid (*Aphis spiraecola* Patch). *Journal of the Agriculture University of Puerto Rico* 31, 257–259.

Aduku, A.O., Dim, N.I. and Hassan, W. (1989) Evaluation of green forages for dry season feeding of rabbits. *Journal of Applied Rabbit Research* 12, 113–115.

Allwood, A.J., Chiranajariwong, A., Drew, R.A., Hanacek, E., Hancock, D., Hengsawad, C., Jinapin, C., Jirasurat, M., Kong Krong, C., Kritsaneepaiboon, S., Leong, C. and Vijaygasegaran, S. (1999) Host plant records for fruit flies (Diptera: Tephritidae) in South-East Asia. *The Raffles Bulletin of Zoology* 7, 99 pp.

Aluja, M. (1993) *Manejo integrado de las moscas de la fruta*. Editorial Tirillas, México, D.F.

Aluja, M., Jiménez, A., Camino, M., Aldana, L., Castrejón, V. and Valdés, M.E. (1994) Determinación de la susceptibilidad de tres variedades de papaya (*Carica papaya*) al ataque de *Toxotrypana curvicauda* (Diptera: Tephritidae). *Folia Entomológica Mexicana* 90, 33–42.

Aluja, M., Jimenez, A., Piñero, J., Camino, M., Aldana, L., Valdes, M.E., Castrejon, V., Jacome, I., Davila, A.B. and Figueroa, R. (1997a) Daily activity patterns and within-field distribution of papaya fruit fly (Diptera: Tephritidae) in Morelos and Veracruz, Mexico. *Journal of Economic Entomology* 90, 505–519.

Aluja, M., Jimenez, A., Camino, M., Piñero, J., Aldana, L., Castrejon, V. and Valdes, M.E. (1997b) Habitat manipulation to reduce papaya fruit fly (Diptera: Tephritidae) damage: orchard design, use of trap crops and border trapping. *Journal of Economic Entomology* 90, 1567–1575.

Andrade, H., Avila, C., García, E., Mora, A., Nieto, D., Téliz, D. and Villanueva, J.A. (1994) La mancha anular del papayo en Veracruz, México, y su manejo integrado. *Séptima reunión científica del sector agropecuario y forestal del Estado de Veracruz.* GIP Grupo Interdisciplinario del Papayo, Veracruz, México, pp. 87–92.

Anonymous (2000a) *Hawaii, Crop Knowledge Master Papaya.* http://www.extento.hawaii.edu/kbase/crop/crops/papaya.htm

Anonymous (2000b) SCALENET. http://www.sel.barc.usda.gov/scalenet

Arenas, L., Avila, C., Cárdenas, E., Etchévers, J., Flores, C., García, E., González, V., Matheis, L., Mora, A., Mora, G., Nieto, D., Riestra, D., Téliz, D., Velázquez, J. and Villanueva, J. (1992) La virosis del papayo en Veracruz: etiología y control. *Quinta Reunión Científica del Sector Agropecuario y Forestal del Estado de Veracruz. Resultados y Avances de Investigación, at Veracruz.* GIP (Grupo Interdisciplinario del Papayo). Veracruz, México, pp. 62–71.

Armstrong, J.W., Hu, B.K. and Brown, S.A. (1995) Single-temperature forced hot-air quarantine treatment to control fruit flies (Diptera: Tephritidae) in papaya. *Journal of Economic Entomology* 88, 678–682.

Baker, R.E.D. (1936) Papaya mosaic disease. *Tropical Agriculture* 16, 159–163.

Bautista, R.C., Harris, E.J. and Lawrence, P.O. (1998) Biology and rearing of the fruit fly parasitoid *Biosteres arisanus*: clues to insectary propagation. *Entomologia Experimentalis Applicata* 89, 79–85.

Becerra L., E.N. (1987) Dinámica poblacional y hospederas de áfidos vectores de virus de la mancha anular del papayo en Veracruz XIV. *Congreso Nacional Fitopatología.* Morelia Michoacán, Mexico, p. 11.

Becerra L., E.N. (1989) Preferencia del color de papaya (*Carica papaya*) y barreras de jamaica (*Hibiscus sabdariffa* L.) Como medio para reducir la transmisión por áfidos de virosis en papaya. *Revista Mexicana Fitopatologia* 7, 218–222.

Ben-Dov, Y. (1993) *A Systematic Catalogue of the Mealybugs of the World (Insecta: Homoptera: Coccoidea: Pseudococcidae and Putoidae) with Data on Geographical Distribution, Host Plants, Biology and Economic Importance.* Intercept, Andover, 686 pp.

Ben-Dov, Y., Miller, D.R. and Gibson, G.A.P. (2000) ScaleNet. http://www.sel.barc.usda.gov/scalenet/scalenet/htm

Bennett, F.D. (1956) Some parasites and predators of *Pseudaulacaspis pentagona* (Targ.) in Trinidad, British West Indies. *Canadian Entomologist* 88, 704–705.

Bennett, F.D. and Brown, S.W. (1958). Life history and sex determination in the diaspine scale, *Pseudaulacaspis pentagona* (Targ.) (Coccoidea). *Canadian Entomologist* 9, 317–324.

Bess, H.A. (1953) Status of *Ceratitis capitata* in Hawaii following the introduction of *Dacus dorsalis* and its parasites. *Proceedings of the Hawaiian Entomological Society* 15, 221–234.

Bhargava, K.S. and Khurana, S.M.P. (1969) Papaya mosaic control by oil spray. *Phytopathologische Zeitshrift* 64, 338–343.

BIOSIS (2000) http://www.york.biosis.org/triton/indexfm.htm

Bird, J. and Adsur, J. (1952) Viral nature of papaya bunchy top. *Journal of the Agriculture University of Puerto Rico* 36, 5–11.

Brown, E.S. (1958a) Revision of the genus *Amblypelta* Stal (Hemiptera: Coreidae). *Bulletin of Entomological Research* 49, 509–541.

Brown, E.S. (1958b) Injury to cacao by *Amblypelta* Stal (Hemiptera: Coreidae) with a summary of food plants of species of this genus. *Bulletin of Entomological Research* 49, 543–554.

Brunner, B.R., Davis, M.J., Ferwerda, F.H., Ramirez, L.M. and Pantoja, A. (1996) Recent advances in papaya bunchy top research. *Annual Meeting of the Puerto Rican Society for Agricultural Sciences,* 22 November 1996. Hato Rey, Puerto Rico.

CABI and EPPO (1997) *Quarantine Pests for Europe.* CAB International, Wallingford, Oxon, UK, 1425 pp.

Campbell, C.W. (1984) Papaya – tropical fruits and nuts. In: Martin, F.W. (ed.) *Handbook of Tropical Food Crops.* CRC Press, Boca Raton, Florida, pp. 246–247.

Campbell, R.J. (1996) South American fruits deserving further attention. In: Janick, J. (ed.) *Progress in New Crops.* ASHS Press, Arlington, Virginia, pp. 431–439.

Castrejón, F. (1987) Aspectos de la biologia y habitos de *Toxotrypana curvicauda* Geastaecker (Diptera: Tephritidae) en condiciones de laboratorio y su distribucion en una plantacion de *Carica papaya* L. en Yutepec, Morelos. BS thesis, Instituto Politecnico Nacional, Mexico, D.F.

Castrejón, F. and Camino, M. (1991) New host plant record for *Toxotrypana curvicauda* (Diptera: Tephritidae). *Florida Entomologist* 74, 466.

Chopra, R.N. (1958) *Indigenous Drugs of India*, 2nd edn. UN Dhur & Sons, Calcutta, India.

Chuman, T., Landolt, P.J., Heath, R.R. and Tumlinson, J.H. (1987) Isolation, identification and synthesis of a male-produced sex pheromone of papaya fruit fly, *Toxotrypana curvicauda* Gerstaecker (Diptera: Tephritidae). *Journal of Chemical Ecology* 13, 1979–1992.

Clavijo, J.A. and Chacín, M.E. (1992) Sphingidae (Insecta: Lepidoptera) reportados como plagas en cultivos venezolanos: clave para las especies. *Boletín Entomológico de Venezuela, Nueva Serie* 7, 119–125.

Collins, F.A. and Whitcomb, W.H. (1975) Natural enemies of the white peach scale, *Pseudaulacaspis pentagona* (Homoptera: Coccidae), in Florida. *Florida Entomologist* 58, 15–21.

Conover, R.A. and Waddill, V.H. (1981). Permethrin as a control for papaya fruit fly. *Proceedings of the Florida State Horticultural Society* 94, 353–355.

Cook, M.T. (1931) New virus disease of plants in Puerto Rico. *Journal of Agriculture of the University of Puerto Rico* 15, 193–195.

Davis, M.J. (1994) Bunchy top. In: Ploetz, R.C., Zentemeyer, G.A., Nishijima, W.T., Rohrback, K.G. and Ohr, H.D. (eds) *Compendium of Tropical Fruit Diseases*. American Phytopathological Society, St Paul, Minnesota, pp. 69–70.

Davis, M.J., Ying, Z., Brunner, B.R., Pantoja, A. and Ferwerda, F.H. (1997) Rickettsial relative associated with papaya bunchy top disease. *Current Microbiology* 36, 80–84.

Duke, J.A. (1983) *Handbook of Energy Crops*. http://www.hort.perdue.edu/newcrop/duke_energy/Carica_papaya.html

Duke, J.A. (1984) *Borderline Herbs*. CRC Press, Boca Raton, Florida.

Duke, J.A. (1985) *Handbook of Medicinal Herbs*. CRC Press, Boca Raton, Florida.

Duke, J.A. (1990) Promissing phytomedicinals. In: Janick, J. and Simon, J.E. (eds) *Advances in New Crops*. Timber Press, Portland, Oregon, USA, pp. 491–498.

Ebesu, R. (1985) The biology of the leafhopper *Empoasca stevensi* Young (Homoptera: Cicadellidae) and its toxicity to papaya. MS thesis, University of Hawaii, 88 pp.

Emeruwa, A.C. (1982) Antibacterial substance from Carica papaya fruit extract. *Journal of Natural Production* 45, 123–127.

Eugene, G., Halliwell, R.S. and Smith, L.R. (1968) Investigación de los virus de la papaya (*Carica papaya* L.) en la Republica Dominicana, con apuntes especiales sobre la asociación de un organismo del tipo mycoplasma con la enfermedad 'bunchy top'. *Instituto Superior de Agricultura. División de Investigaciones Agricolas. Boletin* No. 14.

FAO (2000) FAOSTAT Agricultural Data. http://apps.fao.org

FAO-GPPIS (2000) http://gppis.fao.org/GPPPIS.exe$hostPest?host=382

Fay, H.A., Drew, R. and Lloyd, A. (1997) The eradication program for papaya fruit fly (*Bactrocera papayae* Drew and Hancock) in north Queensland. In: Allwood, A.J. and Drew, R. (eds) *Management of Fruit Flies in the Pacific. ACIAR Proceedings* 76, pp. 259–261.

Fitch, M.M.M., Manshardt, R.M., Gonsalves, D., Slightom, J.L. and Sanford, J.C. (1992) Virus resistant papaya plants derived from tissues bombarded with coat protein gene of papaya ringspot virus. *Biotechnology* 10, 1466–1472.

Fletchman, C.H.W. (1983) *Acaros de importancia agricola*. Livraria Nobel, S.A. São Paulo, Brazil, pp. 112–114.

Flores, R., García, C., Nieto, E., Angel, D., Téliz, D. and Villanueva, J.A. (1995) Integrated management of papaya in Mexico. *Acta Horticulturae* 370, 151–158.

Follett, P.A. (2000) Arthropod pests in Hawaii. *Chronica Horticulturae* 40, 4–10.

Follett, P.A. and Gabbard, Z. (1999) Efficacy of the papaya vapor heat quarantine treatment against white peach scale in Hawaii. *Horticultural Technology* 9, 506.

García, G.B. (1987) Pruebas de transmisión por áfidos (Homoptera: Aphididae) del Virus de la Mancha Anular de la Papaya (*Carica papaya* L.). Tesis de Licenciatura, Facultad de Ciencias Biológicas, U.D.I.C.B.A., Universidad Veracruzana, Xalapa, Veracruz, Mexico.

Gill, R.J. (1997) *The Scale Insects of California*, Part 3, *the Armored Scales (Homoptera: Diaspididae)*. California Department Food and Agriculture, Technical Series Agriculture, Biosystem and Plant Pathology.

González H., Villanueva, J.A. and Miller, D.R. (1999) Parasitoides del piojo harinoso del papayo, *Paracococcus marginatus* Williams y Granara de Willink (Homoptera: Pseudococcidae) en México. *Memorias del XXII Congreso Nacional de Control Biológico*, Oct. 1999, Montecillo, México.

Hancock, D., Hamacek, E., Lloyd, A. and Elson-Harris, M. (2000) The distribution and host plants of fruit flies (Diptera: Tephritidae) in Australia. *Queensland Department of Primary Industries. Information Series* QI99067, 75 pp.

Haque, S.Q. and Parasram, S. (1973) *Empoasca stevensi*, a new vector of bunchy top disease in papaya. *Plant Diseases Reporter* 57, 412–413.

Harkness, R.W. (1967) *Papaya Growing in Florida*. Florida Agricultural Experimental Station Circular S-100.

Harris, E.J. (1989) Pest status in Hawaiian islands and North Africa. In: Robinson, A. and Hooper, G. (eds) *World Crop Pests – Fruit Flies, Their Biology, Natural Enemies and Control*, Vol. 3A. Elsevier, Amsterdam, The Netherlands, pp. 73–80.

Heath, R.R., Epsky, N.D., Jiménez, A., Dueben, B.D., Landolt, P.J., Meyer, W.L., Aluja, M., Rizzo, J., Camino, M., Jerónimo, F. and Baranowski, R.M. (1996) Improved pheromone-based trapping systems to monitor *Toxotrypana curvicauda* (Diptera: Tephritidae). *Florida Entomologist* 79, 37–47.

Hernández, O.E., Luna, C. and Hernández O.A. (1995) Análisis de la disposición espacio-temporal y ácaros-día de *Tetranychus cinnabarinus* Boisd. (Acari: Tetranychidae) en hojas de papayo, en Iguala, Gro., durante 1993. *Memoria XXX Congreso Nacional de Entomología*, 28–31 May 1995, Chapingo, Estado de México.

Hernández, C.E., Riestra, D.D., García, P.E., Ortega, A.L.D. and Mosqueda, V.R. (2000a) Comportamiento del virus de la mancha anular del papayo (VMAP) en cv. 'Maradol Roja' bajo tres sistemas de manejo en Veracruz. *Manejo Integrado de Plagas* (Costa Rica).

Hernández, C.E., Utrera, V., Villanueva-Jiménez, J.A. and Rodríguez-Lagunes, D.A. (2000b) Comportamiento de *Aphis nerii* a extractos acuosos de nim (*Azadirachta indica* A. Juss.) aplicados en plantas de papayo en condiciones de laboratorio. *XXXV Congreso Nacional de Entomología*, Mexico. 6 pp.

Higa, S.Y. and Namba, R. (1971) Vectors of the papaya mosaic virus in Hawaii. *Proceedings of the Hawaiian Entomological Society* 21, 93–96.

Huxham, K. (2000) Annual detections of exotic fruit flies (Tephritidae: Dacinae) in Torres Strait. *International Symposium on Tropical and Subtropical Fruits, Cairns Australia, Symposium Booklet*, p. 95.

Ishii, M. and Holtzmann, O.V. (1963) Papaya mosaic disease in Hawaii. *Plant Disease Reporter* 47, 947–951.

Khurana, P. and Bhargava, K.S. (1971) Three new vectors of papaya viruses. *Journal of Horticultural Science* 46, 209–211.

Knab, F. and Yothers, W.W. (1914) Papaya fruit fly. *Journal of Agricultural Research* 2, 447–454.

Labonne, G., Lupoli, R., Quiot, J.B. and Yvon, M. (1992) The transmissibility of potyvirus by an aphid vector: reliability of measurements under controlled conditions. *Annals of Applied Biology* 120, 269–278.

Landolt, P.J. (1984a) Behavior of the papaya fruit fly *Toxotrypana curvicauda* Gerstaecker (Diptera: Tephritidae) in relation to its host plant, *Carica papaya* L. *Folia Entomologica Mexicana* 61, 215–224.

Landolt, P.J. (1984b) Reproductive maturation and premating period of the papaya fruit fly *Toxotrypana curvicauda* (Diptera: Tephritidae). *Florida Entomologist* 67, 240–244.

Landolt, P.J. (1994a) Fruit of *Morrenia odorata* (Asclepiadaceae) as host for the papaya fruit fly, *Toxotrypana curvicauda* (Diptera: Tephritidae). *Florida Entomologist* 77, 287–288.

Landolt, P.J. (1994b) Mating frequency of the papaya fruit fly (Diptera: Tephritidae) with and without host fruit. *Florida Entomologist* 77, 305–312.

Landolt, P.J. and Heath, R.R. (1988) Effect of age, maturing, and time of day on behavioral response of female papaya fruit flies, *Toxotrypana curvicauda* Gerstaecker (Diptera: Tephritidae), to synthetic sex pheromone. *Environmental Entomology* 17, 47–51.

Landolt, P.J. and Heath, R.R. (1990) Effect of pheromone release rate and time of day on catches of male and female papaya fruit flies (Diptera: Tephritidae) on fruit model traps baited with pheromone. *Journal Economic Entomology* 83, 2040–2043.

Landolt, P.J. and Hendrichs, J. (1983) Reproductive behavior of the papaya fruit fly, *Toxotrypana curvicauda* Gerstaecker (Diptera: Tephritidae). *Annals of the Entomological Society of America* 76, 413–417.

Landolt, P.J. and Reed, H.C. (1990) Behavior of the papaya fruit fly (Diptera: Tephritidae): host finding and oviposition. *Environmental Entomology* 19, 1305–1310.

Landolt, P.J., Heath, R.R. and King, J.R. (1985) Behavioral responses of female papaya fruit flies, *Toxotrypana curvicauda* (Diptera: Tephritidae), to male-produced sex pheromone. *Annals of the Entomological Society of America* 78, 751–755.

Landolt, P.J., Heath, R.R., Agee, Z.H.R., Tumlinson, J.H. and Calkins, C.O. (1988) Sex pheromone-based trapping system for papaya fruit fly (Diptera: Tephritidae). *Journal of Economic Entomology* 81, 1163–1169.

Landolt, P.J., Gonzalez, M., Chambers, D.L. and Heath, R.R. (1991) Comparison of field observations and traping of papaya fruit fly in papaya plantings in Central America and Florida. *Florida Entomologist* 74, 408–414.

Lara, F., Martin, F. and Gonzalez, M. (1989) Studies of several fruit flies and the incidence of papaya (*Carica papaya* L.) infestation in Costa Rica. *Report to Costa Rica Minister of Agriculture*, pp. 1–18.

Liquido, N.J. (1990) Morphological defects in the blossom end of papaya fruits in relation to infestation rates by oriental fruit fly and melon fly (Diptera: Tephritidae). *Journal of Economic Entomology* 83, 1883–1887.

Liquido, N.J. (1991a) Effect of ripeness and location of papaya fruits on the parasitization rates of oriental fruit fly and melon fly (Diptera: Tephritidae) by braconid (Hymenoptera) parasitoids. *Environmental Entomology* 20, 1732–1736.

Liquido, N.J. (1991b) Fruit on the ground as a reservoir of resident melon fly (Diptera: Tephritidae) populations in papaya orchards. *Environmental Entomology* 20, 620–625.

Liquido, N.J. (1991c) Infestation rates of oriental fruit fly (Diptera: Tephritidae) in 'Sunrise' papaya by visual and colorimetric ripeness indices. *Journal of Economic Entomology* 84, 948–953.

Liquido, N.J. (1993) Reduction in oriental fruit fly (Diptera: Tephritidae) populations in papaya orchards by field sanitation. *Journal of Agricultural Entomology* 10, 162–170.

Liquido, N.J., Cunningham, R.T. and Couey, H.M. (1989) Infestation rates of papaya by fruit flies (Diptera: Tephritidae) in relation to the degree of fruit ripeness. *Journal of Economic Entomology* 82, 213–219.

Machain L., M. (1983) *Plagas del papayo en Veracruz y sus depredadores*. Cotaxtla, Veracruz, México: Instituto Nacional de Investigaciones Agrícolas, CIAGOC, Mexico, 23 pp.

Martorell, L.F. (1976) *Annotated Food Plants Catalog of the Insects of Puerto Rico*. University of Puerto Rico, Agricultural. Experimental Station, Rio Piedras, Puerto Rico, 303 pp.

Martorell, L.F. and Adsuar, J. (1952) Insects associated with papaya virus diseases in the Antilles and Florida. *Journal of Agriculture of the University of Puerto Rico* 45, 863–869.

Mason, A.C. (1922) Biology of the papaya fruit fly, *Toxotrypana curvicauda*, in Florida. *United States Department Agricultural Bulletin* 1081.

McPheron, B.A. and Steck, G.J. (1996) *Fruit Fly Pests, a World Assesment of Their Biology and Management*. St Lucie Press, Delray Beach, Florida, 586 pp.

Medina, S. and Franqui, R.A. (1999a) La entomología económica en Puerto Rico. Insectos y ácarinos de los frutales en Puerto Rico VI: Plagas de la papaya, *Carica papaya* L. *Agrotemas de Puerto Rico* 10, 20–22.

Medina, S. and Franqui, R.A. (1999b) La entomología económica en Puerto Rico. Insectos y ácarinos de los frutales en Puerto Rico VII: Plagas de la papaya, *Carica papaya* L. *Agrotemas de Puerto Rico* 10(5), 18.

Medina, S., Franqui, R.A. and Cortes, M. (1999) La entomología económica en Puerto Rico. Insectos y ácarinos de los frutales en Puerto Rico V: Plagas de la papaya, *Carica papaya* L. *Agrotemas de Puerto Rico* 10, 20.

Messing, R. (1999) *Managing Fruit Flies on Farms in Hawaii*. University of Hawaii, College of Tropical Agriculture and Human Resources Publications. IR-4, Sept. 1999.

Miller, D.R., Williams, D J. and Hamon, A.B. (1999) Notes on a new mealybug (Coccoidea: Pseudococcidae) pest in Florida and the Caribbean: the papaya mealybug, *Paracoccus marginatus* Williams and Granara de Willink. *Insecta Mundi* 13, 179–188.

Morton, J.F. (1977) *Major Medicinal Plants: Botany, Culture and uses*. C.C. Thomas Publisher, Springfield, Illinois.

Morton, J. (1987) Papaya, In: Morton, J.L. (ed.) *Fruits of Warm Climates*. Creative Resource Systems, Miami, Florida, pp. 336–346.

Monzu, N., Hardie, D. and Johnston, R. (2000) *Papaya Fruit Fly Identification*. http://www.agric.wa.gov.au:7000/ento/papaya.htm

Nakahara, S. (1982) *Checklist of the Armored Scales (Homoptera: Diaspididae) of the Continuous United States*. USDA, Animal and Plant Health Inspection Service, Plant Protection and Quarantine, Beltsville, Maryland.

Nakahara, S. and Gill, R.J. (1985) Revision of *Philephedra*, including a review of *Lichtensia* in North America and description of new genus, *Metapulvinaria* (Homoptera: Coccidae). *Entomography* 3, 1–42.

Namba, R. and Higa, S.Y. (1981) Papaya mosaic transmission as affected by the duration of the acquisition probe of the green peach aphid – *Myzus persicae* (Sulzer). *Proceedings of the Hawaiian Entomological Society* XXIII(3), 431–433.

Namba, R. and Kawanishi, C.Y. (1966) Transmission of papaya mosaic virus by the green peach aphid. *Journal of Economic Entomology* 59, 669–671.

Nariani, T.K. (1956) Leaf curl of papaya. *Indian Phytopathology* IX, 151–155.

Nguyen, R., Gillis, S. and Harris, D. (1993) 'Solo' sunrise papaya as a host of the Caribbean fruit fly, *Anastrepha suspensa* (Loew), under field and laboratory conditions. *Proceedings of the Florida State Horticultural Society* 106, 118–119.

Nielson, M.W. (1968) The leafhopper vectors of phytopathogenic viruses (Homoptera: Cicadellidae) taxonomy, biology, and virus transmission. Agricultural Research Service. United States Department of Agriculture. *Technical Bulletin* No. 1382.

Nuñez, E. (1982) *Plantas medicinales de Puerto Rico*. Estación Experimental Agrícola de Puerto Rico, Boletín 176. Editorial de la Universidad de Puerto Rico, Río Piedras, Puerto Rico.

Opina, O.S. and Tomines, R.L. (1996) Management of papaya ringspot by isolation and intercropping of papaya. *Philippine Phytopathology* 32, 51–56.

Peck, S.L. and McQuate, G.T. (2000) Field tests of environmentally friendly malathion replacements to suppress wild Mediterranean fruit fly populations. *Journal of Economic Entomology* 93, 280–289.

Peña, J.E. (1986) Feeding behavior of *Toxotrypana curvicauda* (Diptera: Tephritidae) on young papaya seeds. *Florida Entomologist* 69, 427–428.

Peña, J.E. (1993) Pests of mango in Florida. *Acta Horticulturae* 341, 395–406.

Peña, J.E. and McMillan, R.T. Jr (1986) *Verticilliun lecani*, a new fungal parasite of scale *Philephedra tuberculosa* n. sp. (Homoptera: Coccidae) in Florida. *Florida Entomologist* 69, 416–417.

Peña, J.E. and Nagel, J. (1988) Effectiveness of pesticides against two tropical fruit pests. *Proceedings of the Florida State Horticultural Society* 101, 249–251.

Peña, J.E., Baranowski, R.M. and Litz, R.E. (1986) Oviposition of the papaya fruit fly *Toxotrypana curvicauda* Gerstaecker as affected by fruit maturity. *Florida Entomologist* 69, 344–348.

Peña, J.E., Baranowski, R.M. and Litz, R.E. (1987) Life history, behavior and natural enemies of Philephedra tuberculosa n. sp. (Homoptera: Coccidae). *Florida Entomologist* 70, 423–427.

Pérez, J.E., Riestra, D., Mosqueda, R., Rodríguez, D.A., Villanueva-Jiménez, J.A. and García, E. (2000) Extractos acuosos de nim *Azadirachta indica* A. Juss en el manejo integrado del papayo *Carica papaya* L. *Revista Chapingo, Serie Horticultura* 6(1), 81–89.

Petcharat, J. (1997) A small field trial on the release of *Diachasmimorpha longicaudata* Ashmead (Hymenoptera: Braconidae) in an attempt to control the fruit fly *Bactrocera papayae* Drew and Hancock (Diptera: Tephritidae) population. *Khon Kaen Agriculture* 25, 62–65.

Pontis, R.E. (1953) La virosis de la lechosa (*Carica papaya*) en Venezuela I. Transmision del mosaico. *Agronomia Tropical Venezuela* 2, 241–251.

Poulter, N.H. and Cagygill, J.C. (1985). Production and utilization of papain – a proteolytic enzyme from *Carica papaya* L. *Tropical Science* 25, 123–137.

Prasad, V. (1966) Bionomics of the predatory mite, *Phytoseiulus macropilis* (Banks) (Acarina: Phytoseiidae) and its role in the control of the spider mites in Hawaii. PhD dissertation, University of Hawaii.

Quisumbing, E. (1951) *Medicinal Plants of the Philippines*, Bureau of Print, Manila, Philippines.

Rabara, R.C., Valencia, L.D., Sumalde, A.C., Bajet, N.B. and Villegas, V.N. (1996) Correlation between aphid abundance and ringspot virus disease incidence in papaya. *Philippine Journal of Crop Science* 21, 58.

Ramirez, L.M. (1997) Identificación del insecto vector y transmisión del cogollo racimoso de la papaya. Masters Thesis, University of Puerto Rico, Mayagüez Campus, Mayaguez, Puerto Rico.

Roessler, Y. (1989) Control; insecticides; insecticidal bait and cover sprays. In: Robinson, A. and Hopper, G. (eds) *World Crop Pests – Fruit Flies: Their Biology, Natural Enemies and Control*, Vol. 3B. Elsevier, Amsterdam, The Netherlands, pp. 329–336.

Sakimura, K. (1972) *Frankliniella invasor*, new species, and notes on *F. gardeniae* and the *Frankliniella* spp. in Hawaii (Thysanoptera: Thripidae). *Proceedings of the Hawaiian Entomological Society* 21, 263–270.

Schaefers, G.A. (1969) Aphid vectors of the papaya mosaic virus in Puerto Rico. *Journal of Agriculture of the University of Puerto Rico* 53, 1–13.

Schroeder, W.J. and Beavers, J.B. (1997) Citrus root weevils in Florida: identification, biology and control. *Proceedings of the International Society Citriculture* 2, 498–500.

Seelig, R.A. (1970) *Papaya – Fruits and Vegetables: Facts and Pointers*. United Fresh Fruit and Vegetable Association, Washington, DC, p. 7.

Seín, F. and Adsuar, J. (1947) Transmission of the bunchy top disease of papaya (*Carica papaya* L.) by the leaf hopper *Empoasca papayae* Oman. *Science* 106, 130.

Seo, S.T., Farias, G.J. and Harris, E.J. (1982a) Fruit fly: ripening of fruit and its effect on index of infestation of Hawaiian papayas. *Journal of Economic Entomology* 75, 173–178.

Seo, S.T., Farias, G.J. and Harris, E.J. (1982b) Oriental fruit fly: ripening of fruit and its effect on index of infestation of Hawaiian papayas. *Journal of Economic Entomology* 75, 173–178.

Seo, S.T., Tang, C., Sanidad, S. and Takenaka, T.H. (1983) Hawaiian fruit flies (Diptera: Tephritidae): variation of index of infestation with benzyl isothiocyanate concentration and color of maturing papaya. *Journal of Economic Entomology* 76, 535–538.

Sharma, V.C. and Ogbeide, O.N. (1982) Pawpaw as a renewable energy resource for the production of alcohol fuels. *Energy* 7, 871–873.

Sharp, J.L. and Landolt, P.J. (1984) Gustatory and olfactory behavior of the papaya fruit *Toxotrypana curvicauda* Gerstaecker (Diptera: Tephritidae) in the laboratory with notes on longevity. *Journal of the Georgia Entomological Society* 19, 176–182.

Storey, W.B. (1984) Papaya. In: Simmonds, N.W. (ed.) *Evolution of Crop Plants*. Longman, New York, pp. 21–24.

Story, G.E. and Halliwell, R.S. (1969) Association of mycoplasma-like organism with the bunchy top disease of papaya. *Phytopathology* 59, 1336–1337.

Swanson, R.W. and Baranowski, R.M. (1972) Host range and infestation of the Caribbean fruit fly, *Anasthrepa suspensa* (Diptera: Tephritidae), in south Florida. *Proceedings of the Florida State Horticultural Society*, pp. 271–273.

Thompson, F.C. (1998) *Fruit Fly Expert System and Systematic Information Database*. North American Dipterists' Society, Backhuys Publishers, Leiden, The Netherlands

Valencia, G.H. (1975) *Dinámica de población de las principales plagas del papayo en Veracruz. Artículo Número 8*. Memorias del Segundo Congreso del Centro de Investigaciones Agrícolas del Sur Este. Veracruz, México. Instituto Nacional de Investigaciones Agrícolas, Secretariía de Agricultura y Ganadería, 6 pp.

Vargas, R.I., Stark, J.D., Uchida, G.K. and Purcell, M. (1993). Opiine parasitoids (Hymenoptera: Braconidae) of oriental fruit fly (Diptera: Tephritidae) on Kauai Island, Hawaii: island wide relative abundance and parasitism rates in wild and orchard guava habitats. *Journal of Economic Entomology* 22, 246–253.

Vargas, R.I., Walsh, W.A. and Nishida, T. (1995) Colonization of newly planted coffee fields: dominance of Mediterranean fruit fly over oriental fruit fly (Diptera: Tephritidae). *Journal of Economic Entomology* 88, 620–627.

Vargas, R.I., Stark, J.D., Kido, M.H., Ketter, H.M. and Whitehand, L.C. (2000) Methyl eugenol and cue-lure for suppression of male oriental fruit flies and melon flies (Diptera: Tephritidae) in Hawaii: effects of lure mixtures and weathering. *Journal of Economic Entomology* 93, 81–87.

Villanueva, B.J. and Villanueva-Jiménez, J.A. (1994) Transmisibilidad del 'virus de la mancha anular del papayo' (VMAP) entre hospedantes, por medio de afidos (Homoptera: Aphididae). *XXIX Congreso Nacional de Entomología – Southwestern Branch Annual Meeting – Entomology Society of America*. 24–27 April. Monterrey, Nuevo León, México, p. 142.

Villanueva-Jiménez, J.A. and Peña, R. (1991) Afidos (Homoptera: Aphididae) colectados en 'trampas amarillas de agua' en la planicie costera de Veracruz, México. *Agrociencia Serie Protección Vegetal* 2, 7–20.

Waite, G.K. (1990) *Amblypelta* spp. (Hemiptera: Coreidae) and green fruit drop in lychees. *Tropical Pest Management* 33, 353–355.

Waite, G.K. and Huwer, R.K. (2000) Host plants and their role in the ecology of the fruitspotting bugs *Amblypelta nitida* Satl and *Amblypelta lutescens lutescens* (Distant) (Hemiptera: Coreidae). The fruitspotting bug project. http://www.dpi.qld.gov..au/qhi/fsb/content.htm

Webb, R.R. and Davis, M.J. (1987) Unreliability of latex-flow test for diagnosis of bunchy top of papaya caused by a mycoplasmlike organism. *Plant Disease* 71, 192.

Weems, H.V. (1966) The Caribbean fruit fly in Florida. *Proceedings of the Florida State Horticultural Society* 79, 401–405.

Weems, H.V. (1969) Papaya fruit fly *Toxotrypana curvicauda* Gerstaecker (Diptera: Tephritidae). *Florida Department of Agriculture Division of Plant Industry, Entomology Circular* 86, 4 pp.

White, M.I. and Elson-Harris, M.M. (1992) *Fruit Flies of Economic Importance: Their Identification and Bionomics*. CAB International, Wallingford, UK.

Williams, D.J. (1986) Mealybugs on cassava with special reference to those associated with wild and cultivated cassava in the Americas. In: Herren, H.R. and Hennessey, R.N.H. (eds) *Biological Control and Host Plant Resistance to Control the Cassava Mealybug and Green Mite in Africa*. International Institute of Tropical Agriculture (IITA), Ibadan, Nigeria, pp. 49–56.

Williams, D.J. and Granara, M.C. (1992) *Mealybugs of Central and South America*. CAB International, Wallingford, UK, 635 pp.

Wolfengbarger, D.O. (1962) Papaya fruit fly control. *Proceedings of the Florida State Horticultural Society* 75, 381–384.

Wolfengbarger, D.O. and Walker, S.D. (1974) Two major pest problems of papayas. *Proceedings of the Florida State Horticultural Society* 85, 384–385.

Wood, M. and Hardin, B. (2000) Spinosad battles crop pests. *Agricultural Research* 48, 10–12.

Woolcott, G.N. (1933) *An Economic Entomology of the West Indies*. Clay and Sons, Bungay, Suffolk.

Woolcott, G.N. (1948) The insects of Puerto Rico. *Journal of Agriculture of the University of Puerto Rico* 32, 101–145.

Yadava, U.L., Burris, J.A. and McCrary, D. (1990) Papaya a potential annual crop under middle Georgia conditions. In: Janick, J. and Simon, J.E. (eds) *Advances in New Crops*. Timber Press, Portland, Oregon, pp. 364–366.

Yee, W., Akamine, E.K., Aoki, G.M., Hamilton, R.A., Haramoto, F.H., Hine, R.B., Holtzmann, O.V., Ishida, J.T., Keeler, J.T. and Nakasone, H.Y. (1970) Papayas in Hawaii. *University of Hawaii Cooperative Extension Service Circular* 436, 57 pp.

Young, D.A. (1953) Empoascan leafhoppers of the solana group with description of two new species. *Journal of Agriculture of the University of Puerto Rico* 37, 151–160.

Zapata, R. (1994) A review of papaya research in Hawaii, Florida, Guam, Virgin Islands, and Puerto Rico. Cultivo, produccion y manejo de la Papaya. *Proceedings of a Workshop on Papaya*. Colegio Regional de Aguadilla, Agricultural Experiment Station Puerto Rico, 20 May 1994.

6 Pests of Pineapple

Graham J. Petty,[1] Graham R. Stirling[2] and Duane P. Bartholomew[3]
[1]*Institute for Tropical and Subtropical Crops, Agricultural Research Council, Private Bag X1, Bathurst 6166, Republic of South Africa;* [2]*Biological Crop Protection (Pty) Ltd, 3601 Moggill Road, Moggill, Queensland 4070, Australia;* [3]*Department of Agronomy and Soil Science, University of Hawaii at Manoa, 1910 East-West Road, Honolulu, Hawaii 96822, USA*

The pineapple, *Ananas comosus* (L.) Merr., a member of the *Bromeliaceae* family, is a self-sterile herbaceous, perennial, monocotyledonous plant. The commercial cultivars are placed in five distinct groups, i.e. 'Cayenne', 'Spanish', 'Queen', 'Pernambuco' and 'Perolera'. Christopher Columbus and his crew were the first Europeans to discover this fruit when, in 1493, they landed on a West Indian island which they later named Guadeloupe. Since that time the pineapple has been taken to and grown in other countries with tropical and subtropical climates, and a number of pests have gone along with it. Most pests are endemic to the new countries and have adapted to utilize the pineapple to support their life cycles.

This chapter outlines the importance and phenology of pineapple as a commercial crop. It goes on to discuss key pineapple pest biologies and economic thresholds, and the principles and practices of sampling, monitoring and controlling these species. Tables list the pests, worldwide, giving their importance and distribution.

Importance of the Crop

Pineapple ranks after bananas and mangoes as the third largest fruit crop harvested in the tropics and subtropics. Just over 13 million t of pineapple were produced by 76 countries on 704,912 ha of land in 1997 (Food and Agriculture Organization, 1998). Few data are available on the fraction of fruit sent to local markets, processed, or exported (Loeillet, 1997; Barbeau and Marie, 1997; Pinon, 1997; Rougé and N'Goan, 1997). However, Loeillet (1997) reported that 70% of pineapple fruits are consumed within the country of production. Twenty-three countries produce more than 100,000 t (Table 6.1), and production in all countries has increased by 14% in the past 10 years. Pineapple is an important crop in a large number of countries and provides variety in the fruit market and a source of foreign exchange. The low yield for Indonesia and the very high yield for Colombia suggests that some data may not be reliable. However, much of the variation can be explained by differences in climate in the different growing regions, intensity of management, level of inputs, and of relevance to this chapter, the technology available and applied to control pests.

Phenology

The phenology of the pineapple plant begins with the initiation of propagules, all of which

©CAB *International* 2002. *Tropical Fruit Pests and Pollinators* (eds J.E. Peña, J.L. Sharp and M. Wysoki)

Table 6.1. Statistics on pineapple production for countries producing more than 100,000 t in 1997 and the increase in production between 1990 and 1997 (FAO, 1998).

Country	Area (1000 ha)	Production (1000 t)	Yield (t ha^{-1})	Increase or decrease (%, 1990–1997)
Australia	2.67	124.9	46.7	0.0
Bangladesh	13.77	148.5	10.8	9.4
Brazil	55.21	1,936.5	35.1	43.9
China	26.30	899.1	34.2	22.4
Colombia	5.34	329.3	61.6	–3.8
Costa Rica	8.19	260.0	31.7	38.5
Cote d'Ivoire	5.50	260.6	47.4	10.8
Congo	3.30	145.0	25.4	–2.1
Dominican Republic	4.40	109.9	24.7	35.6
Guatemala	3.85	109.8	28.5	38.8
Hawaii	8.05	294.0	36.5	–77.4
India	82.00	1,100.0	13.4	19.9
Indonesia	80.00	537.9	6.7	27.4
Kenya	8.50	290.0	34.1	22.4
Malaysia	7.05	163.0	23.1	–30.7
Mexico	7.69	301.4	39.2	–50.8
Nigeria	100.00	800.0	8.0	4.6
Peru	7.37	125.1	17.0	45.3
Philippines	52.00	1,700.0	32.7	32.0
South Africa	6.10	151.6	24.9	–12.0
Thailand	85.00	2,000.0	23.5	6.7
Venezuela	10.00	189.4	18.9	57.5
Vietnam	24.50	185.0	7.6	–152.9
World	704.90	13,060.2	18.5	14.0

are shoots formed on the stem or fruit of the mother plant. Propagules include ground and stem suckers, hapas, slips, and crown. Ground and stem suckers arise from the stem at its base or along its length. Hapas arise in the transition zone between the stem and fruit peduncle. Slips are borne on vestigal fruits near the top of the peduncle, and the crown develops at the top of the fruit. These propagules consist of a stem and leaves, though large suckers may also have adventitious roots. After planting, development of the pineapple plant is continuous once initiated unless arrested by environmental stress, pests or disease. Some pests attack the propagule while it is still on the plant or in storage, but most problems are encountered after planting.

The first phenological stage after planting is root initiation. The rate of initiation of new roots is determined by water supply and temperature, with little growth occurring if the temperature is below about 15°C. New root development and extension continues at least until the time of flower induction. The presence of white root tips has been used in Hawaii as a diagnostic of root health (Sanford, 1962). Leaf growth resumes at or soon after root initiation begins. New growth of older leaves is readily discerned by an abrupt increase in leaf width at the point where growth resumed. Once root and leaf growth begin, growth is continuous at a rate determined by the prevailing temperature, irradiance, and water and nutrient supply until canopy closure occurs. After the leaf canopy closes in, the degree of mutual shading, which is in part a function of plant population density, reduces the light available to plants and growth is then determined to some extent by the intensity of this shading. In any case, leaf number per plant, and thus plant weight, increases until inflorescence development is environmentally induced or forced with a plant growth regulator (Zhang *et al.*, 1997).

Inflorescence development in commercial practice is initiated by a growth regulator (ethylene gas or ethephon (2-chloroethylphosphonic acid)). Sucker development, the source of the ratoon crop, commences at or soon after the floral initiation process breaks apical dominance in cooler environments such as Australia and South Africa, but in more tropical environments (Cote d'Ivoire, Thailand) may be delayed until after fruit maturation. Circumstantial evidence indicates that stem starch levels at the time of floral initiation determine whether sucker initiation begins at flower induction or much later. Fruitlet initiation and development is temperature dependent (Fleisch and Bartholomew, 1987). In summer in Hawaii fruitlet initiation is completed in 34 days (Bartholomew, 1977) and an average-sized fruit of 'Smooth Cayenne' contains 100 or more fruitlets. Crown development begins once fruitlet development is completed (Bartholomew, 1977). The young inflorescence is protected by a whorl of leaves for the first ±40 days of its development. Anthesis begins 3 to 4 weeks after inflorescence appearance and occurs acropetally over a period of about 2 weeks. Once anthesis is completed and the petals dry, fruit and crown development are essentially continuous until the fruit is mature. Sucker number per plant is highly correlated with plant population density, declining as density increases. Development of suckers may continue after initiation or may be arrested, presumably due to a lack of resources. Those that do develop become the source of the ratoon crop and will grow until the flower initiation process is forced.

Pest injury due to feeding or transmission of disease is to some extent dependent on phenological development. Root pests can attack as soon as root development begins, and pests that feed on leaves can attack at almost any time. Mites can feed on developing fruitlets and propagules as soon as they are accessible and probably before they are visible, and pests can enter fruitlets and feed or transmit diseases at the time of anthesis.

Pests

Nematodes

Plant-parasitic nematodes are present wherever pineapple is grown and are one of the most important constraints to pineapple production worldwide. Nematodes may reduce the length of primary roots to such an extent that plants have poor anchorage, or they may destroy secondary roots and leave infested plants with a poorly developed root system. Such plants are susceptible to moisture and nutrient stress and, because pineapple roots do not regenerate once they are killed back to the stem, nematode-infested plants have little chance of reaching their full yield potential. Although nematodes can cause significant reductions in plant crop yield, they are primarily a problem of ratoon crops. The number, size and vigour of the suckers that produce the ratoon crop is heavily dependent on the health of the initial root system produced by the mother plant. When these roots are damaged by nematodes, ratoon crop yields may be drastically reduced or the crop may fail (Caswell *et al.*, 1990).

More than 100 species of plant-parasitic nematodes have been recorded in association with pineapple roots, and between three and five species are usually found in most pineapple fields. Since this complex of parasitic nematodes must be kept under control if a pineapple plantation is to achieve maximum productivity, this chapter takes a holistic view of nematode management. It concentrates on the general principles which are applicable to all plant-parasitic nematodes and provides only a limited amount of information on particular nematode species.

Key nematode pests

The key nematode pests of pineapple are all endoparasites. They either establish a permanent feeding site within the root and then become sedentary, e.g. root-knot and reniform nematodes, or they live inside roots but remain migratory, e.g. lesion nematodes.

*Root-knot nematode (*Meloidogyne javanica *and* M. incognita*)*

Root-knot nematode is a particularly important pest of pineapple because nematodes invade the tips of primary roots and stop them from elongating (Plate 33). Thus plant anchorage is markedly reduced. A terminal, club-shaped gall is usually produced in response to the developing nematode (Godfrey and Oliveira, 1932), but small, non-terminal galls may also form, causing brooming of the root system (Godfrey, 1936). The disease cycle begins when second-stage juveniles hatch from eggs and establish a feeding site behind the root tip. They then become sedentary and develop through subsequent moults into saccate females. At maturity, each female produces about 1000 eggs which are contained in a gelatinous egg mass. Worm-like adult males are sometimes produced, but they are unimportant because females reproduce by mitotic parthenogenesis. Temperatures of 25–30°C are ideal for nematode multiplication and at these temperatures, the life cycle is completed in 4–5 weeks.

*Reniform nematode (*Rotylenchulus reniformis*)*

In contrast to root-knot nematode, reniform nematode does not prevent primary roots from elongating. Roots infected by reniform nematode will provide the plant with good anchorage, but because secondary root formation is inhibited, root systems tend to be poorly developed. Second-stage juveniles hatch from eggs and develop into males and immature females without feeding. Males do not feed, but immature females enter the root and establish a feeding site. They then become sedentary, developing into kidney-shaped, egg producing adults (Linford and Oliveira, 1940). When soil is moist, root exudates of certain hosts will stimulate eggs to hatch. In dry soil, the nematode survives in the egg stage or as anhydrobiotic juveniles (Apt, 1976; Tsai and Apt, 1979). Reniform nematode is widely distributed in tropical and subtropical regions and soil temperatures in most pineapple-growing regions favour its development. The life cycle is generally completed in about 4 weeks.

*Lesion nematode (*Pratylenchus brachyurus*)*

Lesion nematodes are migratory endoparasites. Adult females lay eggs in root tissue and in soil, but all stages can migrate into and out of roots. Damage is difficult to diagnose in the field, but dark-coloured lesions develop in response to nematodes feeding in cortical tissue. These lesions can be microscopic in size, but they may extend progressively to cover the whole surface of the root. Secondary roots and root hairs are destroyed by the nematode, resulting in a root system composed mainly of poorly developed primary roots. The length of the life cycle is 4 weeks at temperatures of about 30°C (Olowe and Corbett, 1976).

Secondary nematode pests

Spiral nematodes (*Helicotylenchus* spp., *Rotylenchus* spp. and *Scutellonema* spp.), ring nematodes (*Criconemella* spp.), and stubby root nematodes (*Paratrichodorus* spp.) commonly occur in pineapple fields. The most widely distributed species is *H. dihystera*, which is often found at population densities greater than 2000 nematodes per 100 ml soil. Although their pathogenicity has not been studied in detail, it is generally assumed that these ectoparasitic species are not of major economic importance on pineapple. They live in soil, feeding externally on cortical tissues, root hairs and root tips, and the symptoms they produce tend to be relatively non-specific. They can be disregarded as pests unless their population densities are very high.

DISTRIBUTION AND IMPORTANCE Although the key nematode pests of pineapple are widely distributed, management systems, e.g. crop rotation practices, and environmental factors, e.g. soil moisture, soil pH and temperature, can influence distribution at a local level. Thus the distribution and relative importance of various species varies between and within pineapple-growing regions (Table 6.2).

Table 6.2. Distribution and importance of the main nematode pests of pineapple[a].

Nematode	Region	Distribution	Importance
Root-knot nematode (*Meloidogyne javanica*, *M. incognita*)	Africa	Widespread in southern Africa	The key nematode pest in cooler pineapple-growing regions, e.g. southern Africa, Australia, higher elevations in Hawaii. Losses are greatest in sandy soils and well-structured clay loams
	Asia	Significant in Thailand and some areas of the Philippines	
	Australia	Widespread in subtropical areas	
	Hawaii	Significant on a limited hectarage	
	Central and South America	Significant in most areas	
Reniform nematode (*Rotylenchulus reniformis*)	Africa	Generally not present	The most important nematode pest in tropical regions. Favoured by monoculture, low pH soils and situations where there is occasional moisture stress
	Asia	Widespread in the Philippines and some areas of Thailand	
	Australia	Limited to tropical north Queensland	
	Hawaii	Widespread	
	Central and South America	Locally important	
Lesion nematode (*Pratylenchus brachyurus*)	Africa	Widespread in western Africa, limited distribution in southern Africa	Prevalent and of major economic importance in the equatorial tropics. Less important in subtropical areas
	Asia	Widespread	
	Australia	Uncommon, generally occurs at low densities	
	Hawaii	High populations only in limited areas	
	Central and South America	Widespread	

[a]Summaries from Guérout (1975), Keetch (1982), Rohrbach and Apt (1986), Caswell *et al.* (1990), with some additional comments from Dr J.W. Clink (Dole Asia Ltd).

STRATEGIC DECISION MAKING For the last 50 years, nematode pests of pineapple have largely been controlled by pre-plant fumigation. Fumigants were relatively inexpensive and so effective that they were applied routinely as an insurance against crop losses from nematodes. In situations where non-volatile nematicides are needed to suppress resurging nematode populations, they usually have been applied at regular intervals using a calendar-based approach (Rohrbach and Apt, 1986).

With the development of modern concepts of integrated pest management, there has been a move within most sectors of the pineapple industry towards a more strategic approach to nematode management. Nematode infestation levels are determined by sampling fields and decisions on whether action should be taken against the pest are made on the basis of this information. Many pineapple companies regularly sample their fields for nematodes, while a commercial nematode monitoring service is provided for the numerous small growers who comprise the Australian industry.

SAMPLING METHODS Nematodes are never uniformly distributed in pineapple fields. Subtle changes in soil texture, for example, will markedly affect nematode population densities. Thus the only way to obtain reliable information on nematodes and their distribution within a field is to take a number of samples from the field. Since each sample must be representative of the area sampled, it must be a composite of a large number of soil cores. The monitoring service in Australia, for example, is based on a composite sample of 50 cores from a sampling area of 0.5–1 ha. This sampling scheme provides an estimate of nematode density that is within about 25% of the true mean (Stirling and Kopittke, 2000).

MONITORING PROCEDURES Populations of all plant-parasitic nematodes follow the same basic pattern during the pineapple cropping cycle: they are highest during the ratoon crop, they decline during the fallow intercycle period and then re-establish during the plant crop (Fig. 6.1).

Any sampling scheme for nematodes must take these fluctuations into account. In Australia, for example, root-knot nematode is the main pest and samples are collected at four key times during the cropping cycle:

1. Samples collected near the end of the ratoon crop indicate the likely importance of nematodes in the next crop.

2. Samples taken just prior to planting show the impact of the intercycle period on

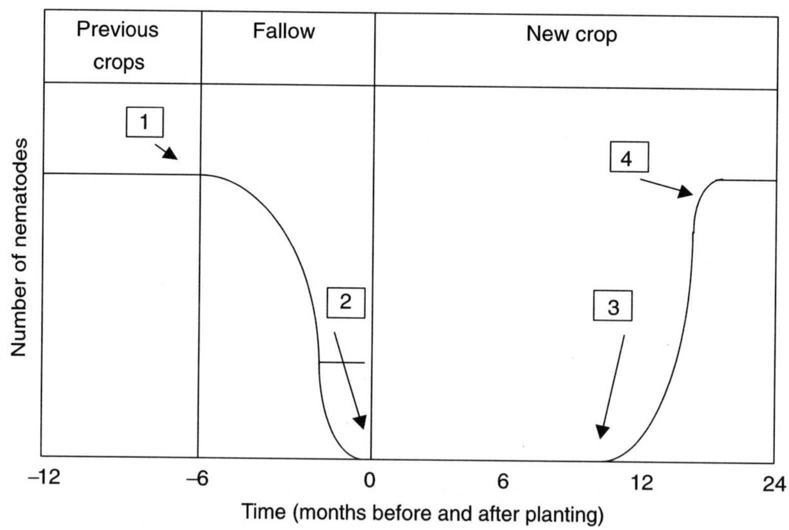

Fig. 6.1. Diagrammatic representation of the dynamics of nematode populations during the pinapple cropping cycle. (The four key times to sample for nematodes are designated by numerals.)

nematode populations. Data from this and the previous sample can be used to determine the need for a pre-plant nematicide.

3. Samples taken about 12 months after planting indicate whether nematode populations have started to increase on the present crop.

4. Samples at plant crop harvest indicate whether the nematode population is high enough to warrant application of a nematicide to the ratoon crop.

Populations of reniform nematode do not decline as rapidly as root-knot nematode during the intercycle period, and they increase more rapidly on the newly planted crop. It may therefore be necessary to modify the above sampling scheme if this nematode is the key pest. Sampling would probably start earlier, and it would continue on a regular basis (perhaps every 3 months) until early in the ratoon crop.

ECONOMIC THRESHOLDS Because of the length of the pineapple cropping cycle, the relatively short generation times of all important plant-parasitic nematodes, and the wide range of factors that can influence the rate of nematode multiplication, it has proved difficult to determine relationships between pre-plant nematode densities and yield. Root-knot nematodes have such a capacity to increase that populations as low as one nematode per 100 ml soil have the potential to cause problems (Keetch, 1982). In Australia, the population density 12 months after planting is considered a more useful indicator of the potential for damage. Populations as high as 200 root-knot nematodes per 100 ml soil have no measurable effect on plant crop yield, but can result in ratoon crop failure. As few as 1–5 nematodes per 100 ml soil at 12 months will reduce ratoon crop yields by 10% (Stirling and Kopittke, 2000).

Damage thresholds for reniform and lesion nematodes on pineapple have not been quantified (Caswell *et al.*, 1991), but in both cases they are likely to be very low. Pre-plant densities of about two reniform nematodes per 100 ml soil can increase to more than 600 nematodes per 100 ml soil within 12 months of planting. When nematicides are applied in this situation, yields in the ratoon crop can be increased by as much as 25% (Sipes and Schmitt 1994a). For lesion nematode, pre-plant populations of less than one nematode per 100 ml soil can substantially reduce fruit weight (Sarah, 1986). Under the environmental conditions of West Africa, population densities of this magnitude can reduce plant crop yields by 30%, and ratoon crops may be uneconomic.

CHEMICAL CONTROL Regardless of where pineapples are grown and which nematode species is the key pest, chemical nematicides are the primary means of control. Ethylene dibromide (EDB), dibromochloropropane (DBCP) and 1,3-dichloropropene (1,3D) were widely used in the pineapple industry until the late 1970s, when the health and environmental problems associated with the fumigant nematicides became apparent (Thomason, 1987). In most countries, 1,3D is now the only registered pre-plant fumigant, and it is generally injected under plastic at a rate of 224–336 l ha^{-1}. Non-volatile organophosphate and carbamate nematicides are also registered as pre-plant treatments in some countries. Commonly used chemicals include fenamiphos, which is formulated as a granule, and oxamyl, which is applied via trickle irrigation. Although they provide an alternative to the soil fumigants, they do not provide the same length of control.

Regardless of which nematicide is used prior to planting, there are many situations where nematodes will increase later in the crop to levels that will cause economic damage. If good ratoon crops are to be produced, the root system must be protected from nematode attack for at least the first 12–15 months, and to achieve this a post-plant nematicide may be needed. When trickle irrigation is available, emulsifiable formulations of fenamiphos and oxamyl may be applied in the irrigation water (Apt and Caswell, 1988; Schneider *et al.*, 1992). Since pineapple is unique in enabling non-volatile nematicides to be translocated systemically from the leaves to the roots (Zeck, 1971), control may also be obtained by spraying these materials on the foliage. The effectiveness of post-plant applications of non-volatile nematicides means that it is possible to monitor nematode populations in pineapple fields and apply control measures before populations increase to damaging levels.

NON-CHEMICAL CONTROL

Fallow. Since plant-parasitic nematodes are obligate parasites of plants, they will die of starvation when fields are kept free of weeds and volunteer pineapple plants. Bare fallows can be achieved by regularly applying herbicides or by cultivation, with the latter having the added benefit of bringing some nematodes to the soil surface where they are killed by the heating and drying effects of the sun. The amount of nematode control achieved with a bare fallow depends on the nematode species involved, the environmental conditions during the fallow and the length of time that the fallow is maintained. The rate of decline in nematode populations is greatest in warm, moist soils because eggs continue to hatch under such conditions and vermiform stages remain metabolically active and will soon exhaust their food reserves. Dryness during the fallow period can reduce its effectiveness as some species, particularly *R. reniformis*, survive well in dry soils (Apt, 1976; Tsai and Apt, 1979). In current systems of pineapple culture, a bare fallow is often maintained for 4–12 months between crop cycles, and this is long enough to reduce nematode populations to non-detectable levels (Guérout, 1975; Stirling and Nikulin, 1993). Bare fallow therefore provides a significant level of control, but has the disadvantage of subjecting soil to the risk of erosion and of reducing populations of beneficial soil microorganisms.

Crop rotation. One method of obtaining the benefits of bare fallow while overcoming its disadvantages is to grow a non-host crop during the intercycle period. A variety of crops has been suggested (Caswell *et al.*, 1990), but some of these are likely to be of limited value because they do not grow vigorously enough to compete with weeds, they do not tolerate the residual herbicides used in many pineapple cropping systems, or they are likely to become weeds in the next pineapple crop. Another problem is that it is difficult to find crops that are poor hosts of all the important nematode pests of pineapple. There are a number of crops that are potentially useful against *Meloidogyne* spp., e.g. grasses such as Rhodes grass (*Chloris gayana*), green panic (*Panicum maximum*), pangola *grass* (*Digitaria decumbens*) and forage sorghum (*Sorghum bicolor* × *S. sudanense*) and legumes such as velvet bean (*Mucuna pruriens*) and forage peanut (*Arachis pintoi*), but all of them are not necessarily poor hosts of *R. reniformis* and *P. brachyurus*. Rhodes grass may also be useful against *R. reniformis* because it is not a host (Caswell *et al.*, 1991), and it was one of a number of crops that increased the rate of decline of reniform nematode in comparison to bare fallow (Ko and Schmitt, 1996).

One plant family which contains crops that have been assessed in recent years for their capacity to suppress nematodes is the *Cruciferae*. When residues of cruciferous plants decompose in soil, glucosinolates present in plant tissue are converted by enzymatic decomposition into isothiocyanates, the active ingredient of the soil fumigant metham sodium. There has therefore been interest in using these crops as green manures to obtain a 'biofumigation' effect. However, results with *Brassica* crops have been equivocal (Mojtahedi *et al.*, 1991, 1993; Johnson *et al.*, 1992), possibly because many *Brassica* spp. are good nematode hosts (Bernard and Montgomery-Dee, 1993; McSorley and Frederick, 1995) or because of low glucosinolate concentrations in cultivars that are currently available. If high glucosinolate cultivars suitable for subtropical and tropical climates are eventually released, it may be worthwhile evaluating them as rotation crops in pineapple cropping systems.

Resistance and tolerance. Because 'Cayenne' is widely accepted in both canning and fresh fruit markets, pineapple plantings throughout the world are dominated by a clone which is both nematode susceptible and intolerant (Py *et al.*, 1984). There have been claims that varieties with nematode tolerance have been identified in the Hawaiian pineapple breeding programme (Caswell and Apt, 1989; Williams *et al.*, 1993) and there are reports of nematode resistance in species other than

Ananus comosus (Caswell *et al.*, 1990). However, this material has never been developed as a commercial cultivar and Sipes and Schmitt (1994b) showed that some of it was not resistant to either *Meloidogyne javanica* or *Rotylenchulus reniformis*.

From a research perspective, one of the challenges of the future is to determine whether modern genetic engineering technologies can improve the nematode resistance of pineapple when conventional selection and breeding programmes have failed. Recent progress with pineapple transformation systems (Graham *et al.*, 1998) suggests that it will eventually be possible to introduce foreign genes into pineapple. The *Mi* gene, which confers resistance to several species of *Meloidogyne*, has been cloned (Kaloshian *et al.*, 1998; Williamson, 1998), while several synthetic resistance systems, e.g. proteinase inhibitors and systems which interfere with giant cell function, are being actively studied in other crops. Less is known about mechanisms of resistance to reniform and lesion nematodes, which means that resistance to these nematodes will be more difficult to achieve.

Alternative cropping systems. In the pineapple cropping system used in most parts of the world, reliance on a monoculture with limited breaks between cycles means that nematode population densities are rarely limited by the absence of a host plant, and some nematodes are always carried over from one crop to the next. Nematode problems are compounded by soil management practices that reduce populations of antagonistic microorganisms to negligible levels (Rohrbach and Apt, 1986). Excessive cultivation, for example, hastens the breakdown of organic matter and indirectly effects the soil biota, while soil fumigation reduces the number and diversity of beneficial soil microorganisms. Thus the typical pineapple-cropping system provides plant-parasitic nematodes with an ample food source and a soil environment in which they can multiply with little competition.

There is a large body of experimental and empirical evidence (Stirling, 1991) to show that soils become more suppressive to nematodes when levels of organic matter are increased. Some of that evidence comes from experiments with Hawaiian pineapple soils in the era before soil fumigation became widespread. Damage caused by root-knot nematode was reduced when chopped pineapple material was incorporated into soil, probably because there was an increase in the activity of naturally occurring parasites and predators of nematodes (Linford, 1937; Linford *et al.*, 1938). The challenge of those interested in developing more sustainable methods of pineapple culture is to develop cropping systems which can harness these natural suppressive forces. Conservation of organic matter will be a key element in such systems and could be achieved with practices such as intercropping, green manuring, minimum tillage, trash retention, mulching and addition of organic amendments.

Integrated pest and disease management. Although the major pests of pineapple are discussed individually in this chapter, it is important to recognize that interactions occur between pests. Nowhere is this more apparent than in soil, where nematodes, white grubs, symphylids and the root rotting fungus, *Phytophthora cinnamomi*, may all act on the same root system. Since damage caused by one pest can influence other components of this root disease complex, it is vital to correctly diagnose such problems and direct control measures at all the causal factors. There is little to be gained, for example, in treating a crop for a chronic *Phytophthora* root rot problem if plant-parasitic nematodes increase on the newly available food source and destroy the new root system.

Arthropods

Arthropods may be defined as invertebrates, with jointed legs and an outer body layer which functions as a rigid protective exoskeleton. The arthropod pests of pineapples include mite, symphylid and insect species. These are listed in Table 6.3, including 102 species from which it may be concluded that the percentage composition for these three taxa is: mites, 4%; symphylids, 3%; insects

Table 6.3. Pineapple pests of the world. Referenced listing of reported locations, on plants and geographically, of arthropod pests (mites, symphylids and insects) of the pineapple, *Ananas comosus* (L.) Merr., with economic importance indicators.

Pests and location	Reference	Plant part affected and economic importance[a]
Acari: Eriophyidae		
Phyllocoptruta sakimurae Keifer		Fruit
Hawaii	Sakimura, 1966	2
Acari: Tarsonemidae		
Steneotarsonemus ananas (Tryon) (pineapple fruit mite/leathery pocket mite)		Floral cavities Young leaf bases
Probably universal	Rohrbach, 1983	
Australia	Tryon, 1898 Waite, 1993	3
Hawaii	Carter, 1967	3
Indonesia	Fauzan, Lampung, 1999, personal communication	2
Ivory Coast	A. Adiko and M. Kèhè, Abidjan, 1999, personal communication	1
Réunion	Vuillaume, 1982	2
South Africa	Petty, 1975	3
Swaziland	Dodson, 1969	3
Tarsonemus sp.		Fruit
Hawaii	Sakimura, 1966	2
Acari: Tenuipalpidae		
Dolichotetranychus floridanus (Banks) (pineapple red mite/flat mite)		Base of leaves
Australia	Waite, 1993	1
Brazil	Sanches, 1980	1
Hawaii	Carter, 1967	1
Indonesia	Fauzan, Lampung, 1999, personal communication	2
Mexico	Linford, 1952	1
Réunion	S. Quilici, Réunion, 1999, personal communication	2
South Africa	Petty, 1978b	1
Most other pineapple growing countries, including Malaysia	Waite, 1993	
Acari: Cryptostigmata (oribatid mites)		Root apex
Lamellobates palustris Hammer		5
Paralamellobates bengalensis Bhaduri & Raychaudhuri		5
Paroppis sp.		5
West Bengal	Sanyal and Das, 1989	
Rostrozetes foveolatus Sellnick		5
Scheloribates curvialatus Hammer		5
Myriapoda: Symphyla (garden centipedes/symphylids)		
Hanseniella unguiculata (Hansen)		Root apex, root hairs
Hawaii	Sakimura, 1966 Carter, 1967	1

Table 6.3. *Continued.*

Pests and location	Reference	Plant part affected and economic importance[a]
Hanseniella spp.		
Australia	Waite, 1993	1
Brazil	N.F. Sanches, Bahia, 1999, personal communication	2
Many other pineapple growing countries	Waite, 1993	
Hanseniella ivorensis Juberthie-Jupeau & Kèhè		Root apex, root hairs
Ivory Coast	Kèhè *et al.*, 1997	1
Scutigerella sakimurai Scheller		Root apex, root hairs
Hawaii	Sakimura, 1966	2
Insects		
Orthoptera: Blaberidae		
Diploptera punctata (Eschscholtz)		Fruit
Hawaii	Sakimura, 1966	2
Orthoptera: Acrididae		
Atractomorpha sinensis Bolivar		Leaves
Hawaii	Sakimura, 1966	2
Taiwan	Hung-Chieh Wen, Taiwan, 1999, personal communication	3
Locusta migratoria Linnaeus		Leaves
Indonesia	Fauzan, Lampung, 1999, personal communication	3
Oxya chinensis (Thunberg)		Leaves
Hawaii	Sakimura, 1966	2
Taiwan	Hung-Chieh Wen, Taiwan, 1999, personal communication	3
Orthoptera: Tettigoniidae		
Conocephalus saltator (de Saussure)		Fruit
Hawaii	Sakimura, 1966	2
Elimae punctifera (Walker)		Fruit
Hawaii	Sakimura, 1966	2
Nastonotus reductus (Brunner von Wattenwyl)		Fruit
Venezuela	F. Leal, Maracay, 1999, personal communication	3
Tarbinskiellus portentosus (Lichtenstein)		Leaves
China	Li Liying, Guangzhou, 1999, personal communication	1
Orthoptera: Gryllidae		
Gryllus bimaculata De Geer (black cricket)		Fruit
Taiwan	Hung-Chieh Wen, Taiwan, 1999, personal communication	3
Teleogryllus mitratus (Burmeister) (oriental garden cricket)		Fruit
Taiwan	Hung-Chieh Wen, Taiwan, 1999, personal communication	3
Teleogryllus oceanicus (Le Guillou)		Fruit
Hawaii	Sakimura, 1966	2

continued

Table 6.3. *Continued.*

Pests and location	Reference	Plant part affected and economic importance[a]
Isoptera: Termitidae		
Macrotermes subhyalinus Rambur		Roots
French Guinea	J.-F. Vayssieres, Ligne Paradis/St Pierre, 1999, personal communication	3
Procornitermes striatus (Hagen)		Leaves, roots
Brazil	N.F. Sanches, Bahia, 1999, personal communication	3
Syntermes obtusus Holmgren		Leaves, roots
Brazil	N.F. Sanches, Bahia, 1999, personal communication	3
Trinervitermes oeconomus (Trägårdh)		Roots
French Guinea	J.-F. Vayssieres, Ligne Paradis/St Pierre, 1999, personal communication	3
Isoptera: Rhinotermitidae		
Coptotermes formosanus Shiraki		Leaves, roots
China	Li Liying, Guangzhou, 1999, personal communication	3
Taiwan	Hung-Chieh Wen,Taiwan, 1999, personal communication	2
Thysanoptera: Thripidae		
Frankliniella fusca (Hinds)		Floral cavities
Hawaii	Sakimura, 1966	3
Frankliniella occidentalis (Pergande)		Floral cavities
Hawaii	Sakimura, 1966	3
Réunion	S. Quilici, Réunion, 1999, personal communication	2
Frankliniella schultzei (Trybom)		Floral cavities
Réunion	S. Quilici, Réunion, 1999, personal communication	2
South Africa	Petty, 1978c	3
Holopothrips ananasi Lima		Leaves
Brazil	N.F. Sanches, Bahia, 1999, personal communication	3
Thrips tabaci Lindeman		Floral cavities, leaves
Hawaii	Sakimura, 1966	1
Philippines	Sérrano, 1935	1
Réunion	Plenet, 1965	2
South Africa	Petty, 1978c	3
Homoptera: Pseudococcidae		
Dysmicoccus brevipes (Cockerell) (pink pineapple mealybug)		Fruit base and peduncle, fruit floral cavities, leaf axils, roots
Worldwide	Carter, 1954 Py and Tisseau, 1965	1
Dysmicoccus neobrevipes Beardsley (grey pineapple mealybug)		Fruit, leaves
Brazil	Sanches, 1981	1
Fiji	Rohrbach, 1983	1
Hawaii	Beardsley, 1959	1
Philippines	Rohrbach, 1983	1
Taiwan	Rohrbach, 1983	1

Table 6.3. *Continued.*

Pests and location	Reference	Plant part affected and economic importance[a]
Very widely distributed, including Micronesia, Malaysia, Mexico, Jamaica	Rohrbach, *et al.*, 1988	
Ferrisia virgata (Cockerell) (striped mealybug)		Fruit, leaves
Hawaii	Sakimura, 1966	4
Taiwan	Hung-Chieh Wen, Taiwan, 1999, personal communication	2
Geococcus coffeae Green		Roots
Hawaii	Sakimura, 1966	4
Phenacoccus solani Ferris		Floral cavities, heart leaves
Hawaii	Carter, 1963	4
South Africa	Willers, 1992	4
Planococcus citri (Risso)		Leaves
Brazil	Sanches, 1980	3
Hawaii	Sakimura, 1966	3
Taiwan	Hung-Chieh Wen, Taiwan, 1999, personal communication	3
Pseudococcus longispinus (Targioni-Tozzetti)		Fruit crown
Hawaii	Rohrbach *et al.*, 1988	3
South Africa	Petty, 1976a	2
Taiwan	Hung Chieh Wen, Taiwan, 1999, personal communication	3
Homoptera: Diaspididae		
Aspidiotus nerii Bouché		Fruit, leaves
Brazil	Sanches, 1980	3
Aulacaspis maculata Cockerell		Fruit, leaves
Unspecified location	Borchsenius, 1966 Recorded on pineapple by Ferald (1903)	5
Diaspis boisduvalii Signoret		Fruit, leaves
Brazil	Sanches, 1980	3
Hawaii	Py *et al.*, 1987	1
Ivory Coast	A. Adiko and M. Kèhè, Abidjan, 1999, personal communication	2
Latin America	Py *et al.*, 1987	1
Sri Lanka	Py *et al.*, 1987	1
Taiwan	Py *et al.*, 1987	1
Diaspis bromeliae (Kerner)		Fruit, leaves
Asia	Petty, 1978a	3
Australia	Petty, 1978a	3
Brazil	Sanches, 1981	3
Hawaii	Sakimura, 1966	1
Ivory Coast	A. Adiko and M. Kehe, Abidjan, 1999, personal communication	2
Pacific Islands	Petty, 1978a	3
Réunion	S. Quilici, Réunion, 1999, personal communication	2
South Africa	Petty, 1978a	3
Taiwan	Hung-Chieh Wen, Taiwan, 1999, personal communication	3

continued

Table 6.3. *Continued.*

Pests and location	Reference	Plant part affected and economic importance[a]
Gymnaspis aechmeae Newstead		Fruit, leaves
Unspecified location	Borchsenius, 1966 Recorded on pineapple by Ferris (1937)	5
Melanaspis smilacis (Comstock) (*M. bromeliae* (Leonardi))		Fruit, leaves
Brazil	Sanches, 1980	3
Hawaii	Sakimura, 1966	2
Pseudaonidia trilobitiformis (Green)		Fruit, leaves
Unspecified location	Borchsenius, 1966 Recorded on pineapple by Balachowsky (1958)	5
Pseudischnaspis anassarum Lindiger (purple scale) This species is a synonym of *Melanaspis smilacis* (Comstock)		Fruit, leaves
Taiwan	Hung-Chieh Wen, Taiwan, 1999, personal communication	2
Homoptera: Ortheziidae		
Orthezia praelonga Douglas		Fruit, leaves
Brazil	N.F. Sanches, Bahia, 1999, personal communication	
Heteroptera: Lygaeidae		
Nysius clevelandensis Evans (grey cluster bug)		Fruit, leaves
Australia	Waite, 1993	2
Nysius vinitor Bergroth (Rutherglen bug)		Fruit, leaves
Australia	Waite, 1993	2
Heteroptera: Coreidae		
Lybindus dichrous (Stal.)		Fruit inflorescence
Argentina	Menezes Mariconi, 1953	3
Brazil	N.F. Sanches, Bahia, 1999, personal communication	3
Lepidoptera: Lycaenidae		
Thecla basilides (Geyer) (*Tmolus echion* (Linnaeus))		Fruit, slips
Brazil	Sanches, 1980	1
Guatemala	Collins, 1960	1
Mexico	Linford, 1952	1
South America	Sakimura, 1966	1
Trinidad	Harris, 1927 Anon., 1995	1
Venezuela	Martinez, 1976	1
Lepidoptera: Castniidae		
Castniomera licus (Drury) (giant/white stem borer)		Stem
Brazil	Sanches, 1980	3
Castnia penelope Schaufuss		Stem
Brazil	Sanches, 1980	2
Venezuela	F. Leal, Maracay, 1999, personal communication	1

Table 6.3. *Continued.*

Pests and location	Reference	Plant part affected and economic importance[a]
Lepidoptera: Nymphalidae		
Dynastor darius (Stichel)		Leaves
Brazil	Sanches, 1980	3
Venezuela	F. Leal, Maracay, 1999, personal communication	3
Lepidoptera: Noctuidae		
Agrotis ipsilon Hufnagel (black cutworm)		Leaf
Taiwan	Hung-Chieh Wen, Taiwan, 1999, personal communication	3
Autographa biloba (Stephens)		Fruit
Hawaii	Sakimura, 1966	2
Elaphria agrotina Guenée		Leaf
Brazil	Sanches, 1981	3
Mythimna convecta Walker (common army worm)		Flowers, developing fruit, leaves
Australia	Waite, 1993	3
Spodoptera exigua (Hübner) (lesser army worm/lesser cotton leaf worm/beet army worm)		Leaf bases
South Africa	Petty, 1990a	3
Spodoptera litura Fabricius (tobacco cutworm)		Leaf
Taiwan	Hung-Chieh Wen, Taiwan, 1999, personal communication	3
Lepidoptera: Coleophoridae		
Batrachedra methesoni Busck		Fruit
Caribbean	Sakimura, 1966	3
Puerto Rico	Perez, 1957, 1959	3
Lepidoptera: Cosmopterygidae		
Pyroderces hemizopha (Meyrick)		Floral cavity (scavenging)
Ivory Coast	Py *et al.*, 1987	4
Pyroderces rileyi (Walsingham) (formerly in the genus *Anatrachyntis*) (pink bud moth)		Floral cavity (scavenging)
Hawaii	Sakimura, 1966	4
South Africa	Petty, 1991	4
Lepidoptera: Tineidae		
Neodecadarchis flavistriata (Walsingham)		Floral cavity (scavenging)
Hawaii	Sakimura, 1966	4
Opogona sacchari (Bojer) (formerly in the genus *Alucita*)		Stem
Réunion	S. Quilici, Réunion, 1999, personal communication	2
Lepidoptera: Tortricidae		
Amorbia emigratella Busck		Fruit
Hawaii	Sakimura, 1966	4

continued

Table 6.3. *Continued.*

Pests and location	Reference	Plant part affected and economic importance[a]
Lepidoptera: Limacodidae		
Darna trima (Moore)		Leaves
China	Li Liying, Guangzhou, 1999, personal communication	1/2
Setothosea asigna (van Eecke)		Leaves
China	Li Liying, Guangzhou, 1999, personal communication	3
Coleoptera: Scarabaeidae (white grubs)		
Melolonthinae		
Antitrogus mussoni (Blackburn) (possible misidentification of *A. consanguineus* (Blackburn) and *A. rugulosus* (Blackburn))	Allsopp, 1993	Roots
Australia	Waite, 1993	3
Asthenopholis subfasciata (Blanchard)		Roots
South Africa	Petty, 1977a	1
Congella valida Péringuey (previously misidentified as *Adoretus tessulatus* Burmeister)		Roots
South Africa	Smith *et al.*, 1995	2
Hoplochelus marginalis (Fairmaire)		Roots
Réunion	Quilici *et al.*, 1992	3
Lepidiota gibbifrons Britton		Roots
Australia	Waite, 1993	3
Lepidiota grata Blackburn		Roots
Australia	Waite, 1993	3
Lepidiota noxia Britton		Roots
Australia	Waite, 1993	3
Lepidiota squamulata Waterhouse		Roots
Australia	Waite, 1993	3
Lepidiota stigma (Fabricius)		Roots
Indonesia	Fauzan, Lampung, 1999, personal communication	2
Macrophylla ciliata (Herbst)		Roots
South Africa	Petty, 1976b	3
Phyllophaga portoricensis Smyth		Roots
Puerto Rico and other Caribbean Islands	Py and Tisseau, 1965	3
Rhopaea magnicornis Blackburn		Roots
Australia	Waite, 1993	3
Trochalus politus Moser		Roots
South Africa	Petty, 1976b	2
Rutelinae		
Adoretus (*Adoretus*) *ictericus* Burmeister		Roots
South Africa	Petty, 1976b	2
Adoretus (*Lepadoretus*) *sinicus* Burmeister		Leaves, roots
Hawaii	Sakimura, 1966	2
Anomala expansa Bates (green beetle)		Roots 1
Taiwan	Hung-Chieh Wen, Taiwan, 1999, personal communication	
Anoplognathus porosus (Dalman)		Roots
Australia	Waite, 1993	3

Table 6.3. *Continued.*

Pests and location	Reference	Plant part affected and economic importance[a]
Blitopertha (*Exomala*) *orientalis* (Waterhouse) (formerly in the genus *Anomala)*		Roots
Hawaii	Sakimura, 1966	1
Popillia bipunctata (Fabricius).		Roots
South Africa	G.J. Petty, Bathurst, 1997, personal observation	4
Popillia mutans Newman		Roots
China	Li Liying, Guangzhou, 1999, personal communication	3
Dynastinae		
Augosoma centaurus (Fabricius)		Fruits, inflorescence
French Guinea	J.F. Vayssieres, Ligne Paradis St Pierre, 1999, personal communication	3
Ivory Coast	Guérout, 1974	2
Heteronychus arator (Fabricius) (black maize beetle)		Subterranean stump
South Africa	Petty, 1977b	1
Strategus jugurtha Burmeister (pineapple rhinoceros beetle)		Roots
Bolivia	Munro, 1954	2
Venezuela	F. Leal, Maracay, 1999, personal communication	2
Strategus validus (Fabricius) (rhinoceros beetle)		Roots
Brazil	N.F. Sanches, Bahia, 1999, personal communication	3
Coleoptera: Curculionidae (Weevils)		
Rhynchophorinae		
Metamasius dimidiatipennis (Jekel) (pineapple black spot beetle)		Fruit, stem
Venezuela	F. Leal, Maracay, 1999, personal communication	1
Metamasius fasciatus (Olivier) (pineapple black beetle)		Fruit, stem
Venezuela	F. Leal, Maracay, 1999, personal communication	1
Metamasius ritchiei Marshall		Fruit, stem
Caribbean	Sakimura, 1966	2
Jamaica	Marshall, 1961	2
Paradiaphorus crenatus (Gyllenhal)		Lower stem
Brazil	N.F. Sanches, Bahia, 1999, personal communication	3
Rhabdoscelus obscurus Boisduval (New Guinea sugar cane weevil)		Fruit
Australia	Waite, 1993	2
Brachycerinae		
Pantomorus cervinus (Boheman) (synonym of *P. godmanni* (Crotch))		Roots
Hawaii	Sakimura, 1966	2

continued

Table 6.3. *Continued.*

Pests and location	Reference	Plant part affected and economic importance[a]
Curculioninae		
Baris sp. (pineapple weevil)		Leaves
Venezuela	F. Leal, Maracay, 1999, personal communication	2
Cholus seabrai Vaurie		Peduncle, slips
Brazil	N.F. Sanches, Bahia, 1999, personal communication	3
Diastethus sp. (formerly *Gladosius* Casey, 1922)	Kuschel, 1983	Leaves
Brazil	N.F. Sanches, Bahia, 1999, personal communication	3
Parisoschoenus ananasi Moure		Leaves
Brazil	N.F. Sanches, Bahia, 1999, personal communication	3
Coleoptera: Colydiidae		
Bitoma sp.		Fruit
Brazil	N.F. Sanches, Bahia, 1999, personal communication	3

[a]1 = major pest; 2 = minor pest; 3 = occasional pest; 4 = minor pest of limited distribution; 5 = economic importance unknown.

93%. A breakdown of the insect species reveals the following composition: Orthoptera, 12% (grasshoppers, 4%; katydids, 4%; crickets, 3%; cockroaches, 1%); Isoptera (termites), 6%; Thysanoptera (thrips), 5%; Homoptera, 15% (mealybugs, 7.5%; scale insects, 7.5%); Heteroptera (bugs), 3%; Lepidoptera, 18% (moths, 16%; butterflies, 2%); Coleoptera (beetles), 35% (white grubs, 24.5%; weevils 9.8%).

It is thus apparent that a tremendous range of ecological niches of the pineapple plant are occupied by species with widely differing biologies – and economic importance (Table 6.3). Due to space constraints it is only possible to deal with a very few species. An attempt has been made to present accounts of the most important ('key') representatives of some of the taxa. Although Table 6.3 lists Isoptera species, it should be noted that they are not obligately pests of pineapples and probably only attack dead or ailing plants. The table does not list species of the following groups, for the reasons given: (i) true ants (Hymenoptera: Formicidae), as these are not pests *per se* but important only by virtue of their association with pineapple mealybugs; (ii) pineapple souring beetles (Coleoptera: Nitidulidae) do not harm sound fruits, and only become a nuisance when subgrade fruits are left to over-ripen in the fields; (iii) fruit flies (Diptera: Tephritidae), e.g. oriental fruit fly, *Bactrocera dorsalis*, melon fly, *Bactrocera cucurbitae* and Mediterranean fruit fly, *Ceratitis capitata*, have been suspected at times, but studies have shown that, if oviposition in fruits does occur, most eggs do not hatch and larvae do not survive (Carter, 1967).

Key arthropod pests

Apart from one insect species, the pink pineapple mealybug, key status does not apply universally to these species, their occurrence and importance in different pineapple producing regions varying considerably. The occurrence of some species may be extremely limiting to crop production in one or more geographical regions but of little or no consequence elsewhere. Classification as 'key' of some of the following arthropod species, should therefore be understood as such, with these species attaining either key or secondary pest status depending on geographical location. Other species given key status may generally be considered as minor pests in specific locations but, because of their universal distribution, their total impact on pineapple production justifies key status.

Pineapple leathery pocket mite/pineapple fruit mite, Steneotarsonemus ananas *(Tryon)*

Steneotarsonemus ananas (Tryon), pineapple leathery pocket mite/pineapple fruit mite, is host-specific to pineapples, occurring worldwide. It is important primarily because of its association with the fruit diseases – inter-fruitlet corking, leathery pocket, and fruitlet core rot/black spot (Plate 34).

BIOLOGY

Description of stages. (M.K.P. Meyer, Pretoria, 1999, personal communication). Egg: oval, opaque white, and large relative to the adult; they are laid singly. Larva: the oval, white active larva has three pairs of legs in the position of the adult's first three pairs of legs. Pupa: mature larvae enter a quiescent pupal stage, which is shiny white and elongate oval. Adult female: elongate oval (150–240 μm wide), light amber to creamy brown; the fourth pair of legs is very thin terminating in long setae (Plate 35). Adult male: the fourth pair of legs of the smaller male is robust and claw-like.

Life cycle. Due to culturing difficulties, very little biological information is available. The cycle may be completed in 7–14 days (Py *et al.*, 1987). Male and female larvae pass through the sessile pupal stage in which transformation to the adult takes place within the bloated larval integument. A fourth pair of legs and genitalia develop distally, appendages are withdrawn from the old integument, and the adult emerges through the dorsally split pupal skin. By parthenogenetic reproduction, males may be produced from haploid eggs. Warm temperatures with very high relative humidity and low light intensity are optimal for the mites' development. Their preferred locations on pineapples are on leaf bases and the inflorescence and developing fruit, especially in blossom cups. Their numbers increase with plant growth, peaking from 6 weeks after forcing (artificially induced flowering) in the warmer months of the year, to 12 or more weeks in the cooler months (G.J. Petty, 1976, unpublished).

FRUIT–DISEASE ASSOCIATION Fruit mites feed on trichome cells on the base of young heart leaves, and on the bracts and sepals of developing fruits (Rohrbach and Apt, 1986). Damage to the latter gives a glossy 'waxy mutant'-like appearance to fruits. These damaged trichomes provide a favourable substrate for development of the fungus, *Penicillium funiculosum* Thom., causal organism of leathery pocket, fruitlet core rot/black spot, and inter-fruitlet corking. Leathery pocket can be particularly severe if fruit maturation occurs under low moisture conditions followed by rainfall during later fruit enlargement, due to cracking of fruit tissues followed by fungus infection (Rohrbach and Apt, 1986). Its incidence is extremely variable and normally low. However, it peaks between January and April (South Africa), and cannery slice recovery has been reduced from a potential 16% to as low as 3% (Petty, 1990b). It was generally found that plants with a high mite infestation after forcing bore fruits with high levels of leathery pocket (G.J. Petty, 1976, unpublished). Mourichon *et al.* (1987) found that fruit mite infestation in pineapple flowers 9 to 10 weeks after forcing could be correlated with the subsequent incidence of leathery pocket. Further evidence for the importance of these mites was provided by research showing that both mites and leathery pocket are very effectively controlled with foliar sprays of endosulfan, which does not control *P. funiculosum* (Petty, 1990c). In the southern hemisphere, relatively moister south-facing plantation slopes are more affected by these fruit diseases than are north-facing slopes.

SAMPLING AND MONITORING Sampling for the monitoring of infestations should take cognizance of the distribution of these mites. Petty (1992) found most mites in the light green zone of leaf bases, i.e. between the soft white basal section and the dark green distal section. On the plant as a whole, total number of mites per leaf increased from the youngest,

A-whorl, leaf to a maximum on 300–400 mm long C-whorl leaves (third spiral from apical meristem), progressively decreasing thereafter to very low numbers lower on the plant where leaf angle to the stem is more obtuse. The lower dirt and trichome contamination of C leaves facilitates mite counting. Because of their small size, a stereomicroscope is required for fruit mite counting – which may be directly on leaves; but the procedure recommended for the pineapple flat mite is advantageous, i.e. removal of mites into liquid suspension, filtration, and total or subsample counting.

ECONOMIC THRESHOLD An economic threshold must be based on fruit mite infestation immediately preceding time of forcing, with the objective of attaining minimal infestation during the following 5 or more weeks. This goal is usually impractical and, therefore, the threshold should be the following situations which are conducive to high leathery pocket incidence: (i) plant (as opposed to ratoon) crop plantations; (ii) plantations on cool, moist aspects; and (iii) forcing dates which ensure harvesting at a time of year historically known for high leathery pocket incidence.

CONTROL

Natural enemies. No effective natural enemies are known, but the thrips, *Podothrips lucasseni* (Krüger), is a predator of fruit mites in Hawaii (Meyer, 1981). However, the host mite may have been incorrectly identified as *S. ananas* (Beer, 1954).

Chemical control. Only when the fruit diseases are potentially of economic importance may it be justifiable to apply control measures for leathery pocket mite and, in fact, measures are seldom applied as these diseases are difficult to predict, depending on weather conditions, and are usually sporadic. Endosulfan is registered for leathery pocket control in South Africa (Krause *et al.*, 1996). Petty (1990c) found that good control may be obtained with foliar sprays of 140 ml endosulfan 350 EC per 100 l water. Five sprays at 4-week intervals starting 4–5 weeks pre-forcing and ending 11–12 weeks post-forcing gave 69.1% control, on average, and up to 90% control. Timing of sprays is critical. Mite infestation should be as low as possible during the 5-week post-force period (Rohrbach and Apt, 1986), and minimal mite infestation following endosulfan treatment requires 5 weeks. In studies on the control of blackspot disease (G.J. Petty, 1999, unpublished), good control resulted from foliar sprays of benomyl (from 1 week before to 11 weeks after forcing) combined with endosulfan (at 5 weeks and 1 week before forcing).

Cultural control. For plant crops on south-facing aspects (southern hemisphere) or north-facing aspects (northern hemisphere) avoid, as far as possible, forcing the plants for crop maturity at a time of year when *P. funiculosum* diseases are known historically to attain maximum severity.

Pineapple flat mite/pineapple red mite, Dolichotetranychus floridanus *(Banks)*

Dolichotetranychus floridanus (Banks), the pineapple flat mite/pineapple red mite, although probably universal (Annecke and Moran, 1982), has been reported in locations including the pineapple growing countries and regions of Central America, the Philippine Islands, Brazil, Hawaii, Cuba, Taiwan, India, South Africa and Australia.

BIOLOGY

Host plants. Their greatest importance is on pineapple, but in Egypt they occur on grass and bamboo species (Wafa *et al.*, 1968–1969), and in India on cardamom spice crops, *Elettaria cardamomum* Maton (Siddappaji *et al.*, 1989).

Description of stages. Egg: light orange and oval; large, relative to mite size. Larva: amber, with distinct red eyes and six legs. Protonymph and deutonymph: orange-yellow, with eight legs. Adult female: orange-red and elongate oval, measuring 450 by 170 μm. Adult male: slightly smaller than females, i.e. 300 by 140 μm, with a more pointed abdomen.

Life cycle. Published information is lacking, but under the hot conditions of late summer and autumn, the cycle may be completed in less than 10 days.

SAMPLING AND MONITORING After digging out a plant sample as large as is practicable, peel back the leaves to determine feeding damage on the bases of older leaves. Older damage appears as dents above leaf bases (Waite, 1993). Feeding sites occur in the narrow space between leaf bases and plant stem, on the whitish adaxial leaf surface (Petty, 1990e). For monitoring of infestation, remove the oldest ten living basal leaves to obtain a mite count. The basal section of each leaf is then excised and immersed in a solution such as ethanol or formaldehyde + acetic acid + ethanol, for some hours. Thereafter, mites may be filtered out of suspension for counting under a stereoscopic microscope. Alternatively, record the mean number of infested leaves per ten leaves for a suitably large plant sample (Petty, 1988).

ECONOMIC THRESHOLDS Plants of all ages are attacked, but the mites are most damaging on new plantings where growth can be severely curtailed, with progressive wilt (Giacomelli and Py, 1981; Waite, 1993). In Australia flat mites are a serious pest, and in Bahia, Brazil, they are widely distributed and important in commercial plantations (Sanches and Zem, 1978). Heavy infestations may kill plants, or reduce their vigour and fruit yield (Meyer, 1981; Meyer *et al.*, 1987; Petty, 1990e).

Damage is primarily to the whitish leaf bases, which develop brown/black necrotic lesions and appear progressively dehydrated, with feeding mites around lesion peripheries, or as orange patches on leaf bases. Infestations may increase on planting material during storage, resulting in damage and rejection, or irregular establishment in the field. Occasionally, crowns are destroyed and fruits damaged before harvesting (Waite, 1993).

The economic threshold of infestation is related to prevailing weather conditions – warmer and drier weather generally lowers the threshold (Petty, 1990e; Waite, 1993) when plant growth and hence leaf elongation is slower, allowing mites to feed for longer periods in specific concealed areas; their numbers increase more rapidly, and initially low infestations may result in economic damage. Under conditions for optimal plant growth, even heavy infestations may not result in significant damage as colonies diminish or disappear. However, if a warm, dry period is anticipated, planting material should be disinfested.

CONTROL

Natural enemies. The phytoseiid mite, *Amblyseius benjamini* Schicha, is commonly found with flat mites in South Africa, their body contents coloured orange by their prey (Meyer, 1981). It also occurs in Australia (Waite, 1993) but cannot suppress severe flat mite infestations. Other South African predatory mites include *Lasioseius* spp. and *Proctolaelaps* spp. (Ascidae), *Anystis baccarum* (L.) (Anystidae), *Cunaxoides* spp. (Cunaxidae) and *Triophtydeus* spp. (Tydeidae) (Petty, 1990e). Mohamed *et al.* (1982) found that the female Egyptian predatory cheyletid mite, *Cheyletus cacahuamilpensis* Baker, consumed 224 flat mites, and the male, 104, during their life spans.

Chemical control. Planting material dips with systemic organophosphate pesticides were effective despite the mites' location in concealed niches (Meyer *et al.*, 1987). Dipping of 'Perola' pineapple slips in 0.11% ethion, 0.03% omethoate or 0.09% vamidithion for 6 min gave better control of flat mites than a number of other pesticides evaluated (Sanches and Zem, 1978). Foliar sprays with systemic organophosphate pesticides gave better control of flat mites than other pesticide groups (Petty, 1990e). For mite control on large 'Queen' pineapple plants, 6 l dimethoate 400 g l^{-1} EC in 2000 l water ha^{-1} gave good control (Petty and Webster, 1981b).

Symphylids

These pineapple associated myriapods are species of *Hanseniella* (Symphyla: Scutigerellidae), e.g. *H. unguiculata*, *H. ivorensis*, and *Scutigerella sakimurai* (Symphyla: Scolopendrellidae). Unidentified species of *Hanseniella* are of economic importance in Martinique, Brazil and Australia, as is *H. ivorensis* in Ivory Coast.

BIOLOGY

Description of stages. Eggs hatch to give the 2 mm long, white, centipede-like symphylids, with six (*Scutigerella* spp.) or seven (*Hanseniella* spp.) pairs of legs. They grow and moult, progressively developing more antennal and body segments and legs so that the 6–10 mm long adults (depending on species) have 12 pairs of legs. They have prominent antennae and well developed cerci.

Life cycle. Ten or 11 eggs are laid in the soil and hatch in about 10 days to give the immature stage, developing into adults which may live for several years. The life cycle may be completed in less than 50 days, under optimum conditions.

Habits and behaviour. They live in soil, especially well structured clay loams, and gravelly sandy and clay loams (Sinclair and Scott, 1997). Stony, coarse grained soil structures are favourable (Waite, 1993). They feed on organic matter and on rootlets and root hairs and, by their pruning action, multiple branching occurs giving roots the typical 'witches' broom' effect (Waite, 1993). Infestations are normally sporadic. Symphylids survive drought conditions by moving deeper into the soil, through naturally formed apertures, but they do not reproduce at these greater than normal depths (Masses, 1979).

MONITORING The presence of these white symphyla may be determined by scattering soil samples on a sheet of black plastic, as they are easily visible on this background. Likewise, shake pineapple root systems over black plastic; the roots should be checked for witches' brooms.

ECONOMIC IMPORTANCE AND THRESHOLD

Symphylids have been a problem to the Hawaiian pineapple industry since about 1925 (Sinclair and Scott, 1997). Known in Australia since 1968, they occur throughout Queensland and south of Mackay are the most important soil arthropod pests; they affect about 35% of the Australian pineapple industry. Only *Hanseniella* spp. is of economic importance.

Symphylids preferentially feed on young, meristematic root tissues, leaving a small cavity in the tip. Resultant symptoms depend on stage of plant development and the intensity and duration of attack. Newly established plants may be unable to develop roots more than a few centimetres in length and the root system becomes a bush of short, branching witches' brooms. A later, less sustained attack will allow the development of longer roots with terminal swellings and secondary roots which, if also damaged, will form witches' brooms (M. Kéhé, 1980, unpublished). Infestations result in plants appearing unthrifty, wilted and near death, with poor ground anchorage and fruit yield (Sinclair and Scott, 1997). Drought conditions exacerbate the problem, and new plantings may be slow to establish under symphylid attack (Waite, 1993). The uniformity of plant damage in plantations, due to symphylids, is positively correlated to the uniformity of soil moisture (Py *et al.*, 1987). Economically important infestations may develop if soil conditions and food supply are optimal, especially with degeneration of residues of soil applied pesticides (Waite, 1993).

In Martinique, experimental symphylid control increased mean fruit mass by 77%, and suckers per plant by 530% (J.J. Lacoeuilhe, 1977, unpublished). In Ivory Coast, similar results were obtained (Py *et al.*, 1987). In Australia, when potted pineapple plants were infested with 12, 24 or 48 symphylids per plant, roots were reduced, in 9 weeks, by 47.7%, 61.7% and 92.8%. The latter failed to establish; the others were severely retarded (Murray and Smith, 1983).

Regarding an economic threshold, Waite (1993) stated that when symphylid numbers in the preceding crop exceed 20 per plant, chemical control is likely to be beneficial.

CONTROL

Natural control. Predation, in Australia, is ineffective (Waite, 1993). In Hawaii, predators include a centipede and a beetle larva (Py *et al.*, 1987).

Cultural control. Thorough intercycle soil preparation will ensure minimal symphylid infestation at time of planting.

Chemical control. Presently, synthetic pesticide application is the only method that can be recommended for symphylid control (Py *et al.*, 1987); pesticides should be applied as pre-plant soil treatments to protect the critical early plant development (Masses, 1979). Good, long lasting control is achieved with lindane at 23 l ha^{-1} of the 20% formulation. Slow release granules of ethoprophos, chlorpyrifos and carbofuran applied as pre-plant, soil incorporated treatments reduced symphylid root damage and significantly increased 'Smooth Cayenne' pineapple yield (Waite, 1996). Post-plant treatments are also usually essential, and must be applied as foliar sprays, or granules at the plant bases (Py *et al.*, 1987).

Integrated pest management. Thoroughly prepare soil during the intercycle period. Prior to planting, determine symphylid infestation on a black plastic sheet (as described above). If numbers exceed the economic threshold, or where soils have a history of high symphylid infestation, ensure that approved pesticides for symphylid control are applied.

Onion thrips or yellow spot thrips, Thrips tabaci *Lindeman*

As a vector of the tomato spotted wilt virus (TSWV) (Sakimura, 1960) from weed host species to pineapples, this thrips is occasionally responsible for severe occurrences of yellow spot (YS) disease of pineapples – first observed in Hawaii in 1926, and subsequently in the Philippines and Australia (Lewcock, 1937), and in 1937 in South Africa (Carter, 1939). More than 80% of fruits in one plantation block in South Africa were discarded due to YS (G. Petty, Eastern Cape, South Africa, 1981, personal observation) (Plate 36). Fortunately, this degree of severity is unusual. Most susceptible to YS are young plantations from crowns; slips are less susceptible. These, especially crowns, are the predominant planting materials for 'Smooth Cayenne' pineapple. Suckers/shoots are least susceptible – perhaps providing less attractive feeding sites for thrips (Collins, 1960). Initial symptoms are small yellow spots on adaxial leaf surfaces, increasing in size up to 15 mm and fusing to form yellow streaks which become brown and necrotic (Collins, 1960). The infection spreads downwards, eventually affecting the apical meristem; with unequal growth the plant assumes a bent growth form, eventually succumbing to the infection. Fruits become infected through the crowns while these are still attached to fruits on the plants. Infected crowns die and dry out, and the internal fruit tissues become necrotic (Pegg, 1993). Infections initiated in floral cavities at flowering destroy the affected fruitlet, and some of those adjoining it, resulting in a dry, blackened cavity – commonly known as 'dead-eye'. There is a tendency for surrounding fruitlets to develop into the cavity and for the whole fruit to assume a curvature towards the dead-eye side (Petty, 1990f) (Plate 37). For onion thrips to become infective they must feed on infected plants as nymphs (not adults), and they remain infective thereafter.

BIOLOGY Both monocotyledons and dicotyledons are included in a wide range of host plants. Cultivated species include cucurbits, crucifers, legumes, garlic, onion, cotton and the tomato family (Annecke and Moran, 1982). Weed hosts, found in pineapple fields, include *Emilia sonchifolia, Emilia sagittata, Bidens pilosa* and *Datura stramonium* (Py *et al.*, 1987).

Description and life cycle. Few males occur in warm climates (Macgill, 1927; Sakimura, 1932) and reproduction, as in glasshouses (Morison, 1957), is mainly parthenogenetic. White, bean-shaped eggs are laid singly, on various parts of the plant, either scattered or in short rows along leaf veins. The eggs are embedded deeply by the ovipositor blades and the incision closes almost completely (Lewis, 1973). At 18°C an average of 80 eggs per female were laid; at higher temperatures, more were laid and they hatched more quickly and more successfully (Lewis, 1973). The amount and quality of food available play an important role in determining egg production. The first of the four nymphal stages emerge after 5–10 days' incubation.

These are pale yellow or brownish, and mature within 30 days. The 1–1.5 mm long adult is slightly darker than nymphs, with yellowish fringed wings. They may be distinguished from *Frankliniella* thrips species occurring in pineapples by ascertaining the following morphological characters (Petty, 1978c). Antennae: seven segments; terminal segment not divided and very short. Pronotum: no prominent setae on front angles. Forevein: distal portion has four setae. At 21°C (Sakimura, 1937) the adult female life span of 59 days includes 3 days pre-oviposition, 50 days oviposition and 6 days post-oviposition.

The method of feeding, which is unique to thrips, has been variously described as: (i) puncturing of tissues and draining cell contents causing their walls to collapse; (ii) piercing the epidermis and rasping the tissues within; or (iii) rasping leaf tissues and sucking the sap. There is a progressive histological degeneration of the leaf as a result of feeding. A glandular secretion from the thrips' mouthparts area is thought to partially predigest their food (Lewis, 1973).

Mixed populations of onion thrips and *Frankliniella* spp. were monitored in a pineapple plantation in the Eastern Cape (South Africa) in 1976 (Petty, 1978c) to determine seasonality, and the effect of a weed infestation (mainly *B. pilosa*). They predominated in the summer and autumn months, and in weed infested areas.

SAMPLING AND MONITORING These actions are not normally taken, as the control measures described below are usually adequate to obviate any thrips-associated problems. However, if monitoring is required, thrips can be satisfactorily trapped (Petty, 1978c) by applying an adhesive coating of polybutene (suitably thinned with paraffin/kerosene) to a sheet of polythene which may then be wrapped around a 5 l cylindrical drum. By affixing the drum to a pipe it can be placed at plant-canopy height and will capture flying adult thrips. The polythene sheet may be removed for the identification and counting, with a stereoscopic dissecting microscope, of thrips adhering thereto.

CONTROL

Natural enemies: predators. According to Lewis (1973), there are many general predators, including anthocorid and mirid bugs, lacewings, dipterous larvae, ladybird beetles and other thrips. Macgill (1939) reported that five or six thrips per day may be killed by mesostigmatid and trombidiid mites, which attach themselves to thrips' intersegmental membranes.

Natural enemies: parasites and pathogens. Species of minute parasitic wasps (Eulophidae) attack the larvae, and Trichogrammatidae and Mymaridae parasitize the eggs. Pathogenic fungi, e.g. *Entomophthora sphaerosperma*, also inflict mortality (Charles, 1941; Stradling, 1968).

Integrated pest control. Onion thrips, and the YS virus disease that they transmit, can be very well controlled with a combination of chemical and cultural practices which maintain plantations free of weeds, especially those species which are virus and thrips hosts. Appropriate herbicides are required. Linford (1943) found that the virus can be carried by thrips from weed hosts for many hundreds of metres. Care should therefore be taken not to disturb such weeds when pineapples are most susceptible to thrips, i.e. in early growth stages and at fruiting.

Pink pineapple mealybug, Dysmicoccus brevipes *(Cockerell)*

Worldwide, pineapple industries regard this mealybug as a major threat to the crop because of its association with the devastating disease, pineapple mealybug wilt (PMW) (Plate 38).

BIOLOGY This species originated in tropical America, where the pineapple itself originated. The mealybug was spread on planting material to plantations around the world (Rohrbach *et al.*, 1988). Its more than 50 host species include sugarcane, perennial grasses, e.g. *Panicum barbinode* and *Tricholaena rosea*, and sisal *Agave sisalana* (Petty, 1976a). The mobile immature crawler stage, with flattened bodies and long hairs, may be wind-dispersed for hundreds of metres within plantations (Rohrbach *et al.*, 1988). The big-headed ant, *Pheidole megacephala* (Fabricius), also plays an important role in its dispersal (Carter, 1932). Pineapple roots and lower leaf axils are predominantly infested (Carter, 1949). On the 'Queen' pineapple cultivar, numbers declined from 3.35 on the oldest living leaf to 0.19 on the tenth leaf, continuing to decline thereafter; 88% of aerial infestation was on the oldest ten basal leaves (Petty and Webster, 1981a). In mature plants, heavy infestations may occur on slip bases, lower fruit surfaces around peduncles, and within fruitlet floral cavities. Planting material infestation, originating from mother plants, may increase during its storage but typically declines if planted in the absence of ants (Carter, 1932).

Reproduction is parthenogenetic and ovoviviparous. However, a bisexual race occurs in Ivory Coast, Madagascar, Dominican Republic and Martinique; apart from this race, pink mealybugs only produce females (Beardsley, 1964). Until about 1959, *D. brevipes* was confused with *Dysmicoccus neobrevipes* Beardsley, the grey pineapple mealybug. Ito (1938) noted biological and behavioural differences, and Beardsley (1959) noted morphological differences as summarized in Table 6.4.

At 23.5°C, developmental stages of the pink mealybug have the following durations (Ito, 1938): three larval (crawler) instars, 34.03 days; adult pre-larva position, 26.58 days; adult larva position, 24.84 days; adult post-larva position, 4.70 days; total adult life span, 56.23 days; total life span, 90 days. On average, 234 yellowish, less than 1 mm long crawlers are produced per reproductive mealybug, with a maximum of 692 recorded.

The 3–4 mm long adult female is elongate-oval, distinctly segmented and coated with a white waxy secretion. Around the lower periphery of their bodies are 34 waxy filaments – lateral filaments are 0.25 × body width; caudal filaments, 0.5 × body width. Mature mealybugs are fairly immobile as body development exceeds leg development.

MEALYBUG–ANT SYMBIOTIC RELATIONSHIP Ant importance in pineapple plantations is solely due to their positive promotion, through symbiosis, of mealybugs and PMW (Reimer *et al.*, 1990). Symbiotically related ant species include *P. megacephala*; fire ant, *Solenopsis geminata* (Fabricius); Argentine ant, *Iridomyrmex humilis* (Mayr); *Technomyrmex albipes* (Smith); *Camponotus friedae* Forel; *Anoplolepis gracilepes* (S. Smith) (formerly known as *A. longipes* Jerdon); and *Araucomyrmex* spp. Caretaking by these species includes the building with debris, and maintenance, of shelters, protecting mealybugs from natural enemies and inclement weather (Rai and Sinha, 1980).

Ants remove mealybug excreted honeydew, preventing an unhygienic build-up of *Capnodium* sp. sooty mould (Beardsley *et al.*, 1982; Duodu and Thompson, 1992). Mealybug leaf infestation of 'Queen' cultivar pineapples decreased by 90% within 20 weeks of *P. megacephala* control with hydramethylnon bait-toxin, and root infestation by almost 100% within 12 weeks (Petty and Tustin, 1993). Under these conditions, a high positive linear correlation prevailed between ant

Table 6.4. Biological and morphological differences between pink and grey mealybugs.

Variable	Pink mealybug	Grey mealybug
Ventral sclerotization of anal lobes	Quadrate	Elongate
Long dorsal abdominal setae	Present	Absent
Reproduction	Parthenogenetic	Bisexual
Mature female colour	Pinkish orange	Greyish
Preferred location on plant	Roots and lower plant	Fruits, crowns, upper plant
Graminaceous plant hosts	Yes	No
Green spotting of leaves	No	Yes

infestation and mealybug leaf infestation ($r = 0.978$, $P < 0.01$), and root infestation ($r = 0.769$, $P < 0.01$). The relationship between ant distribution and mealybug infestation was similar. When uncontrolled, ant numbers did not correlate, i.e. were no longer a limiting factor, and pineapple growth in 52 weeks was reduced by 17%.

PINEAPPLE MEALYBUG WILT PMW commences with cessation of root growth (Carter, 1935), root systems becoming reduced and necrotic. Leaves turn a deep pink, yellowing and wilting; tips die back as edges curl downwards. Fruits may fail to develop, or remain small, fibrous and sour. Symptom expression may take a number of months to develop, and plants may appear to recover (Carter, 1945). PMW is of worldwide distribution, limited only by the climate suitable for the host plant (Carter, 1954). When grown commercially and continuously on the same land, PMW becomes a limiting factor. In Cuba (and many other countries) it is one of the most important pineapple diseases, causing up to 40% crop losses (Borroto *et al.*, 1998). First reported by Hawaiians in 1910 (Rohrbach *et al.*, 1988), it devastated their industry in the 1920s and 1930s. Carter (1933) hypothesized a toxic mealybug secretion aetiology for PMW, later (1963) modifying this hypothesis as subsequent findings indicated a viral aetiology. The existence of a virus in both mealybugs and pineapples with PMW symptoms was proven by Gunasinghe and German (1989), with their finding of double stranded RNA and long, flexuous rod-shaped virus particles. With dimensions of 1200 by 12 nm, and coat protein molecular weight of 23 by 10^3 Da, the subgroup II closterovirus was indicated. This virus was detected with a monoclonal antibody test in mealybugs from PMW plants. It could not be detected in seedling pineapples (Hu *et al.*, 1996). Sether *et al.* (1998) proposed the name 'Pineapple mealybug wilt-associated virus' (PMWaV). PMWaV from diseased Australian pineapples was serologically related to the Hawaiian isolate (Wakman *et al.*, 1995). It appears to comprise a group of closely related closteroviruses (Thomas *et al.*, 1999), and in Australian pineapples a bacilliform virus (PBV) also occurs.

SAMPLING AND MONITORING Mealybug infestation of the roots and basal ten leaves should be assessed (Petty and Webster, 1981a), just prior to environmental conditions favouring their development – plant phenology and seasonal climate being major factors. In South Africa, infestations increase during the drier autumn and winter months.

By assessing symbiotic ant infestations, mealybugs may be indirectly monitored. A peanut butter–soybean oil mixture is highly attractive to big-headed ants, which may be counted on wooden laths with the mixture, placed at suitable intervals around plantation blocks. Even a small number of these ants is sufficient to warrant further investigation and possibly control measures.

ECONOMIC THRESHOLD Occasionally, single mealybugs result in PMW (Carter, 1937). Because of the close big-headed ant–mealybug relationship (Petty and Tustin, 1993), tolerance for both species appears minimal. The relationship between wilt in virus infected plants and mealybug induced stress is presently unknown, and for virus-free plants the effect of mealybug infestation densities. Such information is essential to the formulation of economic thresholds.

CONTROL Once established, wilt is difficult to control because of vegetative propagation and latent symptoms in planting material (Rohrbach and Apt, 1986). Management of the PMW problem requires the integration of monitoring of the species involved and the application of chemical measures designed to maximize the impact of natural and biological controlling factors for mealybugs. Cultural control measures should also be applied if possible.

Mealybug and ant management. From the late 1920s, mealybugs were controlled with the following succession of pesticides: oil emulsions, parathion, malathion, diazinon and dimethoate. The latter are still used as foliar sprays or planting material dips (Krause *et al.*, 1996). Dips materially suppress infestations and their spread in new plantations. However, fumigation with methyl bromide gas is highly effective (Petty, 1984; Willers,

1990; Petty *et al.*, 1991), without phytotoxicity if applied correctly (Petty, 1987). The practice does unfortunately have practical problems and environmental hazards.

From the mid-1940s, ant control pesticides included DD, benzene hexachloride, heptachlor, chlordane, HHDN/dieldrin, Mirex bait toxin, hydramethylnon bait toxin/ Amdro® (American Cyanamid Company, Wayne, New Jersey). Amdro 1500 p.p.m. formulation at 2.24 kg ha^{-1} was shown in 1980 (Su *et al.*) to give big-headed ant control comparable to the then standard treatment, Mirex. Reimer and Beardsley (1990), Petty (1993) and Petty and Manicom (1995) subsequently confirmed its exceptional efficacy. It has full registration in South Africa, at 2 kg ha^{-1}, and is accepted (Samways, 1985) by myrmicine ants (*Pheidole, Myrmicaria* and *Monomorium* spp.) and dolichoderine ants (*Technomyrmex* spp.) but not by formicine species of *Anoplolepis, Acantholepis* and *Camponotus.*

Biological control. A number of potentially effective coccinellid beetles and encyrtid parasitoids have been identified. *Anagyrus ananatis* Gahan (Hymenoptera: Encyrtidae) effectively controls mealybugs in the absence of ants (Gonzalez-Hernandez, 1995). It originates in Brazil, is very host specific, and has a 20-day generation period. *Exochomus concavus* Fursch (Coleoptera: Coccinellidae), under laboratory conditions, consumed 61 and 94 (155 total) mealybugs per larva and adult, respectively (Petty, 1985).

Cultural control. Hot water immersed, PMWaV-infected, 'Smooth Cayenne' crowns lost their infection, and if maintained mealybug-free, PMW did not develop after planting (Ullman *et al.*, 1991). This could not be verified in Australia (Thomas *et al.*, 1999).

Selection of apparently healthy plants in Mexican commercial plantations has resulted in PMW-resistant selections which also have other commercially favourable characteristics (Torres Navarro *et al.*, 1989). The selection process could be speeded up and made more reliable with a technique developed by Hu *et al.* (1997), using specific monoclonal antibodies to the PMWaV in a tissue blotting immunoassay which detects the virus in symptomless plants.

Pineapple scale, Diaspis bromeliae *(Kerner)*

Pineapple plantations worldwide are subject to pineapple scale insect (PSI) infestation (Annecke and Moran, 1982; Bedford *et al.*, 1987; Waite, 1993). Carter (1967) stated that it is the only scale insect of economic importance on pineapples.

BIOLOGY

Hosts, location and damage. Alternative hosts include agave and bromeliads (Waite, 1993) and *Billbergia* spp. (Brimblecombe, 1955). PSI typically infests the lower pineapple leaves, which are more shaded, but where plants are growing in shaded positions PSI are inclined to occur higher on plants. Suckers in densely grown ratoon crops are most subject to damage. Fruits may be very heavily infested, especially ratoon fruits lodged between ridges (walking spaces). PSI infestation of mature fruits in spring is typically of the lower parts (Waite, 1993). Where a PSI has inserted its long, hair-like mouthparts into a leaf for feeding, a yellow spot develops due to its toxic saliva (Py *et al.*, 1987). This may be the full extent of their damage but heavily infested plants assume a greyish scaly appearance, are weakened and stunted, have foliar dieback and small, unsightly, unmarketable fruits. Fruit cracking may result from heavy PSI feeding (Carter, 1967).

Description and life cycle. The adult female PSI secretes and is covered by a hard, circular, waxy, beige coloured shell which at maturity is about 1.3 mm wide (Petty, 1990d). The soft-bodied, yellow insect is roughly triangular in outline; legs and antennae are not apparent and it has long hair-like stylets for mouthparts. A number of oval, translucent-yellow eggs are laid under the scale cover. These hatch in about 7 days. Newly hatched immature 'crawlers' are minute, yellowish and very active. On establishing a feeding site, crawlers shed their legs at the first moult and remain at that site thereafter. The three larval stages, and a pupal stage, are completed in about 60 days in summer (Carter, 1967). Immature male scale insects develop under an elongate, whitish scale cover less than 1 mm long, emerging at maturity as small, winged, orange-coloured insects which mate and die after a short life span.

Population fluctuations. There are up to four, overlapping, generations per year. Eggs are laid throughout the year, in Australia, with peaks in summer and early winter (Waite, 1993). PSI is spread into new plantations mainly on infested planting material when this is not insecticidally treated (Bedford *et al.*, 1987). Crawlers are spread by wind, and on the clothing of farm workers.

SAMPLING AND MONITORING Plants, especially shaded ratoons and where growth is dense, should be visually inspected on a regular basis, especially during dry periods (Petty, 1990d). Where a zinc deficiency in plants occurs, extra vigilance is required as this appears to favour multiplication (Py *et al.*, 1987). Cognizance is required, in monitoring of PSI, of the potential effect of natural enemies, especially hymenopterous parasitoids, on these insects. Less than 50% of female scales were found to be living in most months in Australia (Brimblecombe, 1955) and Murray (1984) ascribed a 40% mortality level to the combined actions of predators and parasitoids. Internal parasitoids, on completing immature development, chew small emergence holes in the scale cover – these are readily detected with the naked eye, or a hand lens.

CONTROL

Natural enemies. The need for chemical control is largely obviated by the existence of a number of effective internal hymenopterous parasitoids. In South Africa these include *Ablerus elegantulus* (Silvestri) and *Encarsia* spp. (Aphelinidae); *Tetrastichus* sp. (Eulophidae) (this species may now have been transferred to a new genus); *Coccidencyrtus ochraceipes* Gahan (Encyrtidae). *C. ochraceipes* was first collected at Hluhluwe, South Africa, by the senior author and identified by Dr G.L. Prinsloo, who noted that it was the first record of the species in the Afrotropical region (Prinsloo, 1985). The female is black with a partially yellow abdomen (lacking in the male), and 0.671 mm in length. They are highly numerous, at times causing heavy scale mortality. At a mean temperature of 24°C, mean development time – egg to adult – was 27 days (G.J. Petty, 1998, unpublished).

In Australia, two small parasitoid wasp species, *Aphytis* sp. and *Aspidiotiphagus* sp., are important, as is the ladybird *Rhyzobius iophanthae* (Coleoptera: Coccinellidae) (Waite, 1993).

Integrated pest management. Pesticides for pineapple mealybug control will also control PSI (Waite, 1993). However, the injudicious use of persistent insecticides may result in a PSI infestation explosion by destroying their natural enemies (Sakimura, 1966). Before reaching a decision to spray insecticides, examine the scale insects to determine if natural enemies are present in significant numbers.

The spread of PSI from heavily infested ratoon crops to new plantations may be controlled either by burning the plants at the end of the ratoon crop cycle, or by sanitizing the planting material ('Queen' suckers or 'Smooth Cayenne' tops and slips). As far as possible, the use of planting material from PSI infested blocks should be avoided. If it must be used it may be effectively sanitized by fumigation with methyl bromide as described by Petty (1984) and Willers (1990). Fumigation of planting material with 40 g m^{-3} methyl bromide gas for 2 h, with a temperature range of 20 to 25°C, gave almost 100% PSI control without plant phytotoxicity (Petty, 1987). Similar results were obtained in Australia (Murray *et al.*, 1979).

Pineapple caterpillar, Thecla basilides *(Geyer)*

The pineapple caterpillar's distribution is limited to the Americas from Mexico to Argentina (Py and Tisseau, 1965), and Trinidad (Harris, 1927).

BIOLOGY

Host plants. In Brazil, on wild pineapple species, e.g. *Ananas ananassoides* (Smith), and other indigenous *Bromeliaceae*; in Honduras, on maize and cacao (Rohrbach, 1983); in Trinidad, on *Heliconia* sp. (Harris, 1927).

Description of stages. The following is according to Harris (1927) and Sanches (1993). Egg: a whitish, reticulated 0.8 mm sphere, flattened basally; a small dorsal spot demarcates the micropyle. Larva: on hatching, the first (of four) instars measures about 1.5 mm, is pale yellow, darkening with age, with head and thoracic region somewhat darker. It is finely pubescent with four rows of long abdominal hairs and four rows of shorter hairs. The second, and subsequent, instars are woodlouse-like and at maturity are about 20 mm long and 6 mm wide. Five pairs of pink longitudinal stripes gives the yellowish body a pink appearance. The head is concealed by the prothorax, the body compressed dorsoventrally, and the terminal abdominal segments are flattened, wedge-like. Pupa: about 13 mm long; pink coloured with dark spots and a characteristic dorsal hump. The imago adult female: a butterfly with a 28–35 mm wingspan, dorsally slate-grey with a darker border and white fringe; posterior wings have two orange spots near a pair of filiform appendages. Ventrally, wings are silvery grey with a number of orange spots. The imago adult male: a somewhat smaller butterfly than the female with a large black spot centrally on the anterior wings.

Habits and life cycle. Female butterflies deposit their eggs singly and widely separated mainly on upper and middle fruitlet bract bases, from prior to anthesis until anthesis is well advanced (Harris, 1927), but also on peduncles and slip bases. The potential egg laying capability is 150 per female. Mature fruits are avoided, as the epidermis is too tough for larvae to penetrate. Already infested fruits are also avoided (Rhainds *et al.*, 1996). Larvae hatch within 5 days and burrow into the bracts, or slips, or directly enter buds and open flowers. After a short feeding period they exit the fruits and again burrow in to form a feeding chamber, up to one-third fruit diameter in depth. Initial symptoms predominate on the fruit's lower half. Within 16 days larvae attach themselves with a fine silk thread to the plant's leaf axils and pupate. Within 11 days the butterfly emerges. Development time, from egg laying to adult, varies from 23 to 32 days, with an adult life span of 22 days, and total life span of about 51 days.

Damage. The larval galleries in fruits fill with a clear gum-like fluid which exudes out of fruits, hardening and darkening to amber and dark brown. These fruits develop an unpleasant smell and taste and are unsaleable. Fusariosis disease may be greatly increased due to creation of ports of entry for *Fusarium subglutinans* fungus infection. Infection is also increased through dissemination of fungus spores by visiting butterflies. Decomposition of the fruit occurs on the plant, with dehydration, shrivelling and blackening. Surveys conducted in Brazilian states have estimated damage at between 30 and 80%. It follows that this species is one of the main pests of pineapple in Brazil, causing great losses to growers (Sanches, 1981). Winter forcings (November–March) in Brazil are most at risk; in Mexico the high risk period commences in April (Py *et al.*, 1987).

MONITORING Slips and developing fruits should be examined for the presence of eggs at the time inflorescences appear. Eggs are easily seen, provided carbide has not been applied for forcing the plants. In areas of low incidence, effective control is easily achieved (Sanches, 1993). Particular vigilance is required at those times of the year when the pest is most active.

ECONOMIC THRESHOLD Plantation densities of 60,000–72,000 plants ha^{-1} suffer more damage than greater or lesser densities (Choairy and Fernandes, 1983). However, the presence of one larva in a fruit is sufficient to render the fruit useless (Sanches, 1981).

CONTROL

Natural enemies. Three hymenopteran natural enemies are known: *Heptasmicra* spp., a chalcid parasite, and *Polistes rubiginosus*, a vespid predator, both in Trinidad (Harris, 1927). The latter was said to be potentially effective. *Metadontia curvidentata* is a hymenopteran predator (Py *et al.*, 1987).

Chemical control. Chemical control is effective if applied from time of flower induction to protect the young developing fruit to beyond anthesis (Py *et al.*, 1987). Organochlorines are advantageous in being less toxic than the organophosphorus pesticides, and give longer residual control. However, their greater persistency is a potential environmental hazard, which encourages greater use of the organohalides and carbamates (Martinez, 1976). Pesticides which have been used to good effect include endrin, heptachlor, chlordane, toxaphene, carbaryl (Py *et al.*, 1987), trichlorfon, malathion, diazinon and parathion (Sanches, 1993). Insecticide sprays applied at 12-day intervals from bud differentiation to the early fruiting stage reduced fruit damage from 35.8 to 0.8% (Suplicy *et al.*, 1966) No reference was found to the use of IGRs (insect growth regulators), which are effectively applied for the control of Lepidoptera in other crops.

Cultural control.

1. The pineapple caterpillar's spread to other geographic regions should be controlled by preventing free movement of its host plants from infested areas.
2. Less/least susceptible pineapple cultivars, e.g. 'Perola' cv. should be grown where possible.
3. Through artificial flower induction, breeding is restricted to the limited time of fruit sensitivity.
4. Flowering may be induced when the pest's numbers are normally minimal.

Pineapple white grubs, Coleoptera: Scarabaeidae

Pineapple white grubs are the larvae of a number of beetle species in the Scarabaeidae. Of a total of 23 species of greater or lesser economic importance in pineapples in various parts of the world, 14 species belong to the Melolonthinae, seven to the Rutelinae and two to the Dynastinae. Most of the problem species occur in Australia and South Africa, and are melolonthids and rutelids (Table 6.3).

BIOLOGY

Host plants. White grubs feed on the roots and subterranean parts of pineapple plants (Smith *et al.*, 1995). Pasture grasses and sugarcane are also hosts and if they precede a pineapple plantation, white grub infestation of the latter crop may result (Waite, 1993).

Description of stages. The beetle subfamilies infesting pineapples are very similar in many respects. The eggs are white, oval and 1–2 mm long producing larvae which, to the untrained eye, are indistinguishable one species from the other. The white or cream coloured larval body is cylindrical and assumes a C-shaped posture, and is soft and wrinkly. It has a reddish brown head capsule with short antennae and prominent, sharply pointed mandibles. Mandibular characteristics, such as the number of notches and a stridulatory area, are diagnostic of the species. Other mouthparts – labrum, maxilla, labium and epipharynx, and the head capsule – provide useful and reliable diagnostic structures, e.g. sutures and setae, for laboratory identification of species. Field identification of larval species is possible by examining, with a hand lens, the setae on the ventral surface of the terminal abdominal segment. The arrangement, number, and structure of these setae (the raster) is diagnostic of different species. The raster area is grey-brown in colour due to the larva's translucent body wall and the soil content of its hindgut.

An individual larva passes through a number of growth stages (instars) and, whereas the body posterior to the head continues to grow in size, in some species up to 35 mm in length, the head capsule – which is a more rigid structure – maintains its size throughout each larval instar and is characteristic and diagnostic of each instar (Plate 39).

Mature larvae pupate in the soil, the pupae gradually darkening to a light brown colour. The pupa of each species has characteristics which, with experience, makes identification possible. Adult beetles developing from the pupae are all of a medium to large size and various shades of brown; some bear white scales or fine hairs. Adult beetles are generally not recognized by pineapple growers as the source of their white grub problem.

Habits and life cycle. Pineapple white grub species have a 1- or 2-year life cycle. A few, e.g. *Lepidiota grata* and *Rhopaea magnicornis*, have either, as unfavourable environmental conditions result in the longer period of development (Allsopp *et al.*, 1997). Beetles lay their eggs in the soil in plant root zones in spring–summer. For the rutelid, *Adoretus ictericus*, the average incubation period at 17–22°C was 19 days (Prins, 1965). The melolonthid, *Macrophylla ciliata*, lays up to 70 eggs and these hatch after about 42 days. Eggs hatch at different times of the year, depending on the species, giving larvae which normally have three instars (commonly designated L.1, L.2 and L.3). For the Australian species (Allsopp *et al.*, 1997) larvae (L.2 or L.3) descend in May/June to greater depths, i.e. 350–600 mm, in soil and hibernate in earthen cells until August/September when they pupate (1-year cycle) or change from L.2 to L.3 (2-year cycle). Pupae mature to adult beetles, emerge from the soil after the first soaking rains between September and November (depending on species) and fly to feeding trees (some do not feed) prior to mating, and egg-laying in the soil. Larvae becoming L.3 in August/September come up again in the soil to feed on roots until April/May of the following year when they again descend to hibernate until September. Development thereafter is as for a 1-year life cycle.

In South Africa, *Asthenopholis subfasciata* is an important melolonthid species. Adult beetles occur from November to January; the males fly actively in December but the females do not appear to fly, and the adults do not appear to feed. Larval peaks in soil occur in April/May (L.1) May/June (L.2) and October/November (L.3). Pupae occur in October/December. Most individuals have a 1-year life cycle; a minority have a 2-year cycle (G.J. Petty, 1990, unpublished).

SAMPLING AND MONITORING Sampling size must compromise between time available and acceptable destruction of plants, on the one hand, and a large enough sample to counteract the uneven distribution of grubs in plantations to give an acceptable variance between samples. Sinclair and Scott (1997) assessed pineapple root volumes for five-plant sample sizes, concluding that this is a practical sample size but insufficient to give a good population estimate. The senior author has found a ten-plant sample per 1500 m^2 plantation block to be adequate; soil in a 300 mm square around each plant to a 150 mm depth is excavated and sifted through to locate and remove grubs. Pre-plant assessments for grub infestation of soil may utilize a large diameter soil auger mounted behind a tractor and operated via the tractor's power-take-off (PTO).

ECONOMIC THRESHOLDS The feeding activities of white grubs – predominantly third instar larvae – result in the progressive destruction of the root system, followed by plants becoming stunted, wilted and chlorotic, a condition aggravated by drought (Petty, 1990 g) (Plate 40). For Australian species with a 1-year cycle, damage occurs from February to May, and from October (year 1) to April (year 2) for 2-year cycle grubs (Allsopp *et al.*, 1997). Heavy infestations result in the subterranean plant stump having a shaved appearance, with feeding cavities. Plants lose their anchorage in the soil and are easily pulled out.

In a lysimeter trial, *A. subfasciata*, L.1 stage larvae were seeded at various infestation levels with growing 'Smooth Cayenne' cv. plants. At the end of a 12-month period, plant growth had been reduced by 5.5% with an infestation density of 0.5 grubs per plant (Petty, 1996).

The probability of infestation by white grubs is greatly determined by the nature and physical structure of soils. Species specific preferences are clearly apparent, e.g. *A. subfasciata* occurs exclusively on red clay–loam soils (Petty, 1996); *Antitrogus consanguineus* on sandy alluvial or yellow podsol soils (Allsopp *et al.*, 1997); and *Lepidiota* spp. only on red volcanic soil (Sinclair and Scott, 1997).

CONTROL

Natural control. In South Africa these include a number of bird species, e.g. *Hagedashia hagedash* and *Numida meleagris*, various mammals, e.g. pigs, rodents, mongooses, and the toad, *Bufo regularis*, on beetles (Petty, 1990h). They are also preyed on by species of wasp, robber fly and wireworm; infected by pathogenic fungi, e.g. the green soil fungus, *Metarhizium anisopliae* and *Paraisaria* sp., and the milky disease bacterium, *Bacillus popilliae* (Allsopp *et al.*, 1997).

Abiotic control. If, after beetle emergence and egg-laying, hot dry weather persists for a lengthy period, mortality of eggs and young hatchlings will result (Allsopp *et al.*, 1997).

Cultural control. Deep pre-plant ploughing in March/April will kill 2-year life cycle L.2 larvae; between January and May, larvae of 1-year cycle grubs will be killed (Allsopp *et al.*, 1997).

Chemical control. Pre-plant chemical soil treatments can be applied only once in more than 4 years. Most presently available pesticides are totally broken down to inactive components in the soil in a far shorter period than 4 years. Fortunately, the most critical period for subterranean plant protection is during the early development of the plant and that can be accomplished with available pesticides. In Australia, the 50% active formulation of chlorpyrifos is applied, pre-plant, at the rate of 5 l ha^{-1}. Annual post-plant booster treatments of 3 l ha^{-1} are applied as high volume foliar sprays (Sinclair and Scott, 1997). Slow/controlled release granular formulations of, *inter alia*, chlorpyrifos gave very promising experimental results for extending this chemical's residual activity in soil, but practical application problems have obstructed its marketing (Sinclair and Scott, 1997).

A promising new strategy to obtain extended control of white grub infestations is the use of carefully timed insecticide sprays to coincide with beetle emergence from the soil and egg laying. By this means, *A. subfasciata* larval infestation was reduced by 78.5% when isazophos was applied in mid-October and again in mid-November (Petty, 1996).

Conclusions

Pineapple is a perennial crop which requires protection from a complex of pests for a period of up to 6 years in cooler subtropical climates. This creates difficulties which do not arise in shorter term crops, in the management of soil dwelling pests, e.g. nematodes, symphylids and white grubs. It is of crucial importance that pineapple root systems be effectively protected during at least the initial 15 months of their development; damage during this time commonly has repercussions mainly in the ratoon crop, on which growers may rely for their profit.

Worldwide, the pineapple industry depends primarily on the use of chemical nematicides for nematode control. These can be very effective but are also costly and hold potential dangers to both user and environment. In order to obtain maximum benefit and minimize the dangers, the need to apply pre-plant and/or post-plant foliar nematicide treatments should be determined by nematode counts for soil, sampled at key times in the natural development cycle of pineapple nematodes.

The efficacy of volatile pre-plant nematicides, e.g. ethylene dibromide, if required, should be ensured by only applying these at soil temperatures and moisture levels which exceed the thresholds for satisfactory soil fumigation; winter temperatures in some southern hemisphere producer countries are frequently sub-optimum. A too rapid loss of fumigant from the soil may be prevented with a plastic mulch. A suitably fine tilth, without large soil clods, promotes fumigation effectiveness.

It is important that forces which are naturally suppressive to nematodes should be encouraged through the development of appropriate cropping systems. In this regard, the organic matter component of soil should be conserved and increased by, for example, green manuring and reincorporation of plants on completion of the ratoon crop cycle. Other strategies in nematode management, which will limit the use of chemical nematicides, are: maintenance of a bare fallow for a number of months; the use of certain nematode-resistant grass and legume species in a system of crop rotation; the use of modern genetic engineering technologies to introduce foreign genes which will impart a degree of resistance or tolerance to the crop.

A knowledge of the previous history of plantation blocks, as regards incidence of soil infestation by pest species, is of great value in developing a strategic management programme. Pest species often have clear preferences for specific soil types and textures, for example, root-knot nematodes on sandy soils and well-structured clay loams; symphylids on, *inter alia*, the latter and on gravelly sandy and clay loams; some white grub species on red clay loams. Pre-plant soil insecticide application for symphylid and white grub control should be made on the basis of previous infestation history and soil type. Preventive control of these pests is required as attempts at post-plant corrective control are generally ineffective. However, certain melolonthid grub species may be effectively controlled by carefully timed adulticide sprays.

The control of fruit and foliar insect and mite pests also tends to be heavily dependent on chemical insecticides/acaricides. To some extent chemical usage can be reduced by the careful selection of planting material from pineries which are essentially infestation free. It is important to start a plantation with propagules (tops or slips) which are minimally infested. If, due to a shortage, infested propagules must be used, they should be disinfested by dipping in a suitable insecticide, or by fumigating the material with an approved fumigant.

Subsequent to the planting of a crop, pest management should be based on sampling and monitoring of plantations for the presence of pests, their natural enemies and symbiotically associated arthropods. The methods, techniques and use of these strategies are discussed in the text above for the different key arthropod species.

When foliar pesticide sprays are deemed necessary, the following should be borne in mind:

1. Spray volume should be adjusted according to the location of the pest on the plant. For protection of fruits and slips, lower volumes are required than for pests on the basal leaves of the plant.

2. Water pH could adversely affect pesticide efficacy, due to alkaline hydrolysis of the active ingredient.

3. Some pesticide mixtures are incompatible – either physically or chemically. Mixtures may also result in phytotoxicity, for example if bromacil weedkiller and dimethoate are applied together, severe plant damage will result.

4. Pineapple pests usually exist in a complex with other pest species. In the selection of a pesticide programme, cognizance should be taken of this – a subject dealt with by Petty and van der Westhuizen (1992). The loss of effective natural control of pests, for example by hymenopterous parasitoids of scale insects, due to injudicious insecticide spray application should be avoided at all costs.

5. To obtain the greatest possible benefit from insecticide sprays, the timing of these should take pest life cycles and behaviour into account.

6. The use of a specific bait toxin to disrupt the symbiotic relationship between ant and mealybug species is more ecologically sound, and effective, than the use of broad spectrum insecticides.

In an integrated pest management programme there should be no weak links which could, for example, promote root growth through disease control only to have the roots destroyed by the nematode component of the total complex.

Acknowledgements

Sincere appreciation is expressed to the following taxonomists of the Plant Protection Research Institute of the Agricultural Research Council of South Africa for their efforts to ensure the correctness of the arthropod nomenclature and classification given in Table 6.3: I. Miller, G.L. Prinsloo, R. Stals, M. Stiller, E.A. Ueckermann and V.M. Uys. Thanks are also due to those who supplied information for the same table, as indicated in the table by 'personal communication'.

References

Allsopp, P.G. (1993) Identity of canegrubs attributed to *Antitrogus mussoni* (Blackburn) (Coleoptera: Scarabaeidae: Melolonthinae). *The Coleopterist's Bulletin* 47(2), 195–201.

Allsopp, P.G., Chandler, K.J., Sampson, P.R., Story, P.G., Brodie, A. and Whittle, P. (1997) Root feeders. In: Agnew, J.R. (ed.) *Australian Sugarcane Pests*. Bureau of Sugar Experiment Stations, Brisbane, pp. 1–21.

Annecke, D.P. and Moran, V.C. (1982) Subtropical fruits, pineapples. In: *Insects and Mites of Cultivated Plants in South Africa*. Butterworths, Durban/Pretoria, p. 91.

Anonymous (1995) Survey on pineapple pests and diseases in the Caribbean. *Tropical Fruits Newsletter* 14, 3–4.

Apt, W.J. (1976) Survival of reniform nematodes in desiccated soils. *Journal of Nematology* 8, 278.

Apt, W.J. and Caswell, E.P. (1988) Application of nematicides via drip irrigation. *Journal of Nematology* (Suppl.)2, 1–13.

Barbeau, G. and Marie F. (1997) La production de l'ananas dans la Caraibe Hispanophone et Anglophone (Pineapple production in the Spanish and English speaking Caribbean). *Acta Horticulturae* 425, 53–62.

Bartholomew, D.P. (1977) Inflorescence development of pineapple (*Ananas comosus* (L.) Merr.) induced to flower with ethephon. *Botanical Gazette* 138(3), 312–320.

Beardsley, J.W. (1959) On the taxonomy of pineapple mealybugs in Hawaii, with a description of a previously unnamed species (Homoptera : Pseudococcidae). *Proceedings, Hawaiian Entomological Society* 17(1), 29–37.

Beardsley, J.W. (1964) Notes on the pineapple mealybug complex, with description of two new species (Homoptera : Pseudococcidae). *Proceedings, Hawaiian Entomological Society* 19(1), 55–68.

Beardsley, J.W., Su, T.H., McEwen, F.L. and Gerling, D. (1982) Field investigations on the interrelationships of the big-headed ant, the gray pineapple mealybug, and the pineapple mealybug wilt disease in Hawaii. *Proceedings of the Hawaiian Entomological Society* 24, 51–67.

Bedford, E.C.G., Bruwer, I.J., de Villiers, E.A. and Petty, G.J. (1987) Hard (armoured) scales. In: Myburgh, A.C. (ed.) *Crop Pests in Southern Africa*, Vol. 2, *Citrus and Other Subtropicals*. Bulletin 411, Department of Agriculture and Water Supply, Republic of South Africa, pp. 27–32.

Beer, R.E. (1954) A revision of the Tarsonemidae of the western hemisphere (Order: Acarina). *The University of Kansas Science Bulletin* 36(16), 1276–1369.

Bernard, E.C. and Montgomery-Dee, M.E. (1993) Reproduction of plant-parasitic nematodes on winter rapeseed (*Brassica napus* ssp. *oleifera*). *Journal of Nematology* 25, 863–868.

Borchsenius, N.S. (1966) *A Catalogue of the Armoured Scale Insects (Diaspidoidea) of the World*. 'Nauka', Moskow, Leningrad, 449 pp. (in Russian).

Borroto, E.G., Cintra, M., González, J. and Borroto, C. (1998) First report of a Closterovirus-like particle associated with pineapple plants (*Ananas comosus* cv. Smooth Cayenne) affected with pineapple mealybug wilt in Cuba. *Plant Disease* 82, 263.

Brimblecombe, A.R. (1955) Pineapple scale investigations. *Queensland Journal of Agricultural Science* 12(3), 81–94.

Carter, W. (1932) Studies of populations of *Pseudococcus brevipes* (Ckl.) occurring on pineapple plants. *Ecology* 13(3), 296–304.

Carter, W. (1933) The pineapple mealybug, *Pseudococcus brevipes*, and wilt of pineapples. *Phytopathology* 23(3), 207–241.

Carter, W. (1935) Studies on the biological control of *Pseudococcus brevipes* (Ckll.) in Jamaica and Central America. *Journal of Economic Entomology* 28, 1037–1041.

Carter, W. (1937) The toxic dose of mealybug wilt of pineapple. *Phytopathology* 27, 971–981.

Carter, W. (1939) Geographical distribution of yellow spot of pineapples. *Phytopathology* 29(3), 285–286.

Carter, W. (1945) Some etiological aspects of mealybug wilt. *Phytopathology* 35(5), 305–315.

Carter, W. (1949) Insect notes from South America with special reference to *Pseudococcus brevipes* and mealybug wilt. *Journal of Economic Entomology* 42, 761–766.

Carter, W. (1954) The insect and the plant diseases. *Journal of Economic Entomology* 47, 210- 215.

Carter, W. (1963) Mealybug wilt of pineapple: a reappraisal. *Annals of the New York Academy of Sciences* 105, 741–764.

Carter, W. (1967) The scale insects. In: *Insects and Related Pests of Pineapple in Hawaii. Manual of the Pineapple Research Institute of Hawaii*, pp. 53–56.

Caswell, E.P. and Apt, W.J. (1989) Pineapple nematode research in Hawaii: past, present and future. *Journal of Nematology* 21, 147–157.

Caswell, E.P., Sarah, J.-L. and Apt, W.J. (1990) Nematode parasites of pineapple. In: Luc, M.,

Sikora, R. A. and Bridge, J. (eds) *Plant Parasitic Nematodes in Subtropical and Tropical Agriculture.* CAB International, Wallingford, UK, pp. 519–537.

Caswell, E.P., DeFrank, J., Apt, W.J. and Tang, C.S. (1991) Influence of non-host plants on population decline of *Rotylenchulus reniformis. Journal of Nematology* 23, 91–98.

Charles, V.K. (1941) A preliminary check list of the entomogenous fungi of North America. *Technical Bulletin, US Department of Agriculture* 21, 707–785.

Choairy, S.A. and Fernandes, P.D. (1983) Plant density in pineapple culture. *Pesquisa Agropecuária Brasileira* 18(9), 985–988.

Collins, J. (1960) *The Pineapple: Botany, Cultivation and Utilization.* Leonard Hill, London, 249 pp.

Dodson, P.G.C. (1969) *Annual Report 1968–1969.* Malkerns Experiment Station, Malkerns, Swaziland.

Duodu, Y.A. and Thompson, W. (1992) Management of ant-mealybug complex in pineapple fields in Guyana. *FAO Plant Protection Bulletin* 40(3), 82–88.

FAO (1998) Food and Agricultural Organization, United Nations. http://apps.fao.org/lim500/nph-wrap.pl?Production.Crops.Primary&Domain=SUA

Fleisch, H. and Bartholomew, D.P. (1987) Development of a heat unit model of pineapple ('Smooth Cayenne') fruit growth from field data. *Fruits* 42, 709–715.

Giacomelli, E.J. and Py, C. (1981) L'ananas au Brésil. *Fruits* 36(11), 645–687.

Godfrey, G.H. (1936) The pineapple root system as affected by root-knot nematode. *Phytopathology* 26, 408–428.

Godfrey, G.H. and Oliveira, J.M. (1932) The development of the root-knot nematode in relation to root tissues of pineapple and cowpea. *Phytopathology* 22, 325–348.

Gonzalez-Hernandez, H. (1995) The status of the biological control of pineapple mealybugs in Hawaii. PhD thesis, The University of Hawaii at Manoa, Honolulu, Hawaii, 128 pp.

Graham, M.W., Hardy, V., Bernard, M., Ko, L., Smith, M., Robinson, S., Buchelli, C., Sawyer, B., O'Hare, T., Jobin, M., Dahler, J. and Underhill, S. (1998) Genetic engineering of pineapple to control blackheart. *Proceedings of the Fourth Asia-Pacific Conference on Agricultural Biotechnology.* Darwin, Australia, pp. 356–358.

Guérout, R. (1974) Attaques d'*Augosoma centaurus* en plantation d'ananas. *Fruits* 29(9), 609–611.

Guérout, R. (1975) Nematodes of pineapple: a review. *Pest Articles and News Summaries* 21, 123–140.

Gunasinghe, U.B. and German, T.L. (1989) Purification and partial characterisation of a virus from pineapple. *Phytopathology* 79(12), 1337–1341.

Harris, W.V. (1927) On a lycaenid butterfly attacking pineapples in Trinidad, BWI. *Bulletin of Entomological Research* 18(2), 183–188.

Hu, J.S., Sether, D.M. and Ullman, D.E. (1996) Detection of pineapple closterovirus in pineapple plants and mealybugs using monoclonal antibodies. *Plant Pathology* 45, 829–836.

Hu, J.S., Sether, D.M., Liu, X.P., Wang, M., Zee, F. and Ullman, D.E. (1997) Use of a tissue blotting immunoassay to examine the distribution of pineapple closterovirus in Hawaii. *Plant Disease* 81(10), 1150–1154.

Ito, K. (1938) Studies on the life history of the pineapple mealybug. *Journal of Economic Entomology* 31, 291–298.

Johnson, A.W., Golden, A.M., Auld, D.L. and Sumner, D.R. (1992) Effects of rapeseed and vetch as green manure crops and fallow on nematodes and soil-borne pathogens. *Journal of Nematology* 24, 117–126.

Kaloshian, I., Yaghoobi, J., Liharska, L., Hontelez, J., Hanson, D. *et al.* (1998) Genetic and physical localization of the root-knot nematode resistance locus *Mi* in tomato. *Molecular and General Genetics* 257, 376–385.

Keetch, D.P. (1982) Nematode pests of pineapple. In: Keetch, D.P. and Heyns, J. (eds) *Nematology in South Africa.* Department of Agriculture and Fisheries, Pretoria, Republic of South Africa, pp. 19–29.

Kèhè, M., Gnonhouri, P. and Adikoko, A. (1997) Evolution des infestations du symphyle *Hanseniella ivorensis* et du nematode *Pratylenchus brachyurus* su ananas en Cote D'Ivoire. *Acta Horticulturae* 425, 465–474.

Ko, M.P. and Schmitt, D. (1996) Changes in plant-parasitic nematode populations in pineapple fields following intercycle crops. *Journal of Nematology* 28, 546–556.

Krause, M., Nel, A. and van Zyl, K. (1996) *A Guide to the Use of Pesticides and Fungicides in the Republic of South Africa,* 37th edn. National Department of Agriculture, Republic of South Africa, Pretoria, 320 pp.

Kuschel, G. (1983) New synonomies and combinations of Baridinae from the neotropic and neoarctic regions (Coleoptera : Curculionidae). *The Coleopterists' Bulletin* 37(1), 34–44.

Lewcock, H.K. (1937) Yellow spot disease of pineapples. *Queensland Agricultural Journal* 48, 665–672.

Lewis, T. (1973) *Thrips, Their Biology, Ecology and Economic Importance.* Academic Press, New York and London, 349 pp.

Linford, M.B. (1937) Stimulated activity of natural enemies of nematodes. *Science* 85, 123–124.

Linford, M.B. (1943) Influence of plant populations upon incidence of pineapple yellow spot. *Phytopathology* 33, 408–410.

Linford, M.B. (1952) Pineapple diseases and pests in Mexico. *FAO Plant Protection Bulletin* 1, 21–25.

Linford, M.B. and Oliveira, J.M. (1940) *Rotylenchulus reniformis*, nov. gen., n. sp., a nematode parasite of roots. *Proceedings of the Helminthological Society of Washington* 7, 35–42.

Linford, M.B., Yap, F. and Oliveira, J.M. (1938) Reduction of soil populations of root-knot nematode during decomposition of organic matter. *Soil Science* 45, 127–141.

Loeillet, D. (1997) Panorama du Marche mondial de l'ananas: l'importance de l'Europe (The world pineapple market: the importance of Europe). *Acta Horticulturae* 425, 37–48.

Macgill, E.I. (1927) The biology of Thysanoptera with reference to the cotton plant, 2. The relation between variations in temperature and the life-cycle. *Annals of Applied Biology* 24, 95–109.

Macgill, E.I. (1939) A gamasid mite *(Typhlodromus thripsi* n. sp.), a predator of *Thrips tabaci* Lind. *Annals of Applied Biology* 26, 309–317.

Marshall, A.K.G. (1961) A new pest attacking pineapple in Jamaica. *Bulletin of Entomological Research* 7, 197–198.

Martinez, N.B. (1976) Preliminary study on control of insects causing 'pineapple gummosis'. *Agronomia Tropical* 26(1), 3–7.

Masses, H. (1979) Lutte contre les symphyles en culture d'ananas dans l'ile de le Martinique. *Congrès sur la lutte contre les insectes en milieu tropical*, 423–438.

McSorley, R. and Frederick, J.J. (1995) Responses of some common Cruciferae to root-knot nematodes. *Journal of Nematology* 27, 550–554.

Menezes Mariconi, F.A. (1953) O 'percevejo do abacaxi'. *Biológico* 19, 155–162.

Meyer, M.K.P. (Smith) (1981) *Mite Pests of Crops in Southern Africa*. South Africa, Department of Agriculture and Fisheries, *Science Bulletin* 397, 92 pp.

Meyer, M.K.P. (Smith), Fourie, P.F. and Petty, G.J. (1987) Plant-eating mites. In: Myburgh, A.C. (ed.) *Crop Pests in Southern Africa*, Vol. 2, *Citrus and other Subtropicals*. South Africa, Department of Agriculture and Water Supply, *Bulletin 411*, pp. 58–70.

Mohamed, M.I., Zaher, M.A. and Hassen, M.F. (1982) Observations on *Cheytelus cacahuamilpensis*, a predator of the tenuipalpid mite *Dolichotetranychus floridanus*. *Entomophaga* 27(3), 343–347.

Mojtahedi, H., Santo, G.S., Hang, A.N. and Wilson, J.H. (1991) Suppression of root-knot nematode populations with selected rapeseed cultivars as green manure. *Journal of Nematology* 23, 170–174.

Mojtahedi, H., Santo, G.S., Wilson, J.H. and Hang, A.N. (1993) Managing *Meloidogyne chitwoodi* on potato with rapeseed as a green manure. *Plant Disease* 77, 42–46.

Morison, G.D. (1957) A review of British glasshouse Thysanoptera. *Transactions of the Royal Entomological Society of London* 109, 467–534.

Mourichon, X., Perrier, X. and Thibaud, B. (1987) Fruitlet core rot and leathery pocket of pineapple in Cote D'Ivoire: research into the main explanatory factors. *Fruits d'Outre Mer* 42(6), 343–352.

Munro, J.A. (1954) Entomology problems in Bolivia. *FAO Plant Protection Bulletin* 2, 97–105.

Murray, D.A.H. (1984) Pineapple scale (*Diaspis bromeliae* (Kerner)) distribution and seasonal history. *Queensland Journal of Agricultural and Animal Sciences* 39(2), 125–130.

Murray, D.A.H. and Smith, D. (1983) Effect of symphyla, *Hanseniella* sp., on establishment of pineapples in south-east Queensland. *Queensland Journal of Agricultural and Animal Sciences* 40(2), 121–123.

Murray, D.A.H., Swaine, G. and Corcoran, R.J. (1979) Methyl bromide fumigation of pineapple scale, *Diaspis bromeliae* (Kerner). *Queensland Journal of Agricultural and Animal Sciences* 36(1), 87–90.

Olowe, T. and Corbett, D.C.M. (1976) Aspects of the biology of *Pratylenchus brachyurus* and *P. zeae*. *Nematologica* 22, 202–211.

Pegg, K.G. (1993) Diseases. In: Broadley, R.H., Wassman III, R.C. and Sinclair, E. (eds) *Pineapple Pests and Disorders*. Queensland Department of Primary Industries, Brisbane, pp. 11–20.

Perez, M.E. (1957) Pineapple gummosis in Puerto Rico, and its control. *University of Puerto Rico Agricultural Experiment Station* 21, 71.

Perez, M.E. (1959) Further experiments on the control of pineapple gummosis in Puerto Rico. *Journal of Agriculture of University of Puerto Rico* 43, 116–127.

Petty, G.J. (1975) Pineapple mites. *Citrus and Subtropical Fruit Journal* 498, 15–18.

Petty, G.J. (1976a) The pineapple mealybug. *Farming in South Africa*, H.15, Department of Agricultural Development, Pretoria, 8 pp.

Petty, G.J. (1976b) The bionomics and control of pineapple white grubs (Coleoptera: Scarabaeoidea) in the Eastern Cape. *Crop Production* 5, 55–59.

Petty, G.J. (1977a) White grub rears head in Eastern Cape. *CSFRI Information Bulletin* 53, 1–2.

Petty, G.J. (1977b) Beetle pests of pineapples: some biological aspects. *Citrus and Subtropical Fruit Journal* 529, 4–7.

Petty, G.J. (1978a) Pineapple scale. *Farming in South Africa*, H.14, Department of Agricultural Development, Pretoria, 3 pp.

Petty, G.J. (1978b) False spider mite of pineapples. *Citrus and Subtropical Fruit Research Institute, Information Bulletin* 75, 8.

Petty, G.J. (1978c) Pineapple pests: thrips. *Farming in South Africa*, H.17, Department of Agricultural Development, Pretoria, 8 pp.

Petty, G.J. (1984) Methyl bromide sterilization of pineapple plants. *Citrus and Subtropical Fruit Research Institute, Information Bulletin* 147, 1–4.

Petty, G.J. (1985) The pineapple mealybug. *Farming in South Africa*, H.15, Department of Agricultural Development, Pretoria, 8 pp.

Petty, G.J. (1987) Control of mealybugs and scale on pineapples by methyl bromide fumigation of planting material. *Phytophylactica* 19(3), 255–258.

Petty, G.J. (1988) Methods for monitoring pineapple insects and mites. *ITSC Information Bulletin* 188, 2–5.

Petty, G.J. (1990a) Minor pests of pineapples. *Farming in South Africa*, H.18, Department of Agricultural Development, Pretoria, 2 pp.

Petty, G.J. (1990b) Leathery pocket in pineapples. *Farming in South Africa*, H.2, Department of Agricultural Development, Pretoria, 2 pp.

Petty, G.J. (1990c) Effect of endosulfan on leathery pocket disease in pineapples, and evidence for the association of the mite, *Steneotarsonemus ananas* (Acarina: Tarsonemidae) with leathery pocket. *Acta Horticulturae* 275, 663–670.

Petty, G.J. (1990d) Pineapple scale. *Farming in South Africa*, H.14, Department of Agricultural Development, Pretoria, 2 pp.

Petty, G.J. (1990e) Pineapple mites. *Farming in South Africa*, H.16, Department of Agricultural Development, Pretoria, 4 pp.

Petty, G.J. (1990f) Thrips in pineapples. *Farming in South Africa*, H.17, Department of Agricultural Development, Pretoria, 4 pp.

Petty, G.J. (1990g) White grubs in pineapples. *Farming in South Africa, Pineapples* H.12, Department of Agricultural Development, Pretoria, 2 pp.

Petty, G.J. (1990h) Black maize beetle in pineapples. *Farming in South Africa, Pineapples* H.13, Department of Agricultural Development, Pretoria 2 pp.

Petty, G.J. (1991) Pineapple bud moth: a new pineapple pest at Hluhluwe. *Institute for Tropical and Subtropical Crops, Information Bulletin* 221, 6.

Petty, G.J. (1992) Leathery pocket mite: distribution on pineapple plants. *Institute for Tropical and Subtropical Crops, Information Bulletin* 244, 6.

Petty, G.J. (1993) Brown house ant control with Amdro. *Institute for Tropical and Subtropical Crops, Information Bulletin* 251, 6.

Petty, G.J. (1996) White grubs in pineapples: a new strategy for management of the problem (Coleoptera : Scarabaeidae). *Pineapple News : ISHS Pineapple Working Group Newsletter* 2(1), 15.

Petty, G.J. and Manicom, B.Q. (1995) Control of the big-headed ant, *Pheidole megacephala*, in pineapple plantations with the proprietary bait Amdro®. *Fruits* 50, 343–346.

Petty, G.J. and Tustin, H. (1993) Ant (*Pheidole megacephala* (F.)) – mealybug (*Dysmicoccus brevipes* (Ckll.)) relationships in pineapples in South Africa. *Acta Horticulturae* 334, 387–396.

Petty, G.J. and van der Westhuizen (1992) Integration of chemical remedies for the control of insect and mite pests in pineapples. *Subtropica* 13(8), 33–41.

Petty, G.J. and Webster, G.S. (1981a) Importance of pineapple mealybug, *Dysmicoccus brevipes* (Ckll.) and its distribution on the pineapple plant. *Subtropica* 2(2), 6–8.

Petty, G.J. and Webster, G.S. (1981b) Importance of spray volume for control of mealybug and flat mite in pineapples. *Subtropica* 2(8), 7–8.

Petty, G.J., Willers, P. and Smart, G. (1991) Methyl bromide fumigation of pineapple planting material. *Subtropica* 12(12), 20–24.

Pinon, A. (1997) L'ananas dans la zone Andine (Pineapple in the Andine area). *Acta Horticulturae* 425, 69–74.

Plenet, A. (1965) Parasites animaux des principales plantes cultivées à la Rèunion. *Congrés de la Protection des Cultures Tropicales*, Chambre de Commerce et d'Industrie, Marseile, 23–25 March 1965.

Prins, A.J. (1965) Notes on the biology and morphology of the wattle chafers, *Monochelus calcaratus* Burm. (Melolthinae), *Hypopholis sommeri* Burm. (Melolonthinae), and *Adoretus ictericus* Burm. (Rutelinae) with some references to natural enemies (Coleoptera: Lamellicornia). *Entomology Memoirs*, Department of Agricultural Technical Services, RSA, Volume 9, 55 pp.

Prinsloo, G.L. (1985) Afrotropical Encyrtidae (Hymenoptera: Chalcidoidea): new records and notes. *Journal of Natural History* 19(2), 277–284.

Py, C., Lacoeuilhe, J.J. and Teisson, C. (1984) *L'ananas: sa culture, ses produits*. Maisonneuve & Larose, Paris, 564 pp.

Py, C., Lacoeuilhe, J.-J. and Teisson, C. (1987) *The Pineapple, Cultivation and Uses*. Maisonneuve & Larose, Paris, 568 pp.

Py, C. and Tisseau, M.A. (1965) *L'ananas*. Maisonneuve & Larose, Paris, 298 pp.

Quilici, S., Vercambre, B. and Bonnemort, C. (1992) Les insectes ravageurs. In: *La Culture du Gèranium Rosat à la Rèunion*. CIRAD/APR/CAHEB/DAF/SAFER. Imp Graphica, St Denis, La Rèunion, pp. 79–90.

Rai, B.K. and Sinha, A.K. (1980) Pineapple: chemical control of mealybug and associated ants in Guyana. *Journal of Economic Entomology* 73(1), 41–45.

Reimer, N.J. and Beardsley, J.W. (1990) Effectiveness of hydramethylnon and fenoxycarb for control of big-headed ant (Hymenoptera: Formicidae), an ant associated with mealybug wilt of pineapple in Hawaii. *Journal of Economic Entomology* 83(1), 74–80.

Reimer, N., Beardsley, J.W. and Jahn, G. (1990) Pest ants in the Hawaiian Islands. In: Vander Meer, R.K., Jaffe, K. and Cedeno, A. (eds) *Applied Myrmecology. A World Perspective*. Westview Press, Boulder, San Francisco & Oxford, pp. 40–50.

Rhainds, M., Gries, G. and Morales, J.L. (1996) Oviposition deterrency in pineapple borer females, *Thecla basilides* (Lepidoptera: Lycaenidae). *Ecological Entomology* 21(1), 105–106.

Rohrbach, K.G. (1983) Pineapple diseases and pests and their potential for spread. *Exotic Plant Quarantine Pests and Procedures for Introduction of Plant Materials* 145–171.

Rohrbach, K.G. and Apt, W.J. (1986) Nematode and disease problems of pineapple. *Plant Disease* 70(1), 81–87.

Rohrbach, K.G., Beardsley, J.W., German, T.L., Reimer, N.J. and Sanford, W.G. (1988) Mealybug wilt, mealybugs and ants on pineapple. *Plant Disease* 72(7), 558–565.

Rougé, B. and N'Goan, M. (1997) L'ananas en Afrique de l'ouest et du centre (Pineapple in West and Central Africa). *Acta Horticulturae* 425, 75–82.

Sakimura, K. (1932) Life history of *Thrips tabaci* on *Emilia sagitata* and its host plant range in Hawaii. *Journal of Economic Entomology* 25, 884–891.

Sakimura, K. (1937) The life and seasonal histories of *Thrips tabaci* Lind. in the vicinity of Tokyo, Japan. *Oyo Dobuts-Zasshi* 9, 1–24.

Sakimura, K. (1960) The present status of thrips-borne viruses. In: Maramorosch, K. (ed.) *Biological Transmission of Disease Agents*. Academic Press, New York and London, pp. 30–40.

Sakimura, K. (1966) A brief enumeration of pineapple insects in Hawaii. *Pineapple Research Institute of Hawaii Technical Paper* No. 315, 7 pp.

Samways, M.J. (1985) Appraisal of the proprietary bait 'Amdro' for control of ants in southern African citrus. *Citrus Journal* 621, 14–17.

Sanches, N.F. (1980) The mites of the pineapple (*Ananas comosus* (L.) Merrill) in the state of Bahia, Brazil, and its insect fauna in the country. MSc thesis, The University of São Paulo, Brazil.

Sanches, N.F. (1981) Entomofauna do abacaxizeiro no Brazil. *Cruz das Almas, BA. EMBRAPA/CNPMF* Documentos 10, 68.

Sanches, N.F. (1993) A broca-do-fruto do abacaxi e seu controle. *Abacaxi em Foco* 77, 2 pp.

Sanches, N.F. and Zem, A.C. (1978) Effectiveness of different acaricides for the control of the mite *Dolichotetranychus floridanus* (Banks, 1900) in Perola pineapple slips. *Anais da Sociedade Entomologica do Brazil* 7(2), 155–161.

Sanford, W.G. (1962) Pineapple crop log-concept and development. *Better Crops Plant Food* 46(3), 32–43.

Sanyal, A.K. and Das, T.K. (1989) Oribatid mites (Acari: Cryptostigmata) associated with pineapple (*Ananas comosus*) root at Kalyani, West Bengal. *Environment and Ecology* 7(4), 971–972.

Sarah, J.L. (1986) Influence of *Pratylenchus brachyurus* on the growth and development of pineapple in the Ivory Coast. *Revue de Nematologie* 9, 308–309.

Schneider, R.C., Zhang, J., Anders, M.M., Bartholomew, D.P. and Caswell-Chen, E.P. (1992) Nematicide efficacy, root growth, and fruit yield in drip-irrigated pineapple parasitized by *Rotylenchulus reniformis*. *Journal of Nematology* 24, 540–547.

Sérrano, F.B. (1935) Pineapple yellow spot in the Philippines. *Philippine Journal of Science* 58, 481–493.

Sether, D.M., Ullman, D.E. and Hu, J.S. (1998) Transmission of pineapple mealybug wilt-associated virus by two species of mealybug (*Dysmicoccus* spp.). *Phytopathology* 88(11), 1224–1230.

Siddappaji, C., Reddy, D.N.R. and Channabasavanna, G.P. (1989) Some observations on the mite pests of cardamom in Karnataka, India. In: Channabasavanna, G.P. and Viraktamath, C.A. (eds) *Progress in Acarology*, Vol. 2. New Delhi, Oxford & IBH Publishing Co., pp. 173–175.

Sinclair, E. and Scott, C. (1997) Soil invertebrates in the Queensland pineapple industry. In:

Allsopp, P.G., Rogers, D.J. and Robertson, L.N. (eds). *Soil Invertebrates in 1997.* Bureau of Sugar Experiment Stations, Brisbane, pp. 134–137.

Sipes, B.S. and Schmitt, D.P. (1994a) Population fluctuations of *Rotylenchulus reniformis* in pineapple fields and the effect of the nematode on fruit yield. *Plant Disease* 78, 895–898.

Sipes, B.S. and Schmitt, D.P. (1994b) Evaluation of pineapple, *Ananus comosus,* for host-plant resistance and tolerance to *Rotylenchulus reniformis* and *Meloidogyne javanica. Nematropica* 24, 113–121.

Smith, T.J., Petty, G.J. and Villet, M.H. (1995) Description and identification of white grubs (Coleoptera: Scarabaeidae) that attack pineapple crops in South Africa. *African Entomology* 3(2), 153–166.

Stirling, G.R. (1991) *Biological Control of Plant Parasitic Nematodes.* CAB International, Wallingford, UK, 282 pp.

Stirling, G.R. and Kopittke, R. (2000) Sampling procedures and damage thresholds for root-knot nematode (*Meloidogyne javanica*) on pineapple. *Australian Journal of Experimental Agriculture* 40.

Stirling, G.R. and Nikulin, A. (1993) Population dynamics of plant parasitic nematodes in Queensland pineapple fields and the effects of these nematodes on pineapple production. *Australian Journal of Experimental Agriculture* 33, 197–206.

Stradling, D.J. (1968) Investigations on the natural enemies of onion thrips (*Thrips tabaci* Lindeman). A work undertaken at the European station during 1968. *Interim Report, Commonwealth Institute for Biological Control,* 8 pp.

Su, T.H., Beardsley, J.W. and McEwen, F.L. (1980) AC-217, 300, a promising new insecticide for use in baits for control of the big-headed ant in pineapple. *Journal of Economic Entomology* 73(6), 755–756.

Suplicy, N., Giacomelli, E.J., Sampaio, A.S. and Orlando, A. (1966) Experiências sôbre o contrôle quimico da broca do fruto do abacaxizeiro. *Biológico* 32(6), 122–126.

Thomas, J.E., Horlock, C.M., Thomson, K.G., Dietzgen, R.G. and Teakle, D.S. (1999) Mealybug wilt of pineapple. *Pineapple News: ISHS. Pineapple Working Group Newsletter* 6, 5–6.

Thomason, I.J. (1987) Challenges facing nematology: environmental risks with nematicides and the need for new approaches. In: Veech, J.A. and Dickson, D.W. (eds) *Vistas on Nematology.* Society of Nematologists, Hyattsville, Maryland, USA, pp. 469–476.

Torres Navarro, H., Lozoya Saldan, A.H. and Uriza Avila, D. (1989) Clonal selection of pineapple, *Ananas cosmosus* (L.) Merr. with resistance to 'marchitez roja' virus and characteristics of commercial value. *Revista Chapingo* 13, 156–160.

Tryon, H. (1898) Fruitlet core-rot of pineapple. *Queensland Agricultural Journal* 3, 462–467.

Tsai, B.Y. and Apt, W.J. (1979) Anhydrobiosis of the reniform nematode: survival and coiling. *Journal of Nematology* 11, 316.

Ullman, D.E., German, T.L., McIntosh, C.E. and Williams, D.D.F. (1991) Effect of heat treatment on a closterovirus-like particle associated with mealybug wilt of pineapple. *Plant Disease* 75(8), 859–861.

Vuillaume, C. (1982) Contribution à l'etude des 'taches noires' sur l'ananas Victoria à la Rèunion. Rèunion annuele IRFA, doc. Interne, 34.

Wafa, A.K., Zaher, M.A. and Yousef, A.A. (1968–1969) Survey of the tenuipalpid mites in UAR (Acarina: Tenuipalpidae). *Bulletin of the Zoological Society of Egypt* 22, 52–59.

Waite, G.R. (1993) Pests. In: Broadley, R.H., Wassman, R.C. and Sinclair, E. (eds) *Pineapple Pests and Disorders.* Department of Primary Industries, Queensland, pp. 21–32.

Waite, G.R. (1996) Substitute chemical control for symphyla in pineapples. *Queensland Department of Primary Industries, Final Report to Horticultural Research and Development Corporation,* 7 pp.

Wakman, W., Teakle, D., Thomas, J.E. and Dietzgen, R.G. (1995) Presence of a closterolike virus and a bacilliform virus in pineapple plants in Queensland. *Australian Journal of Agricultural Science* 46, 947–958.

Willers, P. (1990) Methyl bromide fumigation of pineapple plant material. *Institute for Tropical and Subtropical Crops, Information Bulletin* 217, 6–7.

Willers, P. (1992) *Phenacoccus solani* on pineapple in South Africa. *Institute for Tropical and Subtropical Crops, Information Bulletin* 237, 29–30.

Williams, D.D.F., Fleisch, H., Bartholomew, D.P. and Rohrbach, K.G. (1993) Historical review of pineapple breeding in Hawaii. *Acta Horticulturae* 334, 67–76.

Williamson, V.M. (1998) Root-knot nematode resistance genes in tomato and their potential for future use. *Annual Review of Phytopathology* 36, 277–293.

Zeck, W.M. (1971) A rating scheme for field evaluation of root-knot nematode infestations. *Pflanzenshutz- Nachtricten Bayer* 24, 141–144.

Zhang, J., Malezieux, E. and Bartholomew, D.P. (1997) ALOHA-Pineapple v.2.1: a computer model to predict the growth, development and yield of pineapple. *Acta Horticulturae* 425, 287–296.

7 Pollinators and Pests of *Annona* Species

J.E. Peña,[1] H. Nadel,[2] M. Barbosa-Pereira[3] and D. Smith[4]

[1]*Tropical Research and Education Center, University of Florida, University of Florida 18905 SW 280 Street, Homestead, FL 33031, USA;* [2]*International Centre for Insect Physiology and Ecology (ICIPE), PO Box 30772, Nairobi, Kenya;* [3]*Department of Entomology, ESALQ-USP, Piracicaba, São Paulo, Brazil;* [4]*Department of Primary Industries, Maroochy Horticultural Research Station, Nambour, Queensland 4560, Australia*

Importance of the Crop, Classification and Ecology

The most primitive angiosperm order *Magnoliales* consists of ten families and almost 3000 species. The highly diversified family *Annonaceae* includes about 2300 species, and therefore makes up roughly three-quarters of this order. The genus *Annona* embraces several valuable fruit trees and is centred in the Neotropics, with about 110 species there and only three representatives in the Old World: *Annona senegalensis* Persoon and *A. stenophylla* Engler et Diels are widespread in Africa, but do not occur in America, whereas *A. glabra* L. occurs in both (Safford, 1914; Kessler, 1987). The genus is divided into sections, including the [*Eu-Annona*] *Guanabani* (soursops), *Pilaeflorae* (silky annonas), *Acutiflorae* (custard apples), [*Atta*] *Attae* (custard apples), and *Annonellae* (dwarf annonas). Among these sections two are horticulturally important, the *Guanabani* and *Attae*. The *Guanabani* is characterized by a subglobose, pyramidal flower and broadly imbricate inner petals. The most well-known species within this section are *A. muricata* L., *A. montana* McFadden, *A. glabra* L., *A. salzmannii* A. DC, *A. purpurea* Sesse et Mocino, and *A. senegalensis* Persoon. The *Attae* section is easily distinguished from the other groups by its elongated flowers and triquetrous buds. The inner petals of *Attae* may be absent or minute. The most well-known species within this section are *A. cherimola* P. Miller, *A. squamosa* L., *A. reticulata*, L., and *A. longiflora* Watts. Fruits of *Annona* are fleshy aggregates that arise from coalesced carpels and contain large seeds with reticulate endosperm.

Most species of *Annona* have specific climatic requirements for growth, flowering, and fruit maturation. The origins of the tropical species such as the sugar apple (*A. squamosa*) are the warm lowland regions of Brazil, Guiana, Venezuela, Mexico, and the West Indies. A distinctly subtropical species is the cherimoya (*A. cherimola*), which originates from the cool Andean valleys of Peru and Ecuador at elevations of around 2000 m. The hybrid atemoya (*A. cherimola* × *A. squamosa*) that appeared spontaneously when parent trees were cultivated side by side, and can be produced through manual cross-pollination, exhibits intermediate climatic requirements for growth and fruiting.

Within the family *Annonaceae*, fruits of the cherimoya, sugar apple, atemoya, and soursop (*A. muricata*) have the greatest potential for utilization and export in the American, Caribbean, Asian and Australian countries.

©CAB *International* 2002. *Tropical Fruit Pests and Pollinators*
(eds J.E. Peña, J.L. Sharp and M. Wysoki)

Much of the production of commercially grown species has spread from their indigenous areas to tropical and subtropical parts of Australia, New Zealand, Asia and around the Mediterranean. Thus, production methods vary from more sophisticated systems to less intensive, small farm or backyard type production. For instance, soursop yields in Hawaii can reach 83 kg per tree. At Paramaribo, Surinam, soursop yields of 54 kg per tree are achieved at 278 trees ha^{-1} (Nakasone and Paull, 1998).

Most research on control of *Annona* pest arthropods stemmed from generalized practices already in use on other tropical fruit crops, i.e. intensive application of chemical control measures without knowledge of pest biology or susceptibility, or alternative control methods. The major constraints to high productivity of these crops are not only attack by key pests, but often the lack of pollinating agents. Crop losses due to key pests range from 40 to 90% in heavily infested areas. Natural fruit set in the absence of pollinating agents is lower than 5% of flowers. Survey and manipulation of pollinating agents as well as basic studies on the biology and methods of controlling the pests are prerequisites for successful management of these crops. The purpose of this chapter is to elucidate current knowledge of pollinators and their relationship to fruit set, and to assess studies on key pests of *Annona* crops, including their dynamics, behaviour, and management systems.

Arthropod Fauna

Annonas throughout the world host a broad variety of arthropods. Surveys of arthropods associated with these plants reveal that the number of species found in the Neotropics ranges from a dozen to a few hundred (Ebeling, 1959; Venturi, 1966; D'Araujo *et al.*, 1968; Alata, 1973; Marin, 1973; Gutierrez and Trochez, 1977; Peña *et al.*, 1984; Bennett and Alam, 1985; Anonymous, 1989; Medina-Gaud *et al.*, 1989; Peña *et al.*, 1990; Coto *et al.*, 1995; Peña and Bennett, 1995). The vast majority of these insects are also pests of other well established crops, while some are beneficial pollinators and only two to three are key pests that cause significant crop damage in various production systems.

Pollinators and Pollination

Until the mid-1980s natural pollination was investigated in no more than 20 to 30 species of *Annonaceae* (Gottsberger, 1985), but the number of new studies keeps rising (e.g. Andrade, 1996; Nagamitsu and Inoue, 1997; Webber and Gottsberger, 1997; Momose *et al.*, 1998). Inadequate pollination is implicated as a major factor limiting production of commercial *Annona* fruits in many locations (Gazit *et al.*, 1982). This is attributed, in part, to the temporal separation of female and male function within the flower, which limits its potential to self-pollinate without external factors. The commonest problem is lack of pollinators. This is a direct result of the expansion of plantations into regions outside the native range of the plants and their pollinators, and may be due even to the unnatural conditions imposed by cultivation, regardless of locality.

The majority of *Annonaceae* are pollinated by beetles, although some are pollinated by thrips (e.g. Momose *et al.*, 1998), true bugs (Farre *et al.*, 1997), and even cockroaches (Nagamitsu and Inoue, 1997). Pollination by flies may also occur, though evidence for this is scanty (Gottsberger, 1970). The flowers range from showy to drab, often attracting their pollinators through strong odours. Some species have thermogenic flowers that attain a temperature higher than that of surrounding air, presumably to intensify volatilization of the odoriferous chemicals in the petals (Gottsberger, 1970). Annonaceous flowers produce no nectar, but rewards for pollinators include fleshy edible petals (Gottsberger, 1988), fleshy sterile tissue at the tips of the stamens (Nadel and Peña, 1994), pollen (Gottsberger, 1988; Deroin, 1989), and stigmal exudates (Vithanage, 1984; Gottsberger, 1988). The flowers are used as mating sites by some pollinators (Webber, 1981; Gottsberger, 1988; Deroin, 1989). The activities of beetles in the flowers, including feeding, mating, and quiescence, result in prolonged visits of several hours to a few days while the flowers advance from the female to the male phase.

Annonaceous flowers are protogynously dichogamous, opening as females with receptive stigmas and closed anthers, and later losing stigmal receptivity as the flowers turn into pollen-shedding males (Gottsberger, 1970). This evolutionary adaptation prevents deposition of pollen on to the stigmas in the same flower, and is one of many techniques that plants employ to avoid self-fertilization. Prevention of the transfer of pollen between different flowers on the same plant is another technique used by many *Annonaceae*, and is achieved through synchronization of flowering, where, at any time, open flowers on one plant are functionally of only one sex. These temporal floral traits, commonly enhanced with inherent incompatibility between pollen and ovule from the same plant, render most annonaceous species unable to self-pollinate.

The flowering season of commercial annonas appears to be highly variable but is usually concentrated around the warmer months of the year. The season for sugar apples and atemoyas lasts for 3–5 months. In Florida, USA, flowering of atemoyas begins in April, and sugar apples in May, and continues until early August (Nadel and Peña, 1991b). Likewise, in northeastern India and New Delhi, sugar apples flower approximately from March through August, with a peak in May (Kumar *et al.*, 1977). In Israel, atemoyas and sugar apples flower from June to September (Oppenheimer, 1947; Podoler *et al.*, 1985), whereas in Egypt they flower from May to July (Ahmed 1936a; Rokba *et al.*, 1977). Atemoyas flower from November to the end of January in Queensland, Australia (George *et al.*, 1992).

Cherimoya flowering periods are often shorter. In New Zealand, cherimoyas flower for only 2–9 weeks, during late December through February, with one cultivar reported to produce a second small flush at the end of March (Hopping, 1983). In Chile, peak bloom occurs in January and February (Lopez and Rojas Dent, 1992; Saavedra, 1977). In India, they are reported to flower in August (Thakur and Singh, 1965). However, in California, USA, bloom lasts usually from May to October (Thomson, 1974), and nearly as long in Mexico, from March to July (Castaneda *et al.*, 1997).

Pollination of Commercial Annonas

Natural pollination of commercial *Annona* species was shrouded in mystery for decades. Through a lack of easily observed alternatives, it was assumed to occur through the action of wind, honeybees, or pollen falling on to the stigmas. In the early 1900s, Wester (1910) suggested that the small beetles he occasionally found in sugar apple flowers might be the pollinating agents, but his suggestion went unheeded for the next 60 years. Ahmed (1936a,b) recognized the entomophilous nature of sugar apple flowers, but did not elaborate. Gottsberger's (1970) interest in the evolution of floral traits led him to the Neotropics where he clinched the role of beetles in the pollination of wild *Annona* species. Reiss (1971) was the first to document pollination by nitidulid beetles, to the exclusion of other factors, in cultivated atemoyas and cherimoyas, and Villalta (1988) documented scarab beetle pollination of soursop. Today, natural pollination of *Annona* is believed to be largely restricted to the action of beetles, though some 'self-fertility' is mentioned in much of the literature and awaits elucidation.

Pollination of the larger-flowered *Guanabani* [*Eu- Annona*] section is effected mainly by scarab beetles (Scarabaeidae), and the smaller-flowered *Attae* section by a variety of smaller beetles, mainly sap beetles (Nitidulidae), weevils (Curculionidae), cucujids (Cucujidae), rove beetles (Staphylinidae), and anthicids (Anthicidae). The role of beetles as pollinators was determined through field observations of insects that contact the floral sexual organs and carry pollen, and through examination of fruit set following exclusion or confinement of potential pollinators with the flowers. The story that emerged from areas as widely divergent as South America, Australia, Israel, and the USA revealed that the small-chambered *Atta* flowers share a similar floral and pollination biology, and that it is somewhat different from the large-chambered soursop flowers.

Cherimoya, sugar apple, and atemoya

The pendulous flowers bear three elongate, firm, fleshy petals that are closely appressed at the start of anthesis and gradually spread apart as the flower matures. The flowers are protogynously dichogamous (Ahmed, 1936a; Noonan, 1954; Schroeder, 1956; Thakur and Singh, 1965; Reiss, 1971), spending about 18–25 h in the female phase followed by about 12 h in the male phase (Podoler *et al.*, 1984, 1985; Nadel and Peña, 1994) (Plate 41). The flowers of sugar apple are generally reported to open in early morning, but the shedding of pollen is variously reported to begin either at any time of day (Ahmed 1936a; Kumar *et al.*, 1977), or usually in the afternoon (Wester, 1910), or around midnight (Nadel and Peña, 1994). Atemoyas open in mid- to late afternoon, begin shedding pollen around noon of the next day, and drop the stamens and petals around midnight (Nadel and Peña, 1994). The time of day that they begin to open may differ with climatic conditions (Kumar *et al.*, 1977), species, cultivar, or individual.

Cohorts of flowers open synchronously on a tree (Gazit *et al.*, 1982; Kahn and Arpaia, 1990; Nadel and Peña, 1994), resulting in open flowers that are functionally of the same sex on any tree. A new cohort opens when the petals and stamens of the old cohort drop off, or soon after.

These annonas are most often pollinated by nitidulid beetles (Table 7.1), which breed and feed in decaying fruits or sap flows. The beetles are attracted to the fruity, fermenting odour of *Annona* flowers, especially when hungry (Podoler *et al.*, 1985). Some species occasionally feed in flowers, gnawing tissue at the tip of stamens or the bases of petals (Nadel and Peña, 1994), or feed on pollen grains (Podoler *et al.*, 1985), and possibly on stigmal exudates (Vithanage, 1984). Mating is not known to occur inside the flowers. The beetles generally enter female phase flowers in the morning and remain almost inactive at the base of the petals against the stamens and stigmas, dispersing several hours later after the flowers become male and cover them with pollen (Nadel and Peña, 1994). *Annona* pollen

Table 7.1. Pollinators of commercial annonas.

Commodity	Region	Pollinators	Source
Atemoya, cherimoya	Israel	Nitidulidae: *Carpophilus humeralis*, *C. hemipterus*, *C. mutilatus*, *Haptoncus luteolus*	Gazit *et al.*, 1982
Atemoya	Queensland, Australia	Nitidulidae: *Carpophilus hemipterus*	George *et al.*, 1989
Atemoya	Florida, USA	Nitidulidae: *Carpophilus fumatus*, *C. dimidiatus* spp. cmplx., *Haptoncus luteolus*, *Colopterus posticus*	Nagel *et al.*, 1989
Atemoya, sugar apple	Florida, USA	Nitidulidae: *Carpophilus fumatus*, *C. hemipterus*, *C. humeralis*, *C. marginellus*, *C. mutilatus*, *Colopterus posticus*, *C. truncatus* spp. complex, *Haptoncus luteolus*	Nadel and Peña, 1994
Cherimoya	Quillota, Chile	Nitidulidae: *Carpophilus hemipterus*, *Colopterus* sp.	Lopez and Rojas, 1992
		Nitidulidae: *C. hemipterus*	Lopez and Uquillas, 1997
Cherimoya	Mexico and Michoacan States, Mexico	Cucujidae: *Cryptolestes ferrugineus*, *Silvanus planatus*. Nitidulidae: *Conotelus* sp. Staphylinidae: *Phloenomus* sp.	Castañeda *et al.*, 1997
Cherimoya	California, USA	Staphylinidae: *Eusphalerum* sp.	Kahn, 1997
Cherimoya	Spain	Anthocoridae[a]: *Orius* sp.	Farre *et al.*, 1997
Soursop	Costa Rica	Scarabaeidae: *Cyclocephala amazona*, *C. brittoni*, *C. stictica*	Villalta, 1988

[a]Order Hemiptera; all other pollinators are in the order Coleoptera.

is sticky and remains viable for at least 24 h (Reiss, 1971), sufficiently long for movement between old and new cohorts of flowers.

The number of beetles per flower affects the likelihood of fruit set (Fig. 7.1), and also the quality of the fruit in some cases. All studies provide evidence for increased fruit set as numbers of visiting beetles increase (Gazit *et al.*, 1982; George *et al.*, 1989, 1992; Nadel and Peña, 1994; Lopez and Uquillas, 1997). Fruit symmetry is poor when only 1–3 beetles visit a flower, but is good when at least four beetles visit (Gazit *et al.*, 1982). However, fruit symmetry is sometimes unaffected by the number of pollinators per flower (George *et al.*, 1989; Lopez and Uquillas, 1997). It must be noted, though, that in these studies the number per flower was generally lower (two) than the threshold for good symmetry found by Gazit *et al.* (1982).

The guilds of pollinating beetles in commercial *Annona* species vary geographically, and species may even perform differently in each area. The four major pollinators in Israel, *Carpophilus humeralis*, *C. hemipterus*, *C. mutilatus*, and *Haptoncus luteolus*, are equally effective as pollinators. In Florida, USA, about nine species of native and exotic nitidulids visit the flowers (Plate 42), but *C. mutilatus* is the most important pollinator in terms of efficacy and abundance in flowers, followed by *C. fumatus* and *H. luteolus*. Although *C. humeralis* is very abundant in the annona grove environment, it rarely visits the flowers. When it does visit, it induces a very low rate of fruit set, in contrast with its behaviour in Israel (Nadel and Peña, 1994). The number of species visiting flowers in Ecuador, Colombia and the Caribbean region is similar to that found in Florida (Peña and Bennett, 1995). In other regions the guilds are smaller, and in some cases inadequate for commercial fruit production without the aid of manual pollination.

Because of the dichogamous nature of the flower, the stigmas cannot receive pollen from the same flower under most conditions. Most workers report that the stigmas glisten with moisture when receptive, but turn dry before the stamens split open. However, if humidity is high or temperature moderate, the stigmas are thought to remain moist and receptive until the pollen is released, allowing a small proportion of self-pollination to occur without the aid of insects (George *et al.*,

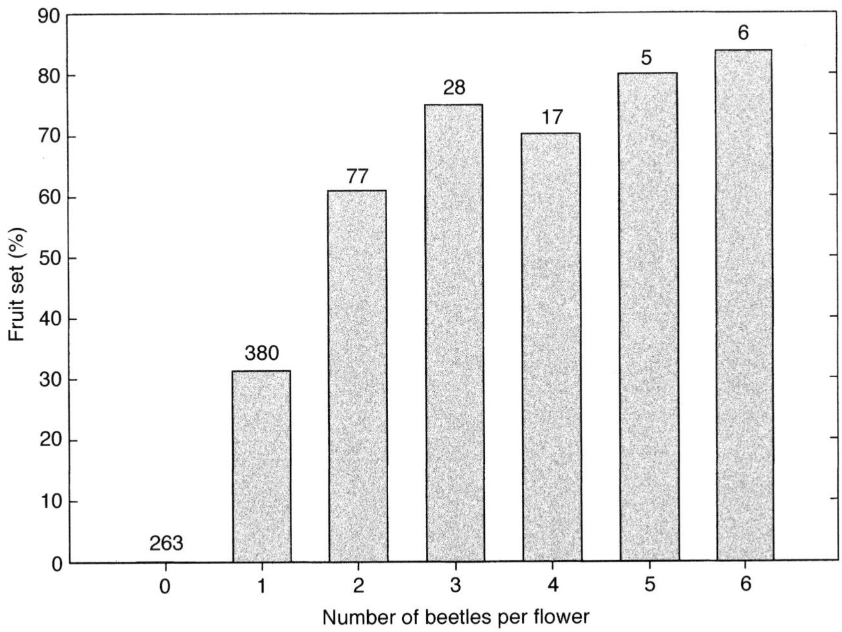

Fig. 7.1. Percentage fruit set as a function of the number of nitidulid beetles per flower. Number of inspected flowers are written above each bar. (Source: Nadel and Peña, 1994.)

1989). Previous workers have noted that *Annona* trees growing near bodies of water or in especially humid climates set higher levels of fruit than those in drier climates (Schroeder, 1943; Sarasola, 1960; Thakur and Singh, 1965; Thomson, 1970). Cultivations in relatively dry areas have low natural fruit set (Thakur and Singh, 1965; Gazit *et al.*, 1982); and in the dry areas of southern California annonas must be manually pollinated to obtain commercial yields (Schroeder, 1943, 1956). However, many of these observations were made before convincing evidence was presented for insect pollination in annonas, and the studies were made outside the native range of the plants and their pollinators. It is likely that climatic factors such as humidity benefit pollinator populations and that natural fruit set in these areas is at least partly due to their activities.

Soursop

Villalta (1988) described the floral morphology, phenology and pollination of *A. muricata* in Costa Rica (Plate 43). The flowers spend about 3 days in the female phase when the outer petals spread apart slightly and a copious viscous exudate appears on the surface of the stigmas. It ends when the tips of the stigmas are shed in the form of a cap-like structure. The ensuing male phase begins at about 0830 h and lasts about 12 h until the stamens and petals fall off. During anthesis the three outer petals spread apart slightly, but the inner three petals remain closed. The flowers attract the beetles one day before the male phase at about 1800 to 2200 h, when they emit a strong odour. Pollen is transported by *Cyclocephala* beetles (Scarabaeidae) on the hairs of the legs and body (Plate 44). The beetles remain in the flowers for 24 h and leave after the pollen is shed. They feed on pollen and on the bases of the petals, and engage in mating. They do not feed on the stigmal exudate, which may contain toxins that protect the stigmas from chewing beetles. Anthesis begins at any time of the day or night, but most often between 1100 and 1500 h; anther dehiscence, however, occurs in synchrony among flowers on a tree and also with other trees.

The petals are stiff and leathery and the inner ones form a large space around the sexual organs that is referred to as the pollination chamber. In the soursop this is very different from the tight space provided for small beetle pollinators in the section *Atta* of *Annona*. Dynastine scarab beetles such as *Cyclocephala* are the most important pollinators of annonas with large pollination chambers, and are generally attracted by nocturnal scents (Gottsberger, 1985, 1988).

In addition to the *Cyclocephala* species that pollinate soursop, *Cyclocephala quatuordecimpunctata* and other scarab beetles are considered likely pollinators for the *Guanabani* group (Gottsberger, 1991). Vidal (1997) reports that the main insects found in *A. muricata* flowers in Mexico are Nitidulidae and Chrysomelidae, while species of Apidae and Formicidae (Hymenoptera) are abundant on the external surfaces of the flowers where contact cannot occur with the flowers' sexual structures.

Villalta (1988) reported between 18% and 24% natural fruit set of soursop in Costa Rica. Some authors believe that entomophilous pollination of *A. muricata* has lower potential than manual pollination in commercial production (Cogez and Lyannaz, 1996).

Pollinator Management

Trials to increase fruit set in atemoya orchards by augmenting nitidulid populations have yielded mixed results. In Israel, one-time additions of decaying apples as an attractant failed to increase the number of beetles in the flowers or fruit set (Galon *et al.*, 1982). Release of 10,000 nitidulids (70% *C. hemipterus* and 30% *C. mutilatus*) in a grove also failed to show any increase in these parameters. Galon *et al.* (1982) suggested that decaying fruit baits compete with the flowers. They found that flowering branches enclosed in mesh bags supplied with nitidulids and decaying apples had only half the percentage of fruit set compared with bags supplied with beetles but no fruit (15% and 29%, respectively). In Australia, success was achieved by adding pineapples in an atemoya orchard throughout the flowering season. This resulted in more beetles in the flowers and 62% higher fruit set compared with an orchard not provided with fruit (George *et al.*, 1992).

Nitidulid pollination in *Annonaceae* can also be improved by using chemical lures (Bartelt *et al.*, 1992; 1994). The effect of nitidulid-pheromone bait stations on sugar apple and atemoya fruit set was determined in southern Florida. Maximum percentage fruit set fluctuated between 10% and 38% during the first 4 weeks after treatment in plots with bait stations, and was significantly higher than in the control plots (Peña *et al.*, 1999).

To the best of our knowledge, no management techniques have been reported for soursop pollinators.

Pests

In the Neotropics 296 species of arthropods are recorded as associated with *Annona* species. The families most frequently observed on *Annona* species are Coccidae (Homoptera), Noctuidae, Oecophoridae (Lepidoptera), and Eurytomidae (Hymenoptera). The most common species in various Neotropical countries are *Bephratelloides cubensis* (Ashmead), *B. pomorum* (F.) (Hymenoptera: Eurytomidae), *Cerconota anonella* Sepp. (Lepidoptera: Oecophoridae), *Parasaissetia nigra* (Neitner), *Saisssetia coeffeae* (Walter), *S. oleae* (Olivier) (Homoptera: Pseudococcidae), and *Cocytius antaeus* (Cramer) (Lepidoptera: Sphingidae). The first three species are considered key pests of *Annona*. They are multivoltine, stenophagous feeders of *Annona* seeds and fruits. *B. cubensis* has been reported from the USA, Caribbean, Central America and South America, whereas *B. pomorum* is listed from Central and South America. Their biology, behaviour and habits are quite similar, except that *B. cubensis* is uniparental. The seasonality of *B. cubensis* on various hosts in a given area can be attributed to the relative fruiting phenologies of the available *Annona* species (Nadel and Peña, 1991b). *Cerconota anonella* has been reported from tropical America and the Caribbean (Peña and Bennett, 1995). Other species are considered secondary and localized. In Asia, the western Pacific and Australasia, the main pests in *Annona* are 2–3 species each of mealybugs, coreid fruit-spotting bugs, *Bactrocera* fruit flies, pyralid fruit-boring moths, and soft scales (De Leon and German, 1917; Estalilla, 1921; Paguirigan, 1951; Galang, 1955; Vinas, 1972; Gabriel, 1975; Cantillang, 1976; Brun and Chazeau, 1980; Coronel, 1983; George and Nissen, 1991; Koesriharti 1991; Smith 1991a,b; Khoo *et al.*, 1991; Waterhouse, 1993).

Key Pests

The *Annona* seed borers, *Bephratelloides* species

Chalcidoid wasps are best known as parasites of other insects. However, many species in the families Torymidae, Eurytomidae, and all Agaonidae and Tanaostigmatidae have evolved the ability to induce galls or feed on plant seeds. Phytophagous eurytomids are considered by some to be the most primitive in the superfamily Chalcidoidea. The Neotropical genus *Bephratelloides* has species known to develop strictly in *Annona* seeds: *B. pomorum*, *B. cubensis*, *B. paraguayensis* (Crawford), and *B. petiolatus* Grissell and Schauff (Grissell and Schauff, 1990); *B. ablusus* Grissel and Foster develops in seeds of the wild *Cymbopetalum*, a close relative of *Annona* in southern Mexico (Grissell and Foster, 1996). *Bephratelloides* species that attack *Annona* species commonly occur in damaging numbers in South and Central America, the Caribbean and southern Florida (Dozier, 1932; Brunner and Acuña, 1967; Brussel and Weidijik, 1975; Peña *et al.*, 1984; Mendes and Pereira, 1997). Hosts recorded for *Annona* seed borers include *A. muricata*, *A. squamosa*, *A. squamosa* × *A. cherimola*, *A. cherimola*, *A. reticulata*, *A. montana* and *A. glabra* (Nadel and Peña, 1991a). Economic damage occurs when the adults chew their way out of the fruit, creating a 2 mm diameter tunnel that provides entry for other insects and decay organisms.

The biology of *B. cubensis* was studied on *A. reticulata* in Cuba by Brunner and Acuña

(1967). They concluded that *B. cubensis* is thelytokous, reproducing without males. It has approximately 4–5 generations year^{-1}. The egg stage lasts 12–14 days, the larval stage 6–8 weeks, the pupal stage 12–18 days, and the adult rarely lives beyond 15 days (Plate 45). One egg is laid per young seed and the larva feeds and pupates within it. Evangelista-Lozano *et al.* (1997) determined that *B. cubensis* has five instars and the average generation time is 62 days. Anjos and Pereira (1997) reported the same number of instars for *B. pomorum* on *A. muricata*. Pereira *et al.* (1997a) determined that the generation time of *B. pomorum* lasts from 43 to 113 days. Sex of *B. pomorum* can be distinguished in the pupal stage by the length of the antennae, which in males extend posteriorly beyond the tips of the wings, and by the elongated segments of the prothoracic tarsi. In the female the antennae do not extend to the tips of the wings and the prothoracic tarsi have rounded segments. Pereira *et al.* (1998) observed that adult females of *B. pomorum* live up to 10 days, while males live for only 2 days. Pereira *et al.* (1998) and Leal *et al.* (1997) have observed an attraction by *B. pomorum* males to caged virgin females and have suggested the possible presence of a female pheromone.

In the atemoya host, *B. cubensis* prefers to oviposit in fruits ranging from 1.5 to 5.5 cm in diameter, which corresponds to fruit ages of 3–7 weeks after bloom (Fig. 7.2). This size range was confirmed by Evangelista-Lozano *et al.* (1997). Although fruits larger than 5.5 cm are probed, usually when *B. cubensis* populations are high, most of these attacks do not result in infestation. Preferred fruit sizes presumably correspond with seeds that have not yet hardened and are easy to penetrate with the ovipositor, while the seeds of older fruits are probably too hard to penetrate. Larger fruits may be less preferred also because the distance from the fruit surface to the seed may exceed the length of the ovipositor. The probes in young sugar apple and atemoya fruits look like dark pinpricks surrounded by a round whitish patch, and are visible for about 2 weeks; in older fruits the whitish patch does not appear, and the probe marks are permanent and often ooze sap.

Oviposition activity by *B. cubensis* in Florida begins at about 0900 h and continues throughout the daylight hours with peaks in activity around 1200–1300 h (H. Nadel and J.E. Peña, unpublished). The wasps appear to spend the night on the underside of leaves on their host trees, and move to the upper surface

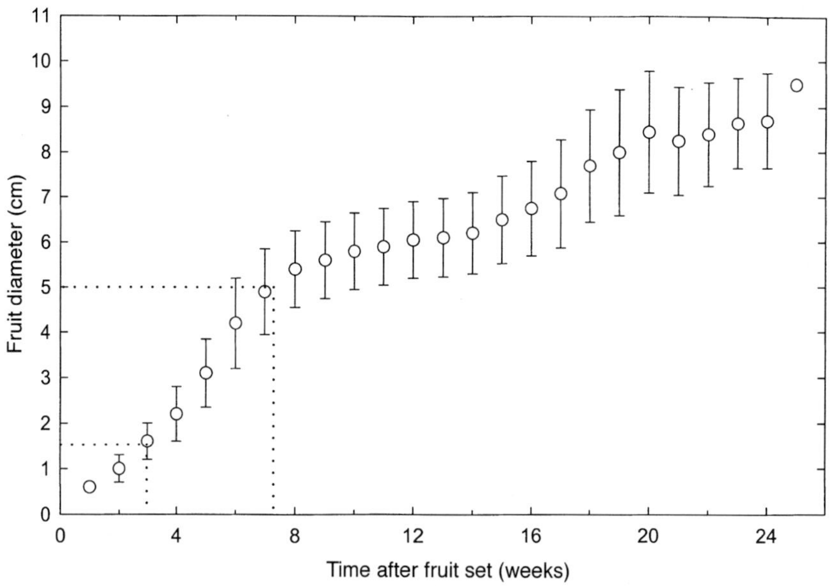

Fig. 7.2. Growth curve for atemoya var. 'Gefner' and fruit size preferred for oviposition by *Bephratelloides cubensis*. Vertical bars represent standard error of the mean. (Source: H. Nadel and J.E. Peña, unpublished.)

at sunrise. Flying individuals can be observed soon afterwards and throughout the day. The females probe fruits repeatedly with the ovipositor, each probe lasting up to about 300 s, but not all result in oviposition. Probes lasting between 100 and 200 s are most likely to result in oviposition (corroborated by seed dissection) (J. Fortier, H. Nadel and J.E. Peña, unpublished).

Pereira *et al.* (1997b,c) reported that the female of *B. pomorum* initially walks on the fruit, using the antennae to locate an oviposition site. However, the purpose of antennation of oviposition sites by chalcidoids is often to detect odours of conspecific competitors, and is also likely to play such a role in this pest genus.

Leal *et al.* (1997) reported that males of *B. pomorum* are strongly attracted to female-baited traps, showing a peak of activity at noon. Lima *et al.* (1997) determined that males are attracted to fruit odours while they search for emerging females. Males are able to locate the region of the fruit where adults will emerge and regularly try to copulate with newly emerged adults of either sex. Females can copulate more than once.

The response of *Bephratelloides* to various *Annona* fruit volatiles has not been investigated extensively. McLeod and Pieris (1981) and Idstein *et al.* (1984) discovered several volatiles characteristic of ripe *Annona* fruits. Among these, methyl hexanoate is significantly attractive to *B. cubensis* but other mixtures of *Annona* compounds such as 3 : 1 hexanoic acid: octanoid acid, or a 2 : 1 : 1 mixture of isoamyl alcohol : butanol : linalool are not attractive. The positive response to the first compound was obtained in an olfactometer but could not be replicated under field conditions (Peña, 1988).

MONITORING AND SAMPLING Nadel and Peña (1991b) monitored the weekly infestation rates of *B. cubensis* in commercially grown sugar apples and atemoyas in Florida. They reported that infestation increased throughout the summer in atemoyas but remained very low in sugar apples. The observed patterns were attributable to the relative fruiting phenologies of the commercial species and of the overwintering host, the bullock's heart (*A. reticulata*). In Florida, *A. reticulata* sets fruit in September–November, and fruits remain on the trees as late as May. Atemoyas set fruit from April to August, and sugar apples from May to August (Nadel and Peña, 1991a). Nadel and Peña (1991b) concluded that *B. cubensis* populations in Florida overwinter mainly in *A. reticulata* and then move to atemoyas that begin to set fruit during April. After a developmental time of 9 weeks, the adults emerge from the early atemoyas and proceed to infest younger atemoya and available sugar apple fruits. A second and third peak of adult emergence and infestation occur in atemoya and sugar apples until young fruits are no longer available and the wasps switch to *A. reticulata* fruits in September. Wasp populations increase throughout the warm months in Florida because of the availability of large concentrations of atemoya trees, and are bottlenecked in winter because *A. reticulata* is grown only sporadically as a garden tree. Low infestations in sugar apples may be due to a lack of incentive for the pests to disperse from plentiful atemoya orchards to other areas. No evidence of diapause was found in mummified stemoya and sugar apple fruits overwintering on trees or on the ground.

Peña *et al.* (1984) observed that peaks of activity of adult *B. cubensis* are observed at 1500 h when the average temperature fluctuates around 31–33°C. While testing the possibility of visual monitoring to assess population levels of this pest, H. Nadel and J.E. Peña (unpublished) observed that 50% of adult seed borers were concentrated in the middle third of the tree's height, on both leaves and fruits. Of these, 67% were concentrated in the outer half of the tree canopy. Peaks in wasp numbers on the trees occurred between 1100 and 1700 h, and numbers were highest during the warmer part of the summer. Flight was observed mostly before 1000 h and after 1600 h.

YIELD LOSSES Injury is caused by the emerging adult as it bores through the seed and fruit pulp (Plate 46). A circular hole is created on the surface that facilitates entry of microorganisms and other insects, and causes a significant necrosis (Korytowski and Peña, 1966; Nadel and Peña, 1991; Pereira, 1996). Yield loss due to attack of *B. pomorum* in soursop ranges from 41% to 80% (Reyes, 1967; Escandon *et al.*, 1982; Zarate 1987).

CONTROL TACTICS Chemical control is widely used, but is seemingly inefficient in some areas (Zarate, 1987). Zarate (1987) recommends the use of organophosphates and carbamates but does not report the results of their application. He also recommends the use of organophosphates mixed with sugar or molasses as attractant. Peña and Nagel (1988) found that the use of malathion plus molasses, fenvalerate, or permethrin provided significant adult mortality of the seed borer; however, counts of infested fruits remained high after treatment. The authors suggested that the poor performance of the insecticides was due to insufficient knowledge of pest dynamics, emergence peaks and pesticide timing. Ramnanan (1996) reported that malathion provided significant damage reduction by *B. pomorum* on soursop. Chemical interventions are made in some countries using full-coverage airblast sprayers, generally at rates of 1000 and 2000 l ha^{-1}, or they are done by backpack sprayers, where coverage is erratic.

Fruit bagging is considered by several researchers (Villalobos, 1987; Zarate, 1987) to be one of the best methods to prevent *B. cubensis* infestation. Ramnanan (1996) recommends sleeving or bagging fruits when the fruit is 5–10 cm in diameter. H. Nadel and J.E. Peña (unpublished) found no infestation in fruits bagged before they reached 2 cm in diameter, compared with unbagged fruits under the following conditions: 39% infested in an unsprayed atemoya orchard, 16% in a chemically treated atemoya orchard, and 11% in an unsprayed sugar apple orchard. In another study (H. Nadel and J.E. Peña, unpublished), atemoya fruits bagged with chlorpyrifos-treated and plain polyethylene bags also remained free of infestation, compared with 8% of infested controls. Bagged fruits were also cleaner and aesthetically better than unbagged fruit. However, bagging encourages the growth of mealybugs on some fruits, probably because natural enemies are excluded. Although the method is effective, a cost comparison must be made between bagging and other potential control methods to determine its economic feasibility.

In some countries, burying infested fruits is advised as the best method to keep the infestation at low levels. Zarate (1987) recommends collecting all infested fruit in the orchard, placing them in a 60 cm deep hole and covering the fruit with $CaCO_3$ and aldrin (!) 2.5%. At present, the decision to take action for the seed borer depends on a combination of: (i) presence of significant numbers of vulnerable fruits that are 1.5–5 cm in diameter; and (ii) visual monitoring to determine the presence of the adult seed borer in the orchard.

It is also prudent to remove (or avoid planting) unpreferred *Annona* species from areas of commercial production if those species allow the wasps to overwinter. Removal of the winter-fruiting *A. reticulata* from southern Florida would probably reduce damage by *B. cubensis* to below economic levels (Nadel and Peña, 1991b).

BIOLOGICAL CONTROL No significant native parasitization or predation of *Bephratelloides* species has been reported. When wasp exit holes are observed in the fruit, ants will invade the tunnels, but their effect on the insects remaining in the seeds is unknown. In Florida, the fungus *Beauveria bassiana* was collected from a *B. cubensis* adult still inside the seed (J.E. Peña, unpublished). To our knowledge this is the only biological control agent identified from *B. cubensis*. The fungus was applied to *B. cubensis* adults under laboratory conditions and provided 90% adult mortality up to 8 days after treatment.

PLANT RESISTANCE Martinez and Cabrera (1997) reported a higher incidence of *B. cubensis* in some soursop cultivars IV-10, VII-14 and 1V-16 compared with infestation registered in the cultivar IV-16 in Puerto Rico. Selection for trees with good quality seedless fruits would provide completely effective control against *Bephratelloides*. Parthenocarpy can be artificially induced on normal trees by plant hormones (Campbell, 1979). Campbell (1979) treated female-phase flowers and then the fruits repeatedly with a spray of gibberellins (1000 p.p.m.), and found good fruit set and production of parthenocarpic, seedless fruits that were, however, smaller than seeded fruits and of poorer flavour.

Lepidopterous annona fruit borers

The annona fruit borers, *Cerconota anonella* (Oecophoridae), *Talponia batesi* Heinrich (Tortricidae) and *Thecla ortyginus* (Lycaneidae), are second in importance as key pests of *Annona*. *Cerconota anonella* is recorded in South America, Central America and the Caribbean. (Fennah, 1937a; Zenner and Saldarriaga, 1969; Lawrence, 1974; Gutierrez and Trochez, 1977; Martinez and Godoy, 1983). The larvae damage the fruit epidermis, pulp and seeds of soursop and cherimoya during feeding (Fennah, 1937a). In Brazil, yield losses can be as high as 70%. *Talponia batesi* is distributed in the Caribbean, Venezuela, Colombia, Brazil and Mexico (Castañeda and Pineda, 1997). *Thecla* spp. are reported from Mexico, the Caribbean region, Central and South America (Fennah, 1937a).

Annona fruit borer, Cerconota anonella

The life cycle of *C. anonella,* as investigated by Bustillo and Peña (1992), averages 36.4 days. The adult moth is uniformly of a pale straw colour, with one spot on the upper surface of the forewing, halfway between the fore- and hindwing margins at approximately the distal two-thirds of the wing (Fennah, 1937a) (Plate 47). The female is approximately 1 cm long with a 2.5 cm wingspan, while males are 0.8 cm long with a wingspan of 1.8 cm (Nuñez and De la Cruz, 1982). They have nocturnal habits, and deposit eggs not only on unripe fruits, but also on the inflorescence (Costa Lima, 1945; Pireira *et al.*, 1991b). Eggs are elongated or ovoid, 0.6 mm in length and 0.25 mm in diameter. They are pale green to translucent in colour, and bear parallel and transverse ridges (Fennah, 1937a; Nuñez and de la Cruz, 1982). After hatching, the neonate larva is off- white in colour, which changes to reddish purple during later instars. The insect undergoes five instars, and the larval stage lasts 18.6 days at room temperature (21°C). Newly hatched *C. anonella* larvae make small traces all over the fruit surface, and penetrate the fruit on the fourth day after eclosion (Ruiz, 1991). While feeding, the larvae periodically clear their tunnel by pushing excreta towards the entrance. The orifice of the tunnel grows larger with larval maturation, and webbing forms across it like a screen, which becomes covered with faecal pellets and particles of fruit. The characteristic symptom of the attack of *C. anonella* is the excrement eliminated by the larvae, and the necrosis caused by opportunistic fungi that enter the fruit through the injury (Fennah, 1937a) (Plate 48). Damage to flowers is observed during spring. Yield losses caused by this pest fluctuate between 70% and 100% (Martinez and Godoy, 1983; Calzavara and Muller, 1987; Ruiz, 1991).

Preliminary observations show that *C. anonella* males are attracted to virgin females placed in cardboard traps; male capture was observed in 45% of the traps tested (Bustillo and Peña, 1992). The same authors observed that adults were attracted to black-light traps, which could be useful in monitoring populations of *C. anonella.*

MONITORING Junqueira *et al.* (1996) recommend placing black light traps (1 trap ha^{-1}) and starting treatment as soon as one moth is captured per trap. Bustillo and Peña (1992) suggest using virgin females placed in small cages as bait in Delta traps, approximately 1.5 m above the ground. Duarte (1947) suggests inspecting the orchard at the onset of blooming and collecting infested fruits in screen cages that restrain the moths but allow the exit of the parasitoids, *Brachymeria* and *Apanteles*.

BIOLOGICAL CONTROL Biological control of *C. anonella* was studied in Colombia and Ecuador. Sampling for parasites of *C. anonella* consisted of rearing collections of *Annona* fruits. Two braconid species, an *Apanteles* sp., and a species in an unknown genus of the subfamily Rogadinae, were identified as natural enemies of *C. anonella* in Colombia and Ecuador. Parasitism by *Apanteles* sp., however, was very low, ranging around 2–5% in Colombia and 2% in Ecuador (J.E. Peña, unpublished). The observations were in contrast to high parasitism levels from other braconids (*Apanteles* spp.) and an ichneumonid, *Xiphosomella* sp., reported from Venezuela by Martinez and Godoy (1983). The pupae are parasitized by the chalcid, *Brachymeria pseudovata* (Costa Lima, 1948; Ruiz, 1991; Pireira *et al.*, 1991a).

CHEMICAL AND CULTURAL CONTROL The effect of various types of protective bags and insecticides was tested in 3–5-year-old *A. muricata* trees. Bagging and use of bags treated with chlorpyrifos resulted in less damaged fruit compared with the control (Bustillo and Peña, 1992). Pesticide-treated bags also reduce infestation by *B. cubensis* in Costa Rica (Villalobos, 1987) and may therefore give an adequate control of both insect species. Ramnanan (1996) suggests that permethrins and cipermethrins provide good control of the damage due to *C. anonella*. In Brazil, Junqueira *et al.* (1996) recommend spraying the small fruits with fenthion, monocrotophos or endosulfan every 15 days.

Thecla ortygnus

This borer is established in Guatemala, Mexico, Costa Rica, Panama, Trinidad and southern Brazil as a pest of soursop (Fennah, 1937b). The adult butterfly is 12 mm long with a 36 mm wingspan. The sexes are differentiated by the colour of the wings, males having blue iridescent wings with well-defined dark marginal areas. The wings in females are smoky blue with black margins. Eggs are deposited in the perianth close to the floral peduncle. Generally, eggs are laid singly, but during high infestations, it is possible to find two eggs per flower. Eggs are hemispherical, 0.9 mm long and 0.5 mm wide (Fennah, 1937b). The newly eclosed larva is 1.3–1.9 mm long, with a characteristic light green colour. When eggs are deposited on flowers, the larva bores through the petals and feeds on stamens and stigmas. When an egg is deposited on a young fruit, the larva penetrates into the pulp. Fully developed larvae reach 17 mm in length. The larval stage has four instars and lasts 12 days; the pupal stage lasts 12–14 days (Fennah, 1937b).

Fruit infested by *Thecla* sp. exhibit irregular holes surrounded by a mass of excrement that the larvae push out from their tunnels (Saldarriaga *et al.*, 1987).

CONTROL Chemical control should be applied at the begining of flowering. Removal of infested flowers is suggested in Trinidad (Fennah, 1937b), while removal or application of malathion when the fruit is small is recommended in Colombia (Saldarriaga *et al.*, 1987).

Talponia batesi

Castañeda and Pineda (1997) determined that *T. batesi* prefers fruit of 1.3 to 8.1 cm diameter, but it can also cause flower injury. The larval stage of *T. batesi* lasts 91 days, while the pupal and adult stages last 15 and 10 days, respectively. In Mexico, infestations of *T. batesi* peak during the summer, but injury to fruit from larval feeding is more noticeable during autumn.

Yellow peach moth Conogethes punctiferalis

Yellow peach moth, *Conogethes punctiferalis* (Guenée) (Pyralidae), occurs throughout South-East Asia and in Japan, Indonesia and Australia (Smith, 1991a; Smith *et al.*, 1997). It is an attractive yellow moth with a wingspan of 25 mm. In eastern Australia, larvae damage the pulp of mature fruit from February to May, usually with losses of less than 5%, but occasionally 50% or higher. The Annona variety Pinks Mammoth is much more susceptible than African Pride.

The scale-like eggs are deposited singly on the fruit surface and the young larvae burrow through the epidermis into the pulp. Activity is betrayed by brown frass extruded to the fruit surface. After about 3 weeks, the last stage larvae pupate in shelters of frass on the fruit surface. The life cycle takes about 6 weeks and there are 5–6 generations each year. This moth infests a wide range of other fruit hosts such as papaya, citrus and stonefruit. Field crops like maize and sorghum are also favoured hosts and can be a source of infestation to fruit crops.

In Australia, tachinid parasitoids such as *Argyrophylax proclinata* Crosskey are important, providing up to 40% parasitism. The adult parasitoid (which is about 8 mm long and similar to a housefly in appearance) attaches its eggs to the host larvae. The fly larva develops within the host, killing the host larva at or near the time of pupation.

The assassin bug, *Pristhesancus plagipennis* Walker, is a significant predator of the larvae in eastern Australia (Smith, 1991a; Smith *et al.*, 1997).

Atis moth borer, Anonaepestis bengalella

The atis moth borer, *Anonaepestis* (*Heterographis*) *bengalella* Raq. (Pyralidae), is the most important fruit boring pest in the Philippines and is found also in India and Indonesia (De Leon and German, 1917; Estatilla, 1921; Galang, 1955; Gabriel, 1975; Coronel, 1983; Koesriharti, 1991). Larvae tunnel in the pulp of small fruit and extrude frass to the surface. Damaged fruit fail to develop and often fall. The mature larva pupates close to the surface.

Other uncommon Lepidoptera

Phycita semilutea Walker (Pyralidae) has been observed on annonas in Thailand (Morokote, 1999, personal communication). The orange fruit borer *Isotenes miserana* (Walker) (Tortricidae) is an uncommon pest in eastern Australia.

Fruit flies

In eastern Australia, the Queensland fruit fly, *Bactrocera tyoni* (Froggatt), is a significant pest of custard apples and soursops, infesting maturing fruit from March to May. Eggs are laid in batches of about a dozen particularly where the stem joins the fruit (Smith, 1991a). The thinner-skinned African Pride variety is more susceptible than the Pinks Mammoth. Infested fruit can be difficult to detect and Cuelure male attractant traps are used to help monitor fly populations, while protein baits are applied for control (Smith, 1991a). The lesser Queensland fruit fly, *B. neohumeralis* Hardy, also occurs in custard apples in eastern Australia. In South-East Asia, the fruit flies *B. dorsalis* (Hendel) and *B. papayae* Drew and Howard can infest custard apple and soursop. In Ecuador and Colombia, infestation of cherimoya by *Anastrepha* spp. and *Ceratitis capitata* is characterized by early ripening and drop of fruit (J.E. Peña, personal observation; Saldarriaga *et al.*, 1987).

Irapuá bee

In northern Brazil, the Irapuá bee, *Trigona spinipes* F. (Hymenoptera: Apidae), damages leaves, shoots, flowers and young fruits (Bastos, 1985; Gallo *et al.*, 1988). Damage to soursop fruits and flowers is characterized by rasping on the epidermis. This insect is brown, 5.0–7.5 mm long, and colonies build nests in trees.

Braga *et al.* (1998) recommend destruction of nests close to soursop groves and weekly monitoring of flowers and fruits. If damage is extreme, chemical control should be applied (Braga *et al.*, 1998). Control, however, must be carefully considered, due to the important pollination service the bees perform on various other flowers in the area.

Coreid spotting bugs

The banana spotting bug, *Amblypelta lutescens lutescens* (Distant), in north Queensland, and both *A. lutescens* and the fruitspotting bug, *A. nitida* Stal, in southern Queensland and northern New South Wales, are serious pests of custard apple.

Adults and nymphs pierce the fruit from near fruit set until harvest, causing round black spots (up to 1 mm in diameter) and damage about 1 cm deep (Smith, 1991a). Adult bugs are yellow-green and 15 mm long. The eggs are pale green, oval and about 2 mm long. The five nymphal instars are ant-like and pinkish with prominent antennae and button-like abdominal scent glands. The life cycle lasts about 6 weeks in summer and there are at least three overlapping generations each year.

Infestations are worst in trees adjacent to eucalypt forests where the bugs breed. There are no significant natural enemies and control is with selective pesticides applied when 2% or more of fruit show fresh damage (Smith, 1991a).

Thrips

In eastern Australia, the red-banded thrips, *Selenothrips rubrocinctus* (Giard), and in

Thailand (Morokote personal communication, 1999) and India, the orchard thrips, *Chaetonaphothrips orchidii* (Bagnall), occasionally cause grey scarring on the fruits.

Minor Pests

Several other species of insects and mites may cause minor losses as foliage, trunk, and branch feeders. In the Neotropics, the foliage feeders most frequently reported in *Annona* are scales, mealybugs, leafhoppers and whiteflies (Homoptera), lace bugs (Hemiptera), and some Lepidoptera and mites. Trunk and branch feeders consist mainly of coleopteran stem borers.

Foliage Pests

Soft scales

About 20 species of soft scales are reported from *Annona* in the Neotropics (Peña and Bennett, 1995). The scales *Parasaissetia nigra*, *Saissetia coffeae*, *S. oleae* and *Philephedra tuberculosa* Nakahara and Gill, cause damage by feeding on sap. Sudden increases in populations of these insects usually coincide with plant stress and absence of effective natural enemies.

In eastern Australia the most common scale pest of custard apples is *P. nigra*. It infests the twigs, leaves and fruit, producing copious honeydew and encouraging heavy sooty mould. Infestations are well attended by ants (*Pheidole* spp. and *Iridomyrmex* spp.) and tend to occur more on young trees up to 2 m high.

In Australia, the pteromalid parasitoid *Scutellista caerulea* (Fonscolombe) normally controls *P. nigra* but is easily disrupted by pesticides or by ants. Larvae of the pyralid moth, *Catoblemma dubbia* (Buttler), are predators.

Other much less common soft scales in eastern Australia are long soft scale, *Coccus longulus* (Douglas), soft brown scale, *C. hesperidum* L., and pink wax scale, *Ceroplastes rubens* Maskell. In Brazil, the scales *S. coffeae* and *Ceroplastes* sp., are considered occasional pests of *Annona* (Junqueira *et al.*, 1996).

Many of these scale species occur on custard apples in New Caledonia (Brun and Chazeau, 1980). Two lac scales also occur on custard apples in India – *Kerria communis* (Mahdibassan) and *Tachardia labata* (Green). *Icerya aegyptiaca* (Douglas), *H. biclavis* and *Ceroplastes floridensis* Comstock are also recorded from India.

Armoured scales

Armoured scales are also uncommon and include *Aspidiotus destructor* Signoret, *Hemiberlesia palmae* (Cockerell), *Pseudaulacaspis pentagona* (Targioni-Tozetti), *Chrysomphalum aonidum* (L.), *C. dictyospermi* (Morgan), *Ischnaspis longirostris* (Signoret), *Howardia biclavis* (Comstock), *Lindingaspis* sp., *Aonidiella orientalis* (Newstead) and *Abgrallaspis cyanophyllis* (Signoret).

Mealybugs

Some species of mealybugs can cause significant crop losses, although they seldom approach major pest status. Phillips *et al.* (1987) regard *Pseudococcus aonidum* (Cockerell) as a minor sporadic pest that, in areas of heavy infestation, may affect the fruit, flowers and leaf growth. Peña and Bennett (1995), in a review of pests of annonas in the Neotropics, list *Dysmicoccus brevipes* (Cockerell), *Ferrisia virgata* (Cockerell), *Nipaecoccus nipae* (Maskell), *Planococcus citri* (Risso), *Pseudococcus longispinus* (Targioni), *P. maritimus* (Ehrhorn) and *Pseudotectococcus anonae* Hempel. Factors that may contribute to increases in mealybug populations include certain ants that reduce effectiveness of natural enemies and transport mealybugs from one place to another. Phillips *et al.* (1987) reported that the Argentine ant, *Iridomyrmex humilis* displaces other ant species in cherimoya orchards in California and interferes with biological control agents associated with the long-tailed mealybug *Pseudococcus aonidum*. Most parasites of these mealybugs are hymenopterans of the family Encyrtidae and the most effective predators are coccinellid beetles. The judicious use of several insecticides can result in reduction of *P. aonidum* (Phillips *et al.* 1987). However, the use of some pesticides (e.g. organophosphates) is limited because they are not currently registered for use in *Annona* crops.

In the Caribbean, a recent introduction of the pink mealybug, also known as the hibiscus or grape mealybug, *Maconellicoccus hirsutus* (Green) (previously reported as *Pseudococcus hibisci* Hall or *Phenacoccus hirsutus* Green), produced severe damage to soursop during the late 1990s. Its original distribution is in the Oriental, Australian, Palaearctic and Ethiopian regions (Mani, 1989). It has also been reported from Pakistan and some islands in the Pacific, including Hawaii and Guam (Beardsley, 1985; Peña, 1998). *Maconellicoccus hirsutus* is an extremely polyphagous species, utilizing at least 144 genera in 74 plant families. Some major hosts include mango, hibiscus, palms, coffee, grape, citrus and *Annona* species. Williams (1985) reported that it causes bunchy leaves on limes in Australia. Tropical fruit crops in the Caribbean that are affected include *Annona* species, carambola, litchi, mango, avocado, *Citrus* sp., bananas and papaya (Etienne *et al.*, 1998).

The development of this species occurs in 3–4 weeks and the females can produce approximately 500 eggs each (Ghose, 1972). Ghose (1972) demonstrated that reproduction in this species is strictly sexual. During late autumn and winter, the female seeks a sheltered position to lay her eggs, usually forming an aggregation of conspecific females. In the summer, females may not seek shelter to lay eggs. Eggs hatch in 3–8 days (Misra, 1920; Ghose, 1972) and the nymphal stage lasts 10–22 days (Mani, 1989).

Infestation symptoms appear first on the growing tips. Shoots become twisted, with shortened internodes, forming bunchy heads of small bushy leaves at the tips, assuming a multiheaded appearance with multiple damage. In heavy infestations, leaves and shoots become compact and crisp. Symptoms in mulberry are known as Tukra disease (Misra, 1920). On mango, infested flowers dry and drop, resulting in fewer, smaller, abnormally shaped fruit that may drop early. *Maconellicoccus hirsutus* is one of the mealybug species known to have toxic saliva that stunts and kills young shoots.

Several factors affect *M. hirsutus* infestations: weather, host, and perennial versus annual host growth. It prefers weak or young growth, and the apical portions of plants. The rate of development increases with increasing temperatures, but slows with rising relative humidity (Babu and Azam, 1987; Mani and Thontadarya, 1988).

Several insecticides have been tested against this mealybug. Beevi *et al.* (1992) reported that the use of neem oil resulted in reduction of egg hatching. However, most researchers agree that chemical and cultural control provide relief for only a short period of time or are ineffective (McComie, 1996).

An integrated pest management programme has been proposed for the Caribbean region, with classical biological control as the main component. One of the candidate natural enemies, the lady beetle *Cryptolaemus montrouzieri* Mulsant (Coleoptera: Coccinellidae), was introduced to Trinidad from India in 1996 (Gautam *et al.*, 1996). Results of this predator introduction are given by McComie (1997). At the time of lady beetle release, the pink mealybug density averaged 19 ovisacs and 14 adults per shoot but declined significantly within 12 weeks (McComie, 1996). Another important natural enemy of *M. hirsutus* is the parasitoid *Anagyrus kamali* Moursi (Hymenoptera: Encyrtidae), while Mani (1989) lists at least 30 species of predators and 16 hymenopterous parasitoids that can be considered candidates for biological control of this mealybug.

The citrus mealybug, *Planococcus citri*, is a serious pest in eastern Australia, infesting the fruits when they are half- to full-grown in December to January. Less common species in this area are *Ferrisia virgata*, *Planococcus pacificus* Cox, *Pseudococcus longispinus* (Targioni-Tozzetti), *Maconellicoccus hirsutus*, and *Pseudococcus lilacinus* (Cockerell).

The citrus mealybug is a cosmopolitan species infesting a wide range of hosts. Eggs (600) are laid in a loose cottony egg sac. There are three moults for female mealybugs and four for males, the whole life cycle taking about 6 weeks in summer. There are 4–6 generations year^{-1}. *P. citri* produces copious honeydew resulting in heavy sooty mould that disfigures the fruit. Low hanging fruit attract heavy ant activity (*Pheidole* spp. and *Iridomyrmex* spp.) that disrupts predators and parasitoids, exacerbating the problem. Tree skirting is practised in winter after each crop to lift branches off the ground and to improve the efficacy of ant sprays applied to the trunk in spring and summer.

The main natural enemies of *P. citri* in eastern Australia are the parasitoid, *Leptomastix dactylopii* Howard, the mealybug ladybird, *C. montrouzieri*, and the lacewing, *Oligochrysa lutea* (Walker) (Smith 1991b). *L. dactylopii* can achieve 50–80% parasitism while up to 40% of mealybug-infested fruit attract activity from *C. montrouzieri* and/or *O. lutea*. *L. dactylopii* is augmentatively released at about 10,000 ha^{-1} in early summer.

In the Philippines, mealybugs, mainly *F. virgata* and *P. lilacinus*, are common pests of custard apples (De Leon and German, 1917; Paguirigan, 1951; Galang, 1955; Vinas, 1972; Cantillang, 1976; Coronel, 1983). *P. lilacinus* (and *P. citri*) occur on a wide range of hosts throughout South-East Asia, and are recorded from custard apple in India, Malaysia, Thailand, Vietnam, Indonesia and Brunei (Khoo *et al.*, 1991; Waterhouse, 1993). They are commonly attended by ants such as *Oecophylla smaragdina* (F.) and *Dolichoderus thoracicus* (Smith). Coccinellids such as *Scymnus* spp., slugs, and parasitoids, such as *Anagyrus* spp, help keep populations in check.

Whiteflies

In Cuba, several species of whiteflies have been observed in soursop, sugar apple and bullock's heart. They include *Aleurocanthus trachoides* (Back), *Bemisia tabaci* (Grenadius), *Aleurodicus dispersus* Russell, and *Aleurothrixus floccosus* (Maskell) (Vazquez *et al.*, 1996). Whitefly populations tend to increase in poorly tended orchards as well as in trees planted too densely. The damage they cause is of two main types. One is caused by the adults while sucking sap from shoots, thus weakening them. The other is caused by the immature stages that suck sap from the underside of the leaves, favouring the development of sooty mould and hindering photosynthetic activity (Vazquez *et al.*, 1996).

The whitefly, *Diauropora decempunctata* (Quaintance and Baker), occurs on custard apples in India; *Aleurodicus destructor* Mackie occurs throughout South-East Asia (Waterhouse, 1993).

Lace bugs

Soursops are not generally troubled with tingids, but local outbreaks of the species *Antiteuchus tripterus* (F.) and *Corythucha gossypii* (F.) may occur. These insects are found on the undersides of leaves and on fruits. Destruction of leaf cells due to the sucking habit of these insects causes yellowish and silver or brownish necrotic areas (Gutierrez and Trochez, 1977). Gutierrez and Trochez (1977) noted that feeding by *A. tripterus* on fruits causes stunting and desiccation, with extensive fruit drop later.

Aphids

The cotton aphid, *Aphis gossypii* Glover, and the black citrus aphid, *Toxoptera aurantii* (Boyer de Fosncolombe), infest young leaf shoots.

Lepidopterous leafminers

Lepidopterous leafminers and other caterpillars ocurr in annonas, but only as minor or localized pests in some areas, and are commonly held in check by parasites. Arevalo (1982) studied damage caused by *Phyllocnistis* sp. (Gracillariidae), a pest of *A. cherimola* in Colombia and Ecuador. Arevalo (1982) reported that 85% of the leaves are attacked by this insect, and described its various life stages. Two common biological control agents of this pest are *Apanteles* sp. and *Horismenus* sp. (Arevalo, 1982).

Coleopteran leafminers

In Brazil, *Prinomerus anonicola* Bondar (Coleoptera: Curculionidae) deposits eggs on leaves of soursop, and the emerging larvae make irregular blotch mines on the upper side of the leaf. This pest is more common in plant nurseries, but can attack field-grown plants (Braga *et al.*, 1998).

Lepidopterous leaf feeders

In eastern Australia, the main leaf feeder is the larva of the pale-green triangle butterfly, *Graphium ecrypylus lycaon* (C. and R. Felder). The smooth, velvety, green, yellow and tan caterpillars eat large holes in the young leaves and occasionally feed on the skin of the fruit. They are heavily parasitized by tachinid flies and are rarely a problem except on young trees less than 2 m high.

In South-East Asia (India, Thailand, Indonesia) the caterpillars of *Meganotron rufescens* (Butler) (Sphingidae), *Papilio agamemnon* L. (Papilionidae), *Archips micaceanus* (Tortricidae), and *Attacus atlas* L. (Satumiidae) are minor pests (Koesriharti, 1991; Waterhouse, 1993). Larvae of two woodboring species, *Zuezera coffeae* Nietner (Cossidae) and *Squamurae disciplaga* (Swinhoe) (Metarbelidae), damage twigs (Vinas, 1972; Gabriel, 1975; Khoo *et al.*, 1991).

Mites

Polyphagotarsonemus latus Banks infestations are mostly observed in greenhouses, inducing smaller leaves to roll towards the lower surface. In the field, damage to small fruits has been observed as bronzing that covers the fruit and prevents their development (Ochoa *et al.*, 1994). *Brevipalpus phoenicis* affects up to 85% of the epidermal surface of unripe soursop fruit. The fruit becomes bronze coloured with darker epidermal cracking and lighter striations, resembling rust. On cherimoya, damage is similar to that of *Brevipalpus pseudostriatus* Ochoa and Salas (Ochoa *et al.*, 1994). Astudillo (1989) reported that *B. chilensis* attacking cherimoya in Chile seldom causes significant injury to the foliage. Ochoa *et al.* (1994) also reported that leaves infested by *Eriophyes annonae* Keifer (= *Aceria annonae* [Keifer]) show initial protuberances on the upper surface corresponding to a fine lint of trichomes on the lower surface. As the leaves develop and the colony increases in size the lint grows to surround the main leaf vein, a symptom called erineum. Young leaves attacked early can shrivel or fold inwards together and galls may be observed on the upper surface.

In eastern Australia, mites are minor pests of custard apple. The tetranychids *Tetranychus urticae* Koch, *T. ludeni* Zacher, and *T. neocaledonicus* (André) occur. In Thailand, *Oligonychus biharensis* (Hirst) and *Brevipalpus* sp. are minor pests (Morokote, 1999, personal communication).

Pests of the Trunk and Branches

Coleopteran stem borers

In eastern Australia, the elephant weevil, *Orthorhinus cylindrirostris* (F.), is a sporadic pest on custard apples. The adults cause fruit drop by chewing pieces of bark from the fruit stalk (Smith, 1991a). A number of beetles occasionally bore into the trunk and main limbs, e.g. *Xyleborus perforans* (Wollaston), *Leptopius setosus* Lea, and *Euthyrrhinus meditabundus* (F.). In the Philippines, the white grub, *Anomala* sp., attacks the roots, sometimes causing wilting (Vinas, 1972; Gabriel, 1975; Coronel, 1983).

In Brazil, the curculionid *Cratosomus bombinus bombinus* is a 22-mm-long weevil that is black to dark grey with transverse yellow stripes on its thorax and elytra (Junqueira *et al.*, 1996) (Plate 49). The female inserts its eggs in branches and the larvae bore through the vascular system, reducing plant growth and vigour (Caloba and Silva, 1995; Junqueira *et al.* 1996). The characteristic symptom of infested branches is the discharge of a dark exudate mainly where the branches fork. This injury serves as a port of entry for the fungi *Lasiodiplodia theobromae* and *Phomosis* sp. (Caloba and Silva, 1995; Junqueira *et al.*, 1996).

Egg incubation lasts approximately 16 to 21 days; the larval stage lasts approximately 100 days. The larva pupates inside the stem and the adult emerges after a 50-day period (Junqueira *et al.*, 1996).

CONTROL In Brazil, it is recommended to cut and burn infested branches, followed by painting the cut with a salve made with lime, copper sulphate, sulphur, Diazinon, sodium chloride, soybean oil and water. Another method is injection of insecticides into the tree or plugging exit holes with beeswax or soap (Junqueira *et al.*, 1996). Planting *Cordia verbenacea* as an adult trap crop is also recommended (Nascimento *et al.*, 1986). An unidentified tachinid has been collected from pupae (Dominguez, 1980).

In Brazil, another curculionid, *Heilipus velamen*, is also considered a serious pest of trunks and branches (Costa Lima, 1956; Pereira *et al.*, 1996). The symptoms of attack from this weevil are similar to those observed when *Cratosomus bombinus bombinus* attacks annona trees; however, the galleries made by *Heilipus* affect only the tree cortex and bark (Plate 50). Junqueira *et al.* (1996) reported that adults and larvae may be infected by the fungus *Metarhizium* sp. while Goncalves (1973) has observed *Agonocryptus* sp. (Ichneumonidae) parasitizing the larvae.

Present Status of Insect Pest Management in the Neotropics

Annona culture in the Neotropics has been experiencing significant changes over the past 20 years, which also bears on insect management. There have been large increases in *Annona* hectarage in the Caribbean region, Central America, South America, and California. In consequence, some advancement has occurred in understanding the pollinator and pest communities in these crops. However, before satisfactory pollinator and pest management programmes are established, more information must be gathered on insect pollination mechanisms, as well as on basic biological information on the key pests, *Bephratelloides* spp. and *Cerconota*. In general, chemical interventions and very timid cultural approaches are the status of arthropod management so far achieved for *Annona* species in the Neotropics.

Bagging of fruit to prevent infestation, sanitation of borer-infested fruit in groves, and general spray schedules against the two major key pests is the norm. Monitoring techniques based on insect behaviour and biology have not been developed. However, surveys of insects attacking these crops provide at least some information on the type of damage and frequency of pests (Gutierrez and Trochez, 1977; Astudillo, 1989). Thus, from the information available at present, sanitation and/or chemical treatment does not suffice for the two key pests. Their basic biology and behaviour, and those of their natural enemies, must also be incorporated into a holistic pest management strategy.

Pest Management in Custard Apples in Eastern Australia

The key pests in Australia are *P. citri*, *A. l. lutescens*, *A. nitida*, *B. tryoni*, *P. nigra* and *C. punctiferalis*. Minor pests include *G. ecrypylus lycoen*, *O. cylindrirostris*, tetranychid mites, thrips, bark-boring weevils and other scales. Because of the size of trees and the difficulty of effectively spraying for mealybugs and scales, an integrated approach to pest control is vital.

Biological control is an important component of management. All pests except the *Amblypelta* spp. and *B. tryoni* have significant natural enemies. *P. citri* is parasitized by the encyrtid wasp, *L. dactylopii*, and is attacked by the mealybug ladybird, *C. montrouzieri*, and the lacewing, *O. lutea* (Walker). *L. dactylopii* provides 50–80% parasitism, and up to 40% of mealybug-infested fruit attract activity from the two predators. *P. nigra* is normally heavily parasitized by the pteromalid wasp, *S. caerulea*. Dusty conditions arising, for example, from roads should be avoided, as dust will disrupt parasitoids. Infestations often develop in young trees especially when there are many ants of the genera *Pheidole* and *Iridiomyrmex*. The fungus *V. lecanii* causes heavy mortality in *P. nigra* and other soft scales during wet weather.

The tachinid fly, *A. proclinata*, is an important parasitoid of *C. punctiferalis*, parasitizing up to 40% of the late instar larvae (Smith, 1991b). The assassin bug *P. plagiopennis* is a common predator of *C. punctiferalis*, *G. ecrypylus* and other caterpillars. Tachinid flies also heavily attack larvae of *G. ecrypylus*.

The larvae of the fruit flies *B. tryoni*, *B. neohumeralis*, *B. dorsalis* and *B. papayae* are parasitized by the braconids *Opius perkinsi* Fullaway, *Fopius deeralensis* (Fullaway), *F. arisanus* (Sonan) and *Diachasmimorpha tryoni* (Cameron), but these have limited impact.

The introduced parasitoid *L. dactylopii* is commercially mass reared and augmentatively released by growers during early summer at about 10,000 ha^{-1}. Adults of *C. montrouzieri* are also released by growers if the species is absent. Spraying for fruitspotting bugs must be coordinated with parasitoid releases. Sprays should be avoided for at least 7 days after a release, and the follow-up spray should not be applied after 3–4 weeks when the next generation of parasitoids are emerging. Alternatively, if a spray has just been applied, 7 days at least should elapse before parasitoid release. To avoid redundancy, *L. dactylopii* should not be released sooner than 1 week before or after a major emergence of conspecifics in the target area.

Ant control is important for pest management. At the end of the growing season, i.e. during winter, the tree skirts are pruned up about 1 m off the ground to minimize contact with the soil the following season. Fruits near the soil are more likely to develop diseases like diplopia rot and cylindrocladium spot and to develop heavy mealybug and ant infestations. The ants disrupt natural enemies of mealybugs and scales. Ant sprays are applied two or three times (in early and mid-season) to the lower trunk.

Attention must be given to orchard hygiene and alternative host plants that can increase pest pressure in the custard apple orchard. Mature fruit infested with yellow peach moth or with fruit fly should be collected and destroyed. Preferred fruit fly hosts like guava and loquat should not be planted in or near the orchard.

Although biocontrol options are available for mealybugs and scales, selective spraying is usually necessary for *Amblypelta* spp. and *B. tryoni*. When fruitspotting bug damage is detected (usually during mid-summer to mid-autumn), selective spraying with a 'soft' insecticide is necessary. Some growers attempt to control the bugs during December–January, before releasing *L. dactylopii*, but sometimes spraying is also necessary later in the season. Selective pesticides disrupt natural enemies of *P. citri* but still allow significant survival of the beneficial species. The fruit fly *B. tryoni* is controlled with weekly applications of protein bait (from the time that the fruit are three-quarters grown to harvest). A patch of foliage about 1 m^2 is sprayed on each tree. Selective insect growth regulators are available for fruit borer control if necessary.

Monitoring is important to minimize the number of sprays. *P. citri* is considered a serious problem if 20% or more of the fruit are infested with one or more large adult females. Ten randomly selected fruit are examined *in situ* on 20 random trees ha^{-1} at fortnightly intervals from January to April. Mealybugs are checked for parasitism and predation.

To determine the level of infestations of *P. nigra*, five 1-year-old lateral branches 1 cm thick are assessed on 20 random trees ha^{-1} once or twice per season. Scales are checked for parasitism.

Fruitspotting bugs are monitored at fortnightly intervals from late December to late March by sampling ten random fruit *in situ* on each of 20 random trees ha^{-1}. Action is necessary if 2% or more have fresh damage. Damaged fruit are removed during sampling.

Yellow peach moth is monitored as for fruitspotting bug, from February to April at fortnightly intervals, and fruits are checked for signs of frass. Action is necessary if 5% or more of fruit are attacked.

References

Ahmed, M.S. (1936a) Pollination and selection in *Anona squamosa* and *Anona Cherimolia*. *Ministry of Agriculture of Egypt, Technical Science Service Horticultural Section Bulletin* No. 57.

Ahmed, M.S. (1936b) The Anonas in Egypt. *Ministry of Agriculture of Egypt, Horticultural Section Booklet* No. 14.

Alata, J. (1973) Lista de insectos y otros animales dañinos a la agricultura del Perú. *Ministerio de Agricultura, Estacion Experimental Agricola de la Molina, Manual* No. 38, 47–48.

Andrade, B.M. (1996) Pollination and breeding system of *Xylopia brasiliensis* Sprengel (Annonaceae) in south-eastern Brazil. *Journal of Tropical Ecology* 12, 313–320.

Anjos, N. and Pereira, M.J.B. (1997) Quantidade de instares larvais em *Bephratelloides pomorum* (Fab.) (Hymenoptera: Eurytomidae). *Resumos 16 Congresso Brasileiro de Entomologia, Salvador, BA, Marco de 1997*, p. 63.

Anonymous (1989) *Prospeccion de Insectos y Acaros Fitofagos Asociados al Chirimoyo*. Universidad Catolica de Valparaiso, Facultad de Agronomia, Chile, 107 pp.

Arevalo de, I.S. (1982) Anotaciones sobre *Phyllocnistis* spp., minador del follaje del chirimoyo. *Revista Colombiana de Entomologia* 8, 15–22.

Astudillo, P. (1989) *Determinacion de Insectos y Acaros Asociados a Frutos de Chiromoyo Annona cherimola Mill. Determinacion y evaluacion de la naturaleza de desecho*. Universidad Catolica de Valparaiso, Facultad de Agronomia, Quillota, Chile, 107 pp.

Babu, T.R. and Azam, K.M. (1987) Studies on biology, host spectrum and seasonal fluctuation of the mealybug *Maconellicoccus hirsutus* (Green) on grapevine. *Indian Journal of Horticulture* 44, 284–288.

Bartelt, R., Dowd, P.F., Vetter, R.S., Shorey, H.H. and Baker, R.C. (1992) Responses of *Carpophilus hemipterus* (Coleoptera: Nitidulidae) and other sap beetles to the pheromone of *C. hemipterus* and other related coattractants in California field tests. *Environmental Entomology* 21, 1143–1153.

Bartelt, R., Vetter, R.S., Carlson, D.G. and Baker, T.C. (1994) Responses to aggregation pheromones for five *Carpophilus* species (Coleoptera: Nitidulidae) in a California date garden. *Environmental Entomology* 23, 1534–1543.

Bastos, J.M. (1985) *Principais plagas das culturas e seus controles*, 3rd edn. São Paulo, Nobel, Brazil, 329 pp.

Beardsley, J.W. (1985) *Maconellicoccus hirsutus* (Green). *Proceedings of the Hawaii Entomological Society* 25, 27.

Beevi, N., Janarthanan, R. And Natarajan, K. (1992) Efficacy of some insecticides against eggs of *Maconellicoccus hirsutus* Green on mulberry *Morus alba*. *Journal of Insect Science* 5, 114.

Bennett, F.D. and Alam, M. (1985) *An Annotated Check List of the Insects and Allied Terrestrial Arthropods of Barbados*. Caribbean Agricultural Research Development Institute, 81 pp.

Braga, R., Oliveira, M.A., Warumby, J. and Moura, J. (1998) Pragas da gravioleira. In: Braga, E., Cardoso, J. and Chagas, F. (eds) *Pragas de fruteiras tropicais de importancia agroindustrial*. Empresa Brasileira de Pesquisa Agropecuaria, pp. 131–141.

Brun, L.O. and Chazeau, J. (1980) *Catalogue des Ravageurs d'Interest Agricole de Nouvelle – Caledonie*. Office de la Recherche Scientifique et Technique Outre Mer, Centre de Noumee, Nouvelle – Caledonie, 120 pp.

Brunner, S.C. and Acuña, J. (1967) *Sobre la biologia de Bephrata cubensis Ashm., el insecto perforador de las frutas Annonaceas*. Academia Ciencias de Cuba. Instituto Agronomico Serie Agricultura 1, 14 pp.

Brussel, E.W. and Wiedijk, F. (1975) Prospects for cultivation of soursop in Surinam with special references to the soursop wasp *Bephrata maculicollis* Cam. *Suriname Landbouw* 21, 48–61.

Bustillo, A.E. and Peña, J.E. (1992) Biology and control of the *Annona* fruit borer *Cerconota anonella* (Lepidoptera: Oecophoridae). *Fruits* 47, 81–84.

Caloba, J.S. and Silva, N.M. da (1995) Insects asociados a graviola, *Annona muricata* L., e *Rollinia mucosa* (Jacq.) Bail no estado do Amazonas. *Anais Sociedade Entomologia Brasileira* 24, 179–182.

Calzavara, B. and Muller, C. (1987) Fruticultura tropical: a gravioleira *Annona muricata* L. Belem. *Empresa Brazileira de Pesquisa Agropecuaria, CNTU, Documento* 47, 36 pp.

Campbell, C.W. (1979) Effect of gibberellin treatment and hand pollination on fruit-set of atemoya (*Annona* hybrid). *Proceedings of the Tropical Region, American Society of Horticultural Sciences* 23, 122–124.

Cantillang, P.L. (1976) Philippine atis – the sweetest fruit. *Greenfields* 6, 26–27.

Carneiro, J. da and Bezerril, E. (1993) Controle das brocas dos frutos (*Cerconota anonella*) e da semente (*Bephratelloides maculicollis*) da graviola no Planlato da Ibiapaba. *Annais Sociedade Entomologia Brasil* 22, 155–160.

Castañeda V., A. and Pineda, S.G. (1997) Identification, description and behavior of cherimola's fruit borer in Coatepec Harina, Mexico. *Memorias Congreso Internacional de Annonaceaes, Chapingo, Mexico*, 39 pp.

Castañeda V., A., Pineda, S.G. and Rebollar, A. (1997) Searching for cherimola pollinator insects in the states of Mexico and Michoacan. I. Identification. *Memorias Fundacion Salvador Sanchez Colin*. Coatepec Harinas, pp. 197–204.

Cogez, X. and Lyannaz, J.P. (1996) Manual of pollination of sugar apple. *Tropical Fruit Newsletter* 19, 5–6.

Coronel, R.E. (1983) *Promising Fruits of the Philippines*. University of the Philippines, Laguna, Philippines, 508 pp.

Costa Lima, A. (1945) Insetos do Brasil, Tomo 5. Lepidopteros 1a parte. *Escola Nacional de Agronomia, Serie Didactica* No. 7. 5, 259–270.

Costa Lima, A. (1948) Entomofagos sul Americanos (Parasitos e predadores) de insetos nocivos a agricultura. *Boletin Sociedade Brasileira da Agronomia* 11, 1–32.

Costa Lima, A. (1956) Insetos do Brasil, Tomo 10: Coleopteros 4a parte. *Escola Nacional de Agronomia, Serie Didactica No. 12*. 10, 94–106.

Coto, D., Saunders, J.L., Vargas, C. and King, A. (1995) Plagas invertebradas de cultivos tropicales con enfasis en America Central. *Centro Agricola Investigacion y Enseñanza, Serie Tecnica* 12, 7–9.

D'Araujo e Silva, A.G., Goncalves, C.R., Galvao, D.M., Goncalves, A.J., Jones, J., Silva, M.D. and DeSimoni, L. (1968) Quarto catalogo dos insectos que vivem nas plantas do Brasil, seus *Parasitos e predadores, Parte II, Tomo I*, 622 p., Tomo II 253 pp.

De Leon, B. and German, J. (1917). Forms of some Philippine fruits. *Philippine Agriculturist and Forester* 5, 251–283.

Deroin, T. (1989) Evolution des modalites de la pollinisation au cours du developpement des axes aeriens chez une Annonacee savanicole soumise aux feux annuels: *Annona senegalensis* Pers. *Comptes Rendus de l'Académie des Sciences Paris serie III*, 308, 307–311.

Dominguez, G. (1980) Insectos perjudiciales de la guanabana (*Annona muricata* L.) en el estado Zulia, Venezuela. *Revista Lationoamericana de Ciencias Agrarias* 15, 43–55.

Dozier, H.L. (1932) Two important West Indian seed-infesting chalcid wasps. *Journal of Agriculture of Puerto Rico* 16, 103–112.

Duarte, F.E. (1947) Insetos holometabolicos. *Agronomia* 6, 178–211.

Ebeling, W. (1959) *Subtropical Fruit Pests*. University of California Press, 389 pp.

Escandon, Q., Lamus, E. and Calderon, J. (1982) Incidencia de *Bephrata maculicollis* Cameron perforador del fruto del guanabano *Annona muricata* L., metodos preliminares de control y rentabilidad de la practica en dos zonas del Valle, Colombia. Facultad de Ciencias Agropecuarias, Palmira, Colombia. 80 pp. (Thesis).

Estatilla, H. (1921) The atis moth borer. *Philippine Agriculture* 10, 169.

Etienne, J., Matile-Ferrero, D. and Lablanc, F. (1998) First record of the mealybug *Maconellicoccus hirsutus* (Green) from Guadeloupe: present state of this pest of crops in the French Caribbean (Hem.: Pseudococcidae). *Bulletin de la Societé Entomologique de France* 103, 173–174.

Evangelista-Lozano, S., Aldana, L. and Valdes, M. (1997) Estudio del barrenador de la semilla de la guanabana *Bephratelloides cubensis* Ashmead (Hymenoptera: Eurytomidae). *Memorias Congreso Internacional de Annonaceaes, Chapingo, Mexico*, 264 pp.

Farre, M., Perez, J.M., Hermoso, M. de los A. and Orgata R., J.M. (1997) Estudio sobre polinizacion de chirimoyo (*Annona cherimola* Mill) en Espana. *Memorias Congreso International de Annonaceas. Chapingo, Mexico*, pp. 43–55.

Fennah, R.G. (1937a) Lepidopterous pests of soursop in Trinidad. *Tropical Agriculture* 14, 175–178.

Fennah, R.G. (1937b) Lepidopterous pests of soursop in Trinidad. *Thecla ortygnus. Tropical Agriculture* 14, 244–247.

Gabriel, B.P. (1975) *Insects and Mites Injurious to Philippine Crop Plants*. Department of Entomology, College of Agriculture, University of Philippines, Los Baños, Laguna.

Galang, F.G. (1955) *Fruit and Nut Growing in the Philippines*. AIA Printing Press, Malabon, Rizal, Philippines.

Gallo, D., Nakano, O., Silvera-Neto, S., Carvalho, R., Batista, G., Berti, E., Parra, J., Zucchi, R., Alves, S. and Vendramin, J. (1988) *Manual de entomologia agricola*. Agronomica Ceres, São Paulo, 649 pp.

Galon, I., Gazit, S. and Podoler, H. (1982) Improvement of natural fruit set of *Annona* by increasing the population of pollinating insects. *Alon Halontea* 9, 611–614.

Gautam, R.D., De Chi, W. and Maraj, C. (1996) A successful journey of ladybirds from India to Trinidad. *Interamerican Institute Ciencias Agricolas, Caraphin News* 14, 1–2.

Gazit, S., Galon, I. and Podoler, H. (1982). The role of nitidulid beetles in natural pollination of *Annona* in Israel. *Journal of the American Society Horticultural Science* 107, 840–852.

George, A.P. and Nissen, R.J. (1991) *Annona cherimola* Miller, *Annona squamosa* L., *A. cherimola* × *A. squamosa*. In: Verheij, E.W.M. and Coronel, R.E. (eds) *Plant Resources of South East Asia. 2. Edible Fruits and Nuts*. Pudoc-DLO Wageningen, the Netherlands and Prozea Foundation, Bogor, Indonesia, 447 pp.

George, A.P., Nissen, R.J., Ironside, D.A. and Anderson, P. (1989) Effects of nitidulid beetles on pollination and fruit set of *Annona* spp. hybrids. *Scientia Horticulturae* 39, 289–299.

George, A.P., Nissen, R.J. and Campbell, J. A. (1992) Pollination and selection in *Annona* species (cherimoya, atemoya, and sugar apple). *Acta Horticulturae* 321, 178–185.

Ghose, S.K. (1972) Biology of the mealybug, *Maconellicoccus hirsutus* (Green) (Pseudococcidae: Hemiptera). *Indian Agriculturist* 16, 323–332.

Goncalves, C.R. (1973) Observacoes sobre hospedeiros de Himenopteros da Familia Ichneumonidae no Brasil. *Annais Sociedade Entomologica do Brasil* 2, 31–36.

Gottsberger, G. (1970) Beitrage zur Biologie von Annonaceen-Bluten. *Osterreichische Botanische Zeitschrift* 118, 237–279.

Gottsberger, G. (1985) Pollination and dispersal in the Annonaceae. *Annonaceae Newsletter* 1, 6–7.

Gottsberger, G. (1988) The reproductive biology of primitive angiosperms. *Taxon* 37, 630–643.

Gottsberger, G. (1991) As annonaceas do cerrado e sua polinizacao. *Revista Brasileira Biologia* 54, 391–402.

Grissell, E.E. and Foster, M.S. (1996) A new *Bephratelloides* (Hymenoptera: Eurytomidae) from seeds of *Cymbopetalum* (Annonaceae) in Mexico. *Proceedings of the Entomological Society of Washington* 98, 256–263.

Grissell, E.E. and Schauff, M.E. (1990). A synopsis of the seed-feeding genus *Bephratelloides* (Chalcidoidea: Eurytomidae). *Procedings of the Entomological Society of Washington* 92, 177–187.

Gutierrez, B.A. and Trochez, A. (1977) Estudio sobre las plagas de las anonaceas en el Valle del Cauca. *Revista Colombiana de Entomologia* 3, 39–47.

Hopping, M.E. (1983) Pollination and fruit set of cherimoya. *Australian Horticulture* 82, 11–14.

Idstein, H., Heeres, W. and Schneier, P. (1984) High resolution gas chromatography, mass spectrometry and gas chromatography Fourier transform analysis of *A. cherimola* volatiles. *Journal of Agricultural and Food Chemistry* 32, 383–389.

Junqueira, N., Cunha, N. and Oliveira, M. (1996) *Graviola para a exportacao: aspectos fitosanitarios.* Empresa Brasileira da Pesquisa Agropecuaria, Brasilia, 67 pp.

Kahn, T.L. (1997) *Pollination Theory and Practice. Cherimoya Handbook.* The California Cherimoya Association, pp. 2–220.

Kahn, T.L. and Arpaia, M.L. (1990) Methods of cherimoya hand pollination. *California Cherimoya Association Newsletter* 3, 1–3.

Kessler, P.A. (1987) Some interesting distribution patterns in the Annonaceae. *Annonaceae Newsletter (Utrecht, The Netherlands)* 6, 7 pp.

Khoo, K.C., Ooi, P.A.C. and Ho, C.T. (1991) *Crop Pests and their Management in Malaysia.* Tropical Press, Kuala Lumpur, Malaysia.

Koesriharti (1991) *Annona muricata* L. In: Verheij, E.W.M. and Coronel, R.E. (eds) *Plant Resources of South East Asia. 2. Edible Fruits and Nuts.* Pudoc-DLO Wageningen, Holland and Prozea Foundation, Bogor, Indonesia, 447pp.

Korytowski, G.C. and Peña, D. (1966) *Bephrata cubensis* Ashm. (Hymenoptera: Eurytomidae): Una nueva especie dañnina a las anonaceas en el Perú. *Revista Peruana de Entomologia* 9, 56–60.

Kumar, R., Hoda, M.N. and Singh, D.K. (1977) Studies on the floral biology of custard apple (*Annona squamosa* Linn.). *Indian Journal of Horticulture* 34, 252–256.

Lawrence, G. (1974) A preliminary list of fruit pests in Trinidad. *Journal of the Agricultural Society of Trinidad and Tobago* 24, 258–265.

Leal, W.S., Moura, J., Bento, J., Vilela, E. and Pereira, P. (1997) Electrophysiological and behavioral evidence for a sex pheromone in the wasp *Bephratelloides pomorum. Journal of Chemical Ecology* 23, 1281–1290.

Lima, I., Do Nascimento, R., Mendoca, C., Duarte, A. and Silva, L. (1997) Mating behavior of the *Annona* seed borer *Bephratelloides pomorum* (Hymenoptera: Eurytomidae) in semi-field conditions. *Memorias Congreso Internacional de Annonaceaes, Chapingo, Mexico,* 272 pp.

Lopez, E. and Rojas Dent, R. (1992) Arthropods associated with cherimoya (*Annona cherimola* Mill.) flowering in Quillota, Chile. *Acta Entomologica Chilena* 17, 101–106.

Lopez, E. and Uquillas, C. (1997) *Carpophilus hemipterus* (Coleoptera: Nitidulidae) as cherimoya (*Annona cherimola* Mill.) pollination agent under controlled conditions. *Acta Entomologica Chilena* 21, 89–99.

Mani, M. (1989) A review of the pink mealybug *Maconellicoccus hirsutus* (Green). *Insect Science and its Application* 10, 157–167.

Mani, M. and Thontadarya, T.S. (1988) Population dynamics of the mealybug *Maconellicoccus hirsutus* (Green) and its natural enemies in grapevine ecosystem. *Journal of Biological Control* 2, 93–97.

Marin, J.C. (1973) Preliminary list of pests of Annonaceae, sapodilla (*Achras zapota* L.) and guava (*Psidium guajava* L.) in Venezuela. *Agronomia Tropical* 23, 205–216.

Martinez, N.B. de and Godoy, F.J. (1983) Natural enemies of the soursop fruit borer,

Cerconota anonella Sepp. *Agronomia Tropical* 33, 155–161.

Martinez, S. and Cabrera, I. (1997) Incidencia de *Bephratelloides cubensis* Ashmead y *Apate monacha* (Fabricius) en seis selecciones de guanabana en Puerto Rico. *Memorias Congreso Internacional de Annonaceaes, Chapingo, Mexico*, p. 37.

McComie, L.D. (1996) Incidence and treatment of the hibiscus mealybug *Maconellicoccus hirsutus* (Green) one year after its discovery in Trinidad. *First Regional Symposium on the hibiscus mealybug* June 1996, St Georges, Grenada, pp. 4–6

Mcleod, A. and Pieris, N.M. (1981) Volatile flavor components of soursop *Annona muricata. Journal of Agricultural and Food Chemistry* 29, 488–490.

Medina-Gaud, S., Bennett, F., Segarra, A.E. and Pantoja, A. (1989) Notes on insect pests of soursop (guanabana), *Annona muricata* L., and their natural enemies in Puerto Rico. *Journal of Agriculture of the University of Puerto Rico* 73, 383–390.

Mendes, J. and Pereira, M.J.B. (1997) Danos de *Bephratelloides pomorum* (FAB) (Hymenoptera: Eurytomidae) em pomares de graviola (*Annona muricata* L.). *Resumos 16 Congresso Brasileiro Entomologico. Salvador-Bahia, Marco 1997*, p. 296.

Misra, C.S. (1920) Tukra disease of mulberry. *Proceedings Third Entomological Meeting, Pusa*, pp. 610–618.

Momose, J., Nagamitsu, T. and Inoue, T. (1998) Thrips cross-pollination of *Popowia pisocarpa* (Annonaceae) in a lowland dipterocarp forest in Sarawak. *Biotropica* 30, 444–448.

Nadel, H. and Peña. J.E. (1991a) Hosts of *Bephratelloides cubensis* (Hymenoptera: Eurytomidae) in Florida. *Florida Entomologist* 72, 207–211.

Nadel, H. and Peña, J.E. (1991b) Seasonal oviposition and emergence activity of *Bephratelloides cubensis* (Hymenoptera: Eurytomidae), a pest of *Annona* species in Florida. *Enviromental Entomology* 72, 208–211.

Nadel, H. and Peña, J.E. (1994) Identity, behavior, and efficacy of nitidulid beetles (Coleoptera: Nitidulidae) pollinating commercial *Annona* species in Florida. *Environmental Entomology* 23, 878–886.

Nagamitsu, T. and Inoue, T. (1997) Cockroach pollination and breeding system of *Uvaria elmeri* (Annonaceae) in a lowland mixed-dipterocarp forest in Sarawak. *American Journal of Botany* 84, 208–213.

Nagel, J., Peña, J.E. and Habbeck, D. (1989) Insect pollination of atemoya in Florida. *Florida Entomologist* 7, 207–211.

Nakasone, H.Y. and Paull, R.E. (1998) *Tropical Fruits*. CAB International, Wallingford, UK, 445 pp.

Nascimento, A., Mesquita, A. and Caldas, R. (1986) Flutuacao populacional e manejo da broca da laranjeira, *Cratosomus flavofasciatus* Guerin, 1844 (Coleoptera: Curculionidae) com Maria Preta *Cordia verbenaceae* (Borraginaceae). *Annals Soceidade Entomologia Brasileira* 15, 45–51.

Noonan, J.C. (1954) Review of investigations of the *Annona* species. *Natural Horticultural Magazine* 33, 219–225.

Nuñez, R. and de la Cruz, J. (1982) Reconocimiento y descripcion de las principales insectos observados en cultivares de guanabana (*Annona muricata* L.) en el departamento del Valle. *Acta Agronomica* 32, 45–51.

Ochoa, R., Aguilar, H. and Vargas, C. (1994) Phytophagous mites of Central America: an illustrated guide. *Centro Agronomico Investigacion y Ensenanza. Serie Technica* 6, 234 pp.

Oppenheimer, C. (1947) The acclimatisation of new tropical and subtropical fruit trees in Palestine. *Bullettin Agricultural Research Station, Rehovot* 44, 94–103.

Paguirigan, D.B. (1951) Outline of Philippine crops: soursop and sugar-apple. *Agricultural Industrial Life* 13, 34–35.

Peña, J.E. (1988) *Toxotrypana* y *Bephratelloides* spp. ejemplos de adaptacion de monofagos a plantas hospederas. *Memorias XV Congreso Socolen, Julio 1987, Bogota, Colombia*, pp. 42–46.

Peña, J.E. (1998) Current and potential arthropod pests threatening tropical fruit crops in Florida. *Proceedings of the Florida State Horticultural Society* 111, 327–329.

Peña, J.E. and Bennett, F.D. (1995) Arthropods associated with *Annona* spp. in the neotropics. *Florida Entomologist* 78, 329–349.

Peña, J.E and Nagel, J. (1988) Effectiveness of pesticides against two tropical fruit pests. *Procedings of the Florida State Horticultural Society* 101, 249–251.

Peña, J.E., Glenn, H. and Baranowski, R.M. (1984) Important insect pests of *Annona* sp. in Florida. *Proceedings of the Florida State Horticultural Society* 97, 337–340.

Peña, J.E., Nadel, H. and Torres, V. (1990) Pests of *Annona* species. *Tropical Fruit World* 1, 121–122.

Peña, J.E., Castiñeiras, A., Bartelt, R. and Duncan, R. (1999) Effect of pheromone bait stations for sap beetles (Coleoptera: Nitidulidae) on *Annona* fruit set. *Florida Entomologist* 82, 475–480.

Pereira, M.J.B. (1996) Biologia de *Bephratelloides pomorum* (Fab.) (Hymenoptera: Eurytomidae), broca da semente de graviola. Tese de Mestrado em Entomologia, Universidade federal de Vicosa, Minas Gerais, 71 pp.

Pereira, M.J.B., Bastos-Neto, E. and Cavalcanti, M. (1991a) Parasitoides de larvas de *Cerconota Anonella* (Sepp. 1830) (Lep: Stenomatidae). *Congresso Brasileiro de Iniciacao Cientifica em Ciencias Agrarias, 10: Lavras. Resumos. Minas Gerais:FEAB, ESAL*, p. 40.

Pereira, M.J.B., Bastos-Neto, E.M., Cavalcanti, Lima, I. and Santana, A. (1991b) Ciclo biologico de *Cerconota anonella* (Sepp. 1830) (Lep: Stenomatidae) em dieta artificial. *Congresso Brasileiro de Iniciacao Cientifica em Ciencias Agrarias, 10, 1991, Lavras. Resumos. Mians Gerais: FEAB, ESAL*, p. 44.

Pereira, M.J.B., Anjos, N. and Picanco, M. (1996) Ocorrencia de *Heilipus velamen* (Coleoptera: Curculionidae) em Minas Gerais. *Congresso Brasileiro de Zoologia, 21, Porto Alegre. Resumos, Rio Grande do Sul: SBZ,UFRGS*, p. 82.

Pereira, M.J.B., Anjos, N. and Picanco, M. (1997a) Ciclo biologico de *Bephratelloides pomorum* (Fab.) (Hymenoptera: Eurytomidae) da broca da semente. *16 Congresso Brasileiro de Entomologia, Salvador, BA, Marco de 1997*, p. 50.

Pereira, M.J.B., Sant'ana, A. and Anjos, N. (1997b). Comportamento de oviposicao de *Bephratelloides pomorum* (Fab) (Hymenoptera: Eurytomidae), broca da semente de graviola. *16 Congresso Brasileiro de Entomologia. Salvador, BA, Marco de 1997*, p. 50.

Pereira, M.J.B., Anjos, N. and Picanco, M. (1997c) Ciclo biologico del barrenador de semillas de guanabana *Bephratelloides pomorum* (Fab.) (Hymenoptera: Eurytomidae). *Agronomia Tropical* 47, 4.

Pereira, M.J.B., Eiras, A. and Anjos, N. (1998) Comportamiento de cortejo y copula de la broca de semilla de guanabana *Bephratelloides pomorum* (Fab.) (Hymenoptera: Eurytomidae). *Revista Biologia Tropical* 46, 105–108.

Philips, P., Bekey, R. and Goodall, G. (1987) Argentine ant management in cherimoyas. *California Agriculture* (March–April 1987) 41, 8–9.

Podoler, H., Galon, I. and Gazit, S. (1984) The role of nitidulid beetles in natural pollination of Annonas in Israel: attraction of nitudulid beetles to *Annona* (atemoya) flowers in Israel. *Oecologia Applicata* 5, 369–381.

Podoler, H., Galon, I. and Gazit, S. (1985) The effect of atemoya flowers on their pollinators: nitidulid beetles. *Acta Oecologica* 6, 251–258.

Ramnanan, N. (1996) Recommendations for the sustainable cultivation of soursop (Part II). *Tropical Fruit Newsletter* 18, 12–14.

Reiss, A. (1971) Pollination and fruit set of *Annona cherimola* and *Annona squamosa*. MSc thesis, The Hebrew University of Jerusalem, Israel.

Reyes, A.J. (1967) Algunas recomendaciones para el control del perforador de las semillas de Annonaceaes (*Bephrata* sp. Orden Hymenoptera). *Revista Agricultura Tropical, Bogota, Colombia* 25, 325–326.

Rokba, A.M., El-Waked, A.T. and Ali, N.H. (1977) Studies and evaluation of some annona varieties. *Agricultural Research Review* 55, 13–21.

Ruiz, R.V. (1991) Manejo de problemas entomologicos en huertos de guanabana. *XXII Foro entomologico plagas de frutales en Colombia y alternatives de manejo. Casos: Guanabana, curuba, citricos*. Universidad Nacional de Colombia, Palmira, pp. 59–92.

Saavedra, E. (1977) Influence of pollen grain stage at the time of hand pollination as a factor on fruit set of cherimoya. *HortScience* 12, 117–118.

Safford, W.E. (1914) Classification of the genus *Annona* with descriptions of new and imperfectly known species. *United States National Herbarium, Smithsonian Institution* 18, 68 pp.

Saldarriaga, A., Polania, I., Cardenas, R., Posada, L. and Garcia, F. (1987) *Guia para el Control de Plagas. Instituto Colombiano Agropecuario. Manual de Asistencia Tecnica No. 1*, 4th edn., 401 pp.

Sarasola, L. (1960) Cherimoya culture in Spain. *California Avocado Society Yearbook* 44, 47–53.

Schroeder, C.A. (1943) Hand pollination studies on the cherimoya. *Proceedings of the American Society Horticultural Science* 43, 39–41.

Schroeder, C.A. (1956) Cherimoyas, sapotes and guavas in California. *California Avocado Society Yearbook* 40, 49–56.

Schroeder, C.A. (1971) Pollination of cherimoya. *California Avocado Society Yearbook* 54, 119–122.

Smith, D. (1991a) Insect pests. In: Sanewski, G. (ed.) *Custard Apple Cultivation and Crop Protection*. Queensland Department of Primary Industries, Information Series Q190031, 103 pp.

Smith, D. (1991b) The use of *Leptomastix dactylopii* Howard (Hymenoptera: Encyrtidae) to control *Planococcus citri* (Risso) (Hemiptera: Pseudococcidae) in custard apples in Queensland. *General and Applied Entomology* 23, 3–8.

Smith, D., Beattie, G.A.C and Broadley, R. (eds) (1997) *Citrus pests and their natural enemies: integrated pest management in Australia*. Queensland Department of Primary Industries, Horticultural Research and Development Corporation Publication, Information Series QI 97030, 272 pp.

Thakur, D.R. and Singh, R.N. (1965) Studies on pollen morphology, pollination and fruit set in some annonas. *Indian Journal of Horticulture* 22, 10–18.

Thomson, P. (1970) The cherimoya in California. *California Rare Fruit Growers Yearbook* 2, 20–34.

Thomson, P. (1974) Cherimoya and pawpaw pollination. *Pomona* 7, 3–6, 24.

Vazquez, L., Jimenez, R. and Iglesia, M. (1996) Whiteflies (Homoptera: Aleyrodidae) inhabiting the principal fruit trees grown in Cuba. *Tropical Fruits Newsletter* 20, 3–4.

Venturi, F. (1966) Insetti e arachnidi delle pieante comuni del Venezuela signalati nel periodo 1938- 1963. Relazione e monografie agrarie subtropicali e tropicali. *Firenze Instuti Agronomico per l'Otremare* 86, 13–16.

Vidal, H.L. (1997) Identificacion y frequencia de insectos polinizadores en guanabana *Annona muricata. Memorias Congreso Internacional de Annonaceaes, Chapingo, Mexico*, p. 36.

Villalobos, E. (1987) Use of endosulfan and polyethylene bags to control *Bephrata* sp. Ashmead, the *Annona* seed borer in (*Annona cherimola* Mill). MSc thesis, National University, Costa Rica.

Villalta, R. (1988) Estudio de la biologia floral e identificacion de agentes polinizadores de la guanabana (*Annona muricata* L.) en la zona atlantica de Costa Rica. Thesis, National University, Costa Rica, 61 pp.

Vinas, R.C. (1972) Atis (*Annona squamosa* L., Annonaceae). In: *Cultural Directions for Philippine Agricultural Crops*, Vol. I, *Fruits*. Public Affairs Office Press, Bureau of Plant Industries, Manila, pp. 31–36.

Vithanage, H.I.M.V. (1984) Pollen–stigma interactions: development and cytochemistry of stigma papillae and their secretions in *Annona squamosa* L. (Annonaceae). *Annals of Botany* 54, 153–167.

Waterhouse, D.F. (1993) *The Major Arthropod Pests and Weeds of Agriculture in Southmost Asia: Distribution, Importance and Origin*. ACIAR Monograph Series, Canberra, ACT, 141 pp.

Webber, A. (1981) Algunos aspectos da biologia floral de *Annona sericea* Dun. (Annonaceae). *Acta Amazonica* 11, 61–65.

Webber, A.C. and Gottsberger, G. (1997) Floral biology and pollination of *Bocageopsis multiflora* and *Oxandra euneura* in Central Amazonia. *Annonaceae Newsletter* 11, 61–66.

Wester, P.J. (1910) Pollination experiments with anonas. *Bulletin of the Torrey Botanical Club* 37, 529–539.

Williams, D.J. (1985) Australian mealybugs. *London British Museum Natural History* 13, 194–196.

Zarate, R.D. (1987) *Principales problemas fitosanitarios del guanabano Annona muricata y medidas preliminares de control*. Universidad Nacional de Colombia, FCA-Palmira, Colombia. 28 pp.

Zenner, I. and Saldarriaga, A. (1969) Fruit borer of annona and soursop *Cerconota* (*Stenoma*) *anonella* Sepp. (Lepidoptera: Stenomiidae). *Agricultura Tropical Bogota* 25, 325–326.

8 Pests and Pollinators of Avocado

Manes Wysoki,[1] Michael A. van den Berg,[2] Gad Ish-Am,[3] Shmuel Gazit,[3] Jorge E. Peña[4] and Geoff K. Waite[5]

[1]Department of Entomology, ARO, Institute of Plant Protection, The Volcani Center, PO Box 6, Bet Dagan 50250, Israel; [2]Institute for Tropical and Subtropical Crops, Private Bag X11208, Nelspruit 1200, South Africa; [3]Department of Horticulture, Faculty of Agricultural, Food and Environmental Quality Sciences, The Hebrew University of Jerusalem, PO Box 12, Rehovot 76100, Israel; [4]Queensland Horticulture Institute, Tropical Research and Education Center, University of Florida, 18905 SW 280 Street, Homestead, FL 33031, USA; [5]Maroochy Research Station, PO Box 5083, SCMC, Nambour, Queensland 4560, Australia

The avocado, *Persea americana* Mill. (Lauraceae), is of Central American origin and well known to the native inhabitants of Mexico, Central America and northern South America (Schieber and Zentmyer, 1987; Zentmyer *et al.*, 1987). It was brought to the West Indies and first reported in Jamaica in 1696. It was taken to the Philippines at the end of the 16th century; to the Dutch East Indies by 1750 and to Mauritius in 1780; it was first brought to Singapore between 1830 and 1840; it reached India in 1892 but never became very popular; it was planted in Hawaii in 1825, and introduced from Mexico into Florida in 1833 and California in 1871 (Morton, 1987). It is grown in many tropical and subtropical countries in areas suitable for cultivation, e.g. Australia, South Africa, Israel and the USA (Florida and California). Today, avocado is grown commercially not only in North America and throughout tropical America and the larger islands of the Caribbean, but also in Polynesia, the Philippines, Australia, New Zealand, Madagascar, Mauritius, Madeira, the Canary Islands, Algeria, tropical Africa, South Africa, southern Spain, southern France, Sicily, Crete, Israel and Egypt.

According to the FAO, the five top avocado-producing countries are: Mexico, Indonesia, the USA, the Dominican Republic and Brazil. World avocado production reached 2 million t in 1999 (FAO, 1999). According to Swirski *et al.* (1997), the main producing countries in Africa and its surrounding islands are South Africa, Zaire, Cameroon, Kenya, Egypt and the Canary Islands.

The objective of this chapter is to review information on pests and pollinators of avocado in its growing areas throughout the world. Insect pests and mites of avocado have also recently been reviewed by Waite and Martinez (2002).

©CAB *International* 2002. *Tropical Fruit Pests and Pollinators* (eds J.E. Peña, J.L. Sharp and M. Wysoki)

PHYTOPHAGOUS ARTHROPODS: PHYLUM ARTHROPODA

Class Arachnida

ORDER ACARI

Tetranychidae

The most important tetranychids affecting avocado are *Oligonychus punicae* (Hirst), *O. perseae* (Tuttle, Baker and Abbatiello), *O. coffeae* (Nietner), *O. yothersi* (McGregor), *Paratetranychus* sp., *Panonychus citri* (McGregor) and *Eotetranychus sexmaculatus* (Riley).

Tea red mite, Oligonychus coffeae *(Nietner)*

In South Africa, the tea red mite was described by Nietner (1861, unpublished) from Sri Lanka (De Villiers, 1998). It is known to attack avocado, cashew nuts, citrus, coffee, guava, mango, tea, cotton, grape, mulberry, protea, rubber, strawberry, and about 30 other plant species (Meyer, 1987). *O. coffeae* prefers the dorsal surface of avocado leaves but will inhabit both sides during severe infestations. In general, these mites favour the older, firmer leaves (Meyer, 1987). Injury to plants is first seen as a yellowish spotting along the midrib and veins of leaves and occasionally on the petioles. As the mites continue to feed, these patches darken, until the entire leaf becomes deeply bronzed and necrotic, often falling from the plant; thus growth is retarded (Meyer, 1987; De Villiers, 1998).

A heavy infestation of tea red mite occurred on young avocado trees in the Nelspruit area after 120 ha of Pinkerton avocados had been sprayed with wettable sulphur (De Villiers, 1998). In Australia, the tea red spider mite is a common pest of avocados in Queensland and northern NSW, usually appearing during the autumn. This may be a result of the numerous sprayings of endosulfan, applied for the control of fruitspotting bugs.

Das (1955) reported that the mite's life cycle on tea in India lasts 9–12 days in summer and about 30 days in winter. Under these conditions, 22 generations can occur annually. These mites are spread by wind, animals, implements and labourers (Meyer, 1987). Many natural enemies, mostly predacious mites, are associated with *O. coffeae*. In South Africa, according to Meyer (2001), the following predacious mites were found: *Anystis baccarum* (L.) (Anystidae), *Agistemus africanus* (Meyer and Ryke), *A. tranatalensis* Meyer (Stigmaeidae), *Eupalopsellus brevipilus* (Meyer and Ryke) (Eupalopsellidae), *Tydeus grabouwi* Meyer and Ryke (Tydeidae), *Amblyseius munsteriensis* Van der Merwe, *A. herbicolus* (Chant), *A. multidentatus* (Swirski and Shechter), *A. transvaalensis* (Van der Merwe and Ryke), *A. tutsi* Pritchard and Baker, and *Typholdromus buccalis* Van der Merwe (Phytoseiidae). In Australia, the main predators of the mite in coastal areas are *Stethorus* spp., whereas in drier inland areas, phytoseiid mites prey on them (Waite and Pinese, 1991). Coccinellids, especially *Stethorus madecassus* Chazeau in the Malagasy Republic, were found feeding on eggs and other stages of *O. coffeae*. The larvae of *Chrysopa* spp. also attack the active stages of this mite (Blommers and Gutierrez, 1975).

In South Africa, chemical control of the tea red mite is for the most part unnecessary since its natural enemies keep its numbers low. In Australia, when necessary, chemical control is practised with sprays of fenbutatin oxide. Avocados can withstand approx. 20% leaf bronzing before action needs to be taken (Waite and Pinese, 1991).

Avocado brown mite,
Oligonychus punicae *(Hirst)*

In Mexico, *Oligonychus punicae* (Hirst) and *Oligonychus perseae* (Tuttle, Baker and Abbatiello) appear to be the most important mite species. *O. punicae* are found on the leaf's upper side, causing leaf bronzing and a reduction in photosynthetic activity. In California, the avocado brown mite, *O. punicae*, and the six-spotted mite, *Eotetranychus sexmaculatus* (Riley), are pests of avocados (McMurtry, 1985a; Bailey and Olsen, 1990a; Aponte and McMurtry, 1997a,b).

In Mexico, different cultivars show different responses to *O. punicae*. One generation can be obtained in 15.4 days when the average temperature is 22°C (Hernandez *et al.*, 2000). *O. punicae* densities are reduced under low temperature and high humidity. The same authors believe that *O. punicae* originated in the south of Mexico and north of Guatemala. *O. punicae* infests 71% of the avocado orchards in the area of Uruapan, Michoacan. Coria-Avalos (1993) recommends applying sulphur and propargite three or four times during the dry season, between December and May. Reyes and Salgado (1994) demonstrated that avocado cultivars 175PLS, 54PLS, 131PLS, 120PLS, 18PLS, 137PLS and 30PLS show some tolerance to *O. punicae*, while Colin and Rubi (1992) indicated that cultivars Colinmex and 148PLS are moderately susceptible. The cultivars Rincon and Fuerte have been considered tolerant to the mite species by Ortega and Ewart (1972; cited in Gallegos, 1983).

The phytoseiid mites, *Amblyseius limonicus* Garman and McGregor and *Typhlodromus floridanus* Muma are considered to have good predatory potential against *O. punicae* (McMurtry and Johnson, 1966; McMurtry, 1985a,b).

Persea mite, Oligonychus perseae *Tuttle, Baker and Abbatiello*

The persea mite, *Oligonychus perseae* Tuttle, Baker and Abbatiello (previously misidentified as *Oligonychus peruvianus* (McGregor)), is found in Mexico and Costa Rica. It was found in California in 1990 damaging the avocado variety Hass, and to a lesser extent the varieties Gwen and Reed, whereas no damage was inflicted on varieties Fuerte, Zutano and Bacon. *O. perseae* feed in colonies under protective webbing along veins and midribs on the undersides of leaves, producing necrotic, purplish, irregular spots (Hoddle, 1998). Recently *O. perseae* was found in avocado orchards of western Galilee, Israel (E. Swirski, personal observation) The mites are covered by a whitish cobweb, where they spend most of their lives. Their damage is characterized by almost geometrical necrosis of areas close to the secondary and mid-veins. Salinas (1992) determined the life cycle of this species to be 20.95 days under laboratory conditions. Ramirez *et al.* (1993) calculated life and fecundity tables for *O. perseae* at 10 and 14°C. Development from the egg to adult emergence required 227 degree-days. The reproductive period ranged from 5 to 14 days at all temperatures tested. The estimated r_m values showed that this species can only grow at 25°C (Ramirez *et al.*, 1993). Martinez (1989) reported that *O. perseae* had developed resistance to organophosphates in the states of Morelos, Mexico and Michoacan.

According to Hernandez *et al.* (2000), no known specific predators of *O. perseae* have been reported to date. Hernandez *et al.* (2000) suggested the removal of fallen foliage as a cultural control method. However, this tactic does not appear to be very practical. Use of hot-pepper extracts did not afford any type of mite control in Mexico (Reyes *et al.*, 1995). Andrade (1988) found the best chemical control with organophosphates and the worst with sulphur.

The following predacious mites were released and established in southern California against the persea mite: *Galendromus annectens* (DeLeon), *G. helveolus* (Chant), *G. pilosus* (Chant), *G. porresi* (McMurtry), *Iphiseius degenerans* (Berlese), *Iphiseiodes zuluagai* Denmark and Muma, *Metaseiulus occidentalis* (Nesbitt), *Neoseiulus californicus* (McGregor) and *Typhlodromus rickeri* Chant. Additionally, the native *Euseius hibisci* Chant was released. *Stethorus picipes* Casey (Coleoptera: Coccinellidae) is important in controlling the avocado brown mite but does not suppress outbreaks of persea mite. In addition, the six-spotted thrips (Thysanoptera: Thripidae), a specialized predator of the spider mite, is commonly found feeding in persea mite nests (Hoddle, 1998).

Avocado red mite, Oligonychus yothersi *(McGregor)*

In Florida, the avocado red mite is a common pest of avocados. Feeding is initially confined to the upper surface of avocado leaves; it is found first along the midrib, then along secondary leaf veins. The areas along the veins become reddish brown. During heavy infestations, leaves can be covered with mites' cast skins. Damage to the leaf area is regularly observed from October through February, causing up to a 30% reduction in leaf photosynthetic activity. Leaves affected by this mite regularly drop earlier (45–60 days after infestation) than their uninfested counterparts. This mite is an occasional pest in some orchards and is seldom observed in others. Periodic inspections are recommended during December, January and February. Control measures may be started when the population reaches six or more mites per leaf (Peña and Johnson, 1999). The duration of the life stages varies from 14 to 15 days. Females are capable of laying 40–50 eggs during their life span (Peña and Johnson, 1999). In Florida, few miticides are registered for use on avocados when fruit is present. Sulphur or oil emulsion sprays are recommended (Peña and Johnson, 1999).

Six-spotted mite, Eotetranychus sexmaculatus *(Riley)*

The six-spotted mite was introduced into Queensland in 1986, apparently on budwood that had been illegally imported from California. When its presence became known, an eradication campaign was mounted and all known infested plantings were treated with miticide. The infestation did not spread and it has not been a problem since. The phytoseiid *Phytoseiulus persimilis* Athias-Henriot was found among mite colonies on mature avocado trees. It is suspected that this predator, along with *Stethorus* spp. and other phytoseiids, has suppressed the populations to undetectable levels (Waite and Pinese, 1991). *E. sexmaculatus* infests avocados in New Zealand, but chemical control is rarely necessary (D. Steven, 1997, personal communication).

Paratetranychus *spp.,* Panonychus citri *(McGregor),* Tetranychus cinnabarinus *Boisduval,* Tetranychus urticae *Koch*

In Peru, *Paratetranychus* spp. and *Panonychus citri* (McGregor) mites affect avocado when it is grown together with citrus trees (Wolfe *et al.*, 1969). In Israel, red spider mites are rarely found in avocado orchards; however, in the early 1980s, the carmine mite *T. cinnabarinus* and the two-spotted mite *T. urticae* were found in some avocado plots adjacent to cotton fields, which were treated aerially with synthetic pyrethroids. Occasionally, the oriental spider mite *Anychus orientalis* Klein causes considerable damage to leaves of variety Pinkerton, especially in drier areas (Swirski *et al.*, 1998).

Eriophyidae

In Brazil, the mite *Tegolophus perseaeflorae* (Keifer, 1969) injures flowers and contributes to flower drop. In Florida, *T. perseaflorae* is observed in developing buds. Peña and Denmark (1996) related the presence of this mite to excessive flower drop and fruit deformation. The mites cause necrotic spots, and subcircular and irregular openings on apical leaves. Feeding by this mite may cause fruit deformation and discoloration. The adult avocado bud mite has a yellowish appearance. Its life cycle has not been determined. Avocado bud mite populations begin to increase from March to May (Peña and Johnson, 1999). Medina *et al.* (1978) recommend chemical applications to reduce their density.

The eriophyid *Calepitrimerus muesebecki* Keifer causes bronzing and virus-like chlorotic damage to avocado foliage in Guatemala; the mite may cause damage while remaining hidden inside the vegetative buds (Velasquez and Santizo, 1992).

Tarsonemidae

The broad mite, *Polyphagotarsonemus latus* (Banks) (Tarsonemidae), sometimes attacks avocado seedlings in glasshouses, causing rolling and browning of the young leaves, severe harm to the foliage of apical buds, development of shoots by lateral buds, as well as dwarfing of the seedlings (Waite and Pinese, 1991; Swirski *et al.*, 1998).

Class Insecta

ORDER ISOPTERA

Termitidae

Africa has a large number of termite species, many of which are present in avocado-growing areas. They live in nests and feed on the cellulose in wood, grass, humus and other plant matter. In avocado orchards, termites destroy mainly the roots and lower stems of young trees, but they may also cause ringbark and kill large trees. Termites are therefore more destructive during the first year after planting, as the trees become less susceptible once they are well established (Webb, 1974).

In Africa, termite mounds are often large and occupy considerable surface area. *Macrotermes* spp. mounds often exceed 4 m in height. In some areas, *Microhodotermes* spp. build mounds which cover up to 25% of the ground surface.

The largest contribution to our knowledge of the taxonomy, biology, distribution and ecology of termites in Africa was made by Coaton and Sheasby (e.g. Coaton, 1971, 1974; Coaton and Sheasby, 1972, 1978) and Skaife (1956). Ruelle (1985) provided a synopsis of our knowledge of the termites of southern Africa.

Aspects of termite control have been studied by a large number of researchers (Skaife, 1956; Nel, 1961; Sands, 1962; Hartwig, 1966; Coaton and Sheasby, 1972). There are three groups of termites that are of economic importance in avocado orchards, all in the family Termitidae.

1. Large fungus growers in the genera *Macrotermes* and *Odontotermes*.
2. Small fungus growers in the genera *Microtermes*, *Allodontotermes* and *Ancistrotermes*.
3. Carton nest builders of the genus *Microcerotermes*. Members of this group are of less importance than those of the other two groups.

Natural enemies of termites include ants, especially of the genera *Dorylus* and *Anoplolepis*. Fungi and parasitic nematodes seem to play an important role in causing termite nests to die out. Anteaters dig the nests open and can totally destroy them; various birds scratch open the protective clay passageways and eat the exposed termites. An exceptionally large variety of animals, including insects, birds and mammals, feed on winged termites that fly out after the rains. This natural mortality plays a large role in limiting the establishment of new nests (Gouse *et al.*, 1998). In South Africa, the following control measures are recommended.

PRETREATMENT OF NEW ORCHARDS It is advisable to find and destroy termite nests in and around the area in which a new orchard is to be established. This should be done before soil preparations commence. Since termites can work up to or even more than 100 m from their nests, it is important that the nests be destroyed in adjacent areas as well.

According to Gouse *et al.* (1998), it is often difficult to find the nests of *Odontotermes badius* (Havil), especially after the soil has been prepared. The nests can occur from 300 mm to more than 3 m deep in the soil (Hartwig, 1966). The termite mounds are usually only slightly raised, 100–300 mm in height and approximately 1.5 m in diameter above ground. The nest of *O. badius* can be detected because when the termites enlarge it, the discarded soil is cemented together to form small compact heaps, close to the nest. During humid weather, numerous small white mushrooms can be seen on these sites for a few days. A product containing carbon bisulphide, cresylic acid and *para*-dichlorobenzene may be used for nest fumigation.

Some nests of *Odontotermes* spp. are characterized by air pipes, or 'chimneys' of clay. *Mactrotermes* spp. make heaps above the ground with a thick wall. *Ancis-trotermes* and *Microtermes* make small underground nests that are spread over a wide area. There are no signs above the ground to betray their nests. The nests of *Microcerotermes* spp. are also completely underground (Gouse *et al.*, 1998).

The nests of small fungus-grower termites are small and spread over a wide underground area. These termites are very common and can probably be found in many of the tropical and subtropical orchards in Africa. The small fungus-grower termites prefer dead plant material. Therefore, their presence is not always a problem, especially where mulching is practised. However, precautions should be taken to ensure that the trees are not exposed to serious stress (Gouse *et al.*, 1998).

The winged adults or alates that fly out of the nests during the late afternoon after spring rains usually come out of passages that open directly into the central nest cavity, making these most suitable for fumigation (Gouse *et al.*, 1998).

In some countries in Africa, carbosulfan is registered to protect newly planted trees. If no treatments have been applied to the soil and the plants are attacked by termites, 200 g l^{-1} imidacloprid SL (Confidor) can be applied by soil drench or trunk application to protect the trees against termite attacks (Gouse *et al.*, 1998).

It has been shown that once the leaves and flush of a young tree have started to wilt as a result of termite damage, subsequent control of this pest will not be able to save that specific tree. However, if chemical control is then applied to the adjacent trees, many of those that have not yet shown wilting symptoms can still be saved.

ORDER HEMIPTERA – SUBORDER HOMOPTERA

Aleyrodidae

The citrus blackfly, *Aleurocanthus woglumi* Ashby, is a minor pest of avocado. According to Bedford (1998b), the original home of the citrus blackfly is India, Ceylon and the Philippines. It also occurs in the West Indies, Costa Rica, Canal Zone, Mexico and Kenya. The citrus blackfly breeds on 75 different plants, including avocado, banana, citrus, coffee, grape, mango and persimmon (Bosman, 1959). The citrus blackfly is an important pest of citrus in many countries. It produces honeydew, on which sooty mould grows and stains the fruit, leaves and branches. Severe infestations stunt tree growth and inhibit blossom formation (Van den Berg and De Villiers, 1987a; Bedford, 1998c).

The egg of the citrus blackfly is creamy white when freshly laid, turning brown with age. It has a short pedicel which attaches it to the leaf. Eggs are laid on the underside of the leaves in characteristic spirals. The nymphs and the pupae are sedentary, dark brown to black. Adults are just over 1 mm in length. The males are light blue and smaller than the females, which are dark blue to black (Clausen and Berry, 1932; Quayle, 1938; Ebeling, 1959). The general colour of the wings is broken by lightish spots on the edges. Ebeling (1959) states that a female lays more than 100 eggs and that there can be 35–50 eggs per spiral. The minimum life cycle from egg to egg is 6–9 weeks (Clausen and Berry, 1932). In South Africa, during the early summer, the egg stage lasts 18 days and the total life cycle about 80 days (Bedford, 1998b). Adult numbers peak in October, with smaller peaks in early December and February/March and considerable overlapping of stages.

The parasitoid, *Eretmocerus serius* Silvestri, has been introduced and successfully established in many countries for the biological control of *A. woglumi*. Up to 72% of the pupae of *A. woglumi* were parasitized in citrus orchards in South Africa (Bedford and Thomas, 1965).

In Mexico. the species *Trialeurodes vaporariorum* (Westwood), *Tetraleurodes* spp. and *Paraleyrodes perseae* Quantaince have little importance in avocado. *Tetraleurodes* is more abundant during the months of June through November. During heavy attacks, leaves from the lower tree canopy become weak and the injury may cause defoliation. *P. perseae* is found close to the mid-vein of the leaves and its feeding causes chlorotic circular spots on them. Heavy damage may be associated with low flower production and defoliation. Peaks are observed between June and November (Coria-Avalos, 1993). In Cuba, *Aleurodicus cardini* (Back.) is recorded as a pest of avocados; its biocontrol agents include *Baccha clavata* (F.), *B. parvicornis* Loew (Syrphidae), *Chrysopa* sp. (Chrysopidae), *Isodromus iceryae* (How.) and *Carthasis distinctus* (Harris) (Nabidae) (Brunner *et al.*, 1975).

In Israel, the Japanese bayberry whitefly, *Parabemisia myricae* (Kuwana), was discovered in 1978, causing heavy damage to avocado and citrus trees (Sternlicht, 1979). Local predacious mites of the family Phytoseiidae, lacewings (Neuroptera), lady beetles (Coccinellidae), predacious bugs of the family Anthocoridae and parasitic wasps were not effective. Thus, the parasitoid *Eretmocerus debachi* Rose and Rosen (Aphelinidae) was imported from California, successfuly established and eventually controlled *P. myricae* (Rose and DeBach, 1992). Other exotic natural enemies, *Eretmocerus* sp., *Encarsia* sp. (Aphelinidae) (from Japan), the lady beetles *Nephaspis amnicola* Wingo and *Delphastus pusillus* (Le Conte) (from Hawaii), the beetle *Cybocephalus binotatus* Grouvelle (Cybocephalidae) and the fungus *Aschersonia aleyrodis* Webber were released in numerous avocado and citrus groves but probably did not become established (Swirski *et al.*, 1987). In Israel, outbreaks of the Japanese bayberry whitefly were recorded in 1992 in some avocado orchards of western Galilee. These outbreaks were probably caused by drift of baits containing insecticides (malathion), which were applied aerially against the Mediterranean fruit fly, *Ceratitis capitata* Wiedemann (Trypetidae), in adjacent groves of citrus, deciduous or subtropical fruit trees. *E. debachi* was re-established and controlled the pest population (Swirski *et al.*, 1998).

Five species of Aleurodidae attack avocado in California: *Aleurodicus dugessii* Cockerell, *Trialeurodes vaporariorum* (Westwood), *Tetraleurodes morii* (Quaintance), *Paraleyrodes mineii* Iaccarino and *Tetraleurodes perseae* Nakahara (Soliman and Hoddle, 1998; Faber and Philips, 2001). The last species was recently found in avocado plantations in Israel.

Aphididae

Cotton aphid, Aphis gossypii *Glover*

A. gossypii has worldwide distribution (Schwarz, 1998; Millar, 1994). In Israel, it is found on young foliage and occasionally on inflorescences of two avocado varieties, Nabal and Horshim (Swirski *et al.*, 1991b). According to Avidov and Harpaz (1969), damage caused by the cotton aphid is characterized by foliar and floral discoloration and deformities caused by the aphid's toxic saliva, resulting in stunted development of infested plants, and growth of extensive layers of sooty mould on the excreted honeydew.

The nymphs are greenish to brownish with a very distinctive appearance, being spotted with an exudation of powdery wax (Van den Berg and De Villiers, 1987a). Winged and apterous females are found, whereas males are absent. Females are yellow-green to green, or almost black depending on their host plant, with a body length between 1.2 and 2.0 mm (Avidov and Harpaz, 1969). Live young aphids are produced by both winged and apterous females. The females may live for 2 to 3 weeks and produce two or more offspring each day (Hill, 1975). Ebeling (1959) refers to a nymphal period of 3–20 days.

The aphelinid, *Aphelinus* sp., attacks *A. gossypii* in South Africa (Prinsloo and Neser, 1994).

Green peach aphid, Myzus persicae *(Sulzer)*

The green peach aphid was described by Sulzer in 1776 (Müller and Schöll, 1958). It is cosmopolitan and was recorded in Africa as early as 1801 (Theobald, 1914). In South Africa it has been recorded on many plant species representing 55 plant families (Millar, 1994). These include its primary host, peach, as well as avocado, granadilla and papaya. *M. persicae* feeds on the flowers and young leaves of avocado. The damage caused by sucking sap and by fouling the plants with cast skins, honeydew and black sooty mould is for the most part insignificant (Daiber, 1992). It also transmits viral diseases in many other plants (Van den Berg and De Villiers, 1987a).

The aphids are soft-bodied, slow-moving insects, 1.5–2.5 mm in length, with long, spindly legs. Winged and apterous, sexual and asexual forms occur. The wingless, asexual females are light green with fairly long, green cornicles. The winged asexual females have a dark head and thorax, a dark patch on the upper side of the abdomen and dark cornicles. The sexual males are winged and superficially similar to the winged asexual females. The sexual females are apterous.

In tropical and subtropical areas, the green peach aphid lives asexually throughout the year but sexual forms develop in areas with cold winters. In the colder areas, some of the winged asexual females leave various vegetable and other hosts and fly to the primary host, which is peach (Daiber, 1992).

Two parasitoid species, *Aphelinus abdominalis* (Dalman) and *Aphelinus asychis*

Walker (Aphelinidae), have been bred from *M. persicae* (Prinsloo and Neser, 1994). According to Daiber (1992), the green peach aphid is attacked by ladybird beetles and by a fungal disease. Control of the green peach aphid on avocado is seldom necessary. In some countries in Africa, mercaptothion is registered for the control of aphids on avocado, granadilla and papaya.

Pseudococcidae

Long-tailed mealybug – Pseudococcus longispinus *(Targioni-Tozzetti),* Pseudococcus nipae *(Maskell),* Planococcus lilacinti *Cockerell*

The long-tailed mealybug, *P. longispinus*, has a cosmopolitan distribution (Ebeling, 1959; Hattingh *et al.*, 1998). In temperate regions it is mostly unable to pass the winter outdoors but may still be a pest in protected environments and glasshouses. In many parts of Africa, the long-tailed mealybug is a potentially serious pest in all areas where avocados are cultivated.

The long-tailed mealybug has a soft, oval, segmented body. The distinctions between the head, thorax and abdomen are not clear. The life cycle of *P. longispinus* lasts about 4 weeks in summer and several months in winter, with three overlapping generations per year (De Villiers *et al.*, 1987d).

Both sexes of the mealybug pass through three nymphal instars. After the third instar, the female transforms into an adult while the male forms a flimsy, cottony cocoon in which the pupa forms and the adult develops. The adult male has one pair of wings and two long waxy anal filaments (Ebeling, 1959). The long-tailed mealybug is viviparous.

All stages of the mealybug, except for the male pupa, are motile and often move slowly from one area to another. Mealybugs are usually held at low numbers by natural enemies. Most outbreaks are associated with chemical disruption of these enemies. Furthermore, ants are attracted to the honeydew that mealybugs excrete. By protecting them against natural enemies, ants may also help cause a mealybug outbreak.

In Israel, during the 1960s and 1970s, avocado orchards adjacent to cotton fields were heavily damaged by the long-tailed mealybug, *P. longispinus*. Drift of broad-spectrum insecticides from aerially sprayed fields killed its natural enemies, thereby interfering with the biological equilibrium and resulting in severe outbreaks of mealybug. Additionally, females of the honeydew moth, *Cryptoblabes gnidiella* Milliere, were attracted to the honeydew of the mealybug and their caterpillars nibbled at the fruit (Ben Yehuda, 1990; Wysoki *et al.*, 1992). The problem was solved by drastically limiting aerial applications of broad-spectrum insecticides within 200 m of avocado orchards, and by releasing the parasitic wasps *Arhopoideus peregrinus* Compere (= *Hungariella peregrina*) in 1954 and *Anagyrus fusciventris* (Girault) introduced from Australia in 1971 (Swirski *et al.*, 1980, 1998).

In Cuba, several biological control agents are used to control the mealybug *Pseudococcus nipae* (Maskell): *Scymnus bahamicus* Csy (Coccinellidae), *Lobodiplosis pseudococci* (Felt), *Pseudoaphycus utilis* (Timberlake), *Verticillium lecanii* (Zimmerman) and *Empusa fresenii* Nowak (Zygomycota).

In the Philippines, *Planococcus lilacinti* Cockerell infests young shoots and fruit peduncles. Heavy infestations may cause fruit to drop but the pest is generally of minor concern (Cendana *et al.*, 1984).

Coccidae

White wax scale, Ceroplastes destructor *Newstead*

C. destructor has been collected in South Africa from avocado (Munro and Fouché, 1936). It has also been recorded on coffee and persimmon from Kenya and Uganda (Hill, 1975). However, the most important host in Africa is citrus. In Australia, the white wax scale *Gascardia destructor* (Newstead), Indian white wax scale *Ceroplastes ceriferus* (Fabricius) and pink wax scale *Ceroplastes rubens* Maskell are minor pests that rarely warrant spraying (Waite and Pinese, 1991).

According to Cilliers (1998), the wax of the adult female *C. destructor* is shiny white with two external bands of pulverized wax leading upwards from the stigmal grooves. The stigmal groove has about 45 spines, of which the middle subconical spine is the largest. The eggs are brick red and smooth when laid. According to Zeck (1932), *C. destructor* lays about 3000 eggs. The crawlers are brick red, and settle along the midribs or main veins of the dorsal surface of the leaves. Very soon after the first moult, *C. destructor* migrates from the leaves to the twigs (Cilliers, 1967). The third moult occurs about 3 months after the crawler has settled. There is a pre-oviposition period of about 8 months.

Observations of *C. destructor* in South Africa are largely restricted to *Melia azedarach*, on which it goes through one generation a year, the eggs being produced from October to December. This unisexual species has three moults on *M. azedarach* (Cilliers, 1967). A total of 24 primary parasitoids attack this wax scale in South Africa (Cilliers, 1967; Snowball, 1969; Annecke and Insley, 1971; Prinsloo, 1984). Two secondary parasitoids have also been recorded. Three important predators of the white wax scale are *Coccothera spissana* Zell. (Olethreutidae), *Eublemma costimacula* Saalm. and *E. scitula* Rambur (Noctuidae) (Cilliers, 1998). In Australia, *Paraceraptrocerus nyasicus* (Compere) provides good control of *C. destructor*.

Soft brown scale, Coccus hesperidum *L.*

The soft brown scale is a cosmopolitan pest of many tropical and subtropical plants. Infestations on citrus are usually the consequence of non-judicious insecticidal usage and/or of uncontrolled ant activity (Annecke and Georgala, 1978). These latter authors indicated that infestations on other plants could have been caused by the same factors. Munro and Fouché (1936) list guava, mango, papaya and many others, including citrus, as host plants. Avocado (Ebeling, 1950) and granadilla (De Villiers and Van den Berg, 1987b) are also attacked. In avocado, the direct loss of fruit due to soft brown scale appears to be small. However, fruit may be stained by sooty mould deposits and consequently downgraded. On English ivy, *Hedera helix*, at 22°C, the parent females are ovoviviparous, producing thin-walled eggs that hatch within a few minutes (Annecke, 1966). Crawlers settle on leaves and twigs. Feeding sites are often selected on the main or subsidiary veins on both the ventral and dorsal surfaces of leaves. The first instar lasts from 15 to 46 days, body length increasing from 0.37 to 0.41 mm (mean 0.39 mm). The second instar lasts from 10 to 57 days and increases from 0.80 to 1.01 mm (0.92 mm). This instar is characterized by the formation of a longitudinal dorsal row of minute ventral columns of glassy wax. The duration of the pre-oviposition period is 27–74 days, the body length increasing from 1.26 to 1.82 mm (1.52 mm) before eggs are produced and from 1.52 to 2.70 mm (2.07 mm) afterwards. No further moult ensues; a pre-oviposition period of 1–2 months intervenes (Annecke, 1966). The sexually mature female may reach a body length of 5 mm (Bodenheimer, 1951; Avidov and Harpaz, 1969). A mean 73 eggs, but up to 200, are produced in the pest's lifetime (Bodenheimer, 1951). Although males have been described by Newstead (1903) from the date palm, *Phoenix dactylifera*, in England, the occurrence of males is unknown in South Africa (Annecke and Georgala, 1978).

In spring, crawlers of soft brown scale tend to move outwards in the tree (Annecke and Georgala, 1978). According to Smith *et al.* (1997), only the egg-producing adults are sedentary, since the nymphs and non-producing adults of the soft brown scale may move from one feeding position to another, especially if feeding conditions deteriorate.

Ants of various species are invariably associated with infestations of this pest. The brown house ant, *Pheidole megacephala* (F.), the pugnacious ant, *Anoplolepis custodiens* (Smith), and related species are commonly encountered. In late summer, the growth of sooty mould, which may cause extensive blackening of the foliage fruit, often pinpoints small populations in a tree (Annecke and Georgala, 1978).

There is evidence of three overlapping generations per year on citrus in South Africa. Surges of crawlers seem to appear in mid-November and early February, and a third one probably in May (Annecke, 1959, 1966).

Annecke (1959, 1964) recorded 22 primary and six secondary hymenopterous parasitoids from soft brown scale. A number of predators of soft brown scale have been recorded (Searle, 1964). Predators seem to be of greater importance at times of soft brown scale outbreaks and may be of lesser value in maintaining the scale at low levels for long periods.

Although ant control is usually not a standard recommendation in orchards of tropical non-citrus and subtropical crops (Annecke and Georgala, 1978; Bedford, 1998b), it is regarded as essential for suppressing soft brown scale in orchards where it occurs. Soft brown scale can be chemically controlled on avocado by using 500 g l^{-1} mercaptothion EC at 250 g 100 l^{-1} water or 250 g kg^{-1} mercaptothion WP at 500 g 100 l^{-1} water (Krause *et al.*, 1996). These latter authors recommend spot-spraying individual trees with large populations of *C. hesperidum*, and following up with ant control.

Pyriform scale, Protopulvinaria pyriformis *(Cockerell)*

Protopulvinaria pyriformis has a cosmopolitan distribution (Wysoki, 1987; Du Toit and De Villiers, 1988) In Israel, *P. pyriformis* was first recorded in 1980 (Ben-Dov and Amitai, 1980), and is now found in almost all areas of the country. In Africa, the pyriform scale (also called the heart-shaped scale), was first recorded on avocado in 1916 at Pietermaritzburg, KwaZulu-Natal (Brain, 1920). *P. pyriformis* is the most damaging insect affecting avocado in Peru (Wolfe *et al.*, 1969) and sometimes causes serious damage to avocado trees in Florida (Ebeling, 1959) and to citrus and avocado trees in Spain (Del Rivero, 1966). A preference for certain avocado cultivars was observed (De Villiers and Van den Berg, 1987b; Wysoki, 1987). 'Nabal' is the most susceptible cultivar to this scale, followed by cultivars. 'Ein Vered', 'Reed', 'Hass' and 'Fuerte', whereas 'Ettinger' is attacked when located close to infested trees of the susceptible cultivars. In laboratory trials, the pyriform scale completed development and reproduced on 'Nabal' and 'Ein Vered' seedlings, and to a lesser extent on those of 'Hass'. Survival of the scale was rather poor on 'Fuerte', 'Horshim' and 'Wurz', but higher on 'Ettinger', 'Pinkerton' and 'Reed' (De Meijer *et al.*, 1989).

P. pyriformis feeds on avocado, guava, *Aphanamixis polystachya* (= *A. moora rohituka*) (De Villiers, 1981) and citrus (Bedford, 1978a). According to De Villiers (1989), a steady increase in the occurrence of the heart-shaped scale was reported from 1986 to 1989 on avocado in parts of South Africa.

The scale secretes excessive quantities of honeydew on which sooty mould grows, staining the fruit leaves (Du Toit and De Villiers, 1988, 1990; Steyn *et al.*, 1994; De Villiers, 2001d). Furthermore, photosynthesis is affected, leading to crop yield reductions (Ray and Williams, 1982). Leaf drop can occur with heavy infestations.

Du Toit *et al.* (1991) defined the following morphological stages: egg, crawler stage (first instar), second instar, third instar, A– female (unsclerotized females without eggs, ring stage), A+ female (sclerotized females with eggs) (Plate 51). The female lays about 200 or 300 eggs which are kept in a white, woolly secretion beneath her body. According to Du Toit *et al.* (1991), the eggs are oval-shaped and light yellow. The first- to third-instar scales are flat, oval-shaped and light green, with mean lengths of 0.35, 0.73 and 1.12 mm, respectively (Ray and Williams, 1982). The adult egg-laying female (A+) is about 3 mm in length and reddish brown with prominent radial stripes dorsally. A white woolly ovisac is visible around the hind margin of the A+ females. No males have yet been observed in South Africa. In Florida, males have been recorded (Ebeling, 1959). In the laboratory, at 27°C and 70% RH, the life cycle of the heart-shaped scale on avocado seedlings (W.J. Du Toit, 1994, unpublished) was as follows: the mean duration of the first instar (which includes the crawler stage) was 10.5 days, the second instar lasted 17.5 days, the third instar 24.5 days, the A– female stage 28.0 days, and the A+ female stage (including the ring stage) 45.5 days. Two generations are completed per year on avocado in Nelspruit, South Africa (De Villiers, 1989), whereas in Israel, two generations appear annually – a winter generation with a peak in November/December and a summer generation with a high level in June/July

(Hadar, 1993). In South Africa, peaks of the first instar were reached in November and March, and peaks of adults in October and March. The winter generation lasted about 7 months from egg to adult; the summer generation lasted 5 months. *P. pyriformis* feeds on the ventral side of the leaves. Fruits are normally not attacked (De Villiers, 2001d). Avocado cultivars favoured by the scale in South Africa in descending order are Hass, Collinson, Ryan, Fuerte and Edranol (De Villiers and Robertson, 1988).

The following natural enemies of the heart-shaped scale have been identified in South Africa (Robertson and De Villiers, 1986): *Metaphycus galbus* Annecke, *Metaphycus helvolus* (Compere), *Metaphycus stanleyi* (Compere) (Encyrtidae), *Coccophagus basalis* Compere, *Coccophagus pulvinariae* Compere (Aphelididae) and *Tetrastichus* sp. (Eulophidae). Two hyperparasitoids, *Cheiloneurus cyononotus* Waterston (Encyrtidae) and *Marietta javensis* (Howard) (Aphelinidae), were also found. Predators included *Chilocorus angolensis* Crotch and *Hyperaspis senegalensis hottentota* Mulsant.

Du Toit *et al.* (1991) found that, on average, 70% parasitism occurs in the adult female stage, 4% in the second instar, 14% in the third instar. *Metaphycus* spp. were dominant on *P. pyriformis* during the late summer and autumn, but were replaced in winter by *Coccophagus* spp. and *Tetrastichus* sp. According to Du Toit *et al.* (1991), *Marietta javensis* (Howard) has a negative influence on the parasitoid complex and, in Israel, attacks the two *Metaphycus* species, *M. stanleyi* Compere and *M. swirskii* Annecke and Mynhardt. Since local and introduced parasitoids are not sufficiently effective at controlling the scale population, selective scalicides such as oil are used, killing the nymphal stages of the pest but not the ovipositing females (Swirski *et al.*, 1988). In Israel, the parasitic wasps *Microterys flavus* (Howard), *Metaphycus flavus* Howard (Encyrtidae), *Coccophagus lycimnia* Walker (Aphelinidae), predators *Chilocorus bipustulatus* L., *Oenopia* (*Synharmonia*) *conglobata* L., *Scymnus flavicollis* Redtenbacher (lady beetles, Coccinellidae), *Anisochrysa* (*Chrysoperla*) *carnea* Stephens (green lacewings, Chrysopidae), spiders and the fungus *Verticillium lecanii* (Zimmerman) (Hadar, 1993) were unable to curb populations of pyriform scale to below economic thresholds. Thus, efforts were exerted to import various natural enemies. *Metaphycus swirskii* (imported from Kenya), initially the most abundant parasite, was soon replaced by *Metaphycus stanleyi* (imported from the USA, South Africa and Spain), which is today the dominant natural enemy of the pyriform scale in Israel. *Metaphycus helvolus* (imported from the USA) is rare in avocado orchards and *M. galbus* (from South Africa) probably failed to become established. Two lady beetles were imported from Spain: *Cryptolaemus montrouzieri* Mulsant is sporadically found in avocado orchards, whereas *Nephus peyerimhoffi* Sicard is very rare. The two secondary parasites *Marietta javensis* (Howard) and *Pachyneuron concolor* (Forster) have an adverse effect on the efficiency of *Metaphycus* spp. In some avocado plots, 70% of the total parasite fauna may consist of these two secondary parasites. Population studies of the active parasitization of the scale by *M. stanleyi* showed an increase in September, with high levels during the winter, a peak in May and a decline in summer. The high rates of encapsulation of *Metaphycus* eggs by the pyriform scale during the summer may interfere with efficient biocontrol of the pest (Blumberg and Blumberg, 1991; Hadar, 1993). In Peru, biological control of this pest can be achieved by releasing the wasp *Microterys flavius*. The site, duration and rate of oviposition, as well as host marking and preference of host stages in the various species of *Metaphycus*, were studied by Hadar (1993). Since local and exotic natural enemies are not sufficiently effective at suppressing the pyriform scale populations, mineral oil – the selective scalicide – is recommended. It kills the three nymphal instars of the pest, but does not affect the ovipositing females. Oil sprays are likely to produce satisfactory results if applied in December/January (or January/February) and in July/August, when the pest populations consist mainly of young nymphal instars (Hadar, 1993). Only heavily infested trees are treated with broad-spectrum scalicides (Wolfe *et al.*, 1969; Swirski *et al.*, 1998).

A survey carried out in a heart-shaped scale-infested avocado orchard near

Nelspruit, South Africa, demonstrated that trees nearest to a dust road had the highest scale infestation (Robertson and De Villiers, 1986; E.C.G. Bedford, 1945, unpublished). In South Africa, buprofezin WP at 30 g 100 l^{-1} water is registered as a full-cover application when first-generation crawlers become active. The addition of an adjuvant, e.g. Agral, is advised (Krause *et al.*, 1996). A thorough wetting of the inside of the tree canopy, especially the ventral sides of leaves, is essential for effective control (Du Toit and De Villiers, 1988).

Diaspididae

Avocado scale, Fiorinia fioriniae *(Targioni-Tozzetti)*

The avocado scale *F. fioriniae* has been found on avocados in Florida, Hawaii, the continental USA and the West Indies (Ebeling, 1959). In South Africa, it is regarded as a minor pest of avocado and other crops (Munro and Fouché, 1936; De Villiers and Van den Berg, 1987b; Erichsen and Schoeman, 1992; Claassens, 1994). On avocado, the avocado scale attacks the leaves and fruit. Small brown spots develop where nymphs feed, which on 'Hass' fruit, fade as the fruit colour develops (Claassens, 1994).

The female scales are oval, shield-shaped, with a thin, translucent covering. They are brownish yellow to orange-brown, 1.0–1.3 mm long (De Villiers and Van den Berg, 1987b). A morphological description and illustrations of the first instar of *F. fioriniae* have been provided by Howell and Tippens (1977).

Latania scale, Hemiberlesia lataniae *(Signoret)*

The latania scale, or palm scale, was originally described from a palm, *Latania borbonica*, in 1869 (Quayle, 1938). *H. lataniae* has been recorded on avocado in countries including Australia, Israel, South Africa and the USA (California and Florida) (Quayle, 1938; Ebeling, 1959; Avidov and Harpaz, 1969; Howard and Oliver, 1985; Swaine *et al.*, 1985; De Villiers and Van den Berg, 1987b). According to Dekle (1976), it has more than 370 host plants in Florida. Davidson and Miller (1990) state that *H. lataniae* feeds on ornamentals from 78 families of 278 genera.

According to Ebeling (1959), the rostralis of *H. lataniae* irritates the flesh of 'Fuerte' avocado and possibly other thin-skinned avocado cultivars. This is indicated by nodules adhering to the inside of the peel with corresponding depressions in the flesh of ripe fruit. The scale attacks the branches or twigs, leaves and fruit of avocado. Infested fruit is degraded or culled. In South Africa, heavy infestations were observed on fruit of cv. Hass, but smooth-skinned cultivars are also susceptible to this scale insect (De Villiers, 2001c). A heavier scale infestation on 'Hass' fruit hanging near the ground than higher up in the tree was reported (Steyn, 2001a).

The adults are variable: on leaves they are grey to white, circular and convex; on stems they are brown and slightly convex. The exuvia is subcentral and yellow-brown (Davidson and Miller, 1990). Males are not always present (Davidson and Miller, 1990).

H. lataniae is unisexual in southern California (McKenzie, 1935) and Israel (Gerson and Zor, 1973), and bisexual in Maryland (Stoetzel and Davidson, 1974). Annually, latania scale completes two generations in Maryland (Stoetzel and Davidson, 1974), three generations in Egypt (El-Minshawy *et al.*, 1971) and four generations in Israel (Gerson and Zor, 1973). From egg to egg-laying female takes 56–65 days in southern California (McKenzie, 1935). In Australia, *H. lataniae* infests leaves, twigs, branches and fruit. Heavy infestations on terminal growth can cause dieback of the growing points. The presence of scale on the fruit presents an aesthetic problem. Rough-skinned cultivars, especially 'Hass', are the worst affected because the scales cannot be brushed off easily in the packing house. Scale infestations are worse where the natural enemy complex is upset by disruptive insecticides such as methidathion, dimethoate and synthetic pyrethroids. Natural enemies include parasitic wasps of the *Aphytis* sp. *proclia* Walker group, *Signiphora perpauca* Girault, *Encarsia citrina* (Craw.), *Signiphora flavella* Girault, the green lacewing *Chrysopa oblatis* Banks, and the coccinellid *Rhizobius satellus* Blackburn.

Chemical control is via methidathion or low viscosity oil (Waite, 1988).

Natural enemies in South Africa include the aphelinid parasitoids *Aphytis lingnanensis* Compere, *Aphytis africanus* Quednau, the encyrtids *Habrolepis obscura* Compere and Annecke and *Comperiella lemniscata* Compere and Annecke. Predators are the coccinellids *Rhyzobius lophanthae* Blaisdell, *Chilocorus* sp., as well as a cybocephalid *Cybocephalus* sp. (De Villiers and Van den Berg, 1987b; Daneel, 1998).

The predatory mites, *Cheletomimus berlesei* Oudemans (Cheyletidae) and the *Hemisarcoptes* spp. (Hemisarcoptidae) feed on diaspidid crawlers such as those of *H. lataniae* (Gerson and Zor, 1973). Coccinellids such as *R. lophanthae* and *Chilocorus* sp. are high density feeders of scale insects, *Dentifibula obtusilobae* Felt is a cecidomyid predator of the latania scale. Evans and Prior (1990) state that armoured scale insects, particularly in the humid tropics, are subject to periodic and often devastating attacks by highly adapted fungal pathogens, such as *Nectria flammea* (Tull.) Dingley (Ascomycota).

In most areas where the scale is present, their natural enemies keep them under biological control under field conditions. Where natural enemies have been suppressed by the misuse of insecticides, scale-infested avocado fruit need to be cleaned before marketing (Ebeling, 1959; Swaine *et al.*, 1985).

Spanish red scale, Chrysomphalus dictyospermi *(Morgan)*

The Spanish red scale attacks a large number of host plants and is also common in greenhouses (Ebeling, 1959). It feeds on the twigs, branches, leaves and fruits of the plants (Van den Berg *et al.*, 1999). Avocado twigs and branches that have been severely infested with *C. dictyospermi* become roughened and crack. Leaf drop may also occur. The Spanish red scale has an almost cosmopolitan distribution which includes Africa, the Bahamas, the Canary Islands, Corsica, the Philippines, Seychelles and the USA.

Spanish red scale is a pest of avocado and has occasionally been found on avocado and citrus in California and Florida (Quayle, 1938; Ebeling, 1959). In South Africa, *C. dictyospermi* has been recorded on avocado, guava and mango, among others (De Villiers, 2001c). Quayle (1938) provided a list of 67 host plants, including avocado, banana, citrus, guava, mango, pecan, rose, apple and tea plant.

The female scale is circular, slightly convex, 1.5–2.0 mm in diameter and light brown or red-brown. The scale covering may be readily separated from the insect, even during the moulting period, distinguishing it from the red scale, *Aonidiella aurantii* (Maskell) (Quayle, 1938).

Scales from Ventura County, California, reproduce parthenogenetically, whereas in those from the New Orleans district, fertilization is necessary for reproduction (Ebeling, 1959). According to Quayle (1938), there can be from three to four generations per year in southern California and from five to six in southern Florida.

Parasitoids recorded from *C. dictyospermi* in the USA include *Aphytis chrysomphali* (Mercet), *Prospaltella aurantii* (Howard), *Thysanus merceti* (Malenotti) and *Thysanus flavopalliata* (Ashmead). The coccinellids *Rhizobius debilis* (Blackburn) and *Lindorus lophantae* (Blaisdell) seem to be of greater importance than the parasitoids (Ebeling, 1959).

Other scales

In Peru, *Selenaspidus articulatus* Morg. and *Unaspis citri* Comstock are common pests of citrus, regularly found attacking avocado when the orchards are located close to citrus orchards. In Mexico, 14 species of diaspidid scales, *Abgrallaspis howardi* (Cockerell), *Acutaspis albopicta* Cockerell, *Aspidiotus spinosus* Comstock, *Chrysomphalum aonidum* (L.), *C. dyctiospermi* (Morgan), *Diaspis cacoccois* Linchtenstein, *Hemiberlesia diffinis* (Newstead), *H. lataniae* (Signoret), *H. rapax* (Comstock), *Melanaspis aliena* (Newstead), *Myxetaspis personata* (Comstock), *Pinnaspis strachani* (Cooley), *Quadraspidiotus sperniciosus* (Comstock), *Velataspis dentate* (Hoke), three Coccidae, *Saissetia oleae* (Oliver) (McGregor and Gutierrez, 1983), *Coccus hesperidum* L., *Pulvinaria simulans* Cockerell, the mealybugs, *Planococcus citri* (Risso), *Dysmicoccus brevipes* (Cockerell), *Ferrisia virgata* (Cockerell) and *Nipaecoccus nipae* (Maskell), are considered the most frequent Coccoidea attacking avocado.

Flatidae

Avocado planthopper, Decipha viridis *(Synave)*

D. viridis is a minor pest of both avocado and macadamia. Nymphs and adults suck plant sap from the leaves and fruit stalks. Large amounts of honeydew are produced which support the growth of sooty mould. This may spoil the appearance of avocado fruit.

The adult *D. viridis* is light green with a blackish spot on the distal edge. When at rest, the wings are kept folded over the abdomen in tent-like fashion. Eggs are laid in oval-shaped masses which are about 5 mm in diameter. The tops of the eggs are white and the sides of the egg masses light brown. Nymphs are light green with white and yellow stripes on their backs. Long white waxy filaments protrude from the tips of their abdomens.

The life cycle of the avocado planthopper lasts about 6 weeks in summer. There may be up to six generations per year. In South Africa, as the populations of *D. viridis* seem to be stable, there are probably sufficient natural enemies to keep them below the economic level.

Acanalonidae

Parathiscia *sp. near* truncata *Walker*

In South Africa, both adults and nymphs feed on avocado and macadamia. The 8 mm adults are mustard green. At rest, the wings are folded over the abdomen, giving them a flattened appearance. The nymphs are green.

Eurybrachidae

Mottled avocado bug, Parapioxys jucundus *Distant*

Both adults and nymphs of *P. jucundus* feed on avocado and macadamia. This species is probably a phloem feeder. It is more abundant on avocado (3.1%) than on macadamia (0.5%). The adult is 15 mm long and 7 mm wide, with a mottled appearance varying from orange-green through green to a brownish colour; ventrally its abdomen is bright red. The nymphs are dark brown with long anal filaments.

Cicadellidae

The leafhopper, *Idona minuenda* Ball (Cicadellidae) in Mexico, is a common pest of avocados grown in more temperate areas with frequent rainfall. Its peaks are observed during the months of July to October. The nymphs feed on the leaves, causing a grey-coloured injury on the leaf's upper side. When the attack occurs on young leaves these turn yellow and remain underdeveloped. This species is suspected of transmitting a virus.

Penthimiola bella (Stol), the citrus leafhopper, is a well-known pest of citrus in South Africa. It has been shown (Du Toit *et al.*, 1993; Bruwer, 1996) that this species is probably one of the causative agents of protrusions, especially on 'Hass' avocado fruit from old trees. Eggs are laid singly through a minute slit and placed in a superficial envelope formed by the tissue of either leaves or fruit. There are five nymphal instars. The adult is largely brownish, mottled with paler markings measuring 3.0–3.9 mm in length. Incubation lasts 9–20 days at 20–27°C and nymphal development takes 25–63 days (N.H.S. Pretorius, 1971, unpublished). The eggs of *P. bella* are attacked by seven Chalcidoidea species (Annecke, 1965) resulting in 50% parasitism (Annecke, 1964b).

Psyllidae

Trioza anceps *Tuthill*

T. anceps is 2–5 mm long and is considered to be widely distributed in Mexico, where it attacks regional non-commercial cultivars. *T. anceps* causes large, finger-like galls on the dorsal surfaces of leaves of avocado trees in Central America (Hollis and Martin, 1997). However, *T. anceps* is very seldom observed on commercial cultivars, such as 'Hass' and 'Fuerte'. A heavy infestation of this stenophagous species may cause defoliation (Hernandez *et al.*, 2000).

Trioza perseae *Tuthill*

T. perseae is reported from Peru (Chavez *et al.*, 1985) and Venezuela. This insect is approximately 2.5 mm long, yellowish with dark bands on the abdomen and thoracic regions, wings which are almost transparent, with reduced venation. Taxonomic characteristics are reported by Chavez *et al.* (1985). The female deposits its eggs on the tender flush, preferring the leaf underside, close to the mid-vein. The eggs have a pedicel; they are whitish and difficult to see with the naked eye. The emerging nymph is yellowish with inconspicuous body segmentation. Most feeding occurs on the leaf underside, but some nymphs and adults can be found on the upper side of the leaf. Hollis and Martin (1997) provide the identification key to different psyllids associated with *P. americana*.

T. perseae galls are formed on the leaf underside, while the adaxial surface shows depressions which are surrounded by yellowish circles. The psyllid is regularly found in those places that are located more than 900 m above sea level. Other psyllid species reported by Hollis and Martin (1997) are *T. aguacatae* Hollis and Martin deforming shoots and leaves, and *T. godoyae* Hollis and Martin causing leaf roll. Chavez *et al.* (1985) reported several unidentified Eulophidae and Pteromalidae as parasitoids of *T. perseae*.

Membracidae

Metcalfiella monogramma *Germar*

Brom (1970) provides a description and life history of *M. monogramma*, a common pest in Mexico. Its injury is characterized by the deposition of eggs into young branches, causing death of the infested tissue. Brom (1970) considers the most important injury to be caused by the feeding of nymphs and adults.

ORDER HEMIPTERA – SUBORDER HETEROPTERA

Coreidae

Many of the coreids that, among others, also damage tropical and subtropical crops are known as tip wilters, as their feeding often causes wilting of new shoots (Brain, 1929; Smit, 1964a; Van den Berg, 1998). A comprehensive study of three members of this family was also performed by Hartwig and De Lange (1978).

The adults and nymphs of tip wilters suck the sap of succulent new growth. In many plant species, the saliva injected by tip wilters into plant tissue causes wilting and dieback of the new shoots, beyond the point of feeding. Because of dieback of the growth tip, secondary axillary buds start to grow, which may cause dense sprouting. Their attacks are most serious during periods of growth flush, especially on young trees.

In South Africa, the most important species attacking some tropical and subtropical crops, including avocado, are: *Anoplocnemis curvipes* F., *Leptoglossus australis* (F.) and *Pseudotheraptus wayi* Brown.

Large blacktip witer,
Anoplocnemis curvipes *F.*

A. curvipes is the most common tip wilter found on tropical and subtropical crops. The adult is a dull grey-brown, about 25 mm in length. The femora of the males are thickened and noticeably bent, with a thorn-like spine.

The eggs are grey-brown, barrel- shaped, and deposited in single rows, each consisting of about 6–12 eggs, on the branches and leaves of their host plants (Smit, 1964b). At an average temperature of 28°C, eggs hatch after 4 to 6 days on mango and nymphal development takes about 4 to 6 weeks. Depending on the availability of new growth, there are from three to six overlapping generations annually in the Nelspruit area (Van den Berg, 1998).

(Scelionidae), *Ooencyrtus caurus* (Encyrtidae) and *Anastatus* sp. (Eupelmidae) may account for up to 90% of eggs but the immigrant adults which result from non-parasitized eggs cause severe damage. Because immigration is constant throughout the season, individual crops may require sprays of endosulfan every 2 weeks. Orchards in more open situations may require only occasional sprayings. Thin-skinned cultivars, especially Fuerte, appear to suffer most, but the thick-skinned Hass can be just as badly affected without showing the severe lesions and cracks that develop on Fuerte (Fay and Huwer, 1993; Waite and Huwer, 1998).

Pentatomidae

The pentatomids dealt with in this chapter are, except for the cosmopolitan *Nezara viridula* L., indigenous to southern Africa. Many of them probably also occur further north in Africa. In Israel, the variegated caper bug, *Stenozygum coloratum* Klug (Pentatomidae), develops generally on wild caper bush, *Capparis spinosa*, sometimes migrating to various cultivated plants, including avocado (mainly cv. Hass, but also Fuerte and Ettinger). The damage to fruit is accompanied by heavy secretion of persein and by the appearance of black spots (Izhar *et al*., 1990b).

Feeding of pentatomids on small developing fruits and nuts and their pedicels causes many of them to drop, while larger fruits and nuts are blemished. Growth points at which feeding takes place may wither and die.

The biology, life history and control of the pentatomids dealt with here are quite similar for the different species. Furthermore, their natural enemies consist of egg parasitoids (Hymenoptera), adult parasitoids (Diptera: Tachinidae) and predators such as ants. The tachinids may occasionally also parasitize nymphs but complete their development in the adult stink bug. In general, the parasitoids of the different pentatomids are, except for species differences, comparable. Their effectiveness at controlling the different species, however, varies greatly from species to species with seasons of the year.

Powdery stink bug, Atelocera raptoria *Germar*

The powdery stink bug is indigenous to southern Africa. It is brown, sprinkled with red-brown and has a whitish wax. The ventral side of the thorax and abdomen is mostly covered with a whitish growth of a waxy substance. The powdery stink bug occurs on many tropical and subtropical crops. It has been reported to be the most abundant stink bug on avocado, and the second most adundant on macadamia.

Grey-brown stink bug, Coenomorpha nervosa *Dallas*

In South Africa, the dorsal side of the adult grey-brown stink bug is brown with grey-black dots forming patterns all over the head, thorax, abdomen and wings. It is up to 14 mm long, the ventral side of the abdomen and thorax is off-white to a whitish brown. One of the characteristics of this species is the width of the abdomen. It sticks out beneath the wings and is also broader than the thorax. The parts of the abdomen that can be seen from the dorsal side have grey-black lines that accentuate each segment.

C. nervosa was recorded as occurring in fairly large numbers on avocado (Joubert and Claassens, 1994). In later surveys, however, this species was found in small numbers on both avocado and macadamia, where feeding probably takes place on fruits, nuts or on the bark (Van de Berg and Greenl, unpublished data).

Spotted stink bug, Farnya *sp.*

The spotted stink bug is green-brown to greyish brown with yellow-brown spots. It is up to 14 mm in length and feeds on avocado and macadamia. Eggs are whitish, turning light brown as development progresses. They are laid in masses averaging 26 eggs per mass. At a constant 25°C and 75% RH, the egg stage lasts on average 6.8 days, and the first to fifth (final) instar nymphs take, respectively, 5.0, 10.4, 7.7, 6.8 and 10.8 days. From egg to adult takes 48 days (Bruwer, 1992).

Egg parasitism by two *Trissolcus* spp. (Scelionidae) and a *Pachyneuron* sp. (Pteromalidae) at Nelspruit accounted, respectively, for 72.6% and 4.5% of the parasitism (Bruwer, 1992).

Yellow-edged stink bug, Nezara pallidoconspersa *Stål.*

Small numbers of the yellow-edged stink bug *N. pallidoconspersa* were collected on avocado and macadamia. Feeding has been recorded on macadamia by De Villiers (1986). The eggs of *N. pallidoconspersa* are yellowish white and laid in clusters averaging about 70 eggs. These are mostly laid on the leaves of their host plants. The final instar nymph has a black background colour with pale orange dots on the thorax. Two brighter orange dots and white markings are also present on the abdomen.

Adult females of *N. pallidoconspersa* are up to 18 mm long, males from 11 to 14 mm. They are green, with abdomen edges that are yellowish with brown lines that accentuate each segment. They also have yellowish hind-wings which are striking when they fly.

N. pallidoconspersa can be found in the hotter areas, such as Nelspruit and Mpumalanga, where they breed throughout the year on the wild-growing castor oil plants, *Ricinus communis*.

Natural enemies of *N. pallidoconspersa* include the egg parasitoid *Trissolcus* sp. (probably not *Trissolcus basalis* Wollaston) and two unidentified tachinid parasitoids that attack adult stink bugs.

Small green stink bug, Nezara prunasis *Dallas*

The small green stink bug is often encountered on subtropical crops. *N. prunasis* was found in relatively large numbers on avocado (Joubert and Claassens, 1994), and pecan (Joubert and Neethling, 1994). Since no nymphs were found, it is not certain whether feeding and breeding take place on these plants. Adults of *N. prunasis* are green and 9–11 mm in length. At the beginning of the winter they congregate in sheltered areas, such as barns or even in the roofs of houses.

Green vegetable bug, Nezara viridula *L.*

The green vegetable bug is a cosmopolitan species which attacks more than 200 cultivated plants, including avocado, citrus, granadilla, macadamia, pecan, tobacco and various vegetables. In South Africa, Dennill and Dupont (1992) reported that feeding by *N. viridula* causes protrusions on avocado fruit. *N. viridula* was reported on avocado, macadamia and pecan (Joubert and Claassens, 1994; Joubert and Neethling, 1994).

The eggs are yellowish white, barrel-shaped and laid in clusters of 60 to over 100, usually on the underside of leaves, but also on the stems (Van Heerden, 1933). A female may lay up to 320 eggs (Van Heerden, 1933). The developing eggs change to orange with darker orange showing through a whitish lid prior to nymphal emergence. The newly emerged nymphs usually remain closely clustered, but do not feed until the second moult (Van den Berg, 1998). There are five nymphal instars; the final instar is green with reddish colouring in the centre of the abdomen and a row of four whitish dots on each side of the red. The adult is a typical shield stink bug, about 15 mm long. When *N. viridula* hibernates, it changes to an orange-brown to bronze colour. In South Africa, the life cycle from egg to adult lasts about 6 weeks (Van Heerden, 1933). There are three full generations per year and stages overlap in summer. In the Cape Province, the adult hibernates in sheltered sites. In the hotter areas like the Mpumalanga Lowveld, some of the adults become more orange-brown to bronze in colour, but at least some individuals do not hibernate since adults and nymphs may be found on wild-growing castor oil plants throughout the winter months (Van den Berg, 1998).

Two tachinid parasitoids of the adult green vegetable bugs, *Bogosia bequaerti* Villeneuve and *Cylindromyia* sp. (Van den Berg, 1998) and the egg parasitoid *Trissolcus basalis* (Wollaston) (Scelionidae) are present in South Africa but are unable to keep the numbers of *N. viridula* acceptably low. A tachinid parasitoid of the adult green vegetable bug, *Trichopoda pennipes* (F.), was first introduced from Hawaii in 1978, and from Florida in 1983 through 1995, but it failed to become established (De Villiers *et al.*, 1980; Van den Berg, 1995). A second parasitoid, *Trichopoda giacomellii* (Blanchard), was introduced from Australia. Establishment of the two *Trichopoda* spp. has not been confirmed.

Two viruses, NVV-1 and NVV-2, were identified from a green vegetable bug (Von Wechmar *et al.*, 1991). When fed to second instar nymphs, 75% reduced longevity, 67% reduced life span and 95% reduced egg production were achieved (Williamson and Von Wechmar, 1995).

The following method is recommended to monitor stink bug populations in avocado and other trees: choose ten trees at random per unit or block (smaller than 5 ha); spread light-coloured cloth beneath the drip areas of the trees; spray the ten trees with dichlorvos (150 ml 100 l^{-1}); count all stink bugs that fall from the trees during the next hour. As a general recommendation, control should be considered if 1.2 or more stink bugs are counted per tree.

Miridae

Avocado bug, Taylorilygus *sp.*

In Africa, the adult avocado bug, *Taylorilygus* sp., is brown to black and up to 3 mm in length. Two dirty white spots are present on the base of the wings. The first two nymphal instars are green. Adults and nymphs feed on avocado flowers, young avocado fruit and presumably also young leaves. Avocado flowers seem to be preferred to fruit. Because of the large numbers of flowers produced, feeding on them is of minor importance. Damage to avocado fruit is caused within the first few weeks after fruit set. This leads to the development of protrusions on larger fruit which are only visible about a month after feeding. The lesions that occur on avocado fruit are in the form of pimply elevations on the fruit surface (Du Toit *et al.*, 1993) and are known as 'Vosknoppe'. This can be very severe, especially on 'Hass' fruit. Upon dissection, brown corky scar tissue can be found in the centres of the lesions: 30–40 protrusions per fruit make them unsuitable for export. If surveys indicate that large populations of the avocado bug are present shortly after fruit set, chemical control should be applied immediately (Van den Berg *et al.*, 1999). In the Philippines, the mirid bugs *Helopeltis bakeri* Pop. and *H. collaris* Donovan attack the shoots and fruit of avocados, causing significant damage (Cendana *et al.*, 1984).

Dagbertus fasciatus *(Reuter),* D. olivaceous *(Reuter),* Rhinacloa *sp.*

In Florida, a number of mirids (*Dagbertus fasciatus* (Reuter), *D. olivaceous* (Reuter) and *Rhinacloa* sp.) feed and insert their eggs on opening buds, leaves, flowers and small fruit. Attacks seem to especially affect flowers and recently set fruit, causing them to drop. Wounds on fruit may serve as a point of entry for decay organisms. These insects are green-brown, comparatively small at 1 cm in length. Mirids usually appear during the bloom and early fruit-setting stages. It is suggested that weeds and grass in the grove be mowed as closely as practicable to reduce places that could harbour mirids. Mirid populations are most common from January through April, when avocado flowers are fully open.

The parasitoid, *Leiophron*, probably *fumipennis* Loan, has been registered in Florida. Sprays during flowering should be applied late in the afternoon to reduce loss of honeybees (Peña and Johnson, 1999).

Tingidae

Avocado lace bug, Pseudacysta perseae *(Heidemann)*

The avocado lace bug was described in 1908 as *Acysta perseae* from Florida specimens and considered a minor pest of avocado for several years. However, persistent population outbreaks of *P. perseae* observed since the mid-1990s in Florida and in the Caribbean region, reveal that *P. perseae* has become one of the most important pests of avocado (Abud-Antum, 1991; Medina-Gaud *et al.*, 1991). *P. perseae* is found in Florida and Georgia in the USA, Bermuda, the Dominican Republic, Puerto Rico and Mexico (Mead and Peña, 1991). Common hosts for this pest, besides avocado, are red bay, *Persea borbonia* (L.) and camphor, *Cinnamomum camphora* (L.). The life cycle of *P. perseae* was reported by Abud-Antum (1991). It requires 22 days from egg stage to adult. The most complete description of adults and late instar nymphs was given by Heidemann (1908).

P. perseae confines its attack to the lower surface of the foliage, causing chlorosis, necrosis and severe defoliation of avocado, reducing yields (Peña *et al.*, 1998) (Plate 52). This bug usually lives in colonies, depositing eggs upright in irregular rows in clusters on the lower leaf surface. This insect opens an avenue of penetration for the leaf anthracnose fungus, *Colletotrichum gloeosporioides* (Mead and Peña, 1991). Since the avocado lace bug was not considered an important pest, it is suggested that in recent times, suitable natural enemies were decimated by application of pesticides or by some other type of environmental disequilibria.

In Florida, avocado lace bug population densities increase during the dry season (November–February), and decline during spring and summer (Peña *et al.*, 1998). The cultivars 'Waldin', 'Booth 8' and 'Loretta' have the highest natural infestation levels. The most susceptible cultivar appears to be 'Booth 8', with average damage levels of 20–28% to the leaf area. Leaf photosynthesis is reduced by 50% when the leaves sustain 40% damage. Cultivars (e.g. 'Simmonds') with 100% of their leaves infested exhibited early leaf drop and an overall reduction in fruit set. By contrast, a West Indies × Guatemala hybrid was scarcely affected by the pest.

The major biological control agents in Florida are two egg parasitoids, *Oligosita* sp. (Hymenoptera: Trichogrammatidae), and an unidentified mymarid wasp; if undisturbed by chemical applications, the green lace wing *Chrysoperla rufilabris* Burmeister and a mirid, *Hyaliodes vitripennis* Say keep avocado lace bug densities from reaching economic levels.

Several pesticides – M-Pede (soap), citrus oil and Mycotrol (*Beauveria bassiana* (Balsamo) Vuillemin) – significantly reduced lace bug densities compared with the untreated control (Peña *et al.*, 1998).

Lygaeidae

In Israel, although the cottonseed bug, *Oxycaraenus hyalinipennis* (Costa) (Lygaeidae), develops on plants of the family Malvaceae, it sometimes feeds on the leaves and stem of avocado seedlings and may even cause some damage (Swirski *et al.*, 1998).

ORDER HYMENOPTERA

Formicidae

In Peru, *Atta sexdens* L. causes very severe damage by cutting new leaf flushes and by consuming leaves. In Mexico, *Atta mexicana* (Smith) is a variably sized ant, 3 to 14 mm long, with polymorphic markings. It is reddish to dark brown in colour. The workers are polymorphic with spines on the pronotum, mesonotum and propodeum. The ant cuts the leaves into irregular pieces, frequently starting the damage at the leaf apex. The central vein of the leaf remains. The severity of the damage depends on the time of year when the infestation occurs.

In South Africa, the driver or red ant, *Dorylus helvolus* L., causes damage to avocado when the fruit hang or lie on the ground under the trees. Parts of the fruit peel are damaged and fruits with such feeding marks are unsuitable for export. Confusion surrounds the identity of the organism causing this damage. It has been ascribed by Dennill and Erasmus (1991) to an unidentified ant species. However, Erichsen and Schoeman (1992) state that it may be due to feeding by a species of slug or snail.

ORDER THYSANOPTERA

Thripidae

Considerable numbers of thrips species infest avocado orchards in different geographical regions of the world. New species have appeared recently (e.g. the avocado thrips, *Scirtothrips perseae* Nakahara in southern California; the orchid thrips, *Chaetanaphothrips orchidii* (Moulton) in Israel), demanding prompt management. These new arrivals may be the result of increased international commerce, but the severity of their damage may have been intensified by changes in horticultural practices, or by the introduction of new cultivars. For instance, in Israel the introduction of new varieties, highly susceptible to thrips, combined with irrigation supplied at shorter intervals, favoured orchid thrips infestation. The Ardit variety of avocado is susceptible to the greenhouse thrips, *Heliothrips haemorrhoidalis* (Bouché) (Thripidae) and varieties 'TX 531' and 'Horshim' to the black vine thrips, *Retithrips syriacus* (Mayet) (Thripidae).

Citrus thrips, Scirtothrips aurantii *Faure*

In Africa, the citrus thrips, *Scirtothrips aurantii* Faure, has been reported from avocado, citrus, guava and litchi (Faure, 1929), and in Egypt (Mound and Palmer, 1981); its presence has been confirmed in the Cape Verde, Yemen, Mauritius and Réunion (Vuillaume *et al.*, 1981; Gilbert and Bedford, 1998). Both the adults and larvae of *S. aurantii* cause lesions when feeding on fruit of avocado, citrus, guava, macadamia and mango (Grové, 2001a,b). *S. aurantii* is a very important pest on citrus (Bedford, 1943; Wentzel *et al.*, 1978) and mango (Grové, 2001b) but it seldom does any economically significant damage on avocado. Some natural enemies were reported by Grouth, 1994; Grouth and Richard, 1992; and Grouth and Stephen, 1995. For more information, see Chapter 3.

Scirtothrips aguacatae *Johansen and Mojica and* Scirtothrips kupae *Johansen and Mojica*

Larvae and adults of *S. aguacatae* and *S. kupae* are found on young leaf flushes, flowers and fruit. Damage to the fruit epidermis results in fruit deformation. To reduce thrips densities, Hernandez *et al.* (2000) recommended treatment during blooming followed by another two chemical treatments when the thrips populations are increasing. Coria-Avalos (1993) suggested applying pesticides three to four times. The first spray should be applied when the trees show 10% bloom, followed by a second spray at full bloom, a third one immediately after bloom and a fourth application when newly formed fruits are observed. Economic thresholds are observed when 7% of fruits, leaves and flowers are infested with thrips. Cultural control is recommended by removing weeds and fallowing around the tree. Reyes and Salgado (1994) demonstrated that leaves of the avocado cultivars '30PLS', '54PLS', 'Rincoatl' '18PLS' and '158PLS' and blossoms of cultivars '18PLS', '44PLC', 'ColinV-101', '175PLS', '158PLS' and 'PV2' are tolerant to *Scirtothrips*. Ebeling *et al.* (1959) reported that the cultivars 'Fuerte' and 'Dickinson' are moderately resistant to *Scirtothrips* spp.

Scirtothrips perseae *Nakahara*

An unknown species of *Scirtothrips* was discovered damaging fruits and foliage of avocado in California in 1997. The species was described later as *S. perseae* (Nakahara, 1997). *S. perseae* feeding damage to foliage is observed on upper and lower leaf surfaces as well as on developing fruit. Fruit is susceptible to damage until it reaches approximately 2.5 cm in diameter. Feeding scars develop from the calyx and fruit scarring results in 'alligator skin' (Hoddle, 1997). Avocado thrips larvae and adults can build to high densities from autumn through spring on young leaves, causing excessive leaf drop. The main source of economic loss attributable to avocado thrips is scarring of immature fruit in the spring (Hoddle and Morse, 1998).

Female *S. perseae* lay eggs singly into soft plant tissue. Eggs are kidney-shaped and whitish yellow in colour. Between 20 and 31 eggs are laid per female, egg to adult development time lasts 16 to 27 days, and eggs hatch within 9 to 14 days depending on the temperature (Hoddle and Morse, 1998). Following egg hatching, developing thrips pass through two actively immature stages. Larvae pupate in leaf litter under trees, in crevices in the bark, or within persea mite nests on leaves. Surveys of leaf litter indicate that 89% of pupating thrips are found in the upper non-decomposed leaf layers. Hoddle and Morse (1998) recommend the use of sabadilla, because of its short residual activity and low toxicity to most natural enemies. In California, *Franklinothrips vespiformis* is considered to have good potential as a predator during high infestations of *S. perseae* (M.S. Hoddle, 2000, personal communication).

Red-banded thrips, Selenothrips rubrocinctus *(Giard)*

The red-banded thrips has been a serious pest of cacao in the French West Indies since 1901 (Russell, 1912). According to Hill (1975), *S. rubrocinctus* has an almost pantropical distribution. In Taiwan, symptoms similar to those caused by *Heliothrips haemorrhoidalis* result from the feeding of *S. rubrocinctus* (Giard). This species can be a problem in California, Florida, South Africa, Australia, Reunion and Taiwan.

In Florida and Australia, infested leaves are spotted on the upper surface with dark reddish brown faecal pellets, or on the lower side of the leaf (Hill, 1975; Waite and Pinese, 1991), damaged areas turn rusty with numerous small, shiny black spots of excreta. The edges of affected leaves curl up. Heavy feeding on fruit causes a russet appearance, with cracking and decay. Greenhouse thrips damage fruit and tend to feed on the larger and more mature ones. They are found most frequently where two fruits are in contact or where a leaf contacts the fruit. Fruits damaged by slight russetting show leathery scarring and cracks on the skin.

Eggs are inserted into the leaf tissue. They are white, kidney-shaped and about 0.25 mm long (Hill, 1975). The first and second nymphal stages are yellow with a bright red band around the base of the abdomen. When fully grown, the second instar is about 1 mm long. The tip of the nymph's abdomen is turned up and carries a drop of excreta on the anal setae (Hill, 1975). The pre-pseudo pupa is yellowish with red eyes, with a red band across the first three abdominal segments. The appearance of the pseudo pupa is similar to that of the pre-pseudo pupa (Hill, 1975; Steyn, 2001b). The adult female is dark brown and just over 1 mm long. Males are smaller and rare (Hill, 1975). Reproduction by the red-banded thrips is parthenogenetic (Avidov and Harpaz, 1969). Females live about 7 weeks and lay an average of 25 eggs, which hatch in 12 to 18 days (Hill, 1975). The nymphal stage lasts 6 to 10 days and the pre-pseudo pupal and pseudo pupal stages last 3 to 6 days.

The predator *Franklinothrips vespiformis* (D.L. Crawford) is considered a predator of *S. rubrocinctus* and *Heliothrips haemorrhoidales* (Bouché); *Leptothrips macro-ocellatus* (Watson) is considered a predator of *S. rubrocinctus* (Johansen and Mojica-Guzman, 1996). No parasitoids or predators of *S. rubrocinctus* have yet been recorded from South Africa. Observations should be made on leaves and fruit for red-banded thrips during summer and autumn for any developing infestation. In Florida, the following materials are labelled for use against thrips: malathion (various labels), permethrin (Pounce, Ambush), pyrethrins + rotenone (Pyrellin) (Peña and Johnson, 1999). In Australia, the pest is generally suppressed by the frequent sprays of endosulfan applied to control fruitspotting bugs (Waite and Pinese, 1991).

Frankliniella chamulae *Johansen and* Frankliniella bruneri *Watson*

In Mexico, *F. chamulae* and *F. bruneri* are found on avocado flowers in the region of Uruapan. There are unconfirmed reports that both species damage the fruit. Coria-Avalos (1993) reports that *Frankliniella* spp., *Scirthotrips aceri* (Moulton) and *Liothrips perseae* are becoming the most important avocado pests in the Michoacan area. However, this author does not specify which of these species injure the fruit, reducing fruit quality.

Western flower thrips, Frankliniella occidentalis *(Pergande)*

In the past, the Western flower thrips appeared infesting avocado inflorescences near infested mango orchards in the south of Israel. Many different thrips species were also found on avocado inflorescences. Among them, high numbers of *Thrips tabaci* Lindeman and *T. major* Uzel (M. Wysoki, R. zur Strassen and W. Kuzlicky, unpublished).

Black vine thrips, Retithrips syriacus *Mayet*

In Israel, the black vine thrips has recently caused damage to avocado fruits and leaves, cultivars 'Horshim', 'TX-531', '4102', '4203' and 'T-142' being the most heavily infested (Plate 53). The black vine thrips can be readily controlled with sabadilla (Izhar *et al.*, 1992).

Greenhouse thrips, Heliothrips haemorrhoidalis *(Bouché)*

The greenhouse thrips was first described in 1833 (Russell, 1909) (Plate 54). It is widely spread throughout the world, attacking a wide range of economically important crops (Rivnay, 1934; Avidov and Harpaz, 1969; Hill, 1975; Skaife, 1979; De Villiers and Van den Berg, 1987a; De Villiers, 1990c; Steyn, 2001a). The greenhouse thrips is an economically important pest of avocados in California (Ebeling, 1959; McMurtry *et al.*, 1991) and Cuba (Cañizares, 1975), and of citrus in Israel (Rivnay, 1934; Bodenheimer, 1951). In Israel, large populations of *H. haemorrhoidalis* usually start on pest-susceptible trees, such as seedlings of the Mexican cultivars, or trees of cultivars 'Ardit', 'Benik', 'Hass', 'Stuart', 'Nahlat' and 'Horshim' (Plate 55). From these foci, the thrips disperse to commercial cultivars 'Fuerte', 'Nabal' and 'Ettinger Heavy' (Izhar *et al.*, 1990a).

The 1.25-mm-long adult greenhouse thrips is black, with the exception of the legs, wings and antennae which are white (De Villiers, 1990c). The sex ratio is female-biased, with parthenogenetic reproduction. The female lives for about 7 weeks and lays an average of 25 eggs (Avidov and Harpaz, 1969). The eggs are white, kidney-shaped and about 0.3 mm long. They are inserted singly into the leaf tissue just beneath the epidermis. They increase in size and become considerably swollen shortly before hatching. The optimum temperature range for oviposition is between 20 and 28°C (Ebeling, 1959).

The first and second nymphal stages are whitish to slightly yellowish, the eyes are red. Temperature and humidity are critical factors during the development of the nymphs. The optimum level of humidity for development is 85%. If humidity falls below 60%, many of the nymphs desiccate. The two nymphal stages last for about 9 to 30 days (Hill, 1975).

The mature second instar nymph moults to form the pre-pseudo pupa, which is yellowish with red eyes. After a day or two, it moults to form the pseudo pupa. The pseudo pupa has longer wings than the pre-pseudo pupa. The pseudo pupa stage lasts for 3–14 days. The antenna of the pseudo pupa is tilted backwards while that of the pre-pseudo pupa is not.

There are four generations per year in most countries but as many as six to seven generations per year have been recorded in California. The life cycle, from egg to adult, lasts about 8 weeks (Hill, 1975). Mortality rate of the thrips is high at relative humidities below 60% and temperatures above 27°C (Rivnay, 1935).

Various parasitoids and predators have been recorded attacking the greenhouse thrips. In South Africa, the pirate bug, *Orius tripoborus* (Hesse) (Hemiptera: Anthocoridae), feeds on nymphs of *H. haemorrhoidalis* on avocado fruits, reducing thrips outbreaks (Dennill and Erasmus, 1992a,b) and *Thripobius semiluteus* Boucek (Hymenoptera: Eulophidae) parasitizes greenhouse thrips nymphs (Steyn, 1996). This parasitoid has been established in some avocado orchards in California, reducing greenhouse thrips densities within 2 years (McMurtry *et al.*, 1991). In California and Israel, the egg parasitoid *Megaphragma mymaripenne* Timberlake and the imported larval parasitoid, *Thripobious semiluteus*, are important natural enemies of *H. haemorrhoidalis* (Wysoki *et al.*, 1997; Faber and Phillips, 2001). The wasp *Thripobius semiluteus* was introduced into California in 1986 and into Israel in 1991. The female endoparasitoid inserts its egg in first or second stage nymphs. In the laboratory, development of the parasite is completed at 23°C within 22–25 days (McMurtry *et al.*, 1991). The following predators of the greenhouse thrips have been recorded in Israel: *Typhlodromus athiasae* Porath and Swirski, *Amblyseius swirskii* Athias-Henriot (Phytoseiidae), *Franklinothrips megalops* Trybon (Thysanoptera) spiders, and anthocorids (Wysoki *et al.*, 1997).

As the greenhouse thrips occurs sporadically for the most part in many African countries, control measures are seldom necessary. Chemicals registered against thrips on avocado include mercaptothion 25% WP at 500 g 100 l^{-1} water or 50% EC at 250 ml 100 l^{-1} water. Mercaptothion has a 7-day safety period. Sulphur 80% WP is also registered at 300 g 100 l^{-1} water or 98% DP at 10–40 kg ha^{-1} (Steyn, 2001a). In the past, in the USA, oil was combined with pyrethrum against eggs

(Ebeling, 1959). In Israel, the initial killing of post-embryonic stages of the thrips by pyrethrum was very high but the eggs hidden under the plant cuticle were not affected and a population build-up was observed after 15 and 22 days. Oil gave similar results, but was not as effective as pyrethrum. In Israel, two treatments of pyrethrum at an interval of 21 days gave good results (Ben-Yehuda *et al.*, 1994). However, in trials carried out in avocado orchards in California in 1985 and 1986, pyrethrum gave inconsistent results against the greenhouse thrips (Goodall *et al.*, 1987).

ORDER LEPIDOPTERA

Metarbelidae

Bark borers, Salagena obsolescens *Hampson and* Salagena transversa *Walker*

In South Africa, the bark borers, *S. obsolescens* and *S. transversa* feed on avocado, guava, litchi, macadamia, pecan, wild fig (*Ficus* sp.), bush willows (*Combretum zeyheri* and *C. collinum*), jacketplum (*Pappea capensis*), water berry (*Syzygium cordatum*) and *Eugenia* sp. (De Villiers, 1973; Pinhey, 1975; De Villiers *et al.*, 1987c). *S. obsolescens* has been recorded from East Africa, Mozambique, South Africa, West Africa and Zimbabwe (Pinhey, 1975).

The eggs are cream-coloured and laid in clusters on the bark (De Villiers, 2001b). The young, light-brown larva feeds on the bark of trees and later makes a tunnel in the wood for shelter. The final instar can be up to 30 mm in length. The larva feeds on the bark of trees under a frass-covered silken web. The older larva tunnels into the wood, usually where branches fork. The larva may kill a branch. The tunnel which serves as a shelter for the larva is up to 70 mm long with a diameter of about 5 mm when the larva is fully grown (De Villiers, 1973).

The pupa is dark brown and develops in the tunnel, its head facing the opening (De Villiers, 2001b). The adult of *S. obsolescens* (Pinhey, 1975) has a variegated pale and dark brown thorax. The abdomen is grey with prominent black or brown dorsal tufts near the base followed by reddish brown ones. The forewing is cream coloured with rows of black rings encircling reddish brown spots. The forewing is 15–23 mm in length. Moths are active at night and a female can lay about 270 eggs (De Villiers, 1973).

Two parasitic wasps of the genus *Genaemirum* sp. (Ichneumonidae) have been recorded. In South Africa, bark borer larvae can be controlled chemically by spraying the lesions with carbaryl 850 g kg^{-1} WP at 250 g 100 l^{-1} water (De Villiers, 2001b).

Avocado leafroller, Homona spargotis *Meyrick; Ivy leafroller,* Cryptoptila immersana *(Walker)*

In Australia, the major caterpillar pests are the avocado leafroller and the ivy leafroller. These pests web leaves together and feed on them, but the more serious damage is caused when they chew into fruit from the shelters that they construct between touching fruit or by webbing leaves to fruit (Waite and Pinese, 1991). Natural enemies recorded for *C. immersana* include the predatory larva of the syrphid fly *Melanostoma agrolas* Walker, as well as the parasitoids *Goniozus* sp. (Bethylidae), *Sympiesis* sp. (Eulophidae) and *Phytodietus celsissimae* Turner (Ichneumonidae). A trichogrammatid also parasitizes the eggs (Waite and Pinese, 1991). Chemical control has been achieved with chlorpyrifos/ddvp, but the IGR tebufenocide (Mimic) is currently pending registration (Waite and Pinese, 1991).

Oecophoridae

The avocado seed moth, *Stenoma catenifer* Walsingham, is considered one of the most important pests of avocado in the Neotropics and has been reported from Mexico to Argentina. However, in Mexico it is not commonly found in the state of Michoacan (Gallegos, 1983). In Peru, *S. catenifer* affects avocado orchards located close to the jungle or near the coast. In Guyana, it also attacks *Chlorocardium rodiei* (Lauracea) (Cervantes *et al.*, 1999). The avocado seed moth can perforate

95% of the fruits, depending on the type of cultivar attacked (Arellano, 1975, 2000).

The adult is 1 cm long, a yellow-ochre moth showing sexual dimorphism. Moths will regularly mate 2–3 days after emergence and can live up to 10 days. One day after mating, the female deposits as many as 240 eggs on the crevices of the fruit. Boscan and Godoy (1984) determined the life cycle of this insect. The egg is 0.6 mm long, oval and light green. The egg ecloses in 5–6 days, the larva penetrates the fruit and consumes the seed in 20 days. There are five instars. Early instars are whitish, while late instars are pink or reddish. The larval stage lasts 16–21 days and the pupal stage lasts 11–19 days. The life cycle lasts between 44 and 49 days with three complete generations per year. The highest damage is observed from May through August.

In Brazil, *Apanteles* spp. are reported as parasites and in Panama, *Xiphosomella stenomae* Cushanan (Gallegos, 1983). Boscan and Godoy (1982) reported that *Apanteles* spp. were responsible for 30% of the parasitism when a high percentage of trees (80%) were infested with *S. catenifer*. *Chelonus* spp. (Braconidae) and *Eudeleboea* sp. (Ichneumonidae) parasitize *S. catenifer* in Guyana (Cervantes *et al.*, 1999). In Mexico, insecticides should be applied at least 12 times in order to be effective. In Peru, the best control is afforded by removal of infested fruit.

Tortricidae

False codling moth, Cryptophlebia leucotreta *(Meyrick)*

The false codling moth was first reported as a pest of citrus in KwaZulu, Natal (Fuller, 1901). It was also reported to attack citrus in other parts of South Africa (Howard, 1909). According to Hill (1975), the false codling moth is distributed in both tropical and southern temperate regions in Ethiopia, Senegal, the Ivory Coast, Togo, Upper Volta, down to South Africa and also in Mauritius and Madagascar (Angelini and Labonne, 1970; Reed, 1974). The false codling moth, previously known as *Carpocapsa* sp. (Fuller, 1901), was described as *Argyroploce leucotreta* by Meyrick (1913). This species was transferred to the genus *Cryptophlebia* by Clarke (1958). Gunn (1921) reported the biology, ecology and control of the false codling moth on citrus, guava and avocado.

According to De Villiers *et al.* (1987b) and Joubert and Du Toit (1993), a tiny hole is visible where the larva penetrates the skin of the avocado fruit. This is usually surrounded by a white powdery substance. A small feeding tunnel is made just beneath the skin but large larvae are seldom found. Kok (in Milne, 1973a,b; Gunn, 1918) mentions that an adult false codling moth has been reared only once, in that case from an avocado seed. 'Pinkerton' avocados which had been interplanted with maize were extensively damaged by false codling moth after the maize was sprayed with cypermethrin (Joubert and Du Toit, 1993).

The false codling moth has a wide range of indigenous host plants from which cultivated crops have been invaded (Catling and Aschenborn, 1978). A. Schwartz (1981, unpublished) listed some 14 indigenous and 21 cultivated host plants in South Africa, including citrus, guava, litchi, macadamia, pecan, mango and, occasionally, avocado. Furthermore, it is also a pest of peaches (Daiber, 1976), acorns, almonds, olives, tea seeds and walnuts (Catling and Aschenborn, 1978). In Israel, it is a pest of macadamia (Wysoki *et al.*, 1986).

Freshly laid eggs are pearl-white and translucent, turning slightly reddish with a black spot shortly before hatching. The 0.77×0.60 mm egg is oval and flattened, with a reticulate sculpture. Eggs are laid singly on fruits and nuts and, according to Gunn (1921), also on guava leaves.

The first instar larva is cream-white with a dark brown head and is about 1.5 mm in length. The full-grown larva is pinkish red with a brown head and it is 12–15 mm in length. The pupa, which is dark brown, is formed in a silken cocoon in the soil. The adult is inconspicuous, with mottled dark grey coloration (Gunn, 1921; Daiber, 1979; A. Schwartz, 1981, unpublished).

According to Newton (1998), the sex ratio of field populations is close to unity. Females mate shortly after emergence and pre-oviposition lasts 5–6 days in the field and 1–2 days in laboratory cultures. Multiple mating takes place in both sexes (A. Schwartz, 1981, unpublished). Daiber (1980) found an average fecundity of 456 eggs at a constant 25°C and 87 eggs per female at 15°C. Five days after moth emergence, a peak of 29 eggs per female was observed at 20°C. Peak egg laying occurs during the first night of the oviposition period, which is normally 2 days after emergence. It was estimated that five generations can develop per year in the Pretoria area (Daiber, 1980). Adults live between 2 and 3 weeks in the field (Ripley *et al.*, 1939). In an unsprayed citrus orchard, the peak flight activity was found to take place in November and February/March (A. Schwartz, 1981, unpublished). Adults are not attracted to light traps (Gunn, 1921; Catling and Aschenborn, 1978).

A female pheromone was isolated by Read *et al.* (1968, 1974) but it was later concluded (Newton and Mastro, 1989; Newton *et al.*, 1993) that true sex pheromones are comprised of several structural optical isomers. However, bioassays of various combinations of pheromones from Malawi and the Ivory Coast have not excluded the possibility of geographical variation (Attygalle *et al.*, 1986; Newton, 1998).

According to Gunn (1921), the life cycle on guavas in the Pretoria area (commencing in January/February) takes an average of 152 days. Incubation takes from 11 to 14 days, larval development from 59 to 71 days, the prepupal stage from 21 to 30 days and the pupal stage from 43 to 66 days. A. Schwartz (1981, unpublished) found that when fed on artificial medium at 25°C, the false codling moth completes its development in 28 days.

Catling and Aschenborn (1974) report that parasitism by *Trichogrammatoidea cryptophlebiae* Nagaraja may increase from January onwards, resulting in between 59% and 89% parasitism for the rest of the season. The egg-larval parasitoid, *Chelonus curvimaculatus* Cameron (Braconidae), has been reported by Searle (1964) and Broodryk (1969). A number of larval parasitoids have been recorded in southern Africa, namely the ichneumonid *Apophua leucotretae* (Wilkinson), the braconids *Agathis bishopi* Nixon, *Agathis leucotreta* (Nixon), *Bassus* sp., *Phanerotoma curvicarinata* Cam., the chalcid *Oxycoryphe edax* Waterston and an unidentified tachinid (Ford, 1934; Ullyett, 1939; Thompson, 1946; G.J. Begemann, 1994, personal communication).

Rhynochorus albopunctatus (Stol) and *Orius* sp. and *Pheidole megacephala* F. have been recorded as predators of the false codling moth (Omer-Cooper, 1939; Steyn, 1954; Newton, 1998). *Beauveria bassiana* (Balsamo) Vuillemin has often been recorded in pupae found on leaf litter (G.J. Begemann, 1994, personal communication). It has also been reported that infection by an unidentified granulosus virus may destroy complete batches of larvae in laboratory cultures (A. Schwartz, 1981, unpublished).

Catling and Aschenborn (1974) recommended augmentative or inundative releases of the egg parasitoid *T. cryptophlebiae*. This method, practised with strict orchard sanitation, brought a very high infestation of false codling moth down to a lower equilibrium (Schwartz *et al.*, 1982). However, other results have been unpredictable and often unsatisfactory (Newton *et al.*, 1986).

Although *T. cryptophlebiae* were seen to disperse up to 1.3 km in a habitat devoid of host plants (Van den Berg *et al.*, 1987), movement within a citrus orchard is limited (Newton, 1988b) to the extent that heterogeneous population densities have a critical impact on the performance of the parasitoid from orchard to orchard.

According to current information on the distribution of natural enemies of the false codling moth, the prospects for classical biological control are poor (CIBC, 1984). Newton (1998) is of the opinion that parasitoids of other *Cryptophlebia* spp. from places like Australia, Fiji, Hawai, India and Taiwan may be sources for the introduction of parasitoids into southern Africa. Orchard sanitation, recommended by Fuller (1901) and Gunn (1921), is still one of the most important methods of reducing damage by the false codling moth. Nothing has been registered for chemical control of the false codling moth on avocado.

Western avocado leafroller, Amorbia cuneana *Walsingham*

The western avocado leafroller is a sporadic pest of avocados in California and Mexico.

The *A. cuneana* larva rolls and feeds on the folded edge of the leaves. The injury can be extended to superficial feeding on fruits, frequently when two fruits are close together, causing economic losses (Bailey and Olsen, 1990b; Coria-Avalos, 1993).

Adult female moths lay light green, oval-shaped, overlapping flat masses of 5 to 100 eggs. Eggs are generally laid on upper leaf surfaces close to the midrib, hatching within 13 to 15 days. Newly hatched larvae are yellowish green and gradually turn darker green as they reach maturity. *Amorbia* larvae roll and tie leaves together with a silken web. This forms a shelter or nest in which the larva feeds on leaves and fruit and is protected from pesticide applications. Larvae go through five instars, and pupate in leaf rolls. The pupal stage lasts about 17 days. Adult moths are bell-shaped, with a wingspan of about 2.5 cm. Adults are active at night. Female moths deposit 400–500 eggs over the course of 2 or 3 weeks (Faber and Phillips, 2001).

Young *Amorbia* larvae typically feed only on the surface of avocado leaves, leaving a thin brown membrane or skeleton of leaf veins. Mature larvae consume the whole leaf. However, mature avocado trees can tolerate considerable leaf damage by the *Amorbia* larvae without severely affecting tree growth or fruit yield. Nevertheless, fruit damage may occur where the larval web leans against fruit, or where webs are made between touching fruit. Larvae may feed on fruit skin and cause scarring, resulting in severe downgrading or culling (Faber and Phillips, 2001). In California, Hoffman *et al.* (1983) tested two components of the sex pheromone of *A. cuneana*, (E,Z)-10,12 and (E,E)-10,12 tetradecadien-1-ol acetate (McDonough *et al.*, 1982) and obtained optimum trap catches with an isomer content of 29 to 82% EZ (as a percentage of EE + EZ) and dosages of 0.06 to 1.7 mg per rubber septum. Later, Bailey *et al.* (1988) tested a combination of the two components – (E,Z)-10,12 and (E,E)-10,12 tetradecadien-1-ol acetate – at a 1 : 1 ratio and at a 9 : 1 ratio to trap populations of *A. cuneana*. The difference in moth response led them to believe that populations represented different species or different races of *A. cuneana*. Faber and Phillips (2001) suggest monitoring larvae in late spring by looking for leaf roll in young foliage and feeding damage in mature fruit. There are no established thresholds for chemical control. Bailey and Olson (1990b) reported Lannate and Orthene to be the most effective pesticides if applied by ground equipment.

The egg, larval and pupal stages of *Amorbia* are parasitized by a variety of beneficial insects. The most important egg parasitoid is *Trichogramma platneri* Nagarkatti. Results of a preliminary study by Oatman and Platner (1985) showed at least 87% parasitization of *Amorbia* eggs. Fleschner *et al.* (1957) reported *Elachertus proteoteraris* Howard, and an ichneumonid parasitizing the larvae of *Amorbia*, and the tachinid, *Phorocera erecta* Coq., parasitizing the pupae.

Apple leafroller, Tortrix capensana *(Walker)*

In South Africa, the apple leafroller was recorded on citrus as early as 1947 (Taylor, 1957). *T. capensana* is a species indigenous to the Ethiopian region (Pinhey, 1975) that occurs throughout Africa (Begemann *et al.*, 1998). *T. capensana* is polyphagous and feeds on trees indigenous to southern Africa, weeds, and fruit trees such as apple, apricot, citrus, peach, pear and plum (Matthew, 1975; Barrow and Bedford, 1998; Begemann *et al.*, 1998) as well as on avocado (Van den Berg *et al.*, 1999). The larva spins leaves or fruit together to form a protective niche where it feeds on those leaves or fruit. In most cases, only the peel of the fruit is fed on, resulting in its downgrading. In a survey carried out in the Nelspruit-Hazyview region (South Africa), Erichsen and Shoeman (1992) found that the apple and citrus leafrollers together had damaged 0.34% of avocado fruit during the previous season. This represented a calculated loss of US$14,000.

Overlapping eggs are laid in oval egg masses on the dorsal leaf surface (Begemann *et al.*, 1998). Freshly laid eggs are a dull yellow-cream which changes to a yellow-orange before hatching. The newly hatched larva is pale yellow with a black head. As the larva ages, the body darkens to a yellow-green. Five instars occur in males and six in females, reaching 20–25 mm in length (Begemann *et al.*, 1998). Pupation takes place in the tree in leaves that have been webbed together. Female pupae are, on average, 11 mm long, males 9 mm (Begemann *et al.*, 1998).

Both male and female adults are brown, with varying patterns of a darker or lighter colour on the forewing. At rest, the female is on average 13 × 7 mm, the larger male form 10 × 5 mm, the small form 6 × 4 mm (Begemann *et al.*, 1998).

Development averages 10 days in summer and 21 days in winter. At 26°C, male larvae complete their development in 18–29 days, female larvae in 18–29 days, and the pupal stage lasts 8–12 days. The duration from egg to egg is 41–73 days (Begemann *et al.*, 1998). Six generations of *T. capensana* occur annually in citrus orchards in South Africa, the first and second being the most important (Begemann *et al.*, 1998).

A considerable number of natural enemies of the apple leafroller have been recorded. *Trichogrammatoidea lutea* Girault was found to parasitize *T. capensana* eggs on citrus (Begemann *et al.*, 1998). Barrow (1977) reared the following parasitoids from *T. capensana* larvae: *Apanteles* sp. (Braconidae), *Goniozus* sp. (Bethylidae) and *Pediobius amaurocoelus* (Westwood) (Eulophidae). Pupal parasitoids include *Nemorilla afra* Curran (Tachinidae), *Theronia* sp. (Ichneumonidae) and *Brachymeria boranensis* Masi (Chalcididae) (Barrow, 1977). Nothing is registered to control leafrollers on avocado in South Africa. However, *Bacillus thuringiensis* var. *kurstaki* Berliner is registered for leafrollers on citrus (Krause *et al.*, 1996).

Citrus leafroller, Archips occidentalis *(Walsingham)*

The citrus leafroller has been recorded as a significant pest of citrus in Swaziland and as an occasional pest on citrus in some parts of South Africa (Newton and Catling, 1998). On avocado, the larva of the citrus leafroller spins leaves or fruit together to form a protective niche, where it feeds. Damage to the fruit (Van den Berg *et al.*, 1999) is characterized by large epidermal lesions which lead to fruit degradation. A survey in the Nelspruit–Hazyview region of South Africa indicated that the citrus and apple leafrollers damaged 0.34% of avocado fruit, representing a calculated loss of 78,700 rand (US$14,000) for the season (Erichsen and Schoeman, 1992).

The eggs of the citrus leafroller are laid on the leaves in compact, oval-shaped masses of between 20 and 176 eggs (Newton and Catling, 1998). Early larvae are pale orange with a darkly pigmented brown head capsule (Newton and Catling, 1998), turning pale green later (Van den Berg *et al.*, 1999). According to Newton and Catling (1998), the insect usually pupates in the rolled leaves; the pupa is dark brown. The adult's anterior wings are overall brown with two oblique bands and a darker area at the apex (Newton and Catling, 1998). The anterior margin is curved (bell-shaped), more strongly in the female than in the male. The posterior wings are bright orange. The female of *A. occidentalis* has a wingspan of 24 mm, whereas the male is considerably smaller.

Working on citrus, Newton and Catling (1998) found that climatic conditions during the summer months, and particularly the abundance of young leaves and fruitlets at the beginning of the season, led to high populations later on. The insect has no dormant stage, but survives through winter at low population levels on citrus and alternative host plants. Avocado fruit is damaged mostly from January to April (Van den Berg *et al.*, 1999).

Parasitoids reported from this pest (Newton and Catling, 1998) include *Brachymeria microlinea* (Walker), *B. boranensis* Masi (Chalcididae), *Apanteles* spp. (Braconidae) and *Pristomerus* sp. (Ichneumonidae) (Prinsloo, 1984; Newton and Catling, 1998). Because the parasitoids of the citrus leafroller keep the pest in check, chemical intervention is usually not necessary.

Other leafrollers

The brown-headed leafrollers, *Ctenopseustis obliquana* (Walker) and *C. herana* (Felder and Rogenhofer), are the major pests of avocados in New Zealand. A complex of six leafroller species, which feed on both foliage and fruit, can cause up to 30% rejection of fruit for export from unsprayed orchards. An average of seven insecticide applications per year are used in most commercial orchards to combat the pests (Stevens *et al.*, 1996, 1998). Frequent applications of *Bacillus thuringiensis* have been shown to provide acceptable control of these pests despite the increased cost associated with the additional sprays: the advantage of being able to pick fruit immediately after spraying outweighs the extra cost (Stevens, 1997, 1999).

Gracillariidae

In Mexico, *Gracillaria perseae* Busk is considered a pest in the states of Oaxaca and Veracruz (Gallegos, 1983). The adult is 2–3 mm in length with whitish cream-coloured wings; the larvae mine the tender leaves, causing deformations; however, very seldom do they cause defoliation. In general, *G. perseae* is not considered an important pest. There is no knowledge of its life cycle or natural enemies (Gallegos, 1983). In Peru, *Phyllocnistis* spp. (Lepidoptera: Gracillariidae) can reduce tree vigour by mining on the leaves, whereas in Florida, *Phyllocnistis* spp. are considered a minor pest problem.

Apanteles n.sp. (Braconidae) and *Sympiesis dolichogaster* (Ashm.) (Eulophidae) have been reported to parasitize *G. perseae* in Cuba.

Geometridae

Citrus looper, Ascotis reciprocaria reciprocaria *(Walker)*

The citrus looper was recorded in South Africa in the early 1940s by Bedford (1943). It was also recorded on avocado in 1974 (De Villiers, 2001a). *A. reciprocaria reciprocaria* has been recorded in equatorial Africa, Cameroon, Malawi, Uga, South Africa, Zaire, Zambia and Zimbabwe. Besides citrus, avocado, coffee (Robusta and Arabica) and macadamia, the citrus looper has a wide range of host plants (Le Pelley, 1968; Hill, 1975; De Villiers, 2001a). According to Schoeman (1998), 78 plant species, including 62 indigenous and 16 exotic species, are hosts of the citrus looper larva. Examples of these are boxwood, kei apple, pepper tree (*Schinus molle*), *Acacia* sp., *Pappea* sp., *Rhus* sp., wattle, willow and *Ziziphus* sp. (Pinhey, 1975). It is considered to be of minor importance on avocado, coffee and macadamia in certain parts of Africa (De Villiers, 2001a).

On avocado, the larvae feed on the skin of young and mature fruit. Young larvae gnaw on the skin of the fruit superficially while on mature fruit the lesions on the skin caused by full-grown larvae are deeper, slightly penetrating the fruit flesh (De Villiers and Van den Berg, 1987a). In South Africa, Erichsen and Schoeman (1994) found the greatest damage by the citrus looper on cultivar 'Edranol', followed by cultivars 'Pinkerton', 'Hass', 'Fuerte' and 'Ryan' (De Villiers, 2001a).

Braconid larval parasitoids include *Rogas* spp., *Cardiochiles* sp. and the ichneumonid *Charops* sp. (Schoeman, 1998). Dipterous parasitoids include *Peribaea cervina* (Mesnil), *Exorista sorbillans* (Wiedemann), *Sturmia imberbis* (Falten), *Pales coerulea* (Jaennike), and *Muscina stabulans* (Falten). An entomopathogenic fungus, *Beauveria bassiana* (Balsamo) Vuillemin, also attacks the larvae (Schoeman, 1998).

Avocado loopers – Anacamptodes defectaria *(Guenée),* Epimeces detexta *(Walker),* Epimeces matronaria *(Guenée),* Oxydia vesulia transponens *(Walker), and* Sabulodes aegrotata *Guenée*

Several loopers, namely *E. detexta*, *E. matronaria*, *A. defectaria* (Guenée) and *O. vesulia transponens*, feed on avocado leaves in Florida, while *S. aegrotata* is reported in California (Bailey and Hoffman, 1979). In Florida, the most common of these is *E. detexta*, a medium-sized grey or greyish white moth. Young larvae are 0.6 cm or less in size, usually grey or greyish black. They grow rapidly to 3–4 cm in length. Older larvae are generally tan or greenish yellow in colour. *E. detexta* and *S. aegrotata* larvae also feed on flower panicles, even fruit, but prefer the tender growth in the upper part of the tree (Bailey *et al.*, 1986). In Florida, looper infestations appear to be somewhat seasonal and are more severe in spring and summer, generally becoming less of a problem in autumn and winter. The adult *E. detexta* is short-lived, and mates and oviposits soon after emergence from the pupa. Eggs are laid in narrow elongated masses, preferentially on needles of Australian pine (*Casuarina* sp.), and they hatch in about 5 days. The larvae grow rapidly and pupate 17–22 days after egg hatch. The pupal stage can last 10 days. Thus a full generation is expected to last between 34 and 37 days. Pupae drop to the ground and the adult emerges in 12 days to start the cycle again. Some avocados are culled because of damage from feeding on the fruit by two or three kinds of small caterpillars (Peña and Johnson, 1999).

Bailey *et al.* (1986) determined that the compound 6,9-nonadecadiene provides maximum trap catch of *S. aegrotata* males. Lannate 90 WP, lannate 1.8 L, permethrin (Pounce 3.2 EC, Ambush 2 EC) and *Bacillus thuringiensis* Berliner preparations are recommended in Florida for control of the avocado looper (Peña and Johnson, 1999).

Native natural enemies of *E. detexta* include the predators *Calleida decora* (Fabricius) and *Podisus maculiventris* (Say). *Alcaerrhynchus grandis* (Dallas), *Parapanteles* sp. and *Trichospilus diatreae* Cherian are natural enemies of *E. matronaria, A. defectaria* and *O. vesulia transponens*, respectively (Peña *et al.*, 1996). Several attempts to introduce exotic biological control agents, e.g. *Telenomus* sp. and *Trichogramma platneri* Nagarkatti, failed (H. Glenn, personal communication).

In Peru, *Sabulodes caberata* Guenée is common on orchards located near the jungle. Even though it may cause some defoliation, its attack is not frequent enough to demand the application of chemicals.

Ectropis sabulosa *Warren*

E. sabulosa can be a serious pest in north Queensland, Australia, where it feeds on leaves. It can completely defoliate trees, exposing fruit and resulting in sunburn. It can also cause severe damage to fruit when it feeds on the skin. Infestations are more severe near windbreaks and good control in an orchard can be obtained by spraying just a few rows near the windbreak, thus sparing the parasitoid *Apanteles* sp. nr. *vitripennis* Curtis which normally suppresses populations and prevents outbreaks (Waite and Pinese, 1991).

Cleora inflexaria *Snellen,* Lophodes sinistraria *(Guenée), and* Eucyclodes pieroides *(Walker)*

In Australia, the grey looper, *Cleora inflexaria* Snellen, the brown looper, *Lophodes sinistraria* (Guenée), and the bizarre looper, *Eucyclodes pieroides* (Walker), as well as the orange fruit borer, *Isotenes miserana* (Walker) (Tortricidae), are all minor pests which occasionally attack the fruit.

The giant looper, Boarmia (Ascotis) selenaria *Denis and Schiffermuller*

In Israel, the giant looper is the most important pest of avocado in regions where cotton is grown. It produces five generations a year, of which the most destructive are the first (spring) and second (early summer) (Wysoki and lzhar, 1986). Its reproductive behaviour was studied by Hadar (1983).

A long list of natural enemies of the giant looper in Israel was compiled in earlier publications (Wysoki and lzhar, 1980; Swirski *et al.*, 1988). Spiders contribute to some extent to biocontrol of the pest (Mansour *et al.*, 1985), whereas *Apanteles cerialis* Nixon (Braconidae) parasitizes young caterpillars, providing the highest parasitization (70%) in October and November (Wysoki and lzhar, 1981). Tachinid flies (Tachinidae), *Compsilura concinnata* Meigen and *Exorista* nr. *sorbillans* Wiedemann attack late instars of the giant looper and parasitization reaches its peak in late summer or autumn. Local egg parasites of the giant looper have not been recorded in Israel. Thus, two exotic wasps, *Ooencyrtus ennomophagus* Yoshimoto (Encyrtidae) and *Telenomus alsophilae* Viereck (Scelionidae), were introduced, unsuccessfully (Wysoki and lzhar, 1980). Approximately 16 million *Trichogramma platneri* Nagarkatti (Trichogrammatidae) were released from 1988 to 1990 in the avocado groves but have not yet been recovered (Wysoki *et al.*, 1988). In Israel, natural enemies suppress giant looper populations in the avocado groves effectively, but in regions where cotton is widely grown, pesticide-induced outbreaks of the looper may occur owing to the disrupted balance with its natural enemies by the drift of pesticides from aerially sprayed nearby cotton fields.

Infestations are controlled, when necessary, by preparations containing *Bacillus thuringiensis* var. *kurstaki*. Timing of the control measures is based on traps baited with virgin females and on scouting the groves for young caterpillars. Since mass production of giant looper virgin females for monitoring purposes is laborious and expensive, efforts were made to replace the virgin females with synthetic pheromone. (Z,Z)-6,9-*cis*-3S, 4R epoxynonadecadiene and (Z,Z,Z)-3,6,9-nonadecatriene were identified as sex pheromone components (Becker *et al.*, 1990). Bioassays performed by electroantennograph (EAG) in a wind tunnel gave positive results, but in field tests males were not sufficiently attracted to these two compounds (Becker *et al.*, 1990). Following experiments involving decapitation of the giant looper and subsequent PBAN injections, a third compound was revealed.

Noctuidae

American (cotton) bollworm, Helicoverpa armigera *(Hübner)*

H. armigera is known in South Africa as the American bollworm. The true American bollworm is the closely related *Helicoverpa zea* (Boddie), which is not found in southern Africa. *H. armigera* has been recorded throughout Africa, Asia, Australia and Europe (CIE, 1967).

Chelonus curvimaculatus Cameron (Braconidae), a *Pales* sp. and a *Drino* sp. (Tachinidae) were found as parasitoids of *H. armigera* on citrus in Rustenburg, South Africa (Vermeulen and Bedford, 1998). Other parasitoids, *Telenomus ullyetti* Nixon (Ichneumonidae), *Trichogrammatoidae lutea* Girault (Trichogrammatidae) (Parry-Jones, 1937) and *Paradino halli* (Curran) (Tachinidae) (Parry-Jones, 1938) were also found from this host feeding on citrus at Mazoe. According to Vermeulen and Bedford (1998), the indigenous natural enemies are not able to control the periodic outbreaks of *H. armigera*.

Psychidae

Oiketicus kirbyi *(Guild.)*

This insect is common in South American avocado orchards. Eggs are deposited within cases, eclosing in 30 days. Female larvae are 60–70 mm long, while male larvae are 45–50 mm. The larva remains within the case and moves around, selecting new feeding places. The pupa develops within the case; when the adults emerge, the apterous female remains within the case where it is fertilized by the male. Alate males are attracted by female sexual pheromones. Later, the female deposits the eggs in the case in which she spent her life and once the egg depositing is finished, she dies (Sanchez, 1983).

Larval feeding on leaves causes almost uniform 1 cm diameter holes. Likewise, *Oiketicus elongatus* Saunders (Psychidae) occasionally defoliates parts of individual trees. The ichneumonid *Chirotica* spp. (Ichneumonidae) is its natural enemy.

Other lepidopteran defoliators

In Mexico, *Copaxa multifenestrata* (Heinrich-Shaffer) (Saturnidae) is considered a specific but minor pest of avocado. It feeds on older leaves, whereas *Papilio garanas garanas* Hubner (Papilionidae) feeds on tender leaves and may cause some economic damage (Hernandez *et al.*, 2000).

Pyrrhopyge chalybea Scudder (Hesperiidae), or the confetti worm, eventually defoliates avocado trees by making holes in the leaves. The adult deposits its eggs on the foliage, eclosing within 20 days; the neonate larva chews a circular hole in the leaf, folding and hiding in the damaged area; the larva is active during the night. Two generations are observed per year, the first occurring in March and the second in August.

The pine emperor, *Imbrasia cytherea* (F.) (Saturnidae), is an important defoliator of *Pinus* spp. It also attacks a few other exotic and a large number of indigenous plants (PPRI, 1970). The subspecies *Imbrasia cytherea clarki* Geertsema feeds on avocado leaves in Africa

(Van den Berg, 1973). In Africa, *I. cytherea clarki* has been found from Zimbabwe to KwaZulu-Natal in the south. The other subspecies, *I. cytherea cytherea* (F.), occurs further south in the Eastern and Western Cape Provinces (Geertsema, 1971). *I. cytherea clarki* larvae are present from March to October, with damage taking place during the driest and coldest months of the year. The egg stage lasts from 26 days at 20°C to up to 84 days at 14°C. In the field, at approximately 15°C, the five larval instars are completed in about 100 days. Pupation takes place in the soil beneath the trees. The pupal stage lasts about 8 months, mainly during the summer (Van den Berg, 1975).

In Brazil, *Papilio scamer scamer* (Boiusduval) and *Saurita cassra* (L.) can cause serious injury to avocado leaves. *P. scamer scamer* is a brown moth, approximately 80 mm in length. *S. cassra* is a 30-mm-long brown moth. The caterpillars of the first species are whitish brown and during their later instars change to a green colour. Larval instars of *S. cassra* are predominantly dark in colour (Medina *et al.*, 1978).

Fornazier *et al.* (1997) suggested that the cultivars 'Ouro Verde', 'Primavera' and 'Briosqui' are the least attacked by an unidentified Geometridae.

Natural enemies of lepidopteran defoliators

Gallegos (1983) reported that in Mexico, *Trichogramma fasciatus* Perkins spp. have been collected parasitizing eggs of *Pyrrhopyge chalybea* Scudder, and the Eupelmidae *Anastatus* spp. have been found parasitizing its larva. The most important natural enemies of *I. cytherea clarki* include the eupelmid egg parasitoid *Mesocomys pulchriceps* Cameron, three other egg parasitoids, and larval and larval-pupal parasitoids. The most effective predator of *I. cytherea clarki* was found to be the Cape chacma (CK SP) baboon, *Papio ursinus* (Kerr). Other important predators include the vervet monkey, *Cercopithecus aethiops* (L.), and birds such as the Cape raven, *Corvultur albicollis* (Latham). Furthermore, viruses and three species of fungi were found to attack the larvae and pupae (Van den Berg, 1974).

ORDER DIPTERA

Tephritidae

Fruit flies – the Mediterranean fruit fly Ceratitis capitata *(Wiedemann), the Natal fruit fly* Ceratitis rosa *Karsch, and the Marula fruit fly* Ceratitis cosyra *(Walker)*

The Mediterranean fruit fly enjoys a cosmopolitan distribution (Kok and Georgala, 1978; Barnes, 1983; White and Elson-Harris, 1992). According to White and Elson-Harris (1992), the Marula fruit fly and the Natal fruit fly are widespread throughout Africa. The Mediterranean fruit fly has been recognized as a problem for avocado in Central America (Mitchell *et al.*, 1977) and appears in Australia and Israel as well (Ebeling, 1959). It has been shown that fruit fly maggots do not develop in the avocado fruit of commercial cultivars under normal orchard practices (De Villiers and Van den Berg, 1987a; Du Toit and De Villiers, 1990). Furthermore, the avocado is not considered an ideal host for fruit fly development (Brink *et al.*, 1997). There are, however, isolated instances in which larvae were found in over-ripe avocado fruit rotting on the ground underneath the trees.

In Hawaii, Armstrong *et al.* (1983) showed that Sharwil avocados could only be infested with *C. capitata* if fruit flies were confined to artificially damaged avocado fruit in cages on the tree. These authors concluded that avocados harvested with the stem attached and protected from postharvest fruit fly infestations present no danger of fruit flies entering another country. However, avocado fruit can be damaged by the ovipositor of the fruit fly when it lays its eggs. These holes develop into typical cracks or star-shaped lesions (Du Toit *et al.*, 1979). Such damage has also been observed on some commercial varieties in Israel.

The first report on *C. capitata* on avocado fruit in Israel was from Rivnay in 1936, with almost all the fruit being infested (Ebeling, 1959). Later, it came to be considered a major pest of cultivated fruits (Avidov and Harpaz, 1969) and avocado was listed as a host among other subtropical fruits by Swirski and Arenstein (1970) and mentioned as a host of

C. capitata by Freidberg and Kugler (1989). Recently, infestations have been reported on cv. TX- 531 and on newly introduced summer varieties (S. Ben Yehuda, personal communication). Adults of these fruit flies have clear wings with light and dark brown dots and brownish veins. The wings are mostly held in a drooping position (Du Toit, 1998).

The Natal fruit fly is the largest of the three species (6.0 mm in length), followed by the Mediterranean fruit fly (4.5 mm) and the Marula fruit fly (4.0 mm). The thorax of the Natal fruit fly is dorsally greyish brown, with small black and white patches posteriorly and a white patch on either side. Natal fruit fly males have black feathery bristles on their middle pair of legs. The thorax of the Mediterranean fruit fly is black with ivory white markings while the thorax of the Marula fruit fly is brown with three relatively large shiny black spots on each side and two smaller ones further back (Van den Berg *et al.*, 1999).

According to Du Toit (1998), the eggs of fruit flies are white, about 1 mm long and banana-shaped. They are usually laid in clumps in the fruit. Larvae are broad and truncate in the rear, tapering anteriorly to a point. The larvae are creamy white and up to 9 mm long. The mature larva has the ability to flex itself and make short jumps. Pupation takes place in the soil. The puparium is brown, cylindrical and up to 6 mm long.

Depending on the presence of hosts and on climatic conditions, up to 15 generations can be completed per year (Du Toit, 1998). Adults overwinter in evergreen shrubs and trees (Ripley and Hepburn, 1930). Adult populations reach peak numbers in the late summer and autumn (Du Toit, 1998). Most of the damage to avocado fruit is done during these periods.

The larval parasitoids *Opius concolor* Szepligeti, *O. humilis* Silvestri (Braconidae) and a pupal parasitoid *Trichopria capensis* Kieffer (Diapriidae) attack *C. capitata* and *C. rosa* (Searle, 1964; Annecke and Moran, 1982). Futhermore, *Tetrastichus* sp. (Eulophidae) was reared from the pupae of both *C. cosyra* and *C. rosa* (Grové, 2001b). *Pheidole megacephala* F. is a predator of *C. capitata* (Searle, 1964).

Fruit flies can only be controlled successfully if a combination of actions is taken (Du Toit, 1998). Du Toit (1998) suggested that plants that serve as alternative host plants for fruit flies be eradicated. These include bug weed, bramble and wild-growing guavas. Secondly, routine orchard sanitation should be carried out. Thirdly, monitoring adult fruit flies using traps baited with sex pheromones would determine when a population build-up occurs. In most countries, baiting is carried out with either mercaptothion or trichlorofon, applied with sugar or a solution of protein hydrolysate.

Bactrocera spp.

In Australia, the Queensland fruit fly, *Bactrocera tryoni* (Froggatt), attacks avocado fruit from the time it is about half-size. The damage from oviposition causes small star-shaped cracks in the skin. This damage is only aesthetic, as the eggs do not hatch in the hard, green fruit. Although they will sting rough-skinned cultivars (e.g. Hass), the damage is usually not visible and is of no consequence. However, in cultivars 'Fuerte', 'Wurtz' and 'Rincon', the scars are quite noticeable and sprays may be required to prevent excessive damage. Cover sprays of dimethoate have been used, but in an integrated system protein autolysate bait sprays are effective and the preferred option (Waite and Pinese, 1991).

Fruit flies of the *Bactrocera dorsalis* complex, probably *B. philippinensis* and *B. papayae*, attack avocado fruit produced in the Philippines (Drew and Hancock, 1994). A variety of thin-skinned types are grown throughout the islands and have been found to be infested with fruit fly larvae (G.K. Waite, 1999, personal communication).

Anastrepha spp.

In Mexico, Enkerlin *et al.* (1993) found that under artificial laboratory conditions, the fruit fly species *Anastrepha serpentina* Wied., *A. ludens* Loew and *A. striata* Schiner infest cv. 'Hass'. However, under field conditions, 'Hass' avocados were not infested by fruit flies. Therefore, it is doubtful that these fruit flies would use 'Hass' avocado as a normal host.

ORDER COLEOPTERA

Chrysomelidae

Avocado beetles – Monolepta apicalis *(Sahlberg),* Monolepta australis *(Jacoby) and* Monolepta bifasciata *Hornst*

Since early 1990, the avocado beetle *M. apicalis* has been a pest of avocado orchards, mostly in the higher lying areas, in the Kiepersol region of South Africa. The infestations were more severe in orchards situated next to blue gum plantations and natural vegetation. *M. apicalis* was also reported in two other areas in the Mpumalanga Province of South Africa (Claassens, 2001). *M. apicalis* occurred as a sporadic pest in South Africa for two seasons. It fed on avocado twigs, leaves, leaf stalks and fruit. In certain instances, the epidermis of the fruit was completely removed, causing a brown blemish covering the entire fruit. Lighter damage appeared as darkish holes of about 2 mm in diameter. Very young leaves were often eaten completely, while older leaves were mainly eaten from the ventral side, resulting in a skeletonized appearance (Claassens, 2001). *M. apicalis* was also recorded from Namibia in Central Africa (G.L. Prinsloo, Plant Protection Research Institute, Pretoria, 1993, personal communication). *M. bifasciata* (Coleoptera: Chrysomelidae) occasionally damages foliage in the Philippines (Cendana *et al.*, 1984). The red-shouldered leaf beetle, *M. australis* (Chrysomelidae) can defoliate trees when swarms descend on an orchard. In Australia, the black swarming leaf beetle and the brown swarming leaf beetle, *Rhyparida* spp., cause similar damage to avocado (Waite and Pinese, 1991). *Bridelia macrantha*, litchi and macadamia leaves have been reported as hosts of *M. apicalis* (Erichsen and Schoeman, 1993). Other recorded hosts are *Hibiscus trionum, Triumfetta rhomboidae, Sida rhomboidea* and *Protea* spp. (Claassens, 2001). The adult *M. apicalis* is bright red with shiny black elytra. The insects vary from 3 to 5 mm. The life history of *M. apicalis* has not been studied in South Africa. In Australia, the closely related *M. australis* (Jacoby) lays eggs in the soil. The larvae feed on grass roots and are about 5 mm long when fully grown. They pupate in the soil. Their life cycle takes about 2 months, with three to four generations occurring annually. Adults emerge after rain (Fay and De Faveri, 1990). *M. apicalis* is active in avocado orchards in the early mornings. During the warmer parts of the day, the insect seeks shade underneath the leaves and on shady stems. When disturbed, *M. apicalis* will hide behind the leaves or drop and fly away.

In Kiepersol, South Africa, *M. apicalis* was noticed in avocado orchards in 1992 and 1993. In a heavily infested orchard, 50–100 insects per young tree were counted. During 1998, the insect was still present, but in very small numbers. The crop loss due to *M. apicalis* also varied on individual trees from about 5% up to 100% (Du Toit and Claassens, 1993; Erichsen *et al.*, 1993). *Monolepta* spp. are especially threatening because they produce an aggregating pheromone and form swarms that can invade an orchard and cause extensive damage in a short period (Fay and De Faveri, 1990).

Insecticides such as carbaryl, dimethoate, endosulfan and trichlorfon have been used effectively in Australia for the related *Monolepta* spp. (G. Waite, 1992, personal communication). Spot sprays may be adequate once swarms are detected. In South Africa, scouting for beetles should begin in November, in the early morning or late afternoon, under cool conditions when the beetles are more likely to be on the upper side rather than on the underside of the leaf (Erichsen, 1993). The beetles aggregate on windbreaks of *Eucalyptus torreliana*. Growers regularly control swarms on the windbreaks before they invade the avocado orchard. Fay and Defaveri (1990) found that numerous individual beetles present at flowering (i.e. not in a swarming phase) were actually beneficial for pollination. However, when swarms invade, there is no alternative but to spray, and action must be quick to prevent damage. Carbaryl and endosulfan are most commonly used.

Blue-green citrus nibbler, Colasposoma fulgidum *Lefèvre*

The blue-green citrus nibbler is indigenous to South Africa and has been collected on a variety of wild trees and shrubs (Webb, 1974). It also feeds on avocado (Bedford, 1998a), guava (Joubert, 1936) and pecan (De Villiers *et al.*, 1987a). Heavy infestations of *C. fulgidum* occur on rare occasions and damage to the young leaves and avocado fruit is seldom serious. However, young fruit can be scarred, reducing the mature fruit's marketability.

Eggs are about 0.9–1.0 mm long, cylindrical and pale yellow. They are laid in egg clusters which are mostly deposited on the extreme tips of leaves (Bedford, 1998a). The egg stage lasts 2–4 weeks. Soon after the larva has hatched, it falls to the ground, and enters the soil (Joubert, 1936). The newly hatched larva is about 1 mm in length (Bedford, 1998a) and 8 mm when mature. The larva has a whitish colour and an irregular translucent mark on the posterior abdominal segments (Joubert, 1936). The larvae live in the soil at a depth of about 75 mm and the larval stage lasts an estimated 9 months (Joubert, 1936). The pupa is formed in a cell constructed of soil particles, and this stage lasts about 1 month (Joubert, 1936). The adults are mostly bright blue-green or metalic green. They are 4.5 to 6.9 mm long and 2.6 to 4.2 mm wide (Joubert, 1936). Soon after the adults emerge from the soil, they congregate on the foliage of indigenous trees such as the bush willow, *Combretum erythrophyllum*. Some of the females lay more than 400 eggs and they live from about 2 to 4 months (Joubert, 1936). Two (or possibly three) unidentified species of parasitoids have frequently been bred out of *C. fulgidum* eggs (Bedford, 1998a). It is normally not necessary to control this minor pest on avocado trees.

Other chrysomelids

In Brazil, *Sternocolaspis quatuordecimcostata* Lefevre is a common pest of other fruit and forest trees. The greenish blue, 8-mm-long adult has longitudinal stripes on the elytra. The female deposits its egg masses near the soil surface. The adults feed on new foliage and green fruits; injury to fruits is characterized by shallow grooves on the fruit epidermis. The major peaks of this pest are observed during the rainy season in the state of São Paulo, between October and February (Medina *et al.*, 1978).

Bostrichidae

Black giant bostrychid, Apate monachus *F. (Bostrichidae)*

The black giant bostrychid is found attacking avocados in different South American and Caribbean countries. The female of *A. monachus* oviposits on dry branches and trunks. The eggs are laid on areas lacking tree cortex, at random, and they eclose in between 7 and 8 days. The number of eggs per female ranges between 64 and 106. Larval development lasts 67 days, the pupa 8 to 9 days, and adults live for up to 40 days. They are active during the night and the two sexes are found copulating within the galleries during daylight hours. The complete cycle lasts 87 to 190 days. In Brazil, *Apate terebrans* (Pallas) and *Acanthoderes jaspidea* (Germar) affect branches, by ovipositing or sometimes by larval boring and adult feeding (Medina *et al.*, 1978). Descriptions of adults are provided by Medina *et al.* (1978). The adults attack healthy plants, they use their mandibles to construct their galleries within branches and the trunk. Injured branches are recognized by the yellowing of their leaves; these branches commonly break down easily.

A Bethylidae (Hymenoptera) and *Clyptodoryctes* (Braconidae) are registered as natural enemies of *A. monachus*.

Curculionidae

Conotrachelus perseae *Barber,* Conotrachelus aguacatae *Barber, and* Conotrachelus serpentinus *(Klug)*

These three species of *Conotrachelus* are known or suspected pests of commercially grown avocado fruits in Mexico and Central America, but are poorly understood. Taxonomic keys and diagnoses are given by Whitehead (1979), expanding the information provided by Barber (1919). Muñiz and Barrera (1958) provide a taxonomic key that includes species of *Conotrachelus* collected from tropical and subtropical America.

C. perseae is a small, shiny, almost black weevil not usually observed in orchards in which *Heilipus lauri* is prevalent. However, it can cause up to 85% fruit loss. It is observed in the areas of Queretaro, Michoacan, Jalisco, Puebla, Morelos and Guanajuato. The adults are 7 mm long, and they can copulate immediately after emergence. Adults are nocturnal, and remain hidden on folded leaves or in any crevices in the trunk (Plate 56). The adults can attack newly formed fruits, and the females deposit one to four eggs per fruit, but sometimes the number can be as high as 70. Eggs are whitish, and eclose in 7–10 days. The larva develops in the seed and lasts 20–35 days. The number of larvae per fruit can be as high as three or four. The larvae leave the fruit to pupate in the soil at a depth of 5 cm. The life cycle from egg to adult lasts 42–75 days. The first generation starts in January and February and lasts approximately 10 weeks. The second generation starts in April and ends in June or July (Anonymous, 1991). The female prefers to oviposit in the stylar area of the fruit; more damage can be found in the lower tree canopy than in the upper part (Hernandez *et al.*, 2000). The latter authors recommend collection of fallen fruits, weed removal and fallowing the soil to expose pupae. The only biocontrol agent observed to date is *Beauveria bassiana* affecting larvae and pupae.

C. aguacatae causes economic losses in the area of Queretaro, Mexico, ranging from 20 to 80%, depending on the control tactics used. Canseco (1971) explains that there is some confusion surrounding the distribution of the species *C. perseae* and *C. aguacatae*. The species *C. aguacatae* can be confused not only with *C. perseae*, but also with *C. sapotae*. However, the major difference is in the shape of the aedeagus. Canseco (1971) provides a detailed description of the different stages of *C. aguacatae*. It is a stout reddish brown univoltine weevil 4 mm in length that exhibits diurnal activity. The female deposits its eggs on tender branches, but it has been observed damaging older branches. The egg stage lasts 10–12 days, while larvae can live up to 117 days. Pupae will develop in approximately 17–19 days andthe adults live for approximately 34–44 days. Overall, one generation lasts 169–192 days (Hernandez *et al.*, 2000). An extensive taxonomic description of different stages of the weevil is provided by Kissinger (1957), Muñiz (1960) and Diaz (1971). Major activity is observed during the months of June and July. However, Garcia *et al.* (1986) demonstrated that in the warm regions of Yautepec, the presence of adults is observed from May to November; eggs could be collected between July and January and pupae between June and February, while in the more temperate area of Tetela del Volcan, adults are found between June and February, eggs are observed in September, larvae are found throughout the year and pupae from May to December. These authors suspect that some of the weevils could be diapausing as larvae from January to April. Canseco (1971) reports that *C. aguacatae* spends the winter as an adult, hiding in the fallen leaves. During the spring, it moves to the foliage in the tree and starts feeding on the leaves. Later on, the copula starts and as soon as the fruits are formed, they start feeding on them. Emergence of the adults coincides with the first rains. Eggs are inserted in the fruit pulp by the females. After each egg deposition, the female seals the site with a gelatinous substance. Each female will deposit 70 individual eggs. The eggs eclose within 7–10 days and the larva bores through the pulp and seed. When it reaches the seed, the galleries within the seed are sinuous. Close to the end of the larval stage, it bores through the pulp again and falls to the soil surface. It pupates inside the soil at 2.5–5 cm from the surface. The pupal stage can last 15–30 days. The number of overlapping generations is estimated at two (Canseco, 1971). Gallegos (1983) reports that most of the infestations are found in small orchards where the regional cultivars are grown without any type of pest management, while Coria-Avalos (1993) indicates that this pest can be found in orchards located in warm climates. Damage is concentrated in the upper half of the tree top (Salazar-Garcia and Bolio-Garcia, 1992). Sticky green or blue traps are more attractive than white, yellow or red traps. Hernandez *et al.* (2000) recommend cutting and removing infested branches, then applying organophosphates. Insecticides can be applied to the soil to control the larva when it falls to the ground or to control adults after their emergence. Aldrin is recommended, as well as parathion (Canseco, 1971). Some chalcidoid wasps have been observed as active biocontrol agents of these weevils. However, the species have not yet been identified. Huerta *et al.* (1990) collected *Camponotus* sp. within the galleries made by *C. aguacatae* and speculated that *Camponotus* sp. could be preying on *C. aguacatae*. The same authors suggested that the species *Oncophanes* (Braconidae), *Euderus* (Eulophidae), *Erythmelus* (Mymaridae) and *Eurydinoteloides* (Pteromalidae) could be associated with *C. aguacatae*. *Steinernema feltiae* (= *bibiones*) Poinar provided 10% larval mortality, whereas *Heterorhabditis heliothidis* Khan, Brooks and Hirschmann provided 45–75% mortality under laboratory conditions (Huerta *et al.*, 1990).

Large borer weevil, Heilipus lauri *Bohemann*

The large borer weevil damages up to 80% of avocado fruits, causing extensive fruit drop (Plate 57). It is distributed in Mexico in the states of Hidalgo, Morelos, Puebla, Veracruz and Guerrero. The dark brown (with two incomplete yellow bands on the elytra) adult is 14–17 mm long, and emerges from the fallen fruits (Ebeling, 1950; Salgado and Bautista, 1993). It can fly and regularly mates 2.5 months after emergence. The female deposits her eggs under the epidermis of the developing fruit, making a 'half-moon' puncture. The small oval eggs are 1 mm long and change from pale green to cream-coloured. The female deposits one to two eggs per fruit and a total of 36 eggs month^{-1}. Twelve to 15 days after oviposition, the legless larva bores through the pulp into the seed, where it feeds and spends its larval and pupal stages (Ebeling, 1950). The larva has five instars that last approximately 54–63 days. The late instar larva measures approximately 2.5 cm (Anonymous, 1984b). Arellano (1975) provides a key for the identification of *H. lauri*. The pupal stage lasts 15 days. Emerging adults feed on the foliage and live up to 4 months.

In the case of fallen fruits, the larvae sometimes leave the fruit and enter the soil to pupate. Because of larval feeding, rotting of the pulp, mainly near its tunnels, and partial or total rotting of the seed occur, eventually resulting in premature fruit drop. The adult feeds on leaves, buds and sprouts, as well as on the fruit (Ebeling, 1950). Further information on development is provided by Salgado and Bautista (1993). In Mexico, two generations are observed. The first occurs from January to August and the second from July to February (Anonymous, 1991).

Heilipus catagraphus *(Germar)*

In Brazil, *H. catagraphus* attacks branches of avocado. According to Lourencao *et al.* (1984), *H. catagraphus* also feeds on the epicarp and avocado pulp without injuring the seed. The larva of this species attacks different plant species within the families *Lauraceae* and *Annonaceae*. An anonymous (1984a) report recommends the use of $CaCo_3$ to prevent *H. catagraphus* egg deposition.

Copturomimus perseae *Hustache*

In Venezuela and Colombia, *C. perseae*, a small greyish weevil, 3.7–5 mm long with a black transversal band on the elytra, causes damage to branches and stems. Both the larva and the adult bore through them. The legless, curculionid larva is 10–12 mm long and whitish. The damage is normally concentrated on the branches; infested branches show white dust exuding from the galleries made by this insect; affected parts exhibit slow dieback and eventually the tree dies (Saldarriaga, 1977; Avilan *et al.*, 1997). Cultural control, collecting and burning infested branches, is the method suggested by Avilan *et al.* (1997).

Scarabaeidae

In Peru, *Oncideres poecila* Bates (Coleoptera: Scarabaeidae) causes girdling of branches. *O. poecila* deposits its eggs on fallen branches; collection of these branches is recommended to reduce the pest problem. *Macrodactylus mexicanus* Burmeister, a univoltine, highly polyphagous insect is associated with avocado and vegetation found in the hot and temperate areas of central and western Mexico.

Scolytidae

Corthylus, Aegeria and Xylosandrus species

In Mexico, *Corthylus* spp. are considered avocado pests. The adults are 1.2–4.3 mm long, with a stout body and colour that varies from light brown or dark brown to black. It displays sexual dimorphism; the females have a concave front with obvious setae, the antennal club is very large, its anterior margin decorated with a long tuft of setae. Eggs are white, translucent and elliptical. The white larvae are legless and C-shaped. Pupae are exarate.

Different species have two or more generations per year, depending on the altitude and latitude conditions where they are found. The males initiate attack by selecting trunks or branches of live or recently cut trees. The male makes an entry gallery and the initial part of one or both transverse oviposition galleries. The female mates with the male and enters the gallery. Here she digs two oviposition galleries that extend perpendicular to the axis of the trunk or branch, extending approximately 5 or 10 mm deep. The female makes a chamber where ambrosia fungi are inoculated. One egg is deposited in each chamber. Upon emerging, the larvae feed on the fungi that develop in the surrounding debris.

Trunks or branches of weakened or felled trees are attacked. The greatest damage is caused by the introduction of staining fungi in the wood. In Taiwan, ambrosia beetles are reported to attack avocado trees in poor health. Maintaining the health of the tree through good agronomic practice helps reduce attack by these pests (Hung and Jong, 1995).

The avocado bark borer, *Aegeria* sp., feeds beneath the bark of avocado trees in the Philippines. Sap may ooze from the wound and branches may be weakened so that they snap in strong winds. The pest is of minor importance (Cendana *et al.*, 1984).

In Florida, ambrosia beetles, *Xylosandrus* sp., burrow into three trunks, stems and branches. Infested trees are regularly stressed before the attack, but frequently the trees appear to be healthy and vigorous. Fungi accompany the beetles and develop mycelia in tree tissues. As a result, the portion of the tree terminal to the burrow entrance usually dies. White crystals of sap around the burrow entrances are evidence of infestation. Beetles are brownish to almost black and small, about 0.05 × 0.13 cm. Eggs and beetle larvae are white and found in the galleries. Larvae feed on the mycelia of 'ambrosia' fungi growing in the galleries (Peña and Johnson, 1999).

Conclusions

Avocado pest management practices are affected not only by the domestic and export fruit market, but also by consumer attitudes towards health concerns and the cosmetic appearance of the fruit. In general, avocado pest management is largely dependent on the use of pesticides. Costs of pest control, and the current lack of information and lack of registration for a new generation of pesticides (Wysoki *et al.*, 1999), is complicating the continuation or development of pest management programmes in avocado. From the information compiled herein, we can deduce that countries such as South Africa, Mexico, Israel, Australia, and the USA are making an effort to maintain and develop avocado pest management programmes. In Israel, the IPM system in avocado orchards (until the introduction of the orchid thrips) was based on large-scale use of insect pathogens, augmentation of local natural enemies, and importation and acclimatization of exotic natural enemies – parasitoids and predators. Other countries, such as New Zealand and Brazil, are either beginning to work towards this goal, or are accumulating basic information on the biology and life histories of the pests and describing damage inflicted to the crop, while no research is being conducted on other aspects of pest management, i.e. economic injury levels, monitoring, sampling, biological and autocidal control.

Some of the large differences in information provided by different countries could be based on economics. For instance, those countries, where avocado is a major source of income for growers, i.e. the USA, Australia, Israel and Mexico, place more emphasis on pest knowledge and responses to grower needs. Those countries in which avocado is still grown on small farms or in backyards and produced for domestic consumption exhibit very little involvement or research aimed at rational pest management. Sometimes, within the same country contradictory attitudes are observed. For instance, Gallegos (1983) reports that in the Michoacan region, there are big differences in avocado pest management. Some orchards may receive at least 12 insecticide applications per year, while others will not receive any pesticides. According to Colin (1990), to avoid important infestations of either mites or thrips, it is necessary to evaluate different pesticides. The information on pesticides, however, is concentrated on the use of organophosphates and sulphur (Colin, 1991).

The usual recommended control tactics against avocado pests in Colombia concentrate on cultural control (removing and burning infested parts) or application of chemical products.

In Florida, where the most important cultivars grown are Guatemalan or West Indian in origin, avocado pests are entirely controlled through the use of chemical products, applied on a calendar schedule. This practice continues despite the presence of several native biological control agents for some insect pests and several attempts to introduce exotic biological control agents and the development of IPM techniques for others, such as mirids (Peña *et al.*, 1996).

The generally followed rules for the improvement of IPM in avocado and in other subtropical fruit crops are: (i) search for and import suitable and effective natural enemies, propagate them, and distribute them among the infested plants; (ii) import plant material and natural enemies with strict enforcement of quarantine instructions to avoid introduction of other pests or hyperparasitoids; (iii) conserve and augment local natural enemies; (iv) carefully monitor the pests, study their biology and behaviour; (v) do not interfere with the biological equilibrium, use insecticides only if absolutely necessary, avoid the use of non-selective insecticides and use only highly selective ones; (vi) avoid drift of insecticides from adjacent plots by limiting sprays and improving spray techniques to prevent any detrimental effect on natural enemies; (vii) plant pest-resistant cultivars and avoid planting susceptible varieties; (viii) conduct surveys for new pests and pest problems in order to find an appropriate solution as soon as possible, before full establishment of the pest or acceleration of the pest problem by augmentation of local natural enemies.

Pollinators

Avocado flowers and flowering

Flower morphology

The avocado flower is circular, about 1 cm in diameter. It is bisexual, having both female and male reproductive organs. The flower consists of one pistil and six trimerous alternate whorls: two whorls of greenish yellow tepals, three of stamens and one of short arrow-shaped staminodes (Fig. 8.1). Each of the nine stamens bears an anther with four pollen sacs. The valves of the pollen sacs turn up while opening, and draw out a pack of pollen grains attached to them. The avocado pollen grain is spherical and is covered with numerous conical spinules. A pair of oval, yellow-orange nectaries is located at the base of each of the three inner stamen filaments. The three yellow-orange staminodes also function as nectaries. The pistil is located centrally, and consists of a greenish, ball-shaped, superior ovary, a hairy, slender, cylindrical style and a cone-shaped stigma, 0.3–0.6 mm in diameter. The stigmatic surface is composed of elongated papilla cells. Usually, it is somewhat depressed in its centre. All cultivars have a similar flower structure, though they may differ slightly with respect to flower size and some other features (Nirody, 1922; Stout, 1933; Bergh, 1969; McGregor, 1976; Davenport, 1986; Ish-Am and Eisikowitch, 1991b).

The flowers are carried on terminal panicles. A typical panicle carries a few hundred to more than a thousand flowers. New flowers open daily during the long flowering period. Each flower opens twice (dianthesis), usually on two successive days, in a dichogamous–

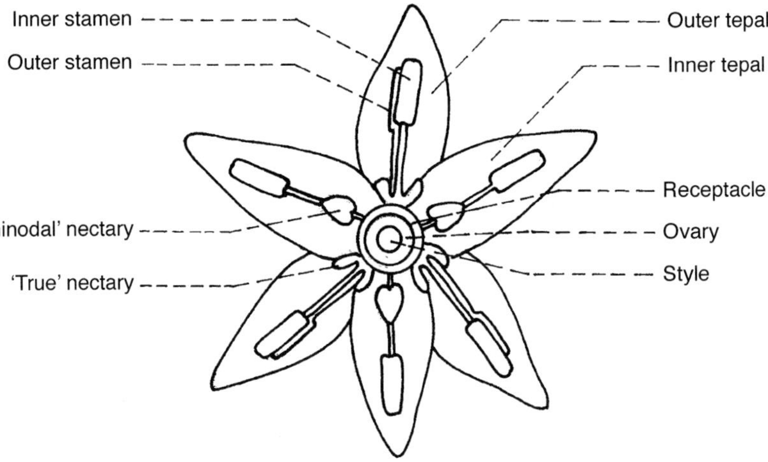

Fig. 8.1. The floral parts of the avocado flower (Ish-Am and Eisikowitch, 1991b).

protogynous rhythm (Fig. 8.2). During the first female (pistillate) opening, the stigma is exposed and white. The stamens with the closed valves are parallel and close to the tepals and nectar is secreted by the three staminodes. At the beginning of the second male (staminate) opening the tepals are 10–20% longer. The three inner stamens move towards the style, and their filaments lengthen until their anthers more or less cover the stigma. The six outer stamens also lengthen, move up and hold a position of about 45° to the pistil. Anther dehiscence takes place about 1 to 2 h later, first by the two lower valves of each anther, and about 1 h later by the two upper valves. Nectar is secreted by the six nectaries, while the three staminodes move closer to the ovary and turn brown. The stigma gradually shrinks, and also turns brown (Nirody, 1922; Stout, 1933; Bergh, 1969; McGregor, 1976; Davenport, 1986; Ish-Am and Eisikowitch, 1991b).

During anthesis, flower shape changes gradually in a regular sequence, which was divided by Ish-Am and Eisikowitch (1991b) into ten distinct morphological stages (Fig. 8.2). These flower stages are similar in all cultivars, and are somewhat influenced by the weather. On hot and dry days, the male flower stigma and staminodes dry upon anthesis, and the tepals bend towards the pedicel. In cool weather, however, both stigma and staminodes of the male stage flower stay fresh during the first dehiscence stages (D_1, D_2, Fig. 8.2) and the tepals spread upright to the style, or even remain only partially open. Under cool conditions, some cultivars open their female flower only partially or not at all (Nirody, 1922; Stout, 1933; Bergh, 1969; Ish-Am and Eisikowitch, 1991b).

Flowering behaviour

All avocados display unique flowering behaviour, which may be termed 'diurnally synchronous dichogamous protogyny, with intermediate closing'. The flower opens twice, first as a female and then as a male. Each flower stage opening and closing occurs nearly synchronously within the tree (and the cultivar). Based on flowering rhythm, all avocado cultivars are divided into two complementary flowering groups. In warm weather, 'Group A' cultivars open female stage flowers from the morning till noon, and male stage flowers during the afternoon. 'Group B' cultivars, on the other hand, open female stage flowers in the afternoon and male stage flowers during the morning hours (Fig. 8.3). This flowering rhythm may be termed 'temporal dioecy' (Clark, 1923; Stout, 1923, 1933; Robinson and Savage, 1926; Bergh, 1969; McGregor, 1976; Papademetriou, 1976b; Davenport, 1986; Ish-Am and Eisikowitch, 1991b).

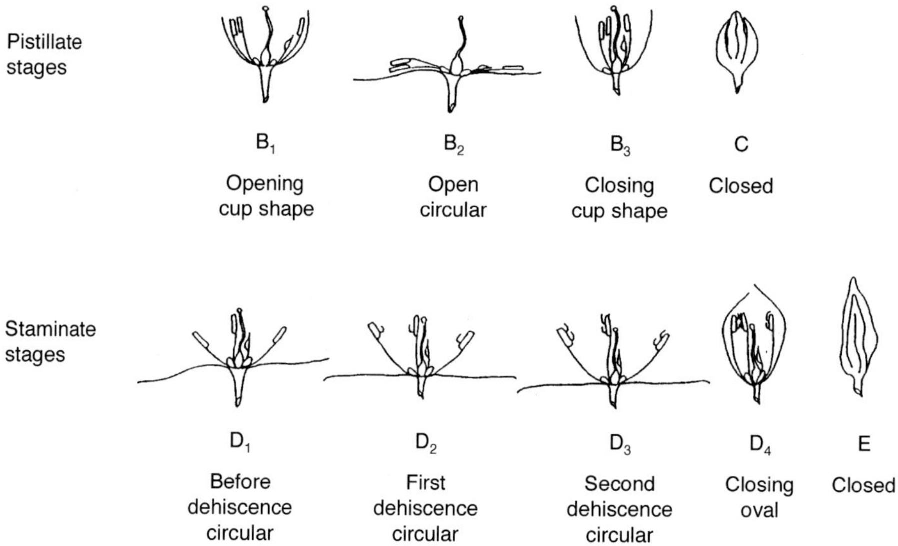

Fig. 8.2. Morphological stages of the avocado flower, at its female (pistillate) and male (staminate) openings and closings (Ish-Am and Eisikowitch, 1991a).

Many avocado cultivars also present a daily phase of overlap, within the tree, of coinciding dehisced male stage and open female stage flowers (Fig. 8.3), which usually takes place for a period of 1–3 h. In some cultivars, this overlap period is almost constant, regardless of temperature, but in others, it gets shorter during warmer weather and may disappear on hot days. In cool weather, there is a significant delay in the whole daily flowering sequence, which may result in complete reversal of the times of day when female and male stage flowers are open (Clark, 1923; Stout, 1923, 1933; Robinson and Savage, 1926; Lesley and Bringhurst, 1951; Gustafson and Bergh, 1966; Bergh, 1969; McGregor, 1976; Papademetriou, 1976b; Sedgley, 1977; Sedgley and Annells, 1981; Sedgley and Grant, 1983; Davenport, 1986; Ish-Am and Eisikowitch, 1991b).

Although the beginning (and the end) of each flower stage manifestation is synchronized within the tree and among trees of a cultivar, the individual flowers do not open and close simultaneously. Rather, they proceed through each stage individually, over a period of 2–3 h. Flowers that open earlier also proceed earlier to the following stages. Therefore, during most of the flowering day, several consecutive flower stages occur concurrently within the panicle (Ish-Am and Eisikowitch, 1991b).

Pollination

Pollination modes

Cross-pollination

Cross-pollination occurs between group B male stage (Plate 58) and group A female stage flowers (Plate 59) in the morning (in warm weather), and vice versa in the afternoon. It may also occur among different cultivars of the same flowering group, when there is a period of overlap between female and dehiscing male openings (Stout, 1923, 1933; Robinson and Savage, 1926; Bergh, 1969; Davenport, 1986; Ish-Am and Eisikowitch, 1991a,b). Cross-pollination effectiveness depends on: the distance between the 'pollenizer' (pollen donor) and the pollinated trees; the effective overlap period between female and male openings; and pollinator mobility, density and effectiveness. In many cases, cross-pollination of a group A cultivar

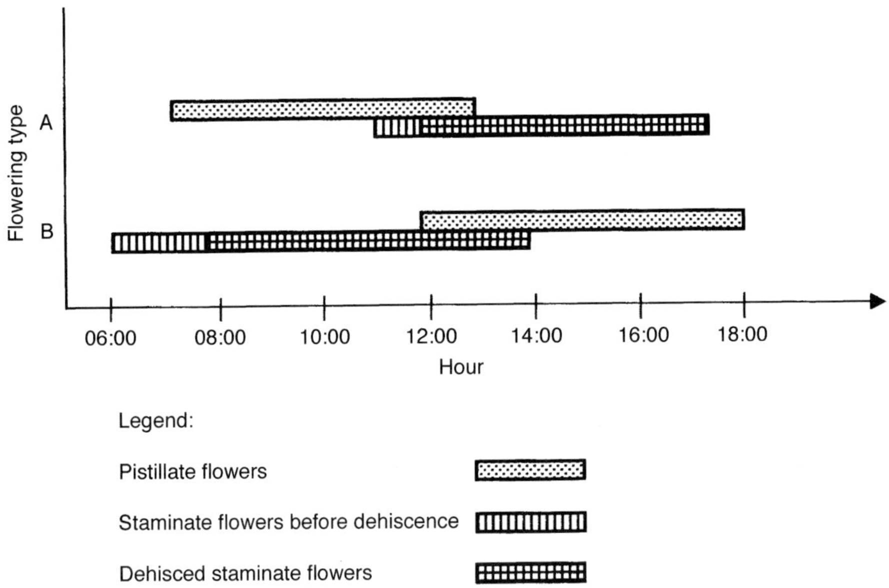

Fig. 8.3. Daily flowering course of A and B type avocados in warm weather.

by a group B is more efficient than the other way around, due to the greater overlap period between group A female and group B male flowering (Stout, 1933; Ish-Am, 1994; Papademetriou, 1976b; Ish-Am and Eisikowitch, 1991b).

Close pollination

Close pollination occurs during the phase of female and dehisced male stage flower overlap within the tree (Stout, 1923, 1933; Robinson and Savage, 1926; Lesley and Bringhurst, 1951; Bergh, 1969; Snir, 1971; Papademetriou, 1976b; Davenport, 1986; Ish-Am and Eisikowitch, 1991a,b). Since close pollination takes place when male and female stage flowers are in close proximity, its efficiency may be very high. It depends on both length and effectiveness of the overlap period within the tree, and on pollinator density and effectiveness. Most Mexican- and Guatemalan-type cultivars, and their hybrids, present a daily effective period of overlap within the tree. Generally, close pollination is more efficient in group A cultivars, since their open female flowers overlap with young pollen-releasing male flowers, and less so in group B types, where newly opened female flowers overlap with old male flowers, which have finished pollen release and are closing (Ish-Am and Eisikowitch, 1991b; Ish-Am, 1994).

Self-pollination

Self-pollination occurs when pollen released at the male stage, reaches the stigma within the same flower. This process does not necessarily demand pollinator involvement, and may be facilitated by wind or gravity. Self-pollination of the male flower is a common phenomenon, but in most cases does not lead to fertilization (Stout, 1923, 1933; Robinson and Savage, 1926; Bergh, 1969; Snir, 1971; Davenport, 1986; Ish-Am and Eisikowitch, 1991a,b). However, Davenport (Davenport, 1985, 1989; Davenport *et al.*, 1994) concluded that in south Florida, spontaneous self-pollination is the predominant means for fruit set.

Pollination requirements

The need for pollination and pollinators

Effective pollination is essential for normal fruit set. Furthermore, in most growing areas self-pollination within the male flower does not result in fruit set. Hence, pollen has to be transferred from dehisced anthers of a male stage flower to a female flower's receptive stigma (Stout, 1923; Peterson, 1955, 1956; Snir, 1971; Sedgley, 1977; Sedgley and Grant, 1983; Shoval, 1987; Robbertse *et al.*, 1997, 1998a). Fruit set is minimal or absent when insect pollinators such as bees and flies are excluded by caging. Therefore, flying pollinators are needed and wind pollination is not effective (Clark, 1923; Lesley and Bringhurst, 1951; Peterson, 1955; Bergh, 1967, 1969, 1975; Gazit, 1976; Du Toit, 1994; Robbertse *et al.*, 1996, 1997). As mentioned earlier, an exception to this rule was found in south Florida by Davenport (Davenport, 1985; Davenport *et al.*, 1994), who reported that spontaneous self-pollination within the male stage flower is the main fruit set mechanism for commercial cultivars.

The need for cross-pollination

As a rule, all avocado types are self-fertile, and may yield good fruit set upon pollination with self-pollen (Stout, 1923; Clark, 1924; McGregor, 1976; Davenport, 1986). Nevertheless, numerous cases of significant decline in yield with increasing distance from pollenizers have been reported (Stout, 1923; Robinson and Savage, 1926; Bergh and Gustafson, 1958; Bergh and Garber, 1964; Bergh *et al.*, 1966; Bergh, 1969, 1975; Goldring *et al.*, 1987; Degani *et al.*, 1989, 1990 1997; Guil and Gazit, 1992; Ish-Am, 1994; Kobayashi *et al.*, 1996; Johannsmeier *et al.*, 1997; Clegg *et al.*, 1998). Pollination experiments under enclosure also demonstrated a clear advantage for cross- vs. self-pollination (Clark and Clark, 1926; Stout, 1933; Peterson, 1955; Gazit, 1976). Using hand pollination, a clear advantage in pollen germination and pollen tube growth was shown for foreign over self-pollen (Robbertse *et al.*, 1996), and most avocado cultivars were found to achieve a higher initial set when cross-pollinated (Papademetriou, 1976b; Gazit and Gafni, 1986; Eisenstein and Gazit, 1989). In addition, isozyme analysis studies revealed selective young fruit drop, with a clear advantage for outcrossed fruits over the selfed ones (Degani and Gazit, 1984; Degani *et al.*, 1986, 1989, 1990, 1997).

In many cases, avocado cultivars have been found to differ significantly in their need for cross-pollination, and in their effectiveness as pollenizers (Gazit, 1976; Degani *et al.*, 1997; Gazit and Gafni, 1986; Eisenstein and Gazit, 1989; Robbertse *et al.*, 1996, 1997). Moreover, the cross-pollination advantage during the initial fruit set and young fruit drop period appears to increase under conditions of stress, such as hot and dry weather. The perceived situation, therefore, is that all avocado cultivars are self-fertile, but, due to cross-pollination advantages during fertilization and throughout young fruit drop, mainly under stress conditions, most of them need cross-pollination to realize their full yield potential.

Pollination as a yield limiting factor

One may assume that pollination rate should not be a yield-limiting factor of avocado, because of the following facts: a medium-sized mature avocado tree produces during the blooming season about 1 million flowers (Lahav and Zamet, 1999); thus, 10,000 to 40,000 female flowers open each day. For all that, a final set of only 400 to 600 flowers per tree is enough to produce a good crop, and only two to three honeybees per tree, during 1 h of overlap between male and female stage flowers, can pollinate this number of flowers. However, most pollinated flowers do not produce a fruit, and when the percentage of pollinated flowers is below 10–20%, yield is usually low (Papademetriou, 1976a; Tzafti, 1981; Eisikowitch and Melamud, 1982; Shoval, 1987). Noticeable initial fruit set occurs only when five to ten bees are observed on a medium-sized tree during the daily female blooming period. At least 1 week at this visitation level is needed for a fair crop, and much more for a good one (Ish-Am, 1994; Ish-Am and Eisikowitch, 1998a).

The honeybee, which serves as the major pollinator for avocado in most countries, tends to abandon avocado orchards when a more attractive bloom is available. In addition, it does not perform efficient cross-pollination, due to its location-constancy behaviour (Stout, 1923, 1933; Lesley and Bringhurst, 1951; Bringhurst, 1952; Schroeder, 1954; Peterson, 1955; Gustafson and Bergh, 1966; Bergh, 1967, 1975; Gazit, 1976; McGregor, 1976; Eisikowitch and Melamud, 1982; Davenport, 1986; Visscher and Sherman, 1998; Bekey, 1989; Vithanage, 1990; Eardley and Mansell, 1993, 1994, 1996; Du Toit, 1994; Hofshi, 1995; Ish-Am and Eisikowitch, 1998a,b,c; Robbertse *et al.*, 1998a,b). Consequently, inadequate pollination is a major yield-limiting factor of avocado in Israel, California and probably also in other countries, mainly under Mediterranean climate conditions.

Avocado pollination ecology

Avocado flowering behaviour exhibits a sophisticated mechanism, which encourages cross-pollination, usually prevents self-pollination (within a flower) and usually enables the existence of close-pollination (between neighbouring flowers within a tree, and cultivar). The flower carries only one ovule, and its P/O (Pollen grains per Ovule) ratio is 5000–10,000, which lies within the range of obligate cross-pollinated species (Cruden, 1977). Moreover, its long blooming period and mass flowering, with only several hundreds of fruits produced, are typical adaptations for cross-pollination. Even though the flowering rhythm (of most cultivars) enables both cross- and close-pollination, selection, which appears during the fertilization and young-fruit drop periods, favours the cross-pollinated flowers. These features may be interpreted as an adaptation to tropical forest conditions, where tree-species specimens are sparse, and cross-pollination probability is very low.

Pollinators and visitors

The honeybee (*Apis mellifera* L.) is apparently the major pollinator of avocado in all countries.

Flower visitation behaviour

Honeybees visiting avocado flowers usually walk among neighbouring flowers, and fly between inflorescences. While visiting the relatively small flower the bee often slips, and grasps other parts of the inflorescence (Davenport, 1986; Vithanage, 1990; Ish-Am and Eisikowitch, 1991a, 1993). Honeybees usually collect avocado nectar, or nectar with pollen and, rarely, only pollen. The nectar-collecting bees (Plate 60), as well as the nectar-with-pollen collectors, visit both female and male stage flowers, and as such may serve as effective pollinators. In contrast, the pollen-only-collecting bees visit almost exclusively male flowers, and do not contribute to pollination (Stout, 1933; McGregor, 1976; Davenport, 1986; Free, 1993; Ish-Am and Eisikowitch, 1993). While visiting a dehisced male flower, the bee's body becomes dusted with pollen, which the bee cleans off after every two to four visits, while hovering or hanging on a leaf. The nectar-and-pollen collectors pack the pollen into their curbiculae, forming pellets, whereas the nectar-only collectors 'deliberately' unload the pollen and throw it down (Ish-Am and Eisikowitch, 1993). The pollen-only collectors' visits are very short: in about 1 s they touch the anthers while hovering, or while landing for an instant. These bees may also perform one to two refuelling nectar collections per ten pollen collections, during which they may visit female flowers, if present in the vicinity (Ish-Am and Eisikowitch, 1993).

The positions of bees visiting female and male stage flowers of equivalent form are very similar (Fig. 8.4). Only limited and defined zones of the bee's body, the 'collection zones' (Fig. 8.5), contact the flower's reproductive organs. While visiting the male flower, the bee touches the exposed pollen on the open valves with its vertex, proboscidal fossa, legs and some ventral regions (Plate 61) (Figs 8.4 and 8.5), collecting large amounts of pollen there. The same 'collection zones' also touch the stigma of the female flower, due to the similarity of the female and male stage flowers, and to the similar location of the stigma and inner stamens' anthers

(Figs 8.2 and 8.4) (Ish-Am and Eisikowitch, 1993).

Constancy for flower stage and close-pollination efficiency

Most honeybees that visit avocado during the period of overlap between female and male openings within a tree, move between the male and female flowers, collect nectar and pollen, or nectar only, and efficiently carry out close pollination (Clark, 1923; Gustafson and Bergh, 1966; Gazit, 1976; Davenport, 1986; Ish-Am and Eisikowitch, 1993).

Mobility among the trees and cross-pollination efficiency

For cross-pollination implementation, honeybees need to move between pollen-releasing male flower trees and female flower trees of a different cultivar. In mature avocado orchards, an average 40% of the bees were found to move between adjacent trees in 10 min (Ish-Am and Eisikowitch, 1998b). However, this constitutes short-range movement, since during foraging most honeybees visit only one to three adjacent trees (Clark, 1923; Stout, 1923, 1933; Free and Spencer-Booth, 1964; Bergh *et al.*, 1966; Gustafson

Fig. 8.4. The position of the honeybee while visiting avocado flowers at various stages (Ish-Am and Eisikowitch, 1993).

and Bergh, 1966; Bergh, 1967; McGregor, 1976; Davenport, 1986; Visscher and Sherman, 1998; Vithanage, 1990; Free, 1993; Hofshi, 1995, 2000). Indeed, scout bees move, while foraging, further throughout the orchard; but, under avocado orchard conditions they were found to comprise only 2–4% of the field bees (Stout, 1933; Ish-Am and Eisikowitch, 1998b). Therefore, honeybees serve as efficient cross-pollinators only in the case of neighbouring trees with complementary flower types, and their efficiency decreases sharply with increasing distance between the pollen source and the female bloom (Ish-Am, 1994; Ish-Am and Eisikowitch, 1998b).

Attractiveness of avocado bloom to honeybees

The attractiveness of the avocado bloom to honeybees under Mediterranean climatic conditions appears to be low compared to numerous other species that are in bloom at the same time, such as citrus, litchi and wild-flower species. Therefore, honeybee foragers from hives placed in the avocado orchard often abandon the avocado and collect pollen and nectar from competing blooms. In Israel, this phenomenon presents a major yield-limiting factor for the early- and medium-blooming cultivars, which flower in March/April, during the blooming season of citrus and many wildflower species (Clark, 1923; Stout, 1923, 1933; Gustafson and Bergh, 1966; Bergh, 1967; Gazit, 1976; McGregor, 1976; Papademetriou, 1976b; Tzfati, 1981; Eisikowitch and Melamud, 1982; Davenport, 1986; Shoval, 1987; Visscher and Sherman, 1998; Vithanage, 1990; Ish-Am and Eisikowitch, 1998a; Hofshi, 2000).

This low attractiveness to honeybees has been attributed to the avocado flower and flowering properties, which are not well suited to this pollinator (Faegri and Van der Pijl, 1979; Kevan and Baker, 1983; Vithanage, 1990; Ish-Am and Eisikowitch, 1993; Visscher and Sherman, 1998; Ish-Am *et al.*, 1999). The avocado flower is greenish yellow, has a slightly bitter smell and its nectar is fully exposed. It has radial symmetry, lacks a landing platform and nectar trails, and is somewhat small for the honeybee, while the inflorescence is too sparse to be visited as a unit (Davenport, 1986; Vithanage, 1990; Ish-Am and Eisikowitch, 1993). Moreover, neither avocado pollen nor its nectar suits fully the honeybee's needs (Ish-Am, 1994).

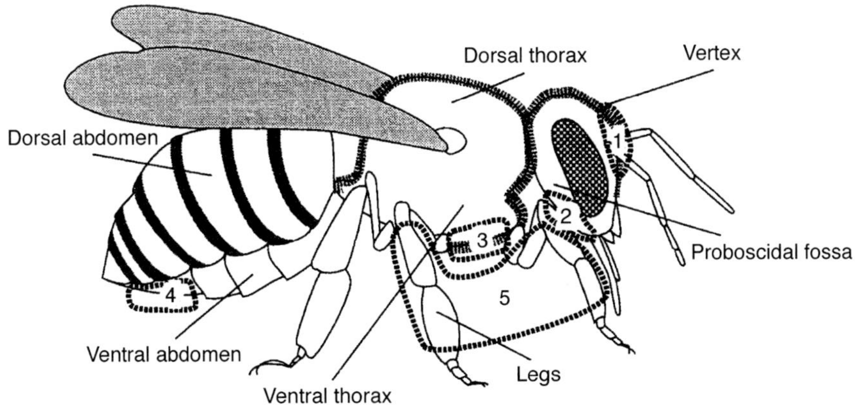

Fig. 8.5. The 'pollinating zones' of avocado pollen on the honeybee body (Ish-Am and Eisikowitch, 1993).

Pollinators and visitors in Central America and Mexico

Visitors

Until recently, little information had been reported on avocado pollination and pollinators in Central America and Mexico. Potential pollinators that have been reported to visit avocado flowers are honeybees, stingless bees, wasps, flies and beetles, and even bats (Nieto, 1984; Crane, 1992; Roubik, 1995). In only a few cases has the effectiveness of these visitor insects as avocado pollinators been studied. In south Puebla, Mexico, Nieto (1984) collected insects visiting avocado flowers. In addition to honeybees, he also found flies of the Syrphidae, Sarcophagidae, Muscidae, Calliphoridae and Tachinidae, wasps of the Vespidae and Ichneumonidae, and beetles of the Scarabaeidae and Lampyridae. He counted visits per 5 min to the avocado flowers, and found: 61 visits for the honeybees, 11 and 8 visits for Calliphoridae and Muscidae flies, respectively, and less than that for the other visitor insects. A recent 5-year study of avocado pollinators has been conducted in Mexico (Castañeda-Vildózola *et al.*, 1999; Ish-Am *et al.*, 1999). In well-maintained orchards, where insecticides are sprayed regularly, the avocado visitor populations were usually small. The species observed included about 45 Diptera species (five families), 20 Hymenoptera species (six families) and five Coleoptera species (five families). The predominant species were the honeybees (*Apis mellifera* L.), and three fly species: *Eristalis tenax* L. (Syrphidae), *Phaenicia mexicana* Macquart (Calliphoridae) and *Palpada mexicana* Macquart (Syrphidae). However, in unsprayed orchards, as well as on backyard trees, much larger insect populations of a greater variety of species were found (Table 8.1). Sometimes, at such locations, hundreds of specimens per tree of tens of species of bees, wasps, flies and others were observed.

Pollinators

For successful pollination, an insect visitor has to exhibit the following behavioural traits (Faegri and van der Pijl, 1979; Ish-Am, 1994): (i) visit both female and male stage flowers; (ii) touch both anthers and stigmas with the same body parts (the 'pollinating zones'); (iii) carry avocado pollen on the 'pollinating zones'; (iv) for cross-pollination – the pollinator visits male and female stage flowers on trees of different cultivars. Indeed, beetle species, which collect pollen and visit only male flowers, hover-flies, which collect nectar while fluttering, without touching anthers or stigmas, and wasps that touch the flower reproductive organs but carry no pollen on their smooth bodies cannot accomplish pollination (Castañeda-Vildózola *et al.*, 1999; Ish-Am *et al.*, 1999). Nevertheless, most of the species visiting avocado flowers do effect its pollination, although sometimes not efficiently. The main effective pollinators observed and collected (Castañeda-Vildózola

Table 8.1. Insects collected on avocado bloom in Mexico. (From Ish-Am *et al.*, 1999.)

Order	Suborder level	Specimens collected (No.)	Species[a] (estimated No.)[b]
Hymenoptera	Meliponinae	444	10
	Other bees	84	16
	Wasps	245	25
Diptera		153	40
Coleoptera		33	10
Heteroptera		44	8
Others		18	6
Total		1021	115

[a]Species identification performed by the SEL (Systematic Entomology Laboratory) of the USDA. Identification of the bee species was carried out by Dr David Roubik of the Smithsonian Tropical Research Institute in Panama.

[b]The number of species is an estimate, since identification had not yet been completed.

et al., 1999; Ish-Am *et al.*, 1999) were: the honeybee, seven to nine species of stingless bees (Apidae: Meliponinae), a few *Bombus* sp. (Apidae: Bombinae), a nectar-collecting social wasp *Brachygastra mellifica* Say (Vespidae) and a blowfly *Chrysomya megacephala* F. (Calliphoridae) (Table 8.2). All of these carry large amounts of avocado pollen on their 'collection zones', which effectively come in contact with the stamens and the stigma, visit many avocado flowers per minute and are attracted to avocado bloom in high numbers.

HONEYBEE (*APIS MELLIFERA* L., APINAE, APIDAE) At present, the feral honeybee population in Central America and Mexico consists of the African race (*A. mellifera scutellata*) and its hybrids with the Italian race. The African race reached Mexico about 15 years ago, and today exists there as a domesticated, as well as a feral honeybee (Roubik, 1998). In most cases, honeybees are the main visitor to the avocado flowers and behave as efficient pollinators. However, sometimes they prefer other blooms, while the local stingless bees and wasp species prefer the avocado (Papademetriou, 1976b; Castañeda-Vildózola *et al.*, 1999; Ish-Am *et al.*, 1999).

STINGLESS BEES (MELIPONINAE, APIDAE) Nine species of this subfamily have been found visiting avocado flowers (Castañeda-Vildózola *et al.*, 1999; Ish-Am *et al.*, 1999) (Table 8.2). These bee species seem to be well adapted for avocado pollination: they are smaller than the honeybee (Plate 62); while visiting avocado flowers they achieve effective contact with both stamens and stigma on the ventral and lateral zones of both the thorax and abdomen. On these zones, they collect large amounts of avocado pollen, which is later transferred to the hind legs and used to build up a pollen load (Plate 63). Six of these species were observed visiting avocado bloom in high density, and moving frequently among male and female stage flowers.

WASPS (HYMENOPTERA) Many wasp species visit avocado flowers (Table 8.1) (Free and Williams, 1976; Papademetriou, 1976b; Castañeda-Vildózola *et al.*, 1999; Ish-Am *et al.*, 1999). The only wasp species found to efficiently pollinate avocados is the 'Mexican honey wasp' (*Brachygastra mellifica* Say) (Table 8.2) (Plate 64). This social wasp collects nectar and stores it as honey for the larvae; it carries large amounts of avocado pollen on its hairy head, thorax and legs, as well as inside unique thoracic cavities. It appears on avocado bloom in high density and visits its flowers at a high rate. This species is found on avocado blooms throughout Mexico. Other wasp species that visit avocado flowers, of the genera *Polistes*, *Mischocyttarus* and others, cannot be considered efficient pollinators, because of low flower visitation rate, small amounts of avocado pollen on the body or small populations on the trees.

FLIES (DIPTERA) Many fly species were collected visiting avocado flowers (Table 8.1). Sometimes they are observed at high densities of tens, or even hundreds per tree (Nieto, 1984; Castañeda-Vildózola *et al.*, 1999; Ish-Am *et al.*, 1999). Some species of the Calliphoridae (Plate 65), Muscidae, Sarcophagidae and Syrphidae also carry large amounts of avocado pollen and make effective contact with stamens and stigmas. However, their pollination efficiency is not high because of a low rate of flower visitation.

Pollination rates

Pollination rates of avocado flowers, at their female stage opening, were determined in a few locations in Mexico. Rates fluctuated widely, from 0 to 52%, in relation to pollinator density (G. Ish-Am *et al.*, 1999, unpublished). In most cases, honeybees, stingless bees, wasps and other pollinators were simultaneously visiting the bloom. However, in one location all honeybees left the avocados for senecio (*Senecio salignus*) bloom, while up to 100 specimens per tree of Meliponinae and *Brachygastra* species remained on the avocado flowers. When only native pollinators were active, pollination rates reached 52% of cross-pollination and daily total pollination. These values are higher than those determined in Israel for a typical honeybee pollination (Ish-Am, 1994).

Table 8.2. Effective pollinators of avocado in Mexico.

Name	Description	Body length (mm)	Collecting		Avocado pollen amount	Density on avocado bloom	Avocado pollination effectiveness
			Nectar	Pollen			
Apis mellifera L.	Social bee	10	+	+	Large	Very high	Very high
Geotrigona acapulconis Strand	Social bee, stingless	6	+	+	Large	Very high	Very high
Trigona nigerrima Cresson	Social bee, stingless	8	+	+	Large	High	Very high
Partamona bilineata Say	Social bee, stingless	6.5	+	+	Large	Very high	Very high
Nannotrigona perilampoides Cresson	Social bee, stingless	4	+	+	Large	High	Very high
Scaptotrigona pectoralis Dalla Torre	Social bee, stingless	6	+	+	Large	Very high	Very high
Trigona (*Frieseomelitta*) *nigra* Provancher	Social bee, stingless	5	+	+	Large	High	Very high
Scaptotrigona mexicana Schwarz	Social bee, stingless	6	+	+	Large	High	Very high
Trigona fulviventris Guérin	Social bee, stingless	7	+	+	Large	Medium	High
Plebeia frontalis Friese	Social bee, stingless	4	+	+	Medium	Medium	Medium
Bombus spp.	Social bees	10–20	+	+	Large	High	Very high
Exomalopsis spp.	Solitary bees	8	+	+	Medium	Low	Medium
Brachygastra mellifica Say	Social wasp	8.5	+	?	Large	High	Very high
Polistes spp.	Social wasps	10–20	+	–	Medium	Medium	Medium
Chrysomya megacephala F.	Blowfly	14	+	–	Medium	Low	Medium

The original avocado pollinators

The avocado flower's sugar amounts are within the typical bee flower range (Cruden *et al.*, 1983; Harder and Barrett, 1992). However, its flower has an open 'general form', with exposed nectar and easily collected pollen. Therefore, it does not fit a specific pollinator species (Faegri and van der Pijl, 1979; Visscher and Sherman, 1998). Since it exhibits mixed features of the typical bee, wasp and fly flowers (Faegri and van der Pijl, 1979; Vithanage, 1990; Roubik, 1995; Visscher and Sherman, 1998; Ish-Am *et al.*, 1999), it is not surprising to find that numerous species of these groups visit the avocado bloom (Table 8.1). Which are the avocado original main pollinators? Evidently not the honeybee, which was only brought to the 'New World' at the beginning of the 16th century by the Europeans (Roubik, 1998). The survey of avocado pollinators that was conducted in Mexico indicates that the original pollinators were stingless bee species and the 'Mexican honey wasp' (Table 8.2). All of these are social hymenopterans, which are adapted to visiting a wide range of flower types year-round, and the avocado bloom was found to be highly attractive to them (Castañeda-Vildózola *et al.*, 1999; Ish-Am *et al.*, 1999). However, since the social hymenopterans may be attracted by competing blooms, the avocado flowers are also available for numerous species of wasps, flies and beetles, which serve as second-order pollinators (Bergh, 1967, 1975; Visscher and Sherman, 1998; Vithanage, 1990; Castañeda-Vildózola *et al.*, 1999; Ish-Am *et al.*, 1999).

All of this aside, the avocado flower may represent a *Meliponinae* flower type, which is mainly adapted for tropical bees, with body sizes of 4–8 mm, and also available for wasp and fly pollinators (Bawa, 1980; Givnish, 1980; Visscher and Sherman, 1998). The synchronized bloom of the mass-flower trees is well suited to the needs and behaviour of the highly social bees (Kubitzki and Kurz, 1984). These insects are well adapted for fast reactions to the trees' blooming changes, exploiting the large amounts of nectar and pollen that are released, thus carrying pollen from male blooming trees to the female ones.

Pollinators and visitors in other tropical climates

Only few reports on avocado pollination and pollinators in the tropics are available. In Jamaica, Free and Williams (1976) found mainly honeybees and *Polistes* wasps visiting the avocado flowers. They found an average of 2710 and 1575 avocado pollen grains per honeybee, and only 580 and 225 pollen grains per *Polistes* wasp, which were collected on dehisced male and female stage flowers, respectively. They also found large numbers of other flower species' pollen grains carried by these insects. Papademetriou (1976b) reported that in Trinidad 'the most abundant species visiting the avocado flowers' are two wasp species of the Vespidae: *Polistes canadensis* and *Metabolybia singulata*. He observed honeybees visiting the avocado for only 2 weeks during one blooming season and almost none during the next one, and assumed that more attractive nectar and pollen plants had drawn the honeybees away. He found more avocado pollen grains on the honeybees than on the wasps. He also saw small numbers of *Musca* sp. and some other fly species visiting the avocado flowers.

In south Florida, Robinson and Savage (1926), Stout (1932), Davenport (1985, 1986, 1989, 1992), and Davenport *et al.* (1994) observed only small numbers of insects visiting the avocado bloom. Most of the time, honeybee visitation to the avocados was very low. Some avocado cultivars, e.g. 'Booth 7' and 'Hardee', were more attractive to honeybees than others. The main visitors to avocado blooms were *Polistes* wasps, flies (Caliphoridae, Muscidae, Tabanidae and Syrphidae), bugs (Miridae, Reduviidae, Phymatidae) (Stout, 1923, 1932; Stout and Savage, 1925; Davenport, 1986, 1989) and many flower thrips (*Frankliniella* spp.). Sometimes half a dozen or more *Frankliniella* per flower were noticed (Davenport, 1986). Although their pollination ability had been considered (Peterson, 1955), they were proven ineffective by comparing enclosed branches where thrips were present to similar enclosed branches treated with insecticides (Davenport, 1992). Pollination rates of the female stage flowers were usually very low, around 1.5% (sometimes up to 4%), with one to two pollen grains per stigma (Davenport, 1986, 1989; Davenport *et al.*, 1994). However, self-pollination within the male stage flower was found to reach 10–80%, according to cultivar and weather. This self-pollination was found to lead to fertilization, and fruit set. Accordingly, it has been concluded that in avocado orchards there is no need for honeybee hives and pollenizer trees, in order to obtain good yields in that region.

Pollinators and visitors in a Mediterranean climate

CALIFORNIA In California, the honeybee is the main avocado pollinator, and other visiting insects are rare (Clark, 1923; Bergh, 1967). Early work revealed the need for honeybee pollination for fruit set, and the need for adjacent foreign pollen donor (pollenizer) trees to maximize yield (Clark, 1923; Stout, 1923 Robinson and Savage, 1926). However, since in some years solid 'Hass' blocks produced well without placing honeybee hives there, growers and researchers concluded that pollination is not a limiting factor. Nevertheless, new pollination studies, and innovative practices involving high densities of both honeybee hives and pollenizer trees, have confirmed the earlier conclusions (Hofshi, 1995, 2000; Kobayashi *et al.*, 1996; Clegg *et al.*, 1998; Visscher and Sherman, 1998).

Numerous other insect visitors to avocado flowers were observed in California, mainly flies and wild bees. Some of these visitors were assumed to contribute to avocado pollination (Bergh, 1967; Schroeder and Hofshi, 1998; Visscher and Sherman, 1998). Ants are common visitors to avocado flowers there, but do not effect pollination (Lesley and Bringhurst, 1951).

ISRAEL In Israel, the honeybee is also the main avocado pollinator and its hives are routinely placed in avocado orchards during the blooming season. However, its effectiveness as an avocado pollinator is often inadequate, constituting a crucial yield-limiting factor (see pages 269 and 270).

Many other insect visitors to avocado flowers have been observed in Israel. Most of them are flies (Calliphoridae, Syrphidae, Muscidae, Bibionidae, Sarcophagidae, Tachinidae and others), bee and wasp species, and more rarely, butterflies, beetles and bugs. Thrips and ant are also observed (Bergh, 1975; Gazit, 1976; Tzfati, 1981; G. Ish-Am, 1994, unpublished). B. Gefen (in Bergh, 1975) reported that the housefly seems to be an efficient transporter of avocado pollen. Ish-Am (1985, 1994) found an average of 200 avocado pollen grains per fly on the body of Muscidae flies, which were collected on dehisced male stage 'Fuerte' flowers. During days of very low honeybee activity (zero to one bee per tree) he found 0% pollinated flowers on days with no fly activity and 10–20% on days with high fly activity. Pollination rates were sometimes even higher, reaching 72%, in a 'Hass' plot that had been covered with fluid manure 2 weeks earlier and had a huge housefly population. However, since fly-pollinated flowers usually carried only one or two pollen grains per stigma, and since after consecutive 'fly days' with no honeybee activity almost no fruit set could be observed, he concluded that the fly's contribution to avocado set is very limited. A beetle pollination experiment was performed by Tzfati (1981). She introduced 'flower beetles' (Scarabaeidae), which had been collected earlier on avocado bloom, into enclosed 'Reed' inflorescences, and after 3 h found 30% pollinated flowers.

Wild bumblebees (*Bombus terrestris* L.), have been observed visiting and pollinating avocado flowers (Ish-Am *et al.*, 2000). Subsequently, a 5-year experiment was conducted by Ish-Am *et al.* (2000) to improve avocado pollination, using hives of domesticated bumblebees of the same species. They compared 3 ha plots in which honeybee hives were placed at different densities (3–10 hives ha^{-1}), to similar plots where 25 bumblebee hives (containing a few hundred bees per hive) per plot had been placed in addition to the honeybees. The best results were obtained in plots of combined honeybee and bumblebee hives. The bumblebees were more effective on cool days and under competitive pollination conditions, and at a distance of more than two rows (10 m) performed cross-pollination more effectively. However, under hot weather conditions and when not attracted by competing blooms honeybees were more efficient.

SOUTH AFRICA In the northern Transvaal, DeMeillon (1979) observed seven *Forcipomyia* and a few *Atrichopogon* species (Ceratopogonidae: Diptera) visiting avocado flowers, whereas honeybees were rarely seen. However, in the eastern Transvaal lowlands, Du Toit and Swart (1993) and Du Toit (1994) found honeybees to be responsible for 81% of the visits. Since other visitor species were very sparse, they concluded that honeybee is the only effective avocado pollinator. Three years later, Johannsmeier *et al.* (1997) studied the role of honeybee pollination in a combined 'Hass' and 'Ettinger' avocado orchard. They found pollination to present a crucial yield-limiting factor, due to both low honeybee visitation rate and low honeybee efficiency in cross-pollination performance at long distances.

In Westfalia, Eardley and Mansell (1993, 1994, 1996) recorded 48 visitor species on avocado bloom. Honeybees alone constituted 85% of the visitors, while six wild bee species and six wasp species made up only 2.8% and 0.4%, respectively. Flies were the second most important visitors: 26 species (Calliphoridae, Muscidae, Syrphidae and others) constituted 8.7% of the visitors. Also recorded were ants, butterflies, bugs and beetles (two, two, two and three species, respectively). Avocado pollen was carried mainly by honeybees and other bee species, as well as by some of the Calliphoridae fly species, whereas the other visitors carried almost no avocado pollen. The authors concluded that in the research area honeybee is the main avocado pollinator, and that a small native carpenter bee species (*Allodape microsticta* Cockerell) is a very efficient avocado pollinator; however, its population is too small to also serve as an effective one. They noted that honeybees were mostly attracted by other flower species, such as litchi, and deduced that the honeybee is not an optimal pollinator for avocado (see also Westerkamp, 1991). Therefore, they called for research aimed at using alternative pollinators from the native fauna to increase avocado productivity.

AUSTRALIA In New South Wales, Vithanage (1990) identified as regular visitors to avocado flowers ('Hass' and 'Fuerte'): honeybees, one wasp and one ant species (Hymenoptera), 11 fly species (Diptera) and one species of each of the orders Coleoptera and Neuroptera. Other species were identified as occasional visitors, and were not recorded. For each of the visitor species, he measured the number of specimens and the number of visited flowers per specimen on ten inflorescences (about 1.0 m^2 of canopy) for each hour of the blooming day. Pooling the data, which were gathered during 3 days $week^{-1}$ from two cultivars at three sites, he calculated the average flower-to-insect ratio for each visitor species. He also determined the average pollen-carrying capacity of each species by washing a sample of six specimens and counting the pollen grains for each. For each visitor species, he calculated the 'pollinator index' as a measure of pollination efficiency. 'Pollinator index' was defined as the product of the average flower-to-insect ratio per visitor species in an hour by the average number of pollen grains per insect. The average number of pollen grains per insect was above 4000 for the honeybees and the wasp species, around 1500 for the beetle and one fly species (*Aphyssura* sp.), 950–150 for another nine fly species, and less than 100 for one fly and the Neuroptera species. The average number of insects per 100 flowers in an hour was above 20 for two fly species (*Simosyrphus gricornis* and *Calliphora* sp.), 16 for the honeybees, 14 and 8 for two other fly species (*Calliphora* sp. and *Chrysomya varipes*, respectively), and two or fewer for the other species. Therefore, the pollination index was more than 25,000 for the honeybees, about 1600 for one calliphorid fly, from 830 to 130 for another four fly species, and less for the other species. Vithanage (1990) concluded that the honeybee was by far the major avocado pollinator in the research area, and that the flies as a group contributed about 14% to the seasonal pollination. He also found that adding honeybee hives to avocado orchards could increase the yield.

Conclusion

The avocado originated and has evolved for millions of years in Central America. Its flowers were pollinated by native pollinators, which co-evolved with the avocado. The avocado flower presents typical generalist features, including large amounts of exposed nectar and pollen. Thus, its rewards are readily available to almost all visitors, from bees, wasps, ants and flies to beetles, bugs and butterflies. However, effective avocado pollinators have to visit both female and male stage flowers and come into contact with the dehisced anthers and the receptive stigma at the same hairy 'pollen collection zones'. Flying insects of small to medium size (3 to 8 mm in length) are especially fit to efficiently collect avocado nectar. While visiting male stage flowers they get pollen at the 'collection zones', which will touch the stigma of female stage flowers.

The main original avocado pollinators apparently are social Hymenoptera species: several small to medium size stingless bees (Meliponinae) and the 'Mexican honey wasp' (*Brachygastra mellifica* Say). Second-order pollinators are numerous species of wasps and flies, and probably also beetles.

The arrival of the honeybee in Central America and the modern agricultural techniques, especially spraying with potent insecticides, changed the ecological conditions there. The honeybee became the major pollinator of many species, including the avocado, and the original pollinators were excluded into the reduced uncultivated areas. In locations where the original pollinator species have survived they are still observed together with the honeybee on avocado bloom, sometimes in large numbers. However, throughout the huge area of sprayed orchards in Michoacán, the honeybee constitutes the only effective pollinator. Over the last two centuries, the avocado has been exported from its region of origin to most tropical and subtropical regions of the world. Since its original pollinator species were never transferred to the new growing regions, it is pollinated there by local species (Valentine, 1987). While in most New World tropics there are local stingless bee and wasp species which are evolutionally related to avocado's original pollinators and may pollinate it effectively, in subtropical countries there are almost no suitable local candidate pollinator species, except for the honeybee. Therefore, in those countries, mostly under a Mediterranean climate, the honeybee serves as the only effective avocado pollinator, whereas many local fly, wasp and bee species visit its flowers, but make only a limited contribution to its pollination. Although honeybee hives are regularly placed in avocado orchards throughout the season, honeybees are frequently attracted to competing local flowers and abandon the avocado bloom. Therefore, inadequate pollination constitutes in those regions an important limiting factor for avocado productivity.

In contrast with the other avocado growing regions, in south Florida insect pollinators were found not to be needed for commercial avocado pollination, since spontaneous self-pollination occurs within the male stage flower. This odd feature may be a product of the local humid and warm climate, combined with human selection for fruitfulness of the local West Indian-related cultivars. In this region there is almost no effective overlap between female and male stage flowers and pollinator activity is very low. In this situation, close- and cross-pollination are inadequate. Hence, apparently the productive cultivars selected there were those that have the rare ability for effective self-pollination at the flower's male stage.

References

Abud-Antum, A. (1991) Presence of avocado lacebug, *Pseudacysta perseae* (Heidemann) (Hemiptera: Tingidae) in Dominican Republic. *Primera Jornada de Proteccion Vegetal, Universidad de Santo Domingo, Dominican Republic*, abstract, p.4.

Andrade, G.S. (1988) Control quimico de la araña cristalina *Oligonychus perseae*, Tutle, Baker and Abatiello en el cultivo del aguacatero. Thesis, Universidad Michoacana de San Nicolas Hidalgo, Uruapan, Michoacan, Mexico.

Angelini, A. and Labonne, V. (1970) Mise au point sur l'etude de *Cryptophlebia* (*Argyroploce*) *leucotreta* (Meyr.) en Côte d'lvoire. *Coton et Fibres Tropicales* 25, 497–500.

Annecke, D.P. (1959) The effect of parathion and ants on *Coccus hesperidum* L. (Coccidae: Hemiptera) and its natural enemies. *Journal of the Entomological Society of Southern Africa* 22, 245–275.

Annecke, D.P. (1964) The encyrtid and aphelinid parasites (Hymenoptera: Chalcidoidea) of soft brown scale, *Coccus hesperidum* Linnaeus (Hemiptera: Coccidae) in South Africa. *Entomology Memoir, Department of Agricultural Technical Services, Republic of South Africa* No. 7, 1–73.

Annecke, D.P. (1965) A new species of *Centroda* Foerster (Hymenoptera: Aphelinidae) parasitic in the eggs of a cicadelid injurious to citrus in South Africa. *South African Journal of Agricultural Science* 8, 1133–1138.

Annecke, D.P. (1966) Biological studies on the immature stages of soft brown scale, *Coccus hesperidum* Linnaeus (Homoptera: Coccidae). *South African Journal of Agricultural Sciences* 9, 205–228.

Annecke, D.P. and Georgala, M.B. (1978) Soft brown scale *Coccus hesperidum* L. In: Bedford, E.C.G. (ed.) *Citrus Pests in the Republic of South Africa*. Science Bulletin Department of Agricultural Technical Services, Republic of South Africa, No. 391, pp. 99–102.

Annecke, D.P. and Insley, H.P. (1971) Catalogue of Ethiopian Encyrtidae and Aphelinidae (Hymenoptera: Chalcidoidea). *Entomology Memoir, Department of Agricultural Technical Services, Republic of South Africa* No. 23, 1–53.

Annecke, D.P. and Moran, V.C. (1982) *Insects and Mites of Cultivated Plants in South Africa*. Butterworth and Co., Durban, 383 pp.

Anonymous (1984a) *Abacticultura paulista*. Boletin tec. CATI, Campinas, Brazil 181, 49–54.

Anonymous (1984b) *Manual Fitosanitario del Aguacate*. Bayer de Mexico, 38 pp.

Anonymous (1991) *Abacate: Cultura, Material Prima e Aspectos Economicos*, 2nd edn. Frutas Tropicais, Instituto de Tecnologia de Alimentos, Campinas, 8, pp. 1–57.

Aponte, G. and McMurtry, J. (1997a) Biology, life table and mating behavior of *Oligonychus perseae* (Acari: Tetranychidae). *International Journal of Acarology* 23, 199–207.

Aponte, G. and McMurtry, J. (1997b) Damage on 'Hass' avocado leaves, webbing and nesting behavior of *Oligonychus perseae* (Acari: Tetranychidae). *Experimental and Applied Acarology* 21, 265–272.

Arellano, G. (2000) El barrenador del fruto del palto *Stenoma catenifer* Walsh y su control natural en Chanchamayo y Satipo. http://lamolina.edu.pe/ciencias/ecologia/revista8.htm.

Arellano, P. (1975) Key for the identification of larvae of avocado stone borers in Mexico. *Folia Entomologica Mexicana* 31, 127–131.

Armstrong, W.J., Mitchell, W.C. and Farais, G.J. (1983) Resistance of 'Sharwil' avocados at harvest maturity to infestation by three fruit fly species (Diptera: Tephritidae) in Hawaii. *Journal of Economic Entomology* 76, 119–121.

Attygalle, A.B., Schwarz, J., Vostrowsky, O. and Bestmann, H.J. (1986) Individual variation in the sex pheromone components of the false codling moth, *Cryptophlebia leucotreta* (Lepidoptera: Tortricidae). *Zeitschrift für Naturforschung* 41c, 1077–1081.

Avidov, Z. and Harpaz, I. (1969) *Plant Pests in Israel*. Israel University Press, Jerusalem, 549 pp.

Avilan, L., Leal, F. and Bautista, D. (eds) (1997) *El aguacatero*. Espasando, Venezuela, 36 pp.

Bailey, J.B. and Hoffmann, M.P. (1979) Omnivorous looper (*Sabulodes aegrotata*) on avocados in California. University of California, Cooperative Extension Service, Berkeley, 3 pp.

Bailey, J.B. and Olsen, K.N. (1990a) Control of two avocado mite pests. *California Agriculture* 44, 31–32.

Bailey, J.B. and Olsen, K.N. (1990b) Chemical control of *Amorbia*, an insect pest of avocado and citrus. *California Agriculture* 44, 10–19.

Bailey, J.B., Olsen, K. and McDonough, L.M. (1986) Field development of a synthetic sex pheromone for omnivorous looper *Sabulodes aegrotata*. *California Avocado Society Yearbook* 70, 147–150.

Bailey, J.B., Hoffmann, M.P. and Olsen, K.N. (1988) Blacklight monitoring of two avocado insect pests. *California Agriculture* 42, 26–27.

Barber, H.S. (1919) Avocado seed weevils. *Proceedings of the Entomological Society of Washington* 21, 53–60.

Barmes, B.N. (1983) Fruit flies – seasonal occurrence, control and monitoring on pome and stone fruits. *Deciduous Fruit Grower* April, 139–146.

Barrow, M.R. (1977) The biology, economic significance and control of *Tortrix capensana* (Wlk.) (Lepidoptera: Tortricidae) on citrus at Zebediela Estate. MSc thesis, University of Natal.

Barrow, M.R. and Bedford, E.C.G. (1998) Apple leaf roller *Tortrix capensana* (Walker). In: Bedford, E.C.G., Van den Berg, M.A. and De Villiers, E.A. (eds) *Citrus Pests in the Republic of South Africa*. Dymamic Ad, Nelspruit, pp. 200–207.

Bawa, K.S. (1980) Evolution of dioecy in flowering plants. *Annual Review of Ecology and Systematics* 11, 15–39.

Becker, D., Cyjon, R., Cosse, A., Moore, I., Kimmel, T. and Wysoki, M. (1990) Identification and enantioselective synthesis of (Z,Z)-6,9-cis-3S,4R-epoxynonadecadiene, the major sex pheromone component of *Boarmia selenaria*. *Tetrahedron Letters* 31, 4923–4926.

Bedford, E.C.G. (1943) The biology and economic importance of the South African citrus thrips, *Scirtothrips aurantii* Faure. *Publications of the University of Pretoria, South Africa, Series 2, Natural Science* 7, 1–68.

Bedford, E.C.G. (1978a) Other scale insects of minor importance. In: Bedford, E.C.G. (ed.) *Citrus Pests in the Republic of South Africa*. Science Bulletin, Department of Agricultural Technical Services, Republic of South Africa No. 391, p. 130.

Bedford, E.C.G. (1978b) Tip wilters and shield bugs on citrus. In: Bedford, E.C.G. (ed.) *Citrus Pests in the Republic of South Africa*. Science Bulletin, Department of Agriculture Technical Services, Republic of South Africa, No. 391, pp. 135–136.

Bedford, E.C.G. (1998a) Blue-green citrus nibbler *Colasposoma fulgidum* Lefèvre. In: Bedford, E.C.G., Van den Berg, M.A. and De Villiers, E.A. (eds) *Citrus Pests in the Republic of South Africa*. Dynamic Print, Nelspruit, pp. 243–244.

Bedford, E.C.G. (1998b) Citrus blackfly *Aleurocanthus woglumi* Ashby. In: Bedford, E.C.G., Van den Berg, M.A. and De Villiers, E.A. (eds) *Citrus Pests in the Republic of South Africa*. Dynamic Print, Nelspruit, pp. 102–104.

Bedford, E.C.G. (1998c) Methods of controlling citrus pests. In: Bedford, E.C.G., Van den Berg, M.A. and De Villiers, E.A. (eds) *Citrus Pests in the Republic of South Africa*. Dynamic Ad, Nelspruit, pp. 8–14.

Bedford, E.C.G. and Thomas, E.D. (1965) Biological control of the citrus blackfly, *Aleurocanthus woglumi* (Ashby) (Homoptera: Aleyrodidae) in South Africa. *Journal of the Entomological Society of Southern Africa* 28, 117–132.

Begemann, G.J., Barrow, M.R. and Bedford, E. (1998) Apple leaf roller *Tortrix capensana* (Walker). In: Bedford, E., Van den Berg, M.A. and De Villiers, E. (eds) *Citrus Pests in the Republic of South Africa*. Dynamic Ad, Nelspruit, 288 pp.

Bekey, R. (1989) To bee or not to be – pollination on avocados. *California Grower* 13, 30–32.

Ben-Dov, Y. and Amitai, S. (1980) The pyriform scale, *Protopulvinaria pyriformis* (Cockerell), in Israel. *Alon HaNotea* 34, 797–798 (in Hebrew).

Ben-Yehuda, S. (1990) Reproductive behavior and phenology of the honeydew moth, *Cryptoblabes gnidiella* Milliere (Lepidoptera: Pyralidae) in avocado orchards. MSc thesis, The Hebrew University of Jerusalem, Israel, pp. 67+111 (in Hebrew).

Ben-Yehuda, S., lzhar, Y., Wysoki, M., Swirski, E. and Eisner, E. (1994) Trials in controlling the greenhouse thrips by insecticides of plant origin and oil. *Alon HaNotea* 48, 114–119 (in Hebrew).

Bergh, B.O. (1967) Reasons for low yields of avocados. *California Avocado Society Yearbook* 51, 161–172.

Bergh, B.O. (1969) Avocado, *Persea americana* Miller. In: Ferwerda, F.P. and Wit, F. (eds) *Outlines of Perennial Crop Breeding in the Tropics*. Veenman and Zonen, Wageningen, The Netherlands, pp. 23–51.

Bergh, B.O. (1975) Avocado research in Israel. *California Avocado Society Yearbook* 58, 103–126.

Bergh, B.O. and Garber, M.J. (1964) Avocado yields increased by interplanting different varieties. *California Avocado Society Yearbook* 48, 78–85.

Bergh, B.O. and Gustafson, D. (1958) Fuerte fruit set as influenced by cross-pollination. *California Avocado Society Yearbook* 42, 64–66.

Bergh, B.O., Garber, M.J. and Gustafson, D. (1966) The effect of adjacent trees of other avocado varieties on Fuerte fruit-set. *Proceedings of the American Society of Horticultural Sciences* 89, 167–174.

Blommers, L. and Gutierrez, J. (1975) Les tétranyques vivant sur agrumes et avocatiers dans la région de Tamatave (Madagascar – est) et quelques-uns de leurs prédateurs. *Fruits* 30, 191–200.

Blumberg, D. and Blumberg, O. (1991) The pyriform scale, *Protopulvinaria pyriformis*, and its common parasitoid, *Metaphycus stanleyi*, on avocado and *Hedera helix*. *Alon HaNotea* 45, 265–269 (in Hebrew, English summary).

Bodenheimer, F.S. (1951) *Citrus Entomology in the Middle East*. Dr W. Junk, The Hague, 633 pp.

Boscan, N. and Godoy, F. (1982) *Apanteles* sp. (Hymenoptera: Braconidae) parasito del Taladrador del fruto del aguacate *Stenoma catenifer* Walsingham (Lepidoptera: Stenomiidae). *Agronomia Tropical* 32, 319–321.

Boscan, N. and Godoy, F. (1984) Observaciones preliminaries sobre la biologia de *Stenoma catenifer* Walsingham (Lepidoptera: Stenomiidae) taladrador del Aguacate (*Persea Americana* Mill). *Agronomia Tropical* 34, 205–208.

Bosman, T. (1959) Blackfly can destroy entire crop. *Farming in South Africa* 35, 6–27.

Brain, C.K. (1920) The Coccidae of South Africa. *Bulletin of Entomological Research* 11, 1–42.

Brain, C.K. (1929) *Insect Pests and Their Control in South Africa*. Cape Town, Nasionale Pers Beperk, 468 pp.

Bringhurst, R.S. (1952) Sexual reproduction in the avocado. *California Avocado Society Yearbook* 37, 210–214.

Brink, T., Steyn, W.P. and De Beer, M.S. (1997) Artificial exposure of different avocado cultivars to fruit flies. *South African Avocado Growers' Association Yearbook* 20, 78–79.

Brom, E. (1970) *El Aguacate*. Comision Nacional de Fruticultura, S.A.G. Mexico, pp. 172–264.

Broodryk, S.W. (1969) The biology of *Chelonus (Microchelonus) curvimaculatus* Cameron (Hymenoptera: Braconidae). *Journal of the Entomological Society of Southern Africa* 32, 169–189.

Brunner, S.C., Scaramuzza, L.C. and Otero, A. (1975) *Catalogo de los insectos que atacan a las plantas economicas de Cuba*, 2nd edn. Academia de Ciencias de Cuba, Instituto de Zoologia, pp. 246–257.

Bruwer, I.J. (1987) Stinkbugs in macadamia orchards. *Institute for Tropical and Subtropical Crops, Information Bulletin* 178, 2–4.

Bruwer, I.J. (1992) The influence of various hemipteran species on macadamia and some factors which can limit nut damage. PhD thesis, University of Stellenbosch, Stellenbosch.

Bruwer, I.J. (1996) Die Hemiptera-plaagkompleks op avokado's in die Kiepersol-gebied. *South African Avocado Growers' Association Yearbook* 19, 33–35.

Cañizares, J. (1975) *Los aguacateros*. Edicion Revolucionaria, Instituto Cubano del Libro, pp. 239–258.

Canseco, F.D. (1971) Biologia y control del barrenador pequeno del hueso del aguacate (*Conotrachelus aguacatae*) Barber en la region de Colima, Queretaro. *Secretaria de Agricultura y Ganaderia, Direccion General de Sanidad Vegetal* 24, 37–48.

Castañeda-Vildózola, A., Equihua-Martinez, A., Valdés-Carrasco, J., Barrientos-Priego, A.F., Ish-Am, G. and Gazit, S. (1999) Insectos polinizadores del aguacatero en los estados de México y Michoacán. Revista Chapingo Serie Horticultura 5 Num. Especial. *Proceedings of Fourth World Avocado Congress, 1999*, pp. 129–136 (in Spanish, English abstract).

Catling, H.D. and Aschenborn, H. (1974) Population studies of the false codling moth, *Cryptophlebia leucotreta* Meyrick, on citrus in the Transvaal. *Phytophylactica* 6, 31–38.

Catling, H.D. and Aschenborn, H. (1978) False codling moth *Cryptophlebia leucotreta* (Meyrick). In: Bedford, E.C.G. (ed.) *Citrus Pests in the Republic of South Africa*. Science Bulletin, Department of Agricultural Technical Services, Republic of South Africa, pp. 165–170.

Cendana, S.M., Gabriel, B.P. and Magallona, E.D. (1984) *Insect Pests of Fruit Crops in the Philippines*. Department of Entomology, University of the Philippines, Los Baños.

Cervantes, L., Lyal, C. and Brown, K. (1999) The stenomatine moth, *Stenoma catenifer* Walsingham: a pre-dispersal seed predator of green heart (*Chlorocardium rodiei* (Schomb) Rowhee, Ritcher and Van der Werff) in Guyana. *Journal of Natural History* 33, 531–542.

Chavez, H., Pereira, C. and Gonzalez, R. (1985) Chicharrita del aguacate *Trioza perseae* Tuthill (Homoptera: Psyllidae) plaga potencial del aguacate en zones altas Del estado de Lara. *Boletin de Entomologia Venezolana* 4, 17–24.

CIBC (Commonwealth Institute of Biological Control) (1984) Possibilities for the biological control of the false codling moth, *Cryptophlebia leucotreta* (Lep.: Tortricidae). *Biocontrol News and Information* 5, 217–220.

CIE (Commonwealth Institute of Entomology) (1967) *Distribution Maps of Insect Pests, Series A (Agricultural)*. Map No. 234.

Cilliers, C.J. (1967) A comparative biological study of three *Ceroplastes* species (Hemiptera: Coccidae) and their natural enemies. *Entomology Memoirs, Department of Agricultural Technical Services, Republic of South Africa*, No. 13, 1–59.

Cilliers, C.J. (1998) Wax scales on citrus. In: Bedford, E.C.G., Van den Berg, M.A. and De Villiers, E.A. (eds) *Citrus Pests in the Republic of South Africa*. Dynamic Ad, Nelspruit, pp. 129–131.

Claassens, V.E. (1994) *Fiorinia*-dopluis op avokadovrugte. *Information Bulletin, Institute for Tropical and Subtropical Crops, Nelspruit* 260, 12.

Claassens, V.E. (2001) Avocado beetle *Monolepta apicalis*. In: Van den Berg, M.A., De Villiers, E.A. and Joubert, P.H. (eds) *Pests and Beneficials of Tropical and Non-Citrus Subtropical Crops in South Africa*. Dynamic Ad, Nelspruit.

Clark, O.I. (1923) Avocado pollination and bees. *California Avocado Association Annual Report* 8, 57–62.

Clark, O.I. (1924) Avocado pollination tests. *California Avocado Association Annual Report* 9, 16–22.

Clark, O.I. and Clark, A.B. (1926) Results of pollination and other experiments on avocados at the orchards of Point Loma Homestead. *California Avocado Association Annual Report* 11, 85–94.

Clarke, J.F.G. (1958) *Catalogue of the Type Specimens of Microlepidoptera in the British Museum (Natural History) Described by Edward Meyrick*, Vol. III. British Museum of Natural History, London, 600 pp.

Clausen, C.P. and Berry, P.C. (1932) *The Citrus Blackfly in Asia and the Importation of its Natural Enemies into Tropical America*. USDA Technical Bulletin No. 320, 59 pp.

Clegg, M.T., Kobayashi, M. and Francis, L. (1998) Outcrossing in avocado: is there a relationship to fruit yield? *Subtropical Fruit News* 6, 24–28.

Coaton, W.G.H. (1971) Five new termite genera from South West Africa (Isoptera: Termitidae). *Cimbebasia Series A* 2(1), 1–34.

Coaton, W.G.H. (1974) Order Isoptera. In: Coaton, W.G.H. (ed.) *Status of the Taxonomy of the Hexapoda of Southern Africa*. Entomological Memoir, Department of Agricultural Technical Services, Republic of South Africa, No. 38, pp. 22–29.

Coaton, W.G.H. and Sheasby, J.L. (1972) Preliminary report on a survey of the termites (Isoptera) of South West Africa. *Cimbebasia Memoir* 2, 1–129.

Coaton, W.G.H. and Sheasby, J.L. (1978) National survey of the Isoptera of southern Africa. 14. The genus *Stolotermes* Hagen (Termopsidae: Stolotermitinae). *Cimbebasia A* 3, 208–213.

Colin, J.G. (1990) Metodos de control de las plagas mas importantes en aguacate. *Centro de investigaciones cientificas y tecnologicas del aguacate en el estado de Mexico, Memoria*, Coatepec Harinas, Mexico, pp. 119–125.

Colin, J.G. (1991) Metodos de control de plagas en el cultivo del aguacate. *Centro de investigaciones cientificas y tecnologicas del aguacate en el estado de Mexico, Memoria*, Coatepec Harinas, Mexico, pp. 121–126.

Colin, J.G. and Rubi, M. (1992) Determinacion de la susceptibilidad de ocho selecciones de aguacate al ataque de araña roja y trips. *Centro de investigaciones cientificas y tecnologicas del aguacate en el estado de Mexico, Memoria*, Coatepec Harinas, Mexico, pp. 149–155.

Coria-Avalos, V. (1993) *Principales plagas del aguacate en Michoacan*. Secretaria de agricultura, Instituto Nacional de Investigaciones Forestales y Agropecuarias, Centro de investigaciones del Pacifico Centro, Michoacan, Mexico, No. 19, 20 pp.

Crane, E. (1992) The past and present status of beekeeping with stingless bees. *Bee World* 73, 29–42.

Cruden, R.W. (1977) Pollen–ovule ratios: a conservative indicator of breeding systems in flowering plants. *Evolution* 31, 32–46.

Cruden, R.W., Hermann, S.M. and Peterson, S. (1983) Patterns of nectar production and plant–pollinator coevolution. In: Bentley, R. and Elias, T. (eds) *The Biology of Nectaries*. Columbia University Press, New York, pp. 80–125.

Daiber, C.C. (1976) A survey of the false codling moth (*Cryptophlebia leucotreta* Meyr.) in peach orchards. *Phytophylactica* 8, 97–102.

Daiber, C.C. (1979) A study of the biology of the false codling moth *Cryptophlebia leucotreta* (Meyr.): the larva. *Phytophylactica* 11, 141–144.

Daiber, C.C. (1980) A study of the biology of the false codling moth *Cryptophlebia leucotreta* (Meyr.): the adult and generations throughout the year. *Phytophylactica* 12, 187–193.

Daiber, C.C. (1992) Insects and nematodes that attack cole crops in Southern Africa – a research review. *Zeitschrift für Pflanzenkrankheiten und Pflanzenschutz* 99, 430–440.

Daneel, M.S. (1998) Palm scale. In: De Villiers, E.A. (ed.) *The Cultivation of Mangoes*. Dynamic Ad, Nelspruit, pp. 154–155.

Das, S.C. (1955) Bionomics of the tea red spider, *Oligonychus coffeae* (Nietner). *Bulletin of Entomological Research* 50, 265–274.

Davenport, T.L. (1985) Avocado flowering, pollination and fruit set. *HortScience* 20, 589.

Davenport, T.L. (1986) Avocado flowering. *Horticultural Review* 8, 257–289.

Davenport, T.L. (1989) Pollen deposition on avocado stigmas in southern Florida. *HortScience* 24, 844–845.

Davenport, T.L. (1992) Pollination habits of avocado cultivars in South Florida. *Proceedings of 2nd World Avocado Congress*, pp. 169–172.

Davenport, T.L., Parnitzki, P., Fricke, S. and Hughes, M.S. (1994) Evidence and significance of self-pollination of avocados in Florida. *Journal of the American Society of Horticultural Sciences* 119, 1200–1207.

Davidson, J.A. and Miller, D.R. (1990) Ornamental plants. In: Rosen, D. (ed.) *Armoured Scale Insects, Their Biology, Natural Enemies and Control*. Vol. 4B, Elsevier, Amsterdam, pp. 603–632.

Degani, C. and Gazit, S. (1984) Selfed and crossed proportions of avocado progenies produced by caged pairs of complementary cultivars. *HortScience* 19, 258–260.

Degani, C., Goldring, A., Gazit, S. and Lavi, U. (1986) Genetic selection during the abscission of avocado fruitlets. *HortScience* 21, 1187–1188.

Degani, C., Goldring, A. and Gazit, S. (1989) Pollen parent effect on outcrossing rate in 'Hass' and 'Fuerte' avocado plots during fruit development. *Journal of the American Society of Horticultural Science* 114, 106–111.

Degani, C., Goldring, A., Adato, I., El-Batsri, R. and Gazit, S. (1990) Pollen parent effect on outcrossing rate, yield and fruit characteristics of 'Fuerte' avocado. *HortScience* 25, 471–473.

Degani, C., El-Batsri, R. and Gazit, S. (1997) Outcrossing rate, yield and selective fruit abscission in 'Ettinger' and 'Ardith' avocado plots. *Journal of the American Society of horticultural Science* 122, 813–817.

Dekle, G.W. (1976) *Florida Armored Scale Insects.* Florida Department of Agriculture and Consumer Services, Gainsville, Florida, 345 pp.

Del Rivero, J.M. (1966) Nota sobre una plaga de agrios y aguacatas. *Boletin Patologia Vegetal Enomologia Agricola* 29, 59–62 (*Review of Applied Entomology* 56 (1968), No. 2118).

De Meijer, A.H., Wysoki, M., Swirski, E., Blumberg, D. and lzhar, Y. (1989) Susceptibility of avocado cultivars to the pyriform scale, *Protopulvinaria pyriformis* (Cockerell) (Homoptera: Coccidae). *Agriculture, Ecosystems and Environment* 25, 75–82.

DeMeillon, W.W. (1979) Subsaharan Ceratopogonidae (Diptera) II. Ceratopogonidae taken on flowers of avocado, *Persea americana*, in South Africa. *Journal of the Entomological Society of South Africa* 42, 187–189.

Dennill, G.B. and Dupont, F.M.A. (1992) An ecological study of the damage done to avocado fruit by Hemiptera in the Hazyview region, South Africa. *South African Avocado Growers' Association Yearbook* 15, 57–61.

Dennill, G.B. and Erasmus, M.J. (1991) A packhouse survey of insect damage to avocados in the Nelspruit/Hazyview area during 1990. *South African Avocado Growers' Association Yearbook* 14, 79–82.

Dennill, G.B. and Erasmus, M.J. (1992a) Basis for a practical technique for monitoring thrips in avocado orchards. *Crop Protection* 11, 89–91.

Dennill, G.B. and Erasmus, M.J. (1992b) The insect pests of avocado fruits – increasing pest complex and changing pest status. *Journal of the Entomological Society of Southern Africa* 55, 51–57.

De Villiers, E.A. (1973) Die voorkoms van basboorders (*Salagena* sp.) op lietsjie en ander vrugtebome. *Citrus and Sub-Tropical Fruit Journal* 477, 17–19.

De Villiers, E.A. (1981) Heart-shaped scale in avocados. *Farming in South Africa*, Leaflet H.4, 2 pp.

De Villiers, E.A. (1986) Bugs on macadamia nuts. *Farming in South Africa. Macadamia* H.1, 3 pp.

De Villiers, E.A. (1989) Seisoensvoorkoms van verskillende stadia van die hartvorminge dopluis. *South African Avocado Growers' Association Yearbook* 12, 58–59.

De Villiers, E.A. (1990a) Bye voordelig vir die vrugteboer. *Information Bulletin Citrus and Subtropical Fruit Research Institute* No. 212, 1.

De Villiers, E.A. (1990b) Coconut bugs on avocados. *Farming in South Africa, Avocado Series* H.5, 2 pp.

De Villiers, E.A. (1990c) Blaaspootjies by avokado's. *Farming in South Africa. Avocado Series* H.3, 1 p.

De Villiers, E.A. (1992) Coconut bug on guavas. *Farming in South Africa, Guavas Series* H.3, 2 pp.

De Villiers, E.A. (1998) Rooiteemyt slaan toe op avokado's. *Information Bulletin. Institute for Tropical and Subtropical Crops*, No. 299, 38.

De Villiers, E.A. (2001a) Citrus looper *Ascotis reciprocaria reciprocaria* (Walker). In: Van den Berg, M.A., De Villiers, E.A. and Joubert, P.H. (eds) *Pests and Beneficials of Tropical and Non-Citrus Subtropical Crops in South Africa.* Dynamic Ad, Nelspruit.

De Villiers, E.A. (2001b) Bark borers *Salagena obsolescens* Hampson and *Salagena transversa* Walker. In: Van den Berg, M.A., De Villiers, E.A. and Joubert, P.H. (eds) *Pests and Beneficials of Tropical and Non-Citrus Subtropical Crops in South Africa.* Dynamic Ad, Nelspruit.

De Villiers, E.A. (2001c) Palm scale *Hemiberlesia lataniae* (Signoret). In: Van den Berg, M.A., De Villiers, E.A. and Joubert, P.H. (eds) *Pests and Beneficials of Tropical and Non-Citrus Subtropical Crops in South Africa.* Dynamic Ad, Nelspruit.

De Villiers, E.A. (2001d) Heart-shaped scale, *Protopulvinaria pyriformis* (Cockerell). In: Van den Berg, M.A., De Villiers, E.A. and Joubert, P.H. (eds) *Pests and Beneficials of Tropical and Non-citrus subtropical Crops in South Africa.* Dynamic Ad, Nelspruit.

De Villiers, E.A. and Robertson, C.M. (1988) Chemiese beheer van hartvormige dopluis op avokado's. *Subtropica* 9(2), 9–12.

De Villiers, E.A. and Van den Berg, M.A. (1987a) Avocado insects of South Africa. *South African Avocado Growers' Association Yearbook* 10, 75–79.

De Villiers, E.A. and Van den Berg, M.A. (1987b) Soft and wax scales. In: Myburg, A.C. (ed.) *Crop Pests in Southern Africa*, Vol. 2, *Citrus and Other Subtropicals.* Plant Protection Research Institute Bulletin 411, pp. 33–37.

De Villiers, E.A. and Wolmarans, M. (1980) Coconut bug, a pest on mango. *Farming in South Africa, Mango Series* H.2, 2 pp.

De Villiers, E.A., Milne, D.L., Du Toit, W.J. and Wolmarans, M.M. (1980) Die invoer van 'n parasitiese vlieg *Trichopoda pennipes* Fabr. en die massateel daarvan op sy gasheer die groenstinkbesie, *Nezara viridula* L. *Information Bulletin, Citrus and Subtropical Fruit Research Institute* 89, 145–151.

De Villiers, E.A., Fourie, P.F. and Petty, G.J. (1987a) Miscellaneous pests. In: Myburgh, A.C. (ed.)

Crop Pests in Southern Africa, Vol. 2, *Citrus and Other Subtropicals*. Plant Protection Research Institute Bulletin 411, pp. 109–111.

De Villiers, E.A., Newton, P.J., Petty, G.T. and Myburgh, A.C. (1987b) Fruit and leaf-eating caterpillars. In: Myburgh, A.C. (ed.) *Crop Pests in Southern Africa*, Vol. 2, *Citrus and Other Subtropicals*. Plant Protection Research Institute, Bulletin 411, pp. 82–90.

De Villiers, E.A., Van den Berg, M.A. and Petty, G.T. (1987c) Borers and rootfeeders. In: Myburgh, A.C. (ed.) *Crop Pests in Southern Africa*, Vol. 2, *Citrus and Other Subtropicals*. Department of Agriculture and Water Supply, Republic of South Africa, Bulletin 411, pp. 91–96.

De Villiers, E.A., Bedford, E.C.G. and Petty, G.T. (1987d) Mealybugs and Australian bug. In: Myburgh, A.C. (ed.) *Crop Pests in Southern Africa*, Vol. 2, *Citrus and Other Subtropicals*. Plant Protection Research Institute, Department of Agriculture and Water Supply, Bulletin 411, pp. 38–42.

Diaz, F. (1971) Biologia y control del barrenador pequeño del hueso del aguacate (*Conotrachelus aguacatae* Barber) en la region de Toliman Qro. *Fitofilo* 66, 37–48.

Drew, R.A.I. and Hancock, D.L. (1994) The *Bactrocera dorsalis* complex of fruit flies (Diptera: Tephritidae: Dacinae) in Asia. *Bulletin of Entomological Research Supplement Series* Supplement No. 2, 68 pp.

Du Toit, A.P. (1994) Pollination of avacados, mangos and litchis. *Plant Protection News* 35, 4–5.

Du Toit, A.P. and Swart, D. (1993) Notes on foraging activity of honeybees in an avocado orchard. *South African Avocado Growers' Annual Yearbook* 16, 37–38.

Du Toit, W.J. (1998) Fruit flies. In: Bedford, E., Van den Berg, M. and De Villiers, E. (eds) *Citrus Pests in the Republic of South Africa*. Dynamic Ad, Nelspruit, 288 pp.

Du Toit, W.J. and Claassens, V.E. (1993) Kewers vreet avokado's. *Information Bulletin, Institute for Tropical and Subtropical Crops* 250, 3.

Du Toit, W.J. and De Villiers, E.A. (1988) Die hartvormige dopluis, *Protopulvinaria pyriformis* (Cockerell), op avokado's. *South African Avocado Growers' Association Yearbook* 11, 79–80.

Du Toit, W.J. and De Villiers, E.A. (1990) Effects of insect growth regulators on the development of heart-shaped scale (Hemiptera: Coccidae) on avocados in South Africa. Tests of Agrochemicals and Cultivars. No. 11. *Annals of Applied Biology* 116(Suppl.), 4–5.

Du Toit, W.J., De Villiers, E.A. and Tuffin, A. (1979) The identification of causes of typical surface lesions on avocado fruit. *South African Avocado Growers' Association Yearbook* 3, 52–53.

Du Toit, W.J., Schutte, M.S. and Steyn, W.P. (1991) Parasitoids of the heart-shaped scale, *Protopulvinaria pyriformis* (Cockerell) (Hemiptera: Coccidae) on avocados in South Africa. *South African Avocado Growers' Association Yearbook* 14, 74–77.

Du Toit, W.J., Steyn, W.P. and De Beer, M.S. (1993) Occurrence of protrusions on avocado fruit and the causative agent. *South African Avocado Growers' Association Yearbook* 16, 100–102.

Eardley, C.D. and Mansell, M.W. (1993) Preliminary report on the natural occurrence of insect pollinators in an avocado orchard. *South African Avocado Growers' Annual Yearbook* 16, 127–128.

Eardley, C.D. and Mansell, M.W. (1994) Report on the natural occurrence of insect pollinators in an avocado orchard: second report. *South African Avocado Growers' Association Yearbook* 17, 117–118.

Eardley, C.D. and Mansell, M.W. (1996) The natural occurrence of insect pollinators in an avocado orchard. *South African Avocado Growers' Association Yearbook* 19, 36–38.

Ebeling, E. (1959) *Subtropical Fruit Pests*. University of California Press, Riverside, 436 pp.

Ebeling, W. (1950) *Subtropical Entomology*. Lithotype Process Co., San Francisco, California, 747 pp.

Ebelling, N., Fleschner, A. and Ricker, D. (1959) Some factors that influence pest populations in avocados. *California Avocado Society Yearbook* 43, 79–82.

Eisenstein, D. and Gazit, S. (1989) Effect of self and cross hand-pollination on initial fruit-set and fruitlet development in avocado. *Alon HaNotea* 43, 339–348 (in Hebrew).

Eisikowitch, D. and Melamud, H. (1982) A preliminary report on the role of honey bees in avocado pollination. *Alon HaNotea* 37, 19–29 (in Hebrew).

El-Minshawy, A.M., El-Sawaf, S.K., Hammad, S.M. and Donia, A. (1971) The biology of *Hemberlesia lataniae* (Sign.) in Alexandria District. *Bulletin of the Entomological Society of Egypt* 55, 461–467.

Enkerlin, W., Reyes, J., Bernabe, A., Sanchez, J., Toledo, J. and Aluja, M. (1993) El aguacate 'Hass' como hospedante de tres especies de *Anastrepha* (Diptera: Tephritidae), en condiciones forzadas y naturals. *Agrociencia, Proteccion Vegetal* 4, 329–348.

Erichsen, C. (1993) Comparison of and scouting for beetle and looper damage to avocado leaves and fruit. *South African Avocado Growers' Association Yearbook* 16, 125–127.

Erichsen, C. and Schoeman, A.S. (1992) Economic losses due to insect pests on avocado fruit in the

Nelspruit/Hazyview region of South Africa during 1991. *South African Avocado Growers' Association Yearbook* 15, 49–54.

Erichsen, C. and Schoeman, A.S. (1993) *Monolepta apicalis* (Sahlberg) (Coleoptera: Chrysomelidae): a new addition to the avocado pest-complex. *South African Avocado Growers' Association Yearbook* 16, 122–124.

Erichsen, C. and Schoeman, A.S. (1994) Identification and potential pest-status of looper on avocado. *South African Avocado Growers' Association Yearbook* 17, 113–116.

Erichsen, C., McGeoch, M.A. and Schoeman, A.S. (1993) Invasion of orchards by the avocado beetle *Monolepta apicalis* (Sahlberg) (Coleoptera: Chrysomelidae): assessment of damage to leaves and fruit. *South African Avocado Growers' Association Yearbook* 16, 118–122.

Evans, H.C. and Prior, C. (1990) Entomopathogenic fungi. In: Rosen, D. (ed.) *Armoured Scale Insects, Their Biology, Natural Enemies and Control,* Vol. 4B, Elsevier, Amsterdam, pp. 3–17.

Faber, B.A. and Phillips, P.A. (2001) *Pests of Avocado. UC Pest Management Guidelines,* University of California. State Wide Integrated Pest Management Project. http://www.ipm.ucdavis.edu/PMG/selectnewpest.avo.html

Faegri, K. and van der Pijl, L. (1979) *The Principles of Pollination Ecology,* 3rd revised edn. Pergamon Press, Oxford, 244 pp.

FAO (Food and Agriculture Organization) (1999) Avocado. *FAO Quarterly Bulletin of Statistics* 12, 120–121.

Faure, J.C. (1929) The South African citrus thrips and five other new species of *Scirtothrips* Shull. *Bulletin of the Faculty of Agriculture, Transvaal University College* 18, 1–18.

Fay, H.A.C. and De Faveri, S.G. (1990) Invasion of flowering avocado trees by *Monolepta australis* (Jacoby) (Coleoptera: Chrysomelidae) and its impact on fruit-set in North Queensland. *Crop Protection* 9, 392–394.

Fay, H.A.C. and Huwer, R.K. (1993) Egg parasitoids collected from *Amblypelta lutescens lutescens* (Distant) (Hemiptera: Coreidae) in north Queensland. *Journal of the Australian Entomological Society* 32, 365–367.

Fleschner, C., Ricker, D. and Johnson, H. (1957) Parasites of amorbia and the omnivorous looper in avocado orchards. *California Avocado Yearbook* 41, 107–117.

Ford, W.K. (1934) Some observations on the bionomics of the false codling moth – *Argyroploce leucotreta,* Meyr. (Family Eucosmidae) – in Southern Rhodesia. *Publications of the British South Africa Co.* 3, 9–34.

Fornazier, M., Pratissoli, D. and Teixeira, C. (1997) Susceptibilidade de cultivares de abacateiro a um geometrideo desfolhador. *Resumos do 16 Congresso Brasileiro de Entomologia, Salvador-BA,* 370 (abstract).

Free J.B. (1993) *Insect Pollination of Crops,* 2nd edn. Academic Press, London, 684 pp.

Free, J.B. and Spencer-Booth, Y. (1964) The foraging behaviour of honey-bees in an orchard of dwarf apple trees. *Journal of Horticultural Science* 39, 78–83.

Free, J.B. and Williams, I.H. (1976) Insect pollination of *Anacardium occidentale* L., *Mangifera indica* L., *Blighia sapida* Koenig and *Persea americana* Mill. *Tropical Agriculture* 53, 125–136.

Freidberg, A. and Kugler, J. (1989) *Fauna Palestina – Insecta IV, Diptera: Tephritidae.* The Israel Academy of Sciences and Humanities, Jerusalem, 212 pp.

Froneman, I.J. and De Villiers, E.A. (1993) Stinkbugs. In: *The Cultivation of Macadamias.* Institute for Tropical and Subtropical Crops, Bulletin 426, pp. 49–52.

Fuller, C. (1901) The Natal codling moth, *Carpocapsa* sp. *First Report of Goverment Entomologist 1899–1900.* Department of Agriculture, Natal, pp. 48–51.

Gallegos, R. (1983) *Algunos aspectos del aguacate y su produccion en Michoacan.* Universidad Autonoma de Chapingo, Grupo Editorial Gaceta, Mexico DF, 311 pp.

Garcia, J.C., Mujica, E., Vazquez, L. and Aranda, E. (1986) Fenologia de *Copturus aguacatae* Kissinger (Coleoptera: Curculionidae) en tres regions del estado de Morelos. *Resumenes, XXI Congreso Nacional de Entomologia,* 16–19 March 1986, Monterrey, Nuevo Leon, Mexico, pp. 61–62.

Gazit, S. (1976) Pollination and fruit set of avocado. In: Sauls, J.W., Phillips, R.L. and Jackson, L.K. (eds) *Proceedings First International Tropical Fruit Short Course: The Avocado.* University of Florida, Gainesville, Florida, pp. 88–92.

Gazit, S. and Gafni, E. (1986) Effect of hand pollination with different pollen donors on initial fruit set in avocado. *Israel Agrisearch* 1, 3–17 (in Hebrew, English abstract).

Geertsema, H. (1971) Description of the stages of two subspecies of the pine emperor moth, *Nudaurelia cytherea* (Fabr.) (Lepidoptera: Saturniidae). *Annals of the University of Stellenbosch,* 37 pp.

Gerson, U. and Zor, Y. (1973) The armoured scale insects of avocado in Israel. *Journal of Natural History* 7, 513–533.

Gilbert, M.J. and Bedford, E.C.G. (1998) Citrus thrips *Scirtothrips aurantii* Faure. In: Bedford, E.C.G.,

Van den Berg, M.A. and De Villiers, E.A. (eds) *Citrus Pests in the Republic of South Africa*. Dynamic Ad, Nelspruit, pp. 164–183.

Givnish, T.J. (1980) Ecological constraints on the evolution of breeding systems in seed plants: dioecy and dispersal in gymnosperms. *Evolution* 34, 959–972.

Goldring, A., Gazit, S. and Degani, C. (1987) Isozyme analysis of mature avocado embryos to determine outcrossing rate in a 'Hass' plot. *Journal of the American Society of Horticultural Science* 112, 389–392.

Goodall, G.E., Bailey, J.R., Phillips, P.A. and Bekey, R.S. (1987) Integrated pest management considerations for greenhouse thrips control in coastal avocado orchards. *South African Avocado Growers' Association Yearbook* 10, 80–82.

Gouse, H.H., DuToit, D.M. and Vorster, A.M. (1998) Termites as pests in citrus orchards. In: Bedford, E.C.G., Van den Berg, M.A. and De Villiers, E.A. (eds) *Citrus Pests in the Republic of South Africa*. Dynamic Ad, Nelspruit, pp. 85–86.

Grout, T.G. (1994)The distribution of phytoseiid mites (Acari: Phytoseiidae) on citrus in southern Africa and their possible value as predators of citrus thrips (Thysanoptera: Thripidae). *Experimental and Applied Acarology* 18, 61–71.

Grout, T.G. and Richards, G.I. (1992) *Euseius addoensis addoensis*, an effective predator of citrus thrips, *Scirtothrips aurantii*, in the eastern Cape Province of South Africa. *Experimental and Applied Acarology* 15, 1–13.

Grout, T.G. and Stephen, P.R. (1995) *Goetheana incerta* parasitizing citrus thrips in southern Africa. *Citrus Journal* 5(4), 30–32.

Grové, T. (2001a) A soft scale *Saissetia* sp. In: Van den Berg, M.A., De Villiers, E.A. and Joubert, P.H. (eds) *Pests and Beneficials of Tropical and Non-Citrus Subtropical Crops in South Africa*. Dynamic Ad, Nelspruit.

Grové, T. (2001b) Citrus thrips *Scirtothrips aurantii* Faure. In: Van den Berg, M.A., De Villiers, E.A. and Joubert, P.H. (eds) *Pests and Beneficials of Tropical and Non-Citrus Subtropical Crops in South Africa*. Dynamic Ad, Nelspruit.

Guil Y. and Gazit S. (1992) Pollination of the Hass avocado cultivar. *Proceedings of 2nd World Avocado Congress*, p. 241.

Gunn, D. (1918) The Bagrada bug (*Bagrada hilaris*). *Department of Agriculture, Union of South Africa, Bulletin* No. 9, 19 pp.

Gunn, D. (1921) The false codling moth (*Argyroploce leucotreta* Meyr.). *Department of Agriculture and Forestry, Union of South Africa, Science Bulletin* No. 21, 1–28.

Gustafson, C.D. and Bergh, B.O. (1966) History and review of studies on cross-pollination of avocados. *California Avocado Society Yearbook* 50, 39–49.

Hadar, D. (1983) Reproductive behavior and control of the giant looper, *Boarmia* (*Ascotis*) *selenaria* Schiff. (Lepidoptera: Geometridae), a pest of avocado in Israel. MSc thesis, The Hebrew University of Jerusalem, Israel (in Hebrew, English summary).

Hadar, D. (1993) Population dynamics of the pyriform scale, *Protopulvinaria pyriformis* Cockerell, and its natural enemies in avocado groves. PhD thesis, The Hebrew University of Jerusalem, Israel (in Hebrew, English summary).

Hall, W.J. and Ford, W.K. (1933) Notes on some citrus insects in Southern Rhodesia. *Mazoe Citrus Experimental Station Publication* 2a, 1–51.

Harder, L.D. and Barrett, S.C.H. (1992) The energy cost of bee pollination for *Pontederia cordata* (Pontederiaceae). *Functional Ecology* 6, 226–233.

Hartwig, E.K. (1966) The nest and control of *Odontotermes lactericius* (Haviland) (Termitidae: Isoptera). *South African Journal of Agricultural Science* 9, 407–418.

Hartwig, E.K. and De Lange, H.C. (1978) Observation on three coreids, *Carlisis wahlbergi* Stål, *Anoplocnemis curvipes* (Fabr.) and *Elasmopoda valga* (L.). Technical Communication, Department of Agricultural Technical Services, Republic of South Africa 141, 12 pp.

Hattingh, V., Cilliers, C.J. and Bedford, E.C.G. (1998) Citrus mealybugs. In: Bedford, E.C.G., Van den Berg, M.A. and De Villiers, E.A. (eds) *Citrus Pests in the Republic of South Africa*. Dynamic Ad, Nelspruit, pp. 112–120.

Heidemann, O. (1908) Two new species of North American Tingidae. *Proceedings of the Entomological Society Washington*, 10, 103–108.

Hernandez, H.G., Johansen, R., Gazca, L., Equihua, A., Salinas, A., Estrada, E., Duran, F. and Yalle, A. (2000) Plagas del aguacate. In: Teliz, D. (ed.) *El Aguacate y su manejo integrado*. Ediciones Mundi-Prensa, Mexico DF, pp. 117–136.

Hill, D. (1975) *Agricultural Insect Pests of the Tropics and Their Control*. Cambridge University Press, Cambridge, 516 pp.

Hoddle, M. (1997) *The Avocado Thrips and Thrips Natural Enemies*. California Avocado Commission, 4 pp.

Hoddle, M. (1998) The biology and management of the persea mite, *Oligonychus perseae* Tuttle, Baker and Abatiello (Acari: Tetranychidae). http://www.biocontrol.ucr.edu/mite1.html

Hoddle, M. and Morse, J. (1998) Avocado thrips: a serious pest of avocados in California. *California Avocado Yearbook* 81, 81–90.

Hoffmann, M.P., McDonough, L.M. and Bailey, J.B. (1983) Field test of the sex pheromone of *Amorbia cuneana* (Walsingham) (Lepidoptera: Tortricidae). *Environmental Entomology* 12, 1387–1390.

Hofshi, R. (1995) Avocado productivity: an overview. *Subtropical Fruit News* 3, 5–8.

Hofshi, R. (2000) Dreaming in reality. *California Avocado Society Yearbook* 82, 137–159.

Hollis, D. and Martin, J. (1997) Jumping plantlice (Hemiptera: Psylloidea) attacking avocado pear trees, *Persea americana*, in the New World, with a review of Lauraceae-feeding among psylloids. *Bulletin of Entomological Research* 87, 471–480.

Howard, C.W. (1909) Orange codling moth (*Encarmonia batrachopa*). *Report of the Department of Agriculture, Transvaal*, 1907–1908, p. 189.

Howard, F.W. and Oliver, A.D. (1985) *Armored Scale Insects (Homoptera: Diaspididae) of Louisiana*. Bulletin, Louisiana Agricultural Experiment Station 762, 125 pp.

Howell, J.O. and Tippens, H.H. (1977) Descriptions of first instars of nominal type-species of eight disputed tribes. *Annals of the Entomological Society of America* 70, 119–135.

Huerta, A., Trujillo, J., Equihua, A. and Carrillo, J. (1990) Enemigos naturals y evaluacion Preliminar de dos nematodos para biocontrol de *Copturus aguacatae* (Coleoptera: Curculionidae), en Atlixco, Puebla, Mexico. *Agrociencia, Proteccion Vegetal* 1, 47–57.

Hung, S.C. and Jong, T.M. (1995) Ambrosia beetle – a weakened avocado tree pest. *Shinung Farm Magazine* 320, 70–71 (in Chinese).

Hung, S.C. and Jong, T.M. (1997) Control of the fruitspotting bug on avocados. *Agriculture World* 163, 54–56 (in Chinese).

Ironside, D.A. (1995) Development of integrated pest management for macadamia in Malawi. *California Macadamia Society Yearbook* 41, 170–173.

Ish-Am, G. (1985) Avocado pollination by honey bees in the 'Fuerte', 'Ettinger' and 'Hass' cultivars. MSc thesis, Tel Aviv University, Israel (in Hebrew, English abstract).

Ish-Am, G. (1994) Interrelationship between avocado flowering and honey bees and its implication on the avocado fruitfulness in Israel. PhD thesis, Tel Aviv University, Israel (in Hebrew, English abstract).

Ish-Am, G. and Eisikowitch, D. (1991a) Possible routes of avocado tree pollination by honeybees. *Acta Horticulturae* 288, 225–233.

Ish-Am, G. and Eisikowitch, D. (1991b) New insight into avocado flowering in relation to its pollination. *California Avocado Society Yearbook* 75, 125–137.

Ish-Am, G. and Eisikowitch, D. (1993) The behaviour of honey bees (*Apis mellifera*) visiting avocado (*Persea americana*) flowers and their contribution to its pollination. *Journal of Apicultural Research* 32, 175–186.

Ish-Am, G. and Eisikowitch, D. (1998a) Low attractiveness of avocado (*Persea americana* Mill.) flowers to honeybees (*Apis mellifera* L.) limits fruit set in Israel. *Journal of Horticultural Science* 73, 195–204.

Ish-Am, G. and Eisikowitch, D. (1998b) Mobility of honey bees (Apidae, *Apis mellifera*) during foraging and agricultural pollination. *Apidologie* 29, 209–219.

Ish-Am, G. and Eisikowitch, D. (1998c) Quantitative approach to avocado pollination. *Proceedings of Third World Avocado Congress*, 1995, pp. 46–51.

Ish-Am, G., Barrientos-Priego, A.F., Castañeda-Vildózola, A. and Gazit, S. (1999) Avocado (*Persea americana* Mill.) pollinators in its region of origin. Revista Chapingo Serie Horticultura 5 Num. Especial. *Proceedings of Fourth World Avocado Congress*, 1999, pp. 137–143.

Ish-Am, G., Regev, I., Peterman, Y., Lahav, E., Degani, C., Elbatzri, R. and Gazit S. (2000) Improving avocado pollination with bumblebees: three seasons summary. *California Avocado Society Yearbook* 92, 119–135.

Izhar, Y., Swirski, E., Ben-Yehuda, S., Wysoki, M. and Hadar, D. (1990a) Susceptibility of the avocado cultivar 'Ardit' to the greenhouse thrips, *Heliothrips haemorrhoidalis* (Bouche) (Thysanoptera: Thripidae) in Israel. *Alon HaNotea* 44, 1009–1014 (in Hebrew, English summary).

Izhar, Y., Wysoki, M., Swirski, E. and Amitai, S. (1990b) The variegated caper bug, *Stenozygum coloratura* (Klug) (Rhynchota: Pentatomidae), and its damage to avocado and persimmon. *Hassadeh* 70, 1244–1245 (in Hebrew, English summary).

Izhar, Y., Ben-Yehuda, S., Swirski, E., Wysoki, M. and Dagan, M. (1992) Occurrence of the black vine thrips, *Retithrips syriacus* Mayet, on avocado in Israel and trials in controlling the pest by insecticides of plant origin. *Alon HaNotea* 46, 523–529 (in Hebrew, English summary).

Jadhav, L.D., Kadam, M.V., Ajri, D.S. and Pokharkar, R.N. (1980) Observations on the biology of leaf-footed plant bug *Leptoglossus membranaceus* Fabricius, a pest of pomegranate. *Bulletin of Entomology* 21, 79–82.

Johannsmeier, M.F., Swart, D.J. and Morudu, T.M. (1997) Honeybees in an avocado orchard: forager distribution, influence on fruit set and colony development. *South African Avocado Growers' Association Yearbook* 20, 39–41.

Johansen, R. and Mojica-Guzman, A. (1996) Reconsideracion del concepto de depredador y parasitoide en Tisanopteros Mexicanos (Insecta) de interes en control biologico natural. *Folia Entomologica Mexicana* 97, 21–38.

Joubert, C.J. (1936) The blue-green citrus nibbler. *Farming in South Africa* 11, 205–206.

Joubert, P.H. (2001) Coconut bug, *Pseudotheraptus wayi* Brown. In: Van den Berg, M.A., De Villiers, E.A. and Joubert, P.H. (eds) *Pests and Beneficials of Tropical and Non-Citrus Subtropical Crops in South Africa*. Dynamic Ad, Nelspruit.

Joubert, P.H. and Claassens, V.E. (1994) Stink bug damage on avocado fruit in the Nelspruit region. *South African Avocado Growers' Association Yearbook* 17, 106–108.

Joubert, P.H. and Du Toit, W.J. (1993) False codling moth damage to avocados. *Institute for Tropical and Subtropical Crops, Information Bulletin* 252, 7.

Joubert, P.H. and Neethling, C. (1994) Kernel damage to pecan nuts in the Nelspruit area of South Africa. *Journal of the Southern African Society for Horticultural Sciences* 4, 42–44.

Keifer, H. (1969) Eriophyid studies. *California Department of Agriculture* 1, 1–20.

Kevan, P.G. and Baker, H.G. (1983) Insects as flower visitors and pollinators. *Annual Review of Entomology* 28, 407–453.

Kissinger, D.G. (1957) Description of a new *Copturus* pest of avocado from Mexico (Coleoptera: Curculionidae: Zygopinae). *Acta Zoologica Mexicana* 2, 1–8.

Kobayashi, M., Henderson, D., Davis, J. and Clegg, M.T. (1996) Outcrossing in avocado: is there a relationship to fruit yield? *California Avocado Society Yearbook* 80, 63–74.

Kok, I. and Georgala, M.B. (1978) Family Tephritidae: fruit flies. In: Bedford, E. (ed.) *Citrus Pests in the Republic of South Africa*. Department of Agricultural Technical Services, Republic of South Africa, Science Bulletin 391, 253 pp.

Krause, M., Nel, A. and Van Zyl, K. (1996) *A Guide to the Use of Pesticides and Fungicides in the Republic of South Africa*. National Department of Agriculture, Pretoria, Government Printer, 320 pp.

Kubitzki, K. and Kurz, H. (1984) Synchronized dichogamy and dioecy in Neotropical Lauraceae. *Plant Systematics and Evolution* 147, 253–266.

Lahav, E. and Zamet, D. (1999) Flowers, fruitlets and fruit drop in avocado trees. Revista Chapingo Serie Horticultura 5 Num. Especial. *Proceedings of Fourth World Avocado Congress, 1999* 95–100.

Le Pelley, R.H. (1968) *Pests of Coffee*. Longmans, Green and Co., London, 590 pp.

Lesley, J.W. and Bringhurst, R.S. (1951) Environmental conditions affecting pollination of avocados. *California Avocado Society Yearbook* 35, 169–173.

Lever, R.G.A.W. (1969) Pests of the coconut palm (*Pseudotheraptus wayi*). *Food and Agricultural Organization* 77, 38–42.

Lourencao, A., Rossetto, C. and Soares, N. (1984) Ocorrencia de adultos de *Heilipus catagraphus* Germar 1824 (Coleoptera: Curculionidae) danificando frutos de abacateiro. *Bragantia Campinas* 43, 245–253.

Mansour, F., Wysoki, M. and Whitcomb, W.U. (1985) Spiders inhabiting avocado orchards and their role as natural enemies of *Boarmia selenaria* Schiff. (Lepidoptera: Geometridae) larvae in Israel. *Acta Oecologica* 6, 315–321.

Martinez, B.R. (1989) El cultivo del aguacate en Mexico, sus problemas fitosanitarios, soluciones y perspectives. *Fruticultura de Michoacan* 2, 7–13.

Matthew, J. (1975) The orange Tortrix, *Tortrix capensana* Walker (Lepidoptera: Tortricidae), as a pest of citrus, with special reference to its significance in orchards under integrated biological control. *Proceedings of the First International Congress of the Entomological Society of Southern Africa*, pp. 221–234.

McDonough, L.M., Hoffman, M., Bierl-Leonhardt, B., Smithisler, C., Bailey, J. and Davis, H. (1982) Sex pheromone of the avocado pest, *Amorbia cuneana* (Walsingham) (Lepidoptera: Tortricidae): structure and synthesis. *Journal of Chemical Ecology* 8, 255–265.

McGregor, S.E. (1976) *Insect Pollination of Cultivated Crop Plants*. USDA Agriculture Handbook 496, Washington, DC, 411 pp.

McKenzie, H.L. (1935) Life history and control of latania scale on avocado. *California Avocado Association Yearbook* 80–82.

McMurtry, J.A. (1985a) Avocado. In: Helle, W. and Sabellis, M.W. (eds) In: *Spider Mites, Their Biology, Natural Enemies and Control. World Crop Pests*, Vol. IB, Elsevier, Amsterdam, The Netherlands, pp. 327–332.

McMurtry, J.A. (1985b) Studies on a *Euseius* species complex in avocado in Mexico, Central America, with a description of new species (Acari: Phytoseiidae). *Acarologia* 26, 107–116.

McMurtry, J.A. and Johnson, H. (1966) An ecological study of the spider mite *Oligonychus punicae* (Hirst) and its natural enemies. *Hilgardia* 37, 363–402.

Lahav, E. and Zamet, D. (1999) Flowers, fruitlets and fruit drop in avocado trees. Revista Chapingo Serie Horticultura 5 Num. Especial. *Proceedings of Fourth World Avocado Congress, 1999* pp. 95–100

McMurtry, J.A., Johnson, H.G. and Newberger, S.J. (1991) Imported parasite of greenhouse thrips established on California avocado. *California Agriculture* 45, 31–32.

Mead, F. and Peña, J.E. (1991) Avocado lace bug, *Pseudacysta perseae* (Hemiptera: Tingidae). *Florida Department Agriculture and Consumer Services, Division of Plant Industry Entomology Circular* 346, 4 pp.

Medina, J.C., Bleinroth, E.W., Sojiro, J. and Leite, W. (1978) *Frutas Tropicais* 1. Abacate. Istituto de Tecnologia de Alimentos, Campinas, Brazil, pp. 63–68.

Medina-Gaud, S., Segarra, A. and Franqui, R. (1991) The avocado lacewing bug *Pseudacysta perseae* (Heidemann) (Hemiptera: Tingidae). *Journal Agriculture University of Puerto Rico* 75, 185–188.

Meyer, M.K.P. (Smith) (1987) *African Tetranychidae (Acari: Prostigmata), with Reference to the World Genera. Entomology Memoirs* No. 69, Department of Agriculture and Water Supply, Republic of South Africa, 175 pp.

Meyer, M.K.P. (Smith) (2001) Tea red mite, *Oligonychus coffeae* (Nietner). In: Van den Berg, M.A., De Villiers, E.A. and Joubert, P.H. (eds) *Pests and Beneficials of Tropical and Non-Citrus Subtropical Crops in South Africa*. Dynamic Ad, Nelspruit.

Meyrick, E. (1913) Description of South African Micro-Lepidoptera. *Annals of the Transvaal Museum* 3, 267–276.

Millar, I.M. (1994) *A Catalogue of the Aphids (Homoptera: Aphidoidae) of Sub-Saharan Africa*. Plant Protection Research Institute Handbook No. 4, 130 pp.

Milne, D.L. (1973a) Avocado pests. In: Milne, D.L. (ed.) *Insect Pests of Subtropical Fruit Crops. Entomological Memoir* No. 33, Department of Agricultural Technical Services, Republic of South Africa, pp. 1–3.

Milne, D.L. (1973b) Bagrada bug, *Bagrada hilaris*. In: Milne, D.L. (ed.) *Insect Pests of Subtropical Fruit, Entomological Memoir* No. 33. Department of Agricultural Technical Services, Republic of South Africa, p. 36.

Mitchell, W., Andrew, C., Hagen, K., Hamilton, R., Harris, E., Maehler, E. and Rhode, R. (1977) The Mediterranean fruit fly and its economic impact on Central American countries and Panama. UC/AIP Pest Manaagement and Related Environmental Protection Project, University of California, Berkeley.

Morton, J.F. (1987) *Fruits of Warm Climates*. J.F. Morton, Miami, Florida, 505 pp.

Mound, L.A. and Palmer, M.G. (1981) Identification, distribution and host-plants of the pest species of *Scirtothrips* (Thysanoptera: Thripidae). *Bulletin of Entomological Research* 71, 467–479.

Müller, F.P. and Schöll, S.E. (1958) Some notes on the aphid fauna of South Africa. *Journal of the Entomological Society of Southern Africa* 21, 382–414.

Muñiz, R. (1960) *Copturus aguacatae* Kissinger: plaga del agucatero (*Persea gratissima* Gaertn.) en Mexico. *Fitofilo* 12, 42–48 (Mexico).

Muñiz, R. and Barrera, A. (1958) Clave para los barrenadores de las ramas del aguacatero (*Persea gratissima* Gaertin) en la America Tropical y Subtropical(Coleoptera: Curculionidae: Zygopinae) *Resumenes, XXI Congreso Nacional de Entomologia*, 16–19 Marzo 1986, Monterrey, Nuevo Leon, Mexico, pp. 61–62.

Munro, H.K. and Fouché, F.A. (1936) *A List of the Scale Insects and Mealy Bugs and Their Host-Plants in South Africa*. Department of Agriculture and Forestry, Union of South Africa, Bulletin 158, 104 pp.

Murray, D.A.H. (1976) Insect pests of passion fruit. *Queensland Agricultural Journal* 102, 145–151.

Nakahara, S. (1997) *Scirtothrips perseae* (Thysanoptera: Thripidae), a new species infesting avocado in southern California. *Insecta Mundi* 11, 189–192.

Nel, J.J.C. (1961) Die Groot Grasdraertermiet in die Oranje Vrystaatstreek. Technical Communication No 12. *Symposium on Entomological Problems*, 17–21 July 1961. Department of Agricultural Technical Services, Pretoria, pp. 104–110.

Newstead, R. (1903) *Monograph of the Coccidae of the British Isles*, Vol II. Ray Society, London, 270 pp.

Newton, P.J. (1988a) Inversely density-dependent egg parasitism in patchy distributions of the citrus pest *Cryptophlebia leucotreta* (Lepidoptera: Tortricidae) and its agricultural efficiency. *Journal of Applied Ecology* 25, 145–162.

Newton, P.J. (1988b) Movement and impact of *Trichogrammatoidea cryptophlebiae* Nagaraja (Hymenoptera: Trichogrammatidae) in citrus orchards after inundative releases against the false codling moth, *Cryptophlebia leucotreta* (Meyrick) (Lepidoptera: Tortricidae). *Bulletin of Entomological Research* 78, 85–99.

Newton, P.J. (1998) False codling moth, *Cryptophlebia leucotreta* (Meyrick). In: Bedford, E.C.G., Van den Berg, M.A. and De Villiers, E.A. (eds) *Citrus Pests in the Republic of South Africa*, Dynamic Ad, Nelspruit, pp. 192–200.

Newton, P.J. and Catling, H.D. (1998) Citrus leaf roller *Archips* (= *Cacoecia*) *occidentalis*

(Walshingham). In: Bedford, E.C.G., Van den Berg, M.A. and De Villiers, E.A. (eds) *Citrus Pests in the Republic of South Africa*. Dynamic Ad, Nelspruit, pp. 207–210.

Newton, P.J. and Mastro, V.C. (1989) Field evaluations of commercially available traps and synthetic sex pheromone lures of the false codling moth, *Cryptophlebia leucotreta* (Meyr.) (Lepidoptera: Tortricidae). *Tropical Pest Management* 35, 100–104.

Newton, P.J., Anderson, T. and Vercuiel, S.W. (1986) Crop losses due to false codling moth, *Cryptophlebia leucotreta* (Meyr.) (Lepidoptera: Tortricidae), on citrus in South Africa from 1983 to 1985. *Citrus and Subtropical Fruit Research Institute* 162, 10–12.

Newton, P.J., Thomas, C.D., Mastro, V.C. and Schwalbe, C.P. (1993) Improved two-component blend of the synthetic female sex pheromone of *Cryptophlebia leucotreta*, and identification of an attractant for *C. peltastica*. *Entomologia Experimentalis et Applicata* 66, 75–82.

Nieto, A.R. (1984) Observacion preliminar de la polinizacion entomofila en aguacate *Persea americana* Mill. *Revista Chapingo* 9, 54–55.

Nirody, B.S. (1922) Investigations in avocado breeding. *California Avocado Association Annual Report* 1921–1922, 65–78.

Oatman, E.R. and Platner, G. (1985) Biological control of two avocado pests. *California Agriculture* 39, 21–23.

Omer-Cooper, J. (1939) *Remarks on False Codling Moth*. Mimeograph, Rhodes University, Grahamstown, pp. 1–17.

Ondieki, J.J. (1975) Diseases and pests of passion fruit in Kenya. *Acta Horticulturae* 49, 291–293.

Papademetriou, M.K. (1976a) Some aspects of the flower behavior, pollination and fruit set of Avocado (*Persea americana* Mill.) in Trinidad. *California Avocado Society Yearbook* 60, 106–153.

Papademetriou, M.K. (1946b) Percentage fruit set in Avocados (*Persea americana* Mill.). *California Avocado Society Yearbook* 59, 135–142

Parry-Jones, E. (1937) The over wintering pupa of *Heliothis armigera* Hübner (*obsoleta* Fabr.) in Southern Rhodesia. *Publication, Mazoe Citrus Experimental Station* 6, 19–36.

Parry-Jones, E. (1938) The biology of a tachinid parasite (*Sturmia rhodesiensis*, sp. n.) of the cotton bollworm (*Heliothis armigera* Hübner) in Southern Rhodesia. *Publication, Mazoe Citrus Experimental Station* 7, 11–34.

Peña, J.E. and Denmark, H. (1996) An eriophyid *Tegolophus perseaflorae* (Acari: Eriophyidae) new to Florida and the USA. *Florida Entomologist* 79, 74–76.

Peña, J. and Johnson, F. (1999) *Insect Management of Avocados. Insect Management Guide*. Department of Entomology and Nematology, Florida Cooperative Extension Service, University of Florida. http://edis.ifas.ufl.edu/Body_IG068

Peña, J.E., Baranowski, R. and Nadel, H. (1996) Pest management of tropical fruit trees in Florida. In: Rosen, D., Bennett, F. and Capinera, J. (eds) *Pest Management in the Subtropics – Integrated Pest Management – a Florida Perspective*. Intercept, Andover, UK, pp. 349–370.

Peña, J.E., Sundhari, S., Hunsberger, A., Duncan, R. and Schaefer, B. (1998) Monitoring, damage, natural enemies and control of avocado lace bug, *Pseudacysta perseae* (Hemiptera: Tingidae). *Proceedings of the Florida State Horticultural Society* 111, 330–334.

Peterson, P.A. (1955) Avocado flower pollination and fruit set. *California Avocado Society Yearbook* 39, 163–169.

Peterson, P.A. (1956) Flowering types in the avocado with relation to fruit production. *California Avocado Society Yearbook* 40, 174–177.

Pinhey, E.C.G. (1975) *Moths of Southern Africa*. Tafelberg Publishers, Cape Town, 273 pp.

PPRI (Plant Protection Research Institute) (1970) *Check List of Insects on Forest Trees and Shrubs in South Africa. Entomological Memoir* No. 21. Department of Agricultural Technical Services, Republic of South Africa, 81 pp.

Prinsloo, G.L. (1984) An illustrated guide to the parasitic wasps associated with citrus pests in the Republic of South Africa. *Department of Agriculture, Republic of South Africa, Science Bulletin* No. 402, 1–119.

Prinsloo, G.L. and Neser, O.C. (1994) The southern African species of *Aphelinus* Dalman (Hymenoptera: Aphelinidae), parasitoids of aphids (Homoptera: Aphidoidae). *Journal of African Zoology* 108, 143–162.

Quayle, H.J. (1938) *Insects of Citrus and Other Subtropical Fruits*. Comstock Publishing Company, Ithaca, 583 pp.

Ramirez, M., Otero-Colina, G., Lopez-Collado, J. and Resendiz, B. (1993) Constantes termicas y tablos de vida de Oligonychus perseae (Trombidiformes: Tetranychidae). *Agrociencia Serie Proteccion Vegetal* 4, 67–80.

Ray, C.H. and Williams, M.L. (1982) Description of the immature stages of *Protopulvinaria pyriformis* (Cockerell) (Homoptera: Coccidae). *Florida Entomologist* 65, 169–176.

Read, J.S., Warren, F.L. and Hewitt, P.H. (1968) Identification of the sex pheromone of the false codling moth (*Argyroploce leucotreta*). *Chemical Communications* 792–793.

Read, J.S., Hewitt, P.H., Warren, F.L. and Myburgh, A.C. (1974) Isolation of the sex pheromone of the moth *Argyroploce leucotreta*. *Journal of Insect Physiology* 20, 441–450.

Reed, W. (1974) The false codling moth, *Cryptophlebia leucotreta* Meyr. (Lepidoptera: Olethreutidae) as a pest of cotton in Uganda. *Cotton Growing Review* 51, 213–225.

Reyes, J.C. and Salgado, M. (1994) Tolerancia de algunas selecciones de agucate al Attaque de araña roja (*O. punicae*) y thrips (*Scirtothrips* sp.). *Memorias, Centro de investigaciones cientificas y tecnologicas del aguacate en el estado de Mexico,* Fundacion Salvador Sanchez Colin, Coatepec Harinas, Mexico, pp. 175–178.

Reyes, J.C., Rubi, M. and Aguilar, J. (1995) Manejo organico en el cultivo del aguacate. *Memorias Fundacion Salvador Sanchez Colin, CICTAMEX,* Coatepec Harinas, Mexico, pp. 29–37.

Ripley, L.B. and Hepburn, G.A. (1930) A menace to the fruit industry: bug-tree harbours fruit-fly. *Farming in South Africa* 72, 1–2.

Ripley, L.B., Hepburn, G.A. and Dick, J. (1939) Mass breeding of false codling moth, *Argyroploce leucotreta* Meyrick, in artificial media. Plant Industry Series No. 53. Entomology. *Department of Agriculture and Forestry, Union of South Africa, Science Bulletin* 207, 1–18.

Rivnay, E. (1934) The biology of the greenhouse thrips, *Heliothrips haemorrhoidales* Bouché in Palestine. *Hadar* VII, 241–246.

Rivnay, E. (1935) Ecological studies of the greenhouse thrips *Heliothrips haemorrhoidalis* in Palestine. *Bulletin of Entomological Research* 25, 267–278.

Robbertse, P.J., Coetzer, L.A. and Johannsmeier, M.F. (1996) Hass yield and fruit size as influenced by pollination and pollen donor – a joint progress report. *South African Avocado Growers' Association Yearbook* 19, 63–67.

Robbertse, P.J., Johannsmeier, M.F. and Morudu, T.M. (1997) Pollination studies in Hass avocado in relation to the small fruit problem. *South African Avocado Growers' Association Yearbook* 20, 84–85.

Robbertse, P.J., Kohne, S., Morudu, T.M. and Johannsmeier, M.F. (1998a) Pollination of avocado in South Africa. *Subtropical Fruit News* 6, 18–20.

Robbertse, P.J., Johannsmeier, M.F. and Morudu, T.M. (1998b) Pollination of Hass avocados. *South African Avocado Growers' Association Yearbook* 21, 63–68.

Robertson, C.M. and De Villiers, E.A. (1986) Parasites of avocado pest bite the dust. *Citrus and Subtropical Fruit Research Institute, Information Bulletin* 168, 9.

Robinson, T.R. and Savage, E.M. (1926) *Pollination of the Avocado.* USDA Departmental Circular 387, 16 pp.

Rose, M. and DeBach, P. (1992) Biological control of *Parabemisia myricae* (Kuwana) (Homoptera: Aleyrodidae) in California. *Israel Journal of Entomology* 25–26, 73–95.

Roubik, D.W. (1995) *Pollination of Cultivated Plants in the Tropics.* Food and Agricultural Organization, Agricultural Services Bulletin 118, Rome.

Roubik, D.W. (1998) The killer bee saga. *Subtropical Fruit News* 6, 13–14.

Ruelle, J.E. (1985) Order Isoptera (Termites). In: Scholtz, C.H. and Holm, E. (eds) *Insects of Southern Africa.* Butterworths, Durban, pp. 53–61.

Russell, H.M. (1909) The greenhouse thrips. *United States Department of Agriculture, Bureau of Entomology, Bulletin* 64(4), 43–59.

Russell, H.M. (1912) The red-banded thrips. *United States Department of Agriculture, Bureau of Entomology, Bulletin* 99, Part 2.

Salazar-Garcia, S. and Bolio-Garcia, J.M. (1992) Damage caused by avocado branch weevil (*Copturus aguacatae* Kiss.) in Fuerte avocado trees in Atlixco, Puebla, Mexico. *Proceedings Second World Avocado Congress,* Orange, California, p. 269 (abstract).

Saldarriaga, A. (1977) Las plagas del aguacate y su control. *Separata Curso Frutales, Compendio,* 20. Instituto Colombiano Agropecuario, 70 pp.

Salgado, M. and Bautista, N. (1993) El barrenador grande del hueso del aguacate (*Heilipus lauri* Bohemann) en Ixtapan de la Sal, Mexico. *Fundacion Salvador Salvador Sanchez Colin, Cictamex, Memorias.* December 1993, pp. 225–231.

Salinas, A.P. (1992) Ciclo biologico de la araña cristalina del aguacatero *Oligonychus perseae* (Tuttle, Baker and Abatiello) (Prostigmata: Tetranychidae). Tesis profesional, Universidad Autonoma de Chapingo, Mexico.

Sanchez, H.M. (1983) *Plagas y enfermedades de los frutales.* Pueblo y Educacion, La Habana, Cuba, 195 pp.

Sands, W.A. (1962) The evaluation of insecticides as soil and mound poisons against termites in agriculture and forestry. *Bulletin of Entomological Research* 53, 179–192.

Schieber, E. and Zentmyer, G. (1987) Exploring for sources of resistence among *Persea americana* and Persea slchiedeana in middle America (1987). *South African Avocado Growers Association Yearbook* 10, 20–21.

Schoeman, O.P. (1998) Citrus looper. In: Bedford, E., Van den Berg, M. and De Villiers, E. (eds) *Citrus*

Pests in the Republic of South Africa. Dynamic Ad, Nelspruit, 288 pp.

Schroeder, C.A. (1954) Some aspects of pollination in the avocado. *California Avocado Society Yearbook* 38, 159–162.

Schroeder, C.A. and Hofshi, R. (1998) Some aspects of pollination and fertilization in subtropical fruit species. *Subtropical Fruit News* 6, 2–5.

Schwartz, A., Marias, A. and Van der Kooij, R. (1982) Beheer van valskodlingmot in Citrusdal deur grootskaalse vrylating van 'n eierparasiet. *Information Bulletin, Citrus and Subtropical Fruit Research Institute* 113, 17–20.

Schwarz, R.E. (1998) Citrus aphids. In: Bedford, E.C.G., Van den Berg, M.A. and De Villiers, E.A. (eds) *Citrus Pests in the Republic of South Africa*. Dynamic Ad, Nelspruit, pp. 106–108.

Searle, C.M.St.L. (1964) A list of the insect enemies of citrus pests in Southern Africa. *Department of Agricultural Technical Services, Republic of South Africa, Technical Communication* No. 30, 1–18.

Sedgley, M. (1977) The effect of temperature on floral behaviour, pollen tube growth and fruit set in avocado. *Journal of Horticultural Science* 52, 135–141.

Sedgley, M. and Annells, C.M. (1981) Flowering and fruit-set response to temperature in the avocado cultivar 'Hass'. *Scientia Horticulturae* 14, 27–33.

Sedgley, M. and Grant, W.J.R. (1983) Effect of low temperatures during flowering on floral cycle and pollen tube growth in nine avocado cultivars (*Persea americana*). *Scientia Horticulturae* 18, 207–213.

Shoeman, O.P. (1998) Citrus looper. In: Bedford, E.C.G., Van den Berg, M.A. and De Villiers, E.A. (eds) *Citrus Pests in the Republic of South Africa*. Dynamic Ad, Nelspruit, pp. 212–216.

Shoval, S. (1987) Pollination rate and pollen tube growth of avocado, in relation to yield. MSc thesis, The Hebrew University, Rehovot, Israel (in Hebrew, English abstract).

Skaife, S.H. (1956) *Dwellers in Darkness: an Introduction to the Study of Termites*. Longmans Green, London, 134 pp.

Skaife, S.H. (1979) *African Insect Life*. C. Struik Publishers, Cape Town/Johannesburg, 278 pp.

Smit, B. (1964a) A further chapter in the history of entomology in South Africa. *Journal of the Entomological Society of Southern Africa* 23, 3–15.

Smit, B. (1964b) *Insects of Southern Africa. How to Control Them*. Oxford University Press, Johannesburg, 399 pp.

Smith, D., Beattie, G.A.C. and Broadley, R. (1997) *Citrus Pests and Their Natural Enemies: Integrated Pest Management in Australia*. Queensland Department of Primary Industries Information series Q 197030, Brisbane, Australia, 272 pp.

Snir, E. (1971) Flowering, pollination and fruit-set in the avocado. MSc thesis, The Hebrew University, Rehovot, Israel (in Hebrew, English abstract).

Snowball, G.J. (1969) Prospects for the biological control of white wax scale (*Gascardia destructor*) in Australia by South African natural enemies. *Journal of the Entomological Society of Australia* 5, 22–33.

Soliman, G. and Hoddle, M. (1998) Biology of the red banded whitefly, *Tetraleurodes perseae* Nakahara (Homoptera: Aleyrodidae). http://www.biocontrol.ucr.edu

Sternlicht, M. (1979) *Parabemisia myricae* (Homoptera: Aleyrodidae) – a new pest in Israel. *Hassadeh* 59, 1830–1831 (in Hebrew).

Stevens, P. (1999) The impact of new pests on New Zealand avocado production. In: Arpaia, M. and Hofshi, R. (eds) *Proceedings of Avocado Brainstorming'99*. California Avocado Commission, University of California, Riverside, p.119.

Stevens, P.S. (1997) Avo-Green – an option for reducing pesticide use. *Proceedings of Australia – New Zealand Avocado Conference*, Rotorua, New Zealand, pp. 112–118.

Stevens, P.S., McKenna, C.E. and Steven, D. (1996) Avocado pest management in New Zealand. *Phytoparasitica* 24, 63.

Stevens, P.S., McKenna, C.E. and Steven, D. (1998) Pest management for avocados in New Zealand. *Proceedings of World Avocado Congress III*. 22–28 October 1995. Tel Aviv, Israel, pp. 429–432.

Steyn, J.J. (1954) The pugnacious ant (*Anoplolepis custodiens* Smith) and its relation to the control of citrus scales at Letaba. *Memoirs of the Entomological Society of Southern Africa* 3, 1–96.

Steyn, W.P. (1996) First record of *Thripobius semiluteus* Boucek (Hymenoptera: Eulophidae), as a parasitoid of pine-tree thrips nymphs in South Africa. *Subtropica* 17(9), 14–16.

Steyn, W.P. (2001a) Greenhouse thrips *Heliothrips haemorrhoidales* (Bouché). In: Van den Berg, M.A., De Villiers, E.A. and Joubert, P.H. (eds) *Pests and Beneficials of Tropical and Non-citrus Subtropical Crops in South Africa*. Dynamic Ad, Nelspruit.

Steyn, W.P. (2001b) Red-banded thrips *Selenothrips rubrocinctus* (Giard). In: Van den Berg, M.A., De Villiers, E.A. and Joubert, P.H. (eds) *Pests and Beneficials of Tropical and Non-Citrus Subtropical Crops in South Africa*. Dynamic Ad, Nelspruit.

Steyn, W.P., Du Toit, W.J. and De Villiers, E.A. (1994) Effect of insect growth regulator, CGA 211446, on the third instar of the heart-shaped scale on avocados. *Subtropica* 15(4), 15–18.

Stoetzel, M.B. and Davidson, J.A. (1974) Biology, morphology and taxonomy of immature stages of 9 species in the Aspidioti. *Annals of the Entomological Society of America* 67, 475–509.

Stout, A.B. (1923) A study in cross-pollination of avocado in southern California. *California Avocado Association Annual Report* 7, 29–45.

Stout, A.B. (1932) Sex in avocados and pollination. *California Avocado Association Yearbook* 17, 172–173.

Stout, A.B. (1933) The pollination of avocados. *Florida Agricultural Experimental Station Bulletin* 257, 1–44.

Stout, A.B. and Savage, E.M. (1925) The flower behavior of avocados with special reference to interplanting. *Proceedings of the Florida State Horticultural Society* 38, 80–91.

Swaine, G., Ironside, A. and Yarrow, W.H.T. (1985) *Insect Pests of Fruit and Vegetables*. Queensland Department of Primary Industries, Information Series Q 183021, Brisbane, 176 pp.

Swirski, E. and Arenstein, Z. (1970) List of common insects found in subtropical orchards. In: *Diseases and Pests of Fruit Trees (Except Citrus) in Israel*. The Volcani Institute of Agriculture Research, Bet Dagan, Israel. Published on the occasion of the 18th International Horticultural Congress, pp. 12–15.

Swirski, E., lzhar, Y., Wysoki, M., Gurevitz, E. and Greenberg, S. (1980) Integrated control of the long-tailed mealybug, *Pseudococcus longispinus* [Hom.: Pseudococcidae], in avocado plantations in Israel. *Entomophaga* 25, 415–426.

Swirski, E., Blumberg, D., Wysoki, M. and lzhar, Y. (1987) Biological control of the Japanese bayberry whitefly, *Parabemisia myricae* (Kuwana) (Homoptera: Aleyrodidae), in Israel. *Israel Journal of Entomology* 21, 11–18.

Swirski, E., Wysoki, M. and lzhar, Y. (1988) Integrated pest management in the avocado orchards of Israel. *Applied Agricultural Research* 3, 1–7.

Swirski, E., lzhar, Y. and Wysoki, M. (1991) Appearance of *Aphis gossypii* Glover and *Aphis spiraecola* Patch (Rhynchota: Aphidoidea) on avocado, persimmon and macadamia. *Alon HaNotea* 45, 413–416 (in Hebrew, English summary).

Swirski, E., Wysoki, M. and Ben-Dov, Y. (1997) Avocado. In: Ben-Dov, Y. and Hodgson, C.J. (eds) *Soft Scale Insects – Their Biology, Natural Enemies and Control*. Elsevier Science Publishers, Amsterdam, pp. 231–240.

Swirski, E., Wysoki, M. and Izhar, Y. (1998) Avocado pests in Israel. *Proceedings of the World Avocado Congress III*. Tel-Aviv, Israel, 22–27 October 1995, pp. 419–428.

Taylor, J.S. (1957) Notes on Lepidoptera in the Eastern Cape Province. Part IV. *Journal of the Entomological Society of Southern Africa* 20, 312–332.

Theobald, F.V. (1914) African Aphididae. *Bulletin of Entomological Research* 4, 313–337.

Thompson, W.R. (1946) A catalogue of the parasites and predators of insect pests. Sec. 1. Parasite host catalogue. Part 8. In: *Parasites of the Lepidoptera (N-P)*. Imperial Agricultural Bureaux, Belleville, Canada, pp. 386–523.

Tzfati, N. (1981) Some aspects of the relation between honeybee activity and pollination rate of avocado in the Jordan Valley. High school final thesis, Beit-Yerach (in Hebrew).

Ullyett, G.C. (1939) Parasites of the false codling moth (*Argyroploce leucotreta* Meyr.) in South Africa. Mimeograph. *Proceedings of the Entomological Conference*, Union Buildings, Pretoria, pp. 54–55.

Valentine, D.H. (1987) The pollination of introduced species, with special reference to the British Isles and the genus *Impatiens*. In: Richards, A.J. (ed.) *The Pollination of Flowers by Insects*. Academic Press, London, pp.117–123.

Van den Berg, M.A. (1973) Host plants of three saturniids and the degree of defoliation they cause to *Pinus patula* Schlechtd. & Cham. *Phytophylactica* 5, 65–70.

Van den Berg, M.A. (1974) Natural enemies and diseases of the forest insect *Nudaurelia cytherea clarki* Geertsema (Lepidoptera: Saturniidae). *Phytophylactica* 6, 39–44.

Van den Berg, M.A. (1975) Bio-ecological studies on forest pests 3. *Nudaurelia cytherea clarki* Geertsema (Lepidoptera: Saturniidae). *Forestry in South Africa* 17, 1–16.

Van den Berg, M.A. (1995) Is suksesvolle vestiging van *Trichopoda pennipes* (F.) (Dipt.: Tachinidae), parasitoïde van stinkbesies bereik? *Macadamia Yearbook* 3, 76–78.

Van den Berg, M.A. (1998) Tip wilters and stink bugs on citrus. In: Bedford, E.C.G., Van den Berg, M.A. and De Villiers, E.A. (eds) *Citrus Pests in the Republic of South Africa*. Dynamic Ad. Nelspruit. pp. 161–163.

Van den Berg, M.A. and De Villiers, E.A. (1987a) Psyllids, aphids and whiteflies. In: Myburgh, A.C. (ed.) *Crop Pests in Southern Africa*, Vol. 2, *Citrus and Other Subtropicals*. Perskor, Pretoria, pp. 43–48.

Van den Berg, M.A. and De Villiers, E.A. (1987b) Leafhoppers and stinkbugs. In: Myburg, A.C.

(ed.) *Crop Pests in Southern Africa*, Vol. 2, *Citrus and Other Subtropicals*. Plant Protection Research Institute Bulletin 411, pp. 49–53.

Van den Berg, M.A., Newton, P.J., Deacon, V.E. and Crause, C. (1987) Dispersal of *Trichogrammatoidea cryptophlebiae* (Hymenoptera: Trichogrammatidae), an egg parasitoid of the false codling moth, *Cryptophlebia leucotreta* (Lepidoptera: Tortricidae), in an empty habitat. *Phytophylactica* 19, 515–516.

Van den Berg, M.A., De Villiers, E.A. and Joubert, P.H. (1999) *Identification Manual for Avocado Pests*. Dynamic Ad, Nelspruit, 53 pp.

Vanderplank, F.L. (1958) Studies on the coconut pest, *Pseudotheraptus wayi* Brown (Coreidae), in Zanzibar. 1. A method of assessing the damage caused by the insect. *Bulletin of Entomological Research* 49, 559–584.

Van Heerden, P.W. (1933) The green stink bug (*Nezara viridula* Linn.). *Annals of the University of Stellenbosch* 11A(7), 1–26.

Velasquez, M. and Santizo, L. (1992) El acaro *Calepitrimerus muesebecki*, plaga del follaje del aguacate. *Nota Tecnico Cientifica* 14. Instituto de ciencia and tecnologia agricola (Guatemala), Disciplina de proteccion vegetal, 2 pp.

Vermeulen, J.B. and Bedford, E.C.G. (1998) American bollworm *Helicoverpa (= Heliothis) armigera* (Hübner). In: Bedford, E.C.G., Van den Berg, M.A. and De Villiers, E.A. (eds) *Citrus Pests in the Republic of South Africa*. Dynamic Ad, Nelspruit, pp. 217–221.

Viljoen, H.M. (1986) Kokosneutstinkbesie-'n potensiële plaag op avokado's. *South African Avocado Growers' Association Yearbook* 9, 72–74.

Visscher, P.K. and Sherman, G. (1998) Insect visitors to avocado flowers. *Subtropical Fruit News* 6, 7–10.

Vithanage, H.I.N.V. (1990) The role of the European honeybee (*Apis mellifera* L.) in avocado pollination. *Journal of Horticultural Science* 65, 81–86.

Von Wechmar, M.B., Williamson, C. and Collett, H. (1991) *Nezara viridula* virus: preliminary characterization of a naturally occurring biocontrol agent. *Journal of the Southern African Society for Horticultural Sciences* 1, 29.

Vuillaume, C., Aubert, B., Viladerbo, B. and Laville, E. (1981) The principal pests of avocado in Réunion. *Fruits* 36, 347–350.

Waite, G.K. (1988) Biological control of latania scale on avocados in south-east Queensland. *Queensland Journal of Agricultural and Animal Sciences* 45, 165–167.

Waite, G.K. and Huwer, R.K. (1998) Host plants and their role in the ecology of the fruitspotting bugs *Amblypelta nitida* Stal and *Amblypelta lutescens lutescens* (Distant) (Hemiptera: Coreidae). *Journal of the Australian Entomological Society* 37, 340–349.

Waite, G.K. and Martinez, B.R. (2002) Insect and mite pests. In: Whiley, A.W., Schaffer, B. and Wolstenholme, B.N. (eds) *Avocado: Botany, Production and Uses*. CAB International, Wallingford, UK.

Waite, G.K. and Pinese, B. (1991) Pests. In: Broadley, R.H. (ed.) *Avocado Pests and Disorders*. Queensland Department of Primary Industries Information Series, Q190013, 74 pp.

Way, M.J. (1951) An insect pest of coconuts and its relationship to certain ant species. *Nature, London* 168, 302.

Way, M.J. (1953) Studies on *Theraptus* sp. (Coreidae); the cause of the gumming diseases of coconuts in East Africa. *Bulletin of Entomological Research* 44, 657–667.

Webb, D.V.V. (1974) *Forest and Timber Entomology in the Republic of South Africa. Entomological Memoir* No. 34. Republic of South Africa, Department of Agricultural Technical Services, 21 pp.

Wentzel, P.C., Georgala, M.B. and Bedford, E.C.G. (1978) Citrus thrips, *Scirtothrips aurantii* Faure. In: Bedford, E.C.G. (ed.) *Citrus Pests in the Republic of South Africa*. Department of Agricultural Technical Services, Pretoria, Republic of South Africa, Science Bulletin 391, pp. 137–141.

Westerkamp, C. (1991) Honeybees are poor pollinators – why? *Plant Systematics and Evolution* 177, 71–75.

White, I. and Elson-Harris, M. (1992) *Fruit Flies of Economic Significance: Their Identification and Bionomics*. CAB International, Wallingford, UK, 601 pp.

Whitehead, D. (1979) Recognition characters and distribution records for species of *Conotrachelus* (Coleoptera: Curculionidae) that damage avocado fruits in Mexico and Central America. *Proceedings Entomological Society Washington* 8, 105–107.

Williamson, C. and Von Wechmar, M.B. (1995) The effects of two viruses on the metamorphosis, fecundity and longevity of the green stinkbug, *Nezara viridula*. *Journal of Invertebrate Pathology* 65, 1–5.

Wolfe, H., Franciosi, R., Van Oordt, E. and Figueroa, R. (1969) El culivo del palto en el Peru. Ministerio de Agricultura y Pesqueria, Estacion experimental Agricola de La Molina, Lima, Peru. *Boletin Tecnico* 73.

Wysoki, M. (1987) A bibliography of the pyriform scale, *Protopulvinaria pyriformis* (Cockerell), 1894 (Homoptera: Coccidae), up to 1986. *Phytoparasitica* 15(1), 73–77.

Wysoki, M. and Izhar, Y. (1980) The natural enemies of *Boarmia* (*Ascotis*) *selenaria* Schiff. (Lepidoptera: Geometridae) in Israel. *Acta Oecologica, Oecologia Applicata* 1, 283–290.

Wysoki, M. and Izhar, Y. (1981) Biological data on *Apanteles cerialis Nixon* (Hymenoptera: Braconidae), a parasite of *Boarmia* (*Ascotis*) *selenaria* Schiff. (Lepidoptera: Geometridae). *Phytoparasitica* 9, 19–25.

Wysoki, M. and Izhar, Y. (1986) Fluctuation of the male population of *Boarmia* (*Ascotis*) *selenaria* (Lepidoptera: Geometridae) estimated by virgin female baited traps. *Acta Oecologica, Oecologia Applicata* 7, 251–259.

Wysoki, M., Chen, C., Klein, Z. and Bar-Zakay, I. (1986) The false codling moth, *Cryptophlebia leucotreta* (Meyrick), a new potential pest in Israel. *Alon HaNotea* 40, 572–576 (in Hebrew).

Wysoki, M., de Jong, M. and Rene, S. (1988) *Trichogramma platneri* Nagarkatti (Hymenoptera: Trichogrammatidae), its biology and ability to search for eggs of two lepidopterous avocado pests, *Boarmia* (*Ascotis*) *selenaria* Schiff. (Geometridae) and *Cryptoblabes gnidiella* (Milliere) (Phycitidae) in Israel. *Trichogramma and Other Egg Parasites. Second International Symposium*, Guanghzou, China, 10–15 Nov. 1986. INRA, Paris (Les Colloques de l'INRAS, No. 43).

Wysoki, M., Ben-Yehuda, S. and Rosen, D. (1992) Reproductive behavior of the honeydew moth, *Cryptoblabes gnidiella* (Lep., Pyralidae). *Alon HaNotea* 47, 2–10 (in Hebrew).

Wysoki, M., Kuzlitzky, W., Izhar, Y., Swirski, E., Ben-Yehuda, S., Hadar, D. and Rene, S. (1997) Successful acclimatization of *Thripobius semiluteus*, a parasitoid of *Heliothrips haemorrhoidalis* (Bouche) in Israel. *Phytoparasitica* 25, 155.

Wysoki, M., Rene, S., Eliahu, M. and Blumberg, D. (1999) Influence of spinosad on natural enemies of avocado and other pests. In: Arpaia, M. and Hofshi, R. (eds) *Proceedings of Avocado Brainstorming '99*, Calfornia Avocado Commission, University of California, Riverside, 124 pp.

Yasuda, K. and Kinjou, T. (1983) Seasonal prevalence of the leaf-footed bug, *Leptoglossus australis* Fabricius (Heteroptera: Coreidae) on fruit of the wild host, *Diplocyclos palmatus* C. Jeffrey, and development of immature stages at various constant temperatures. *Proceedings of the Association for Plant Protection of Kyushu* 29, 89–91.

Zeck, E.H. (1932) Investigations on two white wax scales (*Ceroplastes*) as pests in Australia. *Agricultural Gazette of New South Wales* 43, 611–616.

Zentmyer, G., Schieber, E. and Popenoe, W. (1987) Early history of the avocado during the time of the conquistadores. *South African Growers' Association Yearbook* 10, 11–12.

9 Pests of Guava

W.P. Gould[1] and A. Raga[2]
[1]USDA-ARS, Subtropical Horticulture Research Station, 13601 Old Cutler Road, Miami, FL 33158, USA; [2]Instituto Biologico, Caixa Postal 70, 13.001-970, Campinas, SP, Brazil

Guava, *Psidium guajava* L., belongs to the *Myrtaceae*. This plant family has between 110 and 130 species of trees and shrubs. *Myrtaceae* contains a number of fruit trees of economic importance including rose apple (*Syzygium jambos* (L.)), Surinam cherry (*Eugenia uniflora* L.), and Java plum (*Syzygium cumini* (L.)). Most of the species occur naturally from southern Mexico to South America and the Caribbean. Guava is a small tree that is grown worldwide in the tropics and warm subtropics for its edible fruit.

Tree Phenology, Origin, World Production, Importance

Tree phenology

Guavas can flower and bear fruit continuously in the tropics; however, there are normally two crops a year. In Florida and Puerto Rico (USA) there is a large crop in early to mid-summer (June and July), and a smaller crop in late winter (February) (Morton, 1987). A smaller crop is harvested in Hawaii (USA) in April to May and a heavier crop in September and November (Nakasone and Paull, 1998). In India and Malaysia the main crop is harvested in mid-winter and the lesser crop during the rainy season (July–September) (Morton, 1987; Kwee and Chong, 1990). Through manipulation of water, fertilizer, pruning and defoliation, guavas can be forced to produce a larger crop outside of the normal fruit phenology. In south Brazil, guavas are available in the fresh market all year long. Fruits are mature 90–150 days after flowering, depending on the variety or clone of the fruit and cultural and weather conditions (Nakasone and Paull, 1998). The large white flowers attract a large variety of pollinating insects including many species of bees. In many tropical regions the honeybee, *Apis mellifera* L., is the most important pollinator species.

Origin

The guava is native to the American tropics, and probably originally grew from Peru north to Mexico and the Caribbean (Kwee and Chong, 1990). Early Spanish explorers in the Caribbean observed that guavas were cultivated by the Caribbean Indians. The Spanish brought the guava with them to the Philippines, and from there it spread to most Asian countries. The guava was deliberately spread but it is also dispersed by animals eating the fruits. In some parts of the world it is considered a weed because it invades habitats and displaces native plant species. Today guavas are found in all subtropical and tropical parts of the world.

©CAB *International* 2002. *Tropical Fruit Pests and Pollinators* (eds J.E. Peña, J.L. Sharp and M. Wysoki)

World production, importance

Guavas are either eaten fresh, or used in juice, ice cream, jellies, pastes or preserves. According to Chitarra (1996) guava is considered an important fruit nutritionally because of the high level of ascorbic acid which can be five times greater than in oranges and ten times greater than in tomatoes.

In south Florida, guavas are grown as backyard fruits and 32.4 ha are grown commercially (Murray and Campbell, 1989). Hurricane Andrew, in 1992, did a large amount of damage to south Florida's tropical fruit industry. About 84% of the guava trees survived the hurricane (Crane *et al.*, 1993) but some groves were abandoned by owners put out of business by the storm.

In many parts of the tropics guavas are a major crop. In recent years production for juice processing has increased. The countries which produce the largest crops of guavas are India, Brazil and Mexico (Kwee and Chong, 1990). Brazil is the largest producer of guavas, in 1996 production volume was estimated at 251,264 t, grown on 11,035 ha. About 80% of Brazilian production is in the states of São Paulo (southeast) and Pernambuco (northeast), including fresh market and industry processing (Agrianual, 2000). In 1987, guavas were estimated to be grown on more than 50,000 ha in India, producing 27,319 t annually (Morton, 1987). Kwee and Chong (1990) reported that in 1969 India produced over 200,000 t. Most of this fruit is locally consumed, so production figures are difficult to determine. Other countries with significant production are Colombia, Cuba, Dominican Republic, Egypt, Guyana, Haiti, Jamaica, Kenya, Malaysia, Philippines, South Africa, Taiwan, and USA (primarily Florida, Hawaii and Puerto Rico) (Kwee and Chong, 1990). Exact production figures are often not reported, or multiple fruit crops are reported together. In the USA, Hawaii produces and consumes the bulk of guava juice and papaya juice. Mexico produces guavas commercially in the central states of Aguascalientes and Zacatecas (Aluja and Liedo, 1986). White guava juice is exported from India, and pink juice is exported from South American countries such as Brazil, Venezuela and Ecuador. According to European importers, pink guava juice has a higher usage rate because of its colour and flavour (Galinsky, 2000).

Key Pests

Fruit flies

BIOLOGY Wherever guavas are grown commercially, fruit flies are the key pests (Table 9.1). Guava is almost a universal host for fruit-infesting Tephritidae. The female fruit fly seeks the host fruit using odour and visual stimuli; using a piercing ovipositor she deposits eggs under the surface of the fruit (Plate 66). Some fruit fly species deposit single eggs, and leave oviposition-deterring pheromones to ensure that the larvae will have enough resources, while other species deposit several hundred eggs in one oviposition puncture. Even if eggs are not laid in the fruit or do not hatch, the oviposition punctures or stings can render a fruit unmarketable. After several days the eggs hatch into larvae. The larvae burrow and feed under the surface of the fruit for 7–10 days depending on temperature. The larvae pass through three instars inside the fruit. As the larvae grow, and especially if high populations are present, they may leave the surface region of the fruit and burrow through the central pulp area. During cool periods little or no development occurs, which extends the amount of time the insect spends inside the fruit. Under favourable conditions the larvae emerge from the fruit 1–2 weeks after oviposition. The larvae cut an exit hole from the fruit and drop or 'jump' from the fruit. If the fruit has fallen to the ground, the larvae may pupariate immediately underneath the fruit. They usually burrow several centimetres into the soil before pupariation (Hennessey, 1994). The duration of the puparial stage again depends on temperature. Under favourable conditions adults emerge 7–10 days after pupariation. The adult flies must feed for several days to a week on carbohydrate and protein sources before they can mate and lay eggs. Mating takes place in groups or leks on or near the host tree. Dispersal after emergence from pupariation is common and allows flies to colonize new areas and find new hosts as the original host fruits age.

SAMPLING, MONITORING Trapping is used to detect and identify fruit fly populations. Traps are baited with pheromones used by the flies to attract mates, parapheromones derived originally from plants, and also with protein, cane sugar or food-based odours. The pheromone and parapheromone lures attract primarily males. The protein, cane sugar or food-based odours attract both sexes. Lures are not available for all species. The pheromones and parapheromones are specific for closely related species of fruit flies. Methyl eugenol and cuelure attract many species of *Bactrocera*. Tri-medlure only attracts species closely related to the Mediterranean fruit fly (Medfly), *Ceratitis capitata* (Wiedemann). Developing pheromone lures can be a difficult and an expensive process, so lures exist mainly for the most economically important species. The protein lures are useful in general surveys, but they are not as powerful an attractant as some of the parapheromone lures.

There are a number of widely used trap types. The McPhail trap (McPhail, 1937) is a glass or plastic trap often used with a liquid protein bait (Plate 67). The bait odour travels out through a central hole. Flies attracted to the odour fly upwards into the trap and fall into the liquid. This type of trap has been used to trap nuisance flies for many years in Asia and Europe even before McPhail rediscovered it (Steyskal, 1977). The advantage of this trap is that it can be used to detect many fruit fly species including *Bactrocera*, *Anastrepha* and *Ceratitis*. The disadvantages are that it is heavy, expensive, glass versions are fragile, and the liquid refilling can be a logistical problem.

Another commonly used trap is the cylindrical trap which can be plastic (Steiner trap) or cardboard (Jackson trap). In the Steiner trap an insecticide is placed along with the bait in a cotton wick. In the Jackson trap a pheromone or parapheromone bait is placed inside along with a panel coated with a sticky substance. The attracted flies become stuck on the sticky panel. The Jackson trap is inexpensive and attracts great numbers of flies. If the traps are baited with pheromones they catch only males. Some of the pheromones are very host-specific, others less so. This host specificity is good if only one type of fly is desired, but pheromone or parapheromone baits do not exist for many species.

Less widely used is a fruit mimic trap, a coloured ball coated with sticky substance. These traps often catch non-target organisms and are messy to work with. They are most commonly used for temperate fruit flies, in the genus *Rhagoletis*, in temperate fruit crops. They are also used for the papaya fruit fly, *Toxotrypana curvicauda* Gerstaecker (Landolt *et al.*, 1991, 1992). A variety of these traps are commercially available from integrated pest management (IPM) supply companies. Trapping is the basic tactic for fruit fly IPM because the monitoring can prevent damage, and avoid the establishment of exotic species in new regions.

Another way to sample populations of fruit flies is to collect the host fruits and rear the larvae that emerge. This can be a labour-intensive process, but it gives accurate infestation rates, and also gives information on parasitism. The fruit are simply collected, and held in containers over a pupariation substrate such as sand or vermiculite. The larvae emerging from the fruits can be sifted from the pupariation medium and held for emergence. Souza Filho (2000) obtained an average of 36.9 larvae kg^{-1} of unsprayed guava in the state of São Paulo from eight tephritid species.

ECONOMIC THRESHOLDS Fruit fly population management can be placed in one of two categories; management by individual growers to produce a marketable crop, and management on a large scale by some regulatory programme. In large-scale eradication programmes, the capture of a single fruit fly will trigger increased trapping and mass suppression efforts. The economic threshold for the individual grower will depend upon whether the crop is intended to be used for juice or pulp or for the fresh market. The threshold for the fresh market is very low because a single larva in the pulp is generally not tolerated by the consumer. As a result of this, economic thresholds have not been determined. Unsprayed guavas are almost always infested with fruit fly larvae.

Table 9.1. Fruit fly pests of guava.[a]

Order	Family	Species	Common name	Distribution	Parts attacked	Pest status
Diptera	Tephritidae	*Anastrepha antunesi* Lima		Central and South America, Trinidad	fruit	minor
Diptera	Tephritidae	*Anastrepha bahiensis* Lima		Brazil	fruit	minor
Diptera	Tephritidae	*Anastrepha bistrigata* Bezzi		South America	fruit	minor
Diptera	Tephritidae	*Anastrepha daciformis* Bezzi		South America	fruit	minor
Diptera	Tephritidae	*Anastrepha fraterculus* (Wiedemann)	South American fruit fly	Central and South America, Trinidad and Tobago	fruit	key
Diptera	Tephritidae	*Anastrepha leptozona* (Hendel)		Brazil	fruit	key
Diptera	Tephritidae	*Anastrepha ludens* (Loew)	Mexican fruit fly	Central America	fruit	key
Diptera	Tephritidae	*Anastrepha minensis* Lima		South America	fruit	minor
Diptera	Tephritidae	*Anastrepha obliqua* (Macquart)	West Indian fruit fly	Central America, Caribbean, South America	fruit	key
Diptera	Tephritidae	*Anastrepha ocresia* (Walker)		Caribbean	fruit	minor
Diptera	Tephritidae	*Anastrepha parishi* Stone		Central and South America	fruit	minor
Diptera	Tephritidae	*Anastrepha pseudoparallela* (Loew)		Brazil	fruit	minor
Diptera	Tephritidae	*Anastrepha punctata* Hendel		South America	fruit	minor
Diptera	Tephritidae	*Anastrepha schultzi* Blanchard		Argentina	fruit	minor
Diptera	Tephritidae	*Anastrepha serpentina* (Wiedemann)	Serpentine fruit fly	Central America, South America	fruit	key
Diptera	Tephritidae	*Anastrepha sorocula* Zucchi		Brazil	fruit	key
Diptera	Tephritidae	*Anastrepha striata* Schiner	Guava fruit fly	Central and South America	fruit	key
Diptera	Tephritidae	*Anastrepha suspensa* (Loew)	Caribbean fruit fly	Caribbean, Florida	fruit	key
Diptera	Tephritidae	*Anastrepha turpiniae* Stone		Brazil	fruit	minor
Diptera	Tephritidae	*Anastrepha zenildae* Zucchi		Brazil	fruit	minor
Diptera	Tephritidae	*Bactrocera aquilonis* (May)		Northern Australia	fruit	minor
Diptera	Tephritidae	*Bactrocera breviaculeus* (Hardy)		Northern Australia	fruit	minor
Diptera	Tephritidae	*Bactrocera bryoniae* (Tryon)		Australia, Indonesia, Papua New Guinea	fruit	minor
Diptera	Tephritidae	*Bactrocera caudata* (Fabricius)		Brunei, India, Indonesia, Malaysia, Myanmar, Taiwan, Thailand, Vietnam	fruit	minor
Diptera	Tephritidae	*Bactrocera correcta* (Bezzi)	Guava fruit fly	India, Nepal, Pakistan, Sri Lanka, Thailand	fruit	minor
Diptera	Tephritidae	*Bactrocera cucurbitae* (Coquillett)	Melon fly	Egypt, Kenya, Tanzania, New Guinea, Papua New Guinea, Bangladesh, Cambodia, China, Indonesia, Pakistan, Philippines, Taiwan, Thailand, Myanmar, Malaysia, Hawaii, Guam, Iran	fruit	minor
Diptera	Tephritidae	*Bactrocera dorsalis* (Hendel) (species complex)	Oriental fruit fly	Southern India, South-East Asia, Malaysia, Guam, Hawaii, Taiwan, China, Japan, Myanmar, Thailand	fruit	key

Diptera	Tephritidae	*Bactrocera diversa* (Coquillett)		China, India, Sri Lanka, Burma, Pakistan, Thailand	fruit	minor
Diptera	Tephritidae	*Bactrocera facialis* (Coquillett)		Tonga	fruit	minor
Diptera	Tephritidae	*Bactrocera frauenfeldi* (Schiner)	Mango fly	Australia, New Guinea, South-East Asia, South Pacific	fruit	minor
Diptera	Tephritidae	*Bactrocera jarvisi* (Tryon)		Australia	fruit	key
Diptera	Tephritidae	*Bactrocera kirki* (Froggatt)		South Pacific, Western Samoa	fruit	minor
Diptera	Tephritidae	*Bactrocera melanota* (Coquillett)		South Pacific, Cook Islands	fruit	minor
Diptera	Tephritidae	*Bactrocera neohumeralis* (Hardy)		Australia, Papua New Guinea	fruit	key
Diptera	Tephritidae	*Bactrocera passiflorae* (Froggatt)	Fijian fruit fly	South Pacific, Fiji, Niue Island, Tonga	fruit	minor
Diptera	Tephritidae	*Bactrocera pedestris* (Bezzi)		Malaysia	fruit	minor
Diptera	Tephritidae	*Bactrocera psidii* (Froggatt)		South Pacific, New Caledonia	fruit	minor
Diptera	Tephritidae	*Bactrocera tau* (Walker)		India, Nepal, Cambodia, China, Indonesia, Laos, Malaysia, Philippines, Taiwan, Thailand, Vietnam, Burma, Sri Lanka, Pakistan	fruit	minor
Diptera	Tephritidae	*Bactrocera trivialis* (Drew)		Indonesia, Papua New Guinea	fruit	minor
Diptera	Tephritidae	*Bactrocera tryoni* (Froggatt)	Queensland fruit fly	Australia, New Guinea, South Pacific	fruit	key
Diptera	Tephritidae	*Bactrocera xanthodes* (Broun)		Southern Pacific, Cook Islands, Fiji, Western Samoa, Tonga, Vanuatu	fruit	minor
Diptera	Tephritidae	*Bactrocera zonata* (Saunders)	Peach fruit fly	India, South-East Asia, Indonesia, Laos, Sri Lanka, Thailand, Vietnam, Mauritius	fruit	key
Diptera	Tephritidae	*Ceratitis capitata* (Wiedemann)	Mediterranean fruit fly	Hawaii, Africa, South America, Europe, Central America	fruit	key
Diptera	Tephritidae	*Ceratitis catoirii* Guérin-Méneville	Mascarene fruit fly	Mauritius, Réunion, Seychelles	fruit	minor
Diptera	Tephritidae	*Ceratitis cosyra* (Walker)	Marula fruit fly	Africa, Kenya, Zambia, Zimbabwe, South Africa	fruit	minor
Diptera	Tephritidae	*Ceratitis malgassa* Munroe	Madagascan fruit fly	Madagascar	fruit	minor
Diptera	Tephritidae	*Ceratitis quinaria* (Bezzi)	Five-spotted fruit fly	Africa, Botswana, Malawi, Namibia, South Africa, Sudan, Zimbabwe, Middle East, Yemen	fruit	minor
Diptera	Tephritidae	*Ceratitis rosa* Karsch	Natal fruit fly	Africa, Angola, Ethiopia, Kenya, Malawi, Mali, Mozambique, Nigeria, Rwanda, South Africa, Swaziland, Tanzania, Uganda, Zaire, Zambia, Zimbabwe	fruit	key
Diptera	Lonchaeidae	*Neosilba* spp.		Brazil	fruit	minor

[a]References: Drew *et al.*, 1978; Norrbom and Kim, 1988; Robinson and Hooper, 1989; White and Elson-Harris, 1992; Foote *et al.*, 1993; Kapoor, 1993; Malavasi and Zucchi, 2000.

CONTROL TACTICS Control tactics for large-scale eradication programmes are outside of the scope of this work, so this discussion will centre on control tactics for the individual grower. The primary methods of control have been chemical control and physical barriers.

Biological control has been attempted on an area-wide basis by release of parasites against fruit flies (Plate 68). Due to chemical spraying and consumers' low tolerance for fruit damage, these projects have had limited success. A long-term augmentative release project might be the most successful technique, but high costs of parasite mass-rearing have prevented this from being adopted on a wide scale.

Chemical control is used almost everywhere that guavas are grown commercially. Cover sprays of carbamates, synthetic pyrethroids, and organophosphates are applied on a calendar basis to prevent fruit fly populations from building up in the orchard. These sprays are not effective unless the area sprayed is very large because of the flight ability of the fruit fly. One of the potential negative results of calendar spraying on a large scale is outbreaks of secondary pests. When a large guava orchard in Florida (USA) was sprayed on a regular basis to prevent fruit fly infestation, outbreaks of scale insects became a problem. The mass spraying of crops destroys natural and introduced biological controls (predators and parasites). In south Florida (USA), parasites of the Caribbean fruit fly, *Anastrepha suspensa* (Loew), both naturally occurring and augmentative releases, have been reduced to very low levels by regular pesticide spraying. Only in unsprayed reservoirs such as parks, back yards and natural areas, can any parasites be found (W.P. Gould, unpublished; Souza Filho *et al.*, 2000). In the state of São Paulo, four braconid species reached 3.1% of parasitism of *Anastrepha* spp. where 94.9% were *Doryctobracon areolatus* (Szépligeti) (Souza Filho, 2000). Four species of Figitidae were detected in guavas in Brazil (Guimarães *et al.*, 1999). The importance of Figitidae is probably underestimated in the biological control agent complex.

Few plant resistance studies have been attempted against *Anastrepha* sp. in guavas, but works by Hennessey *et al.* (1995, 1996) and A. Raga (unpublished) have shown that there is potential in this area. In general, fruits which develop more rapidly (are available on the tree for a shorter time), have thicker rind, or have unfavourable chemical environments in the peel, may effectively reduce fly populations (Greany and Shapiro, 1993). Jalaluddin *et al.* (1998a) reported that pre-colour break treatment of guava fruits with 50 p.p.m. gibberellic acid (GA_3) reduced fruit susceptibility to infestation by *Bactrocera correcta* (Bezzi). Jalaluddin *et al.* (1998b) demonstrated that cultivar 'Lucknow 46' was highly resistant to *B. correcta*, whereas cultivars 'AC 10', 'Lucknow 49' and 'Chittidar' were moderately resistant, susceptible and highly susceptible to *B. correcta* attack, respectively.

Autocidal techniques have been used in eradication programmes with great success. This technique is very expensive and requires mass rearing of millions of insects which are then sterilized and released. Many outbreaks of exotic fruit flies have been eradicated using this tool in combination with bait sprays. Bait sprays have not generally been used by growers but have remained primarily a tool of large-scale eradication projects. Eradication projects are extremely expensive and are undertaken by government and international agencies when the potential cost from damage by new invasive species is extreme. Costs for these programmes are high, so they are not attempted against well established or native species. The widespread spraying of pesticide over urban areas has also raised concerns among the public about the desirability of these programmes. The effect of these spray programmes on non-target organisms in the environment is another serious concern.

The second major method of preventing infestation is the use of barriers. Plastic netting, plastic bags and paper bags have been used successfully to prevent fruit flies from ovipositing in guavas (Howard, 1986; Kwee and Chong, 1990) (Plate 69). Fruits must be wrapped early in their development before they are suitable for oviposition by fruit flies. The bags must be adequately ventilated to prevent humidity from allowing establishment of fungi. The bags may also allow build-up of scale insects and mealybugs. The main disadvantage to bagging of fruit is the labour cost (Kwee and Chong, 1990).

Some alternative methods of control are either not practised, or are only practised as parts of large-scale regulatory efforts. Qureshi and Hussain (1993) found that mass trapping reduced fly populations of *Bactrocera dorsalis* (Hendel) and *B. zonata* (Saunders) in guava and mango orchards to subeconomic levels (5.5%) which were considerably lower than those produced by pesticide spraying (20.25%).

It is often recommended that growers remove and destroy all unused fruits in the field, to reduce reservoirs of fruit fly larvae. This practice has been an effective part of large-scale eradication programmes; on a small scale its effectiveness may not be as great (Burton, 1930; Hansen and Armstrong, 1990; Liquido, 1993). Soil drench pesticide sprays have also been used against fruit fly puparia and larvae in large eradication programmes. This method is expensive and has limited effectiveness.

Fruit flies present a considerable phytosanitary or quarantine problem for shippers of fresh market fruits. Unless the guavas are grown in areas certified to be free of fruit flies, a postharvest commodity treatment will need to be applied prior to shipment to areas which do not have that fruit fly species. Few quarantine treatments have specifically been developed for guavas. Currently, hot water, hot air, and irradiation have been developed as treatments for the Caribbean fruit fly (Gould and Sharp, 1992; Arthur *et al.*, 1993; Gould, 1994). Hot water and irradiation treatments have been used to ship Florida guavas to Texas and California within the USA.

Secondary Pests

There are many secondary or minor pests of guava. Most of these species are of minor importance, but population outbreaks can occur locally, making these pests important. Very little is known about many of these pests, and management strategies are also not available for most of them. Common sense dictates that controls be applied only as a last resort, and that the minimum treatment necessary be used to effect control.

Hemiptera: Heteroptera (bugs)

In Malaysia, *Heliopeltis theobromae* Miller (Miridae), a serious pest of cacao, also attacks guavas (Kwee and Chong, 1990). These bugs feed on many parts of the plant including the fruit. Damage to the fruit results in necrotic lesions which render it unmarketable. The symptoms of damage are clear, but the causal insect is often missed by the grower. There are no biological controls known, and control is usually by pesticide applications. In south Florida (USA), South America, and parts of the South Pacific, several species of *Leptoglossus* (Coreidae) attack the fruits, also causing necrotic scab-like lesions when the fruits mature (Nafus and Schreiner, 1999; Peña *et al.*, 1999). These types of pests are usually only a problem to the grower during fruit formation, and treatment is seldom necessary. These bugs often form aggregations on only a few trees so spot treatment may be an option. Cultural control of alternative nymphal hosts such as thistle species or other nearby weeds may lower populations of this pest.

Hemiptera: Homoptera (mealybugs, scale insects)

There are two families of mealybugs found on guavas: Pseudococcidae, which contains most of the species; and Margarodidae, the giant mealybugs (Table 9.2). Many species are found worldwide. Some commonly reported species on guava are *Ferrisia virgata* (Cockerell), *Planococcus citri* (Risso), *Planococcus pacificus* Cox, and *Pseudococcus citriculus* Cox (Kwee and Chong, 1990), *Pseudococcus nipae*, *Planococcus minor* (Maskell) and *Pseudococcus lilacinus* Cock.

Mealybugs are soft-bodied insects that secrete a waxy protective layer that looks like cotton or snow. The first instars or crawlers are the only dispersal stage. They are spread by the wind or they may walk to new hosts. Once established on a host, large populations

Table 9.2. Other insect pests of guava.[a]

Order	Family	Species	Common name	Distribution	Parts attacked	Pest status
Hemiptera	Miridae	*Helopeltis theobromae* Miller	Mosquito bug	Malaysia	fruit, leaves, stems, flower buds	minor
Hemiptera	Miridae	*Helopeltis antonii* Signoret		India	fruit	minor
Hemiptera	Coreidae	*Holhymenia clavigera* (Herbst)		Brazil	fruit	minor
Hemiptera	Coreidae	*Leptoglossus australis* (Fabricius)	Leaf footed bug	South Pacific	fruit	minor
Hemiptera	Coreidae	*Leptoglossus balteatus* (L.)	Leaf footed bug	South Florida	fruit	minor
Hemiptera	Coreidae	*Leptoglossus concolor* (Walker)	Leaf footed bug	South Florida	fruit	minor
Hemiptera	Coreidae	*Leptoglossus gonagra* (Fabricius)	Leaf footed bug, citron plant bug	South Florida	fruit	minor
Hemiptera	Coreidae	*Leptoglossus stigma* (Herbst)		Brazil	fruit	minor
Hemiptera	Coreidae	*Leptoglossus zonatus* (Dallas)		North America, Central America, South America	fruit	minor
Hemiptera	Pseudococcidae	*Dysmicoccus brevipes* (Cockerell)	Pineapple mealybug	Worldwide tropics and subtropics	leaves, stems, fruit	minor
Hemiptera	Pseudococcidae	*Ferrisia virgata* (Cockerell)	Striped mealybug	Worldwide tropics and subtropics	terminal shoots, leaves and fruits	minor
Hemiptera	Pseudococcidae	*Maconellicoccus hirsutus* (Green)	Pink hibiscus mealybug	Old World tropics, Caribbean	leaves, stems, fruit	minor
Hemiptera	Pseudococcidae	*Nipaecoccus nipae* (Maskell)	Spiked mealybug	Worldwide tropics	leaves, fruit	minor
Hemiptera	Pseudococcidae	*Nipaecoccus viridis* (Newstead)	Spherical mealybug	Old World tropics	leaves, fruit	minor
Hemiptera	Pseudococcidae	*Perissopneumon ferox* Newstead		India	leaves, fruit	minor
Hemiptera	Pseudococcidae	*Phenacoccus psidiarum* Cockerell		Mexico	leaves, fruit	minor
Hemiptera	Pseudococcidae	*Planococcus citru* (Risso)	Citrus mealybug	Worldwide	leaves, fruit	minor
Hemiptera	Pseudococcidae	*Planococcus lilacinus* Cockerell	Cacao mealybug	Asia, Indian Ocean Islands, Caribbean	leaves, fruit	minor
Hemiptera	Pseudococcidae	*Planococcus minor* (Maskell)	Passionvine mealybug	India, Japan, Pacific Islands	leaves, fruit	minor
Hemiptera	Pseudococcidae	*Planococcus pacificus* Cox		Malaysia	leaves, fruit	minor
Hemiptera	Pseudococcidae	*Pseudococcus citriculus* Cox		Malaysia	leaves, fruit	minor
Hemiptera	Pseudococcidae	*Pseudococcus longispinus* (Targioni Tozetti)	Long-tailed mealybug	Worldwide	stems, leaves, fruit	minor
Hemiptera	Pseudococcidae	*Rastrococcus iceryoides* (Green)	Mango mealybug	Asia, Africa	stems, leaves, fruit	minor
Hemiptera	Pseudococcidae	*Rastrococcus invadens* Williams	Mango mealybug	Asia, Africa	stems, leaves, fruit	minor
Hemiptera	Margarodidae	*Icerya seychellarum* (Westwood)	Giant mealybug	Asia	leaves, stems	minor
Hemiptera	Margarodidae	*Steatococcus samaraius* Morrison	Steatococcus scale	Southern Pacific	leaves, stems	minor

Hemiptera	Aleyrodidae	*Aleurocanthus woglumi* Ashby	Citrus blackfly	Worldwide tropics	leaves, fruit	minor
Hemiptera	Aleyrodidae	*Aleurodicus dispersus* Russell	Spiralling whitefly	Worldwide tropics	leaves, fruit	minor
Hemiptera	Aleyrodidae	*Aleurodicus maritimus* Hempel		Mexico	leaves, fruit	minor
Hemiptera	Aleyrodidae	*Aleurodicus pulvinatus* (Maskell)	Coconut whitefly	Central and South America, Caribbean	leaves, fruit	minor
Hemiptera	Aleyrodidae	*Aleurothrixus floccosus* (Maskell)	Woolly whitefly	Africa, Asia, Europe, Central and South America, Caribbean	leaves, fruit	minor
Hemiptera	Aleyrodidae	*Aleurotuberculatus canangae* Corbett		Malaysia	leaves, fruit	minor
Hemiptera	Aleyrodidae	*Aleurotuberculatus cherasensis* Corbett		Malaysia	leaves, fruit	minor
Hemiptera	Aleyrodidae	*Aleurotuberculatus psidii* (Singh)		Asia, Malaysia	leaves, fruit	minor
Hemiptera	Aleyrodidae	*Dialeurodes psidii* Corbett		Malaysia	leaves, fruit	minor
Hemiptera	Aleyrodidae	*Parabemisia myricae* (Kuwana)	Bayberry whitefly	Mediterranean, Asia	leaves, stems, fruit	minor
Hemiptera	Aleyrodidae	*Paraleyrodes urichii* Quaintance and Baker		Barbados	leaves, fruit	minor
Hemiptera	Aleyrodidae	*Tetraleurodes truncatus* Sampson and Drews		Mexico	leaves, fruit	minor
Hemiptera	Aleyrodidae	*Trialeurodes floridensis* Quaintance	Avocado whitefly	Mexico, Florida	leaves, stems, fruit	minor
Hemiptera	Aleyrodidae	*Trialeurodes vaporariorum* Westwood	Greenhouse whitefly	Worldwide in greenhouses, Asia	leaves, stems, fruit	minor
Hemiptera	Aphididae	*Aphis craccivora* Koch	Black legume aphid, cowpea aphid, Groundnut aphid	Worldwide	leaves, stems, fruits	minor
Hemiptera	Aphididae	*Aphis gossypii* Glover	Cotton or melon aphid	Worldwide	leaves, stems, fruits	minor
Hemiptera	Aphididae	*Aphis spiraecola* Patch	Spiraea aphid, green citrus aphid	North America, Mediterranean, Africa, Australia	leaves, stems	minor
Hemiptera	Aphididae	*Greenidea ficicola* Takahashi		Asia, Australia, Africa	leaves, stems, fruits	minor
Hemiptera	Aphididae	*Greenidea formosana* Maki		Asia	leaves, stems	minor
Hemiptera	Aphididae	*Myzus ornatus* Laing	Violet aphid	Worldwide	leaves, stems	minor
Hemiptera	Aphididae	*Myzus persicae* Sulzer	Green peach aphid	Worldwide	leaves, stems, flowers, fruits	minor
Hemiptera	Aphididae	*Toxoptera aurantii* (Boyer de Fonscolombe)	Black citrus aphid	Worldwide tropics and subtropics	leaves, stems	minor
Hemiptera	Coccidae	*Ceroplastes janeirensis* Gray	Soft scale	Brazil	leaves, stems, fruits	minor

continued

Table 9.2. *Continued.*

Order	Family	Species	Common name	Distribution	Parts attacked	Pest status
Hemiptera	Coccidae	*Ceroplastes floridensis* Comstock	Soft scale	Worldwide tropics and subtropics	leaves, stems, fruits	minor
Hemiptera	Coccidae	*Ceroplastes rubens* Maskell	Red wax scale	Africa, Asia, Caribbean, Southern Pacific	leaves, stems, fruits	minor
Hemiptera	Coccidae	*Ceroplastes rusci* L.	Fig wax scale	Mediterranean, Africa, Asia, Caribbean, South America	leaves, stems, fruits	minor
Hemiptera	Coccidae	*Cloropulvinaria psidii* Maskell	Green shield scale	Southern Pacific	leaves, stems, fruits	minor
Hemiptera	Coccidae	*Coccus acuminatus* Signoret		Malaysia	leaves, stems, fruits	minor
Hemiptera	Coccidae	*Coccus celatus* De Lotto		Malaysia, Papua New Guinea	leaves, stems, fruits	minor
Hemiptera	Coccidae	*Coccus hesperidum* L.	Brown soft scale	Worldwide	leaves, stems, fruits	minor
Hemiptera	Coccidae	*Coccus viridus* Green	Green scale	Worldwide tropics and subtropics	leaves, stems, fruits	minor
Hemiptera	Coccidae	*Eucalymnatus tessellatus* (Signoret)		Florida, India, Bermuda, Tonga	leaves, stems, fruits	minor
Hemiptera	Coccidae	*Kilifia acuminata* (Signoret)	Acuminate scale	Asia, Southern Pacific, Trinidad and Tobago	leaves, stems, fruits	minor
Hemiptera	Coccidae	*Kilifia deltoides* DeLotto		Southern Pacific	leaves, stems, fruits	minor
Hemiptera	Coccidae	*Parasaissetia nigra* (Nietner)	Nigra scale	Malaysia, Southern Pacific	leaves, stems, fruits	minor
Hemiptera	Coccidae	*Protopulvinaria pyriformis* (Cockerell)	Pyriform scale	Worldwide distribution	leaves, stems, fruits	minor
Hemiptera	Coccidae	*Pulvinaria ficus* Hempel	Guava mealy scale	Caribbean	leaves, stems, fruits	minor
Hemiptera	Coccidae	*Pulvinaria psidii* (Maskell)		Malaysia	leaves, stems, fruits	minor
Hemiptera	Coccidae	*Pulvinaria urbicola* Cockerell		Malaysia, Caribbean	leaves, stems, fruits	minor
Hemiptera	Coccidae	*Saissetia coffeae* (Walker)	Hemispherical scale	Worldwide distribution	leaves, stems, fruits	minor
Hemiptera	Coccidae	*Saissetia miranda* (Cockerell and Parrott)		Florida	leaves, stems, fruits	minor
Hemiptera	Coccidae	*Saissetia neglecta* (DeLotto)	Caribbean black scale	South Pacific	leaves, stems, fruits	minor
Hemiptera	Coccidae	*Saissetia oleae* (Olivier)	Black scale, olive scale	Worldwide distribution	leaves, stems, fruits	minor
Hemiptera	Coccidae	*Vinsonia stellifera* (Westwood)	Stellate scale	Florida, Caribbean, South Pacific	leaves, stems, fruits	minor
Hemiptera	Diaspididae	*Aonidiella aurantii* (Maskell)	California red scale	Worldwide distribution	leaves, stems, fruits	minor
Hemiptera	Diaspididae	*Aonidiella orientalis* (Newstead)	Oriental scale, oriental yellow scale	Worldwide distribution	leaves, stems, fruits	minor
Hemiptera	Diaspididae	*Aspidiotus destructor* Signoret	Coconut scale	Worldwide distribution	leaves, stems, fruits	minor
Hemiptera	Diaspididae	*Chrysomphalus dictyospermi* (Morgan)	Dictyospermum scale, Spanish red scale	Worldwide distribution	leaves, stems, fruits	minor

Hemiptera	Diaspididae	*Chrysomphalus ficus* Ashmead	Florida red scale	Florida	leaves, stems, fruits	minor
Hemiptera	Diaspididae	*Hemiberlesia cyanophylli* (Signoret)	Cynophyllum scale	Florida	leaves, stems, fruits	minor
Hemiptera	Diaspididae	*Hemiberlesia diffinis* (Newstead)		Florida, Mexico	leaves, stems, fruits	minor
Hemiptera	Diaspididae	*Hemiberlesia lataniae* (Signoret)	Latania scale	Worldwide distribution	leaves, stems, fruits	minor
Hemiptera	Diaspididae	*Hemiberlesia palmae* (Cockerell)	Palm scale	India, Southern Pacific	leaves, stems, fruits	minor
Hemiptera	Diaspididae	*Hemiberlesia rapax* (Comstock)	Greedy scale	Worldwide distribution	leaves, stems	minor
Hemiptera	Diaspididae	*Lepidosaphes similis* Beardsley		Southern Pacific	leaves, stems	minor
Hemiptera	Diaspididae	*Lepidosaphes laterochitinosa* Green	Armoured scale	Florida, China, Southern Pacific	leaves, stems	minor
Hemiptera	Diaspididae	*Parlatoria pergandii* Comstock	Chaff scale	Worldwide distribution	leaves, stems, fruits	minor
Hemiptera	Diaspididae	*Pseudaonidia clavigera* (Cockerell)	Camellia mining scale	Florida	bark	minor
Hemiptera	Diaspididae	*Pseudaonidia duplex* (Cockerell)	Camphor scale	Florida, China, Korea	leaves, stems, fruits	minor
Hemiptera	Diaspididae	*Pseudischnaspis bowreyi* (Cockerell)	Bowrey scale	Florida	leaves, stems, fruits	minor
Hemiptera	Diaspididae	*Pseudoparlatoria parlatorioides* (Comstock)	False parlatoria scale	Florida	leaves, stems, fruits	minor
		Selenaspidus articulatus (Morgan)				
Hemiptera	Diaspididae	*Unaspis citri* (Comstock)		Brazil	leaves	minor
Hemiptera	Diaspididae	*Triozoida* sp.	Citrus snow scale	Worldwide distribution	leaves, stems, fruits	minor
Hemiptera	Psyllidae	*Liothrips bondari* Moulton		Brazil	leaves	minor
Thysanoptera	Phlaeothripidae	*Selenothrips rubrocinctus* (Giard)		Brazil	leaves, fruit	minor
Thysanoptera	Thripidae		Red-banded thrips	Worldwide tropics	leaves, flower buds, fruit	minor
		Selenothrips dorsalis Hood				
Thysanoptera	Thripidae			South-East Asia	leaves, flower buds, fruit	minor
		Eupseudosoma involuta (Sepp)				
Lepidoptera	Arctiidae	*Hypercompe icasia* Cramer		Brazil	leaves	minor
Lepidoptera	Arctiidae	*Brachmia* sp.		Lesser Antilles, Puerto Rico	leaves	minor
Lepidoptera	Gelechiidae	*Buzura suppressaria* (Guenée)		Malaysia	shoot tips	minor
Lepidoptera	Geometridae	*Thyrinteina arnobia* (Stoll)		China, India, Sri Lanka	leaves	minor
Lepidoptera	Geometridae	*Phocides polybius* Reakirt		Brazil	leaves	minor
Lepidoptera	Hesperiidae		Guava skipper, gusano cabezón	North, Central, South America	leaves	minor
Lepidoptera	Hesperiidae	*Pyrrhopyge charybdis* Westwood & Hewitson		Brazil	leaves	minor
		Argyresthia eugeniella Busck				
Lepidoptera	Hyponomeutidae		Guava fruit worm	Florida	fruit	minor

continued

Table 9.2. *Continued.*

Order	Family	Species	Common name	Distribution	Parts attacked	Pest status
Lepidoptera	Lasiocampidae	*Eutachyptera psidii* (Salle)	Palomilla mexicana de la seda	Mexico	leaves	minor
Lepidoptera	Lasiocampidae	*Odonestis vita* Moore		Malaysia	leaves	minor
Lepidoptera	Lycaenidae	*Deudorix isocrates* (Fabricius)	Anar caterpiller	India, Bangladesh	leaves	minor
Lepidoptera	Lymantriidae	*Euproctis fraterna* (Moore)	Coffee hairy caterpillar	Malaysia, Bangaledesh, India, Thailand, Myanmar	leaves	minor
Lepidoptera	Lymantriidae	*Lymantria monacha* L.	Nun moth	Europe, Asia	leaves	minor
Lepidoptera	Lymantriidae	*Lymantria xylina* Swinhoe		China, Taiwan		minor
Lepidoptera	Megalopygidae	*Megalopyge defoliata trujillo* Schaus	Borreguillo	Mexico	leaves	minor
Lepidoptera	Metarbelidae	*Indarbela quadrinotata* (Walker)	Bark eating caterpillar	Bangladesh, India, Myanmar, Pakistan, Sri Lanka	bark	minor
Lepidoptera	Mimallonidae	*Mimallo amilia* (Stoll-Cramer)	Sack bearer	Brazil	leaves	minor
Lepidoptera	Noctuidae	*Anua tongaensis* Hampson		South Pacific	leaves	minor
Lepidoptera	Noctuidae	*Eudocima fullonia* (Clerck)	Fruit piercing moth	Africa, Asia, South Pacific	fruits	minor
Lepidoptera	Noctuidae	*Heliothis armigera* Hubner		Malaysia	fruits	minor
Lepidoptera	Noctuidae	*Gonodonta clotilda* (Stoll)		Venezuela	leaves	minor
Lepidoptera	Noctuidae	*Spodoptera littoralis* (Boisduval)	Egyptian cotton leafworm	Europe, Africa, Asia Minor	leaves, fruits	minor
Lepidoptera	Nymphalidae	*Melanitis leda ismene* Cramer	Rice butterfly	Africa, Asia, Australia	leaves	minor
Lepidoptera	Psychidae	*Cremastopsyche pendula* Joannis	Bagworm	Malaysia	leaves	minor
Lepidoptera	Psychidae	*Eumeta minuscula* Butler	Bagworm	Malaysia	leaves	minor
Lepidoptera	Psychidae	*Oiketicus kirbyi* Guilding		Brazil, Colombia, Trinidad, Tobago, Costa Rica	leaves	minor
Lepidoptera	Pyralidae	*Conogethes punctiferalis* (Guenée)	Castor capsule borer	South-East Asia	fruits	minor
Lepidoptera	Saturniidae	*Attacus atlas* L.	Atlas moth	South-East Asia	leaves	minor
Lepidoptera	Saturniidae	*Citheronia laocoon* Cramer		Brazil	leaves	minor
Lepidoptera	Stenomidae	*Timocratica palpalis* (Zeller)		Brazil	branches	minor
Lepidoptera	Tineidae	*Psorosticha zizyphi* (Stainton)	Citrus leafroller	India	leaves	minor
Lepidoptera	Tortricidae	*Amorbia emigratella* Busck		Malaysia	buds, leaves	minor
Lepidoptera	Tortricidae	*Amorbia* spp.		Brazil	fruits	minor

Lepidoptera	Tortricidae	*Cryptophlebia leucotreta* Meyrick	False codling moth	Africa	fruits	minor
Lepidoptera	Tortricidae	*Dudua aprobola* (Meyrick)		South Pacific	leaves	minor
Lepidoptera	Tortricidae	*Strepsicrates ejectana* (Walker)	Guava bud moth	South Pacific		minor
Lepidoptera	Tortricidae	*Strepsicrates rothia* Walker		Malaysia	buds, leaves	minor
Lepidoptera	Tortricidae	*Strepsicrates smithiana indetana* (Dyar)		Malaysia	buds, leaves	minor
Lepidoptera	Tortricidae	*Strepsicrates tetropsis* (Busck)		Trinidad and Tobago	leaves	minor
Coleoptera	Bostrichidae	*Apate monachus* Fabricius	Black borer	Africa, Caribbean	boring in stems, twigs	minor
Coleoptera	Cerambycidae	*Retrachyderes thoracicus* (Oliver)	Trunk borer	Brazil	boring in stems, twigs	minor
Coleoptera	Chrysomelidae	*Costalimaita ferruginea* (Fabricius)	Yellow beetle	Brazil, Venezuela	leaves	minor
Coleoptera	Curculionidae	*Anthonomus irroratus* Dietz		Florida	fruit	minor
Coleoptera	Curculionidae	*Conotrachelus dimidiatus* Champion		Mexico	fruit	key
Coleoptera	Curculionidae	*Conotrachelus psidii* Marshall		Brazil, Venezuela, Bolivia	fruit	key
Coleoptera	Curculionidae	*Diaprepes abbreviatus* L.	Apopka weevil, citrus weevil	Florida, Caribbean	leaves, roots	minor
Coleoptera	Curculionidae	*Diaprepes splengleri* L.	Golden leaf weevil	Caribbean	leaves, roots	minor
Coleoptera	Curculionidae	*Hypomeces squamosus* (Fabricius)	Green weevil	Malaysia	leaves	minor
Coleoptera	Curculionidae	*Pandeleteius vitticollis* Champion		Mexico	leaves	minor
Coleoptera	Curculionidae	*Pantomorus albosignatus* Boheman		Mexico	leaves	minor
Coleoptera	Curculionidae	*Pantomorus cervinus* (Boheman)		Mexico	leaves	minor
Coleoptera	Scarabaeidae	*Adoretus versutus* Harold	Rose beetle	South Pacific, Asia, Indian Ocean	leaves	minor
Coleoptera	Scarabaeidae	*Cyclocephala lunulata* Burmeister	Temolillo	Mexico, Panama	leaves	minor
Coleoptera	Scarabaeidae	*Maladera* sp.		Malaysia	leaves	minor
Coleoptera	Scarabaeidae	*Euphoria* sp.		Mexico	fruits	minor

continued

Table 9.2. *Continued.*

Order	Family	Species	Common name	Distribution	Parts attacked	Pest status
Coleoptera	Scarabaeidae	*Holotrichia rufoflava* Brenske		India	leaves	minor
Coleoptera	Scolytidae	*Hypothenemus eruditus* (Westwood)		India, Korea, Seychelles Islands, Brazil	boring in twigs, stems	minor
Coleoptera	Scolytidae	*Hypothenemus psidii* Hopkins		Philippines	boring in twigs, stems	minor
Orthoptera	Acrididae	*Valanga nigricornis* Sjost	Grasshopper	Malaysia	leaves	minor
Isoptera	Kalotermitidae	*Neotermes connexus* Snyder	Forest tree termite	South Pacific	boring in wood	minor
Acarina	Eriophyidae	*Tegolophus guavae* (Boczek)	Guava mite	Florida	leaves, fruit	minor
Acarina	Tarsonemidae	*Tarsonemus* sp.		Florida	leaves, fruit	minor
Acarina	Tenuipalpidae	*Brevipalpus californicus* (Banks)	Citrus flat mite	Worldwide distribution	leaves, fruit	minor
Acarina	Tenuipalpidae	*Brevipalpus phoenicis* (Geijskes)	False spider mite	Worldwide distribution	leaves, fruit, branches	minor
Acarina	Tetranychidae	*Panonychus* sp.		Malaysia	leaves, fruit	minor
Acarina	Tetranychidae	*Eotetranychus hicoriae* (McGregor)	Pecan leaf scorch mite	Malaysia, India, Republic of Georgia, United States	leaves, fruit	minor
Acarina	Tetranychidae	*Eutetranychus orientalis* Klein	Citrus brown mite	Africa, Asia, Australia	leaves, fruit	minor
Acarina	Tetranychidae	*Oligonychus biharensis* (Hirst)	Red mite	Malaysia, Mexico	leaves, fruit	minor
Acarina	Tetranychidae	*Oligonychus psidium* Estébanes and Baker		Mexico	leaves, fruit	minor
Acarina	Tydeidae	*Tydeus* sp.		Florida	leaves, fruit	minor

[a]References: Silva *et al.*, 1968; Anonymous, 1981; Morton, 1987; Gallo *et al.*, 1988; Medina, 1988; Kwee and Chong, 1990; Kubo and Batista Filho, 1992; Anonymous, 1999; Gonzalez Gaona, 1999; Nafus and Schreiner, 1999; Peña *et al.*, 1999.

are produced around the original colonizer. Mealybugs can be found on twigs, stems, leaves and fruits. Normally these insects do not damage the host severely, but large populations cause the fruit to become misshapen and deformed. Honeydew produced by mealybugs often results in sooty mould which lowers the fruit's market value. Presence of mealybugs on the fruit may also be of phytosanitary importance. Normally parasites and predators control mealybug populations, but misuse or overuse of pesticides may disturb this balance. Oil or soap sprays are often used more effectively since they are safer for biological control agents. Ants are often associated with mealybugs, aphids and other homopterans that produce honeydew. Ants protect mealybugs from predators and parasites. Bagging fruits to protect them from fruit flies actually benefits the mealybugs, protecting them from predators and parasites and forming a favourable microclimate inside the bag.

In Venezuela, the cottony scale *Capulinia* sp. near *jaboticabae* von Ihering (Eriococcidae) has become one of the most destructive guava pests since 1993 (Chirinos, 2000). The adult female measures *c.* 1.36 × 0.86 mm (length × width); the female undergoes two nymphal instars whereas the male undergoes four nymphal instars (Chirinos, 2000). Chirinos (2000) reported that survival of *Capulinia* was lower on cv. 12 (*Psidium friedichsthalianum* × *P. guineense*) and higher in *P. guajava* cv. 'criolla roja'.

Aleyrodidae (whiteflies)

Whiteflies have a life cycle similar to mealybugs. Most of the information given for mealybugs applies to whiteflies as well. One major difference is that adult whiteflies are much more mobile than mealybugs. A number of species have been identified attacking guava (Table 9.2).

In Cuba, Vazquez *et al.* (1996) reported that among fruit trees, guava has the highest frequency of *Aleyrodicus floccosus* Maskell and it also supports the largest number of whitefly species (e.g. *Aleurocanthus woglumi* Ashby, *Aleyrodicus dispersus* Russell, *Aleyroglandulus malangae* Russell, *Bemisia tabaci* (Gennadius) among others).

Aphididae (plant lice, aphids)

There are only two important pest species of aphids attacking guava, and these species are both generalists that attack many crops. Populations seldom build to the level where damage is done or control is necessary. Biological controls of many kinds, parasites and predators, normally keep these aphid populations in check. Their life cycle is similar to other Homoptera, except that they have parthenogenetic forms. This allows for very rapid population increase after colonization. In the sequence of several generations on a new host, reproductive winged forms are produced which then can disperse to new hosts.

Coccidae and Diaspididae (scale insects)

There are two families of scale insects found on guava: Coccidae (soft scales) and Diaspididae (armoured scales). Scales may be found on stems, leaves and fruits. The life cycle of scales is similar to that of mealybugs. Biological control keeps scale populations in check most of the time. Unusual weather conditions like drought, or use of pesticide may cause population outbreaks. Oil or soap sprays are normally effective in controlling scale outbreaks and minimizing damage to biological controls. Some pheromone traps are used to time sprays to coincide with scale outbreaks. With all Homoptera, their extreme reproductive capabilities must be respected. Only apply controls when necessary. Follow-up treatments are needed to control the crawlers which hatched from eggs subsequent to the first treatment. Use the minimum treatment necessary to control the problem. Scale insects may also be a phytosanitary issue, and the coating of scales on the fruits makes them less marketable.

Thysanoptera (thrips)

Thrips are tiny, but important, pests of many crops. The red-banded thrips, *Selenothrips rubrocinctus* (Giard) is a widely distributed pest of tropical crops. It attacks cacao, cashew, mango (Palmer *et al.*, 1989), coffee, rubber, passion fruit, coconuts, avocados as well as guavas. In Brazil it is also recorded in *Eucalyptus* and *Eugenia* (Monteiro, 1994). The economic impact of the damage is hard to assess. The thrips feed on the leaves and fruit and cause a russetting or bronzing of the plant surface. Bronzing on the fruit decreases the marketability and value of the crop. Some predators and parasites are known to attack *S. rubrocinctus* (Bennett and Baranowski, 1982; Dennill, 1992). Chemical control is used in cacao, but is seldom necessary on guava. In South-East Asia, *Scirtothrips dorsalis* Hood also attacks guava (Kwee and Chong, 1990). Thrips are most likely to cause problems when the plant is grown under stress conditions such as drought (Kwee and Chong, 1990) or with nutritional deficiency (Fennah, 1963).

Lepidoptera (butterflies and moths)

There are a number of families of Lepidoptera that have members which can be found on guava. Most Lepidoptera larvae are foliage feeders, but some species bore into the fruits as well. The surface feeding species normally do not cause great damage to the tree. The trees can survive defoliation with little harm to the crop. Frequent severe defoliation can kill small trees. A number of species of Lepidoptera fold the leaves about themselves (leafrollers), so they are difficult to find. Some species feed only at night. The best IPM control method is to use some form of *Bacillus thuringiensis* Berliner, a bacterial pesticide, since it is specific for Lepidoptera and does not harm parasites and predators. Larvae that are protected from direct contact with this natural pesticide, such as those which bore into the fruit, are leafrollers, or bagworms, and are harder to kill. Internal-feeding larvae may also be a phytosanitary problem. Very large larvae like those of the Atlas moth, *Attacus atlas* (Linnaeus), can be hand picked from the trees. The fruit piercing moth, *Eudocima fullonia* (Clerck), is unique in that it is the adult stage that causes damage to the fruit. These moths have a piercing proboscis which they use to feed on the juices of the fruits. This feeding leaves the growing fruit misshapen, and damage is similar to that caused by bugs. Fruit piercing moths are very difficult to control because they can fly great distances in large numbers and cause severe damage. Chemical sprays have little or no effect on the transient adults. The only method to prevent damage is the bagging of the guavas while they are still very small.

Coleoptera (beetles)

The most serious pests of guava outside of the fruit flies are several species of fruit boring weevils. *Conotrachelus psidii* Marshall and *C. dimidiatus* Champion, though limited in distribution, can be a serious problem for growers (Boscán de Martínez and Casares, 1980, 1981). Infestation rates are reported to vary widely among cultivars, and the best control was obtained by timing insecticide application to the emergence of adults from the soil directly beneath the trees. These weevils would be more serious pests if they were widely distributed; but because of their narrow host range, they have not become as widespread and as damaging as the many species of fruit flies.

A number of species of Curculionidae and Scarabaeidae feed as adults on the leaves and cause minor defoliation. Some of the Curculionidae feed on the roots as larvae and may damage nursery trees. Nematodes can be an effective control method for weevils with burrowing larvae. There are several scolytid beetles and a bostrichid beetle that tunnel in the wood of the guava tree (Table 9.2). A heavy infestation of these pests can kill small trees.

Arachnida: Acarina, mites

Mites are ubiquitous on plants, but they are so small they often go unrecognized. Their feeding drains the cell contents, leaving the tissue with a bleached or scalded look. If the feeding extends to the fruits, the fruits may be less marketable. In Costa Rica, colonies of *Oligonychus yothersi* (see Chapter 8) are found on the leaf upperside (Ochoa *et al.*, 1994). In Malaysia, sporadic attacks of *Panonychus* sp., *Eotetranychus* sp. and *Oligonychus biharensis* (Hirst) are observed (Kwee and Chong, 1990). These species are reported feeding on the underside of the leaves, causing leaves to become dull green, and then bronzed. Populations are higher during dry spells and in situations when the plants tend to suffer from water stress, such as when guava is grown in sandy soils (Kwee and Chong, 1990).

In Florida, USA, the mites, *Tegolophus guavae* (Boczek) and *Brevipalpus* spp., cause damage to fruits and tender leaves (Peña *et al.*, 1999). Highest numbers of *T. guavae* are observed in early autumn, through winter and during spring months. The mites are most often observed on the fruit, causing 'pimples' or deformations.

In Florida, USA, *Brevipalpus* sp. densities only increased on the leaves between the summer months and autumn. However, in Central America, *Brevipalpus phoenicis* (Geijskes) have been observed causing fruit to turn brown, smooth and shiny, with epidermal cracking. Symptoms on leaves are characterized by epidermal cracking on the leaves (Ochoa *et al.*, 1994).

In Costa Rica, Ochoa *et al.* (1994) reported that injuries to fruit, i.e. deep striations resembling craters, caused by *Selenothrips rubrocintus* (Thysanoptera: Thripidae), can be used by *B. phoenicis* to form colonies.

Mite populations develop resistance to pesticides very rapidly because they have such a short generation time. Soap or oil sprays may be an alternative control method.

Conclusions

Guava pest management is divided into two parts: (i) dealing with the key pests, usually species of tephritid fruit flies; and (ii) managing the minor pests which are varied. The minor pests seldom need treatment. If a treatment is necessary, the minimum treatment to obtain control should be used. Biological controls and cultural methods should be the first line of defence. Damage to the trees does not usually translate directly into heavy crop losses. An outbreak of scale insects or mealybugs is better tolerated to allow biological controls adequate time to respond.

The fruit flies are a difficult problem. Eradication of exotic species has become a major industry, but it is very expensive and is only undertaken by large governmental organizations. For the control of an established species on guava, bagging the fruits remains the only effective method of control, and the economics of hand labour make it difficult to accomplish on a large scale. Heavy chemical spraying may lead to subsequent outbreaks of secondary pests and should be avoided. The flies are only vulnerable as adults, and the effect of foliar spray is limited. Bait sprays show promise, and as new formulations and baits are invented, these may become key methods of control.

References

Agrianual (2000) *Anuário da Agricultura Brasileira.* Argos Comunicação, São Paulo, 546 pp.

Aluja, M.R. and Liedo, P.F. (1986) Perspectives on future integrated management of fruit flies in Mexico. In: Mangel, M., Carey, J.R. and Plant, R.E. (eds) *Pest Control Operations and Systems Analysis in Fruit Fly Management*, NATO ASI Series G: Ecological Sciences, Vol. 11. Springer Verlag, New York, pp. 9–42.

Anonymous (1981) Plagas de los principales cultivos de Mexico. *Fitofilo* 86, 98–99.

Anonymous (1999) *Crop Protection Compendium Global Module.* CAB International, Wallingford, UK, CD-ROM.

Arthur, V., Leme, M.H.A., Wiendl, F.M., Silva, A.C., Faria, J.T. and Wiendl, J.A. (1993) Desinfestacao de fruto de goiaba infestados pela mosca-da-fruta *Anastrepha obliqua* (Mac., 1835) (Diptera: Tephritidae) através da radiação gama do cobalto-60. *Revista de Agricultura (Brazil)* 68, 207–217.

Bennett, F.D. and Baranowski, R.M. (1982) First record of the thrips parasite *Goetheana parvipennis* (Gahan) (Eulophidae: Hymenoptera) from the Bahamas. *Florida Entomologist* 65, 185.

Boscán de Martínez, N. and Casares, M.R. (1980) El gorgojo de la guayaba *Conotrachelus psidii* Marshall (Coleoptera: Curculionidae). Evaluacion de daños. *Agronomic Tropical (Venezuela)* 30, 77–83.

Boscán de Martínez, N. and Casares, M.R. (1981) Distribucion en el tiempo de las fases del gorgojo de la guayaba *Conotrachelus psidii* Marshall (Coleoptera: Curculionidae) en el campo. *Agronomic Tropical* (*Venezuela*) 31, 123–130.

Burton, R.P. (1930) Citrus grove sanitation. *Proceedings of the Florida State Horticultural Society* 43, 140–142.

Chirinos, D.T. (2000) Biología de la mota blanca del guayabo, *Capulinia* sp., cercana a *jaboticabae* von Ihering (Hemiptera: Eriococcidae) y su potencial de desarrollo de poblaciones sobre varias especies de *Psidium*. Universidad Central de Venezuela, Facultad de Agronomía, Maracay, Venezuela (Thesis).

Chitarra, M.I.F. (1996) Características das frutas de exportação. In: *Goiaba Para Exportação: Procedimentos de Colheita e Pós-colheita*. Ministério da Agricultura, do Abastecimento e Reforma Agrária, Brasília, DF, pp. 9–11.

Crane, J.H., Campbell, R.J. and Balerdi, C.F. (1993) Effect of hurricane Andrew on tropical fruit trees. *Proceedings of the Florida State Horticultural Society* 106, 139–144.

Dennill, G.B. (1992) *Orius thripoborus* (Anthocoridae), a potential biocontrol agent of *Heliothrips haemorrhoidalis* and *Selenothrips rubrocinctus* (Thripidae) on avocado fruits in the eastern Transvaal. *Journal of the Entomological Society of Southern Africa* 55, 255–258.

Drew, R.A.I., Hooper, G.H.S. and Bateman, M.A. (1978) *Economic Fruit Flies of the Pacific Region*. Brisbane, Australia, 137 pp.

Fennah, R.G. (1963) Nutritional factors associated with seasonal population increase of cacao thrips, *Selenothrips rubrocinctus* (Giard) (Thysanoptera), on cashew, *Anacardium occidentale*. *Bulletin of Entomological Research* 53, 681–713.

Foote, R.H., Blanc, F.L. and Norrbom, A.L. (1993) *Handbook of the Fruit Flies (Diptera: Tephritidae) of America North of Mexico*. Cornell University Press, Ithaca, New York, 571 pp.

Galinsky, B. (2000) India, Latin America compete for Europe's tropical juice market. http://www.marketag.com/ma/news/archive/v.35/juice.stm

Gallo, D., Nakano, O., Silveira Neto, S., Carvalho, R.P.L., Batista, G.C., Berti Filho, E.B., Parra, J.R.P., Zucchi, R.A., Alves, S.B. and Vendramim, J.D. (1988) *Manual de Entomologia Agrícola*. Ceres, Piracicaba, São Paulo, 649 pp.

González Gaona, E. (1999) Manejo de plagas en el cultivo del guayabo en Calvillo, Aguascalientes. *Memorias del XXXIV Congreso Nacional de Entomologia*, p. 708.

Gould, W.P. (1994) Heat quarantine treatments for guavas infested with the Caribbean fruit fly. *Proceedings of the Florida State Horticultural Society* 107, 240–242.

Gould, W.P. and Sharp, J.L. (1992) Hot-water immersion quarantine treatment for guavas infested with Caribbean fruit fly (Diptera: Tephritidae). *Journal of Economic Entomology* 85, 1235–1239.

Greany, P.D. and Shapiro, J.P. (1993) Manipulation and enhancing citrus fruit resistance to the Caribbean fruit fly (Diptera: Tephritidae). *Florida Entomologist* 76, 258–263.

Guimarães, J.A., Zucchi, R.A., Diaz, N.B., Souza Filho, M.F. and Uchôa, F., M.A. (1999) Espécies de Eucoilinae (Hymenoptera: Cynipoides: Figitidae) parasitóides de larvas frugívoras (Diptera: Tephritidae e Lonchaeidae) no Brasil. *Anais da Sociedade Entomológica do Brasil* 28, 263–273.

Hansen, J.D. and Armstrong, J.W. (1990) The failure of field sanitation to reduce infestation by the mango weevil *Cryptorhynchus mangiferae* Fabricius (Coleoptera: Curculionidae). *Tropical Pest Management* 36, 359–361.

Hennessey, M.K. (1994) Depth of pupation of Caribbean fruit fly (Diptera: Tephritidae) in soils in the laboratory. *Environmental Entomology* 23, 1119–1123.

Hennessey, M.K., Knight, R.J. Jr and Schnell, R.J. (1995) Antibiosis to Caribbean fruit fly (Diptera: Tephritidae) immature stages in carambola germplasm. *Florida Entomologist* 78, 354–357.

Hennessey, M.K., Knight, R.J. Jr and Schnell, R.J. (1996) Relative resistance of avocado germplasm to Caribbean fruit fly. In: McPheron, B.A. and Steck, G.J. (eds) *Fruit Fly Pests: a World Assessment of Their Biology and Management*. St Lucie Press, Delray Beach, Florida, pp. 339–342.

Howard, F.W. (1986) Protection de goyaves contre la mouch Caraïbe des fruits au moyen de filets. *Fruits* 41, 621–624.

Jalaluddin, S., Katarajan, K. and Sadakathula, S. (1998a) Enhancement of guava resistance to fruit fly. *Fifth International Symposium of Fruit Flies of Economic Importance*. Penang, Malaysia, p. 31.

Jalaluddin, S., Katarajan, K. and Sadakathula, S. (1998b) Relative resistance of guava germplasm to guava fruit fly *Bactrocera correcta* (Bezzi). *Fifth International Symposium of Fruit Flies of Economic Importance*. Penang, Malaysia, p. 32.

Kapoor, V.C. (1993) *Indian Fruit Flies*. International Science Publisher, New York, 228 pp.

Kubo, R.K. and Batista Filho, A. (1992) Ocorrência de *Amorbia* spp. (Lepidoptera: Tortricidae) em frutos de goiabeira (*Psidium guajava* L.) na região de Campinas. *Anais da Sociedade Entomológica do Brasil* 21, 257–260.

Kwee, L.T. and Chong, K.K. (1990) *Guava in Malaysia: Production, Pests and Diseases.* Tropical Press Sdn. Bhd., Kuala Lumpur, Malaysia, 260 pp.

Landolt, P.J., Gonzalez, M., Chambers, D.L. and Heath, R.R. (1991) Comparison of field observations and trapping of papaya fruit fly in papaya plantings in Central America and Florida. *Florida Entomologist* 74, 408–414.

Landolt, P.J., Reed, H.C. and Heath, R.R. (1992) Attraction of female papaya fruit fly *Toxotrypana curvicauda* (Diptera: Tephritidae) to male pheromone and host fruit. *Environmental Entomology* 21, 1154–1159.

Liquido, N.J. (1993) Reduction of oriental fruit fly (Diptera: Tephritidae) populations in papaya orchards by field sanitation. *Journal of Agricultural Entomology* 10, 163–170.

Malavasi, A. and Zucchi, R.A. (2000) *Moscas-das-frutas de Importância Econômica no Brasil – Conheciomento Básico e Aplicado.* Holos Editora, Ribeirão Preto, pp. 277–283.

McPhail, M. (1937) Relation of time of day, temperature and evaporation to attractiveness of fermenting sugar solution to Mexican fruit fly. *Journal of Economic Entomology* 30, 793–799.

Medina, J.C. (1988) Cultura. In: *Goiaba: Cultura, Matéria-prima, Processamento e Aspectos Econômicos.* Instituto de Tecnologia de Alimentos, Campinas, pp. 1–121.

Monteiro, R.C. (1994) Espécies de tripes (Thysanopterea: Thripidae) associadas a algumas culturas no Brasil. MSc thesis. The University of São Paulo, Piracicaba, Brazil.

Morton, J.F. (1987) *Fruits of Warm Climates.* J.F. Morton, Miami, Florida, 505 pp.

Murray, M.J. and Campbell, C.A. (1989) Guava and passionfruit as commercial crops in Florida. *Proceedings of the Florida State Horticultural Society* 102, 212–213.

Nafus, D. and Schreiner, I. (1999) Insect pests of Micronesia. http://www.nmcnet.edu/Lg/aubweb/bugweb/bugroot.htm

Nakasone, H.Y. and Paull, R.E. (1998) *Tropical Fruits.* CAB International, Wallingford, UK, 445 pp.

Norrbom, A.L. and Kim, K.C. (1988) *A List of the Reported Host Plants of the Species* Anastrepha *(Diptera: Tephritidae).* United States Department of Agriculture (Animal Plant Health Inspection Service 81–52), Washington, DC, 114 pp.

Ochoa, R., Aguilar, H. and Vargas, C. (1994) Phitophagous mites in Central America: an illustrated guide. *Serie Tecnica, CATIE, Turrialba, CR* 6, 51–52.

Palmer, J.M., Mound, L.A. and Heaume, G.J. (1989) Guides to insects of importance to man. In: Betts, C.R. (ed.) *Thysanoptera.* CAB International, Wallingford, and British Museum of Natural History, London, 74 pp.

Peña, J.E., Hennessey, M., Duncan, R. and Vasquez, T. (1999) Arthropod seasonality and control of fruit flies on guava in south Florida. *Proceedings of the Florida State Horticultural Society* 112, 206–209.

Qureshi, Z.A. and Hussain, T. (1993) Monitoring and control of fruit flies by pheromone traps in guava and mango orchards. In: Aluja, M. and Liedo, P. (eds) *Fruit Flies: Biology and Management.* Springer Verlag, New York, pp. 375–380.

Robinson, A.S. and Hooper, G. (eds) (1989) *Fruit Flies, Their Biology, Natural Enemies and Control,* Vol. 3a. Elsevier, Amsterdam, 372 pp.

Silva, A.G., Gonçalves, C.R., Galvão, D.M., Gonçalves, A.J.L., Gomes, J., Silva, M.N. and Simoni, L. (1968) *Quarto Catálogo dos Insetos que Vivem nas Plantas do Brasil, seus Parasitos e Predadores.* Ministério da Agricultura, Rio de Janeiro. Tome 1, Part II, 662 pp.

Souza Filho, M.F. (2000) Biodiversidade de moscas-das-frutas (Diptera: Tephritidae) e seus parasitóides (Hymenoptera) em plantas hospeeiras no Estado de São Paulo. MSc thesis. The University of São Paulo, Piracicaba, Brazil.

Souza Filho, M.F., Raga, A. and Zucchi, R.A. (2000) Moscas-das-frutas nos Estados Brasileiros. In: Malavasi, A. and Zucchi, R.A. (eds) *Moscas-das-frutas de Importância Econômica no Brasil – Conheciomento Básico e Aplicado.* Holos Editora, Ribeirão Preto, pp. 277–283.

Steyskal, G.C. (1977) History and use of the McPhail trap. *Florida Entomologist* 60, 11–16.

Vazquez, L., Jiménez, R. and de la Iglesia, M. (1996) Whiteflies (Homoptera: Aleyrodidae) inhabiting the principal fruit trees grown in Cuba. *Tropical Fruits Newsletter* 20, 3–4.

White, I.M. and Elson-Harris, M.M. (1992) *Fruit Flies of Economic Significance: Their Identification and Bionomics.* CAB International, Wallingford, UK, 601 pp.

10 Pests of Minor Tropical Fruits

Peter A.C. Ooi,[1] Amporn Winotai[2] and J.E. Peña[3]
[1]FAO Regional Office for Asia and the Pacific, 39 Phra Atit Road, Bangkok 10200, Thailand; [2]Biological Control Group, Division of Entomology and Zoology, Department of Agriculture, Chatuchak, Bangkok 10900, Thailand; [3]Tropical Research and Education Center, University of Florida, 18905 SW 280 Street, Homestead, FL 33031, USA

Among the tropical fruits, durian, *Durio zibethinus* Murray (*Bombaceae*), mangosteen *Garcinia mangostana* L. (*Clusiaceae*), rambutan, *Nephelium lappaceum* L. (*Sapindaceae*), carambola *Averrhoa carambola* L. (*Oxalidaceae*) and acerola *Malpighia glabra* L. (*Malpighiaceae*) are considered minor crops. Their distribution and cultivation have been restricted to the areas where they originated, perhaps because some of these fruits have a unique eating quality and often require an acquired taste to appreciate them. That is particularly true of durian, the 'king of tropical fruits', whose strong and unmistakable smell makes it the most controversial (Chin and Yong, 1980) among this group (Lim, 1990). Currently, there is much interest in expanding the range where these fruits can be cultivated, particularly in those areas with similar climate. Therefore, it would be worthwhile to recognize the pests recorded on these fruits in their native range so as to minimize their spread to new sites. Moreover, an understanding of the pests will give an idea of their damage potential for other parts of the world. This chapter will attempt to review the literature on pests of durian, mangosteen, rambutan, acerola and carambola in their centre of origin and in some of their adopted growing regions.

Durian (*Durio zibethinus* Murray)

Durian, a common fruit in South-East Asia, originated in Borneo and Sumatra, and is held in high esteem from Sri Lanka and south India to New Guinea. The ripe fruits, or rather the arils which form the edible part, are generally eaten fresh. Its strong and unmistakable smell is present in every market and street stall during the fruiting season. The fruit can be also preserved, deep frozen or consumed in ice creams, cakes and cookies.

Thailand is the largest producer with 444,500 t in 1987, followed by Indonesia. Large fruits of superior cultivars fetch about US$5–10 each in Thailand and Malaysia. The export of fruit from Thailand in 1987 was worth US$9.5 million, with exports to Hong Kong, USA and France.

Crop phenology

The fruit is a globose, ovoid or ellipsoid capsule, up to 25 cm long and 20 cm diameter, green to brownish, covered with numerous broadly pyramidal, sharp, up to 1 cm long spines; valves usually five, thick, fibrous. Seeds are up to 4 cm long and completely

©CAB *International* 2002. *Tropical Fruit Pests and Pollinators* (eds J.E. Peña, J.L. Sharp and M. Wysoki)

covered by a white or yellowish, soft, very sweet aril.

The durian phenology and harvest seasons differ in those countries where it is grown. In Thailand, durian trees flower over a period of 2–3 weeks in March and bear fruit once a year. Fruit development requires 95–105 days after pollination for early cultivars and more than 130 days for late cultivars. In the wet lowland tropics in Malaysia and Indonesia, durian may flower twice a year. In Peninsular Malaysia, the main durian season is between June and August.

Key pests

Yunus and Ho (1980) provided the most comprehensive record of pests of plants in Peninsular Malaysia. These records were based on reports made to the Department of Agriculture since 1920, and 40 insect pests were recorded from durian (Yunus and Ho, 1980). A survey by Waterhouse (1993) listed 20 major pests of durian in South-East Asia. In Malaysia, Khoo *et al.* (1991) described ten insects attacking durian. In eastern Thailand, Sirisingh *et al.* (1994) considered only five or six species as serious pests of durian that required bimonthly insecticide usage. However, Yunus and Balasubramaniam (1981) considered only *Mudaria luteileprosa* Holloway (= *magniplaga*) as the key pest of durian in Malaysia. The more conspicuous insect pests are described in this chapter:

Hypomeces squamosus *F.* *(Coleoptera: Curculionidae)*

This is a polyphagous weevil in which the adult stage feeds on leaves, causing extensive defoliation of young plants (Plate 70). Eggs of the weevil are laid singly in the soil. A female could lay 40–131 eggs (Choonhawong *et al.*, 1982). The egg stage lasts about a week. Larvae, which are yellow with a brown head and black mandibles, feed on roots. The larval stage requires 22–37 days. Full-grown larvae measure 15–20 mm in length and pupation occurs in the same habitat. The pupal stage lasts 10–15 days. Adults are about 15 mm in length.

CONTROL Little is known of the natural biological control of this weevil although unidentified tachinid flies were observed to parasitize it (Choonhawong *et al.*, 1982). When it occurs in large numbers, spot spraying with 0.1% methamidophos or 0.15% acephate solution has been effective.

Allocaridara malayensis *(Crawford)* *(Hemiptera: Psyllidae)*

The durian psyllid is considered to be the most destructive insect pest of durian. Eggs are inserted into young leaves. A cluster of 8–14 eggs forms a yellow or brown circle on the leaf. Full grown nymphs measure about 8 mm long with white, cotton-like, feather shaped particles along its body, especially in the caudal area. Adult longevity lasts about 6 months. Both nymphs and adults live and feed on the underside of the leaf. Injured leaves show yellow spots, followed by leaf curling and leaf fall. The nymphs excrete copious amounts of white honeydew which encourages the development of sooty mould.

CONTROL The natural enemies, *Menochilus sexmaculatus* Fabricius, *Micraspis discolor* (Fabricius), *Coccinella transversalis* (Fabricius), *Chrysopa* sp., spiders, and undetermined encyrtids are associated with durian psyllid (Disthaporn *et al.*, 1996).

Conogethes punctiferalis *(Guenée)* *(Lepidoptera: Pyralidae)*

The highly polyphagous durian fruit borer, previously known in older references as *Dichocrocis punctiferalis* (Guenée), damages both young and mature fruits by boring through the pulp (Plate 71). Infested fruits can be distinguished by the presence of frass covering the infested surface area. Adults can be found on undersides of leaves during the day. Wing width measures 2.0–2.3 cm; the forewing is yellow with scattered black spots. Newly hatched larvae feed on the fruit surface. Larvae are greenish brown with several spots. Full-grown larvae measure 1.5–1.8 cm in length with black spots scattered along the body. Pupation occurs in soil and sometimes on fallen leaves. When reared on castor beans, the larval stage is 12–13 days, and the pupal stage requires 7–9 days. Adult male longevity is 14 days, while female longevity is about 16 days. Highest insect densities occur between April and August in Malaysia and between February and June in Thailand. *Conogethes punctiferalis* uses longan, rambutan and pomegranate as alternative hosts.

CONTROL Fallen fruits should be collected and burnt or buried. Spraying with cyhalothrin-L plus a surfactant at 20-day intervals has been recommended. Little is known regarding biological control of this insect.

Mudaria luteileprosa *Holloway* *(Lepidoptera: Noctuidae)*

The durian seed borer has been misidentified as *Mudaria magniphaga* (Walker) (Kuroko and Lewvanich, 1993) (Plate 72). The females oviposit on the durian fruit spines. Newly hatched larvae feed initially on the skin of the fruit and later bore into the husk and then into the seeds. The full-grown purple-red larvae are *ca.* 4 cm in length. The larvae emerge from the fruit to pupate in damp soil. An exit hole measuring about 5–8 mm in diameter surrounded by white-orange excreta can be observed on the surface of the fruit. All stages of durian fruit are susceptible to *Mudaria* attack. Sirisingh *et al.* (1994) considered that this insect is the most important postharvest pest in Thailand. Losses vary according to locality but generally 50% of the fruits are unmarketable if it is not controlled. Usually only one larva can be found per fruit.

CONTROL Chemical treatment, light traps and baits have been recommended. Spraying should be stopped 2 weeks before harvesting. Fenthion, dimethoate, deltamethrin, methamidophos, and methidathion have been recommended in Malaysia, while carbaryl, methamidophos, endosulphan, and phosalone are recommended in Thailand. Fruit thinning at an early stage can also reduce attack by the pest. Light trapping reduces the moth population and thereby reduces fruit infestation. Ripe bananas mixed with some insecticides are used as bait. Although bagging young fruit provides physical protection against this borer, it is not generally utilized.

Planococcus minor *(Maskell)*, Planococcus lilacinus *(Cockerell)* *(Hemiptera: Pseudococcidae)*

Adults of *Planococcus minor* and *P. lilacinus* are flat and oval, approximately 3 mm in length, pale yellow and covered with wax. One female can lay an average of 100–200 eggs per cluster. Capacity for egg laying ranges from 600 to 800 eggs in a period of 14 days, and eggs hatch within 6–10 days. Crawlers move around until they find suitable feeding sites or until they complete their nymphal development. The female mealybugs have three nymphal stages. Regularly, two or three generations are produced per year. *Planococcus* spp. feed on branches, inflorescences and fruits. Infestations increase though ant tending and damaged plants become stunted and covered with sooty mould. Although mealybugs attack only the fruit epidermis and their proboscis rarely, if ever, penetrates the flesh, infested fruits are considered to have low quality and are non-marketable. *Planococcus* feeds on roots of grasses when no suitable host plants are available. In eastern Thailand, *Planococcus* can be found attacking durian after fruit set during the beginning of the hot and dry season starting in March. This infestation continues until the fruits mature in mid July.

CONTROL Water or oil sprays are considered effective control methods. Indirect control by banding the trunk or branches with pesticides or petroleum oil to keep out ants is also reported to be effective. *Cryptolaemus montrouzieri* Mulsant, *Scymnus* sp. and *Nephus* sp. (Coleoptera: Coccinellidae), *Ankylopteryx octopunctata* (Fabricius) (Neuroptera: Chrysopidae), and *Sphenolepis* sp. (Hymenoptera: Eulophidae) have been reported to be natural enemies of *Pseudococcus* sp. (Disthaporn *et al.*, 1996).

Eutetranychus africanus *(Tucker)* *(Acarina: Tetranychidae)*

Charanasri and Kongchuensin (1991) and Anonymous (2000) reported the African red mite *Eutetranychus africanus* (Tucker) as one of the most important mite pests of durian. Severe infestation often causes yield reduction and increases the amount of pesticides used, with resulting market problems. African red mites feed on the upper side of the leaf especially along the mid-vein, causing withering spots that later spread over the whole leaf. In the eastern part of Thailand where severe damage has been reported, growers sprayed large amounts of acaricides, causing the development of acaricide-resistant mite strains.

BIOLOGY Female African red mite is flat and oval, about 0.42 mm in length and 0.35 mm wide. It is dark brown in colour without an empodium at the end of the legs. A pair of red dots is located at the anterior end. The short setae on the dorsal part of the body are spatulate, whereas the ventral setae are slender. The male African red mite reportedly completes its life cycle within 9.2 days. Generally, the egg stage has been observed to last 4.8 days, but eggs of female mites were reported to eclose within 4.5 days. The mites pass through three nymphal stages of 1.6, 1.3 and 1.6 days, respectively. The unmated male adult has a longevity of 4.7 days. The male is slightly smaller and more elongated than the female. Female longevity lasts 9.3 days. The pre-oviposition period has been 4.8 days. The average number of eggs laid per female is 14. Longevity for unmated females is 8 days while for mated females it is 6.5 days. In Thailand, populations of red mites have been reported to be high from September until October, depending on rainfall and relative humidity. Population peaks in December and January, then declines, increasing again in March and April. In general, the population is usually high during long dry periods. Several species of host plants are damaged by this mite. They include fruit trees, vegetables, legumes and ornamental plants.

CONTROL Charanasri and Kongchuensin (1991) report the phytoseiids *Amblyseius syzygii* Gupta, *Amblyseius cinctus* Corpuz et Rimando, *Amblyseius deleoni* Muma et Denmark, and *Amblyseius largoensis* (Muma) as predators of African red mites, while Anonymous (2000) suggests that the dominant predatory species of the African red mite in Thailand is *Amblyseius largoensis* (Muma). Other unidentified species of Stigmaeidae, Canaxidae and Bdellidae have also been reported. *A. largoensis* develops and multiplies quickly, and the female lays 2–3 eggs day^{-1}. The life cycle from egg stage to adult is 4 days. During its life span, a single *A. largoensis* feeds on 27 African red mite eggs.

Lawana conspersa *(Walker)* *(Hemiptera: Flatidae)*

The white moth cicada has been reported on cocoa and *Bauhinia* spp. Damage is caused by oviposition in the leaf and fruit stalks (Ibrahim and Ibrahim, 1989). Both the adult and nymphs feed on plant sap, causing dehydration of plant parts, and loss in photosynthesis due to the presence of sooty moulds caused by excretions of the insect.

Eggs are laid into young twigs or into midribs of leaves. Nymphs feed on young shoot tips, young leaves and flowers. The nymphs are usually covered with white, waxy material. Adults are 2 cm long with two orange stripes on the basal portion of the forewings (Dammerman, 1929).

CONTROL No chemical control of *L. conspersa* is recommended. Natural biological control by the entomogenous fungus, *Metarrhizium anisopliae* (Metschnickoff) Sorokin var. *anisopliae* Tulloch, has been reported as effective.

Cataenococcus hispidus *(Morrison)* *(Hemiptera: Pseudococcidae)*

This mealybug can be found in large numbers on durian fruits and peduncles, reducing the cosmetic value of the fruits. Its host plants include durian, rambutan, guava and cocoa

(Plate 73). The insect has both sexual and parthenogenetic reproduction. Females are ovoviviparous. Females pass through three nymphal instars which last 8, 6 and 7 days before becoming adults. The males undergo two nymphal stages, one prepupal and one pupal stage, lasting 7, 8, 2 and 4 days, respectively. Adult longevity is 23 days under laboratory conditions.

CONTROL No chemical control is necessary against this insect. Ho and Khoo (1997) and Na (1988) provide an extensive study on the close mutualistic relationship between the ant *Dolichoderus thoracicus* (Smith) (Hymenoptera: Formicidae) and *C. hispidus* and reiterate the necessity to manipulate both organisms for control of the mealybug. Control of the ants associated with the mealybugs may be recommended if any intervention is required. Spraying the basal part of tree trunk with gamma-HCH (0.1% a.i.) or chlorpyrifos (0.1% a.i.) will reduce the ant populations.

Chalcocelis albiguttatus *(Snellen)* *(Lepidoptera: Limacodidae)*

Chalcocelis albiguttatus or gelatine grub is a polyphagous insect recorded on tea, oil palm, durian, rambutan, Malay apple (*Eugenia malaccensis* L.) and candle nut (*Aleurites triloba* Forst.); it is collected regularly in forests in Sabah, Malaysia (Chey *et al*., 1997). This gelatine grub is recorded as an important pest of coconut in Malaysia and Indonesia. Damage caused by larvae scraping off the leaf tissue serves as a portal of entry for plant pathogens.

Eggs are generally flat, scale-like, and laid on the leaf underside. Egg stage is 2–3 days. Total larval stages range from 59 to 70 days. The larva is about 17 mm long; it is characterized by its spines and appears to be enclosed in a gelatinous cover. Full-grown larvae form cocoons attached to leaves or stems. The pupal stage lasts 21–31 days.

CONTROL This insect is usually kept under control by natural biological agents including microbial diseases. However, if chemical control is needed, 0.01% cypermethrin solution can be used for spot spraying.

Daphnusa ocellaris *Walker* *(Lepidoptera: Sphingidae)*

Daphnusa ocellaris is considered to be an endemic pest, with low importance. For example, this sphingid is not recorded as a serious pest of durian (Yunus and Ho, 1980; Waterhouse, 1993). However, Ramasamy (1980) reported a large outbreak of *D. ocellaris* in the Malaysian state of Perak. These opposite reports suggest that not much is known of the potential of this endemic pest, and much less about its natural enemies.

The life cycle of *D. ocellaris* is completed in 32–44 days (Ramasamy, 1980). The cream and pink eggs are laid in groups of 6–20, each measuring 1.5–2 mm and hatch in 4–6 days. The larval stages are completed in 17–21 days. Larvae are voracious defoliators. They are usually green with a distinct dorsal red stripe. Full-grown larvae measure 6–6.5 cm in length (male caterpillars) while female larvae measure 7.5–8 cm long. Pupation occurs in the soil and pupae of male moths are smaller (3.5 cm long) than pupae of female moths (4 cm long). Pupation period varies between 12 and 17 days.

CONTROL Ramasamy (1980) recorded two tachinids from pupae of *D. ocellaris*, namely *Blepharipa sugens* (Wiedmann) and *Exorista* sp. (Diptera: Tachinidae). Ramasamy (1980) determined that light trapping could reduce moth density.

Mangosteen (*Garcinia mangostana* L.)

Mangosteen is known only as a cultivated species, although there have been occasional wild specimens in Malaysia. The mangosteen is probably the most highly praised tropical fruit (Verheij, 1992). The scarcity of mangosteen orchards, limited fruit supply and the fruit's short shelf life are major marketing problems. Much of the limited fruit supply is due to the long juvenile period of 10–15 years that discourages commercial production. Thailand is the largest producer of mangosteen in South-East Asia. The mangosteen fruit is a globose and smooth berry, turning dark purple at ripening. It is a crop of the humid tropics, often found in association with durian. The fruit has a diameter of 4–7 cm. The edible part is the sweet white aril that envelops the seeds within the pericarp (Verheij, 1992).

Generally, the mangosteen season coincides with that of the durian. Floral initiation to anthesis takes about 25 days and the fruit ripens 100–120 days later. The main harvest season in Thailand is May/June and July. In Peninsular Malaysia, fruits are available from June to August, while in Sarawak, fruits are available from November through January.

Key pests

A total of 25 insects were recorded from mangosteen over a period of *c.* 60 years (Yunus and Ho, 1980). A survey by Waterhouse (1993) listed only three major pests of mangosteen in South-East Asia. Khoo *et al.* (1991) described two insects that attack mangosteen.

Hyposidra talaca *(Walker) (Lepidoptera: Geometridae)*

Little is known of the biology of this highly polyphagous insect. *Hyposidra talaca* is found in tropical lowlands and highlands. Besides mangosteen, *H. talaca* has been recorded on cacao, cinchona, coffee, tea and other fruit trees (Entwistle, 1972). The larvae are typical loopers, brown with dorsal transverse rows of white spots. Full-grown larvae drop to the ground by silken threads and pupate about 2–4 cm deep in the soil. Eggs of *H. talaca* are iridescent and laid in clusters. *H. talaca* develops within 2.5–3.5 months in the highlands of Indonesia when recorded on cinchona (Kalshoven, 1981).

CONTROL To our knowledge there is no information on chemical control for this insect. Entwistle (1972) reports that in cocoa undetermined parasitoids reduce *H. talaca* densities.

Phyllocnistis citrella *Stainton (Lepidoptera: Phyllocnistidae)*

This insect is better known as a citrus leafminer. Mining of young leaves cause deformation and often leads to early leaf fall (Plate 74). Its damage is serious on young plants.

Studies of this moth, particularly those on citrus, showed that eggs are laid singly on young leaves. Larvae mined the leaf epidermis, resulting in characteristic serpentine mines lined with excrement. Full-grown larvae would grow up to 3 mm in length. Pupation takes place on the leaf. The complete life cycle in Indonesia was found to be 16 days (Kalshoven, 1981). (For more information, see Chapter 3.)

CONTROL Many natural enemies keep this insect in check in Indonesia. The most important parasitoid is *Ageniaspis* sp. (Hymenoptera: Encyrtidae) (see Chapter 3).

Stictoptera cucullioides *Guenée (Lepidoptera: Noctuidae)*

Stictoptera cucullioides is reported to feed voraciously on young flushes of mangosteen. Two other species have also been recorded, namely, *S. columba* (Walker) and *S. signifera* (Walker). As with many insects that have occasional outbreaks, little is known of the egg and early larval stages. The full-grown, 30 mm long larvae have a distinct orange coloured head capsule. Pupation takes place in the soil. As the insect occurs infrequently, no prophylactic control measures are necessary. Further ecological studies will probably explain the role of biological control in keeping the insect in check. In cases of occasional outbreaks, carbaryl was found to be effective.

Rambutan (*Nephelium lappaceum* Blume)

Rambutan is a tropical relative of litchi, with a distribution that ranges from southern China through the Indo-Chinese region,

Malaysia, Indonesia to the Philippines (Stone, 1992). As the name suggests (the Malay word 'rambut' meaning hair) the fruit is glabrose, resembling a 'burr' (Chin and Yong, 1980). Rambutan, while highly prized in its region of origin, remains a minor fruit internationally. The fruits are consumed fresh, or canned. The canning industry has boosted the cultivation of rambutan in its native range with Thailand as the largest producer. The rambutan fruit is an ellipsoid to subglobular schizocarp, up to 7 cm × 5 cm, usually consisting of one nutlet. The skin colour varies from yellow to purplish red and is usually glabrous. The seed is usually covered by a thick, sweet, juicy, white to yellow, translucent sarcotesta which is the edible part (Stone, 1992). The main flowering period occurs during the dry season. Fruits ripen about 110 days after bloom. In Thailand, insect pests of rambutan are not serious and not much is known of their ecology. The situation may change if some of these pests are transported to other tropical countries where these fruits are being planted (e.g. Costa Rica).

Key pests

A total of 127 insects were recorded from rambutan over a period of *c.* 60 years (Yunus and Ho, 1980). A survey by Waterhouse (1993) listed 28 major pests of rambutan in South-East Asia. Khoo *et al.* (1991) described 15 insects that attack rambutan. Some of the key pests are described in detail below.

Conopomorpha (= Acrocercops) cramerella (Snellen) (Lepidoptera: Gracillariidae)

Conopomorpha (= *Acrocercops*) *cramerella* (Snellen) attained notoriety as the most serious pest of cocoa beans in South-East Asia in the last two decades. The insect is native to the region, as is rambutan (Plate 75). However, for it to switch to a fruit from the New World some 200 years ago (Ooi *et al.*, 1990) is unique. Because of its new found status, much is known of its biology and advances have been made in the management and biological control of this pest. Lim *et al.* (1987) provide an extensive literature review of this insect.

Much of the information on the biology of this insect is from studies made with cocoa. Eggs of *C. cramerella* are laid singly on the surface of the fruit. Each egg is less than 0.5 mm, flat and rather oval in shape; they hatch within a week. The newly hatched larvae bore straight through the base of the egg into the fruit. The larvae tunnel in the fruit until they reach the fruit stalk. In the case of cocoa fruits, the larvae tunnel between the beans causing them to stick together, undersized and poor in quality. In the case of rambutan fruits, the damage is often less obvious and would not usually damage the sarcotesta. In cocoa, the larval stage is completed in 14–18 days. Four to six larval instars have been recorded and a full-grown larva measures 12 mm long; the larva leaves the fruit and pupates on the leaf surface or among leaf debris on the ground. Pupation occurs inside an oval-shaped cocoon. The pupal stage lasts between 6 and 8 days.

As noted above, this insect would tend to damage only the part of the fruit next to the fruit stalk (Plate 76). Hence, its damage to rambutan is often ignored. The current low population densities may suggest that some mortality factors and an early fruiting season may keep *C. cramerella* from reaching higher density peaks.

CONTROL Walker and Huddleston (1987) reported *Chelonus chailini* Walker and Huddleston (Hymenoptera: Braconidae) as an egg–larval parasitoid of *C. cramerella*. *Goryphus* sp. (Hymenoptera: Ichneumonidae), *Ceraphron aguinaldoi* Dessart (Hymenoptera: Ceraphronidae), *Xanthopimpla* sp. (Hymenoptera: Ichneumonidae), *Brachymeria* sp. (Hymenoptera: Chalcididae), *Ooencyrtus ooii* Noyes (Hymenoptera: Encyrtidae), *Trichospilus pupivorous* Ferriere (Hymenoptera: Eulophidae) and *Paraphylax* sp. (Hymenoptera: Ichneumonidae) have been reported from *C. cramarella* attacking rambutan in Peninsular Malaysia (Ooi, 1987; Noyes, 1991).

Adoretus (= Lepadoretus) compressus *(Weber) (Coleoptera: Scarabaeidae)*

Adoretus (= *Lepadoretus*) *compressus* is a highly polyphagous cockchafer beetle that is active at night. This species is the most common of about seven species within the genus. While damage to older rambutan plants may be tolerated, damage to developing seedlings may lead to a significant setback in growth. Characteristically, adults feed on the interveinal areas close to the centre of the leaf. Eggs of this scarabaeid beetle are laid in the soil usually under grasses or weeds. Grubs feed on decaying plant materials and roots. Five larval instars have been recorded and the full-grown larva would grow to 14 mm in length. Pupation takes place in the soil. When large numbers of chafer beetles are observed feeding on rambutan plants, especially young ones, the use of stomach poisons such as trichlorfon at 0.1% a.i. is advised (Rao and Suppiah, 1971). Little is known of the biological control of this insect.

Apogonia cribricollis *Burmeister (Coleoptera: Scarabaeidae)*

Apogonia cribicollis feeds on more than 30 plant species and is known from Singapore and Malaysia. Its damage to young plants is more significant than to mature plants (Entwistle, 1972). Like other cockchafers, eggs of *A. cribricollis* are laid in the soil at the base of plants. The elongated eggs are white in colour, measuring 1.0 × 1.3 mm. These are laid some 2.5–5 cm deep in the soil in groups of 5–35 eggs. The mean incubation period has been recorded to be around 9 days (Lever, 1953). Larvae grew from 3 to 15 mm in 10 to 11 weeks, feeding on the roots of plants. A pupal period of 7–10 days is preceded by a prepupal period of 2–3 days. Beetles are chestnut brown in the first 2 weeks, turning black later. Adults are active at night, feeding on foliage from the leaf margin inwards, leaving an untidy look on the leaves (Entwistle, 1972).

CONTROL Little is known of the biological control of this cockchafer. Often when severe leaf damage is observed, spraying with synthetic pyrethroids such as cypermethrin (0.01% a.i.) has been recommended.

Chalcocelis albiguttatus *(Snellen) (Lepidoptera: Limacodidae)*

See pests of durian.

Parasa lepida *(Cramer) (Lepidoptera: Limacodidae)*

This blue-striped nettle caterpillar is polyphagous and causes severe defoliation of rambutan trees. It is known throughout South and South-East Asia. Desmier de Chenon (1982) provided an extensive study of the insect as a pest of coconut. The egg stage of *P. lepida* lasts 5–7 days. Eggs are laid in groups of 15–40. Upon hatching, caterpillars are gregarious until fully grown. The male larva has seven instars and often eight are recorded for females. The larva has many protuberances on the body. Young larvae are yellow and in the third instar a blue band is observed on the dorsum. Larval development requires an average of 40 days (32–46 days). Pupation is also gregarious and the pupal stage lasts about 22 days (21–24 days).

CONTROL Cock *et al.* (1987) provided a comprehensive list of natural enemies of *P. lepida*. The study by Desmier de Chenon (1982) listed *Apanteles parasae* Rohwer (Hymenoptera: Braconidae), *Chaetexorista javana* Brauer and Bergenstamm (Diptera: Tachinidae), *Sarcorohdendorfia* (*Sarcophaga*) *antilope* (Bottcher) (Diptera: Calliphoridae), and *Buysmania oxymora* (Tosquinet) (Hymenoptera: Ichneumonidae) as the most frequently recorded parasitoids.

Often populations of *P. lepida* are decimated during the rainy season which sets off epizootics of a viral disease. In general, this insect is kept in check by its numerous natural enemies. However, in the rare event of outbreaks, spraying with synthetic pyrethroids or *Bacillus thuringiensis* at 1 kg ha^{-1} has been recommended.

Hyperaeschrella insulicola *(Kiriakoff)* *(Lepidoptera: Notodontidae)*

Occasionally, outbreaks of green caterpillars have been reported from isolated rambutan orchards (Ooi, 1978) (Plate 77). The large numbers of caterpillars quickly defoliate whole trees leaving a 'withering' effect. As quickly as it developed, the outbreaks subside during the dry season. Little is known of the biology and ecology of this insect. Egg and larval stages are found on the tree while pupation occurs among the leaf litter at the base of the tree. Besides rambutan, the insect is also reported to feed on cashew.

Conogetes punctiferalis *(Guenée)* *(Lepidoptera: Pyralidae)*

See pests of durian.

Attacus atlas *L. (Lepidoptera: Saturniidae)*

Attacus atlas L. is one of the largest moths known. Larvae of this moth are polyphagous and it has 37 different host plants (Yunus and Ho, 1980). *A. atlas* causes occasional to severe defoliation on rambutan.

One or several eggs are laid on the leaf lower surface. The eggs develop better under high humidity conditions. The egg stage lasts 10–13 days. Larvae are white with characteristic soft tubercles on the body. Fully grown larvae measure about 150 mm. The larval stage lasts 28–38 days. Pupation occurs within a cocoon constructed from leaves. The pupal stage lasts from 23 to 28 days.

The rather rare outbreaks suggest that the insect is often under natural biological control. It has been reported that insectivorous birds use the caterpillars as prey. Besides predators, seven species of parasitoids have been recorded (Yunus and Ho, 1980).

Adoxophyes privatana *Walker* *(Lepidoptera: Tortricidae)*

Like many insects feeding on rambutan, this webworm is a polyphagous herbivore recorded on 13 species of agricultural and ornamental plants (Yunus and Ho, 1980). It is recorded as an important pest of young flushes of cocoa. Larvae are usually found feeding inside a flush of young leaves held together by the silk produced.

Eggs of this insect are laid on the young leaves or shoots. Upon hatching, the caterpillar feeds on the leaves and soon pulls leaves together with silk. The larval period lasts about 6 days. Pupation occurs within the folded leaves.

Usually damage is not serious, although the plant may look untidy. It is suspected that natural biological control often takes care of the population. A chalcid, *Brachymeria* sp. nr. *apicicornis* (Cameron), has been recorded (Khoo *et al.*, 1991).

Carambola (*Averrhoa carambola* L.)

Known as carambola, star apple, or five corner, this fruit tree is a 5–12 m high evergreen tree native to southern Asia. Leaves are imparipinnate, with flowers in axillary or cauliflorous panicles, pentamerous. The fruit is a large berry, ovoid to ellipsoid in outline, with five pronounced ribs, stellate in cross-section (Samson, 1992). The tree grows well in tropical or subtropical lowland conditions (Sedgley, 1984) and flowers abundantly.

Flowers need to be pollinated. Carambola has been introduced since the 19th century into different areas in the western hemisphere and is now one of the most popular crops in Guyana, and Surinam (Ramsammy, 1989). In Malaysia carambola is produced in mixed family holdings with an average size of less than 2 ha. Estimated production in Malaysia fluctuates around 24,000 t and in the Philippines trees yield approximately 2150 t (Abdul *et al.*, 1989; Samson, 1992). Florida is the only place in continental USA where carambola is grown commercially, with 179 ha of commercial orchards, 89% of which are 4 years old or younger (Campbell, 1989). Most of the information on pests of carambola comes from South-East Asia and Australia. In Florida, the insects that have been recorded from carambola include: *Morganella longispina* (Morgan) (Homoptera: Diaspididae), *Diaprepes abbreviatus* L., *Nezara viridula* (L.),

Ecpantheria scribonia (Stoll) (Lepidoptera: Arctiidae) feeding on leaves and *Platynota rostrana* (Walker) feeding on fruit (Peña and Duncan, 1999).

Key pests

Bactrocera carambolae *(Diptera: Tephritidae)*

The carambola fruit fly, *Bactrocera carambolae* (Drew and Hancock) was reported for the first time in the western hemisphere in Surinam in 1975, although it was properly identified in 1986 (Plate 78). This species is endemic to Indonesia, Malaysia and southern Thailand (Van Sauers-Mueller, 1991). In Surinam, the major hosts are carambola and the Curaçao apple, *Syzygium samarangese*. Minor hosts include: West Indian cherry, *Malpighia punificolia*, mango *Mangifera indica*, sapodilla, *Manilkara achras*, guava *Psidium guajava*, Indian jujube *Ziziphus jujuba*, *Citrus* spp. and cashew *Anacardium occidentale* L., as occasional hosts. Export losses of infested countries are estimated to be US$25.3 million (Midgarden and Fleurkens, 1998).

BIOLOGY The life cycle of *B. carambolae* is typical of other fruit flies. From egg to mature adult takes about 22 days under good conditions (26°C and 70% RH) (Midgarden and Fleurkens, 1998). Eggs take 1–2 days to hatch. The larval stage lasts 6–9 days, and pupation 8–9 days. Adults are 3.5–5 mm, yellowish black with a brown tinge, especially on the abdomen, head and legs, ovipositor of female is knife-shaped (Midgarden and Fleurkens, 1998). Adults become sexually mature 8–10 days after emergence. The minimum period of time for one generation is approximately 30 days (Midgarden and Fleurkens, 1998). Before laying eggs, the adult female fly feeds for a week on protein, e.g. on bacteria growing on fruit and plant surfaces, on bird faeces and on sugars, e.g. honeydew and nectar, and spoiled fruit. Mature adults copulate after groups of males gather and perform a courtship dance in the early evening, just before sunset. Females puncture the skin of green or mature fruit and lay eggs in groups of 3–5 just under the skin (Midgarden and Fleurkens, 1998). Males and females are strong fliers and will fly long distances if they cannot find a good source of food or a site to lay eggs. Data from *B. dorsalis* have shown that the adults can fly over 50 km from the emergence site (Midgarden and Fleurkens, 1998). Adults may live 30–60 days in nature. Females can lay more than 1000 eggs over their lifetime. Eggs are white, banana-shaped and 1 mm long, shining white to milky when ready to hatch. Larvae have three instars inside the fruit where they feed on the pulp and make tunnels in the fruit. Larvae are elongate and pointed at the head. Length of larvae is 1 mm just after hatching to 7–8 mm just before pupation. The colour is white or the same colour as the fruit pulp (Midgarden and Fleurkens, 1998). At the end of the third instar, the larvae leave the fruit and burrow 2–7 cm into the soil to pupate. Pupae are dark reddish brown, barrel-shaped and about 4–5 mm long. Since its detection, the fruit fly has spread throughout the coastal areas of Surinam and French Guiana (Van Sauers-Mueller and Vokaty, 1996).

CONTROL An effective control method called the male annihilation technique has been developed for members of the *Bactrocera dorsalis* complex. Baits are impregnated with a combination of lure and insecticide. Methyl eugenol, a parapheromone lure, is also used to attract male fruit flies before they become sexually mature. Surveillance methods are accomplished using delta-shaped Jackson traps impreganated with methyl eugenol and an insecticide. Another surveillance method performed in the infested areas consists of collecting fruit and rearing the larvae in screened cages with sand or sawdust.

Morganella longispina *(Morgan)* *(Homoptera: Diaspididae)*

Plumose scale, *Morganella longispina*, has been found in Florida since 1980 (Hamon, 1981). The female armour of the scale is circular to oval (1–1.5 mm × 1.0 mm) convex, dull black, with a thick and opaque texture. The male armour is similar in colour, smaller, slightly convex and elongate. The scale is known from Algeria, South Africa, China, India, Sri Lanka, Sandwich Islands, Tahiti, Hawaiian

Islands, Mauritius and most of the Caribbean Islands, including Barbados, Dominican Republic, Trinidad, Haiti, Jamaica, Puerto Rico, and from Guyana and Brazil in South America (Hamon, 1981). Its polyphagous habits include the following host plants: *Citrus* sp., *Ligustrum* sp., *Nerium* sp., *Severinia* sp., *Averrhoa carambola*, *Eucalyptus*, *Carica papaya*, *Coffea* sp., *Ficus*, *Persea americana*, *Mangifera indica*, and macadamia among others (Hamon, 1981).

In Florida, the number of *M. longispina*-infested branches peaked during November 1997 and was reduced between the months of June and July. The cultivar with highest plumose scale infestation was 'B-10', followed by 'Kajang', 'Wai Wei', 'Cheng Chui', 'Meis' and 'Kary' (Peña and Duncan, 1999).

CONTROL The products pymetrozine, pyriproxifen and imidacloprid were effective for reducing plumose scale density compared to the control (Peña and Duncan, 1999). Zimmerman (1948) reported *Archenomus perkinsi* (Fullaway) and *Prospatella koebelei* Howard as parasitoids of *M. longispina* in Hawaii.

Nezara viridula *(L.) (Homoptera: Pentatomidae) and* Leptoglossus *spp. (Homoptera: Coreidae)*

In Florida, the plant bugs, southern stink bug *Nezara viridula* (L.) and the leaf-footed bug *Leptoglossus* spp., move from vegetables planted during the autumm and winter to carambola orchards that have mature and ripe fruit. Adults and nymphs insert their piercing–sucking mouthparts into the fruit, extracting fruit fluids leaving a small puncture. With time, the area around this puncture or scar becomes soft and shows decay. *Leptoglossus* spp. were only observed in February and October 1998; however, the highest number of punctures per fruit were detected between January and February. Lesions on fruit were also observed during July, August and November (Peña and Duncan, 1999). In Florida, this plant bug is also considered a pest of citrus, where its feeding on ripening fruit causes premature colour break and fruit drop (Mead, 1971). In Florida, the carambola cultivar with the highest number of punctures was 'Pasi', followed by 'Sri kembagen' and 'Cheng Chui' (Peña and Duncan, 1999).

Toxoptera aurantii *(Boyer de Fonscolombe) (Homoptera: Aphidae)*

In Florida, the black citrus aphid, *Toxoptera aurantii* (Boyer de Fonscolombe) is found in large colonies on flowering terminals feeding on flower peduncles of carambola (Plate 79). The black citrus aphid is dark reddish to brown. Wingless forms (apterous) are about 2 mm long and by production of honeydew, sooty mould fungi develop around the infested areas. Aphids are a problem in carambola during the flowering season. The major problem could be excessive flower drop, followed by sooty mould on leaves and fruits. Black citrus aphids were most common during the month of November 1997 at the peak of carambola flowering. Other insects such as *Planococcus citri* (Risso) (Homoptera: Pseudococcidae) and *Daghbertus* spp. (Hemiptera: Miridae) were observed on the inflorescences. In Florida, the highest aphid infestation peaks were observed on 'B-10', followed by 'Pasi', 'Wai Wei', 'Cheng Chui', 'Kary' and 'Erlin' (Peña and Duncan, 1999).

Gonodonta *spp. (Lepidoptera: Noctuidae) and* Eudocima *spp.*

Fruit piercing moths, *Gonodonta* spp. *Eudocima* spp. (Lepidoptera: Noctuidae), feed at night by piercing the skin of the ripe or ripening fruit with their strong proboscis, the damage results in crop loss or unmarketable fruit. The internal injury to the fruit resembles a honeycomb. Secondary rots develop at the puncture site and the fermented fruits are frequently visited by other insects, such as sap beetles (Nitidulidae). The fruit piercing moths are dark brown, with orange or dark yellow coloured hindwings. In Florida, the larvae do not feed on carambola, but have been collected from leaves of atemoya (Peña and Duncan, 1999). In Queensland, Australia, fruit piercing moth larvae feed on native vines isolated from the orchard environment (Fay, 2000).

CONTROL Control for fruit piercing moths with insecticides is problematic, and alternative methods have been sought in Australia to combat these pests. Netting trees and bagging fruits can be effective, but are not options for most crops unless other significant pests (e.g. birds, flying foxes) are also controlled. Trapping using black light traps can reduce damage by 60–70%, while a new baiting system has provided 75–85% security in citrus trials (Fay, 2000) For information on fruit piercing moths, see Chapter 11 on the litchi.

Brevipalpus phoenicis *(Geijskes) (Acarina: Tenuipalpidae)*

Tenuipalpids, possibly *Brevipalpus phoenicis* (Geijskes), is also an important pest of citrus, papaya, guava and more than 50 other plant species (Jeppson *et al.*, 1975). This mite appears to be responsible for bronzing of carambola fruit. It prefers crevices and fruit angles. In Florida, *B. phoenicis* peak densities were observed between August and October 1998 (Peña and Duncan, 1999).

Barbados cherry or acerola (*Malpighia glabra* L.)

Barbados cherry or acerola, *Malpighia glabra* (L.) (= *punicifolia* L.), is a tropical fruit native to the West Indies, Central America and South America (Stahl *et al.*, 1955; Phillips, 1991). The genus *Malpighia* is present from south Texas to Peru (Asenjo, 1980). Recently, the Barbados cherry has received attention because its fruits are an exceptionally high natural source of ascorbic acid (vitamin C). Its cultivation has extended throughout the subtropics and tropics (Ledin, 1958) and some of the largest plantings are in Brazil (Anonymous, 1996). Estimated commercial hectarage in the Caribbean region is > 160 ha with a potential crop value of several million US dollars (Gonzalez-Ibañez, 1983). In Florida, Barbados cherry is grown in the southern part of the state in homeowners' yards and as a small commercial crop. Flowering and fruit set occur almost continuously from April through November in Florida, and fruits mature in approximately 30 days (Stahl *et al.*, 1955; Ledin, 1958).

Key pests

Acerola weevil, Anthonomus macromalus *Gyllenhal (Coleoptera: Curculionidae)*

The most important insect pest of Barbados cherry in Florida is the acerola weevil, *Anthonomus macromalus* Gyllenhal (= *A. flavus*, = *A. bidentatus*, = *A. malpighia*) (Coleoptera: Curculionidae). This weevil is small (2.1–3.4 mm long) and dark brown (Plate 80). There is a distinct chevron pattern of light and dark scales on the elytra (Clark and Burke, 1985). Members of this species are oligophagous within the botanical family *Malpighiaceae*. This weevil appears to be native to the Neotropics, with reports from Florida (USA) and from many of the islands of the Caribbean region (Clark and Burke, 1985). The first report in Florida was in 1972 (Stegmaier and Burke, 1974).

Stegmaier and Burke (1974) and Balloff (1993) reviewed the biology of *A. macromalus*. Adults deposit eggs on the anthers of flowers, stem terminals and in immature fruits. The larvae develop in the flowers and fruit, causing extensive damage to floral reproductive structures and to the flesh of the fruit, which reduces yield. Adult weevils feed on immature, expanding leaves which causes a 'shot-gun hole' appearance to the leaves as they expand (Hunsberger *et al.*, 1998).

SEASONALITY In Florida, the adult population peaks in late June through July. Infested trees typically showed leaf damage. Adult weevil densities sampled from the upper and middle strata of branches and lower branches were statistically equal.

The peak months for adult emergence from fruit were from late June through September. In June, the percentage fruit infestation was estimated at 45–75%. By August, the respective estimates were 65–75%. A 1-month delay in peak adult emergence from fruit followed the peak number of adults in the field (Hunsberger *et al.*, 1998).

Both eggs and larvae were found in flowers. Eggs were laid before flower opening. The female eats a hole into the flower bud and deposits the eggs on to the anthers. Larvae were observed eating the floral ovary when the flower was fully open. Larvae were found tunnelling inside the new growth of branches (Plate 81). External evidence of larval feeding was sudden wilt and premature death of infested shoots (Hunsberger *et al.*, 1998).

BIOLOGICAL CONTROL Surveys for parasitoids of *A. macromalus* have yielded a single parasitoid, *Catolaccus hunteri* Crawford (Hymenoptera: Pteromalidae) (Hunsberger and Peña, 1997). No other parasitoids were found in Florida.

PHEROMONES There is some evidence that *A. macromalus* can produce aggregation pheromones. Adult weevils have been found in aggregations in orchards (A.G.B. Hunsberger and J.E.Peña, personal observations) and individuals were attracted to crushed *A. macromalus* in Petri dish arenas (Hunsberger *et al.*, 1998). Furthermore, aggregation pheromones have been identified from two *Anthonomus* spp. Tumlinson *et al.* (1968) identified the aggregation pheromone (grandisol) from the cotton boll weevil (*Anthonomus grandis grandis* Boh.) and Eller *et al.* (1994) identified the aggregation pheromone from the pepper weevil, *A. eugenii* Cano.

CHEMICAL CONTROL The appropriate time to apply adulticides appears to be in the spring, before weevil densities start to increase. Also, because of the uneven and overlapping fruiting phenology, fruits bear different weevil generations, allowing a rapid reinfestation of the orchard.

Other pests

In Brazil, losses caused by pests are not quantified, and the information on pest biology and integrated pest management is very limited (Boaretto and Brandao 1995). In Brazil, several species of aphids (*Toxoptera citricidus* Kirkaldy, *Aphis tavaresi* DelGuercio; *Aphis tavaresi argentiniensis* Blanchard, *Toxoptera aurantii*, *Aphis spiraecola*) are mentioned as pests of acerola (Couceiro, 1985; Araujo and Minami, 1994), of which the most important is considered to be *T. citricidus*, appearing often in very large numbers on the apical buds and fruit peduncles. Batista *et al.* (1994) reported that *T. citricidus* can produce an average of 49.7 nymphs when the female is feeding in acerola. During dry weather infested plants show leaf curling, stunting, and damage to trees and fruit by the discharge of honeydew. In Brazil, Braga *et al.* (1998) recommend careful applications of organophosphates.

Other common pests of acerola in Brazil are: *Orthezia praelonga* Douglas, *Coccus viridis* and *C. hesperidium* L. The recommended control is the use of emulsifiable oils.

Fruit flies (Diptera: Tephritidae)

In Brazil, fruit flies of the genus *Anastrepha* and the Mediterranean fruit fly, *Ceratitis capitata* Wiedemann, are pests of acerola (Malavasi *et al.*, 1980). These flies cause premature fruit maturity, and reduce fruit quality.

Leptoglossus *spp. (Homoptera: Coreidae);* Crinocerus *spp. (Homoptera: Coreidae); and* Nezara viridula *L. (Homoptera: Pentatomidae)*

The coreid *Leptoglossus phyllopus* Herbst and the pentatomid *Nezara viridula* L have been observed attacking acerola fruits in Florida, USA (Ledin, 1958). In Brazil, the coreids *Crinocerus sanctus* Fabr. and *Leptoglossus stigma* Herbst. cause fruit deformation by inserting their proboscis into the fruit, and are probably responsible for early fruit rotting (Braga *et al.*, 1998).

Minor pests

In Brazil, *Trigona spinips* Fabr. (Hymenoptera: Apidae), a pest of citrus and banana, opens galleries on acerola fruits, reducing fruit quality (Braga *et al.*, 1998). (See pests of banana, in Chapter 2.)

Conclusions

We selected these fruit species as examples of minor fruits grown in the world. Very little information can be acquired for management of their pests. Moreover, production of other tropical fruit is becoming more extensive and propagation of several of them, e.g. borojo, *Borojoa patinoi*, bread fruit *Artocarpus* spp., sapodilla, *Manilkara sapota* L., cirvela, *Spondias cytherea* Sonnerat, is becoming more frequent in tropical areas. On the other hand, emigration of ethnic groups from Asia, Africa and the Americas and their settlement in Europe, Australia, USA and Canada, has produced a demand for fruits that have been ignored in the major world markets. Thus, if rules for sanitation and food security are followed in the importing countries, there will be a need to develop IPM programmes for these fruit species.

References

Abdul, W., Ahmad, I. and Hassan, A. (1989) Carambola production, processing and marketing in Malaysia. *Proceedings of the Inter-American Society of Tropical Horticulture* 33, 30–43.

Anonymous (1996) Acerola. California Rare Fruit Growers. www.crfg.org/pubs/ff/acerola.html

Anonymous (2000) Durian pests. http://www.oae.gi.th./projects /durian/thai/news

Araujo, P. and Minami, K. (1994) Acerola. *Campinas: Fundacao Cargil*, 81 pp.

Asenjo, C.F. (1980) Acerola. In: Nagy, S. and Shaw, P.E. (eds) *Tropical and Subtropical Fruit: Composition, Properties, and Uses.* AVI Publishing Company, Connecticut, pp. 341–374.

Balloff, S. (1993) *Anthonomus malpighiae* Clark and Burke (Coleoptera: Curculionidae) a pest of the Barbados cherry in Guadeloupe. MSc thesis, Ecole Nationale Superieure Agronomique de Montpellier, Guadeloupe, 37 pp.

Boaretto, M. and Brandao, A. (1995) Pragas da cultura da acerola. In: Sao Jose, A. and Alves, R. (eds) *Acerola do Brasil: Producao e Mercado.* Vittoria da Conquista: Universidad de Estadual do Sudoeste da Bahia, Brazil, pp. 58–61.

Braga, R., Luna-Batista, J., Campos, L. and Warumby, J. (1998) Pragas da aceroleira. In: Braga, R., Cardoso, J. and das Chagas (eds) *Pragas de Fruteriras Tropicais de Importancia Agroindustrial.* Empresa Brasileira de Pesquisa Agropecuaria, Servicio de Producao de Informacao, Brasilia, pp. 33–40.

Campbell, C.W. (1989) Carambola production in the United States. *Proceedings of the Inter-American Society of Tropical Horticulture* 33, 47–54.

Charanasri, V. and Kongchuensin, M. (1991) African red mite, *Eutetranychus africanus* (Tucker), an important pest of durian in Thailand. *Entomology and Zoology Bulletin* 13, 62–68 (in Thai).

Chey, V.K., Holloway, J.D. and Speight, M.R. (1997) Diversity of moths in forest plantations and natural forests in Sabah. *Bulletin of Entomological Research* 87, 371–385.

Chin, H.F. and Yong, H.S. (1980) *Malaysian Fruits in Colour.* Tropical Press Sendirian Berhad, Kuala Lumpur, 129 pp.

Choonhawong, W., Kongkanjana, A., Wongkobrat, A., Chawanapong, M., Jumroenma, K. and Meksongsee, B. (1982) Study on biology and host plants of leaf eating weevil (*Hypomeces squamosus* Fabricius). In: *Annual Report 1982, Corn and other Field Crops Entomology Research, Entomology and Zoology Division, Department of Agriculture, Bangkok*, pp. 34–38 (in Thai).

Clark, W.E. and Burke, H.R. (1985). Revision of the *venustus* species group of the weevil genus *Anthonomus* Germar (Coleoptera: Curculionidae). *Transactions of the American Entomological Society* 111, 103–170.

Cock, M.J.W., Godfray, H.C.J. and Holloway, J.D. (eds) (1987) Slug and nettle caterpillars. In: *The Biology, Taxonomy and Control of the Limacodidae of Economic Importance on Palms in South-East Asia.* CAB International, Wallingford, UK, 270 pp.

Couceiro, E.M. (1985) Curso de extensao sobre a cultura da acerola: *UFRP, Recife*, 45 pp.

Dammerman, K.W. (1929) *The Agricultural Zoology of the Malay Archipielago.* J.H. de Bussy, Amsterdam, 473 pp.

Desmier de Chenon, R. (1982) *Latoia* (*Parasa*) *lepida* (Cramer) (Lepidoptera: Limacodidae), a coconut pest in Indonesia. *Oleagineux* 37, 177–183.

Disthaporn, S., Kraturuek, C. and Sutthiarrom, S. (1996) *Manual for Integrated Pest Management in Durian.* Thai–German Project IPM in selected fruit crops, Department of Agriculture and Department of Agricultural Extension, Thailand, 97pp.

Eller, F.J., Bartelt, R.J., Shasha, B.S., Schuster, D.J., Riley, D.G., Stansly, P.A., Mueller, T.F., Shuler, K.D., Johnson, B., Davis, J.H. and Sutherland, C.A. (1994) Aggregation pheromone for the pepper weevil, *Anthonomus eugenii* Cano (Coleoptera: Curculionidae): identification and field activity. *Journal of Chemical Ecology* 20, 1537–1555.

Entwistle, P.F. (1972) *Pests of Cocoa*. Longmans, London, 779 pp.

Fay, H.A. (2000) Fruitpiercing moths and fruit-spotting bugs: intractable pests of tree fruits in an insecticide-reduction environment. *International Symposium on Tropical and Subtropical Fruits*. Cairns Australia, 26 November–1 December 2000, Symposium Booklet, p. 27.

Gonzalez-Ibañez, J. (1983) Evaluation of herbicides for container-grown acerola (*Malpighia punicifolia* L.) under greenhouse conditions. *Puerto Rico University Journal of Agriculture* 67, 112–116.

Hamon, A.B. (1981) Plumose scale, *Morganella longispina* (Morgan). Florida Department of Agriculture and Consumer Services. *Division of Plant Industry, Entomology Circular* 226, 2 pp.

Ho, C.T. and Khoo, K.C. (1997) Partners in biological control of cocoa pests: mutualism between *Dolichoderus thoracicus* (Hymenoptera: Formicidae) and *Cataenococcus hispidus* (Hemiptera: Pseudococcidae). *Bulletin of Entomological Research* 87, 461–470.

Hunsberger, A.G.B. and Peña, J.E. (1997) *Catolaccus hunteri* (Hymenoptera: Pteromalidae), a parasite of *Anthonomus macromalus* (Coleoptera: Curculionidae) in south Florida. *Florida Entomologist* 80, 301–304.

Hunsberger, A.G.B., Peña, J.E., Giblin-Davis, R., Gries, G. and Gries, R. (1998) Biodynamics of *Anthonomus macromalus* (Coleoptera: Curculionidae) a weevil pest of Barbados cherry in Florida. *Proceedings of the Florida State Horticultural Society* 111, 334–338.

Ibrahim, A.G. and Ibrahim, D.K. (1989) Biology of *Lawana conspersa* Walker (Homoptera: Flatidae) a minor pest of cocoa in Malaysia. *Malaysian Applied Biology* 18, 93–98.

Jeppson, L.R., Keifer, H. and Baker, E. (1975) *Mites Injurious to Economic Plants*. University of California Press, Berkeley, 614 pp.

Kalshoven, L.G.E. (1981) *Pests of Crops in Indonesia*. (Revised and translated by P.A. van der Laan and G.H.L. Rothschild.) P.T. Ichtiar Baru-Van Hoeve, Jakarta, 701 pp.

Khoo, K.C., Ooi, P.A.C. and Ho, C.T. (1991) *Crop Pests and their Management in Malaysia*. Tropical Press Sendirian Berhad, Kuala Lumpur, 242 pp.

Kuroko, H. and Lewvanich, A. (1993) *Lepidopterous Pests of Tropical Fruit Trees in Thailand* (in Thai). Japan International Cooperation Agency, Tokyo, Japan, 132 pp.

Ledin, R.B. (1958) The Barbados or West Indian cherry. *Florida Agricultural Experiment Station, Bulletin* 594, 15–17.

Lever, R.J.A.W. (1953) Cockchafer pests of cocoa and other crops. *Malaysian Agricultural Journal* 36, 89–113.

Lim, G.S., Ho, C.T., Md. Jusoh Mamat, Teoh, C.H., Khoo, K.C., Chan, L.G. and Ooi, P.A.C. (1987) *The Cocoa Pod Borer* (an introduction and abstracts). Malaysian Plant Protection Society, Kuala Lumpur, Malaysia, 103 pp.

Lim, T.K. (1990) *Durian Diseases and Disorders*. Tropical Press Sendirian Berhad, Kuala Lumpur, 95 pp.

Malavasi, A., Morgante, J.S. and Zucchi, R.A. (1980) Biologia das moscas das frutas (Diptera: Tephritidae) I. Lista de hospedeiros e-ocorrencia. *Revista Brazileira de Biologia* 40, 9–16.

Mead, F. (1971) Leaffooted bug, *Leptoglossus phyllopus* (Linnaeus) (Hemiptera: Coreidae). *Florida Department of Agriculture, Entomology Circular* 107, 2 pp.

Midgarden, D. and M. Fleurkens. (1998) *Carambola Fruit Fly Programme*. http://www.carambolafly.com/English/english.htm

Na, A.B. (1989) The cocoa black ant–mealybug relationship: artificial establishment of *Cataenococcus hispidus* (Homoptera: Pseudococcidae) on cocoa. PhD thesis, Universiti Putra Malaysia, Malaysia.

Noyes, J.S. (1991) A new species of *Ooencyrtus* (Hymenoptera: Encyrtidae) from Malaysia, a prepupal parasitoid of the cocoa pod borer, *Conopomorpha cramerella* (Snellen) (Lepidoptera: Gracillariidae). *Journal of Natural History* 25, 1617–1622.

Ooi, P.A.C. (1978) An important new pest of rambutan. *MAPPS Newsletter* 2, 5

Ooi, P.A.C. (1987) Advances in the biological control of cocoa pod borer. In: Ooi, P.A.C. *et al.* (eds) *Management of the Cocoa Pod Borer*. Malaysian Plant Protection Society, Kuala Lumpur, pp. 103–117.

Ooi, P.A.C., Day, R.K. and Mumford, J.D. (1990). Evolution of the cocoa pod borer. *MAPPS Newsletter* 14, 37–38.

Peña, J.E. and Duncan, R. (1999) Seasonality and control of arthropods on carambola cultivars in southern Florida. *Proceedings of the Florida State Horticultural Society* 112, 210–212.

Phillips, R.L. (1991) Barbados cherry. University of Florida, Institute Food Agricultural Sciences, *Fact Sheet FC-28*, Gainesville, 2 pp.

Ramasamy, S. (1980) An outbreak and some aspects of the biology of the hawk moth, *Daphnusa ocellaris* Walk. on durian in Perak. *Malaysian Agricultural Journal* 52, 213–218.

Ramsammy, P. (1989) The carambola in Guyana. *Proceedings of the Inter-Amercian Society of Tropical Horticulture* 33, 12–25.

Rao, B.S. and Suppiah, K. (1971) Cockchafers as plantation pests in Malaysia and their control. In: Wastie, R.L. and Wood, B.J. (eds) *Crop Protection in Malaysia.* Incorporated Society of Planters. Kuala Lumpur, pp. 173–186.

Samsom, J.A. (1992) Averrhoa L. In: Verheij, E.W.M. and Coronel, R.E. (eds) *Plant Resources of South-East Asia No. 2 Edible Fruits and Nuts.* PROSEA, Bogor, Indonesia, 443 pp.

Sedgley, M. (1984) Oxalidacea. In: Page, P.E. (ed.) *Tropical Tree Fruits for Australia.* Queensland Department of Primary Industries, Brisbane, series QI83018, pp. 125–128.

Sirisingh, S., Ithnin, B. and Unahawutti, C. (1994) Entomology and pests. In: Nanthachai, S (ed.) *Durian: Fruit Development, Postharvest Physiology, Handling and Marketing in ASEAN.* ASEAN Food Handling Bureau, Kuala Lumpur, Malaysia, 156 pp.

Stahl, A.L., Kaplow, M. and Nelson, R. (1955) The present and future possibilities of Barbados cherry. *Proceedings of the Florida State Horticultural Society* 68, 138–143.

Stegmaier, C.E. Jr and Burke, H.R. (1974) *Anthonomus flavus* (Coleoptera: Curculionidae) a fruit-infesting weevil of the Barbados cherry, *Malpighia glabra* (Malpighiaceae), new to North America. *Florida Entomologist* 57, 81–90.

Stone, B.C. (1992) *Nephelium lappaceum* L. In: Verheij, E.W.M. and Coronel, R.E. (eds) *Plant Resources of South-East Asia: No. 2 Edible Fruits and Nuts.* PROSEA, Bogor, Indonesia, 443 pp.

Tumlinson, J.H., Hardee, D.D., Minyard, J.P., Thompson, A.C., Gast, R.T. and Hedin, P.A. (1968) Boll weevil attractant: isolation studies. *Journal of Economic Entomology* 61, 470–474.

Van Sauers-Mueller, A. (1991) An overview of the carambola fruit fly *Bactrocera* species (Diptera: Tephritidae), found recently in Surinam. *Florida Entomologist* 74, 432–440.

Van Sauers-Mueller, A. and Vokaty, S. (1996) Carambola fruit fly projects in Surinam and Guyana. *Tropical Fruits Newsletter* 18, 6–8.

Verheij, E.W.M. (1992) *Garcinia mangostana* L. In: Verheij, E.W.M. and Coronel, R.E. (eds) *Plant Resources of South-East Asia: No. 2 Edible Fruits and Nuts.* PROSEA, Bogor, Indonesia, 443 pp.

Walker, A.K. and Huddleston, T. (1987) *Chelonus chailini* sp. n. (Hymenoptera: Braconidae) from Malaysia, parasitizing gracillariid moths (Lepidoptera). *Bulletin of Entomological Research* 77, 437–440.

Waterhouse, D.F. (1993) *The Major Arthropod Pests and Weeds of Agriculture in Southeast Asia: Distribution, Importance and Origin.* The Australian Centre for International Agricultural Research (ACIAR), Canberra, ACT. Monograph 21, 141 pp.

Yunus, A. and Balasubramaniam, A. (1981) *Major Crop Pests in Peninsular Malaysia,* 2nd edn. Bulletin No. 138. Ministry of Agriculture, Peninsular Malaysia.

Yunus, A. and Ho, T.H. (1980) *List of Economic Pests, Host Plants, Parasites and Predators in West Malaysia (1920–1978).* Bulletin No. 153. Ministry of Agriculture, Malaysia, 538 pp.

Zimmerman, E. (1948) *Insects of Hawaii,* Vol. 5, *Homoptera: Sternorhyncha.* University of Hawaii Press, Honolulu, 464 pp.

11 Pests of Litchi and Longan

G.K. Waite[1] and J.S. Hwang[2]

[1]*Queensland Horticulture Institute, Maroochy Research Station, PO Box 5083, SCMC, Nambour, Queensland 4560, Australia;* [2]*Taiwan Agricultural Chemicals and Toxic Substances Research Institute, 189 Chung-Cheng Road, Wufeng, Taichung Hsien, Taiwan 413, Republic of China*

The litchi (*Litchi chinensis* Sonn.) and longan (*Dimocarpus longan* Lour.) are closely related species belonging to the family Sapindaceae. Of South-East Asian origin, they thrive in subtropical areas with cool dry winters and warm wet summers (Menzel, 1983), but may be grown in tropical areas at high elevation. Around the world, litchis have been successfully grown commercially in latitudes from 15° to 35° (Menzel and Simpson, 1990). Some cultivars are grown in warmer areas in Thailand and Indonesia but the fruit is generally of poor quality. Longans are often cultured in the same geographical areas as litchis but, except in China and Thailand, are usually less popular.

The litchi is a traditional fruit in China, and it occupies a special place in Chinese culture. The first official record dates back to the 2nd century BC, while unofficial records date to 1766 BC. Indo-China is the centre of origin for this species and many old specimen trees have been identified. Several trees in Guangdong and Hainan Island are believed to be at least 2000 years old. Similarly, longan occupies a special place in the culture of Thailand, but in contrast it is a relatively recent arrival there.

Litchi and longan fruits are best when eaten fresh, but they can also be processed by canning, juicing or drying to litchi nuts. In Asia they are usually harvested as whole panicles in order to maintain freshness, and are sold in street markets or at roadside stalls within a few days of harvest. Frequent sprinklings with water help to maintain the colour of the skin and overall freshness. If allowed to dry out, the skin turns brown and becomes brittle, lowering consumer appeal. Although South African and Israeli litchis sent to European markets are treated with sulphur dioxide to retain attractive fruit colour, internal quality is not preserved. This marketing issue needs to be overcome before world trade can expand.

Importance of the Crop

Litchi

World production of litchis is about 1 million t with the bulk of the crop grown in China and India (Partridge, 1997). Though much of this production is consumed as fresh fruit, a large proportion is processed in the form of canned fruit and juice. Over 700,000 t of fresh litchis are estimated to be consumed in South-East Asia and India annually. Because of the special significance of litchis to the Chinese people, demand outstrips production, and fruit is imported from Vietnam and Thailand during the northern season. Countries such as South Africa, Mauritius,

©CAB *International* 2002. *Tropical Fruit Pests and Pollinators* (eds J.E. Peña, J.L. Sharp and M. Wysoki)

Madagascar, Réunion and Australia have relatively low populations and must export part of their crop to maintain profitability. In addition, their crop matures when South-East Asian litchis are unavailable, especially around Chinese New Year. Exports to Singapore, Hong Kong and mainland China can be quite lucrative, especially for suppliers of high quality fruit from Australia. Fruit from African countries, treated with sulphur, is less popular.

Apart from China and India, large litchi industries have been developed in Taiwan, Thailand, Vietnam, Madagascar, South Africa and Réunion. Smaller but expanding industries exist in Australia, Bangladesh, Mauritius, Mexico, the Seychelles, Spain and the USA (California, Florida and Hawaii).

Longan

In general, longans can be successfully grown wherever litchis are cultivated. Although both species flower at about the same time, longans take longer than litchis to mature. China (400,000 t), Thailand (150,000 t) and Vietnam (20,000 t) are the major world producers.

Litchi and Longan Phenology

Litchi and longan are adapted to the warm subtropics and produce the best crops when winters are short, dry and cool, but frost-free. Such climatic conditions initiate the development of flower panicles. The inflorescences emerge during late winter and the flowers open in early spring. Most cultivars set fruit far in excess of what an individual tree can carry through to maturity and will shed the excess at various times during fruit development. Longan flowers about 2 to 3 weeks later than litchi, and matures 4 to 8 weeks later. Harvesting fruit as whole panicles has the effect of pruning and stimulates new leaf growth after harvest. It also reduces tree size. Ideally, trees should produce one or two vegetative flushes after harvest. The aim is to have the second or third flush commence in winter. If conditions are cool during the early part of the flush, development of the new growth will be floral. However, if warm weather is encountered, the new growth will produce leaves (Menzel and Simpson, 1995). Litchi and longan are often cultivated in the same geographical areas and many pests are common to both crops. Tan *et al.* (1998) found that in Guanxi Province, China, more than 90% of the insect species recorded for each were common and suggested that this might be used to advantage in area-wide management of pests.

Pests

Fruit borers (Lepidoptera: Gracillariidae)

Conopomorpha sinensis *Bradley*

DISTRIBUTION AND BIOLOGY The major pest of lychees and longans in China, Taiwan and Thailand is *Conopomorpha sinensis* Bradley, known as the litchi stem-end borer in China and the litchi fruit borer in Thailand (Plate 82). Bradley (1986) described three previously unrecognized species congeneric with the cocoa pod borer, *Conopomorpha cramerella* Snellen, previously known as *Acrocercops cramerella*, which was thought to be the damaging species. Two of the new species, *C. sinensis* and *Conopomorpha litchiella* Bradley attack litchis, while the host range of *C. cramerella* is restricted to rambutan and cocoa (Bradley, 1986). The confusion regarding these true identities and hence the actual species causing certain damage, is demonstrated in papers from India (Singh, 1975; Butani, 1977; Lall and Sharma, 1978), China, (Anonymous, 1978) and Taiwan (Hwang and Hsieh, 1989). Since Bradley's taxonomic study, Yao and Liu (1990), Huang *et al.* (1994, 1997) and Zhang, Z. (1997) have recognized the differences and have described the pest status of each accordingly.

C. sinensis lays yellow, scale-like eggs, 0.4×0.2 mm long, on the fruit any time after fruit set, as well as on new leaves and shoots of litchi and longan. The eggs hatch in 3–5 days and the larva immediately penetrates the fruit, leaf or shoot. One or more eggs may be laid on a fruit but only one larva per fruit survives. Mature larvae are 6–10 mm in length and brownish in colour, or green if they have fed on leaves. After 8 –12 days they leave the feeding site to pupate on or under mature leaves in cream-coloured oval cocoons. The light green pupae change to dark brown just before eclosion, and the moth emerges after 5–7 days. The moth is very small with long filiform antennae and narrow, fringed forewings measuring 8–11 mm when expanded. The moths live for about 5–8 days. In Taiwan, the pest may complete 4–5 generations during the litchi and longan fruiting seasons. It appears to continue to develop over the winter period, either in alternative hosts or in the terminals of the main hosts, litchi and longan. During the off-season, when fruit is not available to the pest, it can survive by feeding on young leaves and shoots, similar to *C. litchiella*.

BIOLOGICAL CONTROL Most of the natural enemies of *C. sinensis* in Thailand are microhymenopterous parasitoids. Most of them also parasitize the litchi leafminer, *C. litchiella* (Dolsopon *et al.*, 1997a). *Phanerotoma* sp., *Colastes* sp. and *Pholestesor* sp. (Braconidae) and *Goryphus* sp. (Ichneumonidae) all attack the larva, while the latter may also attack the pupal stage. The ichneumonid, *Paraphylax* sp., has been reared from pupae, but it is apparently a hyperparasitoid. The braconids are recognized as being the most effective of the complex. Pests present in fallen fruit may have been parasitized either before the fruit has fallen from the tree, or after it has fallen. For this reason, it is recommended that all fallen fruit should be left under trees in order to permit parasitoid populations in the orchard to increase.

Two species of hymenopterous pupal parasitoids, *Phanerotoma* sp. and *Apanteles* sp., have been recorded from *C. sinensis* in Taiwan. *Tetrastichus* sp. and *Elasmus* sp. have been found attacking the larvae, and the egg parasitoid, *Trichogrammatoidea bactrae fumata* Nagaraja, has been introduced to Taiwan from Thailand, with unknown results.

In Bihar, India, *Mesochorus* sp., *Chelonus* sp., *Bracon* sp. and *Apanteles* sp. were found to parasitize *C. cramerella* (most likely *C. sinensis*) (Sharma and Agrawal, 1988). Huang *et al.* (1994) recorded five hymenopterous parasitoids, of which *Phanerotoma* sp. was the most important, causing up to 22% mortality of larvae in fallen fruit.

DAMAGE *C. sinensis* shows a preference for litchis over longans but in both crops, damaged fruit may fall from the tree. Huang *et al.* (1994) found that the pest had damaged 87.9–99.0% of fallen fruit and 16.0–86.5% of fruit remaining on the tree in sprayed orchards. Corresponding damage rates in unsprayed orchards were 96.1–100% and 41.5–96.7%, respectively.

MONITORING AND CONTROL In Thailand, fruit are inspected weekly immediately after fruit set to detect eggs of *Conopomorpha*, which are very small and almost invisible to the naked eye. The use of a ×20 hand lens facilitates monitoring. Infested fruit should be picked as it is found and destroyed, at infestation levels of 1–2%. When the pest becomes more active, permethrin is applied at weekly intervals, ceasing at least 2 weeks before harvest. Assessment of natural enemy activity can be done by inspecting fruit that has fallen from the tree. If parasitism is significant, allowance for this must be made in deciding when to spray. Recently, assessment of moth populations based on the catch of males at sticky traps using pheromones as the bait, has been used. Fruit may also be bagged to exclude the pest. This does not provide complete protection, but fruit colour and quality are improved in both litchi and longan. Because litchis mature earlier than longans, the trees are pruned after harvest to remove any remaining fruit that might harbour pupae of the pest and so present a threat to the longan crop (S. Dolsopon, Bangkok, 1999, personal communication).

The actual sex pheromone of *C. sinensis* has not been determined, but that of *C. cramerella* has been used as a lure in male traps for *C. sinensis* in Taiwan. More moths are caught in traps that are placed inside the canopy than in traps outside, and trapping can be used to monitor the abundance of the pest. The application of cypermethrin, deltamethrin, carbofuran or fenthion during the early fruit set period is recommended to prevent excessive damage during fruit maturation. Moths can be excluded by enclosing the fruit panicles in nylon mesh bags, but the economics of this procedure may preclude its widespread use. If parasitoids are not active, orchard sanitation through the removal and destruction of fallen, infested fruit helps to reduce the build-up of moths.

Conopomorpha litchiella *Bradley*

DISTRIBUTION AND BIOLOGY All stages of the litchi leafminer, *Conopomorpha litchiella* Bradley, are similar to those of the fruit borer. The female moth lays the eggs on a new shoot. The small, light-yellow egg hatches 3–5 days later. The newly hatched larva is creamy white and it bores into the shoot. Larvae also mine in the leaf blade. Mature larvae prefer to feed on the midrib and veins of young leaves. The larval stage occupies 10–14 days after which the larva pupates on mature leaves. The pupal stage lasts 7–10 days and moths live for about a week.

NATURAL ENEMIES The same species of parasitoids that attack the fruit borer effectively suppress *C. litchiella* in Thailand (Dolsopon *et al.*, 1997a).

DAMAGE The leafminer is attracted to leaf flushes of litchi and longan, especially during the rainy season from June to October in Thailand. Litchi is more heavily infested than longan. Damage caused to the shoots causes them to wilt. Dolsopon *et al.* (1997b) found that 75% of litchi shoots were destroyed compared to 50% in longan. In addition, the number of infested shoots was highly correlated with the number of pupae found on leaves.

MONITORING AND CONTROL Bearing trees should be inspected for the presence of leafminer during major flushes in both litchi and longan. Eggs laid on the shoots can be detected by using a ×20 hand lens. Damaged leaves should also be randomly checked. Sprays, if they are necessary, must be applied to the very young leaves of any important flushes, especially the second flush. If 30–40% of larvae are parasitized, spraying is not recommended. Young, non-bearing trees do not need to be sprayed. This allows the parasitoids to multiply. Sticky traps that are used for fruit borer can be used to help monitor for the presence of adult moths, particularly during major flushes in June–October.

Fruit borers (Lepidoptera: Tortricidae)

DISTRIBUTION AND BIOLOGY The tortricid genus, *Cryptophlebia*, contains several species that attack litchis and longans throughout the world. In South Africa and the Indian Ocean islands of Madagascar, Mauritius, Seychelles and Réunion, *Cryptophlebia peltastica* Meyr. is a major pest of litchis (Quilici *et al.*, 1988; De Villiers and Stander, 1989; Quilici, 1996). *Cryptophlebia leucotreta* Meyr. also attacks litchis in South Africa but the frequency of attack is insignificant compared to that of *C. peltastica* (Newton and Crause, 1990). *Cryptophlebia bactrachopa* Meyr. is recorded attacking litchis in Malawi (La Croix and Thindwa, 1986).

The insect originally referred to as *Argyroploce illepida* Butler (= *Cryptophlebia carpophaga* Walsingham) as a pest of litchis in India (Butani, 1977), is actually *Cryptophlebia ombrodelta* Lower (Bradley, 1953). This species also occurs in Thailand, China, Japan, Taiwan, Hawaii and Australia (CIE, 1976), but only in the latter two areas is it regarded as a significant pest of litchis and longans. *Cryptophlebia illepida* is found only in Hawaii where it is a pest of litchi and macadamia (Jones, 1994). All of these species utilize similar alternative host plants such as *Bauhinia galpinii* and other *Bauhinia* spp., *Delonix regia* (flamboyant), *Caesalpinia pulcherima*, *Cassia* spp., *Macadamia integrifolia* and *Acacia* spp. (Ironside, 1974; Newton and Crause, 1990).

Namba (1957) detailed the biology of *C. illepida*, and that of *C. ombrodelta* was described by Ironside (1974). The creamy white eggs are oval and flat, have a reticulate surface and are about 1.0 × 0.8 mm in size. They may be laid singly or in groups of up to 15, on the fruit. As incubation progresses, the eggs turn pink and then red, before finally appearing black as the head capsule of the fully developed larva becomes visible. Under normal summer temperatures in Queensland, eggs hatch in 4–5 days. On hatching, the larva commences to feed on the skin of the fruit, and it then tunnels towards the seed. The mature larva is pinkish in colour with a brown head capsule, and is about 20 mm long. The prepupal period occupies 4 days during which the larva chews an exit hole for its emergence as a moth. When the moth is ready to emerge, the pupa wriggles through the prepared opening to expose about two-thirds of its length and facilitate the moth's emergence. Female moths live for 7–10 days and may lay up to 250 eggs.

DAMAGE All *Cryptophlebia* species that are pests of litchis and longans are fruit borers. The eggs are laid on the surface of the fruit. If the fruit are immature, the young larva will bore directly into the seed, which is completely eaten. One larva may damage two or even three fruit, if very small fruit are attacked. However, there is a preference for more mature fruit with larger seeds, especially as the fruit begins to colour (Rogers and Blair, 1981; Newton and Crause, 1990). At this stage, newly emerged larvae generally tunnel in the skin before attempting to get to the seed. If the route taken to the seed commences adjacent to the peduncle, then the larva usually survives, but if it attempts to bore through the mature flesh, it usually drowns in the juice, which seeps into the wound caused by its feeding. In either case, the fruit will not be marketable. In addition, the juice that oozes from borer wounds may stain neighbouring fruit on a panicle and cause further losses (Waite, 1992a). This is particularly so in cultivars such as Bengal, which produces large panicles bearing many fruit in a tight cluster.

BIOLOGICAL CONTROL The various species of *Cryptophlebia* are known to be attacked by complexes of egg, larval and pupal parasitoids that are not always capable of preventing economic damage in orchard crops. In India the egg parasitoid, *Trichogrammatoidea fulva* Nagaraja, was considered to be the most important of a number of natural enemies of *C. ombrodelta*, parasitizing up to 68% of eggs (Anonymous, 1976). Larval parasitoids recorded in the same study were *Goniozus* sp. nr. *triangulus* Keiff. (Bethylidae), *Agathis* sp. (Braconidae), *Apanteles* sp. (*ater* group) (Braconidae) and *Phanerotoma dentata* Panz. (Braconidae). None of these attained high levels of parasitism. *T. fulva* was introduced into Mauritius from India in 1973 to control *C. leucotreta* on litchi, but its establishment has never been confirmed (Quilici *et al.*, 1988).

C. ombrodelta in Queensland is parasitized by *Apanteles briareus* Nixon, *Apanteles* sp. (*ater* group), *Apanteles* sp. (*myoecenta* group), *Bracon* sp. (Braconidae), *Brachymeria pomonae* (Cameron) (Chalcididae), *Euderus* sp. (Eulophidae), *Echthromorpha insidiator* Smith. F., *Gotra bimaculata* Cheeseman (Ichneumonidae) and *Thelariosoma* sp. (Tachinidae). The reduviid, *Pristhesancus maculipennis* Stål, also occasionally preys on larvae (Ironside, 1974; Sinclair, 1979). The egg parasitoid, *Trichogrammatoidea bactrae* Nagaraja, has been recorded from eggs of *C. ombrodelta* collected from litchis, longans, macadamias and *Bauhina galpinii* in southeast Queensland and in 1998/99, *T. cryptophlebiae* Nagaraja was found for the first time parasitizing a significant proportion of *C. ombrodelta* eggs laid on these same hosts (G.K. Waite, 1999, unpublished) (Plate 83).

The eulophid ectoparasitoid, *Elachertus* sp. nr. *lateralis*, parasitizes *C. ombrodelta* larvae tunnelling in the seed pods of *Bauhinia purpurea* in Guangzhou, China, where parasitism levels of up to 67% have been recorded (G.K. Waite, 1988, unpublished). *C. ombrodelta* has been considered an important pest of litchis in Guangdong Province (Anonymous, 1978) but, in reality, its impact seems to be minor. This is in contrast to Queensland where, although the insect is known as the macadamia nut borer, it is a more serious problem in litchis and longans. This consideration led to the importation of *Elachertus* sp. into Queensland and its release in 1993. The parasitoid has been recovered from the field but no measure of its effect is available.

In South Africa, the egg parasitoid *T. cryptophlebiae* has been found to be particularly effective and it has been used extensively for inundative release into citrus for the control of *C. leucotreta* (Newton and Odendaal, 1990). It has been recorded parasitizing 63% of the eggs of *C. peltastica* in litchis, comparable to the result achieved against *C. leucotreta* in citrus under normal conditions (Newton and Crause, 1990).

In Hawaii, numerous parasitoids attacking *C. illepida* were identified by Namba (1957). These were *Cremastus hymeniae* Vierick, *Horogenes chilonus* (Cushman), *Coccygomimus sanguipes* (Cresson), *C. punicipes* (Cresson), *Pristomerus hawaiiensis* Perkins (Ichneumonidae), *Brachymeria obscurata* (Walker) (Chalcididae), *Bracon mellitor* Say (Braconidae), *Omphale metallicus* Ashmead (Eulophidae), *Perisierola emigrata* Rohwer (Bethylidae) and *Sierola cryptophlebiae* Fullaway (Dryinidae).

MONITORING AND CONTROL Newton and Crause (1990) found that there was no difference in the distribution of eggs laid on litchi fruit by *C. peltastica* and *C. leucotreta* with respect to height on the tree, or aspect. Relatively low oviposition rates were recorded on immature fruit, but the number of eggs laid increased rapidly as fruit approached maturity. Monitoring for infestations should concentrate on maturing fruit, with samples taken from any part of the tree. Similar recommendations apply for *C. ombrodelta* in Queensland litchis (Waite, 1992a). Jones (1995) found that the presence of eggs, either hatched or unhatched, was not a good predictor of larval infestation of litchi fruit in Hawaii. Only 12% of fruit with *Cryptophlebia* spp. eggs, exhibited larval damage. A sequential sampling procedure was developed, which indicated that a maximum 220 fruit sample was required to make a decision with respect to chemical control, but if damage was greater than 10%, the sample could be reduced to fewer than 50 fruit.

De Villiers (1992a) recommended the application of the insect growth regulator triflumuron as a single, full cover spray 40 days before harvest, or two sprays of teflubenzuron a fortnight apart, commencing when the fruit is 10 mm in diameter. An alternative recommendation is to cover the fruit panicles with paper bags to exclude the pest. This also improves fruit colour and overall quality. In Queensland, carbaryl and azinphos-methyl have been used with varying success. Several sprays commencing at fruit colouring have generally been applied, usually on a calendar basis rather than as a result of monitoring for the presence of eggs (Waite, 1992b). Newer insecticides including the insect growth regulator tebufenozide (Mimic®) promise to provide better control with less disruption of natural enemies.

Fruit piercing moths (Lepidoptera: Noctuidae)

DISTRIBUTION AND BIOLOGY Fruit piercing moths attack a range of fruit throughout South-East Asia, the South Pacific and Australia (Banziger, 1982; Fay and Halfpapp, 1993). Fruits such as mango, carambola, citrus, mangosteen and stonefruits, as well as litchi and longan, are damaged by a complex of moth species.

The larvae of fruit piercing moths develop on a variety of host plants. In the Pacific Islands, *Eudocima* (*Othreis*) *fullonia* (Clerck) breeds on *Erythrina* spp., commonly known as coral trees. However, in Australia and Thailand, this species breeds only on vines of the family *Menispermaceae* (*Legnephora, Stephania, Fawcettia, Tinospora, Carronia, Sarcopetalum, Pleogyne* and *Hypserpa* spp.), even though *Erythrina* spp. are common throughout its range in these countries (Sands and Schotz, 1991). *Eudocima salaminia* (Cramer) larvae utilize *Stephania japonica* Miers and its subspecies as host plants. Fay and Halfpapp (1999) recorded *E. fullonia, Eudocima salaminia* (Cramer) and *Eudocima jordani* (Holland) as primary feeders on litchi fruit on the wet tropical coast of Queensland. Populations of fruit piercing moths that appear in the subtropical southeast of the state in late summer and autumn, originate in breeding areas of the tropical north, with seasonal migration bringing them south (Sands and Brancatini, 1995).

DAMAGE Unlike most lepidopterous pests, of which the larva is the damaging stage, in this case it is the adult that causes the damage through its feeding on the fruit. The moths possess a proboscis that is able to drill a neat hole in the skin of a range of fruit, including litchi and longan. Considering the thickness of the skin of some cultivars, this is no easy task. The moths suck juice from the fruit, leaving an opaque area of flesh that is drier than the surrounding tissue. A day or so after feeding by a moth the flesh of the fruit commences to ferment. The fermentation is initiated by yeasts and bacteria that are introduced to the wound on the proboscis of the moth (Sands and Schotz, 1989). *Drosophila* spp., which are attracted to the fermenting fruit, hasten the process of fruit deterioration. Within a few days a frothy exudate begins to seep from the hole and this may stain undamaged fruit on the panicle, causing further losses.

BIOLOGICAL CONTROL Several egg parasitoids of *E. fullonia* have been identified in Papua New Guinea and the western Pacific region. Those originating in Papua New Guinea were thought to provide a good opportunity to achieve biological control of the pest in the Pacific Islands. However, only one of the introduced species, *Telenomus* sp., established well and contributed significantly to egg mortality. Two parasitoids from Papua New Guinea, *Ooencytus papilionis* Ashmead (Encyrtidae) and *Telenomus lucullus* Nixon (Scelionidae), have been identified as potentially useful biological control agents for fruit piercing moths in Australia. They have not yet been introduced because of fears that they will attack non-target species (Sands, 1996).

PEST MONITORING AND CONTROL In Australia and Thailand, farmers go to their orchards at night with spotlights and attempt to manually remove as many moths as possible. This may be achieved by catching the moths as they feed on the fruit, or by catching them in a net or swatting them with a tennis racquet as they fly from tree to tree. When moths are numerous, this becomes a futile exercise.

Australian farmers also make traps by draping shade cloth loosely over a frame of wire and baiting it with fermenting fruit such as citrus and bananas. The moths are attracted to and feed on the fermenting fruit, and become entangled in the folds of shade cloth when they fly off. They are then killed manually each morning. Large numbers of these traps are required to protect an orchard and, even then, substantial damage is sustained. In Thailand, ripe fruit of banana and pineapple are dipped in insecticide and hung in the trees to poison the feeding moths. In some countries, panicles of fruit are covered with paper bags to protect them from a range of pests. These are effective against fruit piercing moths and are recommended in Thailand.

In recent times, parrots and fruit bats have become a severe problem for litchi and longan growers in eastern Australia. To prevent total loss of crops from these pests, many growers have erected protective nets. These have ranged from nets draped over individual trees, to tunnel nets covering entire rows, or complete enclosures that surround the entire orchard. Although these have a high capital cost, they last for years and the entire cost is usually saved through the prevention of damage in one season. These same nets exclude fruit piercing moths and, if the mesh is fine enough, may also keep out some other pests. Nets that are merely draped over trees offer poor protection, since the fruit upon which they rest is still accessible to the various pests.

Loopers (Lepidoptera: Noctuidae)

DISTRIBUTION AND BIOLOGY *Oxyodes scrobiculata* F. and *Oxyodes tricolor* Guen. occupy similar niches in Thailand and Australia, respectively. In Australia, *O. tricolor* attacks litchis in southeast Queensland but is not a pest in north Queensland. The castor oil looper, *Achaea janata* (L.), is a voracious leaf feeder in Queensland and often infests trees in large numbers at the same time as *O. tricolor*.

The white eggs of *O. scrobiculata* are laid on the leaves of litchi, longan and rambutan, with litchi being preferred. The larvae grow to 40–50 mm in length and may be greenish yellow to brown when mature. The larval stage takes 14–17 days to complete and the small, light-brown moths emerge about 10 days after pupation. The main period of activity is during the postharvest vegetative flush.

DAMAGE The caterpillars feed on the foliage of litchi trees and can cause severe defoliation. Although they will eat leaves of any age, they prefer the younger leaves. The whole leaf is consumed, leaving only

Icerya sp., resulting in an outbreak of that pest (Vayssieres, 1997). Systemic insecticides that will control the larvae in the soil without harming the beneficial species are being tested.

Scarab beetles (Coleoptera: Scarabaeidae)

DISTRIBUTION AND BIOLOGY The elephant beetle, *Xylotrupes gideon* (Linnaeus), damages litchi fruit in all production areas of Australia. The larvae develop in the soil or in composting organic matter where they feed on plant roots and the humus material in the compost. They grow to a length of about 50–70 mm and pupate in the soil or compost. The large, heavily sclerotized and sexually dimorphic adults emerge in spring.

DAMAGE The beetles breed in litchi orchards or close by, and are attracted to the trees just prior to harvest by ripe fruit that has split or has been damaged by other pests, especially by parrots or fruit bats. Once they have infested a tree and start feeding on the damaged fruit, they inevitably start to feed on sound fruit and may inflict severe damage.

BIOLOGICAL CONTROL Natural enemies of the beetles are not known.

MONITORING AND CONTROL Regular inspection of orchards when the fruit is ripening is necessary to detect the presence of the beetles. They are tolerant of many of the routine sprays applied for other pests and high dosage rates of carbaryl or chlorpyrifos are required to kill them. Alternatively, they can be manually removed by knocking them into buckets with a stick, but in Australia at least, this is costly because of the labour involved.

Leaf feeding beetles (Coleoptera: Chrysomelidae)

DISTRIBUTION AND BIOLOGY The small black or brown leaf-eating beetles belonging to the genus *Rhyparida* often attack the new leaf flush of litchis and longans in north Queensland. The most serious problem is caused by *Rhyparida discopunctulata* Blackburn, which emerges in swarms after spring rains and proceeds to strip the leaves from host trees in their flight path. The red-shouldered leaf beetle, *Monolepta australis* (Jacoby), may do the same, especially in southern Queensland. The beetles lay their eggs in the soil and the larvae feed on the roots of a range of weeds and grasses. Swarms may emerge at any time of the year, but most swarms emerge after spring storms.

DAMAGE Individual beetles are commonly seen on litchi and longan trees, usually attracted to split fruit. Even when they are numerous but not swarming, no damage results. However, once swarms form and a feeding frenzy develops, extensive damage may occur, with whole trees being stripped of their leaves.

PEST MONITORING AND CONTROL Most growers are aware of the potential for leaf-feeding beetle swarms to cause severe and rapid damage and they monitor orchards regularly for the incidence of the pest. Neighbours will also assist one another in this, and if swarms are detected on windbreak trees, they are sprayed with carbaryl or endosulfan before they can move on to the orchard trees (Waite and Pinese, 1989).

Soft scales (Homoptera: Coccidae)

DISTRIBUTION AND BIOLOGY *Pulvinaria* (*Chlorpulvinaria*) *psidii* (Maskell), the green shield scale, infests litchi trees in China, Taiwan, Australia, Florida and India (Plate 84). It reproduces parthenogenetically (Swirski *et al.*, 1997) and has three to four generations a year in Taiwan. In Queensland, crawlers are produced in early spring by adult scales that infest the leaves and twigs. Some of these crawlers move onto developing flower panicles and later onto the young fruit. The life cycle takes about 5–6 weeks (El-Minshawy and Moursi, 1976) and the progeny of these early colonists often cover the surface of the fruit. The female scales are sometimes mistaken for mealybugs by farmers because the egg masses, which are covered in waxy filaments, cover the end of the scale.

Soft brown scale, *Coccus hesperidum* Linnaeus, is an occasional pest of litchis in Queensland where its parasitoids have been disrupted by chemical sprays or where it is protected by ants (Waite, 1986). *Parasaissetia nigra* (Nietner) and *Saissetia coffeae* (Walker) infest litchis in India, but along with *C. psidii* they are not important (Butani, 1977).

Pink wax scale, *Ceroplates rubens* Maskell, occurs in most litchi and longan producing countries but is a pest only on longans in Queensland, Australia. *Ceroplastes ceriferus* (Fabricius), Indian white wax scale, and *Nipaecoccus vastator* (Maskell) are common on litchis in Taiwan, but are only minor pests. In Réunion, only *Icerya seychellarum* West. and a *Pulvinaria* sp. have been recorded on litchis, the former often attended by the ant *Solenopsis* sp. (Vayssieres, 1997).

The lac insect, *Kerria lacca* Kerr, was introduced into Taiwan from Thailand in 1940 for the production of shellac. Modern synthetic substances have made the insect obsolete for that purpose, but it lives on as a pest of many fruit and flower species, among them litchi and longan. Female scales may be ovoviviparous or produce eggs. Two generations occur each year, in December–February and May–June (Hwang and Hsieh, 1981).

DAMAGE *P. psidii* is not considered to be important in India (Butani, 1977). The scales cause no damage as they feed, but when significant populations develop on the fruit, as they often do in Florida (Butcher, 1954), China, Taiwan and Australia (Waite and Elder, 1996), these become unmarketable because of the presence of the scales. The scales also produce honeydew, which supports the growth of sooty mould not only on infested fruit and panicles, but on those below. This discoloration reduces their appeal to consumers and results in downgrading or rejection of the fruit in the market place. *C. hesperidum* causes similar problems.

C. rubens infests the leaves of longan trees, giving rise to heavy films of sooty mould which probably affect photosynthesis. In addition, on heavily infested trees, every fruit is spoiled by sooty mould deposits.

Heavy infestations of *K. lacca* may cause twigs to wilt and die, affecting flowering and fruiting. The heavy deposits of honeydew which are produced encourage the growth of sooty mould (Hsieh and Hwang, 1981).

BIOLOGICAL CONTROL The mealybug ladybird, *Cryptolaemus montrouzieri* Mulsant, and the green lacewing, *Mallada signata* (Shneider), are the most effective predators of *P. psidii* in Queensland, the egg masses being particularly attractive to them. Numerous parasitoids have been introduced to certain areas over the years to control the pest on ornamentals. *Microterys kotinskyi* (Fullaway) (Encyrtidae) is regarded as an important parasitoid of the scale on ornamentals in Burma (Swirski *et al.*, 1997). No significant parasitism has been noted in the scale on litchis in Queensland, but in China, *Anicetus ceroplastis* Ishii parasitizes nymphs and female scales. An unidentified coccinellid and several predatory mites are also reported to feed on the scales there, and during the wet season, an unspecified entomophagous fungus may cause significant mortality (Anonymous, 1978).

Pink wax scale is often well controlled in coastal Queensland by the parasitoid *Anicetus beneficus* Ishii and Yasumatsu, but such control may be patchy. The parasitoid is known to have poor dispersal abilities and infestations of pink wax scale may develop unchecked in longan orchards because infested trees have been isolated from sources of the parasitoid. In addition, sprays applied to control the macadamia nut borer may suppress developing parasitoid populations.

Several parasitoids and predators are mentioned as attacking *I. seychellarum* in Réunion. These include *Rodolia chermesina* Mulsant (Coccinelidae), *Borniochrysa squamosa* (Tjeda) (Chrysopidae), *Aprostocetus* sp. (Eulophidae) and *Cryptochaetum* sp. (Cryptochaetidae) (Vayssieres, 1997). Parasitoids of *K. lacca* in Taiwan include *Eupelmus tachardiae*, *Tachardiaephagus tachardiae*, *Tachardiaephagus* sp., *Tetrastichus purpureus* and *Phycus* sp. Several predators have also been recorded, but even so, natural control is ineffective (Hwang and Hsieh, 1981).

PEST MONITORING AND CONTROL In Queensland, it is recommended that green shield scale populations be monitored just before and during the emergence of flower panicles to determine infestation levels on the twigs. If a significant proportion of twigs is infested with scales, an oil spray should be applied to prevent infestation of the flowers and subsequently, the fruit (Waite and Elder, 1996). Pink wax scales and the sooty mould they produce are easily seen on the leaves of longans, and corrective sprays of low viscosity oil or methidathion should be applied to limit fruit contamination. Proper timing of sprays with insecticides, such as dimethoate and fenthion, at the time when nymphs are being produced, provides good control of *K. lacca* (Hsieh and Hwang, 1983).

Armoured scales (Homoptera: Diaspididae)

DISTRIBUTION AND BIOLOGY The diaspidid scales, *Hemiberlesia lataniae* (Signoret) and *Fiorinia* sp. nr. *nephelii* Maskell, occasionally infest litchis in Queeensland but they seldom cause problems (Waite, 1992a). Similarly, *Fiorinia nephelii, Parlatoria pseudopyri* Kuwana, *P. cinerea* Danne and Hadden and *Aulacaspis* spp. are recorded from the crop in India, but are of no consequence (Butani, 1977).

DAMAGE The hard scales generally infest twigs and if allowed to multiply unchecked they can kill the terminal growth, and in the case of young trees may threaten their vitality.

BIOLOGICAL CONTROL Microhymenoptera undoubtedly suppress hard scale populations in the absence of disruptive chemical sprays, but the identity of those operating on the above species has not been determined, except for *H. lataniae* in Queensland. There, latania scale attacking avocados has been shown to be controlled by a complex of natural enemies which include *Aphytis* sp. *proclia* group and *Encarsia citrina* (Craw.) (Aphelinidae), as well as *Signiphora flavella* Girault and *S. perpauca* Girault (Signiphoridae). *Chrysopa oblatis* Banks (Chrysopidae) and *Rhizobius satellus* Blackburn were also found to feed on the scales (Waite, 1988).

Bugs (Hemiptera: Tessaritomidae)

DISTRIBUTION AND BIOLOGY Several bugs belonging to the family Tessaritomidae attack litchis and longans throughout China, South-East Asia and Australia. *Tessaritoma papillosa* Drury occurs in southern China, Vietnam, Thailand, Burma, the Philippines and India (Anonymous, 1978), although Butani (1977) notes that *Tessaritoma javanica* Thunberg and *Tessaritoma quadrata* Distant are the species found on litchi in India. In Australia, *Lyramorpha rosea* Westw. is known as the litchi stink bug, but it is a rare visitor to litchi orchards and is never a problem.

In China, apart from utilizing litchi and longan as hosts, *T. papillosa* attacks citrus, pomelo, castor oil, pomegranate, eucalyptus, canna, loquat and rose flowers (Anonymous, 1978). The bug has one generation per year and both the adults and nymphs feed on the terminals, flowers and fruit. Adults tend to aggregate and overwinter mostly on litchis and longans, but may also be found on other hosts, in areas out of the wind but with adequate sunshine. In spring the female bugs are especially attracted to trees with many flowers and new terminals, where they mate and lay up to 14 egg masses, each containing about 14 eggs. These are usually attached to the back of leaves. Peak egg laying occurs in late March in Guangdong, but continues throughout the summer until September (Anonymous, 1978). The eggs, which are round and creamy white in colour, turn red just before hatching takes place after about 13 days at 25°C (Anonymous, 1978; Unahawutti, 1990). Newly hatched nymphs are reddish in colour and the individuals from each egg mass remain aggregated for several hours after hatching, before they disperse. They develop through five instars to the light-brown coloured adult stage in about 80 days. When disturbed, or during the heat of the day, first and second instar nymphs may drop to the ground, returning to the tree later when temperatures have cooled. Nymphs are able to withstand periods of up to 12 days without feeding. Both adults and nymphs are able to expel smelly defensive odours when disturbed.

The first nymphs mature in June while there are still old adults in the trees. These old adults may have lived for up to 370 days, and they die during July/August. The new adults do not mate immediately, since their reproductive organs are not mature. They overwinter and recommence the cycle in spring.

DAMAGE In litchis and longans, adults and nymphs feed on terminals, which may be killed, and also on flowers and fruit, causing these to fall. Zhang, D.P. (1997) studied the physiology of stink bug stings in detail, and found that the type of damage caused resulted in more immature fruit than mature fruit falling from the tree. Liu and Lai (1998) claimed that 20–30% of fruit is damaged by litchi stink bug and that chemical control has not given satisfactory results.

BIOLOGICAL CONTROL In Guangdong, the main natural enemies of *T. papillosa* are the egg parasitoids *Encyrtus* (*Ooencyrtus*) sp., *Anastatus* sp. and *Blastophaga* sp. These may parasitize 70–90% of eggs late in the season, with *Encyrtus* sp. being the most effective (Anonymous, 1978). Similar parasitism levels were recorded by Liu and Lai (1998) when parasitized egg cards were hung in trees during March. Liu and Gu (1998) found that in an orchard under integrated pest management, combined parasitism rates by *Anastatus* sp. and *Ooencyrtus* sp. were 41.9–47.1% in June, but only 0–2.8% in an orchard that relied on chemical control. During the 1970s, biological control of the litchi stink bug in Guangdong was initiated using the egg parasitoid, *Anastatus japonicus* Ashmead, the flat venter wasp, after field trials in the late 1960s had demonstrated its value (Huang *et al.*, 1974). Since early season parasitism rates were of the order of only 10% in April when most of the eggs are laid, natural control did not prevent severe damage to the crop. Boosting parasitoid numbers in the orchard at this time of the year through mass releases of wasps was found to increase parasitism levels to 90% and give good control. The wasps have traditionally been reared in the eggs of the castor silkworm, *Samia cynthia ricini* (Drury), but Liu *et al.* (1988) reported that *Anastatus* sp. had been successfully reared on artificial host eggs. Xin and Li (1989) found that the average rate of parasitism of the artificial eggs was 40–44% while emergence was 94–96%. After continuous rearing for three generations in the artificial eggs, the viability of *A. japonicus* was better than when they were reared in the eggs of *Antheraea pernyi* (Guénerin-Méneville).

In Guangxi Province, Zhou and Xian (1994) report that the main parasitoids of *T. papillosa* eggs are *Ooencyrtus* sp. and *Anastatus* sp. and that these species parasitized an average of 66.8% of eggs. Mass releases of *Ooencyrtus* sp. into litchis and longans in Guangxi have given significant levels of control.

In Thailand, native species of egg parasitoids, *Anastatus* sp. nr. *japonicus* and *Ooencyrtus phongi*, operate in a similar manner to their counterparts in China, i.e. low levels of control during the critical fruit production period early in the season, building up to good levels later. Mass rearing of the local parasitoids in the wild silk worm *Philosamia ricini* Hutt. and releasing them early in the season, produced results similar to those achieved in China, with *Anastatus* sp. and *O. phongi* parasitizing 79% and 21% of eggs, respectively (Nanta, 1992).

MONITORING AND CONTROL

Manual control. When cheap communal labour was available in China in the 1970s, manual control was combined with mass releases of the egg parasitoid to minimize damage caused by the litchi stink bug. It was recommended that adults be destroyed during the winter period by shaking or beating trees to dislodge the adult bugs, which could then be collected and destroyed. Alternatively, they could be caught using nets wetted with human or bovine urine, presumably to induce them to release their grasp on the tree. The burnt remains of dead bugs could be used in the same way. Egg rafts, which are easily detected, could also be removed manually and destroyed, with any parasitized eggs determined by their black colour placed in receptacles for rearing of the parasitoids for later release. Nymphs were swept from the trees using straw dipped in human or bovine urine attached to bamboo poles. A glue made from resin and castor oil and painted on to the tree trunks prevented the nymphs from climbing back up (Anonymous, 1978).

Chemical control. Mass releases of egg parasitoids are effective in controlling litchi stink bug and, at least during the time that people's communes were in place in China, were a cheaper form of control than the use of chemicals (Huang *et al.*, 1974). After the break-up of the communes, chemical insecticides seemed to be the favoured method of control for this pest, although biological control using egg parasitoids is once again being considered. Timing of sprays is especially critical for success with the litchi stink bug because it has been found to vary in its susceptibility at different times of the year to trichlorfon, the commonly used chemical, depending on the body fat content and its nature. Maximum resistance was recorded during the winter when fat levels reached 20–25%, and the proportion of unsaturated fat was high. Maximum susceptibility occurred from February until June/July when fat levels were 4–12% and the proportion of unsaturated fat was low. On this basis, one spray was applied in early March, when resistance was low and before the first eggs had been laid. In late April and early May, adult resistance is still low and most of the new generation nymphs are in instars one and two, which are also susceptible. This is a good time to apply a second spray if necessary, before the nymphs develop to the third instar, which is difficult to kill (Anonymous, 1978). Good coverage, especially of old, very large trees, is essential.

Biological control. The annual life cycle of *T. papillosa* is tied to seasonal conditions, and activity commences in spring, coinciding with the production of litchi and longan flower panicles. Therefore, releases of the egg parasitoid *A. japonicus* are timed for this period in both China and Thailand. The parasitoids are reared in the eggs of silkworm moths which are in turn reared on castor oil leaves. About 1500–2000 silkworm eggs are exposed to 600 mated female parasitoids for 2 days. Female parasitoids are parthenogenetically arrhenotokous. To prevent the production of excess males, fresh host eggs are used, the temperature is maintained at 25–30°C, and humidity at 70–80%, and adequate light is provided. Adult parasitoids emerge after about 20 days at 26–28°C. If necessary, in order to accumulate sufficient numbers of parasitoids that emerge at the same time, the host eggs can be stored at 10–12°C for up to 6 months when the parasitoid larvae are in the first or second instar (Huang *et al.*, 1974). For field release, the parasitized eggs are glued on to cards and attached to the trunks and branches of the trees.

Fruitspotting bugs (Hemiptera: Coreidae)

DISTRIBUTION AND BIOLOGY *Amblypelta nitida* Stål, the fruitspotting bug, and *Amblypelta lutescens lutescens* (Distant), the banana spotting bug, are major pests of most of the tropical and subtropical tree crops grown in Queensland (Waite, 1990). *A. nitida* is more common in southern Queensland, while *A. l. lutescens* dominates in north Queensland.

The adult bugs overwinter on citrus or non-crop hosts, which may be native plants or exotic ornamentals. In spring, when the litchi and longan trees start to flower, the bugs commence to move into the orchards. However, they prefer to feed on green fruit, and most fruitspotting bugs migrate into the trees just after fruit set. Orchards located adjacent to the natural rainforest breeding areas of the bugs are particularly susceptible to attack (Waite and Huwer, 1998). Female bugs lay individual eggs, mostly on leaves. These take about 7 days to hatch and the nymphs pass through five instars before maturing after approximately 42 days.

In South Africa, the coconut bug, *Pseudotheraptus wayi* Brown, has become a problem in a variety of crops over the last 20 years (De Villiers, 1990b). It is not currently recognized as a pest of litchis in that country, but its history of adopting new hosts there suggests that it may eventually become a problem in the crop. *Paradasynus longirostris* Hsiao has recently been described as an important pest of longans in Fujian, China. The adults and nymphs attack the fruit (Huang *et al.*, 2000).

DAMAGE *Amblypelta* spp. feed on the developing seed within the fruit and this initiates abscission with the fruit falling a couple of days later (Plate 85). The puncture mark is invisible on the surface of the fruit and the only way to distinguish bug-damaged from naturally shed fruit is to dissect them to detect the typical tan lesion that is produced on the dark brown seed testa. In south Queensland, *A. nitida* was shown to have damaged an average of 95% of fallen fruit on unsprayed litchis over three seasons, and in north Queensland, *A. l. lutescens* had damaged 98% of fallen fruit (Waite, 1990). Comparable data are not available for longans but observations indicate that a similar situation exists, with the susceptible period extending up to harvest (G.K. Waite, 1990, unpublished). In litchis, adults may occasionally feed on mature fruit, but usually only nymphs, which are trapped on the trees because they are flightless, can be found feeding on the fruit after it has commenced to colour.

BIOLOGICAL CONTROL Fruitspotting bugs have few natural predators apart from spiders, particularly those belonging to the family Thomisidae. However, egg parasitoids often have a significant effect. In north Queensland, the egg parasitoids, *Ooencyrtus* sp., *Anastatus* sp. and *Gryon* sp. account for 30–60% of the eggs of *A. l. lutescens* (Fay and Huwer, 1993). In south Queensland, *Anastatus* sp. and *Gryon meridianum* have been found to parasitize eggs of *A. nitida* and *A. l. lutescens* to a similar extent (Waite and Petzl, 1997).

MONITORING AND CONTROL In most crops attacked by fruitspotting bugs, the insects themselves are extremely difficult to detect, being well camouflaged because of their colour and having good sensory perception, which enables them to quickly escape predation. In litchis and longans, where the fruit is borne on the outside of the tree, the bugs can be easily seen. However, reliance on finding the bugs to determine when to spray is often misleading, particularly if monitoring is carried out irregularly. Because damaged fruit fall to the ground, dissection of this fruit is the best way to sample for the presence of bugs. Litchis and longans usually set more fruit than the tree can mature, and a period of natural fruit drop occurs every season. Inspection of the seed of the fruit which has fallen is the only way to determine if that fruit is part of the natural drop or has been damaged by fruitspotting bugs. If 10% or more of the fallen fruit has been damaged by bugs then a spray of endosulfan should be applied. Usually, a maximum of two sprays of endosulfan applied 2 weeks apart during the first 6 weeks after fruit set is sufficient to protect the crop from the bugs. In longans, which are subject to attack over a longer period, further sprays may be necessary but those applied for the control of *C. ombrodelta* also control fruitspotting bugs.

Mites (Acari: Eriophyidae)

DISTRIBUTION AND BIOLOGY The litchi erinose mite, *Aceria litchii* (Keiffer), also known in China as litchi hairy mite, hairy spider, or dog ear mite (Anonymous, 1978), occurs throughout China and Taiwan (Huang *et al.*, 1990), India (Puttarudriah and Channa Basavanna, 1959; Prasad and Singh, 1981), Pakistan (Alam and Wadud, 1963), Hawaii (Nishida and Holdaway, 1945) and Australia (Pinese, 1981; Waite, 1986). It does not occur in litchis in South Africa, Israel, Canary Islands, Mauritius, Madagascar and Réunion.

Female mites lay eggs singly on the leaf surface among the erineum induced by their feeding. The eggs are small (0.032 mm in diameter), spherical and translucent. They hatch in 3–4 days, producing a slow-moving protonymph. This stage is followed by two further nymphal stages or deutonymphs (Plate 86). The adult stage is reached about 13 days after oviposition (Alam and Wadud, 1963) and 13–15 overlapping generations are produced each year in India and China (Prasad and Singh, 1981; Zhang, Z., 1997). The mites are minute, being 0.13 mm in length and pinkish white in colour. All stages possess only four legs. However, they are quite mobile and are able to move easily from old infested leaves to the new leaves of fresh growth flushes, where their feeding stimulates the production of the erineum among which they shelter and feed. Population abundance follows the flushing activity of the trees, with numbers booming during the summer flush, and moderating during winter when lower temperatures slow both tree growth and mite reproduction. Alam (1959) claimed a reduction in activity during the rainy season in Pakistan, while Thakur and Sharma (1990) found that maximum mite numbers were correlated with maximum temperatures, wind velocity and sunshine during May, in Bihar, India.

Trees may be infested with erinose mites from the time they are planted out if they have been produced as marcots from infested parent trees. Otherwise, for infestation to occur, the mites must move directly between touching trees (Prasad and Singh, 1981) or be either physically transported from tree to tree by human activity or other agents, or carried on the wind (Wen *et al.*, 1991). While wind transport may be the most common method of movement, they may be transported between trees via honeybees at flowering (Waite and McAlpine, 1992; Waite, 1999).

The cause of the development of erineum as a result of feeding by *A. litchii* has been questioned by Somchoudhury *et al.* (1989) and Sharma (1991), who proposed that the erineum does not arise from stimulated leaf cells, but is in fact formed by the thalli of the alga, *Cephaleuros viriscens* Kunze, with the alga and mite sharing a symbiotic relationship. On the basis of this, Somchoudhury *et al.* (1989) proposed a name change for *A. litchii* to 'litchi algal mite'. Saha *et al.* (1996) studied the symptoms associated with the presence of *C. virescens* and *A. litchii* at a fundamental leaf level and agreed that the alga was involved in the production of erinose. While *C. virescens* commonly grows on litchi trunks, branches and leaves, where it may be a problem in its own right, it is doubtful whether it is responsible for the erinose associated with *A. litchii*, since chemical applications that kill the mites allow new foliage to develop without erinose symptoms (Waite and Elder, 1996; Gupta *et al.*, 1997). In addition, erinose symptoms are not present on litchis grown in countries that have never recorded the mite, but where the alga is present. The development of erinea and galls on leaves as a result of feeding by eriophyiids is discussed by Westphal and Manson (1996).

DAMAGE The mites attack the new growth, causing a felt-like erineum to be produced on the leaflets. This may form as small blisters, but if the infestation is severe, it may eventually cover the entire leaflet, causing it to curl. Whole terminals may be deformed. The erineum is at first silver-white, changing as it ages to light brown and then dark reddish brown (Nishida and Holdaway, 1945). Very old erinose is almost black. Maximum populations are present in light brown, verging on dark brown, erinose, and Nishida and Holdaway (1945), quoting Misra (1912), state that mites are not present after the erinose has turned dark brown. This is only partly true, and some mites do remain there until a new vegetative flush is produced, unless seasonal conditions or tree health prevent that. In such cases the infestation may die out before new leaves become available for colonization (G.K. Waite, 1990, unpublished; Wen *et al.*, 1991).

Many leaves may fall if infestations become very severe. While established trees can tolerate substantial infestations without detriment to their growth, that of young trees may be restricted. At flowering, if leaves immediately below a flower panicle are infested, that panicle will also be affected, with attack on the florets preventing fruit set or producing malformed fruit (Plate 87). Even after fruit has set and has developed to half the final size, the mites can colonize it, producing erinose on the skin that detracts from its appearance, and which may make it unmarketable as a quality product.

BIOLOGICAL CONTROL Numerous species of mite have been recorded in association with *A. litchii*. Not all have been proven to be predators. In India, Lall and Rahman (1975) reported that *Phytoseius intermedius* Evans and Macfarlane, *Phytoseius* sp., *Typhlodromus fleschneri* Chant and *Cunaxa setirostris* were found in association with *A. litchii*. Somchoudhury *et al.* (1987) added a further six species, namely *Amblyseius largoensis* Muma, *Amblyseius syzygii*, *Amblyseius herbicolus* (Chant), *Typhlodromus sonprayagenis*, *Typhlodromus homalii*, and *Agistemus* sp. Thakur and

Sharma (1989) added *Amblyseius coccineae* Gupta, *Amblyseius finlandicus* Oudemans, *Amblyseius purni* Gupta, and *Amblyseius paraaerialis* Muma.

In east Pakistan (Bangladesh), Alam and Wadud (1963) claimed that along with an unidentified species of predatory mite, two different spider mites of the genus *Tetranychus* preyed on *A. litchii*. This observation was surely erroneous!

Wu *et al.* (1991) recorded the phytoseiid mites found on litchis in China and the predator guilds associated with *A. litchii* in Australia and China were compared by Waite and Gerson (1994). They found that nine phytoseiid mites, *Amblyseius herbicolus* (Chant), *Amblyseius largoensis* (Muma), *Amblyseius barkeri* (Hughes), *Amblyseius nambourensis* Schicha, *Amblyseius neomarkwelli* Schicha, *Phytoseius hawaiiensis* Prasad, *Phytoseius rubiginosae* Schicha, *Okisieus morenoi* Schicha and *Typhlodromus haramotoi* Prasad, and six other mite species belonging to five families (Anystidae, Ascidae, Cheyletidae, Cunaxidae and Stigmaeidae), were associated with litchi erinose mite in Queensland. A cecidomyiid fly larva, *Arthrocnodax* sp., was also a common predator. The limited census carried out in China over a shorter period recorded nine phytoseiid species – *Amblyseius eharai* Amitai and Swirski, A. *herbicolus, A. largoensis, Amblyseius ovalis* (Evans), *Amblyseius cantonensis* Schicha, *Amblyseius okinawanus* Ehara, *P. hawaiiensis, Phytoseius fujianensis* Wu and *Okisieus subtropicus* Ehara. On advice provided by entomologists at the Guangdong Entomological Institute that *A. eharai* was the most effective predator of *A. litchii*, it was imported into Australia in 1993 for host specificity testing. The mite was never released following a general natural reduction in the severity of erinose mite infestations in commercial orchards around that time. This coincided with the promotion of an integrated approach to controlling the complex of litchi pests, and an increased awareness by farmers of the need to reduce toxic pesticide applications for pests such as fruitspotting bugs and the macadamia nut borer (G.K. Waite, 2001, unpublished).

Agistemus exsertus Gonzalez (Stigmaeidae), has been used to control *A. litchii* in Guangdong, Guangxi and Fujian Provinces in China (Ren and Tian, 2000).

MONITORING AND CONTROL Trees should be monitored specifically for erinose mite incidence on a monthly schedule. Additionally, constant vigilance during the conduct of routine orchard operations will assist in the early detection of mite infestations.

Mechanical. During the non-fruiting period, particularly during postharvest pruning, infested branches should be cut off and burned (Anonymous, 1978; Wen *et al.*, 1991; Zhang, Z., 1997).

Chemical. The critical time for treatment with chemical sprays is when the trees are about to flush (Pinese, 1981; Waite, 1992a; Waite and Elder, 1996). Monitoring for the presence of erinose mite on litchi trees is simply a matter of carrying out regular inspections of the foliage to detect erinose symptoms around the time the trees are expected to flush, although not all the trees in an orchard will do so at the same time. The mites themselves are virtually invisible to the naked eye. However, to determine whether leaves carrying erinose are active, they should be picked and left to desiccate overnight. As the leaves dry, the mites move to the surface of the erinose where they are easily visible the next morning with the aid of a hand lens or microscope. Populations in excess of 100,000 per leaflet have been assessed using this method in association with a washing and centrifuge technique (Waite, 1992a). In Queensland, three sprays of dimethoate or wettable sulphur, applied at 2–3 week intervals during terminal emergence and leaf expansion, protect the new flush from infestation by mites migrating upwards from infested leaves below. If this operation is carried out during the postharvest flush, the mite population on the tree will be minimal when the flower panicles emerge. If infested leaves remain below the emerging flower panicle, a similar series of chemical applications should be made to prevent mites moving up and damaging the flowers and fruit. Sprays applied at times other than flushing often give poor control because of the protection provided to the mites by the erineum in which they shelter.

In India, Prasad and Bagle (1981) found that dicofol gave the best control of litchi erinose mite, but that monocrotophos, carbaryl, cyhexatin and chlordimeform also performed satisfactorily. Sharma and Rahman (1982) confirmed that dicofol gave good control, but dimethoate also provided acceptable results. Sprays of wettable sulphur have been recommended for the control of erinose mite in Australia (Waite, 1992b), China (Anonymous, 1978), Hawaii (Nishida and Holdaway, 1945), and Taiwan (Wen *et al.*, 1991). Other chemicals recommended for use in China include dichlorvos, dimethoate (Anonymous, 1978), dicofol, chlorpyrifos, omethoate and isocarbophos (Zhang, Z., 1997).

Longan erinose mite (Acari: Eriophyidae)

DISTRIBUTION AND BIOLOGY The longan erinose mite, *Aceria longana* Boczek and Knihinicki, is recorded as a sporadic but major pest of longans in Thailand (Chai-ai and Visitpanich, 1997). *A. longana* is specific to longan, severely affecting the terminals and flowers. The mite is microscopic, creamy white, and lives among the erineum produced on the leaves, similar to *A. litchii*, or in the terminal buds. Boczek and Knihinicki (1998) describe this species as being very similar to *A. dimocarpi*.

DAMAGE Mites feeding on the growing points and flowers cause malformation and stunting of these plant parts and severely infested trees may stop growing. Leaves may exhibit curling due to erinose development, and damage to buds produces a witches' broom effect. There appears to be some uncertainty as to whether the witches' broom is always caused by the mite, or whether a phytoplasma is also involved, in which case it is suggested that the mites may be vectors of the disease (S. Dolsopon, Thailand, 1999, personal communication). Increased activity of the mite is noticed during hot, dry weather which often prevails during the February–April period in Thailand. Similar symptoms occur on longans in China where a different mite and a phytoplasma have been implicated.

BIOLOGICAL CONTROL No natural enemies are recorded for the mite, but it is assumed that there is a suite of predatory mites associated with it, similar to that found in association with the litchi erinose mite in Queensland and China (Waite and Gerson, 1994).

MONITORING AND CONTROL A monitoring procedure similar to that recommended for litchi erinose mite should be undertaken and severely infested branches removed and destroyed. Wettable sulphur should be applied according to the flushing activity of the trees.

Longan gall mite (Acari: Eriophyidae)

DISTRIBUTION AND BIOLOGY The longan gall mite, *Aceria dimocarpi* (Kuang), is associated with longans in China where it is recorded as causing erineum on leaves (Kuang, 1997) and also witches' broom symptoms (He *et al.*, 2000).

DAMAGE Witches' broom symptoms have been reported from the provinces of Fujian, Guangxi, Guangdong and Hainan as well as from Taiwan, though the cause of these symptoms has only recently been attributed to *A. dimocarpi* by He *et al.* (2000). As well as relating mite presence to witches' broom symptoms they were able to reduce the incidence of those symptoms in affected orchards through a combination of pruning and the application of miticides. Others claim that the symptoms are caused by a phytoplasma infection that can be transmitted by litchi stink bug and longan psylla and via dodder weed (*Cuscuta tempestris*) and in bud wood (Chen *et al.*, 2000). Typically, infested young leaves fail to expand, and roll up, and shoots form compact clusters. Inflorescences are malformed and flowers fail to produce fruit.

BIOLOGICAL CONTROL Natural enemies of the mite have not been recorded but it would be expected that predatory mites would be associated with infestations in unsprayed situations.

MONITORING AND CONTROL Longan trees should be inspected regularly and when witches' broom symptoms are found, an appropriate miticide should be applied. In Guangdong, the application of a mixture of omethoate, dicofol and colloidal sulphur reduced the incidence of affected shoots by up to 92% (He *et al.*, 2000).

Gall flies (Diptera: Cecidomyidae)

DISTRIBUTION AND BIOLOGY The litchi leaf midge, *Dasyneura* sp., is regarded as one of the major pests of litchi in China (Zhang, Z., 1997). *Litchiomyia chinensis* Yang and Luo was described from specimens reared from galls collected on litchi leaves in Guangdong (Yang and Luo, 1999). The larvae of the midges overwinter in the galls that form on the leaves as a result of their feeding. They pupate in the soil and the adult flies emerging in March/April, initiate the first of seven to eight overlapping generations for the year. The midge prefer trees that have dense foliage and where the environment is damp.

DAMAGE The adult midge lays its eggs, often in lines, on young leaves. When the larvae hatch they mine in the leaf, causing 'watery dots' which, as they grow, become galls. The galls turn brown and eventually drop out, giving the leaf a 'shot-hole' appearance and reducing the area of photosynthetic tissue.

BIOLOGICAL CONTROL No natural enemies are mentioned in the literature.

MONITORING AND CONTROL In susceptible orchards, monitoring is not an option and preventive procedures are adopted. As with the litchi erinose mite, infested leaves can be removed after harvest, and destroyed. Later in the spring, 75 kg ha^{-1} of 2.5% methyl parathion can be distributed on the ground under the tree, or isofenphos (0.001%) can be sprayed on the ground just prior to the expected emergence of the adult flies. In autumn, isocarbophos (0.001%) should be sprayed twice over a period of 14 days during the leaf flush (Zhang, Z., 1997).

Fruit flies (Diptera: Tephritidae)

DISTRIBUTION AND BIOLOGY *Ceratitis capitata* (Weidemann) and *Ceratitis* (*Pterandrus*) *rosa* Karsch are recorded from litchi orchards in South Africa and Réunion and *C. capitata*, *Bactrocera dorsalis* Hendel and *Bactrocera cucurbitae* (Coquillett) from Hawaii. In Queensland, *Bactrocera tryoni* (Froggatt) is occasionally found attacking litchis.

Female fruit flies lay their eggs through the skin of the litchi fruit, often utilizing cracks and wounds made by other pests (Vayssieres, 1997). Although the eggs may hatch, the larvae rarely survive (De Villiers, 1990a, 1992b; Vayssieres, 1997) probably because of the amount of juice present in mature litchis, which drowns the larvae.

DAMAGE All of the above-mentioned species are capable of ovipositing through the skin of litchis, although in some cultivars the thickness of the skin may prevent successful oviposition. Only in South Africa do fruit flies appear to be regarded as a problem in litchis, despite the relatively low damage levels recorded. *C. rosa* is thought to be responsible for most of this damage (Grové *et al.*, 1999a,b). The physical damage thus caused, as well as some damage inflicted on the flesh by early larval instars, allows the development of microorganisms that ferment and rot the fruit (De Villiers, 1992b). Most often, the presence of fruit fly eggs and the occasional larva in litchi fruit can be attributed to oviposition through prior damage.

BIOLOGICAL CONTROL Numerous parasitoids, especially *Opius* spp. and *Biosteres* spp., have been recorded attacking the fruit fly species that damage litchis. Despite the activity of such parasitoids, fruit flies continue to flourish and they often require chemical control in susceptible crops.

MONITORING AND CONTROL In South Africa, the use of pheromone-baited traps is recommended for monitoring fruit fly populations around litchi orchards. Control is achieved with bait sprays of protein hydrolysate mixed with trichlorfon or mercaptothion. Alternatively, fruit panicles may be protected with paper bags applied just after the November fruit drop (De Villiers, 1988). Grové *et al.* (1999c) found that quarantine cold treatment was effective in preventing survival of *C. rosa* in litchi fruit. In other countries, no specific action is recommended because the problem is not serious and is not recognized as a quarantine risk. In Queensland, insecticide sprays applied for the control of the macadamia nut borer help to suppress fruit fly populations in the orchard.

Bark borers (Lepidoptera: Metarbelidae)

DISTRIBUTION AND BIOLOGY The dark-brown, slightly hairy larvae of *Salagena* sp. feed on the bark and wood of litchi trees in the northeastern Transvaal and Natal south coast areas in South Africa. The cream-coloured eggs, which are laid on the bark of the tree, take a few weeks to hatch. On hatching, the larvae bore into the wood, forming tunnels up to 70 mm long and 5 mm in diameter, usually in the crotches of branches. They cover the feeding site with a mass of frass which is held together by webbing. Pupation occurs in the tunnel and moths emerge from November to January. The pest also attacks pecan, macadamia, guava, avocado, lavender tree (*Heteropyxis natalensis*), marula (*Sclerocarya birrae*), wild fig (*Ficus* spp.), bush willow (*Conbretum zeyheri* and *C. collinum*), wild plum (*Pappea capensis*) and water berry (*Syzygium cordatum*) (De Villiers, 1983a).

In India, *Indarbela quadrinotata* Walker and *Indarbela tetraonis* Moore are also polyphagous, attacking trees such as aonla (*Emblica officinalis*), ber (*Zizyphus* spp.), citrus, falsa (*Grewia asiatica*), guava, jack fruit, loquat, mango, mulberry and pomegranate as well as litchi. Eggs are laid in April/May and the caterpillars are fully grown by December. Pupation is delayed until March/April and the insect has one generation per year (Butani, 1977). *Lepidarbela dea* Swinhoe attacks litchi trees in Guangdong. Trials over large areas of orchard showed that the application of the entomopathogenic nematode, *S. carpocapse*, gave an average of 94.3% control of the pest (Xu and Xie, 1997).

DAMAGE The larvae of all species feed on the bark and bore in the wood. If bark damage extends right around branches, they are ringbarked and die. The wounds may also allow infection by fungi, which can cause dieback.

BIOLOGICAL CONTROL Two unidentified parasitoids have been recorded from the pupae of *Salagena* sp. and they are considered to assist in the suppression of the pest.

MONITORING AND CONTROL During routine orchard operations, the frass-covered webbing associated with the workings of the borers can be seen on the branches. Unspecified insecticide applications applied to areas of activity, denoted by a covering of fresh frass, are effective against *Salagena* sp. (De Villiers, 1983a). In India, it was recommended that the frass and webbing be cleared away and the holes plugged with cotton wool soaked in petrol, chloroform, formalin, etc. These chemicals were considered to be too expensive, and Shah (1946) proposed the use of hot water injected by syringe. Later, Khurana and Gupta (1972) suggested using dichlorvos, trichlorfon and endosulfan in the same way.

Fruit borers (Lepidoptera: Lycaenidae)

Deudorix spp. are recorded as minor pests of litchis and longans in India (*Deudorix epijarbas* Moore), China, Thailand (*Deudorix epijarbas amatius*) and Australia (*Deudorix epijarbas diovis* Hewitson). The adults are pretty, grey-blue moths with black, orange and white markings and filamentous tails on the hindwings. Single eggs are laid on the fruit and the larva bores inside, completely destroying the flesh and seed. Unlike *Cryptophlebia* spp., whose larvae usually attack only one fruit, *Deudorix* larvae move from fruit to fruit, damaging three or four in the process. A neat round hole is chewed in the skin of the fruit and the larva plugs this with its flattened rear end, as it feeds inside. The larvae also apparently produce a substance attractive to ants, since these insects can often be seen in attendance. In Queensland, longans are more severely attacked than litchis.

Leaf feeders (Lepidoptera: Saturniidae)

The larvae of *Gonimbrasia belina* Westwood, commonly known as mopane worms, were recorded defoliating litchis in South Africa by van den Berg (1995). It completed two generations in a year and was attacked by four egg parasitoids and one tachinid larval parasitoid. Rather than applying an insecticidal spray to control the pest, it was recommended that larvae be removed by hand and destroyed, or used as a food source.

Branch borers (Lepidoptera: Cossidae)

The coffee leopard moth, *Zeuzera coffeae* Nietner, has two generations a year in mainland China and Taiwan. Eggs are laid in groups of 20–30 in crevices in the bark of litchis and longans. On hatching, the larvae bore into the branches, which may be killed. They pupate just below the surface of the bark, and the empty pupal case remains extruding from the exit hole when the moth emerges. The pest is of minor importance, mainly in abandoned orchards.

Bark borers (Lepidoptera: Yponomeutidae)

Comoritis albicapilla Moriuti is a new pest in Guangdong Province, China, where it damages the bark of litchi trees. The larvae are protected by a double layer of webbing that incorporates the larval faeces, making chemical control very difficult. Entomopathogenic nematodes provided variable levels of control. Application during rainy spring weather was most effective (Xu *et al.*, 2000).

Mealybugs (Homoptera: Pseudococcidae)

The citrus mealybug, *Planococcus citri* (Risso), is widely distributed throughout the world on numerous hosts. It sporadically attacks litchis in Taiwan where it may contribute to the production of sooty mould.

Thrips (Thysanoptera)

In India, *Dolichothrips indicus* Hood and *Megalurothrips distalis* Karny, were reported by Ananthakrishnan (1971) to have attacked the flowers and leaves of litchi respectively. The damage caused by these species was not quantified, but phosphamidon and dimethoate were recommended for their control (Butani, 1977).

The tea yellow thrips, *Scirtothrips dorsalis* Hood, infests litchis and longans in China. All life stages feed on the shoots, causing malformation of the leaves in the new flush. The crimped, yellowish leaves have a mosaic appearance, and eventually lose their sheen and may fall. The thrips are most numerous from August to October, especially when seasonal conditions are dry. Sprays of isocarbophos, omethoate or dimethoate are recommended to control the pest, along with orchard management procedures that encourage uniform flushing (Zhang, Z., 1997).

The plague thrips, *Thrips imaginis* Bagnall, often infests litchi flowers in Queensland. As its name implies, it may be present in extraordinary numbers and it appears to feed on the litchi florets. However, experimental data have indicated that the insect has little, if any, effect on fruit set (Waite, 1992a).

Mites (Acari: Tetranychidae and Tarsonemidae)

The tea red spider mite, *Oligonychus coffeae* (Nietner) (Tetranychidae), occasionally infests litchi leaves in Queensland, but is never a problem.

In Queensland, *Polyphagotarsonemus latus* (Banks) (Tarsonemidae), may occasionally damage individual terminals on orchard trees. However, it is most often seen on nursery trees where it can be easily controlled with sprays of dicofol, sulphur or endosulfan.

Beetles (Coleoptera: Bostrychidae and Scolytidae)

Vayssieres (1997) discussed the occurrence of several species of boring beetles in litchis in Réunion. The species involved were not identified. The bostrychids were all secondary invaders of dead branches, but the scolytids were suspected of being able to attack live wood.

Bugs (Hemiptera: Miridae, Lygaeidae, Rhopalidae)

Five species of mirid bugs were found infesting litchi in Réunion, but no damage could be attributed to their presence (Vayssieres, 1997).

In Queensland, the Rutherglen bug, *Nysius vinitor* Bergroth, invades many crops during spring in some years. The bugs migrate from the weed hosts on which they breed as these dry off. Litchis are often infested at flowering or early fruit set, but the true pest status of the bugs is uncertain, despite 74% of fallen fruit up to 15 mm long having been fed upon by *N. vinitor* in an outbreak year. Much of this fruit may have fallen regardless of being damaged, since it coincided with the annual natural fruit thinning (Waite, 1992a).

The rhopalids, *Leptocoris rufomarginata* (Fabricius) and *Leptocoris tagalica* Burm., infest litchi trees in Queensland on a sporadic basis. They have been shown to cause green fruit to fall, and may sometimes be as important as the fruitspotting bugs, *Amblypelta* spp., in this respect (Waite, 1992a). The damage symptoms are characteristic and easily distinguished from those of *Amblypelta* spp. in that *Leptocoris* spp. leave only a 'pin-prick' on the seed compared to the extensive lesion produced by the former. These bugs are generally controlled by sprays applied for the control of fruitspotting bugs, but if necessary, specific sprays of endosulfan may be applied, based on monitoring and damage assessment carried out by dissecting fallen green fruit.

Pollination

Du Toit and Swart (1995) concluded that litchi has a limited ability for self-pollination and that insect pollination of litchi flowers seems to be necessary to ensure optimum fruit set in the crop. In India, 16 species of bees, flies, wasps and other insects were recorded visiting litchi flowers by Singh and Chopra (1998). Most common were the honeybees, *Apis cerana* Fabricius, *Apis dorsata* Fabricius, *Apis mellifera* Linnaeus and *Apis florea* Fabricius, and the syrphid flies *Melanostoma univittatum* Wiedemann and *Episyrphus balteatus* (De Geer). Pollination by these insects resulted in a 387% increase in fruit set and a 505% increase in fruit retention. Of the bees, *A. dorsata* has been considered the most important, although *A. mellifera* is also a regular visitor to litchi flowers and most likely has a significant impact on pollination (Abrol, 1999; Kitroo and Abrol, 1996; Kumar *et al.*, 1996). In Thailand, *A. mellifera* is preferred as a large-scale producer of honey and for the pollination of longans, while *A. cerana* is preferred for small-scale honey production and the pollination of litchis, rambutans and mangoes (Wongsiri and Chen, 1995). *Trigona iridipennis* Smith may also contribute to pollination in both longans and litchis in Thailand and India (Boonithee *et al.*, 1991; Kumar *et al.*, 1996). Eardley and Mansell (1996) concluded that of 38 insect species visiting litchi flowers near Ofcolaco in South Africa, most were ineffectual as pollinators and that honeybees and several species of indigenous bees such as *Plebeina denoiti* (Vachal), *Meliponula erythra junodi* (Friese), *Ctenoceratina moerenhouti* (Vachal), *Ctenoceratina rufigaster* (Cockerell) and *Braunsapis facialis* (Gerstaecker) contribute significantly to litchi pollination. In contrast, litchis at Tzaneen were visited mostly by honeybees and very few other insects.

When litchi flowers are bagged to deny access to pollinating insects, fruit set is reduced. In India, Kumar *et al.* (1996) recorded an average of 1.4 fruits on panicles when insects were excluded, 8.9 on panicles caged with *A. mellifera* and 14.9 on panicles with free access to pollinators. An average of six fruits per panicle were set in South Africa when honeybees were allowed free access to flowers compared with two per panicle when panicles were bagged (Du Toit, 1994).

Integrated Pest Management (IPM) in Litchis and Longans

As with all crops, the ultimate aim in the protection of the litchi and longan crops from insect pests is to implement a viable integrated pest management (IPM) system. This system will have been developed for each country's growing regions, accounting for the variation in geography and latitude that determines the timing of different phenological processes in the trees, and the local insect and mite fauna.

The tactics adopted against each pest depend on how well they fit in with the overall management strategy for an orchard. Some pests may be effectively controlled by natural enemies, and the options considered for the control of other pests must always take into account the possible side effects on these, and the possible induction of pest problems caused by the injudicious use of insecticides. It should be noted, however, that when a damaging pest infests the crop, it should be controlled, but with consideration of the impact of the chosen method on the rest of the orchard ecosystem.

In China, where litchis have been cultivated for thousands of years, the need to control pests to produce quality fruit has been well recognized over the years. There is a strong belief in the concept of biological control combined with sound orchard management practices, especially the pruning out of infested branches or leaves, as the basis for the IPM system. In addition, many studies have been conducted on the effect of orchard floor management on pest incidence. As well as providing a suitable habitat for natural enemies of some pests, suitable ground cover may also provide beneficial effects through modification of the orchard environment (Liang and Huang, 1994; Liu and Tan, 1999). During the years when farm communes existed, such management systems were widely used. However, since the demise of the communes and the wider availability of more effective insecticides and spray application equipment, adherence to the more environmentally friendly approach has waned.

Zhang, Z. (1997) has divided the year into phenological stages so that the currently recommended management strategies can be implemented as the various growth characteristics, which favour particular pest activity, develop. A disease management system is also included. Although various biological controls are detailed in the preamble, the commercial recommendations, listed below, do not include them:

1. Winter flushing period – suppression of leaf flushes at this time of the year not only promotes flowering, but also suppresses some of the insect and mite pests which overwinter there.
2. Spring flush/flower panicle production period – a spray of trichlorfon for the control of litchi stink bug and suppression of erinose mite and leaf gall midge, in combination with chlorbenside for downy blight disease, is applied at this time. Two sprays of the latter chemical may be required during prolonged rainy periods at flowering.
3. Fruitlet period – the pest targets are stem-end borer and litchi stink bug, which are controlled with a mixture of cypermethrin or chlorpyrifos plus trichlorfon. Ridomil-MZ® or Sandofan® may be applied for downy blight control.
4. Maturing fruit period (varies according to the cultivar) – targets are stem-end borer

and downy blight. Further sprays of the above pesticides are applied at the relevant times.

5. Autumn flush period – to protect this flush from stem-end borer, erinose mite, leaf midge and various lepidopterous caterpillars, two sprays, 10–15 days apart, of isocarbophos or acephate are recommended.

Presumably, releases of *Anastatus* sp. could be substituted for the stink bug sprays, provided they are correctly timed and integrated with the stem-end borer sprays.

The use of *A. japonicus* in combination with *O. phongi* has been adopted in Thailand, where the bug is very well controlled on litchis and longans by mass releases of these egg parasitoids at flowering. If neceassary, carbaryl can be applied at fruit set to control excessive numbers of nymphs if egg parasitism has been poor. The subsequent control of *C. sinensis* using permethrin has no effect on the biological control, since the oviposition period for the bugs, and hence the critical protection period for the parasitoids, has passed by the time the small fruit become susceptible to the borer. In Thailand also, pruning to remove infested leaves and fruit, and to restrict tree size so that necessary sprays can be applied effectively, is recommended. Fruit bagging, while it may not always give perfect pest control, provides the bonus of enhanced fruit colour in both litchi and longan.

Waite (1992a) detailed the procedure that should be adopted for implementing IPM in Queensland litchis. Basically, this system follows the phenological cycle of the trees and by monitoring at least once monthly during autumn/winter and weekly during the flower/fruit production period in spring and summer, the more important pests can be detected and timely controls applied. In this way, conservation biological control is employed for pests such as the soft scales and erinose mite, although the latter may still require occasional chemical control. Leaf-eating loopers are tolerated until excessive foliage is consumed and *Bacillus thuringiensis* (Bt) applied if necessary. In the past, carbaryl has been applied up to five times or azinphos-methyl up to three times during the fruit maturation phase, to control the macadamia nut borer. The imminent introduction of tebufenozide promises the opportunity for better control of this pest, with fewer sprays and less disuption to the natural enemy complex.

The complex of pests attacking litchis and longans throughout the world invariably includes at least one species of fruit borer. In the absence of completely effective natural enemies of these pests, some chemical applications will remain a necessary part of the IPM systems implemented for each region. These will generally be the key determinants not only of the eventual fruit yield and its quality, but also of the viability of the whole IPM system.

Acknowledgements

Dr J.F. Vayssieres, Département des productions fruitières et horticoles, CIRAD-FLHOR, Réunion; Dr M. Van den Berg, Citrus and Subtropical Fruits Research Institute, Nelspruit, South Africa; and Supatra Dolsopon, Chiang Rai Research Centre, Chiang Rai, Thailand.

References

Abrol, D.P. (1999) Pollination importance value of bees. *Insect Environment* 5, 140–141.

Alam, M.Z. and Wadud, M.A. (1963) On the biology of litchi mite, *Aceria litchii* Keifer (Eriophyidae: Acarina) in east Pakistan. *Pakistan Journal of Science* 18, 232–240.

Alam, Z. (1959) On the biology and control of litchi mite, *Aceria* sp. (Eriophyidae: Acarina) in east Pakistan. *Scientist Pakistan* 3, 13–18.

Ananthakrishnan, T.N. (1971) Thrips (Thysanoptera) in agriculture, horticulture and forestry – diagnosis, bionomics and control. *Journal of Scientific and Industrial Research* 30, 113–146.

Anonymous (1976) Report of investigations on natural enemies of *Cryptophlebia ombrodelta* (Lower) (Lep.: Tortricidae) in India. *Commonwealth Institute of Entomology Special Report,* Indian Station, Bangalore, 16 pp.

Anonymous (1978) Lychee pest and disease control. In: Academy of Guangdong Agricultural Science (eds) *Monograph of Guangdong Lychee.* Guangdong Science and Technology Press, pp. 119–146 (in Chinese).

Banziger, H. (1982) Fruit-piercing moths (Lep., Noctuidae) in Thailand: a general survey and some new perspectives, *Mitteilungen der Schweizerischen Entomologiscchen Gesellschaft* 55, 213–240.

Boczek, J. and Knihinicki, D. (1998) Studies on eriophyoid mites (Acari: Eriophyoidea). XXVII. *Bulletin of the Polish Academy of Sciences, Biological Sciences* 46, 141–146.

Boonithee, A., Juntawong, N., Pechhacker, H. and Huttinger, E. (1991) Floral visits to select crops by four *Apis* species and *Trigona* sp. in Thailand. In: van Heemert, C. and de Ruijter, A. (eds) *Proceedings of the Sixth International Symposium on Pollination, Tilburg, Netherlands, 1990*, pp. 74–80.

Bradley, J.D. (1953) Some important species of the genus *Cryptophlebia* Walsingham, 1899, with descriptions of three new species (Lepidoptera: Olethreutidae). *Bulletin of Entomological Research* 43, 679–689.

Bradley, J.D. (1986) Identity of the South-East Asian cocoa moth, *Conopomorpha cramerella* (Snellen) (Lepidoptera: Gracillariidae), with descriptions of three allied new species. *Bulletin of Entomological Research* 76, 41–51.

Butani, D.K. (1977) Pests of litchi in India and their control. *Fruits* 32, 269–273.

Butcher, F.G. (1954) Insect problems in litchi production. In: *Florida Litchi Growers' Association Yearbook and Proceedings*. Second Annual Meeting, Winter Haven, Florida, pp. 15–16.

Chai-ai, P. and Visitpanich, J. (1997) Life cycle of four-legged mite, *Aceria longana* (Acarina: Eriophyidae), the major pest of longan. *Journal of Agriculture, Chiang Mai University* 13, 88–89 (in Thai).

Chen, J.Y., Chen, J. and Xu, X. (2000) Advances in the research of longan witches' broom disease. *1st International Symposium on Litchi and Longan, Guangzhou, China*. Program and Abstracts No. E-03, p.74.

Commonwealth Institute of Entomology (CIE) (1976) Pest: *Cryptophlebia ombrodelta* (Lower). *Distribution Maps of Pests*, Map No. 353.

De Villiers, E.A. (1983a) Bark borers in litchi trees. *Litchis H.2 Farming in South Africa*.

De Villiers, E.A. (1983b) The litchi moth. *Litchis H.3 Farming in South Africa*.

De Villiers, E.A. (1988) Fruit fly. In: *The Cultivation of Litchis*. Institute of Tropical and Subtropical Crops, Nelspruit. *Agricultural Research Council Bulletin* No. 425, 56–58.

De Villiers, E.A. (1990a) Die beheer van vrugtevlieë by lietsjies. *South African Litchi Growers' Association Yearbook* 3, 29–30 (in Afrikaans).

De Villiers, E.A. (1990b) Coconut bugs on avocados. *Farming in South Africa, Avocados* H.5, 2 pp.

De Villiers, E.A. (1992a) Control of the litchi moth, *Argyroploce peltastica* Meyr. *South African Litchi Growers' Yearbook* 4, 8.

De Villiers, E.A. (1992b) Fruit fly. In: *The Cultivation of Litchis*. Agricultural Research Council, South Africa. *Bulletin 425*, 56–58.

De Villiers, E.A. and Stander, G.N. (1989) Proefresultate met trilumuron vir die beheer van die lietsjiemot. *South African Litchi Growers' Yearbook* 2, 20–21 (in Afrikaans).

Dolsopon, S., Dasanoda, M. and Warapitirangsri, W. (1997a) Effectiveness of larval parasitoids to control litchi leafminer, *Conopomorpha litchiella* (Lepidoptera: Gracillariidae) in the field. *Journal of Thai Agricultural Research* 15, 35–44 (in Thai).

Dolsopon, S., Suphakamnuad, N. and Dasanoda, M. (1997b) Crop loss assessment caused by litchi fruit borer, *Conopomorpha sinensis* (Lepidoptera: Gracillariidae) and its parasitoids. *Chiangrai Horticultural Research Centre Annual Report*, Horticultural Research Institute, Department of Agriculture, Thailand, pp. 76–91 (in Thai).

Du Toit, A.P. (1994) Pollination of avocados, mangoes and litchis. *Inligtingsbulletin Institut vir Tropiese en Subtropiese Gewasse* 262, 7–8.

Du Toit, A.P. and Swart, D.J. (1995) Pollination and fruit set in Bengal litchi during the 1993 flowering season at Politsi: second report. *Yearbook of the South African Litchi Growers' Association* 7, 31–32.

Eardley, C.D. and Mansell, M.W. (1996) The natural occurrence of insect pollinators in a litchi orchard. *South African Litchi Growers' Association Yearbook* 8, 27–29.

El-Minshawy, A.M. and Moursi, K. (1976) Biological studies on some soft scale-insects (Hom.: Coocidae) attacking guava trees in Egypt. *Zeitschrift für Angewandte Entomologie* 81, 363–371.

Fay, H.A.C. and Halfpapp, K.H. (1993) Non-odorous characteristics of lychee (*Litchi chinensis*) and carambola (*Averrhoa carambola*) pertaining to fruitpiercing moth susceptibility. *Australian Journal of Experimental Agriculture* 33, 327–331.

Fay, H.A.C. and Halfpapp, K.H. (1999) Activity of fruit-piercing moths, *Eudocima* spp. (Lepidoptera: Noctuidae), in north Queensland crops: some effects of fruit type, locality and season. *Australian Journal of Entomology* 38, 16–22.

Fay, H.A.C. and Huwer, R.K. (1993) Egg parasitoids collected from *Amblypelta lutescens lutescens* (Distant) (Hemiptera: Coreidae) in north

Queensland. *Journal of the Entomological Society of Australia* 32, 365–367.

Grové, T., de Beer, M.S., Dreyer, S. and Steyn, W.P. (1999a) Monitoring fruit flies in litchi orchards. *South African Litchi Growers' Association Yearbook* 10, 34–36.

Grové, T., Steyn, W.P. and de Beer, M.S. (1999b) Exposure of different litchi cultivars to Mediterranean fruit fly, *Ceratitis capitata* (Weidemann) and the Natal fruit fly, *Ceratitis rosa* Karsch. *South African Litchi Growers' Association Yearbook* 10, 37–38.

Grové, T., Steyn, W.P. and de Beer, M.S. (1999c) Exposure of Natal fruit fly, *Ceratitis rosa* Karsch, to litchi export conditions. *South African Litchi Growers' Association Yearbook* 10, 39–40.

Gupta, D., Bhatia, R., Sharma, N.K., Chandel, J.S. and Sharma, R. (1997) Incidence and management of red rust of litchi in lower hills of Himachal Pradesh. *Pest Management in Horticultural Ecosystems* 3, 70–74.

He, D., Zeng, M., Zhou, B., Lin, S., Li, J. and Peng, J. (2000) The cause of witches' broom in longan – discovery of a new gall-mite species. *1st International Symposium on Litchi and Longan*. Guangzhou, China. Programme and Abstracts No. E-04, p.75.

Higgins, J.E. (1917) The litchi in Hawaii. *Hawaii Agricultural Experiment Station Bulletin* No. 44.

Ho, D.P., Liang, H.W., Feng, Z.W. and Zhao, X.D. (1990) A study of the biology and control methods of the long horn beetle *Aristobia testudo* (Voet). *Natural Enemies of Insects* 12, 123–128.

Hsieh, F.K. and Hwang, J.S. (1981) The Indian lac insect in Taiwan, once a beneficial insect, now a nuisance. *Chinese Journal of Entomology* 1, 119–120 (in Chinese).

Hsieh, F.K. and Hwang, J.S. (1983) Further studies on the control of the lac insect, *Kerria lacca* (Kerr.). *Plant Protection Bulletin* 25, 31–40 (in Chinese).

Huang, B., Huang, J. and Chen, L. (2000) Studies on *Paradasynus longirostris* Hsiao and its control (Hemiptera: Coreidae). *1st International Symposium on Litchi and Longan*. Guangzhou, China. Abstract E-12.

Huang, C.C., Chang, K.S. and Chu, Y.I. (1994) Damage and population fluctuation of the litchi fruit borer, *Conopomorpha sinensis* Bradley, in Chia-Nan district, Taiwan. *Plant Protection Bulletin Taipei* 36, 85–95 (in Chinese).

Huang, C.Q., Wu, H.Q., Lin, Y.W., Xie, Y.D., Huang, J. and Huang, B.K. (1997) A review of shoot and fruit borers and two species of gracillariid moths attacking litchi and longan. *Wuyu Science Journal* 13, 125–130 (in Chinese).

Huang, M.D., Mai, S.H., Wu, W.N. and Poo, C.L. (1974) The bionomics of *Anastatus* sp. and its utilization for the control of litchi stink bug, *Tessaritoma papillosa* Drury. *Acta Entomologica Sinica* 17, 362–375 (in Chinese).

Huang, T., Huang, K.W. and Horng, I.J. (1990) Two species of eriophyid mites injurious to litchi trees in Taiwan. *Chinese Journal of Entomology* (Special Publication) 3, 57–64.

Hwang, J.S. and Hsieh, F.K. (1981) Bionomics of the lac insect in Taiwan. *Plant Protection Bulletin* 23, 103–115 (in Chinese).

Hwang, J.S. and Hsieh, F.K. (1989) The bionomics of the cocoa pod borer, *Conopomorpha cramerella* (Snellen), in Taiwan. *Plant Protection Bulletin* 31, 387–395.

Ironside, D.A. (1974) Biology of macadamia nut borer (*Cryptophlebia ombrodelta* (Lower)). *Queensland Journal of Agricultural and Animal Sciences* 31, 201–212.

Jones, V.P. (1994) Oviposition patterns of koa seedworm and litchi fruit moth (Lepidoptera: Tortricidae) on macadamia and litchi in Hawaii. *Journal of Economic Entomology* 87, 1278–1284.

Jones, V.P. (1995) Sampling plans for *Cryptophlebia* spp. (Lepidoptera: Tortricidae) attacking macadamia and litchi in Hawaii. *Journal of Ecomomic Entomology* 88, 1337–1342.

Khurana, A.D. and Gupta, O.P. (1972) Bark-eating caterpillars pose a serious threat to fruit trees. *Indian Farmers' Digest* 5, 51–52.

Kitroo, A. and Abrol, D.P. (1996) Studies on pollen carrying capacity and pollination efficiency of honey bees visiting litchi flowers. *Indian Bee Journal* 58, 55–57.

Kuang, H. (1997) Four new species of Eriophyidae (Acari: Eriophyoidea) from China. *Entomotaxonomia* 19, 74–78 (in Chinese).

Kumar, R., Chaudhary, O.P. and Lenin, J.K. (1996) Studies on pollination of litchi (*Nephilium litchii* Camb.) in Dehradunn India. *Indian Bee Journal* 58, 9–12.

La Croix, E.A.S. and Thindwa, H.Z. (1986) Macadamia pests in Malawi. III. The major pests. The biology of bugs and borers. *Tropical Pest Management* 32, 11–20.

Lall, B.S. and Rahman, M.F. (1975) Studies on the bionomics and control of the erinose mite *Eriophyes litchii* Keifer (Acarina: Eriophyidae). *Pesticides* 9, 49–54.

Lall, B.S. and Sharma, D.D. (1978) Studies on the bionomics and control of the cacao moth *Acrocercops cramerella* Snellen (Lepidoptera: Gracillariidae). *Pesticides* 12, 40–42.

Liang, W.G. and Huang, M.D. (1994) Influence of citrus orchard ground cover plants on arthropod communities in china. *Agriculture, Ecosystems and Environment* 50, 29–37.

Liu, D.G. and Tan, B.L. (1999) Studies on the parasitoid community in a complex litchi orchard. *Natural Enemies of Insects* 21, 134–139 (in Chinese).

Liu, X.D. and Lai, C.Q. (1998) Experiment of control of litchi stink bug by using *Anastatus japonicus* Ashmead. *South China Fruits* 27, 31.

Liu, Y.F. and Gu, D.X. (1998) Control of litchi stink bug by egg parasitoids. *Chinese Journal of Biological Control* 14, 145–147 (in Chinese).

Liu, Z.C., Wang, Z.Y., Sun, Y.R., Liu, J.F. and Yang, W.H. (1988) Studies on culturing *Anastatus* sp. a parasitoid of litchi stink bug, with artificial host eggs. *Colloques de l'INRA* No. 43, 353–360.

Menzel, C.M. (1983) The control of floral initiation in lychee: a review. *Scientia Horticulturae* 21, 201–215.

Menzel, C.M. and Simpson, D.R. (1990) Performance and improvement of lychee cultivars: a review. *Plant Varieties Journal* 44, 197–215.

Menzel, C.M. and Simpson, D.R. (1995) Temperatures above 20°C reduce flowering in lychee (*Litchi chinensis* Sonn.). *Journal of Horticultural Science* 70, 981–987.

Misra, C.S. (1912) Litchi leaf curl. *Agricultural Journal of India* 7, 286–293.

Namba, R. (1957) *Cryptophlebia illepida* (Butler) (Lepidoptera: Eucosmidae) and other insect pests of the macadamia nut in Hawaii. *Proceedings of the Hawaiian Entomological Society* XVI, 285–297.

Nanta, P. (1992) In: *Biological Control of Insect Pests.* Biological Control Branch, Entomology and Zoology Division, Department of Agriculture, Bangkok, Thailand, 206 pp. (in Thai).

Newton, P.J. and Crause, C. (1990) Oviposition on *Litchi chinensis* by *Cryptophlebia* species (Lepidoptera: Tortricidae). *Phytophylactica* 22, 365–367.

Newton, P.J. and Odendaal, W.J. (1990) Commercial inundative releases of *Trichogrammatoidea cryptophlebiae* (Hym.: Trichogrammatidae) against *Cryptophlebia leucotreta* (Lep.: Tortricidae) in citrus. *Entomophaga* 35, 545–556.

Nishida, T. and Holdaway, F.G. (1945) The erinose mite of lychee. *Hawaii Agriculture Experiment Station Circular* No. 48.

Partridge, I. (1997) Tropical fruit becomes major industry. *Rural Research* 174, 9–12.

Pinese, B. (1981) Erinose mite – a serious litchi pest. *Queensland Agricultural Journal* 107, 79–81.

Prasad, V.G. and Bagle, B.G. (1981) Evaluation of acaricides for the control of litchi mite, *Aceria litchii* Keifer (Acarina: Eriophyidae). *Pesticides* 1, 22–23.

Prasad, V.G. and Singh, R.K. (1981) Prevalence and control of litchi mite, *Aceria litchii* Keifer in Bihar. *Indian Journal of Entomology* 43, 67–75.

Puttaruddriah, M. and Channa Basavanna, G.P. (1959) A preliminary account of polyphagous mites of Mysore. In: *Proceedings of the First All India Congress on Zoology, Bangalore,* Vol. 2, 530–539.

Quilici, S. (1996) Enquête sur les ravageurs et auxiliaires sur litchi à la Réunion. *Réport au Département des Productions Fruitières et Horticoles, CIRAD-FLHOR, Ile de la Réunion,* 13 pp.

Quilici, S., Verbizier, B., Trahais, B. and Manikom, R. (1988) Note sur les ravageurs du litchi à la Réunion. *Fruits* 43, 459–464.

Ren, S. and Tian, M. (2000) Litchi pests and their natural enemies. *1st International Symposium on Litchi and Longan*. Guangzhou China. Abstract No. E-02.

Rogers, D.J. and Blair, A.D. (1981) Assessment of insect damage to litchi fruit in northern Queensland. *Queensland Journal of Agricultural and Animal Sciences* 38, 191–194.

Saha, K., Somchoudhury, A.K. and Sarkar, P.K. (1996) Structure and ecology of *Cephaleuros virescens* Kunze and its relationship with *Aceria litchii* Keifer (Prostigmata: Acari) in forming litchi erineum. *Journal of Mycopathological Research* 34, 159–171.

Sahoo, A.K. and Maiti, B. (1992) Control of litchi leaf roller. *Environment and Ecology* 10, 553–554.

Sands, D. (1996) Natural enemies and prospects for biological control of fruit piercing moth. In: Welch, T. and Ferguson, J. (eds) *Proceedings of the Fourth National Lychee Seminar Including Longans, Yeppoon, Queensland.* Australian Lychee Growers' Association, pp. 110–117.

Sands, D.P.A and Brancatini, V.A. (1995) Country Report – Australia. *Proceedings of the Workshop on Fruit Piercing Moths.* CSIRO Division of Entomology, Long Pocket Laboratories, Brisbane, Australia, pp. 47–48.

Sands, D.P.A. and Schotz, M. (1989) Advances in research on fruit piercing moths of subtropical Australia. In: Batten, D. (ed.) *Proceedings of the Fourth Australasian Conference on Tree and Nut Crops.* Lismore, New South Wales, Australia, pp. 378–383.

Sands, D.P.A. and Schotz, M. (1991) Ecology of fruit piercing moths in subtropical Australia. In: Magallona, E.D. (ed.) *Proceedings of the 11th International Congress of Plant Protection.* Manila, Philippines, pp. 229–232.

Shah, R. (1946) An effective and inexpensive method for the control of stem borers in fruit trees with special reference to santra trees in C.P. and Berar. *Current Science, Bangalore* 15, 135.

Sharma, D.D. (1983) New record of predatory mites on litchi mite (*Aceria litchii* Keifer) in Bihar. *FAO Plant Protection Bulletin* 31, 36.

Sharma, D.D. (1991) Occurrence of *Cephaleuros virescens*, a new record of leaf-curl in litchi (*Litchi chinensis*). *Indian Journal of Agricultural Sciences* 61, 446–448.

Sharma, D.D. and Agrawal, M.L. (1988) Studies on the biology and immature stages of litchi fruit borer *Conopomorpha cramerella* (Snellen) (Lepidoptera: Gracillariidae). *Journal of Research, Rajendra Agricultural University* 6, 84–87.

Sharma, D.D. and Rahman, M.F. (1982) Control of litchi mite *Aceria litchii* (Keifer) with particular reference to evaluation of pre-bloom and post-bloom application of insecticides. *Entomon* 7, 55–56.

Sinclair, E.R. (1979) Parasites of *Cryptophlebia ombrodelta* (Lower) (Lepidoptera: Tortricidae) in Southeast Queensland. *Journal of the Australian Entomological Society* 18, 329–335.

Singh, H. (1975) *Acrocercops cramerella* Snell (Gracillariidae: Lepidoptera) as a pest of litchi in Uttar Pradesh and its control. *Indian Journal of Horticulture* 32, 152–153.

Singh, R. and Chopra, S.K. (1998) Flower visitors of litchi (*Litchi chinensis* Sonn.) and their role in pollination and fruit production. *Pest Management and Economic Zoology* 6, 1–5.

Somchoudhury, A.K., Sarkar, P.K., Singh, P. and Mukherjee, A.B. (1987) Bioefficacy of some pesticides against *Aceria litchii* (Keifer) (Acari: Eriophyidae). *First National Seminar on Agricultural Acarology, Kalyani*, 48 pp.

Somchoudhury, A.K., Singh, P. and Mukherjee, A.B. (1989) Interrelationship between *Aceria litchii* (Acari: Eriophyidae) and *Cephaleuros virescens*, a parasitic alga in the formation of erineum-like structure on litchi leaf. In: Basavanna, G.P. and Viraktamath, C.A. (eds) *Progress in Acarology* 2, 147–152.

Storey, R.I. and Rogers, D.J. (1980) Lepidopterous pests of the litchi in north Queensland. *Queensland Journal of Agricultural and Animal Sciences* 37, 207–212.

Swirski, E., Ben-Dov, Y. and Wysoki, M. (1997) 3.3.5 Guava. In: Ben-Dov, Y. and Hodgson, C.J. (eds) *Soft Scale Insects – Their Biology, Natural Enemies and Control (7B)*, pp. 255–263.

Tan, S., Wei, J. and Lan, R. (1998) Analysis of the similarity of the structure of the litchi and longan pest communities. *Guanxi Science and Technology of Tropical Crops* 69, 4–10 (in Chinese).

Thakur, A.P. and Sharma, D.D. (1989) New records of predatory mites on mite pests of economic crops in Bihar. *Biojournal* 1, 155–156.

Thakur, A.P. and Sharma, D.D. (1990) Influence of weather factors and predators on the populations of *Aceria litchii* Keifer. *Indian Journal of Plant Protection* 18, 109–112.

Unahawutti, C. (1990) Insect pests of fruit trees. In: *Proceedings of the Fifth Workshop on Insect Pests and Their Control*. Entomology and Zoology Division, Department of Agriculture, Bangkok, Thailand, 121 pp. (in Thai).

van den Berg, M.A. (1995) Life history of the mopane emperor moth *Gonimbrasia belina* (Lepidoptera: Saturniidae). *Inligtingsbulletin Instituut Tropiese en Subtropiese Gewasse* No. 270, pp. 10–12 (in Afrikaans).

Vayssieres, J.F. (1997) Enquête sur les arthropodes ravageurs et auxiliaires du litchi à la Réunion. *Réport au Département des Productions Fruitières et Horticoles*. CIRAD-FLHOR, Ile de la Réunion, 40 pp.

Waite, G.K. (1986) Pests of lychees in Australia. In: Menzel, C.M. and Greer, G.N. (eds) *The Potential of Lychee in Australia. Proceedings of the First National Seminar*. Sunshine Plantation, Nambour, Queensland, Australia, pp. 41–50.

Waite, G.K. (1988) Biological control of latania scale on avocados. *Queensland Journal of Agricultural and Animal Sciences* 45, 165–167.

Waite, G.K. (1990) *Amblypelta* spp. and green fruit drop in lychees. *Tropical Pest Management* 36, 353–355.

Waite, G.K. (1992a) *Pest Management in Lychees. Final Report, DAQ 81A*, Australian Rural Industries Research and Development Corporation, Canberra, Australia, 42 pp.

Waite, G.K. (1992b) Integrated pest management in lychee. In: Fullelove, G.D. (ed.) *Proceedings of the Third National Lychee Seminar*, Bundaberg, Queensland, Australia, pp. 39–43.

Waite, G.K. (1999) New evidence further incriminates honey-bees as vectors of lychee erinose mite *Aceria litchii* (Acari: Eriophyidae). *Experimental and Applied Acarology* 23, 145–147.

Waite, G.K. and Elder, R.J. (1996) Lychee/longan insect pests and their control. In: Welch, T. and Ferguson, J. (eds) *Proceedings of the Fourth National Lychee Seminar Including Longans*. Yeppoon, Queensland. Australian Lychee Growers' Association, pp. 102–109.

Waite, G.K. and Gerson, U. (1994) The predator guild associated with *Aceria litchii* (Acari: Eriophyidae) in Australia and China. *Entomophaga* 39, 275–280.

Waite, G.K. and Huwer, R.K. (1998) Host plants and their role in the ecology of the fruitspotting bugs *Amblypelta nitida* Stål and *Amblypelta lutescens lutescens* (Distant) (Hemiptera: Coreidae). *Australian Journal of Entomology* 37, 340–349.

Waite, G.K. and McAlpine, J.D. (1992) Honey bees as carriers of lychee erinose mite *Eriophyes litchii* (Acari: Eriophyidae). *Experimental and Applied Acarology* 15, 299–302.

Waite, G.K. and Petzl, N. (1997) A search for egg parasites of *Amblypelta* spp. (Hemiptera: Coreidae) in south-east Queensland. In: Menzel, C. and Crosswell, A. (eds) *Maroochy Research Station Report No. 7*, pp. 13–14.

Waite, G.K. and Pinese, B. (1989) Pests. In: Broadley, R.H. (ed.) *Avocado Pests and Disorders*. Queensland Department of Primary Industries, Brisbane, Australia, pp. 8–24.

Wen, H.C., Lee, H.S. and Lin, C.C. (1991) Field studies on litchi erineum mite (*Eriophyes litchii* Keifer) in southern Taiwan. *Journal of Agricultural Research China* 40, 298–304 (in Chinese).

Westphal, E. and Manson, D.C.M. (1996) Feeding effects on host plants: gall formation and other distortions. In: Lindquist, E.E., Sabelis, M.W. and Bruin, J. (eds) *Eriophyoid mites – Their Biology, Natural Enemies and Control*. Elsevier Science, Amsterdam, pp. 231–242.

Wongsiri, S. and Chen, P. (1995) The effects of agricultural development on honey bees in Thailand. *Bee World* 76, 3–5.

Wu, W.N., Lan, W.M. and Liu, Y.H. (1991) Phytoseiid mites on litchis in China and their application. *Natural Enemies of Insects* 13, 82–91 (in Chinese).

Xin, J.C. and Li, L.Y. (1989) Observation on the oviposition behaviour of *Anastatus japonicus* (Ashmead) and the results of its continuous rearing in vitro. *Natural Enemies of Insects* 11, 61–64 (in Chinese).

Xu, J.L. and Xie, R.C. (1997) Mass application of entomogenous nematode against litchi stem borer. *Acta Phytophylactica Sinica* 24, 150–154 (in Chinese).

Xu, J.L., Han, R.H., Liy, X.L., Cao, L. and Yang, P. (1995) The application of codling moth nematode against the larvae of the litchi longicorn beetle. *Acta Phytophylactica Sinica* 22, 12–16 (in Chinese).

Xu, J.L., Yang, P. and Xie, R.C. (2000) Studies on the applications of some entomopathogenic nematodes against litchi yponomeutid *Comoritis albicapilla* Moriuti. *Acta Phytophylactica Sinica* 27, 27–31 (in Chinese).

Yang, C.K. and Luo, Q.H. (1999) A new genus and species of gall midge (Diptera: Cecidomyiidae) infesting litchi from China. *Entomotaxonomia* 21, 129–132 (in Chinese).

Yao, Zheng-Wei and Liu, Siu-King (1990) Two gracillariid insect pests attacking litchi and longan. *Acta Entomologica Sinica* 33, 207–211 (in Chinese).

Zhang, D.P. (1997) A study of the pericarp of litchi fruit and the injury caused by litchi stinkbug, *Tessaritoma papillosa* (Hem.: Pentatomidae). *Wuyu Science Journal* 13, 198–303.

Zhang, Z. (ed.) (1997) *Litchi Pictorial Narration of Cultivation*. Pomology Research Institute, Guangdong Academy of Agricultural Sciences, 189 pp.

Zhou, Z.H. and Xian, X.Y. (1994) Investigation on the morphology and parasitism of *Ooencyrtus* sp. *Plant Protection* 20, 41–42 (in Chinese).

12 Passion Fruit

Elen L. Aguiar-Menezes,[1] Euripedes B. Menezes,[2]
Paulo Cesar R. Cassino[2] and Marco A. Soares[2]
[1]*Empresa Brasileira de Pesquisa Agropecuária, Centro Nacional de Pesquisa de Agrobiologia, BR 465, Km 7, Caixa Postal 74505, Seropedica, RJ 23890-000 Brazil;*
[2]*Universidade Federal Rural do Rio de Janeiro, Centro Integrado de Manejo de Pragas 'Cincinnato Rory Gonçalves', BR 465, Km 7, Seropedica, RJ 23890-000 Brazil*

Introduction

Passion fruits belong to *Passiflora* L. (family *Passifloraceae*) which has a wide genetic base. While some species are undomesticated, others are cultivated as ornamental plants, for nourishment and for medical purposes. The majority of *Passiflora* species are indigenous to the tropical and subtropical regions of South America; Brazil is the centre of diversity of the *Passifloraceae* (Cunha, 1996; Manica, 1997). Of the 400 known species of *Passiflora*, about 50 or 60 bear edible fruits. The majority of these species are unknown outside their centre of origin (Martin and Nakasone, 1994). A few species are economically important e.g. *Passiflora edulis* botanical form *flavicarpa*, the yellow passion fruit, whose juice and pulp are used extensively as ingredients of beverages, salads, fruit cocktails and desserts (Donadio, 1983). *Passiflora edulis* f. *flavicarpa* Deneger, *P. edulis* Sims. (purple passion fruit) and *P. alata* Dryand (sweet passion fruit) are the main species cultivated in the world. The major producers of passion fruit are found in South America, mainly Brazil, Colombia, Peru and Ecuador (Ruggiero *et al.*, 1996). Commercial plantations of passion fruit are also found in Australia, Hawaii, USA, India, New Guinea, Kenya, South Africa, Sri Lanka and Costa Rica (Kluge, 1998). Other important *Passifloraceae*, such as *Passiflora ligularis* Juss. (granadilla) and *P. quadrangularis* L. (badea, parcha granadina, tumbo) are cultivated in Central America and in the Andean regions of South America (Kluge, 1998).

Commercial production of passion fruits is currently increasing due to industrialization of the processed passion fruit products (Akamine *et al.*, 1954; Pires and São José, 1994). Although the passion fruit crop has great economic potential, its establishment and expansion have been hindered by various problems. For example, a wide host range of diseases, insects and mites attack passion fruit. Some pest species cause significant crop losses, reaching the status of key pests or secondary pests. Another limiting factor is the low sexual self-compatibility. Increased fruit set depends on effective cross-pollination. Therefore, hand cross-pollination is the second most expensive production cost of passion fruit culture. Knowledge of effective pollinating agents might be useful to secure maximum fruit production at lower cost.

Flowering and Fruit Setting

The period of flowering of passion fruit varies among species and among regions. For example, in Hawaii, USA, the passion fruit

©CAB *International* 2002. *Tropical Fruit Pests and Pollinators* (eds J.E. Peña, J.L. Sharp and M. Wysoki)

flowers for 8–9 months a year with two distinct periods of flowering and fruit setting. The first period of flowering occurs in late winter or early spring (April and May), and the fruit matures in midsummer. The second period of flowering occurs in late summer (July and August), and the fruit matures in midwinter (Free, 1993).

According to Akamine *et al.* (1954), there is little or no overlapping of the functional periods of the flowers so that not much crossing takes place between the purple and yellow types. In Hawaii, USA, and Brazil, the flowers of the purple passion fruit open early in the morning, usually around dawn, and close before noon. The flowers of the yellow passion fruit open about noon and close about 2100 or 2200 h (Akamine *et al.*, 1954; Free, 1993; Teixeira, 1994).

In New South Wales, Australia, passion fruit flowers start to open in the night or early morning and start to close at about midday the following day, but the stigmas are fully receptive on the morning of the first day only. Anthers of the most flowers do not dehisce until the afternoon (Cox, 1957).

In Jaboticabal and Botucatu, SP (Brazil), the sweet passion fruit flowers for 12 months a year with two flowering peaks, one in January and February and the other in September and October. Its flowers open at about 0400 to 0500 h and close at 1800 to 2000 h (Vasconcellos, 1991; Ruggiero *et al.*, 1996).

Characteristics of Passion Fruit Flowers and Their Pollination

Recent interest in commercial production of passion fruit has prompted several studies on the pollination ecology. When the flowers first open, the stamens hang down, and the anthers dehisce on the undersides, exposing the pollen, the style remains erect, and there is no stickiness on the stigmatic surfaces. Eventually, the erect styles curve downwards and outwards, and when the process is completed, they are more likely to be touched by insects collecting nectar and pollen. However, for the first hour of flowering, a visiting insect is likely to receive pollen on its body but is unlikely to deposit it on the stigma. When the flower begins to close, the styles return to an upright position. The process of recurvature requires about 1 h. However, the styles of some flowers do not curve downward as much as others, and because there is a greater distance between anther and stigma for a pollinator to bridge, they are less likely to be pollinated. This applies particularly to those flowers whose styles always remain upright, many of which are infertile as female flowers (Akamine *et al.*, 1954; Free, 1993). In Brazil, no fruit set is obtained on flowers of yellow passion fruit pollinated when their styles are upright. On flowers with styles partially curved and totally curved, 13% and 45% fruit set is obtained, respectively (Ruggiero *et al.*, 1976). Studies of Vasconcellos (1991) showed that the percentage of fruit set for sweet passion fruit was 0, 44.19 and 73.47% on flowers with styles upright, partially curved and totally curved, respectively. Under natural conditions, the stigma remains receptive only on the day of flower opening, and the pollen loses its viability after 24 h (Akamine and Girolami, 1959; Ruggiero *et al.*, 1976; Vasconcellos, 1991).

The flowers of passion fruit are fragrant when open. Nectar is secreted in a groove at the base of the gynophore, and the pollen is heavy and sticky. These features, in conjunction with the position of the anthers when the pollen is exposed and the functional position of the stigmas, indicate that flowers of passion fruit are adapted to pollination by insects rather than by wind. Wind is not important in cross-pollination (Akamine *et al.*, 1954; Akamine and Girolami, 1959; Nishida, 1963; Semir and Brown, 1975; Free, 1993), and this was confirmed in studies of caged plants that prevented access to insects; no fruit set occurred although plants flowered profusely (Akamine and Girolami, 1957).

Corbet and Willmer (1980) reported nectar sugar concentration of *P. edulis* f. *flavicarpa* to be about 45–50%, which varies little throughout the day. They calculated that the mean volume of the nectar chambers of yellow passion fruit is 180 μl and that nectar secretion continues throughout the afternoon. The hypothesis in this study is that in order to support large bee pollinators, nectar sugar concentrations will be low, and nectar production rates will be high.

In Hawaii, the principal insects visiting passion fruit flowers include honeybee, *Apis mellifera* L. (Apidae) and carpenter bee, *Xylocopa varipuncta* Patton (Anthophoridae) (Akamine *et al.*, 1954). It is doubtful whether the honeybee is effective for pollinating the flowers because of its small size. However, according to Hammer (1987), the foraging habits of honeybees, not their size, may cause less than expected fruit set. The carpenter bee, on the other hand, is large enough so that, in moving around the flower to obtain nectar, its body brushes along the anthers and stigmas, transferring pollen from one organ to the other (Plate 88).

Approximately, 700 species of bees belong to *Xylocopa* Latreille (Anthophoridae). They are found throughout the tropical regions of the world (Hurd and Moure, 1963; Hurd, 1978) with two generations and activity peaks during the periods of December to March and July to September (Camillo, 1978a; Camillo and Garófalo, 1982), coinciding with the flowering peaks of the passion fruit. Pope (1935) stated that carpenter bee, *X. varipuncta*, certain moths and hummingbirds were large enough to transfer pollen from the stamens to the stigmas of the passion flowers. Nishida (1958, 1963) reported insect species within Diptera, Hymenoptera, Coleoptera, Thysanoptera and Orthoptera visiting flowers of passion fruit in Hawaii, with *X. varipuncta* and hover fly, *Eristalis arvorum* (Syrphidae), being the most abundant species. Nishida (1958, 1963) stated that thrips and midges were too small to transfer the relatively large pollen grains of flowers of passion fruit.

The insects most commonly visiting passion flowers in El Salvador are *Bombus* spp., *Trigona* spp., and *Xylocopa* spp. (Free, 1993). In São Paulo (Brazil), the most common species visiting passion fruit flowers are *Xylocopa* spp., *Epicharis* spp. and *Apis mellifera scutellata* (Nishida, 1963; Ruggiero *et al.*, 1976).

Akamine and Girolami (1959) noticed that hover fly *E. arvorum* and long-horned grasshopper *Conocephalus saltator* (Tettigoniidae) feed on pollen of passion flowers, but considered that their value as potential pollinators outweighed their potential pest status.

In the West Indies, the successful pollinators of passion flowers included three species of hummingbird, and in higher numbers, *Xylocopa mordax* Smith (Corbet and Willmer, 1980). While collecting nectar, *X. mordax* moves around the flower while facing inward, often making at least one complete circuit. Only the tip of the galeae can be inserted past the operculum at the mouth of the nectary and into the nectar groove. Because of the depth of the nectar groove, at least 13 μl of nectar remained inaccessible to the bees. The collected nectar load has about 50–70 μl volume, and 45–49% sugar concentration. *Xylocopa mordax* spends about 8.5 s per flower visit. On sunny days each flower received a mean of four visits in the morning and two in the afternoon; overcast conditions reduced the visits. Besides nectar, *X. mordax* collected pollen on the dorsum when in contact with the fully recurved stigmas in different passion fruit species (Free, 1993). According to Corbet and Willmer (1980), most yellow passion flower pollination by *X. mordax* occurs between 1330 and 1500 h (Fig. 12.1) when the stigmas have curved downward. The daily percentage of flowers pollinated ranged from 25% on overcast days to 94% on bright sunny days. Flowers on lower branches were less likely to receive a visit by *X. mordax* and are less likely to set fruit than flowers on higher branches.

In Brazil, Ruggiero *et al.* (1975, 1976) observed that three species of *Xylocopa* and Africanized honeybees were the most abundant pollinators of passion fruit, but the pollinating efficiency of the honeybee was low compared with *Xylocopa* spp. (3% and 75% set, respectively). *Xylocopa* bees were more efficient pollinating flowers whose styles were totally and partially curved and less efficient on rainy days. Camillo (1978b, 1980) also found that *Xylocopa suspecta* Moure & Camargo is a more effective pollinator of yellow passion flowers than *X. frontalis* (Olivier) in Brazil. The author cited other insects, i.e. *Epicharis rustica* (Friese) (Anthophoridae), *Bombus morio*, *B. atratus*, *Apis mellifera*, *Scaptotrigona postica*, *Geotrigona* sp. (Apidae), and *Oxaea flavescens* Klug (Oxaeidae) visiting flowers of yellow passion fruit. However, *S. postica* and *A. mellifera* usually collect pollen while

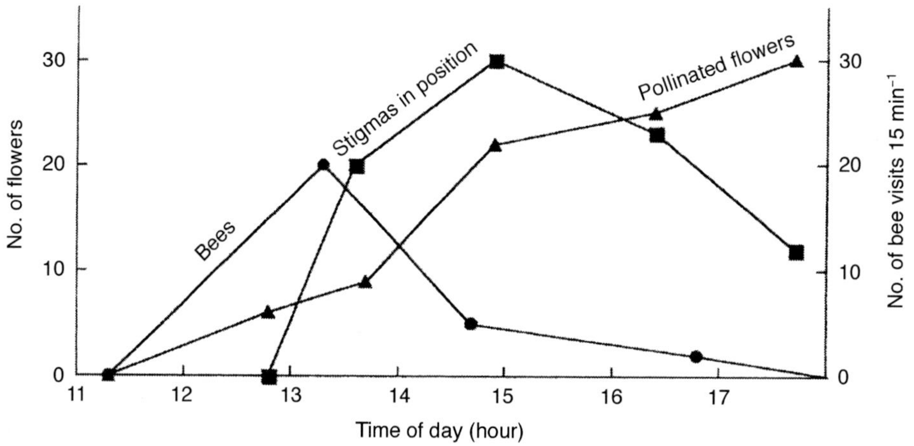

Fig. 12.1. Pollination of *Passiflora edulis flavicarpa*, yellow passion fruit. Changes throughout a day in the number of bee visits, number of flowers with at least one stigma curved downward to another level, and the number of pollinated flowers (those with pollen on at least one stigma) (after Corbet and Willmer, 1980).

E. rustica, *B. morio*, and *O. flavescens* collect nectar.

Sazima and Sazima (1989) also reported that *X. suspecta* and *X. frontalis* were effective pollinators of passion fruit in Ribeirão Preto, SP (Brazil), but *Trigona spinipes* (Fabricius) (Apidae) collected nectar and pollen without pollinating the flowers. When *T. spinipes* was numerous, their visits depleted the flowers of nectar, thereby diminishing foraging by *Xylocopa*. Moreover, *Trigona* attacked and repelled *Xylocopa* when the latter attempted to visit passion fruit flowers, resulting in 6–25% fruit set reduction. The deleterious effect of *Trigona* colonies is likely to be more serious in small plantations.

Hoffmann and Pereira (1996) found the following species of bees visiting flowers of yellow passion fruit in Campos dos Goytacazes, RJ (Brazil): *A. mellifera*, *Xylocopa ordinaria*, *X. frontalis*, *Eulaema nigrita*, and *E. cingulata* (Apidae). Most flower pollination by *Xylocopa* bees occurred between 1400 and 1700 h. Species of *Eulaema* were observed only during the morning.

In Malaysia, Mardan *et al.* (1991) observed that carpenter bee, *Platynopoda latipes*, was the most important pollinator of passion fruit. They suggested that honeybees (*Apis cerana* and *A. dorsata*) were detrimental to fruit set by removing pollen before effective pollination by *P. latipes* could occur.

Three carpenter bees, *Xylocopa mordax* Smith, *X. scutellata* Moure, and *X.* (*Megaxylocopa*) *fimbriata* Fabricius, were the most important pollinators of passion fruit in east and southeast of Lake Maracaibo (Venezuela). *Xylocopa* nests were observed in wooden trellis supporting passion fruit plants. Most flower visits occurred between 1500 and 1800 h (Dominguez-Gil and McPheron, 1992).

Ways of facilitating pollination by *Xylocopa* have been advocated. Nishida (1958) advised that either the area of passion fruit should not exceed the pollinating capacity of the insects present, or the number of pollinators on the crop should be increased. Different ways to increase the population of pollinators have been suggested (Nishida, 1954; Akamine and Girolami, 1959; Cobert and Willmer, 1980). The carpenter bee builds its nest in wood or plant stems, and thus its presence as a pollinating agent can be encouraged by placing wooden posts throughout the passion fruit plantation. The post may be redwood, kukui, or some other suitable soft wood. Abundance of nesting sites might reduce time spent searching for nests and diminish competition between adult females (Akamine *et al.*, 1954; Free, 1993). Studies of Camillo and Garófalo (1982) showed that eucalyptus was the wood preferred by *Xylocopa* bees given a choice between nine types of wood. Hoffmann (1997) recommended the use of posts about 20 cm diameter have been 60–70 cm length in which two holes of 1.5 cm diameter have been made. The first hole is longitudinal, 20 cm deep, and the second one is perpendicular to the first at 15 cm high, allowing them to meet. The first hole is closed by a woody piece or a cork, and the second is used as the nest entrance.

Camillo (1996) increased the pollination of passion fruit by introducing into plantations *Xylocopa* spp. in Holambra, São Paulo (Brazil). Before the placement of *Xylocopa* nests, the natural pollination resulted in 3.2% fruit set. With the introduction of 49 nests of *X. frontalis* and *X. grisescens* into 1.5 ha of passion fruit, the percentage of fruit set increased to 25%.

Carpenter bee populations can be increased by providing seasonal spreads of nectar and pollen sources, thereby reducing competition with other nectar sources while passion fruit is flowering (Akamine *et al.*, 1954). In Brazil, *Hibiscus* spp., *Cassia* spp., *Ipomoea purpurea*, and *Crotalaria juncea* have flowers that are very attractive to carpenter bees (Ruggiero *et al.*, 1996).

Evaluation of the need to either increase numbers of carpenter bees or perform hand pollination is by counting the number of flowers dropped, because this may be caused by lack of fertilization, indicating low population density of insect pollinators. To evaluate whether the pollinators occur in the crop at a suitable level, Ruggiero *et al.* (1996) recommended that three opening flowers per plant should be labelled on a sunny day. For 2–3 ha, this operation must be repeated with > 34 plants, labelling 100 flowers in total. If the area is greater, the quantity of labelled flowers should be increased proportionally. Four days later, the fruit set on the labelled flowers is counted. Of the 100 labelled flowers, Ruggiero *et al.* (1996) observed that 40–50% developed into fruit, meaning that the population density of carpenter bees was adequate. However, values < 30% indicated there was a lack of pollinators and, in this case, flowers must be hand pollinated.

According to Akamine *et al.* (1954), Akamine and Girolami (1957, 1959), and Nishida (1958), honeybees may actually cause unfruitfulness of passion fruit. In areas where honeybees are plentiful, it has been observed that they visit the flowers as soon as flowers have opened. The bees remove the pollen from the anthers and carry it back to the hive. By the time the styles have moved into position where their stigmas can be pollinated by carpenter bees, the pollen may be entirely gone. Unless the carpenter bees have some pollen remaining on their bodies from early visits to the flowers, pollination is entirely precluded and fruit setting fails to occur.

In Hawaii, Nishida (1958) found that in two localities, flowers bagged and pollinated by hand had about the same percentage set as unbagged flowers. However, in two other localities, bagged and hand pollinated flowers had a greater fruit set, indicating that the local insect pollinators were too few in these sites (Fig. 12.2). His results also showed that the fruit set from cross-pollinating, bagged flowers varied from 50 to 100%, depending on the locality. So the maximum fruit set in some localities was limited by factors other than pollinators. The percentage of fruit set from hand pollination and the difference in set between hand and natural pollination also varied within a season. Akamine and Girolami (1959) reported that natural fruit set was less than that achieved by hand pollination. Corbet and Willmer (1980) confirmed that bagged flowers only set fruit following cross-pollination (mean of 77% fruit set). Furthermore, manually cross-pollinating flowers exposed to insect pollinators increased fruit set from 27 to 73%, indicating that natural pollination was inadequate.

Nishida (1963) found that the abundance of honeybees on the crop and the proportion of pollen gathered varied greatly from one month to the next and was probably associated with the presence of competing sources of forage. He observed that sometimes honeybees were so abundant that nearly all pollen was removed soon after the flowers opened. A few honeybees were present on the crop when it was not flowering and in the mornings before the flowers opened; these bees were collecting from the extrafloral nectaries on the leaf petioles. He found that when the density of honeybees present increased to a certain level there was a tendency for the fruit set to decrease. However, there was a positive correlation between the increase in fruit set and the increase in the population of *X. varipuncta* that was present (Fig. 12.3). Nishida (1963) suggested that *X. varipuncta* is a more efficient pollinator than the honeybee, because it is larger, works faster and carries

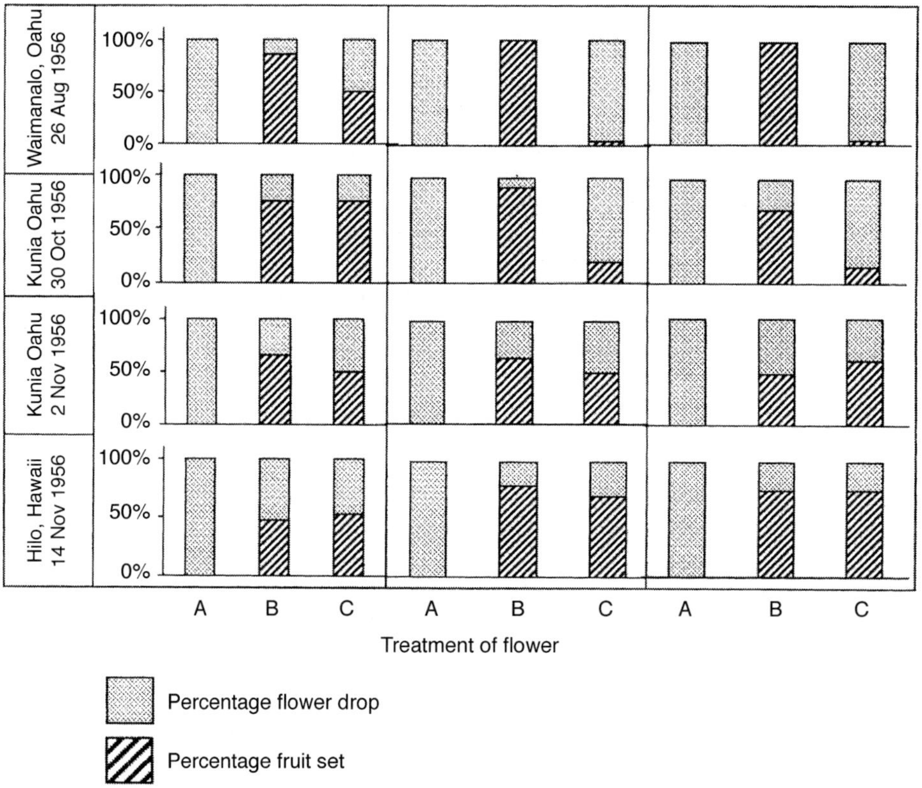

Fig. 12.2. An experiment to determine the efficiency of natural pollination of *Passiflora edulis*, passion fruit, in three sites in each of four localities: A, flowers bagged; B, flowers bagged and hand pollinated; C, flowers not bagged nor hand pollinated (after Nishida, 1958).

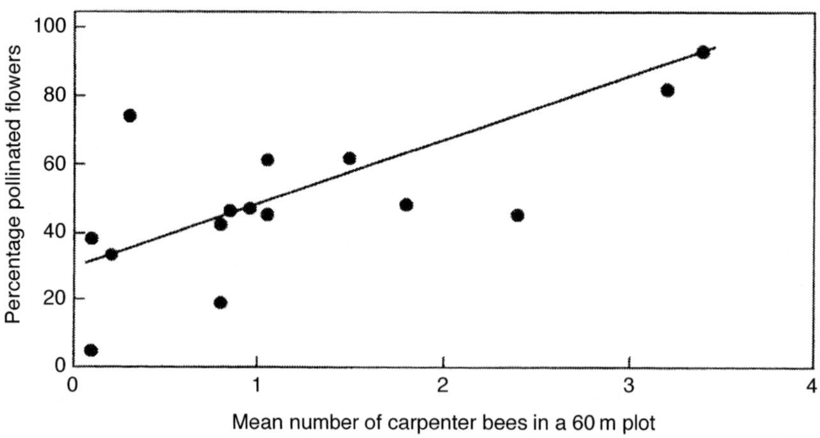

Fig. 12.3. Relationship between the number of *Xylocopa varipuncta*, carpenter bees, in *Passiflora edulis*, passion fruit, fields and percentage of flowers pollinated (after Nishida, 1963).

larger loads. Ruggiero *et al.* (1976) confirmed this by experiments in which caged plants resulted in better fruit set when confined with *Xylocopa* spp. than with honeybees (75% and 45% fruit set, respectively).

On the other hand, Free (1993) suggested that honeybees as pollinators are important because they forage in the lowlands and in the humid uplands, and may be easily manipulated. Cox (1957) reported that honeybees are abundant and effective pollinators of passion fruit in Australia. In Florida, USA, where honeybees are the sole pollinators of *P. edulis*, up to 25% of its flowers produce fruit (Hardin, 1986).

As pointed out by Gilmartin (1958), it has not been determined that cross-pollination between flowers of different clones or varieties is necessary for maximum fruit setting. It is advisable to plant several selected varieties in an orchard to enhance the possibility of cross-pollination and to minimize crop losses which might occur from planting with a variety strain that proved to be highly self-incompatible. Akamine and Girolami (1959) found that any cross that involved variety 'C 39' (Table 12.1) was compatible, but crosses between other varieties were nearly completely incompatible. They suggested that plants of compatible clones, which flower at the same time, should be distributed in the field to ensure the maximum possibility of cross-pollination.

Table 12.1. Reciprocal crosses between varieties of passion fruit, and the percentage of fruit set following cross-pollination. (From Akamine and Girolami, 1959.)

Reciprocal crosses	No. flowers pollinated	Percentage of fruit set
C 39 and C 37	258	92
C 39 and C 77	250	97
C 39 and C 80	167	97
C 37 and C 77	83	2
C 37 and C 80	106	2
C 77 and C 80	80	4

Conclusions

Plants of the *Passifloraceae* depend on cross-pollination to set fruit because their flowers present characteristics that make it difficult for self-pollination, such as presence of stigmas above the level of the anthers and stigma receptivity and low self-fertility. Thus, the passion fruit depends largely on mutualistic relationships, with insects as pollinators. In fact, the flowers of passion fruit are large, attractive, colourful, fragrant, and produce plentiful nectar and pollen that facilitate insect cross-pollination. Most of the studies dealing with pollination of the *Passifloraceae* support the theory that carpenter bees, mainly *Xylocopa* spp., are the most effective pollinators. However, other hymenopterans (e.g. *Apis, Bombus, Epicharis, Oxaea*) and orthopterans, dipterans, etc., visit the flowers of passion fruit, and may pollinate, even though they may be less effective than *Xylocopa*. Thus, as stated by Price (1997) studies of pollination ecology, co-evolution between plant and pollinator, energetic relationships, demographics of plant and pollinator, reproductive strategies, population dynamics and community ecology, are still warranted and needed for *Passifloraceae*.

Pests

Although passion fruit is attacked by several pest species of insects and mites that feed upon all parts of the plant, a limited number of species are clearly of major economic importance. Few have key pest status, while some species are secondary pests because they are sporadic or occur at low population levels, and therefore do not require control strategies. Insect and mite pests that are frequently associated with passion fruit are described below, including their description, life history, behaviour, hosts, damage and control (Santo, 1931; Lordello, 1952b; Correa *et al.*, 1977; ICA, 1987; Dominguez-Gil *et al.*, 1989; Figueiro, 1995; Lima and Veiga, 1995).

Lepidopterous defoliators

Three heliconiine species, *Dione juno juno* Cramer, *Agraulis vanillae vanillae* Linnaeus, and *Eueides isabella huebneri* Ménétries (Nymphalidae), are the most common lepidopterans feeding upon foliage of passion fruit (Dominguez-Gil and McPheron, 1992). *Dione juno juno* is the key pest of passion fruit in Brazil and causes severe damage to the plants (De Bortoli and Busoli, 1987; Gravena, 1987). *D. juno* is distributed in the southern USA, the Antilles, Guyana, Surinam, French Guiana, Trinidad, and from Colombia to Argentina (Toledo, 1991), while *A. vanillae* occurs in South America and over a large part of the southeastern USA (Carter, 1992). *Eueides isabella* is found in Venezuela and south of Brazil (Silva *et al.*, 1968; Brown Júnior and Mielke, 1972; Dominguez-Gil and McPheron, 1992; Boiça Júnior *et al.*, 1993).

DESCRIPTION AND LIFE HISTORY Adults of *D. juno* have orange wings with black borders and venation. The wingspan is about 60 mm. Eggs are laid in groups of 40–70 on the leaf underside, near the border. The 0.9 × 0.6 mm egg is light yellow when first laid, and darkens just before hatching. Eggs hatch in 6–7 days. The larvae pass through four or five instars, requiring 19–27 days to reach full growth. When fully grown, the larva is about 29–35 mm in length, 3–5 mm in width, dark brown, with small yellow spots, and covered with black setae arranged in rows. The larvae are gregarious and, when disturbed, raise their head and thorax, and stand on their pseudolegs. The chrysalis is suspended by the cremaster, is obtect, cream in colour, and about 16–25 mm in length. The pupal stage requires 7–9 days (D´Almeida, 1944; Lordello, 1954; Silva, 1979; Chacón and Rojas, 1984; De Bortoli and Busoli, 1987; Toledo, 1991; Dominguez-Gil and McPheron, 1992; Dominguez-Gil, 1998).

The *A. vanillae* butterfly has red-orange wings, with black markings and venation, and silver spots on the underside (Plate 89). Wingspan varies from 60 to 75 mm (Carter, 1992). The female lays eggs singly on leaves or stems. The eggs are light yellow when recently oviposited, spindle shaped, and about 1 mm in length. The eggs hatch in 3 days (Lordello, 1952a). The newly hatched larva, approximately 3 mm long, is creamy white. The full-grown larva is about 35–40 mm in length, the dorsum has orange, blue and white stripes and six longitudinal lines of branched black spines (Plate 90). The larvae has five instars, and lasts about 17 days (Lordello, 1952a). The chrysalis is suspended by the cremaster, which is usually attached to a branch of the host plant. The chrysalis is about 25 mm in length, sharply angled, and creamy white and lasts about 7 days.

Two-thirds of the forewing of *Eueides isabella huebneri* is dark brown, almost black, with irregular yellow spots, and one-third is orange with black stripes. The hindwings are orange with black borders and a central stripe. The wingspan is 70–80 mm. Females oviposit single eggs on leaves or stems. The eggs are whitish yellow when recently laid, and are darker close to hatching, which occurs 4–7 days after oviposition. The newly hatched larva is 1–3 mm long, white with black head and body hairs. The full-grown larva is 30 mm long. The dorsal surface of its body is black with transversal narrow white stripes, and the dorsal surface of the eighth and ninth abdominal segments is orange. The larvae require five instars. The chrysalis has spines on the thorax and abdomen, differing from the other heliconiines.

HOST PLANTS Caterpillars of *D. juno* feed on all *Passiflora* species, except *P. foetida* (Echeverri *et al.*, 1991; Carter, 1992). According to Boiça Júnior *et al.* (1993), *P. alata*, *P. setacea* and the hybrid *P. alata* × *P. macrocarpa* are more resistant to attack by *D. juno* than *P. edulis*, *P. cincinnata*, *P. caerulea* and the hybrid *P. edulis* × *P. alata*.

INJURY The three species of heliconiine defoliators reduce leaf area, thereby indirectly reducing yield. *Dione juno* usually causes damage that is more serious because of its gregarious behaviour (Fancelli and Mesquita, 1998). During the first instar, the caterpillars scrape the leaf epidermis of young leaves, leaving small holes in the leaves, while older larvae devour both young and older leaves (Lordello, 1954; Chacón and Rojas, 1984). Besides defoliation, the caterpillars may feed on the apical buds, flowers or stems (De Bortoli and Busoli, 1987).

NATURAL ENEMIES Several predators and parasitoids have been reported from these heliconiids. However, these natural enemies are not considered to be effective. For instance, Silva *et al.* (1968) reported *Alcaeorrynchus grandis* (Dallas) and *Apateticus mellipes* Bergroth (Pentatomidae) as predators of *D. juno*. In Argentina, *A. vanillae* was recorded to be parasitized by *Pteromalus caridei* Brèthes (Pteromalidae). Silva (1979) reported *Spilochalcis* spp. (Chalcididae) parasitizing *D. juno*. In the state of Pernambuco (Brazil), Lima and Veiga (1995) found *Spilochalcis* spp., *Polistes* sp. (Vespidae), *Paratrechina longicornis*, *Crematogaster* sp., *Pseudomyrmex gracilis* (Formicidae), and *Forcipomuia* sp. (Ceratopogonidae) as natural enemies of *D. juno*. Ruggiero *et al.* (1996) recorded the hymenopteran wasps, *Polistes* spp. and *Polybia* spp. (Vespidae), as predators of passion fruit heliconiids in Brazil. In Lake Maracaibo (Venezuela), Dominguez-Gil and McPheron (1992) reared two specimens of *Spilochalcis* sp. from field-collected larvae and chrysalis of *A. vanillae* and *E. isabella*.

Lima and Veiga (1993) verified the occurrence of nuclear polyhedrosis virus (NPV) infesting *D. juno* caterpillars in Pernambuco. In Lake Maracaibo, larvae of *D. juno* were also infected by NPV. Once infected, larvae became sluggish, and their cuticles become discoloured and fragile. Chacón and Rojas (1984) estimated that NPV kills 100% of the *D. juno* population in Colombia. A NPV epizootic occasionally reduced populations of *A. vanillae* in plantations located east of Lake Maracaibo where NPV was very abundant during January and February for 3 consecutive years (Dominguez-Gil *et al.*, 1989).

CONTROL In small areas, cultural control during periodic crop inspection includes hand picking and destruction of eggs and caterpillars (Rossetto *et al.*, 1974). However, these methods require considerable time and labour and are often impractical for a large-scale cultivation. In this case, injurious populations of defoliating caterpillars infesting passion fruit must be controlled with insecticidal sprays. Action thresholds have not been defined. Growers spray the foliage, often starting with appearance of the pest, and continue at regular intervals until the crop is harvested. In passion fruit it is very important to protect pollinating insects by timing insecticidal treatments when pollinators are not present in the field. Choosing an insecticide that is selective for the pest and less toxic to pollinators, predators and parasitoids is important in these agroecosystems. *Bacillus thuringiensis* (Bt) Berliner, which effectively controls a variety of caterpillars and has little or no effect on natural enemies, is commonly recommended. Menezes *et al.* (1989) performed a laboratory experiment with different strains of Bt to control *D. juno*. Figueiro (1995) demonstrated that suspensions of *Baculovirus dione* at concentrations of 10, 20, 40 and 80 g of larvae per 500 l of water, were highly pathogenic and efficient at controlling larvae of *D. juno* under laboratory conditions. Studies of Moura *et al.* (2000) on selectivity of insecticides to vespid predators of *D. juno*, showed that the deltamethrin was highly selective to *Polybia scutellaris* and *Polybia fastidiosuscula*, and showed intermediate selectivity to *Protonectarina sylveirae*. Cartap was moderately selective to all three species of predatory wasps. Malathion was selective to *P. sylveirae* and showed intermediate selectivity to *P. fastidiosuscula*.

Coreid bugs

Many species of bugs attack passion fruit and the majority belongs to the Coreidae (leaf-footed bugs). In passion fruit producing areas, three main species of coreids are reported: *Diactor bilineatus* Fabricius, *Leptoglossus* spp. and *Holhymenia* spp. *D. bilineatus* is the most common species, and is known as the passion fruit bug because it feeds only on fruit of *Passiflora* spp. Among the *Holhymenia*, *H. clavigera* (Herbst.) and *H. histrio* (Fabricius) are the most common species attacking passion fruit. The bugs *Leptoglossus*, *L. gonagra* Fabricius and *L. australis* Fabricius, usually cause damage to passion fruit.

D. bilineatus is considered the most important pest of passion fruit in Brazil by Mariconi (1952) and Fancelli and Mesquita (1998). However, Dominguez-Gil and McPheron (1992) consider that *Diactor bilineatus* Fabricius, *Leptoglossus* spp. and *Holhymenia* spp. are the second most significant phytophagous group of pests in passion fruit plantations in the Lake Maracaibo region (Venezuela).

DESCRIPTION AND LIFE HISTORY Adults of *D. bilineatus* are orange on the ventral face of the head, and the dorsal face is dark metallic green with two orange longitudinal lines that continue on the prothoracic tergum and the scutellum, both of which are dark metallic green. The hind legs have the tibia expanded and leaf-like, dark green in colour with some orange markings. Males are about 20 mm in length, and females, 21.5 mm. When disturbed, they walk or make short flights, mainly in cold periods. A female lays a batch of six to nine eggs on the underside of the leaves. They are about 3 mm long, light yellow, bright, elipsoid and flattened in the base. The incubation period for eggs is 13–16 days. The nymphs, which pass through five instars, require about 43–46 days to reach the adult stage (Mariconi, 1952). The first instar nymph has an orange head that turns dark blue in the following instars. The thorax is orange, and a white stripe surrounds the femurs and tibiae. The hind legs are characterized by the expanded leaf-shaped tibia. The abdomen is orange with six pairs of lateral, dark blue expansions from the third to the eighth abdominal segments (Dominguez-Gil, 1998).

The adult body of *Holhymenia* spp. is black with orange spots. The legs are reddish orange. The head, the prothoracic tergum and the scutellum are black with white spots (De Bortoli and Busoli, 1987; Brandão *et al.*, 1991; Dominguez-Gil, 1998).

The adult *L. gonagra* (Plate 91) is about 15–19 mm in length and dark brown in colour. The head colour ranges from dark brown to almost black, with two yellow longitudinal lines. The prothoracic tergum is brown with a yellow transverse line. The antennae are brown with the second, third and distal two-thirds of the forth segments light yellow in colour. The hind tibiae are expanded and leaf-like. The longevity of adults is about 37 days. Eggs are 1.4 mm in length, dark brown with triangular cross-section. The eggs hatch in 8 days. Newly hatched nymphs are reddish with black legs and antennae. The bugs pass through five nymphal instars in about 55 days (Chiavegato, 1963, De Bortoli and Busoli, 1987). The adult passion vine bug, *L. australis*, is elongate, approx. 20 mm in length, and dull black in colour with orange spots on the underside of the body. In Hawaii, passion vine bugs migrate from surrounding scrub to infest passion fruit plantations. Neglect of vines may allow populations of the bug to build up. Feeding usually occurs on flowers or green-mature fruit. Nymphs often cluster on fruit when feeding. Damage to mature fruit is not pronounced; however, young fruit develops dimple-like surface blemishes at the feeding sites (Murray, 1976).

HOST PLANTS Silva *et al.* (1968) listed several species of *Passiflora* as being hosts of *D. bilineatus*. Mariconi (1952) verified that *P. quadrangularis* is seriously damaged by this pest. Besides passion fruit, *H. clavigera* feed on guava (Fancelli and Mesquita, 1998). Silva *et al.* (1968) mentioned *P. edulis* as host of this species. *L. gonagra* feeds on a large number of host plants, including passion fruit, chayote, citrus, tobacco, guava, sunflower, cucumber, grape, pomegranate, São Caetano melon (*Cayaponia espelina*), bixa (*Bixa orellana*), araçazeiro (*Psidium araca*), and *Anisosperma passiflora* (Chiavegato, 1963).

NATURAL ENEMIES Silva *et al.* (1968) reported that *D. bilineatus* eggs were parasitized by *Hadronotus barbiellinii* Lima (Scelionidae). Eggs of *H. clavigera* were reported to be parasitized by *Hexacladia smithii* Ashmead (Encyrtidae) (Silva *et al.*, 1968).

INJURY Both immature and adult bugs injure the crop, piercing stems, leaves, fruits and flowering buds, and sucking plant juices. However, the nymphs prefer to feed on flowering buds and young fruits, usually resulting in excessive dropping. The adults may also attack leaves, stems and fruits at any stage of ripening. If larger fruits are fed upon, they wilt and show a wrinkled surface. Fruits may also develop dimple-like blemishes at the feeding sites on the fruit surface (Murray, 1976). *Leptoglossus gonagra* often causes misshaping or dropping of young fruits (Chiavegato, 1963).

CONTROL In small passion fruit producing areas, hand picking and destruction of eggs, nymphs and adults is recommended (Mariconi, 1952). Chiavegato (1963) suggests the removal of the alternative cucurbit host, 'São Caetano melon', a preferred host of *L. gonagra*, or to avoid the cultivation of chayote and *Anisosperma passiflora* in adjacent areas as tactics to reduce pest densities. In southeast Queensland, Australia, regular inspections are recommended during the summer months to detect any build-up of *L. australis* (Murray, 1976).

Stem weevil

In Brazil, the stem weevil *Philonis* spp. (Curculionidae) was first reported in Alagoas in 1972. Currently, the infestation has expanded to several states in Brazil (Warumbi and Veiga, 1978; Leão, 1980; Torres Filho and Araújo, 1981; Oliveira and Busoli, 1983; Cruz *et al.*, 1993; Racca Filho *et al.*, 1993; Boaretto *et al.*, 1994). It is commonly found on borders of young plantations (Fancelli, 1992a). Severe outbreaks of this pest have caused the eradication of 250 ha in Brazil (Rossetto *et al.*, 1978). Racca Filho *et al.* (1993) reported the occurrence of *P. passiflorae* O'Brien and *P. obesus* Champion in Rio de Janeiro. The species that occurs in São Paulo was identified as *P. crucifer* (Piza Júnior and Kavati, 1995).

DESCRIPTION AND LIFE HISTORY Adults of *P. passiflorae* are about 7 mm in length, brown with whitish elytra with two brown stripes. Adults of *P. crucifer* are 4 mm in length, brown with black markings. They are nocturnal (Piza Júnior and Kavati, 1995). According to Santos and Costa (1983) and Boaretto *et al.* (1994), females lay eggs on young or old stems. The eggs hatch in 8–9 days. The larvae are white and legless. The full-grown larva is about 8 mm long. The larval stage is 53–64 days, and the pupal period is 14–35 days (Costa *et al.*, 1979). All stages of development of this species occur inside the stem. Pupae and recently emerged adults are frequently found in cocoons spun by the full-grown larva (De Bortoli and Busoli, 1987).

HOST PLANTS In Brazil, yellow passion fruit is susceptible to attack by *Philonis* spp. while *Passiflora alata*, *P. maliformis*, *P. serrato digitada* and *P. caerulea* are not infested by this pest (Oliveira and Busoli, 1983). Cruz *et al.* (1993) observed that yellow passion fruit is very susceptible to *Philonis obesus* attack, but *P. alata* and *P. giberti* show some plant resistance.

INJURY Larvae of *Philonis* spp. feed within the stems, opening longitudinal galleries inside stems that prevent plant development. The attacked stems are easily identified by the presence of excrement and sawdust (Santos and Costa, 1983). As the larva develops, infested stems become weak and frail, and die (Fancelli, 1992a). According to De Bortoli and Busoli (1987), the simultaneous attack of several larvae is characteristic of weevil infestations, which causes hypertrophy in stems where the pupal cell will be constructed (Rossetto *et al.*, 1978; Oliveira and Busoli, 1983; Racca Filho *et al.*, 1993). Attack by the stem weevil also causes fruit drop before maturation (Costa *et al.*, 1979).

CONTROL Periodic inspections of the crop are essential for an early detection of weevil-infested stems (Fancelli, 1992a). When infestation symptoms are detected on the crop, affected stems should be pruned and burned (De Bortoli and Busoli, 1987). According to Leão (1980) and Costa *et al.* (1979), a contact insecticide (e.g. decamethrin at 25% (5–10 g a.i. ha^{-1})) should be applied during early afternoon hours for stem weevil control, at the time of adult emergence. After 4–5 days, systemic insecticides for control of future stem infestations should be used.

Flies

Some fly species feed upon the fruits of *Passiflora* spp., and others attack flowering buds. In Brazil, Lordello (1954), Santos and Costa (1983) and Teixeira (1994) reported the following genera of flies damaging passion fruits: *Anastrepha* Schiner (Tephritidae) and *Lonchaea* Fallén (Lonchaeidae). *A. consobrina* (Loew), *A. curitis* Stone, *A. dissimilis* Stone,

A. fraterculus (Wiedmann), *A. kuhlmanni* Lima, *A. lutzi* Lima, *A. pseudoparallela* (Loew), *A. striata* Schiner, and *A. xanthochaeta* Hendel are reported as being the most common species associated with passion fruit in Brazil (Santos and Costa, 1983; Teixeira, 1994; Zucchi, 1988, 2000). *Anastrepha pallidipennis* Guerne was reported on yellow passion fruit in Colombia (Chacón and Rojas, 1984). The oriental fruit fly, *Bactrocera dorsalis* (Hendel), melon fly, *Bactrocera cucurbitae* Coquillett, and the Mediterranean fruit fly, *Ceratitis capitata* Wiedmann, are known to attack the passion fruit vines in Hawaii, USA (Back and Pemberton, 1918); however, the relative importance of each species appears to vary with respect to location of the vineyard (Akamine *et al.*, 1954). The Queensland fruit fly, *Bactrocera tryoni* (Froggatt), is the most important insect pest of passion fruit in Australia (Murray, 1976).

Neosilba pendula (Bezzi) and *Dasiops* sp. (Lonchaeidae) are the most common species attacking flowering buds of passion fruit (Rossetto *et al.*, 1974; Silva, 1982; Fancelli and Mesquita, 1998). *Dasiops* sp. attacking flowering buds was reported in Rio de Janeiro, Brazil (Silva *et al.*, 1968; Vasconcellos *et al.*, 1995). The species of *Dasiops* known to attack flowering buds or fruits of *Passiflora* spp. in the Americas are *D. curubae* Steykal, *D. inedulis* Steykal, and *D. passifloris* McAlpine (Steyskal, 1980).

Other flies may also feed upon flowering buds, such as *Lonchaea cristula* McAlpine (Lonchaeidae) and *Zapriothrica salebrosa* Wheeler (Drosophilidae) (Chacón and Rojas, 1984). Hernández *et al.* (1985) observed that *L. cristula* is more common in curuba (*Passiflora molissima*) when this crop is cultivated near areas where other host fruits grow. The larvae of these fly species destroy pollen by boring into anthers, and may cause intensive dropping of infested buds (Chacón and Rojas, 1984).

DESCRIPTION AND LIFE HISTORY *Anastrepha* adults are 6.5–8.0 mm in length, predominately yellow in colour, with brown and yellow markings on the wings. The adult Medfly is a smaller colourful insect with yellow and black markings on the body and black and orange markings on the wings. Adult flies emerge from puparia buried up to 2–5 cm deep in the soil, and crawl to the surface. They feed on juice of ripening fruits (Plate 92) and on honeydew excreted by aphids, mealybugs, and soft scale insects. Females deposit their eggs mainly in ripening fruit, depositing two to six eggs in the cavity beneath the skin. After 2 or 3 days, the whitish eggs hatch. The cream coloured larvae bore into the fruit pulp and contaminate it with bacteria and fungi, which cause the fruit to decay. Large fruits may contain as many as 100 larvae. Under favourable conditions, larvae complete development in about 9–13 days. They exit the fruit while it is hanging on the tree or after it has fallen to the ground. In the tropical areas, the pupae complete development in 10–14 days. In temperate areas, the pupae complete development in 7–11 days during the summer, but in winter may remain dormant for several months. Flies of the genus *Anastrepha* produce a variable number of generations depending on the inhabited region (Orlando and Sampaio, 1973; Morgante, 1991).

The adult of *Bactrocera tryoni* is wasp-like in appearance, about the size of a house fly, with transparent wings bearing a dark band on the front margin. Bright yellow patches interrupt the general reddish brown body colour. The female lays several pale cream, elongate eggs beneath the skin surface of the fruit. Creamy coloured maggots may emerge from the eggs in 2 or 3 days and tunnel within the fruit while feeding. During the warmer weather the larval stage is completed in about 2 weeks. The mature larvae then leave the fruit to burrow into soil to pupariate for an additional 2 weeks, after which adults emerge from puparia. Very few eggs laid in immature passion fruit produce adult flies. The development of woody tissue around eggs in the rind of the fruit prevents some eggs from hatching, or when hatching occurs, causes high mortality of young larvae. Egg hatching in ripe fruit is mostly unaffected since the fruit has ceased growing and does not form the woody tissue around the eggs (Murray, 1976). In Queensland (Australia), *B. tryoni* invades passion fruit vines from alternative host plants and is most active from September to April (Swaine *et al.*, 1985).

The adult *Dasiops curubae* is blackish blue. The wings are hyaline and slightly smoky yellowish, while the calypters and wing fringes are pale yellowish (Steyskal, 1980).

The adult *Dasiops inedulis* is bright metallic dark blue with hyaline wings; the calypters and wing fringes are yellowish to nearly white (Steyskal, 1980). Peñaranda *et al.* (1986) reported that the length of the life cycle of this species takes 22.8 days under laboratory conditions. The period of incubation requires 2–3 days; the larval stage, 4–9 days; and the pupal stage, 10–17 days (Peñaranda *et al.*, 1986).

Dasiops passifloris was described by McAlpine (1964), who recorded its distribution in Florida. Adults are metallic blue-black, and females have a long ovipositor resembling those species in the Otitidae and Tephritidae. They oviposit one to four eggs in the pulp of immature fruit. After hatching, larvae bore through the immature fruit and feed on developing seeds. The maturing larvae then begin feeding on the fruit pulp immediately beneath the skin. In about 12 days the larvae mature and drop to the ground, where they pupariate within the soil or possibly beneath some refuse. The pupal stage lasts 14 days (Stegmaier, 1973).

The adult *Neosilba pendula* is about 4 mm long, bright metallic dark blue, with hyaline wings (Rossetto *et al.*, 1974).

HOST PLANTS The highly polyphagous *Anastrepha* spp. infest approximately 270 plant species in 41 families, and are considered to be the major fruit pests of tropical and subtropical America. *Passiflora* has been reported as a host of the larvae of two groups of *Anastrepha* (*chiclayae* and *pseudoparallela*) (Norrbom and Kim, 1988; Stefani and Morgante, 1996). *Anastrepha chiclayae* Greene has been found associated with *Passiflora* spp. in Mexico (Hernández-Ortiz, 1992). Larvae of *A. limae* Stone feed upon fruits of *P. quadrangularis* (Stone, 1942; Caraballo, 1981). Lordello (1954) observed infestations by *Anastrepha* and *Lonchaea* species on *Passiflora quadrangularis* and *P. macrocarpa*. The Medfly attacks 253 kinds of fruits, nuts, and vegetables; many of which are tropical plants. *Neosilba pendula* is known as the key pest of cassava, and is a secondary pest of several fruits, especially tangerine (Rossetto *et al.*, 1974). *Dasiops curubae* damages flowers of curuba (*P. mollissima*) (Steyskal, 1980; Causton, 1993). *Dasiops inedulis* is reported in Panama to be a serious pest of purple granadilla, *P. edulis* (Steyskal, 1980). This species has been implicated in 21–65% loss of flowering buds of passion fruit in the Cauca Valley (Colombia) (Peñaranda *et al.*, 1986). *Dasiops passifloris* attacks fruits of *P. suberosa* (syn., *P. pallida*) (Steykal, 1980).

INJURY Fruit fly adult damage is caused by oviposition in green fruits, causing disfigurations of the fruit surface. The larvae damage the fruit by feeding on its pulp, contaminating it with bacteria and fungi (Plate 93), and causing premature fruit drop (Medina *et al.*, 1980; Santos and Costa, 1983; Morgante, 1991). According to Akamine *et al.* (1954), the oriental, melon, and Mediterranean fruit flies puncture the fruit while the rind is still tender. As the fruit enlarges, a woody area (callus) develops around the puncture. If the fruit is small and undeveloped, the damage may be sufficient to cause it to shrivel and fall from the plant. If the fruit is well developed, it may continue to maturity. At the time of ripening, the area around the puncture has the appearance of a small, woody crater, which disfigures the outer appearance of the fruit but does not impair pulp quality. Although oviposition scars are present on ripening fruits, they generally do not contain living larvae. Larvae appear to be able to develop better in immature than in mature fruit.

Oviposition by *B. tryoni* in immature green fruit also results in the formation of calluses in the skin of the fruit at the puncture site. Punctured fruits may persist on the plant to maturity but are not acceptable for fresh market sale because of the damage (May, 1953; Hargreaves, 1979).

According to Murray (1976), passion fruit increase rapidly in size during the first 10–15 days after fruit set. During this period the skin of the fruit is turgid and easily punctured by the ovipositor. Infested immature fruit shows characteristic skin blemishes. The woody tissue, which forms around the eggs, develops a hard raised area around the puncture mark. Egg laying or puncture often cause young fruit to shrivel and drop. Puncture marks are difficult to detect on ripe fruit. A few days after larval infestation, mature fruit will show wrinkling and breakdown.

Anastrepha pseudoparallela lays eggs in unripe fruits of *P. alata*, and the larvae develop by feeding on the seeds. Cyanogenic compounds are present in all parts of *Passiflora* plants, including seeds of unripe fruits. These glycosides protect the plant by preventing feeding by herbivore species. Thus, the use of these resources by *A. pseudoparallela* for larval breeding is probably associated with its ability to tolerate these chemical defences and suggests a high degree of specialization (Stefani and Morgante, 1996).

The larvae of flies that attack the flowering buds and immature fruits cause premature fruit drop (Brandão *et al.*, 1991). Immature fruits infested by *D. passifloris* become dirty, whitish green in colour, while infested ripe fruits become bluish white (Stegmaier, 1973). *Dasiops inedulis* larvae bore into the anthers, and the ovary, causing flowering bud drop (Peñaranda *et al.*, 1986).

NATURAL ENEMIES Most species of tephritid fruit flies are attacked by a complex of larval parasitoids while egg and pupal parasitoids are much less common (Bateman, 1972). *Doryctobracon* Enderlein, *Diachasmimorpha* Viereck, *Opius* Wesmael, *Psyttalia* Walker and *Utetes* Foerster (Braconidae) are the most common larval parasitoids of tephritid fruit flies (Wharton, 1996). *Pachycrepoideus vindemiae* (Rondani) and *Spalangia endius* Walker (Pteromalidae) are pupal parasitoids of Medfly (Back and Pemberton, 1918). In Colombia, *Opius* sp., *Zelus rubidus* (Reduviidae), and spiders of Thomisidae were reported as natural enemies of *D. inedulis* (Peñaranda *et al.*, 1986).

According to Silva *et al.* (1968), larvae of *N. pendula* are parasitized by *Alysia lonchaeae* Lima, *Ganaspis carvalhoi* Dettmer, *Tropideucoila weldi* Lima (Cynipidae), and *Opius* sp. and preyed upon by *Belonuchus rufipennis* (Fabricius) (Staphylinidae).

CONTROL Akamine *et al.* (1954) argued that one of the most important steps in controlling fruit flies is the elimination of over-ripe papaya, tomato, and other fruits in which the flies breed and on which the adults feed. Santos and Costa (1983) recommended that passion fruit must be planted far away from coffee plantations and wild host plants that grow adjacent to the passion fruit crop should be removed. Fruit flies may be controlled using bait sprays composed of molasses (7%) or protein hydrolysate (1%), and an insecticide. The bait is sprayed over 1 m^2 of the plant canopy, using 100–200 ml of bait per plant (Santos and Costa, 1983). The bait should be applied during the night (Rossetto *et al.*, 1974). Boaretto *et al.* (1994) reported that bud flies may be controlled by insecticide baits composed of fenthion, molasses and water. The bait is applied at the beginning of the flowering peak, and usually three applications spaced at 8–10 days are necessary. The authors also suggested burying the attacked buds and planting trap crops, such as sweet pepper.

Mites

Several species of mites have been reported from passion fruit (Sanches, 1996). *Brevipalpus phoenicis* (Geijskes) (Tenuipalpidae), the red spider mites *Tetranychus mexicanus* (McGregor) and *T. desertorum* Banks (Tetranychidae) are known to infest passion fruit plants. Warm temperature and low precipitation favour development of these species (Haddad and Millán, 1975; Oliveira, 1987; Brandão *et al.*, 1991). On the other hand, *Polyphagotarsonemus latus* (Banks) (Tarsonemidae) prefers high temperatures and > 80% relative humidity (Oliveira, 1987; Brandão *et al.*, 1991).

DESCRIPTION AND LIFE HISTORY *Brevipalpus phoenicis* mites are quite small, e.g. 0.3 mm in length, and pass through five stages in their life cycle: egg, larva, protonymph, deutonymph, and adult. Adults are bright red, depositing bright red, oval eggs of about 0.1 mm long, on the underside of leaves or in crevices on the stems (Swaine *et al.*, 1985). The length of the cycle from egg to adult varies from as little as 18 days (30°C) to as long as 49 days (20°C) (Oliveira, 1987). In Queensland (Australia), the life cycle takes about 6 weeks. According to Flechtmann (1989), this mite develops on both upper and lower leaf surface, but prefers the lower leaf surface. Large numbers congregate in leaf axils, along grooves in the terminal shoots and leaf stalks, and along the main veins of leaves (Swaine *et al.*, 1985). Severely infested leaves are completely webbed by spider mites (Oliveira, 1987).

Female spider mites are < 0.5 mm long, and red. Males are slightly smaller than females, and greenish yellow. The life cycle of spider mites comprises five stages: egg, larva, protonymph, deutonymph, and adult. The length of female and male life cycles is about 18 and 20 days, respectively (Oliveira, 1987). The period of incubation is 5–6 days. The larva of *T. mexicanus* is light yellow with three pair of legs, and the larval stage requires 4–7 days. Its protonymph is reddish yellow with four legs and develops in 4–5 days. The deutonymph completes its development in 2–4 days (Dominguez-Gil, 1998).

Polyphagotarsonemus latus females are about 0.2 mm in length. The body varies in colour from white to yellowish. Males are smaller than females, and hyaline (Brandão *et al.*, 1991). The entire cycle from egg to adult takes about 3–5 days. The species develops rapidly through four stages: egg, larva, pupa, and adult. This mite attacks young leaves, and its colonies are localized on the lower leaf surface (Oliveira, 1987).

HOST PLANTS A wide variety of host plants are attacked by the mites. *Brevipalpus phoenicis* feeds on citrus, coffee, cashew, papaya, banana, guava, pomegranate, apple, loquat, peach, pear, grape, grevillea, and various weeds (Oliveira, 1987). *Tetranychus desertorum* occurs on cotton, sweet potato, bean, papaya, passion fruit, strawberry, peach, tomato, grape, and certain ornamentals. *Tetranychus mexicanus* feeds upon cotton, citrus, apple, papaya, passion fruit, pear, peach, cacao, walnut, and ornamentals (Flechtmann, 1989). Hosts of *P. latus* are bean, potato, cotton, coffee, citrus, apple, pumpkin, walnut, grape, peach, pepper, rubber plantation, and various weeds (Oliveira, 1987).

INJURY *Brevipalpus phoenicis* is responsible for general discoloration of the leaves, and necrosis, culminating in leaf drop. Attacked young stems dry from the extremity to the base and eventually die (Flechtmann, 1989). In Queensland, Australia, *B. phoenicis* infestations are usually most damaging during the summer and autumn. Heavy infestation may result in defoliation (Swaine *et al.*, 1985). In Hawaii, *B. papayensis*, known as red mite, is one of the most troublesome pests of passion fruit, but it is usually most damaging in areas of low rainfall and during prolonged dry weather. Its effects are yellowing, shrivelling, and falling of the leaves. With heavy and prolonged infestation, leaf fall increases and the vine has the appearance of dying back. At the same time, developing fruit may begin to shrivel and fall prematurely from the plant. Close examination reveals the presence of mites as scattered reddish patches on the surface of the fruit, particularly around the stem end, along the midrib and veins of the leaf, especially on the under-surface. If red spider mites are left uncontrolled, the plant may eventually die (Akamine *et al.*, 1954).

Red spider mites cause a general weakening of the plants. Initial damage to foliage appears as fine silver speckling on the lower surface of the leaves, which turn brownish on the upper side as mites continue to feed. If a large number of mites are present, entire leaves or plants turn yellow and necrotic (Oliveira, 1987). Photosynthesis and transpiration of the plants are suppressed. Dense populations of spider mites produce silken webs that cover the leaves. Heavy infestations cause leaves to drop and plants to lose vigour (Oliveira, 1987).

P. latus induces malformations in developing leaves, which later dry and drop. It may attack flowering buds, causing a reduction in the number of flowers, and in turn, of fruits produced per plant (Oliveira, 1987; Flechtmann, 1989).

NATURAL ENEMIES Important natural enemies of spider mites are predacious mites belonging to Phytoseiidae. The life history of these predators is closely related to that of their host. Larvae and adults of *Stethorus* sp. (Coccinellidae) were also observed as predators of *T. mexicanus* in passion fruit plantations in Lake Maracaibo (Venezuela) when spider mites reached high population densities (Dominguez-Gil and McPheron, 1992).

CONTROL Periodic inspections of the orchard and other adjacent hosts, including weeds, are essential to verify the occurrence and first symptoms of mite attacks (Oliveira, 1987; Brandão *et al.*, 1991). Mites have also become resistant to many of the organophosphate miticides. Selective miticides, dosages, timing, and refining application techniques may be useful in an integrated mite management system. The four principal requirements for a practical operation are: (i) presence of predacious mites in the orchard; (ii) knowledge of the appearance and habits of plant-feeding and predacious mites; (iii) careful examination of relative numbers of predators and plant-feeding mites, particularly during a period when rapid population changes are occurring; and (iv) knowledge of pesticides to use, how to use them, and what pesticides to avoid, in order to conserve predators.

Flechtmann (1989) recommended the use of sulphur that is not toxic to pollinating insects. According to Piza Júnior (1992), fenthion, propargite, chlorfentezine, and avermectin are effective miticides.

Secondary Pests

Secondary pests include various species of insects that may become abundant, and occasionally damage the passion fruit crop. The insects in this group are either associated frequently with a particular environmental condition, or else occur within limited geographical areas.

Aphids

Aphids (Aphidae) are known to attack passion fruit vines, although they seldom cause serious damage. Nevertheless, at least three species of aphids, *Myzus persicae* (Sulzer), *Aphis gossypii* (Glover), and *Macrosiphum solanifolii* Ashmead (= *M. euphorbiae* (Thomas)), must be regarded as potentially important pests of passion fruit.

DESCRIPTION AND LIFE HISTORY The characteristic of *Myzus* is the presence on the head of tubercles at the base of the inner side of antennae. Because of this, the frons shows an outline that is scooped out in the middle. The apterous form of *M. persicae* is 1.2–2.3 mm long, and frequently pale green in colour, but populations also occur that are yellowish or tending to reddish. The cornicles are long and cylindrical, sometimes with a slight swelling of the distal part, which is often blackish. The cauda is subtriangular, shorter than the cornicles. The length of the antennae is a little less than that of the body. The alate form is about 1.2–2.2 mm in length and green in colour; head, antennae and thorax are brown or blackish (Barbagallo *et al.*, 1997).

The apterous form of the cotton aphid, *A. gossypii*, has an ovoid body shape, and is medium to small in size (1.0–1.8 mm in length). Colour is variable, from ochreous brown to mottled, more or less dark green or even bluish tinged. Antennae are brown with the middle part cream coloured; cornicles and cauda are blackish brown. The alate form has head and thorax blackish, as are the antennae, cornicles and cauda; the abdomen has the same variation in colour as does the apterous form. Length of body is 1.2–2.0 mm. The cotton aphid has a nearly cosmopolitan world distribution. This species is of greater seriousness in warm-temperate regions and in the intertropical zone. It is typically anholocyclic, remaining active during the whole year with uninterrupted generations of parthenogenetic females. However, there are recorded cases of the appearance of sexual forms, with the subsequent deposition of resistant eggs on various plants (Barbagallo *et al.*, 1997).

M. solanifolii is about 2.6–4.0 mm in length, green, with wax secretions on the body of immature forms (Barbagallo *et al.*, 1997).

HOST PLANTS Peach is the preferred primary host of *M. persicae*. It may infest other *Prunus* species, in particular almond and plum. Its secondary host plants include numerous wild and cultivated plants, such as passion fruit (Barbagallo *et al.*, 1997). *Aphis gossypii* infests numerous species of dicotyledonous plants, including passion fruit. Favoured hosts are in the *Malvaceae* (cotton, hibiscus, etc.) and

Cucurbitaceae (pumpkin, cucumber, watermelon, melon) (Barbagallo *et al.*, 1997). *M. solanifolii* is a very polyphagous species, showing preference for the *Solanaceae*, i.e. potato, tomato, etc. (Barbagallo *et al.*, 1997).

INJURY Although these aphids cause malformation in foliage, they are more important as disease vectors. *Myzus persicae* and *A. gossypii* transmit a virus disease that causes a disease associated with hardening of fruits (Brandão *et al.*, 1991; Piza Júnior and Resende, 1993). *Myzus persicae* and *M. solanifolii* are vectors of the passion fruit woodiness virus in Australia. In Hawaii, however, where these two species are present, this virus does not occur (Akamine *et al.*, 1954).

NATURAL ENEMIES Naturally occurring predators and parasites are effective against aphids. The Coccinellidae are effective against cotton aphids, and in particular the larval stage of *Scymnus*. Other predators include the Chrysopidae (*Chrysoperla*), Cecidomyiidae (*Aphidoletes*) and Syrphidae (*Syrphus*). Parasitism by *Lysiphlebus* sp. (Aphidiidae) has been reported (Barbagallo *et al.*, 1997). According to Grasswitz and Paine (1993), *Lysiphlebus testaceipes* (Cresson) parasitizes *Myzus*, *Aphis*, and *Macrosiphum*. Silva *et al.* (1968) reported parasitism of *M. solanifolii* by *Aphidius platensis* Brèthes, *A. brasiliensis* Brèthes, *Diaeretiella rapae* (McIntosh) (Aphidiidae), and predation by *Bacha clavata* (F.) (Syrphidae), *Coccinella ancoralis* Germar, *Cycloneda sanguinea* (L.), and *Eriopis connexa* (Germar) (Coccinellidae). They also reported parasitism of *M. persicae* by *Aphelinus mali* (Hald.) (Aphelinidae), *A. platensis*, and *D. rapae* in Argentina and Uruguay. Cabbage aphid is the primary host of *D. rapae* which is commercially available for release against a wide range of aphids, especially *Myzus* and *Brachycaudus* spp. (Hsieh and Allen, 1986).

Caterpillars

Caterpillars of *Azamora penicillana* (Walker) (Pyralidae) are defoliators of passion fruit (Santos and Costa, 1983; Fancelli, 1992b; Fancelli, 1993). *Peridroma saucia* (Hübner) (Noctuidae) attacks the floral structure, and may reduce fruit production (Chacón and Rojas, 1981). *Pyrausta perelegans* (Hampson) (Pyralidae) is also associated with passion fruit flowers. In Colombia, this species is one of the most important pests of curuba, and may infest 70% of the crop. Caterpillars of *Aepytus* (*Pseudodalaca*) *serta* (Schaus) (Hepialidae) and *Odonna passiflorae* Clarke (Oecophoridae) are passion fruit stem borers (Chacón and Rojas, 1984).

DESCRIPTION AND LIFE HISTORY Caterpillars of *A. penicillana* lodge in flowering buds or in a nest made from leaves, which are joined by silk webs produced by the insect. The caterpillar is whitish, and when fully grown, 25 mm long. The adult is a small pale greyish moth (Santos and Costa, 1983; Fancelli, 1992b; Fancelli, 1993). *P. saucia* is a moth with a 39 mm wingspan. Forewings are crimson red, hindwings are whitish. Females lay their eggs on the underside of leaves, usually in clusters of 60–244. Eggs are about 0.7 mm in diameter. Eggs are creamy white when first laid, but turn reddish blue when close to hatching, after 8–10 days, depending on temperature. The newly hatched larva, approximately 0.97 mm long, is reddish brown; full-grown larvae are about 40 mm, and brownish grey. The larva has six instars and becomes fully grown in 31–38 days. The pupa is about 18.1 mm in length and dark brown. The pupal stage lasts from 18 to 22 days (Chacón and Rojas, 1981).

Moths of *A. serta* are pale brown and have a wingspan of about 45 mm. Eggs, each about 0.65 mm in diameter, are light yellow and laid on the bark of stems near the ground. The larva is cream in colour with a dark brown head. The full-grown larva is about 38 mm. Pupae are light brown and about 39 mm long, and 34 mm in length for females and males, respectively (Chacón and Rojas, 1984).

The adult moth of *O. passiflorae* has a wingspan of 24–30 mm. The head and thorax are greyish, and the dorsal of the abdomen is olive-green anteriorly, and greyish posteriorly. The full-grown larva is 18–21 mm long with a light brown head and cream coloured body, with several setae. The pupa is dark yellow and about 9.5–13.0 mm long. The pupal stage occurs inside galleries constructed by the larvae (Chacón and Rojas, 1984).

The adult *P. perelegans* is a pale coloured microlepidopteran with a wingspan of about 35 mm. The wings are yellow semi-hyaline. The borders of the forewings are dark rose. The young larva is about 1.95 mm long and green. The full-grown larva is about 23.8 mm long, and the pupa is dark brown and about 13.24 mm long (Chacón and Rojas, 1984).

HOST PLANTS *A. penicillana* was reported damaging a wild species of passion fruit (*Passiflora cincinnata*) in Brazil (Fancelli, 1993). *P. saucia* damages and causes reduction in fruit production of curuba (*Passiflora mollissima*). It is a polyphagous insect, feeding on potato (*Solanum tuberosum*), oak (*Quercus suber*), *Calendula officinallis*, cotton, tobacco, bean, tomato, lucerne, soybean, and beet (Chacón and Rojas, 1981).

INJURY Although the caterpillars of *A. penicillana* cause defoliation, the most serious damage is caused by the phytotoxic effects of the fluid secreted by the caterpillar on the leaves and young stems. Heavy infestations cause leaves to dry and drop, and passion fruit plants lose vigour and bear fewer flowers. In Bahia, Brazil, the population peak of this pest occurs during the rainy season (April to June) (Santos and Costa, 1983; Fancelli, 1992b, 1996).

P. saucia larvae feed upon floral structures of *P. mollissima*. Young larvae migrate from leaves to the flowers where they feed on the floral tube, nectary and gynophore, causing flower dropping. The sixth instar larvae may occasionally continue feeding on the young fruit, or drop onto the soil to pupate. In Colombia, *P. saucia* infested 64% of the flowers during the summer (July to September) (Chacón and Rojas, 1981).

Larvae of *A. serta* bore into roots located near the surface, and occasionally bore into stems. Stem injury is characterized by the presence of sawdust. A single larva is regularly found in 1-year-old plants, while in 6–8-year-old plants, up to five larvae may develop (Chacón and Rojas, 1984).

The damage of *O. passiflorae* caterpillars is characterized by the presence of sawdust outside the principal and lateral stems. Several larvae in different stages of development attack simultaneously at the same point of the stem, and cause cellular hypertrophy. They form galleries in different directions, resulting in total destruction of the stem.

The caterpillars of *P. perelegans* infest 6-month-old plants, and remain during the whole vegetative period. They attack the buds and developing flowers, feeding on nectaries, gynophores, and young fruits (Chacón and Rojas, 1984).

NATURAL ENEMIES Naturally occurring predators and parasites are particularly effective against *P. saucia* in Colombia. A tachinid fly, *Incamyia* sp., is an important factor for reducing the population of *P. saucia* caterpillars. Another dipterous parasitoid is *Megaselia scalaris* (Phoridae). Adults of the predator *Anisotarus* sp. (Carabidae) feed on caterpillars and prepupae. Some caterpillars may also be infected by bacteria (*Bacillus cereus*, *Pseudomonas aeruginosa* and *Streptococcus* spp.) and nematodes (Pseudodiplogasteridae). The larval stage of *O. passiflorae* is infected with the fungus *Beauveria bassiana* and is parasitized by the hymenopteran *Neotheronia* sp. (Ichneumonidae). *Sathon* sp. (Braconidae) and *Enytus* sp. (Ichneumonidae) parasitize larvae of *P. perelegans*. The former has gregarious behaviour, and on average 11 adult wasps may emerge from one larva (Chacón and Rojas, 1984).

CONTROL According to Chacón and Rojas (1984), the infestation of *A. serta* depends on the wood used to made the trellises. The authors suggest the use of resistant wood such as mangrove (*Rhizophora mangle*). Wood of Barbados cherry (*Malpighia glabra*) and *Cassia tomentosa* are susceptible to attack by *A. serta*, and are not recommend for trellises.

Mealybugs

Citrus mealybug, *Planococcus citri* Risso, and the passion vine mealybug, *Planococcus pacificus* Cox (Pseudococcidae), are pests of lesser importance on passion fruit.

DESCRIPTION AND LIFE HISTORY Citrus mealybug, *P. citri*, is a small, oval-shaped sucking insect commonly found on passion fruit in Hawaii, USA. A white, mealy powder covers the upper surface of the insect body. Wax strands radiate from the body with slightly longer strands posteriorly. Females are wingless and vary in length from 3 to 4 mm. The males are fragile, with two wings, and 2 mm in length with two long white filaments extending from the end of the abdomen. The female is active and feeds throughout its life. The male feeds only during the first stage (Murray, 1976). After mating, the female deposits up to 500 yellowish eggs in a loose cottony mass or ovisac and then dies. Crawlers emerge from the eggs in 3–9 days and moult several times until the adult stage is reached. There are three moults in the female and four in the male. Approximately 4 weeks are required for completion of the cycle during warm weather (Murray, 1976).

The females of passion vine mealybug, *P. pacificus*, are white, oval and about 3–4 mm long. Eggs are laid in a loose, cottony mass and hatch to produce crawlers 3–9 days later. Development from egg to adult takes about 4 weeks during summer. In Queensland (Australia), these species are most common in late summer and autumn (Swaine *et al.*, 1985).

HOST PLANTS Citrus mealybug infests citrus and many greenhouse and indoor plants. Other plants recorded as its hosts include avocado, pineapple, pumpkin, cotton, rice, sweet potato, potato, cacao, coffee, sugarcane, chayote, tobacco, guava, mango, rose, pomegranate, etc. (Silva *et al.*, 1968).

INJURY Mealybugs characteristically aggregate on the plant, especially at leaf nodes and under dead leaves and trash. Aggregation may also occur under dried flower bracts. Secretion of a sugary solution from the mealybugs promotes growth of a black fungal mould on the fruits and leaves. Ants are often found tending mealybugs for this secretion and interfere with the natural control of the mealybugs by parasites and predators. If a severe infestation occurs, loss of vigour, leaf drop, and fruit malformation may occur. Unchecked, an infestation may cause death of the plant (Murray, 1976; Swaine *et al.*, 1985).

NATURAL ENEMIES Lady beetles (Coccinellidae), especially mealybug lady beetle, *Cryptolaemus montrouzieri* Mulsant, and maculate lady beetle, *Harmonia octomaculata* (Fabricius), substantially reduce mealybug numbers. Of secondary importance are small wasp parasitoids such as *Leptomastidea abnormis* (Girault) (Encyrtidae) and *Ophelosia* sp., and lacewing larvae (*Oligochrysa lutea* (Walker)) (Murray, 1978; Swaine *et al.*, 1985).

Silva *et al.* (1968) reported several species of parasitoids and predators of *P. citri* in Argentina. It is parasitized by *Apanteles paraguayensis* Brèthes (Braconidae), *Coccophagus caridei* (Brèthes) (Aphelinidae), *Anagyrus coccidivorus* Dozier, *A. pseudococci* (Girault), *Leptomastidea abnormis* (Girault), *Leptomatrix dactylopii* Howard (Encyrtidae), and *Pachyneuron* sp. (Pteromalidae). *Leptomastix dactylopii* is commercially available. It is a yellowish brown wasp that lays its eggs in late instar nymphs and adult mealybugs. *Leptomastix* prefers hosts in warm, sunny, humid environments. It may complete one generation in 2 weeks at 30°C or in 1 month at 21°C (Fisher, 1963).

CONTROL Clusters of mealybugs under dead leaves are well protected from the insecticide sprays, and little control can be achieved unless vines are cleaned thoroughly to allow spray penetration. Pruning may enhance the effectiveness of the spray; however, this is often impractical, as laterals to be pruned are generally bearing fruits (Murray, 1976).

According to Murray (1976), occasional outbreaks of this pest are best controlled by two sprays of 1 : 60 oil or methidathion 0.05% combined with 1 : 100 oil, 2 weeks to 1 month apart. The 1 : 60 oil is preferred, as methidathion is highly toxic to the mealybug's natural enemies. For good control, thorough coverage is essential.

Scales

Soft brown scale (*Coccus hesperidum* Linnaeus) (Coccidae) may occasionally infest leaves and stems of passion fruit. California red scale, *Aonidiella aurantii* (Maskell) (Diaspididae), is most common on older passion fruit vines (Swaine *et al.*, 1985).

DESCRIPTION AND LIFE HISTORY Adults of soft brown scale are approximately 3 mm long, and covered by a brown, oval, dome shaped scale. A sweet, sticky secretion produced by this insect promotes growth of sooty mould on the fruit and leaves. Ants also tend the scale for this secretion (Murray, 1976).

California red scale is a small, flattened, reddish orange scale. The dull-red female produces living young or crawlers that shelter under the parent scale for some time. After leaving the protection of the parent scale, the crawlers quickly settle on the vine or fruit and then moult – twice if a female, three times if a male – before reaching the adult stage. The life cycle from egg to adult takes about 6 weeks in summer (Swaine *et al.*, 1985).

HOST PLANTS California red scale infests citrus and many ornamental plants (Forster *et al.*, 1995). Silva *et al.* (1968) cited several other host plants of this species such as pumpkin, coconut, papaya, rose, mulberry, etc.

Soft brown scale uses various plants as host, such as avocado, sapodilla, plum, mulberry, coconut, gladiolus, papaya, laurel, salvia, maté, pear, rose, grape, etc. (Silva *et al.*, 1968).

INJURY Soft scales and diaspidids injure plants by sucking sap, and when numerous can kill the plant. They sometimes heavily encrust the leaves, fruits, twigs or branches. Mealybugs may be found on almost any part of the host plant from which they suck the sap (Murray, 1976; Swaine *et al.*, 1985).

NATURAL ENEMIES Parasitic wasps are important to control *A. aurantii*, mainly *Comperiella bifasciata* (Howard), and *Aphytis chrysomphali* (Mercet) (Aphelinidae) (Murray, 1976; Swaine *et al.*, 1985). In Argentina, this species was reported to be parasitized by the aphelinids, *A. chrysomphali*, *A. maculicornis* Masi, and *Aspidiotiphagus citrinus* (Crawford) (Silva *et al.*, 1968).

Azya luteipes Mulsant, *Coccidophilus citricola* Brèthes, and *Pentilia egena* Mulsant have been recorded as predators of California red scale. Two species recorded as pathogenic fungi of this scale are *Nectria coccophila* and *Myriangium duriaei* (Silva *et al.*, 1968).

According to Forster *et al.* (1995), *Aphytis melinus* is the most important parasitoid attacking California red scale. The adult is a tiny yellow wasp. The female *A. melinus* feeds on and oviposits in immature scales, preferring the virgin adult female scale. The solitary, ectoparasitic larva leaves a flat and dehydrated scale body beneath the scale cover, where the parasitoid's cast skin and faecal pellets (meconia) may be observed. The parasitoid's short life cycle (10–20 days) results in two or three parasitoid generations for each scale generation. *Comperiella bifasciata* is an important encyrtid that parasitizes California red scale. Adult parasitoids are black, with two white stripes on the female's head. One parasitoid generation requires about 3–6 weeks to develop, with faster development occurring on larger (later instar) hosts and at warmer temperatures.

Parasitoids of *C. hesperidum* in Argentina are *Aneristus coccidis* Blanchard, *Coccophagus caridei* (Brèthes), *Ablerus ciliatus* De Santis (secondary parasitoid) (Aphenilidae), *Aphycus flavus* Howard, *A. luteolus* (Timberlake), and *Cheiloneurus longisetaceus* De Santis (Encyrtidae). Among the predators is *Azya luteipes* Mulsant (Coccinellidae) (Silva *et al.*, 1968).

CONTROL Chemical control is often not required since parasitization by small wasps substantially reduces populations. Should chemical control be necessary, a 1 : 60 oil spray is satisfactory (Murray, 1976).

Termites

Termites are increasingly common in passion fruit plantations, but losses attributable to them have not been quantified. Three termite species, *Heterotermes convexinotatus* (Snyder), *Amitermes foreli* Wasmann, and *Microcerotermes arboreus* Emerson, were observed to feed on roots and stems of 2–4-year-old passion fruit plants in Venezuela (Dominguez-Gil and McPheron, 1992).

DESCRIPTION AND LIFE HISTORY *Heterotermes* (Rhinotermitidae) is a widespread genus. It is characterized by the soldier which has a long and rectangular head. It does not have teeth on the interior curvature of the mandibles. The pronotum is wide and flat (Hadlington, 1987). It has subterranean habits, and does not construct an exposed nest. The colony always remains in contact with the soil through the galleries (Dominguez-Gil, 1998).

Amitermes (Termitidae) is cosmopolitan in distribution and especially conspicuous in the tropics and in the warmer areas of the temperate zones (Krishna and Weesner, 1970). The members of this genus are essentially subterranean in habit. The nest is usually situated in the soil (Krishna and Weesner, 1970). Its soldiers have the mandibles curvated, thin, not too long, and with a prominent tooth and not clearly rectangular (Hadlington, 1987).

Subterranean soldiers of *Microcerotermes* (Termitidae) have long, rectangular mandibles, which are serrated on the interior face (Hadlington, 1987). A queen of *M. arboreus*, measuring 21 mm in length, may deposit 1680 eggs in 24 hours (Krishna and Weesner, 1970).

INJURY Termites penetrate and excavate the roots and continue upwards within the stems. The plant often dies, and death may be associated with the presence of soil pathogens, which usually cause rotting, including *Fusarium* spp. and *Phytophthora* spp. (Dominguez-Gil and McPheron, 1992; Piza Júnior, 1992).

CONTROL The use of tillage operations to reduce populations of soil-inhabiting insects may work in several ways. In the case of termites, it may change the physical condition of soil and expose the colony to the sun. Piza Júnior (1992) recommended that after tillage, the soil should be treated with Thiodan 350 CE (endosulfan) at 100–500 cm^3 per 100 l of water. The soil must be treated when it is wet to allow the penetration of the insecticidal solution. When the crop is already established, the insecticidal solution must be applied to the soil around the plants in large quantities to reach a depth of 35 cm.

Bees

In some passion fruit growing areas, the honeybee *Apis mellifera* L. (Apidae) is considered a pest since it robs the pollen from the carpenter bees, thereby causing a reduction of fruit set (Akamine *et al.*, 1954). Adults of *Trigona spinipes* Fabricius (Apidae), known as irapuá or arapuá in Brazil, attack leaves, stems, trunk, developing buds, developing fruits, and fruit peduncles of several plant species (Puzzi, 1966; Bastos, 1985; Teixeira *et al.*, 1996). It may be found from northern to southern Brazil (Silva *et al.*, 1968; Bleicher and Melo, 1993). Carvalho *et al.* (1994) reported serious damage caused by *T. amazonensis* to yellow passion fruit in Acre (Brazil).

DESCRIPTION AND LIFE HISTORY *Trigona spinipes* is about 5–6.5 mm in length, black with transparent wings and without an ovipositor (Santos and Costa, 1983). It constructs its nests on trees, usually between their branches, or in abandoned termite nests. It uses fibrous filaments of plant material and agglutinative elements, mainly resin. Like honeybees, they exist in large colonies with a queen, without corbiculae, and with thousands of workers (Riek, 1979).

INJURY *Trigona spinipes* causes malformation of foliage and dropping of flowers, resulting in a reduction in the number of fruits produced per plant. It also attacks developing flowering buds (Fancelli and Mesquita, 1998).

HOST PLANTS *Trigona spinipes* damages various plant species, especially flowering buds and leaves, of sapodilla, mulberry, banana, citrus, coconut, mango, rose, pine, and fig (Silva *et al.*, 1968).

NATURAL ENEMIES Silva *et al.* (1968) reported the parasitism of larvae of *T. spinipes* by *Pseudohypocera nigrofascipes* Borgn. & Schn. (Phoridae).

CONTROL Recommendations to prevent honeybees from robbing passion fruit flowers of their pollen have been made. One of them is to plant more attractive plant species such as eucalyptus and basil in adjacent areas to

Table 12.2. Phytophagous arthropods associated with passion fruit (*Passiflora* spp.) around the world.

Class/Order	Family	Scientific name	Feeding habits	Damaging stage	References
ARACHNIDA					
Acari	Tarsonemidae	*Tarsonemus stammeri* Schaarschmidt	Fruit sucker	Larva, nymph and adult	Dominguez-Gil and McPheron,
		Polyphagotarsonemus latus (Banks)	Leaf sucker	Larva, nymph and adult	1992; Teixeira, 1994
	Tetranychidae	*Tetranychus mexicanus* (McGregor)	Leaf sucker	Larva, nymph and adult	Brandão *et al.*, 1991;
		Tetranychus desertorum (Banks)	Leaf sucker	Larva, nymph and adult	Dominguez-Gil and McPheron,
		Tetranychus urticae (Koch)	Leaf sucker	Larva, nymph and adult	1992; Teixeira, 1994
	Tenuipalpidae	*Brevipalpus phoenicis* (Geijsks)	Leaf sucker	Larva, nymph and adult	Teixeira, 1994
INSECTA					
Thysanoptera	Thripidae	*Frankliniella auripes* Hood	Feed on pollen	Adult	Chacón and Rojas, 1984
Coleoptera	Chrysomelidae	*Trichaltica bogotana* Harold	Leaf chewer	Adult	Lordello, 1952b; Silva *et al.*,
		Euryscopa cingulata Latreille	Leaf chewer	Adult	1968; Corrêa *et al.*, 1977;
		Cacoscelis marginata (F.)	Leaf chewer	Adult	Chacón and Rojas, 1984;
		Cacoscelis famelica (F.)	Leaf chewer	Adult	Dominguez-Gil and McPheron,
		Cacoscelis melanoptera (Germar)	Leaf chewer	Adult	1992; Teixeira, 1994
		Cacoscelis orphelia Bechyné	Leaf chewer	Adult	
		Cacoscelis walteriana Bechyné	Leaf chewer	Adult	
		Diabrotica poss. *undecinpunctata* M.	Leaf and flower chewer	Adult	
		Maecolaspis sp.	Leaf chewer	Adult	
		Monomarca sp.	Flower-bud chewer	Adult	
		Acalymma sp.	Seedling chewer	Adult	
	Curculionidae	*Litostylus diadema* (F.)	Leaf chewer	Adult	Dominguez-Gil and McPheron,
		Philonis passiflorae O'Brien	Stem borer	Larva	1992; Racca Filho *et al.*, 1993;
		Philonis obesus Champion	Stem borer	Larva	Piza Júnior and Kavati, 1995
		Philonis crucifer	Stem borer	Larva	
	Scolytidae	*Chramesus bispinus* Wood	Stem borer	Larva and adult	Chacón and Rojas, 1984
	Cerambycidae	*Stenygra conspicua* (F.)	Stem borer	Larva	Chacón and Rojas, 1984;
		Lepturges sp.	Stem borer	Larva	Teixeira, 1994
		Stizocera sp.	Stem borer	Larva	
	Meloidae	*Epicauta atomaria* Germ.	Leaf chewer	Adult	Teixeira, 1994
	Scarabaeidae	*Cyclocephala melanocephala* (F.)	Flower and leaf chewer	Adult	Rossetto *et al.*, 1974; Chacón
		Leucothyreus sp.	Flower and leaf chewer	Adult	and Rojas, 1984; De Bortoli
					and Busoli, 1987
	Anthribidae	*Araecerus fasciculatus* Degeer	Calyx and fruit chewer	Larva and adult	Chacón and Rojas, 1984

Hymenoptera	Apidae	*Trigona amalthea* (Olivier)	Flower chewer	Adult	Carvalho *et al*., 1994; Sazima
		Trigona spinipes F.	Flower chewer	Adult	and Sazima, 1989;
		Trigona amazonensis	Flower chewer	Adult	Dominguez-Gil, 1998
	Formicidae	*Crematogaster* sp.	Feed on stem	Adult	Dominguez-Gil and McPheron,
		Solenopsis sp.	Feed on stem	Adult	1992
Isoptera	Rhinotermitidae	*Heterotermes conexinotatus* (Snyder)	Feed on root and stem	Adult (workers)	Dominguez-Gil, 1998
	Termitidae	*Microcerotermes arboreus* Emerson	Feed on root and stem	Adult (workers)	Dominguez-Gil and McPheron,
		Amitermes foreli Wasmann	Feed on root and stem	Adult (workers)	1992
Lepidoptera	Nymphalidae	*Eueides isabella huebneri* Ménétries	Leaf chewer	Larva	Santo, 1931; Lordello, 1952a,
		Eueides isabella dianosa (Hübner)	Leaf chewer	Larva	1954; Silva *et al*., 1968; Brown
		Eueides aliphera aliphera (Godart)	Leaf chewer	Larva	Júnior and Mielke, 1972;
		Dione juno juno (L.)	Leaf chewer	Larva	Chacón and Rojas, 1984; De
		Dione glycera (C. & R. Felder)	Leaf chewer	Larva	Bortoli and Busoli, 1987;
		Agraulis vanillae vanillae (L.)	Leaf chewer	Larva	Carter, 1992; Dominguez-Gil
		Agraulis vanillae maculosa S.	Leaf chewer	Larva	and McPheron, 1992
		Dryadula phaetusa (L.)	Leaf chewer	Larva	
		Philaethria dido (L.)	Leaf chewer	Larva	
		Philaethria wernickei wernickei (R)	Leaf chewer	Larva	
		Heliconius silvana robigus (Weymer)	Leaf chewer	Larva	
		Heliconius sara apseudes (Hübner)	Leaf chewer	Larva	
		Heliconius ethilla narcaea (Godart)	Leaf chewer	Larva	
		Heliconius erato phyllis (F.)	Leaf chewer	Larva	Chacón and Rojas, 1984
	Geometridae	*Sabulodes* sp. poss. *matrica* Druce	Leaf chewer	Larva	Chacón and Rojas, 1984
	Hepialidae	*Aepytus* (P.) *serta* (Schaus)	Stem borer	Larva	Silva *et al*., 1968; Chacón and
	Noctuidae	*Peridroma saucia* (Hübner)	Flower chewer	Larva	Rojas, 1984
		Copitarsia consueta (Walker)	Flower chewer	Larva	
		Spodoptera ornihogalli (Guenée)	Flower chewer	Larva	Chacón and Rojas, 1984
	Oecophoridae	*Odonna passiflorae* Clarke	Stem borer	Larva	Santos and Costa, 1983;
	Pyralidae	*Pyrausta perelegans* (Hampson)	Bud, flower, fruit borer	Larva	Chacón and Rojas, 1984;
		Azamora penicillana (Walker)	Leaf chewer	Larva	Fancelli, 1992b
		Pococera sp.	Calyx and fruit chewer	Larva	Chacón and Rojas, 1984
	Cossidae	*Langsdorfia* sp.	Stem borer	Larva	
Homoptera	Coccidae	*Coccus hesperidium* L.	Stem sucker	Nymph and adult	Murray, 1976; Chacón and
		Ceroplastes sp.	Stem sucker	Nymph and adult	Rojas, 1984
	Aleyrodidae	*Hexaleurodicus* sp.	Leaf, stem sucker	Nymph and adult	Chacón and Rojas, 1984

continued

Table 12.2. *Continued.*

Class/Order	Family	Scientific name	Feeding habits	Damaging stage	References
	Diaspididae	*Selenaspidus articulatus* (Morgan)	Leaf, fruit, stem suckers	Nymph and adult	Murray, 1976; Swaine *et al.*, 1985; ICA, 1987
		Aonidiella auranti (Maskell)		Nymph and adult	
	Aphididae	*Myzus persicae* (Sulzer)	Leaf sucker	Nymph and adult	Akamine *et al.*, 1954; Brandão *et al.*, 1991; Piza Júnior and Resende, 1993
		Aphis gossypii (Glover)	Leaf sucker	Nymph and adult	
		Macrosiphum solanifolii Ashmead	Leaf sucker	Nymph and adult	
	Pseudococcidae	*Planococcus citri* Risso	Leaf sucker	Nymph and adult	Murray, 1976; Swaine *et al.*, 1985
		Planococcus pacificus Cox	Leaf sucker	Nymph and adult	
Hemiptera	Coreidae	*Holhymenia histro* (F.)	Leaf, fruit sucker	Nymph and adult	Mariconi, 1952; Chiavegato, 1963; Silva *et al.*, 1968; Murray, 1976; Dominguez-Gil, 1998; Chacón and Rojas, 1984; De Bortoli and Busoli, 1987; ICA, 1987; Dominguez-Gil and McPheron, 1992
		Holhymenia clavigera (Herbest.)	Leaf, fruit sucker	Nymph and adult	
		Veneza zonatus (Dallas)	Leaf, fruit sucker	Nymph and adult	
		Leptoglossus gonagra (F.)	Leaf, fruit sucker	Nymph and adult	
		Leptoglossus conspersus (Stal)	Leaf, fruit sucker	Nymph and adult	
		Anisoscelis foliacea (F.) *marginella*	Leaf, fruit sucker	Nymph and adult	
		Anisoscelis flavolineata (F.)	Leaf, fruit sucker	Nymph and adult	
		Diactor bilineatus (F.)	Leaf, fruit sucker	Nymph and adult	
Diptera	Tephritidae	*Anastrepha pallidipennis* Guerne	Feed on fruit	Larva	Akamine *et al.*, 1954; Lordello, 1954; Murray, 1976; Santos and Costa, 1983; Chacón and Rojas, 1984; Zucchi, 1988; Teixeira, 1994
		Anastrepha consobrina (Loew)	Feed on fruit	Larva	
		Anastrepha ethalea (Walker)	Feed on fruit	Larva	
		Anastrepha grandis (Macquart)	Feed on fruit	Larva	
		Anastrepha kuhlmanni Lima	Feed on fruit	Larva	
		Anastrepha lutzi Lima	Feed on fruit	Larva	
		Anastrepha pseudoparallela (Loew)	Feed on fruit	Larva	
		Ceratitis capita (Wiedemann)	Feed on fruit	Larva	
		Dacus curcubitae Coquillett	Feed on fruit	Larva	
		Bactrocera dorsalis (Hendel)	Feed on fruit	Larva	
		Dacus tryoni (Froggatt)	Feed on fruit	Larva	
	Drosophilidae	*Zapriothrica salebrosa* Wheeler	Flower-bud chewer	Larva	Chacón and Rojas, 1984
	Lonchaeidae	*Lonchaea cristula* McAlpine	Feed on flower bud	Larva	Rossetto *et al.*, 1974; Steyskal, 1980; Chacón and Rojas, 1984
		Silba pendula Bezzi	Feed on flower bud	Larva	
		Dasiops curubae Steykal	Feed on flower bud	Larva	
		Dasiops inedulis Steykal	Feed on flower bud	Larva	
		Dasiops passifloris McAlpine	Feed on fruit	Larva	

passion fruit. Collection of wild swarms is also recommended (Boaretto *et al.*, 1994). The control strategies recommended for *T. spinipes* include the destruction of nests near the crop, and weekly inspections to verify the occurrence of this pest on flowers. In exceptional cases, chemical control is recommended.

Conclusions

Several different species of arthropods have been reported in passion fruit. Other pests doubtless occur, and new ones will appear in the future. Fortunately the majority of these species are not injurious. The species listed in Table 12.2 are generally accepted as being responsible for most of the insect and mite damage wherever passion fruit grows.

For control of insect and mite pests which attack passion fruit, we must consider two basic problems: (i) creation and preservation of conditions favourable to carpenter bees, whose function in pollination is of vital importance for fruit set; (ii) suitable control of insects and mites that damage the plant. Additionally, a latent problem is the conservation of natural enemies, which is complicated because beneficial and noxious insects and mites are closely associated with the plant. The timing of spraying is critical, so that applications are not made when passion fruit flowers are open and the carpenter bees are active. The choice of a selective pesticide, with low toxicity to predators and parasites, is important to maintain not only natural control but also pollinators.

References

Akamine, E.K. and Girolami, D.G. (1957) Problems in fruit set in yellow passion fruit. *Hawaii Farm Science* 5, 3–5.

Akamine, E.K. and Girolami, D.G. (1959) *Pollination and Fruit Set in Yellow Passion Fruit*. Hawaii Agricultural Experiment Station, Honolulu, Technical Bulletin 39, 32 pp.

Akamine, E.K., Hamilton, R.A., Nishida, T., Sherman, G.D. and Storey, W.B. (1954) *Passion Fruit Culture*. University of Hawaii, Extension Circular 345, 23 pp.

Back, E.A. and Pemberton, C.E. (1918) The Mediterranean fruit fly in Hawaii. *United States Department of Agricultural Bulletin* 536, 1–118.

Barbagallo, S., Cravedi, P., Pasqualini, E. and Patti, I. (1997) *Aphids of the Principal Fruit-Bearing Crops*. Bayer, Milan, 123 pp.

Bastos, J.A.M. (1985) *Principais Pragas das Culturas e Seus Controles*, 3rd edn. Nobel, São Paulo, 329 pp.

Bateman, M.A. (1972) The ecology of fruit flies. *Annual Review of Entomology* 17, 493–518.

Bleicher, E. and Melo, Q.M.S. (1993) *Artrópodes Associados ao Cajueiro no Brasil*. EMBRAPA/CNPAT, Fortaleza, Documentos 9, 33 pp.

Boaretto, M.A.C., Brandão, A.L.S. and São José, A.R. (1994) Pragas do maracujazeiro. In: São José, A.R. (ed.) *Maracujá, Produção e Mercado*. DFZ/UESB; Vitória da Conquista, pp. 99–107.

Boiça Júnior, A.L., Lara, F.M. and Oliveira, J.C. (1993) Resistência de genótipos de maracujazeiro ao ataque de *Dione juno juno* Cramer, 1779 (Lepidoptera: Nymphalidae). In: *Resumos do 14 Congresso Brasileiro de Entomologia*. SEB, Piracicaba, p. 143.

Brandão, A.L.S., São José, A.R. and Boaretto, M.A.C. (1991) Pragas do maracujazeiro. In: São José, A.R., Ferreira, F.R. and Vaz, R.L. (eds) *A Cultura do Maracujá no Brasil*. FUNEP, Jaboticabal, pp. 136–168.

Brown Júnior, K.S. and Mielke, O.H.H. (1972) The heliconians of Brazil (Lepidoptera: Nymphalidae). Part II. Introduction and general comments, with a supplementary revision of the tribe. *Zoologia* 57, 1–40.

Camillo, E. (1978a) Aspectos reprodutivos de algunas espécies de *Xylocopa* (Hymenoptera: Anthophoridae). *Ciência e Cultura* 30, 594–595.

Camillo, E. (1978b) Polinização do maracujazeiro. In: *Anais do 2 Simpósio Sobre a Cultura do Maracujazeiro*, Jaboticabal, pp. 32–39.

Camillo, E. (1980) Polinização do maracujazeiro. In: Ruggiero, C. (ed.) *Cultura do Maracujazeiro*. FCAN, Jaboticabal, pp. 47–53.

Camillo, E. (1996) Utilização de espécies de *Xylocopa* (Hymenoptera: Anthophoridae) na polinização do maracujá amarelo. In: *Anais do 2 Encontro Sobre Abelhas*. Ribeirão Preto, pp. 141–146.

Camillo, E. and Garófalo, C.A. (1982) On the bionomics of *Xylocopa frontalis* (Olivier) and *Xylocopa grisescens* (Lepeletier) in southern Brazil. I. Nest construction and biological cycle. *Revista Brasileira de Entomologia* 42, 571–582.

Caraballo, J. (1981) Las moscas de las frutas del género *Anastrepha* Schiner, 1868 (Diptera: Tephritidae) de Venezuela. MSc thesis, University Central de Venezuela, 210 pp.

Carter, D. (1992) *Butterflies and Moths*. Dorling Kindersley, London, 304 pp.

Carvalho, E.F., Calixto, A.R.Y., Silva Filho, J.R. and Morato, E.F. (1994) Avaliação da polinização artificial na redução dos danos causados por *Trigona amazonensis* na cultura do maracujá amarelo. In: *Anais do 13 Congresso Brasileiro de Fruticultura*. SBF, Salvador, pp. 798–799.

Causton, C.E. (1993) Una mosca del genero *Dasiops* (Diptera: Lonchaeidae) atacando la curuba (*Passiflora mollissima*) en el Edo. Merida, Venezuela. *Boletín de Entomología Venezolana* 8, 146.

Chacón, P. and Rojas, M. (1981) Biologia y control natural de *Peridroma saucia*, plaga de la flor de la curuba. *Revista Colombiana de Entomologia* 7, 47–53.

Chacón, P. and Rojas, M. (1984) Entomofauna asociada a *Passiflora mollissima, P. edulis* f. *flavicarpa* y *P. quadrangularis* en el Departamento del Valle del Cauca. *Turrialba* 34, 297–311.

Chiavegato, L.G. (1963) *Leptoglossus gonagra* – praga do maracujá. *O Agronômico*, 15, 31–36.

Corbet, S.A. and Willmer, P.G. (1980) Pollination of the yellow passionfruit: nectar, pollen and carpenter bees. *Journal of Agricultural Science* 95, 655–666.

Corrêa, L.S., Ruggiero, C. and Oliveira, J.C. (1977) Ocorrência de *Acalymma* sp. (Coleoptera: Chrysomelidae) sobre mudas de maracujá-amarelo (*Passiflora edulis* f. *flavicarpa* Deg.). *Científica* 5, 229–230.

Costa, J.M., Correa, J.S., Santos, Z.F.A. F. and Ferraz, M.C.V.D. (1979) *Estudos da Broca do Maracujazeiro na Bahia e Meios de Controle*. EPABA, Salvador, Comunicado Técnico 37, 10 pp.

Cox, I.E. (1957) Flowering and pollination of passion fruit. *Agricultural Gazette of New South Wales* 68, 573–576.

Cruz, C.A., Graça, D. and Lima, A.F. (1993) Ocorrência de *Philonis obesus* em maracujazeiro no Estado do Rio de Janeiro. In: *Resumos do 14 Congresso Brasileiro de Entomologia*. SEB, Piracicaba, p. 16.

Cunha, M.A.P. (1996) Recursos genéticos e modificações em métodos de seleção para produtividade em maracujá. *Revista Brasileira de Fruticultura* 18, 413–423.

D´Almeida, R.F. (1944) Estudos biológicos sobre alguns lepidópteros do Brasil. *Archivos de Zoologia do Estado de São Paulo* 4, 44.

De Bortoli, S.A. and Busoli, A.C. (1987) Pragas. In: Ruggiero, C. (ed.) *Cultura do Maracujazeiro*. Legis Summa, Ribeirão Preto, pp. 111–123.

Dominguez-Gil, O.E. (1998) Fauna fitófaga de parchita maracuyá (*Passiflora edulis* f. *flavicarpa*) en las regiones oriental y suroriental de la cuenca del Lago de Maracaibo, Venezuela: características morfológicas. *Boletín del Centro de Investigaciones Biológicas* 32, 1–66.

Dominguez-Gil, O.E. and McPheron, B.A. (1992) Arthropods associated with passion fruit in western Venezuela. *Florida Entomologist* 75, 607–612.

Dominguez-Gil, O.E., Vera, H., Socarras, G. and Carvajal, R. (1989) Posible epizootia de tipo viral en larvas de *Eueides isabella* Ménétries y *Dione juno juno* (Cramer) defoliando plantaciones de parchita maracuyá en el Estado Zulia. In: *Resúmenes del 11 Congreso Venezolano de Entomologia*. Sociedad Venezolana de Entomología, Maracaibo, p. 33.

Donadio, L.C. (1983) Frutíferas tropicais de valor potencial. In: Donadio, L.C. (ed.) *Fruticultura Tropical*. FCAV/UNESP, Jaboticabal, pp. 20–30.

Echeverri, F., Cardona, G., Torres, F., Pelaez, C., Quiñones, W. and Renteria, E. (1991) Ermanin: an insect deterrent flavonoid from *Passiflora foetida* resin. *Phytochemistry* 30, 153–155.

Fancelli, M. (1992a) *A Broca da Haste do Maracujazeiro*. EMBRAPA/CNPMF, Bahia, Maracujá em Foco 53, 2 pp.

Fancelli, M. (1992b) *A Lagarta de Teia do Maracujazeiro*. EMBRAPA/CNPMF, Bahia, Maracujá em Foco 54, 2 pp.

Fancelli, M. (1993) Ocorrência de *Azamora penicillana* (Walk.) (Pyralidae: Chrysauginae) em maracujá silvestre. *Anais da Sociedade Entomológica do Brasil* 22, 121–124.

Fancelli, M. (1996) Insetos-pragas do maracujazeiro e controle. In: Lima, A.A., Borges, A.L., Santos Filho, H.P., Fancelli, M. and Sanches, N.F. (eds) *Instruções Práticas Para o Cultivo do Maracujazeiro*. EMBRAPA/CNPMF, Bahia, Circular Técnica, 20, 44 pp.

Fancelli, M. and Mesquita, A.L.M. (1998) Pragas do maracujazeiro. In: Braga Sobrinho, R., Cardoso, J.E. and Freire, F.C.O. (eds) *Pragas de Fruteiras Tropicais de Importância Agroindustrial*. EMBRAPA/SPI, Brasília, pp. 169–180.

Figueiro, C.L.M. (1995) Aspectos biológicos e controle da lagarta *Dione juno juno* (Lepidoptera: Heliconiinae) do maracujazeiro (*Passiflora edulis*) por *Baculovirus dione*. MSc thesis, UFPa, Brazil, 73 pp.

Fisher, T.W. (1963) *Mass Culture of* Cryptolaemus *and* Leptomastix *– Natural Enemies of Citrus Mealybug*. University of California Agricultural Experiment Station Bulletin, No. 797.

Flechtmann, C.H.W. (1989) *Ácaros de Importância Agrícola*. Nobel, São Paulo, 189 pp.

Forster, L.D., Luck, R.F. and Grafton-Cardwell, E.E. (1995) *Life Stages of California Red Scale and its*

Parasitoids. University of California, Division of Agricultural Natural Research Publication, 21529.

Free, J.B. (1993) *Insect Pollination of Crops*, 2nd edn. Academic Press, London, pp. 404–409.

Gilmartin, A.L. (1958) Post-fertilization seed and ovary development in *Passiflora edulis* Sims. *Tropical Agriculturist* 35, 41–58.

Grasswitz, T.R. and Paine, T.D. (1993) Influence of physiological state and experience on the responsiveness of *Lysiphlebus testaceipes* (cresson) (Hymenoptera: Aphidiidae) to aphid honeydew and host plants. *Journal of Insect Behaviour* 6, 511–528.

Gravena, S. (1987) Perspectiva do manejo integrado de pragas. In: Ruggiero, C. (ed.) *Cultura do Maracujazeiro*. Legis Summa, Ribeirão Preto, pp. 134–145.

Haddad, G.O. and Millán, F. (1975) *La Parchita Maracuyá* (Passiflora edulis *f.* flavicarpa *Degener*). Fondo de Desarollo Frutícola, Caracas, Boletim Técnico 2, 82 pp.

Hadlington, P. (1987) *Australian Termites*. University Press, New South Wales, Australia.

Hammer, L.H. (1987) The pollinators of yellow passionfruit – do they limit the success of *Passiflora edulis* f. *flavicarpa* as a tropical crop? In: *Proceedings of the Annual Meeting of the Florida State Horticulture Society* 100, 283–287.

Hardin, L.C. (1986) Floral biology and breeding systems for the yellow passion-fruit, *Passiflora edulis* f. *flavicarpa*. In: *Proceedings of the Inter-American Society of Tropical Horticulture* 30, 35–44.

Hargreaves, J.R. (1979) Damage of passion-fruit by the Queensland fruit fly, *Dacus tryoni* (Froggatt). *Queensland Journal of Agricultural and Animal Sciences* 36, 147–150.

Hernández, M.R., Rojas, M. and Trochez, A. (1985) Evaluacion de la proteina hidrolizada de maiz para la captura de *Lonchaea cristula* (Diptera: Lonchaeidae), plaga de la curuba en el Valle del Cauca. *Revista Colombiana de Entomologia* 11, 47–50.

Hernández-Ortiz, V. (1992) *El Genero* Anastrepha *Schiner (Diptera: Tephritidae) en México, Taxonomia, Distribucion y Sus Plantas Huespedes*. Instituto de Ecología, Veracruz, 162 pp.

Hoffmann, M. (1997) Polinização do maracujá amarelo *Passiflora edulis* f. *flavicarpa*. In: São José, A.R.S., Bruckner, C.H., Manica, I. and Hoffmann, M. (eds) *Maracujá: Temas Selecionados (1), Melhoramento, Morte Prematura, Polinização e Taxonomia*. Cinco Continentes, Porto Alegre, 5, pp. 58–70.

Hoffmann, M. and Pereira, T.N.S. (1996) Polinização do maracujá-amarelo (*Passiflora edulis* f *flavicarpa* Deg.) na região de Campos dos Goytacazes, RJ. In: *Anais do 2 Encontro Sobre Abelhas*. Ribeirão Preto, p. 330.

Hsieh, C.Y. and Allen, W.W. (1986) Effects of insecticides on emergence, survival, longevity, and fecundity of the parasitoid *Diaretiella rapae* (Hymenoptera: Aphidiidae) from mummified *Myzus persicae* (Homoptera: Aphididae). *Journal of Economic Entomology* 79, 1599–1602.

Hurd, P.D. (1978) *An Annotated Catalog of the Carpenter Bees (Genus* Xylocopa *Latreille) of the Western Hemisphere (Hymenoptera: Anthophoridae)*. Smithsonian Institution Press, Washington, 106 pp.

Hurd, P.D. and Moure, J.S. (1963) *A Classification of the Large Carpenter Bees (Xylocopini) (Hymenoptera: Apoidea)*. University of California Publications in Entomology, 366 pp.

ICA (1987) *Guia para el Control de Plagas*, 4th edn. Ministerio de Agricultura, Instituto Colombiano Agropecuário/SOCOLEN, Manual de Asistencia Tecnica 1, pp. 232–237.

Kluge, R.A. (1998) Maracujazeiro *Passiflora* sp. In: Castro, P.R.C. and Kluge, R.A. (ed.) *Ecofisiologia de Fruteiras Tropicais*. Nobel, São Paulo, pp. 32–47.

Krishna, K. and Weesner, F.M. (1970) *Biology of Termites*. Academic Press, New York, 643 pp.

Leão, J.A.C. (1980) Considerações sobre as principais doenças e pragas do maracujá no Nordeste do Brasil. In: *Anais do 1 Encontro Estadual da Cultura do Maracujá*. EMATER-SE, Aracajú, pp. 67–71.

Lima, M.F.C. and Veiga, A.F.S.L. (1993) Ocorrência do vírus da poliedrose nuclear (PNV) em lagartas de *Dione juno juno* (Cramer, 1779) sobre maracujazeiro, em Recife – Pernambuco. In: *Resumos do 14 Congresso Brasileiro de Entomologia*. SEB, Piracicaba, p. 143.

Lima, M.F.C. and Veiga, A.F.S.L. (1995) Ocorrência de inimigos naturais de *Dione juno juno* (Cr.), *Agraulis vanillae maculosa* S. e *Eueides isabella dianosa* (Hüb.) (Lepidoptera: Heliconiidae) em Pernambuco. *Anais da Sociedade Entomológica do Brasil* 24, 631–634.

Lordello, L.G.E. (1952a) Insetos que vivem sobre o maracujazeiro. I – Notas bionômicas acerca de *Dione vanillae* (L., 1758) (Lep.: Nymphalidae). *Revista de Agricultura* 29, 23–29.

Lordello, L.G.E. (1952b) Insetos que vivem sobre o maracujazeiro. II – Contribuição ao conhecimento de *Cascoscelis famelica* (F., 1787) (Col.: Chrysomelidae). *Dusenia* 3, 387–393.

Lordello, L.G.E. (1954) Insetos que vivem sobre maracujazeiro. III – Notas acerca de *Dione juno* (Cramer) (Lep.: Nymphalidae) e relação de alguns outros insetos habitualmente coligidos de *Passiflora* spp. *Revista de Agricultura* 31, 23–29.

Manica, I. (1997) Maracujazeiro: taxonomia, anatomia e morfologia. In: São José, A.R.S., Bruckner, C.H., Manica, I. and Hoffmann, M. (eds) *Maracujá: Temas Selecionados (1), Melhoramento, Morte Prematura, Polinização e Taxonomia.* Cinco Continentes, Porto Alegre, pp. 7–24.

Mardan, M., Yatim, I.M. and Khalid, M.R. (1991) Nest biology and foraging activity of carpenter bee on passion fruit. In: *The 6th International Symposium on Pollination.* Tilburg, pp. 127–132.

Mariconi, F.A.M. (1952) Contribuição para o conhecimento do *Diactor bilineatus* (Fabricius, 1803) (Hemiptera: Coreidae), praga do maracujazeiro (*Passiflora* spp.). *Arquivos do Instituto Biológico* 21, 21–42.

Martin, F.W. and Nakasone, H.Y. (1994) The edible species of *Passiflora.* In: Schaffer, B. and Andersen, P.C. (eds) *Handbook of Environmental Physiology of Fruit Crops, Subtropical and Tropical Crops,* Vol. 2. CRC Press, Boca Raton, pp. 101–113.

May, A.W.S. (1953) Queensland host records for the Dacinae. *Queensland Journal of Agricultural Science* 10, 36–79.

McAlpine, J.F. (1964) Descriptions of new Lonchaeidae (Diptera). *Canadian Entomologist* 96, 661–700.

Medina, J.C., Garcia, J.L.M., Lara, J.C.C., Tocchini, R.P., Hashizume, T., Moretti, V.A. and Canto, W.L. (1980) *Maracujá, da Cultura ao Processamento.* ITAL, Campinas, Frutas Tropicais 9, 207 pp.

Menezes, E.B., Rabinovitch, L., Ferreira, I.T., Moura, V.R.A., Pereira, A.C. and Guaycurus, T.V. (1989) Ação da biomassa de *Bacillus thuringiensis* isolados no Brasil: uma aproximação preliminar contra lepidópteros. In: *Resumos do 12 Congresso Brasileiro de Entomologia.* SEB, Belo Horizonte, p. 232.

Morgante, J.S. (1991) *Moscas-das-frutas (Tephritidae): Características Biológicas, Detecção e Controle.* SENIR, Brasília, Boletim Técnico 2, 19 pp.

Moura, M.F., Picanço, M., Gonring, A.H.R. and Bruckner, C.H. (2000) Seletividade de inseticidas a três Vespidae predadores de *Dione juno juno* (Lepidoptera: Heliconidae). *Pesquisa Agropecuária Brasileira* 35, 251–257.

Murray, D.A.H. (1976) Insect pests on passion fruit. *Queensland Agricultural Journal* 102, 145–151.

Murray, D.A.H. (1978) Effect of fruit fly sprays on the abundance of the citrus mealybug, *Planococcus citri* (Risso) and its predator, *Cryptolaemus montrouzieri* Mulsant, on passionfruit in southeastern Queensland. *Queensland Journal of Agricultural and Animal Science* 35, 344–346.

Nishida, T. (1954) Entomological problems of the passion fruit. *Hawaii Farm Science* 3, 3–7.

Nishida, T. (1958) Pollination of the passion fruit in Hawaii. *Journal of Economic Entomology* 51, 146–149.

Nishida, T. (1963) *Ecology of the Pollinators of the Passion Fruit.* Hawaii Agricultural Experiment Station, Honolulu, Technical Bulletin 55.

Norrbom, A.L. and Kim, K.C. (1988) *A List of the Reported Host Plants of Species of* Anastrepha *(Diptera: Tephritidae).* USDA-APHIS, Washington, Miscellaneous Publications 8, 52 pp.

Oliveira, C.A.L. (1987) Ácaros. In: Ruggiero, C. (ed.) *Cultura do Maracujazeiro.* Legis Summa, Ribeirão Preto, pp. 104–110.

Oliveira, J.C. and Busoli, A.C. (1983) *Philonis* sp. (Coleoptera, Curculionidae), nova praga do maracujazeiro em Jaboticabal, SP. In: *Resumos do 8 Congresso Brasileiro de Entomologia.* SEB, Brasília, p. 281.

Orlando, A. and Sampaio, A.S. (1973) 'Moscas das frutas', notas sobre o reconhecimento e combate. *O Biológico* 39, 143–150.

Peñaranda, I.A., Chacón, P. and Rojas, M. (1986) Biología de la mosca de los botones florales del maracuyá *Dasiops inedulis* (Diptera: Lonchaeidae) en el Valle del Cauca. *Revista Colombiana de Entomologia* 12, 16–22.

Pires, M.M. and São José, A.R. (1994) Custo de produção e rentabilidade da cultura do maracujazeiro. In: São José, A.R. (ed.) *Maracujá, Produção e Mercado.* DFZ/UESB; Vitória da Conquista, pp. 223–233.

Piza Júnior, C.T. (1992) *Pragas e Doenças do Maracujá.* CATI, Campinas, Comunicado Técnico 96, 9 pp.

Piza Júnior, C.T. and Kavati, R. (1995) *7th Ciclo de Debates Sobre a Cultura do Maracujá.* CATI, Campinas, Comunicado Técnico 124, 6pp.

Piza Júnior, C.T. and Resende, J.A.M. (1993) Virose do endurecimento dos frutos do maracujazeiro. CATI, Campinas, Comunicado Técnico 103, 5 pp.

Pope, W.T. (1935) *The Edible Passion Fruit in Hawaii.* Hawaii Agricultural Experiment Station, Honolulu, Technical Bulletin 74, 22 pp.

Price, P.W. (1997) Pollination ecology. In: Price, P.W. (ed.) *Insect Ecology.* John Wiley & Sons, New York, pp. 239–266.

Puzzi, D. (1966) *Pragas dos Pomares de Citros do Estado de São Paulo.* Instituto Biológico de São Paulo, São Paulo, Publicações 116, 58 pp.

Racca Filho, F., Guajará, M.S. and Lima, A.F. (1993) O gênero *Phinolis* Champion, 1906 no Estado do Rio de Janeiro (Coleoptera, Curculionidae).

In: *Resumos do 14th Congresso Brasileiro de Entomologia*. SEB, Piracicaba, p. 166.

Riek, E.F. (1979) Hymenoptera (wasps, bees, ants). In: *The Insects of Australia*. Melbourne University Press, Carlton, pp. 867–959.

Rossetto, C.J., Cavalcante, R.D., Grisi Júnior, C. and Carvalho, A.M. (1974) *Insetos do Maracujazeiro*. Instituto Agronômico, Campinas, Circular Técnica 39, 12 pp.

Rossetto, C.J., Longo, R.S., Rezende, J.A.M. and Branco, E.M.C. (1978) Ocorrência de *Philonis* sp. (Coleoptera: Curculionidae) como praga de maracujazeiro. *Ciência e Cultura* 30, 9.

Ruggiero, C., Lam-Sanchez, A. and Miguel, S. (1975) Estudos da incompatibilidade em flores do maracujá amarelo (*Passiflora edulis* f. *flavicarpa* Deg.). In: *Anais do 3rd Congresso Brasileiro de Fruticultura*. SBF, Rio de Janeiro, pp. 491–495.

Ruggiero, C., Banzatto, D.A. and Lam-Sanchez, A. (1976) Studies on natural and controlled pollination in yellow passion fruit (*Passiflora edulis* f. *flavicarpa* Deg.). *Acta Horticulturae* 57, 121–124.

Ruggiero, C., São José, A.R., Volpe, C.A., Oliveira, J.C., Duringan, J.F., Baumgartner, J.G., Silva, J.R., Nakamura, K., Ferreira, M.E., Kavati, R. and Pereira, V.P. (1996) *Maracujá Para Expotação: Aspectos Técnicos da Produção*. EMBRAPA/SPI, Brasília, Publicações Técnicas FRUPEX 19, 64 pp.

Sanches, N.F. (1996) Ácaros do maracujazeiro. In: Lima, A.A., Borges, A.L., Santos Filho, H.P., Fancelli, M. and Sanches, N.F. (eds) *Instruções Práticas Para o Cultivo do Maracujazeiro*. EMBRAPA/CNPMF, Bahia, Circular Técnica 20, 44 pp.

Santo, E. (1931) *Inimigos e Doenças das Fruteiras*. Biblioteca Agric. D'O Campo, Rio de Janeiro, 80 pp.

Santos, Z.F.A.F. and Costa, J.M. (1983) *Pragas da Cultura do Maracujá no Estado da Bahia*. EPABA, Salvador, Circular Técnica 4, 10 pp.

Sazima, I. and Sazima, M. (1989) Mamangavas e irapuás (Hymenoptera: Apoidea): visitas, interações e consequências para polinização do maracujá (Passifloraceae). *Revista Brasileira de Entomologia* 33, 109–118.

Semir, J. and Brown, K.S. (1975) Maracujá: flor da paixão. *Revista Geografica Universal* 2, 41–47.

Silva, A.G.A., Gonçalves, C.R., Galvão, D.M., Gonçalves, A.J.L., Gomes, J., Silva, M.N. and Simoni, L. (1968) *Quarto Catálogo dos Insetos que Vivem nas Plantas do Brasil, Seus Parasitos e Predadores*, Part II, Vol. 1. Ministério da Agricultura/DDIA, Rio de Janeiro, 622 pp.

Silva, C.C.A. (1979) Morfologia e biologia de *Dione juno juno* (Cramer, 1779) (Lepidoptera: Heliconiidae). MSc. thesis, UFPR, Curitiba, Brazil, 99 pp.

Silva, L.M.S. (1982) O controle das pragas do maracujazeiro e a preservação dos seus inimigos naturais. In: *Anais do 2nd Encontro Estadual Sobre a Cultura do Maracujá*, pp. 54–58.

Stefani, R.N. and Morgante, J.S. (1996) Genetic variability in *Anastrepha pseudoparallela*: a specialist species. In: McPheron, B.M. and Steck, G.J. (eds) *Fruit Fly Pests, a World Assessment of Their Biology and Management*. St Lucie Press, Delray Beach, pp. 277–280.

Stegmaier, C.E. (1973) *Dasiops passifloris* (Diptera: Lonchaeidae), a pest of wild passion fruit in south Florida. *Florida Entomologist* 56, 8–10.

Steyskal, G.C. (1980) Two-winged flies of the genus *Dasiops* (Diptera: Lonchaeidae) attacking flowers or fruit of species of *Passiflora* (passion fruit, granadilla, curuba, etc.). In: *Proceedings of Entomological Society of Washington* 82, 166–170.

Stone, A. (1942) *The Fruit Flies of the Genus* Anastrepha. United States Department of Agriculture, Washington, Miscellaneous Publications 439, 112 pp.

Swaine, G., Ironside, D.A. and Yarrow, W.H.T. (1985) *Insect Pests of Fruit and Vegetables*. Queensland Department of Primary Industries, Brisbane.

Teixeira, C.A.D., Aviles, D.P. and Fernandes, L.M. (1996) Danos de *Trigona* sp. (Hymenoptera: Apidae) em flores de maracujazeiro (*Passiflora edulis*) em Porto Velho. In: *Resumos do 8th Encontro de Pesquisadores da Amazonia*. UNIR/PIUAL, Porto Velho, p.13.

Teixeira, C.G. (1994) Maracujá. I – cultura. In: Teixeira, C.G., Castro, J.V., Tocchini, R.P., Nishida, A.L.A.C., Turatii, J.M., Leite, R.S.S., Bliska, F.M.M. and Garcia, E.B. (eds) *Maracujá, Cultura, Matéria-Prima, Processamento e Aspectos Econômicos*, 2nd edn. ITAL, Campinas, pp. 3–131.

Toledo, Z.D.A. (1991) Fauna del noroeste argentino. Contribucion al conocimiento de los lepidopteros argentinos. IX. *Dione juno* (Cramer) (Lepidoptera: Rhopalocera: Heliconiidae). *Acta Zoológica Lilloana* 40, 109–117.

Torres Filho, J. and Araújo, G.C. (1981) Broca do caule *Philonis* sp., nova praga do maracujazeiro, *Passiflora edulis* Sims. no Estado do Ceará. *Fitossanidade* 5, 50–51.

Vasconcellos, M.A.S. (1991) Biologia floral do maracujá doce (*Passiflora alata* Dryand) nas condições de Botucatu – SP. MSc thesis, UNESP, Jaboticabal, SP, Brazil, 98 pp.

Vasconcellos, M.A.S., Aguiar, E.L., Bicalho, A.C., Menezes, E.B. and Prado, A.P. (1995)

Ocorrência de *Dasiops* n. sp. (Diptera: Lonchaeidae) em maracujazeiro, no município de Itaguaí, RJ. In: *Resumos do 15 Congresso Brasileiro de Entomologia*. SEB, Caxambu, p. 335.

Warumbi, T.T. and Veiga, A.F.S.L. (1978) Uma nova praga no maracujá no nordeste do Brasil. In: *Resumos do 3 Congresso Latinoamericano de Entomologia*. CEPLAC, Ilhéus, p. 20.

Wharton, R.A. (1996) Parasitoids of fruit-infesting Tephritidae – how to attack a concealed host. In: *Abstracts of XX International Congress of Entomology*. Firense, Italia, p. 665.

Zucchi, R.A. (1988) Moscas-das-frutas (Dip., Tephritidae) no Brasil; taxonomia, distribuição geográfica e hospedeiros. In: *Anais do 1 Encontro Nacional Sobre Moscas-das-Frutas*. Fundação Cargill, Campinas, pp. 1–10.

Zucchi, R.A. (2000) Espécies de *Anastrepha*, sinonímias, plantas hospedeiras e parasitóides. In: Malavasi, A. and Zucchi, R. A. (eds) *Moscas-das-frutas de importância econômica no Brasil, conhecimento básico e aplicado*. Holos, Ribeirão Preto, pp. 41–48.

13 Quarantine Treatments for Pests of Tropical Fruits

Jennifer L. Sharp[1] and Neil W. Heather[2]

[1]USDA-ARS, 13601 Old Cutler Road, Miami, FL 33158, USA; [2]School of Land and Food, The University of Queensland, Gatton College, Queensland 4345, Australia

---

Multitudes of new pests regularly threaten agricultural commodities and significantly impact agricultural businesses that market the commodities on a global scale, especially with expanding trade and tourism. Primary pests of major concern include members of the phyla Arthropoda and Mollusca. These insect, mite, slug and snail pests attack many commodities, including the tropical fruits discussed in this volume. The risk of pest introduction and means to stop the establishment of these pests in new areas continues to be a major concern of regulatory officials. Without the use of quarantine intervention by these authorities, pests could be transported over the globe, rapidly become established in new areas, devastate tropical fruits, and cause economic havoc to members of the agricultural communities and to consumers, who demand high quality tropical fruits grown and marketed without the use of chemicals. Means to protect tropical fruits include the use of quarantine, which is intended to prevent the establishment of exotic pest species in places where they are not already found.

Prevention of pest entry by inspection, with or without pretreatment, and early detection of incursions through active monitoring are the first lines of defence employed by regulatory officials who must use these methods to avoid expensive programmes required to eradicate quarantine pests once they become established. For example, officials dealing with various exporters and importers handling commodities originating from countries with quarantine pests must satisfy the importing countries' rules, regulations, and requirements before the commodities are permitted entry into that country. Numerous strategies commencing with a Pest Risk Analysis (AQIS, 1991; Shannon, 1994; Anonymous, 1996) have been developed to ensure quarantine security before any pest host is permitted entry. Quarantine security, a degree of statistical probability and confidence (Chew, 1994; Robertson *et al.*, 1994), may require quarantine treatments including operational systems to ensure that tropical fruits do not contain any unwanted target pests capable of establishment on arrival.

When the use of plant quarantine to prevent the movement of exotic pests first became important for fresh horticultural produce early in this century, disinfestation treatments, such as fumigation, heat or cold, were virtually the sole means by which effective control could be exercised over the risk of infested produce. For almost all pests, and especially for fruit flies, the earliest treatments were the application of heat or cold. These were supplanted for a time by fumigants (e.g. ethylene dibromide, methyl bromide) and sometimes by residual chemicals (e.g.

©CAB *International* 2002. *Tropical Fruit Pests and Pollinators* (eds J.E. Peña, J.L. Sharp and M. Wysoki)

dimethoate). Consumer preference is now identified clearly as being for the use of non-chemical and residue-free physical treatments with heat/cold, modified atmospheres and, in an increasing number of countries, irradiation (Heather, 1994). Modern developments in electronic temperature control and greater physiological understanding of fruit tolerance are enabling the increasing use of heat disinfestation, which takes only a few hours and is much faster than cold, which often requires many days or weeks. Heat is particularly useful for many tropical fruits which, not unexpectedly, tend to be intolerant of cold. When commodities cannot tolerate heat or cold because the treatments may cause damage, other treatments are available and will be discussed in this chapter. An increasing number of countries are using radiation as an alternative treatment to fumigation and other treatment methods to control tropical fruit pests.

Successful disinfestation treatments must achieve the quarantine security needed to control the pest specified by authorities without causing damage to the product, which would render it unacceptable to consumers. The required efficacy needed to reach the security goal varies from country to country. For the United States of America (USA) the required efficacy has usually been 99.9968% (probit 9) demonstrated at the 95% confidence level with no survivors from 100,000 treated insects (Baker, 1939; Couey and Chew, 1986). For Japan, efficacy is determined as no survivors from a minimum of 30,000 treated pests, and New Zealand has a concept of Maximum Pest Limit, currently of allowing five surviving flies in 1,000,000 pieces of fruit for critical fruit fly species such as Mediterranean fruit fly, *Ceratitis capitata* (Wiedemann) and Queensland fruit fly, *Bactrocera tryoni* (Froggatt) (Baker *et al.*, 1990). It is not unusual for a prospective market (e.g. Japan and New Zealand) to require a treatment to be demonstrated repetitively on each cultivar to be imported. Each efficacy requirement forms part of a disinfestation research procedure or protocol, aspects of which may be unique to each market. Increasingly, these traditional standards are being questioned, as most would have been set arbitrarily or based on anecdotal evidence. Their origins probably lie more in the efficacies achievable from treatments available at the time than what was required by the risk. Disinfestation treatments are likely to remain the most important means of obtaining market access where there are pest quarantine impediments.

The US Department of Agriculture, Animal and Plant Health Inspection Service, *Treatment Manual* provides compilations of treatments presented as schedules for all commodities requiring quarantine action before the commodities are imported into the USA (APHIS, 1998). The schedules are arranged in sections and cover a wide variety of applications including fumigation, heat, cold, combination treatments, and hot water treatments. The manual is available by writing to: Printing and Distribution, US Department of Agriculture, Animal and Plant Health Inspection Service, Methods and Development, 4700 River Road, Room 1A01, Riverdale, MD 20737. Several states within the USA, such as Florida, Texas, Arizona, California, and Hawaii, have separate quarantine treatment manuals and directives, with procedures and treatment schedules that must be followed to satisfy each state's regulatory requirements. Other countries have similar internal and international schedules which list proclaimed pests and conditions of entry for commodities, including treatments before or after entry. In Chapter 1 of this volume, J. Peña discusses the pests associated with tropical fruits. Many of these pests require quarantine regulatory action. Quarantine pests of tropical fruits include numerous species of tephritid fruit flies, leaf and fruit miners, other moths and flies, weevils and other beetles, thrips, scales and mealybugs, mites, snails and slugs.

Quarantine pests can be restricted to the surface of fruits, under a calyx, in residual structures such as in the navel of some citrus cultivars, or they can be present internally in the pulp, inside seeds, or simply present as hitch-hikers on fruits and in packaging.

This chapter discusses various quarantine treatment measures that may be applied to keep tropical fruits free from exotic pests. The quarantine measures have been divided arbitrarily into three broad categories which are: measures to eliminate need for treatment;

single treatments; and combination treatments. Emphasis is given to studies published since the mid-1990s. Comprehensive book reviews for quarantine treatments have been presented by Sharp and Hallman (1994), Paul and Armstrong (1994), and Champ *et al.* (1993).

Measures that Eliminate the Need for Treatments

Pest risk analysis (PRA)

There is an international standard for pest risk analysis (PRA) endorsed by the 1995 Food and Agriculture Organization Conference (Anonymous, 1996). PRA is typically done in the recipient country but relies largely on information provided by the producer country. For the discussion herein, it divides conveniently into probability of arrival and probability of establishment. The probability of pest arrival is more complex, if not more difficult, to assess than the second, and includes the host status of the commodity, seasonal pest incidence, field pest management and handling procedures after harvest, all of which influence pest survival. It is essential that any PRA be done with absolute objectivity.

Pest-free area

Regulatory officials may declare areas within a state, municipality, or district to be free of certain injurious pests. The area must meet certain criteria according to lawful codes and regulations. In most countries, a pest-free area or zone status needs to rely on the long-term climatic unsuitability of production areas to the pest, islands separated by sea distance, mountainous areas, and major desert zones with climatic extremes. Production areas granted pest-free status need to be protected by quarantine prohibitions and monitored by trapping or sampling to enable early detection and eradication of incursions by the pest. Ongoing protection can also be supported by mass sterile release programmes if the pest lends itself to this type of strategy (e.g. fruit fly). Many examples exist that make use of the pest-free area concept: honeydew melons from an area in Brazil declared to be free of *Anastrepha grandis* (Macquart) can be imported into the USA (Federal Register, 1993a); municipalities within the State of Sonora, Mexico, are proclaimed free of Mediterranean fruit fly and *Anastrepha* species (Federal Register, 1994a); and some geographical subdivisions of the Riverina and Sumaysia districts of Australia are proclaimed free of Mediterranean fruit fly, Queensland fruit fly, and other exotic members of the Tephritidae (Federal Register, 1995b).

Production-based pest management assurance systems

Ideally, an export quarantine security programme should be based on a holistic pest management system such as the model of Jang and Moffitt (1994). Alone, a quality control system structured in this way may be adequate to meet the quarantine security required by an importing country; but this can be difficult to demonstrate conclusively. Quarantine security assurance of this kind works best when groups such as producers, packers, exporters, and importers are all dedicated to a market. If the contribution of pest management systems to quarantine security is recognized, for the groups which have the potential to enable less severe treatment schedules, it can lead to a longer time that these tropical fruits may remain in the marketplace available for consumers.

Non-host status

Tropical fruits that are never attacked by pests or are resistant to infestation by them are termed non-hosts of those pests. However, the condition can be difficult to determine, and even more often difficult to understand. Reviews that discuss the term and its concepts have been provided by Armstrong (1994) and Greany (1994). The non-host status of a commercially produced tropical fruit to a pest needs to be assessed in terms of the probabilities, both that it might become infested and that viable adults capable of initiating a further generation could emerge from the exported product. The susceptibility of tropical fruits to attack by a quarantine pest can differ because of several factors including the type of fruit, its physical characteristics, and its stage of physiological maturity and ripeness at the time of harvest. Perhaps the widest use of non-host status is for trade of bananas harvested at the stage of maturity and ripeness commonly known as hard green. Bananas at the preclimacteric hard-green stage are recognized internationally as non-hosts of fruit flies of concern, although ripening bananas may be susceptible (Armstrong, 1994). Another example is 'Solo' papaya from the Costa Rica provinces of Guanacaste, San Jose, and Puntarenas, which can be imported into continental USA, Puerto Rico, and the US Virgin Islands (Federal Register, 1992a). Rigid quality control in the packing house ensures that the exact fruit physiological conditions required are met in order to ensure non-susceptibility. If tropical fruit can be harvested at an early stage of ripeness that is not subject to infestation by a pest, no treatment is required. If physical characteristics of tropical fruit, such as having a thick and hard peel, preclude attack by a pest, then the fruit may be able to be exported without treatment. Gould *et al.* (1996) showed that Florida grown litchi, *Litchi chinensis* Sonn., cultivars 'Brewster' and 'Mauritius', and longan, *Dimocarpus longan* (Lour.), cultivar 'Kohala', were non-hosts of Caribbean fruit fly, *Anastrepha suspensa* (Loew), as long as the peel was intact and without any cut or opening.

Sampling and inspection

Statistical sampling theory has a long history of usage in industry as well as quarantine. Guidelines are available to determine sample size for combinations of tolerance and statistical confidence limits (Couey and Chew, 1986). Inspection of a sample of a consignment of produce is part of most quarantine security systems. It is done at export, import, and, often, one or more times when tropical fruits are packed for shipment to markets. Inspection is done to find and remove pests not normally found with the tropical fruits. Although not strictly tropical, stone fruits from Chile present the risk of introducing various pests which do not normally feed on stone fruits but which may be present in shipments of the fruit as hitch-hiking pests (Federal Register, 1989). Given that all assumptions underlying a sampling plan can be met, inspection is a very powerful tool. Although normally labour-intensive and hence costly, it can provide a means by which trade can proceed when no treatment is available. Sampling and inspection need to be recognized more widely both as a bona fide treatment and as a component of an integrated or holistic quarantine security system. The USA often uses sampling and inspection and, for example, permits the importation of avocado fruits from New Zealand subject to inspection at the port of first arrival by US plant quarantine inspectors. If pests are found, then the avocados would be treated to kill the pests (Federal Register, 1991). If the avocados are never infested by a pest or never used as a source of transportation (e.g. hitch-hiker), then no need exists for a treatment, and the avocados may be imported. Use of a preclearance programme, involving inspection by US or Japanese inspectors in the country of origin, is another example of eliminating a need for treatment. Other ways to eliminate the requirement for a treatment include the establishment of pest-free areas (Riherd *et al.*, 1994), elimination of the pest from an area by eliminating host materials, and use of sterile insect releases. 'Hass' avocado fruits from Mexico may enter the USA without treatment if the avocados are shipped to and consumed in Alaska (Federal Register, 1993b). Also, avocados from Hawaii may be imported into Alaska without treatment. The US Department of Agriculture has concluded that climatic conditions in Alaska ensure that pests of avocados in Mexico and Hawaii will not pose a risk to agriculture in that state and that the possibility that some of the avocados might move from Alaska to other states is almost nil (Federal Register, 1993b, 1994b). When quarantine measures such as pest risk analysis, the establishment and maintenance of pest-free areas, proving non-host status, sampling and inspection are not successful, other means may be required, such as the use of a statistically proven successful single treatment, combinations of single treatments, and systems approaches.

Single Treatments

Individual treatments that provide quarantine security when they are used alone are termed single treatments. Residue data as appropriate should be included when a treatment uses a chemical. Data must show that the treatment residues do not exceed a Maximum Residue Level (MRL), an amount based on good agricultural practice and usually much lower than any health safety level. Commodity tolerance to the treatment should form part of the data required to establish the treatment. Selected examples of single treatments used to control pests of tropical fruits are presented by Heather (1993a).

Insecticide application after harvest

Dimethoate and fenthion applied as dips or in-line flood sprays have proven to be highly effective disinfestation treatments to control Queensland fruit fly in tomatoes and mangoes and *Bactrocera cucumis* (French), in members of the *Cucurbitaceae* (Heather, 1993a, 1994). Each insecticide is used at a concentration of 400 mg l^{-1}, and fruit is dipped or flood sprayed for 1 min. Despite its effectiveness to control pests, insecticide applications after harvest pose problems with residue, operator exposure, consumer acceptance, and are not approved treatments in countries such as the USA and Japan, which differentiate between before and after harvest application in their proclamation of MRLs. However, they are economical, logistically simple, and have fewer problems of product injury due to phytotoxicity than physical treatments or fumigants, and they confer residual protection against any perceived risk of reinfestation subsequent to treatment.

Fumigation

A fumigant is a toxic chemical released in the gaseous phase and dispersed in toxic concentrations so that it reaches pest species that could be found either on the surface of the tropical fruit, under the calyx, in the naval areas of the tropical fruit, in the pulp tissues, and inside the seed(s). Concentration of the fumigant, time of exposure ($C \times T$ product), temperature and relative humidity, and load factor (percentage of the chamber occupied by the commodity versus the size of the chamber) are critical to the efficacy of the treatment. Fumigants that control pests of tropical fruits include methyl bromide, methyl iodide, ethylene dibromide, phosphine, and hydrogen cyanide. Fumigation practices and fumigants used for disinfestation purposes have been reviewed by Stark (1994) and Yokoyama (1994).

Methyl bromide

Methyl bromide is a colourless, odourless gas, non-flammable in air, stable, and not corrosive. It has been used since the 1930s to disinfest agricultural warehouses, museums and homes, freight containers, processed commodities, fresh fruits and vegetables, ornamental foliage and cut flowers, and bulk grain. It is the principal fumigant used for the postharvest treatment of a wide range of food and non-food commodities by countries around the globe (APHIS, 1998). However, methyl bromide has been identified as a significant ozone-depleting chemical and its production and use by the USA are being restricted (Federal Clean Air Act, 1990; Anonymous, 1992). The US Congress recently amended the Federal Clean Air Act of 1990 to completely phase out the use of methyl bromide by 1 January 2005 (Anonymous, 1998).

Methyl iodide

Also known as iodomethane, methyl iodide is a colourless, transparent liquid having a boiling point at 42.5°C. It is a stronger methylating agent than methyl bromide, rapidly destroyed by ultraviolet light, and probably not likely to deplete the stratospheric ozone as does methyl bromide. It has been tested successfully as a soil fumigant to control fungi, nematodes, and weeds (Ohr *et al.*, 1996; Zhang *et al.*, 1997). In preliminary studies, Sharp and King (1997) showed that it controlled eggs and larvae of Caribbean fruit fly *in vitro* and in grapefruit and guavas. Guavas infested with eggs and larvae of the fly were fumigated for 2 h with doses of 16, 32, and 48 g m^{-3}. Treated eggs and larvae did not survive treatment at any of these doses. Treatment of guavas infested in a similar manner and exposed to 16 g m^{-3} of methyl iodide for 20, 30, and 40 min resulted in 54%, 62%, and 64% mortality, respectively. Caribbean fruit fly eggs, 24 h old, obtained from a laboratory colony and exposed to 2, 4, and 8 g m^{-3} for 30 min resulted in 24%, 49%, and 90% mortality, respectively. This study was the first to demonstrate that methyl iodide could be used to control a fruit fly quarantine pest. The fumigant, however, is not registered for use to control quarantine pests that attack tropical fruits.

Ethylene dibromide

Ethylene dibromide is a colourless, nonflammable liquid. Until 1983, the chemical was used extensively to control pests in many commodities. Beginning 28 September 1983, actions were taken by the US Environmental Protection Agency to remove ethylene dibromide from the marketplace because it was found to be carcinogenic to mammals (Federal Register, 1983). It cannot be used to treat fruit for consumption in the USA. It is currently approved in some countries, including Australia, but under conditions of greatly reduced MRLs, which restrict its usage. Because of the impending loss of methyl bromide, questions have been raised about the possibility of reinstatement of ethylene dibromide for some uses.

Phosphine

Phosphine (hydrogen phosphide) is a colourless, flammable gas usually generated from an aluminium phosphide tablet or pellet preparations activated when exposed to moisture in the air at temperatures above 15°C. Magnesium phosphide as plates was developed to generate phosphine under conditions of high relative humidity. It is also available as phosphine in cylinders for direct application to a fumigation space. Phosphine acts slowly and requires 5–12 days to control most pests. It is corrosive to metals, especially copper electrical wiring. It is not currently approved for controlling pests of fresh tropical fruits because of the severe injury caused at dosages required to kill quarantine pests. It may also cause chromosome rearrangement among operators (Garry *et al.*, 1989).

Hydrogen cyanide

Hydrogen cyanide was one of the earliest quarantine fumigants used in the 20th century. It is a colourless gas or liquid formulated for fumigation as sodium cyanide, calcium cyanide, or potassium cyanide. It is very soluble in water and unsafe to use on moist materials. A solution of hydrogen cyanide in water is a dilute acid, which renders tropical fruits unpalatable, perhaps hazardous for human consumption, and unmarketable due to burn, wilt, and discoloration. Under very controlled conditions, hydrogen cyanide is used in California to kill surface pests such as scales that attack fruits (Fiskaali, 1989).

Hot water immersion

Some tropical fruits can be immersed in water heated to about 46°C for a time, usually 1–2 h, to control pests without unacceptable loss of market quality. Fruit injury can be minimized by preconditioning some fruits such as mangoes to 37°C for 12 h before immersion (Joyce and Shorter, 1994). Hot water immersion is the preferred method to control tephritid fruit flies in mangoes in continental USA. Fruit size is limited to 700 g for each mango imported into the USA because data on the temperature and time mortality relationship required to control fruit fly eggs and larvae have not been developed for larger mangoes (APHIS, 1998). This is not a major problem since most retail marketed mangoes are in the vicinity of 450–500 g. Nevertheless, some mango producers in Caribbean countries and Mexico have requested that the US Department of Agriculture develop a hot water immersion quarantine treatment that controls fruit fly pests in mangoes weighing more than 700 g. Studies are underway to determine the time of exposure in water at 46.7°C required to control fruit flies such as West Indian fruit fly, *Anastrepha obliqua* (Macquart), eggs and larvae present in 'Keitt' mangoes weighing up to 950 g. McGuire and Sharp (1997) reported that 'Keitt' mangoes weighing 701–950 g immersed in water at 48°C without preconditioning were damaged at the exposure times of 180 min, a time of exposure believed necessary to control fruit fly eggs and larvae that might be present in the larger mangoes.

Due to the success of the hot water immersion treatment to control fruit fly eggs and larvae that attack mangoes, the number of commercial hot water treatment facilities in Mexico, Caribbean countries, Central and South America could soon exceed 100 (P. Whiterell, USDA-APHIS, North Carolina, 1998, personal communication). The major supplier of hot water treatment equipment has been the USA. Mangoes treated with hot water benefit from the treatment because postharvest decay-forming organisms are reduced or killed, thus extending the shelf life of the fruit, and the fruit are also cleaned by the treatment. Australia uses locally manufactured hot water–benlate dips set at around 52°C for 5–10 min to control diseases of mango fruit after harvest (Johnson *et al.*, 1997). The longer times necessary for fruit fly control have resulted in fruit damage when attempted commercially on the predominant 'Kensington' variety (Johnson and Heather, 1994). Hot water is a preferred method used to control fruit fly pests in a variety of tropical fruits that tolerate the treatment (Smith, 1992; Corcoran *et al.*, 1993; Sharp, 1994; Waddell *et al.*, 1997).

Vapour heat

Heated air that is saturated with water vapour has been used successfully to control fruit flies in mango, papaya, and citrus in the USA since around 1930. Vapour heat treatments are approved for mango from South-East Asia and Australia for the markets in Japan and for citrus, papaya, and pineapple for markets in the USA (APHIS, 1998). Vapour heat, depending on temperature and air flow rate, is not as efficient a heat conductor as hot water. Application of vapour heat as a quarantine treatment was first used to control Mediterranean fruit fly in grapefruit in Florida in 1929 (Hallman and Armstrong, 1994). The application was crude but effective. As with hot water, time of exposure and target temperature needed to control pests are critical factors that influence market quality of treated fruits. Vapour heat treatment temperatures usually range between 40 and 47°C and require several hours of exposure. Time of exposure needed to reach a target temperature can be controlled depending on the temperature and the flow rate of the saturated air as it passes over the fruit as well as the type of fruit treated, size, and density of the load.

Forced air heating

A system of heating tropical fruits using forced air that is heated, humidified, and circulated around fruit was developed in Japan and termed a differential pressure–vapour heat process (Sugimoto *et al.*, 1983). Unfortunately this was abbreviated to vapour heat and has become confused with the original vapour heat treatment developed in the USA. The Japanese vapour heat treatment system is used for exports of mango and papaya from the Philippines, Hawaii, Thailand, and Australia to Japan and is accepted primarily by world markets located in South-East Asia, Malaysia, and Australia. New Zealand is another country currently developing forced air heating–disinfestation treatment equipment. Typical fruit fly disinfestation schedules for mangoes treated with forced air heating are based on fruit temperatures at the seed surface 46–47°C held for up to 15 min (Heather *et al.*, 1997).

Air that is heated but humidified below the level at which condensation occurs on the treated fruit has been termed heated air (Sharp *et al.*, 1991) or high temperature forced air

(Armstrong *et al.*, 1989). Recently developed treatments first use low then high humidity air within the same treatment cycle to reduce fruit injury without loss of efficiency (Unahawutti *et al.*, 1992). Thermal heating with air becomes less efficient as the humidity is decreased, but some fruits tolerate treatment better with this system compared with immersion in heated water. Fruits such as citrus do not tolerate hot water immersion treatment, but the market quality of citrus is not adversely affected by forced heated air. Sharp (1993a), Mangan and Ingle (1994) and Sharp and McGuire (1996) demonstrated that tephritid fruit flies are controlled in citrus with hot air without damaging the fruit. Also, Armstrong *et al.* (1995a) reported that Mediterranean fruit fly, oriental fruit fly, *Bactrocera dorsalis* (Hendel), and melon fly, *Bactrocera cucurbitae* Coquillett in Hawaii were controlled in papaya with forced hot air without fruit injury. The US Department of Agriculture approved the use of forced air heating to control pests of mango and grapefruit from Mexico imported into the USA (Federal Register, 1995a), and for tangerines, oranges (except for navel oranges), and grapefruit from Mexico and areas of the USA in Texas and California (Federal Register, 1997b). Also, Gould (1996) reported that papaya fruit fly, *Toxotrypana curvicauda* Gerstaecker, eggs and larvae were controlled in papaya exposed to forced air heat at 48°C for 167 min.

Cold temperature

Cold storage at temperatures above 0°C is used to disinfest different fruits (Gould, 1994). Treatments can require up to 3 weeks of cold storage after the commodity has reached the temperature to achieve total disinfestation of quarantine pests. For fruit flies the time required for cold storage varies greatly from species to species, but this is probably related more to the disjunct origins of the underlying research. For example, applicability of the method depends on the cold tolerance of the pest relative to that of the host commodity, the temperature, and the storage time. Citrus, litchi, guava, and carambola can be treated with cold temperature, but mangoes are damaged at temperature and time combinations required to control pests. Hill *et al.* (1988), Jessup *et al.* (1993), and Heather *et al.* (1996) reported disinfestation treatments of Australian-grown oranges, lemons, and mandarins that controlled Mediterranean fruit fly and Queensland fruit fly with storage times of 16 or 14 days at 1°C. Armstrong *et al.* (1995b) demonstrated that the cold temperature and exposure time combinations needed to control Mediterranean fruit fly, oriental fruit fly, and melon fly in Hawaiian-grown carambolas were about the same as those shown by Gould and Sharp (1990).

Controlled (modified) atmospheres

A modified atmosphere maintained with almost no variation in gas compositions is called a controlled atmosphere. Examples of controlled or modified atmospheres used to control pests of tropical fruits are elevated concentrations of carbon dioxide, reduced oxygen levels, combinations of the two, or the incorporation of nitrogen gas at different concentrations. The gaseous mixtures are contained in a confined space where the tropical fruits are exposed to different temperatures and time periods. Each fruit cultivar varies in its response to a particular modified atmosphere with respect to fruit quality. Usually, however, modified atmospheres extend the market life of the fruits and control pests. Exposure of the tropical fruits usually requires several days of treatment to particular mixes of gasses at elevated temperatures to control pests.

Carpenter and Potter (1994) reviewed the literature on controlled atmosphere disinfestation. Whiting *et al.* (1995) showed that disinfestation of fruit to control Tortricidae in a modified atmosphere was accelerated at a temperature of 40°C. Whiting and van den Heuvel (1995) controlled diapausing adult two-spotted spider mites, *Tetranychus urticae* Koch, with various controlled atmosphere treatments applied at 20 and 40°C.

Coatings and wraps

Atmospheres also may be altered by use of film wraps and fruit coatings (Hallman *et al.*, 1994). Coatings trap gases inside the fruits and restrict gaseous exchange. Methyl cellulose and shellac, when applied to the surface of fruits, can cover eggs and trap surface pests and seal them, resulting in the death of the pest presumably by suffocation. Internal feeding pests are believed to be controlled by atmospheres that are modified inside coated fruits. Usually, carbon dioxide levels are increased and oxygen levels are decreased in the pulp tissue.

Coatings such as Prima Fresh 31 (S.C. Johnson and Son, Racine, WI) contain amine fatty acid soap, waxes, and food-grade shellac. When applied as a coating, Prima Fresh 31 reduced the number of immature Caribbean fruit fly in grapefruit and guava but was not adequate as a single treatment (Hallman *et al.*, 1994, 1995). Prima Fresh 31, however, killed all exposed eggs, nymphs and adults of Chilean false spider mite on cherimoya from Chile that were immersed in the wax material (Thompson, 1990; Undurraga and Lopez, 1992).

Incorporation of insect growth regulators such as 20% methoprene with wax when applied to papaya was reported by Saul and Seifert (1990) to control Mediterranean fruit fly, oriental fruit fly, and melon fly eggs and larvae at the 99.9968% efficacy level.

Irradiation

Irradiation is a proven method of disinfesting tropical fruits and many other agricultural commodities of quarantine pests (Komson *et al.*, 1987; Heather, 1993b; Burditt, 1994). The American Society for Testing and Materials published a standard (guide) that explains dosimetry in radiation research on food and agricultural products (ASTM, 1997). Irradiation treatment of the fruit is done by exposure to an ionizing energy source which can be gamma rays, X-rays, or accelerated electrons. Such radiations are referred to as ionizing radiations because their energy is great enough to dislodge electrons from atoms and molecules and convert them to ions. Gamma rays and X-rays form part of the electromagnetic spectrum. Gamma rays are produced by the spontaneous disintegration of radionuclides. Cobalt-60 and caesium-137 are the radionuclides used exclusively to sterilize food and medical products which form the basis of a commercial industry. Cobalt-60 is produced by neutron bombardment of the metal cobalt-59 inside a nuclear reactor, then double encapsulated in stainless steel pencils to prevent any leakage during its use in a radiation plant. Cobalt-60 has a half-life of 5.3 years. Caesium-137 is produced by reprocessing used nuclear fuel elements and is not in widespread commercial use. Caesium-137 has a half-life of 30 years. Neither of these sources gives rise to radiation levels in the treated commodity which could cause radioactivity, due to the inherently low energy levels.

Machines capable of producing high energy electron beams by accelerating electrons are machine sources of radiation. A beam of accelerated electrons from a linear accelerator is used to directly irradiate the commodity to be disinfested or to bombard a metal target and produce X-rays to which the commodity is exposed. Electron machines capable of accelerating electrons produce high energy electron beams, but electron beams cannot penetrate deeply into fruits and are limited to a maximum of about 4 cm or fruit 8 cm in diameter. Cobalt-60 and caesium-137 gamma rays will penetrate more than a metre through bulk commodities, although their strength is attenuated requiring rotation to achieve a uniform dosage at the outside and reduced dosage at the centre. The difference is known as the minimum–maximum ratio (expressed as D_{min} and D_{max}) and varies with the commodity and geometry of the irradiator. Machine sources may be preferred by some users because no radioactive materials are used, and when the radiation process is completed, the power to the machine is turned off. Their disadvantage lies in their very high capital cost and mechanical complexity compared with cobalt source irradiators, which is the type most widely used by the industry. The process cannot increase the normal radioactivity level of the fruit, regardless of the exposure time or the dose used to treat the fruit. Radiation affects rapidly dividing cells in arthropods and living plant products and slows the development of colour and ripening of fruits and vegetables. Some types of living plant structures, such as fruits, tolerate radiation treatment better than others, thus allowing them to be treated to control pests.

Radiation dose is the amount of radiation energy absorbed by the fruit as it passes through the radiation field (Federal Register, 1996a). The Gray (Gy) is currently the accepted Système International (SI) unit. One Gy is equal to 0.001 kilogray (kGy), 100 rad, or 0.1 kilorad, terms that were often used in the past. A working group of the International Consultative Group on Food Irradiation (ICGFI) concluded that a minimum absorbed dose of 150 Gy would provide quarantine security against all species of Tephritidae and 300 Gy would provide quarantine security against most other arthropod pests (ICGFI, 1994). The US Department of Agriculture has accepted irradiation as a quarantine treatment and developed schedules of irradiation doses against fruit flies of from 150 to 250 Gy to treat commodities for movement from Hawaii to the mainland USA and to allow carambola and litchi to be moved from Hawaii with a minimum irradiation treatment dose of 250 Gy (Federal Register, 1996a,b, 1997a). The same agency has reviewed the literature concerning irradiation for pests of tropical fruits and recommended a minimum dose to control seven major economically important species of tephritid fruit flies with doses ranging from 150 to 250 Gy depending on the species. These required doses far exceed research results on some species. For example, Heather *et al.* (1991) demonstrated a maximum requirement for > 99.9968% efficacy of < 100 Gy for both Queensland fruit fly and Jarvis fruit fly, *Bactrocera jarvisi* (Tryon). Use of the lowest dose that will meet the minimum efficacy required by an importing country will reduce the possibility of product injury.

The USA recently assumed leadership in using irradiation as a quarantine pest disinfestation treatment, largely because of the objections raised by activist consumer groups. However, with a better informed public and the loss of highly effective chemicals such as ethylene dibromide and methyl bromide used as fumigants, irradiation is becoming a more accepted type of quarantine treatment. The US Food and Drug Administration regulates the use of irradiation in the USA and permits consumption there of fresh fruits treated with irradiation doses up to 1 kGy. For spices and herbs, 30 kGy is allowed (Federal Register, 1986). Many publications present literature reviews for irradiation as a quarantine treatment (Moy, 1985; IAEA, 1992; Anonymous, 1993). The International Atomic Energy Agency, Vienna, Austria, published a booklet in 1991 that discussed facts about food irradiation (Anonymous, 1991). The booklet contains 14 fact sheets that explain the food irradiation process and is available by writing to: The ICGFI Secretariat, Joint FAO/IAEA Division of Nuclear Techniques in Food and Agriculture, Wagramerstrasse 5, PO Box 100 A-1400, Vienna, Austria.

Microwave energy

Microwaves are radio frequency waves (not heat energy). However, microwave energy is converted to heat by interaction with charged particles and polar molecules. The agitation is evidenced as heat. Microwaves in the electromagnetic spectrum ranging from 10 kHz to 100 GHz have been used to heat objects by converting electromagnetic energy to heat energy and are a potential quarantine treatment method to control some insect pests and also to study the biology of treated pests due to various heating rates (Hallman and Sharp, 1994; Sharp *et al.*, 1999). Microwave energy impinging on spherical and cylindrical products is focused toward their centres (Buffler, 1993). Sharp (1993b, 1996) studied heating profiles in different fruits by exposing them to various heating rates with a research-quality microwave oven and developed time and temperature data at various powers, which could be used to control pests. Subsequently, Sharp *et al.* (1999) studied the effects of rapid heating on the mortality of mature larvae of Caribbean fruit fly and found that rapid heating failed to control the larvae at temperatures normally lethal to the larvae.

The studies also showed that at controlled power levels larvae can be killed, perhaps without significant damage to the treated fruit. Until cost efficient ways can be found to obtain precise consistent temperatures, the microwave method is unlikely to be adopted commercially.

Combination Treatments

Two or more single treatments used simultaneously or sequentially to provide quarantine security are combination treatments (Mangan and Sharp, 1994). Examples of such treatments are modified atmosphere followed by fumigation, modified atmosphere plus temperature, and refrigeration followed by fumigation. Grapes grown in Australia may be imported into the USA after refrigeration in transit and then fumigation at the port of arrival (Federal Register, 1990). Another example of a combination treatment is soapy water immersion followed by immersion in wax to control Chilean false spider mite, *Brevipalpus chilensis* (Baker), on cherimoya and limes from Chile (Federal Register, 1992b; APHIS, 1998), and use of warm soapy water and brushing to control external pests such as scales and mealybugs on durian (APHIS, 1998).

Systems approach

Jang (1996) discussed a systems approach to quarantine security to control Mediterranean fruit fly, oriental fruit fly, and melon fly in 'Sharwil' avocado fruits grown in Hawaii.

Components of systems approaches include seasonal pest incidence determined by surveys, trapping for basic levels of incidence in the production area, sampling, pest management in the field, inundative sterile insect releases, cultural practices, host resistance, safeguards after harvest, inspection, and quality control (Jang and Moffitt, 1994; Jang, 1996). An example of the systems approach is the importation into the USA of papayas from Brazil and Costa Rica. Conditions of entry into the USA include requirements for growing, treating, packing, and shipping, field sanitation, fruit fly trapping, and issuance of a phytosanitary certificate (Federal Register, 1998). Also, treatments must be consumer safe, environmentally friendly, and easy to regulate and enforce. 'Magic bullet' treatments such as ethylene dibromide and methyl bromide fumigations are no longer available for use in many parts of the world. The difficult challenge is to develop successful new strategies following the loss of these chemicals.

Conclusion

Until tropical fruits are produced in pest-free enclosures and remain free of exposure to pests or until fruits are genetically altered to be resistant to pests, quarantine treatments and methodologies will be required to control pests that attack tropical fruits. The risk that new pest introductions will be encountered increases, for example, as world populations increase, fruit production and international trade increase, and as new markets are created to provide fresh fruits to a growing population of consumers over the globe. The expansion of air travel allows rapid shipment of fruits within hours from one side of the globe to the another. Contraband commodities carried by travellers who fail to heed quarantine requirements are arguably the greatest risk of introduction of new pests to an area. Access for commercial produce under conditions of low quarantine security risk removes much of the temptation for travellers to carry it as high risk contraband.

To keep the cost of tropical fruits low for consumers, quarantine treatments must be affordable. New treatments must be easily adapted by industry and be economical to use. Average costs to build facilities that use hot water, forced hot air, vapour heat, and irradiation currently range from US$200,000 to US$3,000,000 (Anonymous, 1991; US Environmental Protection Agency, 1996). Estimated costs for using controlled atmospheres are US$5000 per container per shipment (US Environmental Protection Agency, 1995). All costs are expected to increase.

References

Anonymous (1991) *Facts About Food Irradiation.* A Series of Fact Sheets from the International Consultative Group on Food Irradiation, IAEA/PI/A33E, 91–05549. International Atomic Energy Agency. Vienna, Austria, 38 pp.

Anonymous (1992) *United Nations Environmental Programme.* Methyl bromide: atmospheric science, technology and economics. UN Headquarters, Ozone Secretariat, Nairobi, Kenya.

Anonymous (1993) Cost–benefit aspects of food irradiation processing. In: Proceedings of Symposium. Aix-En-Provence, 1–5 March. International Atomic Energy Agency, Vienna, 505 pp.

Anonymous (1996) *International Standards for Phytosanitary Measures. Part 1 – Import Regulations. Guidelines for Pest Risk Analysis.* Secretariat of IPPC, FAO of UN, Rome, 21 pp.

Anonymous (1998) *Changes to Methyl Bromide Phaseout in the United States.* http//:www.epa.gov/ozone/mbr/harmoniz/html

APHIS (Animal and Plant Health Inspection Service) (1998) *Plant Protection and Quarantine Treatment Manual.* US Department of Agriculture. US Government Printing Office, Washington, DC.

AQIS (Australian Quarantine and Inspection Service) (1991) *The Application of Risk Management in Agricultural Quarantine Import Assessment.* AGPS, Canberra, 33 pp.

Armstrong, J.W. (1994) Commodity resistance to infestation by quarantine pests. In: Sharp, J.L. and Hallman, G.J. (eds) *Quarantine Treatments for Pests of Food Plants.* Westview Press, Boulder, Colorado, pp. 199–211.

Armstrong, J.W., Hansen, J.D., Hu, B.K.S. and Brown, S.A. (1989) High-temperature forced-air quarantine treatment for papaya infested with tephritid fruit flies (Diptera: Tephritidae). *Journal of Economic Entomology* 82, 1667–1674.

Armstrong, J.W., Hu, B.K.S. and Brown, S.A. (1995a) Single-temperature forced hot-air quarantine treatment to control fruit flies (Diptera: Tephritidae) in papaya. *Journal of Economic Entomology* 88, 678–682.

Armstrong, J.W., Silva, S.T. and Shishido, V.M. (1995b) Quarantine cold treatment for Hawaiian carambola fruit infested with Mediterranean fruit fly (Diptera: Tephritidae) eggs and larvae. *Journal of Economic Entomology* 88, 683–687.

ASTM (American Society for Testing and Materials) (1997) *Standard Guide for Dosimetry in Radiation Research on Food and Agricultural Products.* Designation: E1900–97, 9 pp.

Baker, A.C. (1939) The basis for treatment of products where fruitflies are involved as a condition for entry into the United States. *US Department of Agriculture Circular 551.*

Baker, R.T., Cowley, J.M., Harte, D.S. and Frampton, E.R. (1990) Development of a maximum pest limit for fruit flies (Diptera: Tephritidae) in produce imported into New Zealand. *Journal of Economic Entomology* 83, 13–17.

Buffler, C.R. (1993) *Microwave Cooking and Processing: Engineering Fundamentals for the Food Scientist.* Van Nostrand Reinhold, New York, 166 pp.

Burditt, Jr, A.K. (1994) Irradiation. In: Sharp, J.L. and Hallman, G.J. (eds) *Quarantine Treatments for Pests of Food Plants.* Westview Press, Boulder, Colorado, pp. 101–117.

Carpenter, A. and Potter, M. (1994) Controlled Atmospheres. In: Sharp, J.L. and Hallman, G.J. (eds) *Quarantine Treatments for Pests of Food Plants.* Westview Press, Boulder, Colorado, pp. 171–198.

Champ, B.R., Highley, E. and Johnson, G.I. (eds) (1993) *Postharvest Handling of Tropical Fruits. Proceedings of an International Conference.* Chiang Mai, Thailand, 19–23 July 1993. Australian Centre for International Agricultural Research Proceedings No.50, 500 pp.

Chew, V. (1994) Statistical methods for quarantine treatment data analysis. In: Sharp, J.L. and Hallman, G.J. (eds) *Quarantine Treatments for Pests of Food Plants.* Westview Press, Boulder, Colorado, pp. 33–46.

Corcoran, R.J., Heather, N.W. and Nimmo, A.L. (1993) Comparative heat response of eggs of the Queensland fruit fly, *Bactrocera tryoni* and Mediterranean fruit fly, *Ceratitis capitata.* In: *Proceedings, Australasian Postharvest Conference.* 20–24 September 1993. The University of Queensland, Gatton College, Lawes, Queensland 4343, Australia, pp. 303–307.

Couey, H.M. and Chew, V. (1986) Confidence limits and sample size in quarantine research. *Journal of Economic Entomology* 79, 887–890.

Federal Clean Air Act (1990) Public Law 101–549, enacted 15 November 1990. Washington, DC.

Federal Register (1983) Environmental Protection Agency. Notices. Tuesday, 11 October 1983. OPP-30000/25D; PH-FRL 2450-1. Ethylene dibromide; intent to cancel registrations of pesticide products containing ethylene dibromide; determination concluding the rebuttable presumption against registration; availability of position document. *Federal Register* 48 (197), 46234–46248.

Federal Register (1986) Department of Health and Human Services. Food and Drug Administration, Part I11. Friday 18 April 1986.21 CFR Part 179. Irradiation in the production, processing, and handling of food; final rule. *Federal Register* 51 (75), 13376–13399.

Federal Register (1989) Department of Agriculture. Animal and Plant Health Inspection Service. Proposed Rules. Thursday, 13 July 1989.7 CFR Part 319. Docket No. 88-176. Apricots, nectarines, peaches, and plums from Chile. *Federal Register* 54(133), 29566–29569.

Federal Register (1990) Department of Agriculture. Animal and Plant Health Inspection Service. Rules and Regulations. Tuesday, 26 June 1990. 7 CFR Parts 300 and 319. Docket No.90-039. Importation of grapes from Australia. *Federal Register* 55(123), 25950–25954.

Federal Register (1991) Department of Agriculture. Animal and Plant Health Inspection Service. Rules and Regulations. Wednesday, 15 May 1991.7 CFR Part 319. Docket No.91-039. Importation of avocados from New Zealand. *Federal Register* 56(94), 22295–22297.

Federal Register (1992a) Department of Agriculture. Animal and Plant Health Inspection Service. Proposed Rules. Friday, 3 January 1992. 7 CFR Part 319. Docket No.91-068. Importation of papayas from Costa Rica. *Federal Register* 57(2), 217–219.

Federal Register (1992b) Department of Agriculture. Animal and Plant Health Inspection Service. Rules and Regulations. Monday, 30 November 1992.7 CFR Part 319. Docket No.91-160-2. Importation of cherimoyas from Chile. *Federal Register* 57(230), 56434–56437.

Federal Register (1993a) Department of Agriculture. Animal and Plant Health Inspection Service. Rules and Regulations. Tuesday, 25 February 1994.7 CFR Part 319. Docket No.92-126-2. Honeydew melons from Brazil. *Federal Register* 58 (36), 11363–11364.

Federal Register (1993b) Department of Agriculture. Animal and Plant Health Inspection Service. Rules and Regulations. Tuesday, 27 July 1993. 7 CFR Part 319. Docket No. 92-111-3. Hass avocados from Mexico. *Federal Register* 58(142), 40033–40037.

Federal Register (1994a) Department of Agriculture. Animal and Plant Health Inspection Service. Rules and Regulations. Monday, 28 February 1994.7 CFR Part 319. Docket No.93-095-2. Importation of apples, apricots, peaches, persimmons, pomegranates, and citrus for Sonora. *Federal Register* 59(39), 9381–9382.

Federal Register (1994b) Department of Agriculture. Animal and Plant Health Inspection Service. Rules and Regulations. Wednesday, 28 December 1994.7 CFR Part 318. Docket No. 93-088-2. Avocados from Hawaii. *Federal Register* 59(248), 66638–66642.

Federal Register (1995a) Department of Agriculture. Animal and Plant Health Inspection Service. Rules and Regulations. Monday, 6 February 1995.7 CFR Parts 300 and 319. Docket No. 93-028-5. Grapefruit and mangoes from Mexico; addition of treatment. *Federal Register* 60(24), 6957–6958.

Federal Register (1995b) Department of Agriculture. Animal and Plant Health Inspection Service. Proposed Rules. Monday, 11 September 1995.7 CFR 319. Docket No. 93-119-1. Importation of citrus fruits from Australia. *Federal Register* 60(175), 47101–47103.

Federal Register (1996a) Department of Agriculture. Animal and Plant Health Inspection Service. Rules and Regulations. Wednesday, 15 May 1996.7 CFR Part 319. Docket No. 95-088-1. The application of irradiation to phytosanitary problems. *Federal Register* 61(95), 24433–24439.

Federal Register (1996b) Department of Agriculture. Animal and Plant Health Inspection Service. Proposed Rules. Tuesday, 23 July 1996.7 CFR Part 318. Docket No. 95-069-1. Papaya, carambola, and litchi from Hawaii. *Federal Register* 61(142), 38108–38114.

Federal Register (1997a) Department of Agriculture. Animal and Plant Health Inspection Service. Rules and Regulations. Thursday, 10 July 1997. 7 CFR Parts 300 and 318. Docket No.95-069-2. Papaya, carambola, and litchi from Hawaii. *Federal Register* 62(132), 36967–36976.

Federal Register (1997b) Department of Agriculture. Animal and Plant Health Inspection Service. Proposed Rules. Tuesday, 30 December 1997. 7 CFR Parts 300 and 301. Docket No.96-069-1. High-temperature forced-air treatments for citrus. *Federal Register* 62(249), 67761–67764.

Federal Register (1998) Department of Agriculture. Animal and Plant Health Inspection Service. Rules and Regulations. Friday, 13 March 1998. 7 CFR Part 319. Docket No.96–046–5. Importation of fruits and vegetables; papayas from Brazil and Costa Rica. *Federal Register* 63(49), 12383–12396.

Fiskaali, D.A. (ed.) (1989) *California Commodity Treatment Manual, Vol 1. Treatments.* State of California Department of Food and Agriculture, 342 pp.

Garry, V.F., Griffith, J., Danzl, T.J., Nelson, R.L., Whorton, E.B., Krueger, L.A. and Cervenka, J. (1989) Human genotoxicity: pesticide applicators and phosphine. *Science* 246, 251–255.

Gould, W.P. (1994) Cold storage. In: Sharp, J.L. and Hallman, G.J. (eds) *Quarantine Treatments for Pests of Food Plants*, Westview Press, Boulder, Colorado, pp. 119–132.

Gould, W.P. (1996) Mortality of *Toxotrypana curvicauda* (Diptera: Tephritidae) in papayas exposed to forced hot air. *Florida Entomologist* 79, 407–413.

Gould, W.P. and Sharp, J.L. (1990) Cold-storage quarantine treatment for carambolas infested with the Caribbean fruit fly (Diptera: Tephritidae). *Journal of Economic Entomology* 83, 458–460.

Gould, P., Nguyen, R., Hennessey, M.K., Sharp, J.L. and Crane, J. (1996) Infestation studies on lychees and longans to determine their host status in relation to the Caribbean fruit fly. *Proceedings of the Florida State Horticultural Society* 109, 276–277.

Greany, P.D. (1994) Plant host status and natural resistance. In: Paull, R.E. and Armstrong, J.W. (eds) *Insect Pests and Fresh Horticultural Products. Treatments and Responses*. CAB International, Wallingford, UK, 360 pp.

Hallman, G.J. and Armstrong, J.W. (1994) Heated air treatments. In: Sharp, J.L. and Hallman, G.J. (eds) *Quarantine Treatments for Pests of Food Plants*. Westview Press, Boulder, Colorado, 149–163.

Hallman, G.J. and Sharp, J.L. (1994) Radio frequency heat treatments. In: Sharp, J.L. and Hallman, G.J. (eds) *Quarantine Treatments for Pests of Food Plants*. Westview Press, Boulder, Colorado, pp. 165–170.

Hallman, G.J., Nisperos-Carriedo, M.O., Baldwin, E.A. and Campbell, C.A. (1994) Mortality of Caribbean fruit fly (Diptera: Tephritidae) immatures in coated fruits. *Journal of Economic Entomology* 87, 752–757.

Hallman, G.J., McGuire, R.G., Baldwin, E.A. and Campbell, C.A. (1995) Mortality of feral Caribbean fruit fly (Diptera: Tephritidae) immatures in coated guavas. *Journal of Economic Entomology* 88, 1353–1355.

Heather, N.W. (1993a) Quarantine disinfestation of tropical fruits: non-chemical options. In: Champ, B.R., Highley, E. and Johnson, G.I. (eds) *Postharvest Handling of Tropical Fruits*. The Australian Centre for International Agricultural Research (ACIAR) Proceedings No.50, ACIAR, Canberra, pp. 272–279.

Heather, N.W. (1993b) Irradiation as a quarantine treatment of agricultural commodities against arthropod pests. *Management of Insect Pests: Nuclear and Related Molecular and Genetic Techniques*. International Atomic Energy Agency, Vienna, Austria, pp. 627–640.

Heather, N.W. (1994) Pesticide quarantine treatments. In: Sharp, J.L. and Hallman, G.J. (eds) *Quarantine Treatments for Pests of Food Plants*. Westview Press, Boulder, Colorado, pp. 89–100.

Heather, N.W., Corcoran, R.J. and Baños, C. (1991) Disinfestation of mangoes against two Australian fruit flies (Diptera: Tephritidae). *Journal of Economic Entomology* 84, 1304–1307.

Heather, N.W., Whitfort, L., McLaughlan, R.L. and Kopittke, R.A. (1996) Cold disinfestation of Australian mandarins against Queensland fruit fly (Diptera: Trypetidae). *Postharvest Biology and Technology* 8, 307–315.

Heather, N.W., Corcoran, R.J. and Kopittke R.A. (1997) Hot air disinfestation of Australian 'Kensington' mangoes against two fruit flies (Diptera: Tephritidae). *Postharvest Biology and Technology* 10, 99–105.

Hill, A.R., Rigney, C.J. and Sproul, A.N. (1988) Cold storage of oranges as a disinfestation treatment against the fruit flies *Dacus tryoni* (Froggatt) and *Ceratitis capitata* (Wiedemann) (Diptera: Tephritidae). *Journal of Economic Entomology* 81, 257–260.

ICGFI (1994) *Irradiation as a Quarantine Treatment of Fresh Fruits and Vegetables*. A Report of the Working Group Convened by ICGFI (1994). *ICGFI Document* No. 17, 40 pp.

International Atomic Energy Agency (1992) *Use of Radiation as a Quarantine Treatment of Food and Agricultural Commodities. Proceedings of the Final Research Coordination Meeting*. Kuala Lumpur, Malaysia, 27–31 August 1990. IAEA, Vienna, Austria, 220 pp.

Jang, E.B. (1996) Systems approach to quarantine security: postharvest application of sequential mortality in the Hawaiian grown 'Sharwil' avocado system. *Journal of Economic Entomology* 89, 950–956.

Jang, E.B. and Moffitt, H.R. (1994) Systems approaches to achieving quarantine security. In: Sharp, J.L. and Hallman, G.J. (eds) *Quarantine Treatments for Pests of Food Plants*. Westview Press, Boulder, Colorado, pp. 225–237.

Jessup, A.J., de Lima, C.P.F., Hood, C.W., Sloggett, R.F., Harris, A.M. and Beckingham, M. (1993) Quarantine disinfestation of lemons against *Bactrocera tryoni* and *Ceratitis capitata* (Diptera: Tephritidae) using cold storage. *Journal of Economic Entomology* 86, 795–802.

Johnson, G.I. and Heather, N.W. (1994) Disease and pest control in tropical fruits. In: Champ, B.R. and Highley, E. *Postharvest Technology for Agricultural Products in Vietnam*. ACIAR Proceedings No.60, ACIAR, Canberra, pp. 100–126.

Johnson, G.I., Sharp, J.L., Milne, D.L. and Oosthuyse, S.A. (1997) Postharvest technology and quarantine treatments. In: Litz, R.L. (ed.) *The Mango: Botany, Production and Uses*. CAB International, Wallingford, UK, pp. 447–507.

Joyce, D.C. and Shorter, A.J. (1994) High-temperature conditioning reduces hot water treatment injury of 'Kensington Pride' mango fruit. *HortScience* 29, 1047–1051.

Komson, P., Sutantawong, M., Smitasiri, E., Lapasatukul, C. and Unahawutti, U. (1987) Irradiation as a quarantine treatment for the oriental fruit fly *Dacus dorsalis* Hendel in mangoes. *Modern Insect Control: Nuclear Techniques and Biotechnology*. IAEA Vienna, pp. 319–324.

Mangan, R.L. and Ingle, S.J. (1994) Forced hot-air quarantine treatment for grapefruit infested with Mexican fruit fly (Diptera: Tephritidae). *Journal of Economic Entomology* 87, 1574–1579.

Mangan, R.L. and Sharp, J.L. (1994) Combination and multiple treatments. In: Sharp, J.L. and Hallman, G.J. (eds) *Quarantine Treatments for Pests of Food Plants*. Westview Press, Boulder, Colorado, pp. 239–247.

McGuire, R.G. and Sharp, J.L. (1997) Quality of colossal Puerto Rican 'Keitt' mangoes after quarantine treatment in water at 48°C. *Tropical Science* 37, 154–159.

Moy, J.H. (1985) *Radiation Disinfestation of Food and Agricultural Products. Proceedings of an International Conference*. Honolulu, Hawaii, 14–18 November 1983. Hawaii Institute of Tropical Agriculture and Human Resources. University of Hawaii at Manoa, Honolulu, Hawaii, 424 pp.

Ohr, H.D., Sims, J.J., Grech, N.M., Becker, J.O. and McGiffen, M.E. Jr (1996) Methyl iodide, an ozone-safe alternative to methyl bromide as a soil fumigant. *Plant Disease* 80, 731–735.

Paul, R. and Armstrong, J.W. (1994) *Insect Pests and Fresh Horticultural Products: Treatments and Responses*. CAB International, Wallingford, UK, 360 pp.

Rigney, C.J. and Wills, P.A. (1986) Efficacy of gamma irradiation as a quarantine treatment against Queensland fruit fly. In: Moy, J. H. (ed.) *Radiation Disinfestation of Food and Agricultural Products*. University of Hawaii at Manoa, Honolulu, Hawaii, pp. 116–120.

Riherd, C., Nguyen, R. and Brazzel, J.R. (1994) Pest free areas. In: Sharp, J.L. and Hallman, G.J. (eds) *Quarantine Treatments for Pests of Food Plants*. Westview Press, Boulder, Colorado, pp. 213–223.

Robertson, J.L., Preisler, H.K., Frampton, E.R. and Armstrong, J.W. (1994) Statistical analyses to estimate efficacy of disinfestation treatments. In: Sharp, J.L. and Hallman, G.J. (eds) *Quarantine Treatments for Pests of Food Plants*. Westview Press, Boulder, Colorado, pp. 47–65.

Saul, S. and Seifert, J. (1990) Methoprene on papayas: persistence and toxicity to different developmental stages of fruit flies (Diptera: Tephritidae). *Journal of Economic Entomology* 83, 901–904.

Shannon, M.J. (1994) APHIS. In: Sharp, J.L. and Hallman, G.J. (eds) *Quarantine Treatments for Pests of Food Plants*. Westview Press, Boulder, Colorado, pp. 1–10.

Sharp, J.L. (1993a) Hot-air quarantine treatment for 'Marsh' white grapefruit infested with Caribbean fruit fly (Diptera: Tephritidae). *Journal of Economic Entomology* 86, 462–464.

Sharp, J.L. (1993b) Microwaves as a quarantine treatment to disinfest commodities of pests. In: Champ, B.R., Highley, E. and Johnson, G.I. (eds) *Postharvest Handling of Tropical Fruits. Proceedings of an International Conference*. Chiang Mai, Thailand, 19–23 July 1993, No.50. Australian Centre for International Agricultural Research, Canberra, Australia, 500 pp.

Sharp, J.L. (1994) Hot water immersion. In: Sharp, J.L. and Hallman, G.J. (eds) *Quarantine Treatments for Pests of Food Plants*. Westview Press, Boulder, Colorado, pp. 133–147.

Sharp, J.L. (1996) Microwave heating of agricultural products. In: Vijaysegaran, S., Pauziah, M., Mohamed, M.S. and Ahmad Tarmizi, S. (eds) *Proceedings of the International Conference on Tropical Fruits*, Vol. II, Kuala Lumpur, Malaysia, 23–26 July 1996. Malaysian Agricultural Research and Development Institute, Malaysia, pp. 329–343.

Sharp, J.L. and Hallman, G.J. (1994) *Quarantine Treatments for Pests of Food Plants*. Westview Press, Boulder, Colorado, 290 pp.

Sharp, J.L. and King, J.R. (1997) Methyl iodide as a quarantine treatment for Caribbean fruit fly. In: Stanley, D. (ed.) *Methyl Bromide Alternatives Newsletter*. US Department of Agriculture 3, p. 11.

Sharp, J.L. and McGuire, R.G. (1996) Control of Caribbean fruit fly (Diptera: Tephritidae) in navel orange by forced hot air. *Journal of Economic Entomology* 89, 1181–1185.

Sharp, J.L., Gaffney, J.J., Moss, J.I. and Gould, W.P. (1991) Hot-air treatment device for quarantine research. *Journal of Economic Entomology* 84, 520–527.

Sharp, J.L., Robertson, J.L. and Preisler, H.K. (1999) Mortality of mature third instar Caribbean fruit fly (Diptera: Tephritidae) exposed to microwave energy. *Canadian Entomologist* 131, 1–7.

Smith, E.S.C. (1992) Fruit fly disinfestation in mangoes by hot water dipping. In: Ooi, P.A.C.,

Lim, G.S. and Teng, P.S (eds) *Proceedings of the Third International Conference on Plant Protection in the Tropics*. Kuala Lumpur, Malaysia. Malaysian Plant Protection Society. No. 6, 188–195.

Stark, J.D. (1994) Chemical fumigants. In: Paull, R.E. and Armstrong, J.W. (eds) *Insect Pests and Fresh Horticultural Products: Treatment and Responses*. CAB International, Wallingford, UK, pp. 69–84.

Sugimoto, T., Furusawa, K. and Mizobuchi, M. (1983) Effectiveness of vapor heat treatment against the oriental fruit fly, *Dacus dorsalis* Hendel, in green pepper and fruit tolerance to the treatment. *Research Bulletin Plant Protection Service* 19, 81–88 (in Japanese).

Thompson, A. (1990) Efecto de la aplicatión de cera y aceie en frutos de chirimoya (*Annona cherimola* Mill.) cv. Concha Lisa, sobre el control de *Brevipalpus chilensis* Baker y evaluación del comportamiento de la fruta en post-cosecha. Thesis in agronomy, Universidad Católica de Valparaiso, Chile, 92 pp.

Unahawutti, U., Chettananachitara, C., Poomthong, M., Komson, P., Smitasiri, E., Lapasathukool, C., Worawissithumrong, W. and Intarakumheng, R. (1992) Evaluation of vapor heat treatment for control of the oriental fruit fly and the melon fly in 'Nan Klarngwan' mango. *Technical Report, Department of Agriculture, Bangkok, Thailand*, 106 pp.

Undurraga, M.P. and Lopez, L.E. (1992) Use of wax and soapy water, alone or combined as a quarantine method of control of *Brevipalpus chilensis* Baker on cherimoyas (*Annona cherimola* Mill.) (Report) Universidad Católica de Valparaiso, Valparaiso, Chile, 22 pp.

United States Environmental Protection Agency (1995) Office of Air and Radiation (6205-1). EPA430-R-95-009. July 1995. Stratospheric Ozone Protection. Alternatives to Methyl Bromide. Ten Case Studies – Soil, Commodity, and Structural Use.

United States Environmental Protection Agency (1996) Office of Air and Radiation (6205-1). EPA430-R-96-021. December 1996. Stratospheric Ozone Protection. Alternatives to Methyl Bromide. Vol. 2 – Ten Case Studies. Soil, Commodity, and Structural Use.

Waddell, B.C., Clarke, G.K. and Maindonald, J.H. (1997) Comparative mortality responses of two Cook Island fruit fly (Diptera: Tephritidae) species to hot water immersion. *Journal of Economic Entomology* 90, 1351–1356.

Whiting, D.C. and van den Heuvel, J. (1995) Oxygen, carbon dioxide, and temperature effects on mortality responses of diapausing *Tetranychus urticae* (Acari: Tetranychidae). *Journal of Economic Entomology* 88, 331–339.

Whiting, D.C., O'Connor, G.M., van den Heuvel, J. and Maindonald, J.H. (1995) Comparative mortalities of six tortricid (Lepidoptera) species to two high-temperature controlled atmospheres and air. *Journal of Economic Entomology* 88, 1365–1370.

Yokoyama, V. (1994) Fumigation. In: Sharp, J.L. and Hallman, G.J. (eds) *Quarantine Treatments for Pests of Food Plants*. Westview Press, Boulder, Colorado, pp. 67–87.

Zhang, W.M., McGiffen, M.E. Jr, Becker, J.O., Ohr, H.D., Sims, J.J. and Kallenbach, R.L. (1997) Dose response of weeds to methyl iodide and methyl bromide. *Weed Research* 37, 181–189.

Index

Browse Read and Buy

www.cabi.org/bookshop

ANIMAL & VETERINARY SCIENCES
BIODIVERSITY CROP PROTECTION
HUMAN HEALTH NATURAL RESOURCES
ENVIRONMENT PLANT SCIENCES
SOCIAL SCIENCES

Subjects
Search
Reading Room
Bargains
New Titles
Forthcoming

CABI Publishing
A division of CAB International

Online BOOK SHOP

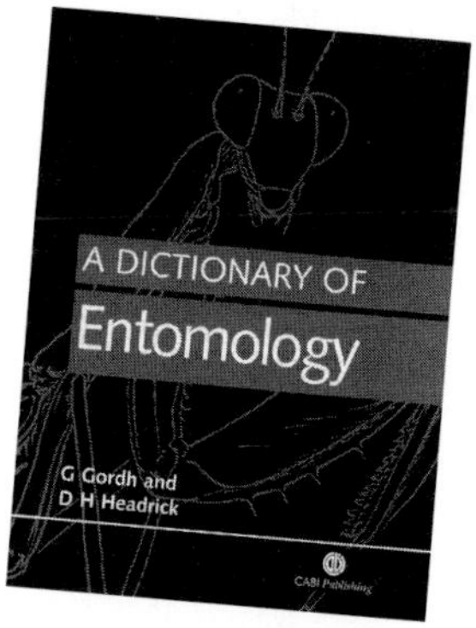

Order & Pay Online!

MasterCard
VISA
AMERICAN EXPRESS

★ FULL DESCRIPTION ★ BUY THIS BOOK ★ BOOK OF THE MONTH

Tel: +44 (0)1491 832111 Fax: +44 (0)1491 829292